Smart Innovation, Systems and Technologies

Volume 222

The Smart Innovation, Systems and Technologies book series encompasses the topics of knowledge, intelligence, innovation and sustainability. The aim of the series is to make available a platform for the publication of books on all aspects of single and multi-disciplinary research on these themes in order to make the latest results available in a readily-accessible form. Volumes on interdisciplinary research combining two or more of these areas is particularly sought.

The series covers systems and paradigms that employ knowledge and intelligence in a broad sense. Its scope is systems having embedded knowledge and intelligence, which may be applied to the solution of world problems in industry, the environment and the community. It also focusses on the knowledge-transfer methodologies and innovation strategies employed to make this happen effectively. The combination of intelligent systems tools and a broad range of applications introduces a need for a synergy of disciplines from science, technology, business and the humanities. The series will include conference proceedings, edited collections, monographs, handbooks, reference books, and other relevant types of book in areas of science and technology where smart systems and technologies can offer innovative solutions.

High quality content is an essential feature for all book proposals accepted for the series. It is expected that editors of all accepted volumes will ensure that contributions are subjected to an appropriate level of reviewing process and adhere to KES quality principles.

Indexed by SCOPUS, EI Compendex, INSPEC, WTI Frankfurt eG, zbMATH, Japanese Science and Technology Agency (JST), SCImago, DBLP.

All books published in the series are submitted for consideration in Web of Science.

More information about this series at http://www.springer.com/series/8767

Amaresh Chakrabarti · Ravi Poovaiah ·
Prasad Bokil · Vivek Kant
Editors

Design for Tomorrow—Volume 2

Proceedings of ICoRD 2021

Springer

Editors
Amaresh Chakrabarti
Centre for Product Design
and Manufacturing
Indian Institute of Science
Bengaluru, Karnataka, India

Prasad Bokil
IDC School of Design
Indian Institute of Technology Bombay
Mumbai, India

Ravi Poovaiah
IDC School of Design
Indian Institute of Technology Bombay
Mumbai, India

Vivek Kant
IDC School of Design
Indian Institute of Technology Bombay
Mumbai, India

ISSN 2190-3018 ISSN 2190-3026 (electronic)
Smart Innovation, Systems and Technologies
ISBN 978-981-16-0118-7 ISBN 978-981-16-0119-4 (eBook)
https://doi.org/10.1007/978-981-16-0119-4

This Springer imprint is published by the registered company Springer Nature Singapore Pte Ltd.
The registered company address is: 152 Beach Road, #21-01/04 Gateway East, Singapore 189721, Singapore

Conference Organisation

Steering Committee

Balan Gurumoorthy, Indian Institute of Science, Bengaluru, India
Cees de Bont, Loughborough University, UK
John Gero, University of North Carolina, Charlotte, USA
Kota Harinarayana, National Aeronautical Laboratories, India
Panos Papalambros, University of Michigan, USA
Steve Culley, University of Bath, UK

Advisory Committee

S. Saleem Ahmed, Vellore Institute of Technology, Vellore, India
Tracy Bhamra, Loughborough University, UK
K. S. Bhanumathi, Kovair, India
Henri Christiaans, Ulsan National Institute of Science and Technology, South Korea
Anjan Das, Confederation of Indian Industry, India
U. B. Desai, Indian Institute of Technology Hyderabad, India
Kees Dorst, University of Technology Sydney, Australia
Srinivasan Dwarakanath, Airbus, India
Shankar Prem Goel, National Institute of Advanced Studies, Indian Institute of Science, Bengaluru, India
Anil K. Gupta, Honey Bee Network, SRISTI and GIAN, India
Imre Horvath, Delft University of Technology, Netherlands
Jitendra Jadhav, CSIR-National Aerospace Laboratories, Bengaluru, India
Arun Jaura, Michelin, India
Deep Kapuria, The Hi-Tech Gears, India
Yong Se Kim, Sungkyunkwan University, South Korea
Ananda Kumar, Elado Consulting Private Limited, India
Larry Leifer, Stanford University, USA

Banshildar Majhi, Indian Institute of Information Technology, Design and Manufacturing, Kancheepuram, India
Udayant Malhoutra, Dynamatic Technologies Limited, Bengaluru, India
K. Rajalakshmi Menon, Centre for Airborne Systems, Defence Research and Development Organisation, India
Arabinda Mitra, Office of PSA to GOI India
T. S. Mruthyunjaya, Chairman Emeritus, CPDM, Indian Institute of Science, India
M. Muruganant, Indus University, India
Nagahanumaiah, Central Manufacturing Technology Institute, India
Shailesh Nayak, National Institute of Advanced Studies, Indian Institute of Science, Bengaluru, India
Srinivas Nidamarthi, ABB Ltd, Michigan, USA
Sudarsan Rachuri, Department of Energy, USA
P. Radhakrishnan, PSG Institute of Advanced Studies, India
V. Bhujanga Rao, National Institute of Advanced Studies, Indian Institute of Science, Bengaluru, India
B. Ravi, Indian Institute of Technology Bombay, India
Sofia Ritzén, KTH Royal Institute of Technology, Sweden
Rajkumar Roy, City, University of London, UK
Al Salour, The Boeing Company, USA
Raman Saxena, USID Foundation, India
Reena Sharma, Defence Research and Development Organisation, India
Vijay Srinivasan, National Institute of Standards and Technology, USA
B. N. Suresh, Indian Institute of Space Science and Technology, India
Kristin L. Wood, Singapore University of Technology and Design, Singapore

Conference Chair

Ravi Poovaiah, Indian Institute of Technology Bombay, India

Programme Chair

Amaresh Chakrabarti, Indian Institute of Science, Bengaluru, India

Vice-Chairs

Prasad Bokil, Indian Institute of Technology Bombay, India
Vivek Kant, Indian Institute of Technology Bombay, India

Co-Chairs

Lucienne Blessing, Singapore University of Technology and Design, Singapore
Monica Bordegoni, Politecnico Di Milano, Italy
Debkumar Chakrabarti, Indian Institute of Technology Guwahati, India
John Clarkson, University of Cambridge, UK
Benoit Eynard, University of Technology of Compiègne, France
Tim Mcaloone, Technical University of Denmark, Denmark
Yukari Nagai, Japan Advanced Institute of Science and Technology, Japan
Luciana Pereira, Federal University of ABC, Brazil
Raghu Prakash, Indian Institute of Technology Madras, India
Sandro Wartzack, University of Erlangen-Nuremberg, Germany

International Programme Committee

Shakuntala Acharya, Indian Institute of Technology Guwahati, India
Sofiane Achiche, Polytechnique Montréal, Canada
Rajat Agrawal, Indian Institute of Technology Roorkee, India
Janet Allen, University of Oklahoma, USA
Manish Arora, Indian Institute of Science, Bengaluru, India
Sandipan Bandyopadhyay, Indian Institute of Technology Madras, India
Niccolò Becattini, Politecnico Di Milano, Italy
Alessandro Bertoni, Blekinge Institute of Technology, Sweden
Marco Bertoni, Blekinge Institute of Technology, Sweden
Mehul Bhatt, Örebro University, Sweden
Ritwij Bhowmik, Indian Institute of Technology Kanpur, India
Braj Bhushan, Indian Institute of Technology Kanpur, India
Abhijit Biswas, Indian Institute of Science, Bengaluru, India
Pradipta Biswas, Indian Institute of Science, Bengaluru, India
Jeremy Bonvoisin, University of Bath, UK
Yuri Borgianni, Free University of Bozen-Bolzano, Italy
David C Brown, Worcester Polytechnic Institute, USA
Francesco Buonamici, University of Florence, Italy
Felician Campean, University of Bradford, UK
Gaetano Cascini, Politecnico Di Milano, Italy
Philip Cash, Technical University of Denmark, Denmark
Amaresh Chakrabarti, Indian Institute of Science, Bengaluru, India
Jayanta Chatterjee, Indian Institute of Technology Kanpur, India
Yong Chen, Shanghai Jiao Tong University, China
Peter Childs, Imperial College London, UK
François Cluzel, University of Paris-Saclay, France
Nathan Crilly, University of Cambridge, UK

Umberto Cugini, Politecnico Di Milano, Italy
Pranab K. Dan, Indian Institute of Technology Kharagpur, India
Amarendra Kumar Das, Indian Institute of Technology Guwahati, India
Suman Devadula, MIT World Peace University, India
Debayan Dhar, Indian Institute of Technology Guwahati, India
Jay Dhariwal, Indian Institute of Technology Delhi, India
J. E. Diwakar, Indian Institute of Science, Bengaluru, India
Alex Duffy, University of Strathclyde, UK
Raghu Echempati, Kettering University, USA
Claudia Eckert, The Pennsylvania State University, USA
Boris Eisenbart, Swinburne University of Technology, Australia
Yvonne Eriksson, Mälardalen University, Sweden
Georges Fadel, Clemson University, USA
Mario Fargnoli, Sapienza University of Rome, Italy
Francesco Ferrise, Politecnico Di Milano, Italy
Susan Finger, Carnegie Mellon University, USA
Kikuo Fujita, Osaka University, Japan
L. S. Ganesh, Indian Institute of Technology Madras, India
Georgi V. Georgiev, University of Oulu, Finland
Detlef Gerhard, Ruhr-University Bochum, Germany
Kilian Gericke, University of Rostock, Germany
Ashitava Ghosal, Indian Institute of Science, Bengaluru, India
Philippe Girard, University of Bordeaux, France
Milane G. Goncalves, Delft University of Technology, Netherlands
Ravindra Goonetilleke, The Hong Kong University of Science and Technology, Hong Kong
Lapo Governi, University of Florence, Italy
Serena Graziosi, Politecnico Di Milano, Italy
Matthias Guertler, University of Technology Sydney, Australia
Bala Murali Gunji, Vellore Institute of Technology, Vellore, India
Sean Hanna, University College London, UK
Bisheshwar Haorongbam, National Institute of Design Assam, India
Peter Hehenberger, University of Applied Sciences Upper Austria, Austria
Ben Hicks, University of Bristol, UK
Stephan Husung, Em Engineering Methods AG, Darmstadt, Germany
David Inkermann, Technical University of Clausthal, Germany
William Ion, University of Strathclyde, UK
Ashok Iyer, Manipal Academy of Higher Education - Dubai Campus, UAE
Santosh Jagtap, Blekinge Institute of Technology, Sweden
R. Jayaganthan, Indian Institute of Technology Madras, India
Camille Jean, Arts et Métiers, France
Shiva Ji, Indian Institute of Technology Hyderabad, India
Anirudha Joshi, Indian Institute of Technology Bombay, India
Purba Joshi, Indian Institute of Technology Bombay, India
Iestyn Jowers, The Open University, UK

Tero Juuti, University of Tampere, Finland
Pratul Chandra Kalita, Indian Institute of Technology Guwahati, India
Udo Kannengiesser, Johannes Kepler University Linz, Austria
Sougata Karmakar, Indian Institute of Technology Guwahati, India
Sonal Keshwani, Indian Institute of Science, Bengaluru, India
Mohammed Rajik Khan, National Institute of Technology Rourkela, India
Maaike Kleinsmann, Delft University of Technology, Netherlands
Edwin Koh, Singapore University of Technology and Design, Singapore
Christian Köhler, Saarland University of Applied Sciences, Germany
Saptarshi Kolay, Indian Institute of Technology Roorkee, India
Lauri Koskela, University of Huddersfield, UK
Srinivas Kota, Birla Institute of Technology and Science, Pilani, India
Dieter Krause, Hamburg University of Technology, Germany
Vinayak Krishnamurthy, Texas A&M University, USA
Rishikesha Krishnan, Indian Institute of Management Indore, India
Tore Kristensen, Copenhagen Business School, Denmark
Peter Krogh, Aarhus University, Denmark
Petter Krus, Linköping University, Sweden
Jyoti Kumar, Indian Institute of Technology Delhi, India
Udaya Kumar, Indian Institute of Technology Guwahati, India
Senthil Kumaran, Indian Institute of Information Technology, Design and Manufacturing, Kancheepuram, India
Guillaume Lamé, University of Paris-Saclay, France
Tobias Larsson, Blekinge Institute of Technology, Sweden
Sang Leigh, Massachusetts Institute of Technology, USA
Torben Lenau, Technical University of Denmark, Denmark
Rungtai Lin, National Taiwan University of Arts, Taiwan
Ying Liu, Cardiff University, UK
Ying Liu Chieh, Chang Gung University, Taiwan
Joaquim Lloveras, Polytechnic University of Catalonia, Spain
Jianxi Luo, Singapore University of Technology and Design, Singapore
N Madhusudanan, Corporate Technology, Siemens, India
Johan Malmqvist, Chalmers University of Technology, Sweden
Monto Mani, Indian Institute of Science, Bengaluru, India
M Manivannan, Indian Institute of Technology Madras, India
Nicolas Maranzana, Arts et Métiers ParisTech, France
Dorian Marjanović, University of Zagreb, Croatia
Christian Marxt, ETH Zürich, Switzerland
Yoshiyuki Matsuoka, Keio University, Japan
Deepak John Matthew, Indian Institute of Technology Hyderabad, India
Jason Matthews, University of the West of England, UK
Christopher Mccomb, The Pennsylvania State University, USA
Alison Mckay, University of Leeds, UK
Gavin Melles, Swinburne University of Technology, Australia

Dirk Schaefer, University of Liverpool, UK
Vinodh Sekar, National Institute of Technology, Tiruchirappalli, India
Harivardhini Selvakumar, PES University, India
Dibakar Sen, Indian Institute of Science, Bengaluru, India
Joy Sen, Indian Institute of Technology Kharagpur, India
Anshuman Sharma, TCS, India
Apurbba Kumar Sharma, Indian Institute of Technology Roorkee, India
Arlindo Silva, Singapore University of Technology and Design, Singapore
Inderdeep Singh, Indian Institute of Technology Roorkee, India
Sumer Singh, Indian Institute of Technology Delhi, India
Vishal Singh, Indian Institute of Science, Bengaluru, India
Ricardo Sosa, Auckland University of Technology, New Zealand
Mario Storga, University of Zagreb, Croatia
Eswaran Subrahmanian, Carnegie Mellon University, USA
Shankar Subramanian, Indian Institute of Technology Madras, India
Abinash Swain, Indian Institute of Technology Roorkee, India
Puneet Tandon, Indian Institute of Information Technology, Design and Manufacturing, Jabalpur, India
Asokan Thondiyath, Indian Institute of Technology Madras, India
Anshuman Tripathy, Indian Institute of Management Bengaluru, India
Yasushi Umeda, University of Tokyo, Japan
Paul Varghese, Trysquare Engineers and Builders, India
Nilesh Vasa, Indian Institute of Technology Madras, India
Gokula Vasantha, Edinburgh Napier University, UK
Srikanth Vedantam, Indian Institute of Technology Madras, India
Srinivasan Venkataraman, Indian Institute of Technology Delhi, India
Alok Verma, Old Dominion University, USA
Nikola Vukasinovic, University of Ljubljana, Slovenia
Fedja Vukic, University of Zagreb, Croatia
Parag Vyas, GrauBär Design Studios, India
Robert Wendrich, University of Twente, Netherlands
Andrew Wodehouse, University of Strathclyde, UK
Pradeep G. Yammiyavar, Indian Institute of Technology Guwahati, India
Yong Zeng, Concordia University, Canada

Organising Committee

Shakuntala Acharya, Indian Institute of Technology Guwahati, India
Rupa Agarwal, National Institute of Fashion Technology, India
Prasad Bokil, Indian Institute of Technology Bombay, India
Vivek Kant, Indian Institute of Technology Bombay, India
Ajanta Sen, Indian Institute of Technology Bombay, India
Avinash Shende, Indian Institute of Technology Bombay, India

Web Admin

Roopa Narayan Sahoo, IDC, Indian Institute of Technology Bombay, India

Student Organising Committee

Abdul Mohsin Ali, Indian Institute of Technology Guwahati, India
Shivalika Bansal, Indian Institute of Technology Bombay, India
Apoorv Naresh Bhatt, Indian Institute of Science, Bengaluru, India
Kaushik Bhattacharya, Indian Institute of Science, Bengaluru, India
Debajit Brahma, Indian Institute of Technology Bombay, India
Spandan Das, Indian Institute of Technology Guwahati, India
Deepshikha Dash, Indian Institute of Technology Bombay, India
Arshiya Gahlot, Indian Institute of Technology Bombay, India
Felnunmoi Gangte, Indian Institute of Technology Bombay, India
Anubhav Gangwar, Indian Institute of Technology Bombay, India
Divya Gaur, Indian Institute of Technology Bombay, India
Kiran Ghadge, Indian Institute of Science, Bengaluru, India
Bhavana Ghanta, Indian Institute of Technology Bombay, India
Puneeth K. S., Indian Institute of Science, Bengaluru, India
Anirvina Karnala, Indian Institute of Technology Bombay, India
Ishaan Kaushal, Indian Institute of Science, Bengaluru, India
Shreya Kondvilkar, Indian Institute of Technology Bombay, India
John Kujur, Indian Institute of Technology Bombay, India
Anurag Kumar Singh, Indian Institute of Technology Bombay, India
Siddharth Kumar, Indian Institute of Technology Guwahati, India
Anubhab Majumder, Indian Institute of Science, Bengaluru, India
Saikat Mondal, Indian Institute of Technology Bombay, India
Deblina Pandit, Indian Institute of Technology Bombay, India
Arivukkarasu Periyasamy, Indian Institute of Technology Bombay, India
Stuti Prakashkumar, Indian Institute of Technology Bombay, India
Subin Raj, Indian Institute of Science, Bengaluru, India
Priyanka Rawal, Indian Institute of Technology Bombay, India
S. Madhuri, Indian Institute of Technology Bombay, India
Nishath Salma, Indian Institute of Science, Bengaluru, India
Pratiti Sarkar, Indian Institute of Technology Bombay, India
Rounak Sengupta, Indian Institute of Technology Bombay, India
Namrita Sharma, Indian Institute of Technology Guwahati, India
Shivangi, Indian Institute of Technology Bombay, India
Sanjay Singh, Indian Institute of Science, Bengaluru, India
Paridhi Subbaian Kaliamoorthy, Indian Institute of Science, Bengaluru, India
JayaSurya Sudakaran, Indian Institute of Technology Bombay, India
Charu Tripathi, Indian Institute of Science, Bengaluru, India

Sponsors

Centre for Product Design and Manufacturing, Indian Institute of Science, Bengaluru, India
Indian Institute of Science, Bengaluru, India
Springer

Endorsers

The Design Society, UK
Design Research Society, UK

About the Conference

Design is ubiquitous; it pervades all spheres of life and has been around as long as life has taken up the task of purposefully changing the world around it. Research into design and the emergence of a research community in this area have been relatively new. Its development has been influenced by the multiple facets of design (human, artefact, process, organisation, the micro- and macro-economy and the ecology by which design is shaped) and the associated diversification of the community depending on the facets of focus or that of their applications. Design is complex, balancing the needs of multiple stakeholders and requiring a multitude of areas of knowledge to be utilised, with resources spread across space and time.

ICoRD'21 is the eighth in a series of conferences intended to be held every two years in India to bring together the international community from diverse areas of design practice, education and research. It aims to showcase cutting-edge research about design to the stakeholders; aid the ongoing process of developing and extending the collective vision through emerging research challenges and questions; and provide a platform for interaction, collaboration and development of the community in order for it to take up the challenges to realise the vision. The conference is intended for all stakeholders of design and, in particular, for its practitioners, researchers, pupils and educators.

The collection of papers in these two book volumes constitutes the Proceedings of the Eighth International Conference on Research into Design (ICoRD'21) held on virtual platform during 7 to 10 January 2021 at the IDC School of Design, IIT Bombay, Powai, Mumbai.

ICoRD series was initiated by Centre for Product Design and Manufacturing (CPDM) at Indian Institute of Science (IISc), Bengaluru, in 2006. Since then, it has been hosted in 2009 and 2011 (both at IISc), 2013 (at IIT Madras), 2015 (IISc), 2017 (IIT Guwahati) and 2019 (IISc). CPDM has pioneered design research in India for the last two decades. IISc is one of India's leading science and technology institutions and is one of the Institutes of Eminence decreed by MHRD, Government of India.

ICoRD'21 has been organised jointly by IDC, IIT Bombay, and CPDM, IISc Bengaluru. ICoRD'21 has been hosted on virtual platform by IDC School of Design, Indian Institute of Technology Bombay, Mumbai, India. IDC School of Design has overseen the inception of design in a free India for the past 50 years. IIT Bombay is one of India's leading technological institutions and is decreed by the Government of India as an Institute of National Importance. As a city, Mumbai serves as a home to a variety of technological and service sectors.

The theme of ICoRD'21 has been 'Design for Tomorrow'. The world as we know it in our times is increasingly becoming connected. In this interconnected world, design has to address new challenges of merging the cyber and the physical, the smart and the mundane, and the technology and the human. As a result, there is an increasing need for strategising and thinking about design for a better tomorrow. Our theme for ICoRD'21 serves as a provocation for the design community to think about rapid changes in the near future to usher in a better tomorrow.

The conference contained:

- Keynote presentations from eminent (inter)national experts and practitioners
- Presentations of refereed papers as podium presentations
- Panel discussions to present perspectives on topics of general interest
- A series of workshops on topics of special interest
- Networking sessions for young researchers.

Preface

Design is ubiquitous; it pervades all spheres of life and has been around ever since life has been engaged in purposefully changing the world around it. While some designs have transcended time, most designs are in a perpetual process of being evolved. Research into design and the emergence of a research community in this area have been relatively new. Its development has been influenced by the multiple facets of design (human, artefact, process, organisation, ecology, micro- and macro-economy by which design is shaped and which it shapes in turn) and the associated diversification of the community depending on the facets of focus or that of their applications. Design is complex, balancing the needs of multiple stakeholders and requiring a multitude of areas of knowledge to be utilised, with resources spread across space and time.

The collection of papers in these two book volumes constitutes the Proceedings of the Eighth International Conference on Research into Design (ICoRD'21) held at Indian Institute of Technology Bombay, India (this time on virtual platform), during 7–10 January 2021. ICoRD'21 is the eight in a series of biennial conferences held in India to bring together the international community from diverse areas of design practice, teaching and research. The goals are to share cutting-edge research about design among its stakeholders; aid the ongoing process of developing a collective vision through emerging research challenges and questions; and provide a platform for interaction, collaboration and development of the community in order for it to address the global and local challenges by forming and realising the collective vision. The conference is intended for all stakeholders of design and, in particular, for its practitioners, researchers, teachers and students.

Five hundred and fifty-four abstracts were submitted to ICoRD'21, from which 529 were accepted for full paper submission. A total of 295 full papers were submitted, which were reviewed by experts from the ICoRD'21 International Programme Committee comprising 238 members from over 149 institutions or organisations from 27 countries spanning five continents. Finally, 234 full papers, authored by 460 researchers (460 unique authors, actually 606 authors' entries in 236 papers) from 21 countries spanning five continents, were selected for presentation at the conference and for publication as chapters in this book. ICoRD has steadily grown over the last seven editions, from a humble beginning in 2006 with 30 papers and

60 participants, through 75 papers and 100 participants in ICoRD'09, 100 papers and 150 participants in ICoRD'11, 114 papers and 170 participants in ICoRD'13, 118 papers and 200 participants in ICoRD'15, 178 papers and 230 participants in ICoRD'17 and 169 papers and 352 participants in ICoRD'19.

All papers were presented in ICoRD'21 in the podium mode. It had keynotes from prominent researchers and practitioners from around the world such as Steve Eppinger, MIT, USA; Punya Mishra, Arizona State University, USA; Armand Hatchuel, Mines ParisTech, France; Tetsuo Tomiyama, International Professional University, Tokyo, Japan; Paul Hekkert, TU Delft, The Netherlands; Tomas Ramos, Nova University, Lisbon, Portugal; Dibakar Sen, Indian Institute of Science, Bengaluru, India, etc.

The chapters in the three book volumes together cover all three major areas of products and processes: functionality, form and human factors. The spectrum of topics ranges from those focusing on early stages such as creativity and synthesis, through those that are primarily considered in later stages of the product life cycle, such as safety, reliability or manufacturability, to those that are relevant across the whole product life cycle, such as collaboration, communication, design management, knowledge management, cost, environment and product life cycle management. Issues of delivery of research into design, in terms of its two major arms: design education and practice, are both highlighted in the chapters of the book volumes. Foundational topics such as the nature of design theory and research methodology are also major areas of focus. It is particularly encouraging to see in the chapters the variety of areas of application of research into design—aerospace, health care, automotive, biomedical and so on.

The theme of ICoRD'21 is 'Design for Tomorrow'. The world as we know it in our times is increasingly becoming connected. In this interconnected world, design has to address new challenges of merging the cyber and the physical, the smart and the mundane, and the technology and the human. As a result, there is an increasing need for strategising and thinking about design for a better tomorrow. Our theme for ICoRD'21 serves as a provocation for the design community to think about rapid changes in the near future to usher in a better tomorrow. ICoRD'21 is organised jointly by IDC, IIT Bombay, and CPDM, IISc Bengaluru. IDC School of Design has overseen the inception of design in a free India for the past 50 years. IIT Bombay is one of India's leading technological institutions and is decreed by the Government of India as an Institute of National Importance. As a city, Mumbai serves as a home to a variety of technological and service sectors.

On behalf of the Steering Committee, Advisory Committee, Organising Committees and Co-Chairs, we thank all the authors, delegates, institutions and organisations that participated in the conference. We also thank the members of the International Programme Committee for their support in reviewing the papers for ICoRD'21, which is essential for maintaining the quality of the conference, and for their support in putting this book together.

We are thankful to the Design Society and Design Research Society for their kind endorsement of ICoRD'21. We thank Indian Institute of Technology (IIT) Bombay and its IDC School of Design, and Indian Institute of Science (IISc), Bengaluru, and its Centre for Product Design and Manufacturing (CPDM), for their support of this event. We also wish to place on record and acknowledge the enormous support provided by the Student Organising Committee in managing the review process, in preparation of the conference programme and this book and the conference as a whole. In particular, we wish to thank Mr. Apoorv Bhatt, Mr. Roopa Narayan Sahoo and Ms. Nishath Salma for their valuable contributions. We also thank the large and dedicated group of student volunteers of IIT Bombay, IISc Bengaluru, IIT Guwahati and other institutions in the organisation of the conference. Finally, we thank Springer, especially its Editor Ms. Swati Meherishi and its editorial support team, for their wonderful support, including their sponsoring of Springer book vouchers for the Winners of Distinguished Paper Awards at ICoRD'21.

Bengaluru, India — Amaresh Chakrabarti
Mumbai, India — Ravi Poovaiah
Mumbai, India — Prasad Bokil
Mumbai, India — Vivek Kant

Contents

About the Editors

Prof. Amaresh Chakrabarti is a Senior Professor and current Chairman for the Centre for Product Design and Manufacturing, Indian Institute of Science (IISc), Bengaluru. He has B.E. in Mechanical Engineering from Indian Institute of Engineering Science and Technology, Shibpur, M.E. in Design of Mechanical Systems from Indian Institute of Science, Bengaluru, and Ph.D. in Engineering Design from the University of Cambridge, UK. He led the Design Synthesis group at the EPSRC funded Engineering Design Centre at the University of Cambridge before joining IISc as an Associate Professor in 2001. His interests are in synthesis, creativity, sustainability, and informatics. He has published 16 books, over 300 peer-reviewed articles, and has 11 patents granted/pending. He has been on the Advisory Board and the Board of Management of the Design Society, UK; member of the CII National Committee on Design India; member of the Jury of India Design Mark; and member of the CII Smart Manufacturing Council India. He founded IDeASLab—India's first Design Observatory. He is the Founding Programme chair for International Conference Series on Research into Design (ICoRD), Conference Chair for the 22nd CIRP Design Conference 2012 and the 3rd International Conference on Design Creativity 2015 (3rd ICDC) and vice-Chair for AI in Design and Design Computing and Cognition Conferences. In 2014, he co-initiated India's first Smart Factory R&D platform.

Prof. Ravi Poovaiah is senior faculty member at the Industrial Design Centre (IDC), Indian Institute of Technology (IIT) Bombay. He has backgrounds in Mechanical Engineering, Product Design and Graphic Arts Education having studied respectively at IIT Madras and Bombay and at the Rhode Island School of Design (RISD), Providence, USA. His current pedagogic as well as research and design interests are in fields related to Interaction Design, New Media Design, Visual Design and Product Design and his research interests are in areas related to Visual Language, Information Visualisation, Visual Narratives, Way Finding Systems, Interaction Devices, Collaborative Social and Learning Environments and Designing for Children. He is involved with building digital resources related to 'Design Learning', 'Folk Tales', 'Interactive Systems', 'Designing for Children', 'Design of Way-finding Systems' and 'Design in India' with free and open access to networked information. He is also the co-director of a research project named COSMIC on experimenting with Social Media at IIT Bombay along with collaborative Institutes NTU and NUS in Singapore. Prof. Poovaiah has worked on projects with major industry leaders such as Microsoft, Yahoo, Google India, Motorola India, Indian Oil Corporation, Bharat Petroleum Corporation, Bharat Electronics Limited, Siemens India, among others.

Dr. Prasad Bokil is a mechanical engineer and communication designer with Master's and Ph.D. in communication design from IDC School of Design, Indian Institute of Technology (IIT) Bombay, India. He teaches courses in typography and graphic design at undergraduate as well as postgraduate levels. His research interests are visual language, visual narratives and game design. He is currently working towards 'design for wellbeing'. Dr. Bokil also worked as a Research Assistant at IDC, IIT Bombay, where he worked with a team of artisans at studio under guidance of Prof. A. G. Rao to enhance the packaging of some of the products. He is involved in several funded projects, including the Kanimuni project funded by MHRD under the Design Innovation Centre at IIT Guwahati. This project aims to develop educational games to build interest and confidence to learn English, Mathematics, Science and 21st century skills in school students. Dr Bokil has (co-)organised many conferences and workshops, including Typography Day 2013, Himalayan Languages Symposium 2016, ICoRD'17 in 2017, Type Design Workshop 2013, Game Design Workshop 2015, Design Research Workshop 2015, Workshop on Software product development process in Agile environment March 2015, Informatica 2016, Kanimuni: DIC Game Design Workshop 2016, etc.

Dr. Vivek Kant is an Assistant Professor at IDC School of Design, Indian Institute of Technology (IIT) Bombay, India. He is an active researcher in the field of systems design, human factors (cognitive ergonomics/cognitive engineering), sociotechncial systems, and history and philosophy of design engineering. Dr. Kant pursued his B.S. in Electrical Engineering from Dayalbagh Educational Institute, India in 2004, M.S. in Cognitive Science from University of Allahabad, India in 2008, M.A. in Psychology from University of Connecticut, USA in 2011, and Ph.D. in Systems Design Engineering from University of Waterloo, Canada in 2015. He was a Postdoctoral Fellow at the Institute of Catastrophe Risk Management, Nanyang Technological University (NTU), Singapore, and the Division of Sociology, School of Humanities and Social Sciences, NTU along with Future Resilient Systems, Singapore-ETH Centre. He has also worked as an Assistant Systems Engineer for Tata Consultancy Services, India.

Part I
Design of Systems, Services and Product Service Systems

Chapter 1
The Future of Home Service: Integration of User Behavior and Scenario Planning in the Domestic Plumbing Service Design

Sachin Shivaji Jadhav and Pratul Chandra Kalita

Abstract In the present era, with rapidly changing measures and uncertainty, it is beneficial for researchers, organizations, as well as for the government to anticipate future events. The substantial economic boost in the on-demand service sector leads to the development of home services in India. As increasingly products and services sold over the Internet, it becomes all the more essential to build up knowledge of design in planning and anticipating future events. This paper proposes user behavior analysis and plausible futuristic scenarios with the practical case of domestic plumbing services. The research aims to identify the interrelationships and significance of service aspects in domestic plumbing. The study also aims to understand the future in this context through scenario development. We conducted in-depth interviews, including exploratory surveys. Structured questionnaires were developed to study various aspects of stakeholders of the system viz. plumbing service provider, plumber, and users. Further, insights from the survey result were incorporated into scenario planning. Scenario planning was used to discover new strategic options for the future and to gain a deeper level of foresight. This study provides an integrative approach of user behavior study and scenario planning for service design with special emphasis on domestic plumbing. It acts as a road map for service design in general, primarily to similar service sectors viz. electrical maintenance, domestic appliances maintenance, and carpentry. The study will benefit students and practitioners in the field of service design. It would finally contribute to design for development.

1.1 Introduction

Service design, as a design practice and research area, has developed into a design-led approach to service innovation [1]. Service innovation is more about improved

S. S. Jadhav (✉) · P. C. Kalita
Department of Design, Indian Institute of Technology, Guwahati, Guwahati, Assam 781039, India
e-mail: sjsachin@iitg.ac.in

P. C. Kalita
e-mail: pratulkalita@iitg.ac.in

A. Chakrabarti et al. (eds.), *Design for Tomorrow–Volume 2*, Smart Innovation, Systems and Technologies 222, https://doi.org/10.1007/978-981-16-0119-4_1

customer experience, processes, and actions [2, 3]. Also, it involves human actors, physical resources/technologies, and processes [4, 5]. Within design communities, service design conceptualized as design-centered contributions to service innovation based on a human-centered perspective and creative methods [6, 7]. Also, service design was influenced by emotional design, design thinking, and contextual design. Researchers [6] argues that design impacts on product innovation are generally related to the attributes of physical objects, while design impacts on service innovation require different dimensions. Service design has also been examined as a set of collaborative and cross-disciplinary activities for service innovation [8].

Technological advancements and a competitive business environment have increased the growth of on-demand services [9]. A robust economic boost in the on-demand service sector and growth of startups have a smooth the way for the development of home services in India. Significant sectors of home services include installation and repair, cleaning, health and wellness, shifting of furniture, events organization, and business services. In the present era, with rapidly changing measures and uncertainty, it is beneficial for researchers, organizations, as well as for the government to anticipate future events. The amount of trade conducted electronically has grown extraordinarily since the spread of the Internet [10, 11]. More recently, forecast by the Indian Brand Equity Foundation [12] mentioned that Internet users in India are expected to increase from 445.96 million in 2017 to 829 million by 2021. There is a lack of studies on user behavior and scenarios in the context of home services. Therefore, it could be a valuable addition to the design field and business organizers for developing and anticipating plausible scenarios in the home service sector.

1.1.1 Purpose of the Article

As increasingly products and services sold over the Internet, it becomes all the more essential to build up knowledge of design in planning and anticipating future events. The research aims to identify the interrelationships and significance of service aspects in domestic plumbing. The study also aims to understand the future in this context through scenario development.

1.2 Methodology: Integrating User Behavior and Scenario Planning

In the literature, researchers presented well build-upon and established research methods on user behavior and scenario planning. This research paper integrates existing research methods to conduct user behavior analysis and develop plausible futuristic scenarios with an emphasis on design innovation in domestic plumbing services. We conducted in-depth interviews, including exploratory surveys. The purpose of the survey was to study the plumbing tools and service aspects in domestic plumbing. Structured questionnaires were developed to study various aspects of stakeholders of the system viz. plumbing service provider, plumber, and users. During the survey, we critically observed the possible requirement and relationships of plumbing tools, service providers, and users. Further, insights from the survey result were incorporated into scenario planning. Scenario planning started with identifying key focal issues derived from user behavior study. From the secondary data, the driving forces of change and trends were identified. Further, analysis of each trend was carried out to understand the impact and uncertainty in domestic plumbing. These trends were mapped on impact vs. uncertainty with the scale low to high. Finally, two scenarios were generated considering the current processes of services, communication between stakeholders, technological aspects, and demographic locations.

1.3 Theoretical Background—Foundational Concepts in User Behavior and Scenario Planning

1.3.1 User Behavior

Customer behavior data is required to enhance the design activities, such as target customer identification, idea generation, and information content generation [13]. Today, the trend of increased mobile networking allows marketers to understand and interact with customer behavior [14]. Moreover, this trend changed the role of customers from isolated to connected, from unaware to informed, and from passive to active [15]. Customer behavior includes several factors viz. gender, age, occupation, living environment, and cultural background [16]. These factors determine the uncertainty of customer behavior. Therefore, it is beneficial to understand customer behavior to deliver and fulfill needs. Studies on customer behavior use quantitative and qualitative methods such as surveys, in-depth interviews, and focus groups.

1.3.2 Scenario Planning

With rapidly changing measures and uncertainty, it is beneficial for decision-makers and designers to be able to adapt and anticipate future events. The scenario approach is one of the tools that support such decision making. Scenario planning viewed as an alternative to forecasting [17, 18], where scenario planning focuses on unexpected but plausible outcomes that represent a break from the past. However, forecasting emphasis on continuing trends, most likely pathways to estimate uncertainties, as in the recent past. The journals 'Futures' and 'Long Range Planning' publish a wide range of academic and practitioner articles on the application of scenario planning toward the problem-solving area of management research [19]. The roots of scenario planning are in military strategy studies [20]. In the 1960s, Herman Kahn was an early founder of scenarios, and he promoted the idea of 'thinking the unthinkable' and used scenarios as a tool for business predictions. The scenarios are beneficial for strategy development, innovation, risk management, visioning, executive learning, and used in both new business and old business [21]. Scenarios usually serve one of two functions: one is risk management, where scenarios enable strategies and decisions to be tested against possible futures, while the other is creativity and sparking new ideas [22].

1.3.3 Integrating User Behavior and Scenario Planning Approaches

From literature, studies have shown that utilizing user behavior data could improve the design activities for service organizations [23, 24]. User behavior supports identifying demographic data. Also, enhance the customers' experience, contexts, requirements, and customization [16, 23]. By understanding and analyzing these data, service organizations can transform insightful information for strategic market planning [25]. Ordenes et al. [23] indicated that customer feedback could be utilized to gain information on customers' experience. Lim et al. [24] used customer data to generate information contents and design service concepts.

Scenario planning is a strategic planning method to anticipate uncertainties in business environments [17, 21]. The objective of scenario planning is to create plausible future events. Lena Bo et al. [26] indicated that the scenario development process consists of generating ideas and gathering data, integrating, and checking the scenario's consistency. The standard approach to scenario planning is Intuitive logics, and the development process includes eight stages [17]. These eight stages are viz. setting the agenda, determining the driving forces, clustering the driving forces, defining the cluster outcomes, impact/uncertainty matrix, framing the scenarios, scoping the scenarios, and developing the scenarios. There are different approaches to create scenarios [19, 22. However, they all have a common starting point, i.e., identifying the key focal issue, processes, time scale, and driving forces of change and trends. User behavior can provide essential information on customers' experience, requirements, contexts, and demographic data to define focal issues and processes in this context.

1.4 Results and Discussions

1.4.1 User Behavior on Domestic Plumbing Services

Structured questionnaires were administered to a sample size of 160 respondents. The data for this study was collected through the survey, which was a paper-based questionnaire. The rationale for using a survey is primarily because these data did not currently exist. Also, a better source for insights into plumbing service aspects is from respondents themselves. Plumbing service aspects were categorized into the groups viz. corrective maintenance, preventive maintenance, the time required to fix plumbing issues, servicemen responsiveness, and mode of contact. Figure 1.1 illustrates the demographic data of domestic plumbing services.

The chi-square independence test is a procedure for testing if two categorical variables are related in some populations. Thus, chi-square tests were conducted for a significance level of 0.05 to observe the variance between the frequency of corrective maintenance, preventive maintenance, the time required to fix plumbing issues, the responsiveness of servicemen, and the mode of contact made with variables viz. residential typology, locality, occupation, age group, and gender. Table 1.1 shows the sig. of the Pearson chi-square test. The study shows that there is a significant difference in the frequency of preventive maintenance with residential typology (sig. = 0.001); servicemen responsiveness with residential typology (sig = 0.019); time required to fix plumbing issues with occupation (sig. = 0.044) and age group (sig. = 0.003).

Cross-tabulation for the frequency of preventive maintenance and residential typology indicates that 47.3% of respondents with the detached house have never

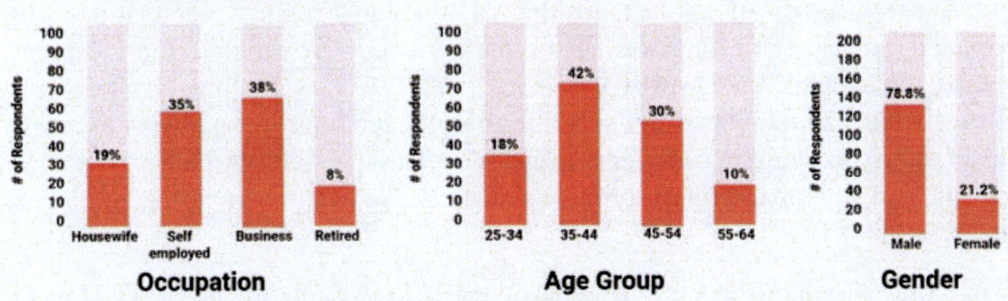

Fig. 1.1 Demographic data of domestic plumbing services

Table 1.1 Significance of Pearson chi-square test

Variables	Residential typology	Locality	Occupation	Age group	Gender
Corrective maintenance	0.461	0.245	0.589	0.634	0.681
Preventive maintenance	0.001	0.297	0.924	0.073	0.358
Time required to fix plumbing issues	0.514	0.086	0.044	0.003	0.190
Servicemen responsiveness	0.019	0.101	0.637	0.741	0.395
Mode of contact	0.120	0.079	0.659	0.642	0.514

called for preventive maintenance. Also in attached house (26.4%), apartment blocks (20.0%), and residential buildings (6.4%), we see a decline in percentage. 36.0% of respondents with apartment blocks have called 1 or 2 times for preventive maintenance in a year. 32.0% of respondents with residential buildings have called more than three times for preventive maintenance in a year. Cross-tabulation for the frequency of servicemen responding to customer plumbing issues and residential typology shows that 37.5% of respondents with the attached house disagree with servicemen responding to customer plumbing issues, also similar responses found in the detached house (21.9%), apartment blocks (28.1%), and residential buildings (12.5%). Cross-tabulation for the frequency of time required to fix plumbing issue and occupation results that 48.1% of respondents with business occupation have said the time taken to fix plumbing issues takes 4–8 h. 38.8% of respondents with self-employed occupation have said the time taken to fix plumbing issues takes more than 24 h. Cross-tabulation for the frequency of time required to fix plumbing issue and age group results that 59.2% of respondents with 35–44 age group have said the time taken to fix plumbing issues takes more than 24 h. 54.5% of respondents with 45–54 age group have said that the time taken to fix plumbing issues takes 4–8 h.

One-way-ANOVA test was conducted for a significance level of 0.05 to observe the variance of service frequency of dripping faucets issue, water pressure issue, running toilet issue, slow/ clogged drain and leaky pipe issue with corrective maintenance, preventive maintenance, the time required to fix plumbing issues, servicemen responsiveness, mode of contact, residential typology, locality, occupation, age group, and gender. The ratings of the service frequency are measured in a 7-point semantic differential scale where $1 = $ never and $7 = $ very frequently. Statistical analyses helped us to identify and describe complex relationships between the variables of domestic plumbing services. Table 1.2 shows the ANOVA means, standard deviation, and significance of descriptive statistics. The analysis of variance showed that

- The effect of occupation on dripping faucet issue was significant, $F(3,156) = 4.375$, $p = 0.005$.

Table 1.2 Descriptive statistics of one-way ANOVA

Dependent variable	Independent variable	Mean	SD	Sig
Dripping faucet issue	Occupation	4.14	1.635	0.005
Dripping faucet issue	Residential typology	4.14	1.635	0.039
Running toilets issue	Locality	3.26	1.771	0.036
Leaked pipes issue	Residential typology	2.83	1.855	0.004

- The effect of resident typology on dripping faucet issue was significant, F (3,156) $= 2.859$, $p = 0.039$.
- The effect of locality on running toilet issue was significant, F (2,157) $= 3.402$, $p = 0.036$.
- The effect of resident typology on leaky pipes issue was significant, F (3,156) $= 4.160$, $p = 0.004$.

Further, multiple comparisons of post-hoc tests were conducted to observe the significant differences within the groups of respondents. The Tukey HSD test reflects that

- There is a statistically significant difference in dripping faucet issues between the groups of occupation, i.e., the housewife and the retired ($p = 0.003$), the self-employed, and the retired ($p = 0.013$), and the business and the retired ($p = 0.013$).
- There is a statistically significant difference in dripping faucet issues between the groups of residential typology, i.e., the detached house and the apartment blocks ($p = 0.021$). However, there were no differences between the residential buildings and the detached house ($p = 0.709$), residential buildings and the attached house ($p = 0.998$), and residential buildings and the apartment blocks ($p = 0.645$)
- There is a statistically significant difference in running toilet issues between the groups of locality, i.e., the very old and the new ($p = 0.028$). However, there were no differences between old and very old ($p = 0.657$), as well as old and the new ($p = 0.452$).
- There is a statistically significant difference in leaky pipe issues between the groups of residential typology, i.e., the residential buildings and the detached house ($p = 0.002$), residential buildings, and the attached house ($p = 0.014$). However, there were no differences between the apartment blocks and the detached house ($p = 0.646$), apartment blocks and the attached house ($p = 0.922$), and apartment blocks and the residential buildings ($p = 0.058$)

1.4.2 *Scenario Planning on Domestic Plumbing Services*

The scenario planning started with a key focal issue of services toward domestic plumbing. Likely, issues of services such as preventive maintenance, corrective maintenance, plumber's responsiveness to the customer, and strategic management to the service provider. Plausible scenarios were created for a time frame of 5–10 years. Scenario planning accounts for the effect of multiple drivers of change, trends, and delivers significant possibilities, risks, and opportunities. In this study, we identified the driving forces of change and trends for domestic plumbing from four perspectives viz. political, economic, social, and technological. The political drivers of change are Skill India, Digital India, Pradhan Mantri Kaushal Vikas Yojana, and National Skill Development Corporation. The economic drivers of change are gig economy, family income, demand, and supply. The social drivers of change are buying habits, lifestyle, education, and online platforms. The technological drivers of change are video in real-time, mixed reality, and drone delivery. The aforementioned driving forces of change and trends were collected from government reports, research articles, and market surveys. The impact vs. uncertainty matrix was conducted for each identified driving force of change. Figure 1.2. depicts the impact vs. uncertainty mapping for the plumbing service system. A brief discussion on driving forces of change and trends are as follows.

Political. The government plays a role as a regulator and influence the adoption process of business plans for organizations. Also, promote initiatives and schemes to suit the needs of the industry and enable a decent quality of life to its population. For instance, the government had approved the skill certification scheme, Pradhan Mantri Kaushal Vikas Yojana. The objective of skill certification scheme was to earn a livelihood, and employable for Indian youth based on skill training. Also, the report

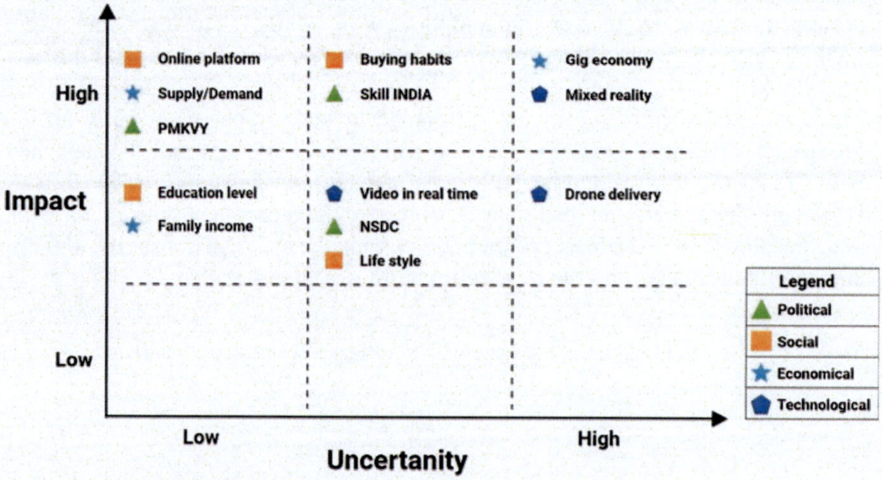

Fig. 1.2 Impact versus Uncertainty mapping of domestic plumbing services

indicates that there is an incremental human resource requirement of 103.4 million during 2017–2022.

Economic. Standard skills and knowledge are the driving forces of economic growth for any country. To drive sustainable economic growth and generate vast scale employment opportunities, the government initiated 'Startup India' to nurture innovations and startups. Also, the gig economy in the country attracts the ability to work from anywhere at flexible hours.

Social. Nowadays, social media are emerging and new ways of marketing and increasing online business. The impact of customer reviews and feedback directly affects the way of doing online business. Consumers have a much wider choice of products/services available in the cyber market. Consumers can compare products, features, prices, and even look up reviews before they select what they want. Consumers enjoy more extensive access to assistance and advice from experts and peers. Consumers also avail of fast services and delivery of products/services.

Technological. Technology may transform jobs of the future and lean more toward tech-enabled. Digital services—from E-learning to online payments, E-health to broadcasting preventive measures, news, online video streaming are all seeing a boom. There could be a rise in technological advancement and consumers' adoption of technologies as they embrace technology while working from home. The aid of drones was beneficial during the Covid-19 crisis—for example, medical deliveries, spraying disinfectants, surveillance, and monitoring public places.

Two scenarios were created for the plumbing service system as follows,

Scenario 1: With Indian cities witnessing significant growth, plumbing services providing online will expand sufficiently to tier II and tier III cities. Figure 1.3 illustrates the existing online plumbing services in urban areas and why to choose on-demand home services. In this scenario, the extent to which online ordering for

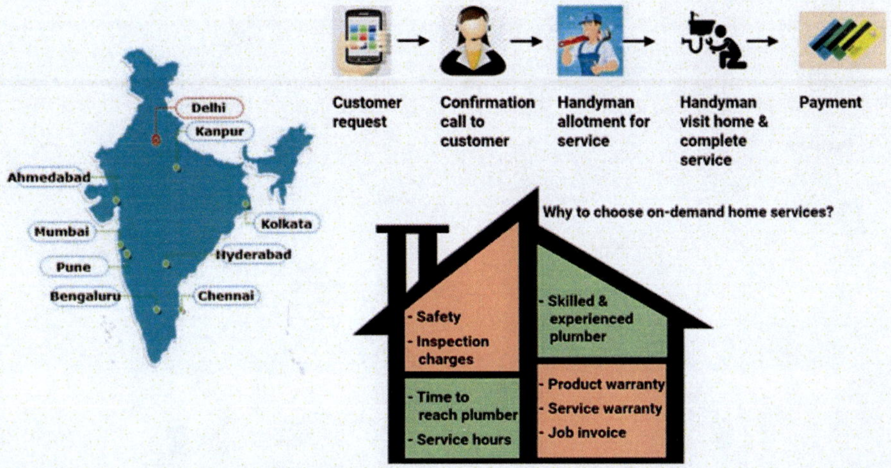

Fig. 1.3 Scenario of plumbing service system

plumber service is limited to urban cities. The number of plumber service requests could be high, but most consumers request plumber service through a phone call, neighbors' recommendation, and walk-in. The Indian plumbing association initiated the process of professionalization. However, plumbing apprentices work as helpers in urban cities with a practicing plumber [27]. These apprentices learn the plumbing skills as they worked on practical jobs. As such, plumbers concentrated more on practical approach and lack of training, plumbing codes, and certification programs. Therefore, the service provided by the plumbers were subjective and challenging to regulate. The strategic implications for service providers and consumers in this scenario are challenging, particularly in terms of identifying certified plumbers. The strategic actions for on-demand home services need to ensure.

- Encourage the apprentices to take up certification programs to improve labor status and income through professionalization.
- Uniform guidelines on health and safety measures. Policies on monetary benefits and job security for plumbers.
- Refocus on business models and areas of potential growth in the future.
- The advantage of delivering first of its kind in services to attract more customers.
- Invest in research and development for new products and services.
- The designer focused online and Web site platforms for better user navigation, user interactions, secure payment methods, and transparency in pricing details.

Scenario 2: Potential customers are likely to solve minor plumbing issues with the aid of specific instruction through online streaming. Figure 1.4 depicts the scenario of the plumbing service system. In this scenario, the stakeholders involved are viz. customers, service providers, delivery agents, and plumber. Customers can access the website for resolving minor plumbing issues with proper instruction and specific

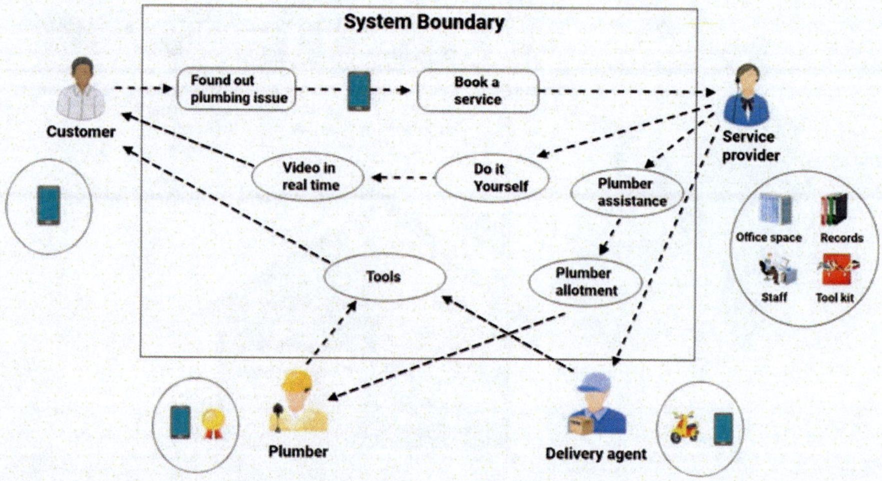

Fig. 1.4 Scenario of plumbing service system

images to particular plumbing issues. This design solves the problem of long waiting times for the arrival of plumbers. From the survey data, minor plumbing issues include dripping faucets/taps, low water pressure, clogged drain/toilets, and leaked pipes. Also, customers can compare and purchase the tools/spare parts for a particular plumbing issue through the information obtained from the Web site. This design solves the problem of transparency in the cost of tools/spare parts. Further, customers can make direct contact with the plumbers through the internet for major issues of on-demand services. This design solves the search and availability of plumbers in the market. Also, the support process such as easy access to the Web site, inventory of products, and availability of plumbers must be added to the solution, so that the quality of the whole operational process, including onstage and backstage process, can be enhanced. Service providers could have opportunities to attract new customer segments with specific content that delivers value. The strategic implication for business organizers in this scenario would.

- To consider factors such as mobility, privacy risk, and assessment of the service provider, which may affect the adoption of home services.
- Managing trade-offs and balances between customers and plumbers in need of urgency.
- Design management strategies in service quality, easy access, availability, and customization in service processes.
- To develop a new business model with an emphasis on design thinking approaches.

1.5 Conclusion

This study proposes an integrative approach of user behavior and scenario planning for service design with special emphasis on domestic plumbing. The study revealed aspects of customer service requirements and related components viz. corrective maintenance, preventive maintenance, operation time, service frequency, replacement of spare parts, consumables, and fittings.

For the case study, user behavior analyses reveal complex relationships between the variables of domestic plumbing services. In particular, chi-square test results show a significant difference of frequency of preventive maintenance with residential typology (sig. = 0.001); servicemen responsiveness with residential typology (sig = 0.019); time required to fix plumbing issues with occupation (sig. = 0.044) and age group (sig. = 0.003). Also, from the One-way ANOVA test, it has been observed that the frequency of dripping faucets issues significantly varies with occupation (sig = 0.005) and residential typology (sig = 0.039); Running toilet issue significantly vary with locality (sig = 0.036), and leaky pipes issue with residential typology (sig = 0.004). Scenario planning accounts for the effect of multiple drivers of change, trends, and delivers significant possibilities, risks, and opportunities. In this study, we identified the driving forces of change and trends for domestic plumbing from four perspectives viz. political, economic, social, and technological. The political drivers of change are Skill India, Digital India, Pradhan Mantri Kaushal Vikas Yojana, and

National Skill Development Corporation. The economic drivers of change are gig economy, family income, demand, and supply. The social drivers of change are buying habits, lifestyle, education, and online platforms. The technological drivers of change are video in real-time, mixed reality, and drone delivery.

The research contributes to the early stages of development for service design. It acts as a road map for service design in general, primarily to similar service sectors viz. electrical maintenance, domestic appliances maintenance, carpentry, etc. The study will benefit students and practitioners in the field of service design. It would finally contribute a design for the development of domestic plumbing services. Among other practical home service cases, this study has limited consideration of only a domestic plumbing service. Future research could be conducted with the practical instances of electrician, carpentry, repair, and installations of home service. Also, research may be undertaken in developing and measuring the readiness of domestic plumbing.

References

1. Yu, E., Sangiorgi, D.: Service design as an approach to implement the value cocreation perspective in new service development. Service Res. **21**(1), 40–58 (2018)
2. Fließ, S., Kleinaltenkamp, M.: Blueprinting the service company—managing service processes efficiently. J. Bus. Res. **57**(4), 392–404 (2004)
3. Lee, C.H., Chen, C.H., Trappey, A.J.C.: A structural service innovation approach for designing smart product service systems: case study of smart beauty service. Adv. Eng. Inform. **40**, 154–167 (2019)
4. Chen, H.R., Cheng, B.W.: Applying the ISO 9001 process approach and service blueprint to hospital management systems. TQM J. **24**(5), 418–432 (2012)
5. Costa, N., Patrício, L., Morelli, N., Magee, C.L.: Bringing service design to manufacturing companies: integrating PSS and service design approaches. Des. Stud. **55**, 112–145 (2018)
6. Yu, E., Sangiorgi, D.: Exploring the transformative impacts of service design: the role of designer–client relationships in the service development process. Des. Stud. **55**, 79–111 (2018)
7. Wetter-Edman, K., Sangiorgi, D., Edvardsson, B., Holmlid, S., Grönroos, C., Mattelmäki, T.: Design for value co-creation: exploring synergies between design for service and service logic. Serv. Sci. **6**(2), 106–121 (2014)
8. Patrício, L., Fisk, R.P., e Cunha, J.F., Constantine, L.: Multilevel service design: from customer value constellation to service experience blueprinting. J. Serv. Res. **14**(2), 180–200 (2011)
9. van der Burg, R.J., Ahaus, K., Wortmann, H., Huitema, G.B.: Investigating the on-demand service characteristics: an empirical study. J. Serv. Manage. **30**(6), 739–765 (2019)
10. Laudon, K.C.: E-commerce 2017, 13th edn. Pearson Publication Inc. USA (2017)
11. Lee, M.-C., Han, M.-W.: E-Business model design and implementation in supply-chain integration. In: 2009 International Symposium on Web Information System and Applications Proceedings, vol. 8, pp. 1–4. Academy Publisher China (2009)
12. E-COMMERCE. India Brand Equity Foundation, July 2019
13. Kim, M.J., Lim, C.H., Lee, C.H., Kim, K.J., Park, Y., Choi, S.: Approach to service design based on customer behavior data: a case study on eco-driving service design using bus drivers' behavior data. Serv. Bus. **12**(1), 203–227 (2018)
14. Rust, R.T., Huang, M.H.: The service revolution and the transformation of marketing science. Mark. Sci. **33**(2), 206–221 (2014)
15. Blazevic, V., Lievens, A.: Managing innovation through customer coproduced knowledge in electronic services: an exploratory study. J. Acad. Mark. Sci. **36**(1), 138–151 (2008)

16. Zhao, J., Wang, X., Zhou, Y.: Study and implementation of user behaviour analysis. In: Int. Conf. on Advanced Communication Technology, vol. 1, pp. 692–695. IEEE (2010)
17. Derbyshire, J., Giovannetti, E.: Understanding the failure to understand New Product Development failures: mitigating the uncertainty associated with innovating new products by combining scenario planning and forecasting. Technol. Forecast. Soc. Chang. **125**, 334–344 (2017)
18. Amer, M., Daim, T.U., Jetter, A.: A review of scenario planning. Futures **46**, 23–40 (2013)
19. Page, S.J., Yeoman, I., Connell, J., Greenwood, C.: Scenario planning as a tool to understand uncertainty in tourism: the example of transport and tourism in Scotland in 2025. Curr. Issues Tourism **13**(2), 99–137 (2010)
20. Bradfield, R., Wright, G., Burt, G., Cairns, G., Van Der Heijden, K.: The origins and evolution of scenario techniques in long range business planning. Futures **37**(8), 795–812 (2005)
21. Lindgren, M., Bandhold, H.: Scenario planning: the link between future and strategy, 1st edn. Palgrave Macmillan, New York (2003)
22. Bishop, P., Hines, A., Collins, T.: The current state of scenario development: an overview of techniques. Foresight **9**(1), 5–25 (2007)
23. Ordenes, F.V., Theodoulidis, B., Burton, J., Gruber, T., Zaki, M.: Analyzing customer experience feedback using text mining: a linguistics-based approach. J. Serv. Res. **17**(3), 278–295 (2014)
24. Lim, C.H., Kim, M.J., Heo, J.Y., Kim, K.J.: Design of informatics-based services in manufacturing industries: case studies using large vehicle-related databases. J. Intell. Manuf. **29**(3), 497–508 (2018)
25. Huang, M.H., Rust, R.T.: IT-related service: a multidisciplinary perspective. J. Serv. Res. **16**(3), 251–258 (2013)
26. Bo, L., Ho, M.: Scenario types and techniques : towards a user's guide. Futures **38**, 723–739 (2006)
27. Ranganathan, A.: Professionalization and market closure: the case of plumbing in India. Ind. Labor Relat. Rev. **66**(4), 902–932 (2013)

Chapter 2
India Post Services—A Speculative Service Design for Behaviour Change

Shivani Ganwani, Radhika Verma, Ravi Mahamuni, and Varnika Naik

Abstract Service design has proved its potential in creating effective and innovative solutions, desirable by users as well as service provider organizations. Service design, which follows holistic and human-centred approach, exhibits an enormous scope to address complex, large-scale problems like designing citizen services that are experiential, efficient and effective. Citizen services involve complexity due to the diverse user base and their varying needs; rapid technological advancements and increasing demands of users. The involvement of multiple stakeholders, socio-political and environmental changes across the globe makes it more complicated. In order to explore the effectiveness of the service design approach, methods and tools for citizen services, we undertook a speculative case study of Indian Postal Services. The study was broadly divided into research, analysis, ideation and conceptualization stages. A mix of quantitative and qualitative methods was used for user research, and derived insights were taken forward through ideation workshops using tools provided by CraftChange—service design for behaviour change framework. Ideas were clustered to create a service ecosystem comprising of peripheral as well as core services for India Post. This study emphasizes on adopting differentiating practices, tools and methods while designing citizen services, to achieve greater adoption and sustained usage of the services. Service design for behaviour change approach towards citizen services seems to be promising to have more holistic, valuable and sustainable services for the users, service providers, associated organizations and the society and environment.

S. Ganwani · R. Verma (✉) · R. Mahamuni · V. Naik
Tata Consultancy Services Ltd, Mumbai, India
e-mail: radhika.verma@tcs.com

S. Ganwani
e-mail: shivani8.s@tcs.com

R. Mahamuni
e-mail: ravi.mahamuni@tcs.com

V. Naik
e-mail: varnika.naik@tcs.com

A. Chakrabarti et al. (eds.), *Design for Tomorrow–Volume 2*, Smart Innovation, Systems and Technologies 222, https://doi.org/10.1007/978-981-16-0119-4_2

2.1 Introduction

India, the world's second-largest populated country, has remarkable diversity across geography, religion, climate, language, culture and economy. India is the origin of four of the major religions in the world with 121 languages, Hindi being the most widely used [1, 2]. India is also a blend of several different cultures. Different social strata exist in India based on lower, middle and higher economic classes [3]. Apart from such diversity, infrastructural support, and access to various products and services also vary across regions within the country. Due to such variations and diversity, the Indian consumer market is considered as complex.

With such a large, diverse population in India and the ever-growing technological era, there is a massive pile of expectations from postal services to satisfy growing citizen needs. The Department of Posts (DoP), trading as India Post, established in 1854, is one of the oldest citizen services. It is the world's most widespread postal network, having over 1.55 lakh post offices [4]. Mail service is the core of India Post services apart from several additional services like banking and insurance, retail services like bill collection and distributing railway tickets. Post offices are also utilized as Passport Seva Kendra, called as Post Office Passport Seva Kendra. It has established a strong network of services in every corner of the nation, significantly in the rural sectors along with an international presence also.

Complexities within citizen services—World view. Across the world, government postal service business has been declining because of technology and market savvy private competitors, and electronic diversion along with other interlinked factors like demographic changes, pricing, evolving security and privacy concerns [5, 6]. Hence, the need for reforming or restructuring the postal services is recognized across the world. Postal services in many countries are additionally providing financial, banking, insurance services to their citizens, providing online services, exploring innovative ways to generate revenues [7].

In general, for all citizen services, governments around the globe have been looking for new ways to tackle the new scale of problems indicating the need for change [8]. Commonly, governments providing citizen services often encounter internal issues such as Disconnects between analysis of problems and creation of solutions; Poor understanding of citizen needs; Government departments working in silos; Designing for the average, leading to missing out on the extremes; and Incremental solutions by patching up seemingly speedy solutions to meet urgent needs [9]. It is also essential to train and educate the staff as well as citizens to deliver and use the services effectively. Moreover, it has become challenging for the governments to fulfil the diversifying and ever-increasing needs of the citizens, while also controlling the budget deficits. Today's citizens are experiencing dissatisfaction with government services due to reasons such as slow speed, complex processes and the effort required to navigate through processes [10].

To summarize, rising and diversified expectations of citizens; rapid technological advancements and various constraints faced by the government lead to complexity in designing large-scale citizen services. Changes are required across government policies and processes, infrastructure, staff providing the service to the citizens and the citizens who are users of the services, to develop successful citizen services working effectively for citizens as well as the government. Therefore, there is a profound need to borrow more modern and innovative approaches.

Adopting a service design approach. Design thinking and related approaches have been found effective in the private sector. They are also capable of meeting the immediate needs of the citizens (as 'users' of the government services) as well as to help governments achieve their broader aims (better quality of life for its citizens) while taking care of the constraints. '…design thinking does sit within the broader gamut of citizen-centred approaches, it is more about empowering passive citizens and understanding their experiences of government policy and services' [11]. Many governments across the globe have established Innovations Labs that adopt design approaches at the core, the UK being a pioneer in this [12].

A relatively new field, service design, has been adopted in various domains such as banks, retail, transport and even public sector services like education and healthcare [13]. Service design, due to its characteristics [14] such as human-centred, iterative, co-creative and holistic, seems to be the natural fit towards designing effective citizen services. Considering the whole lifecycle and the entire ecosystem, not focusing on single-point interactions, service design transforms the whole service and brings out changes in policies, processes, infrastructures and interactions. In the broader context, it has the potential to effectuate changes in human behaviours, society and the environment.

With a broader aim of exploring service design approach and its methods for designing citizen services in mind, we undertook a representative case study of India Post considering the diversity and variations found in India, the complexity of the service and larger scale. While there are a plethora of challenges in designing citizen services, we believe that this study acts as a promising step towards the evaluation of service design approach, processes and methods in this context.

2.2 Understanding the Current Context

Governments that are managing multiple public services have realized the importance of adapting themselves with the changing time. Specific trends have been observed, such as integration of artificial intelligence and related technologies that can specifically cater to a vast volume of data, among government services. Developing economies like India have identified to use AI for healthcare, education, smart city initiative as a few core areas in their national strategy [15]. Developing a unique digital identity of the citizens, to provide better citizen services, is another priority for the government. Enrolment of 1.2 billion citizens in Aadhaar Card in India is an example of this model.

Additionally, nudging citizens by using fundamental principles of behavioural science has enabled governments to make informed decisions in policymaking and process changes [15]. There is also a massive shift towards inculcating a sense of smartness within cities, by creating smart ecosystems, data and connectivity, smart communities, transportation and more. With the emergence of digitization and smart cities, a key initiative that governments are emphasising on is citizen experience.

With excessive effort being assigned to these initiatives in silos, collectively these initiatives can build a service ecosystem that can be leveraged by government-driven public services to enhance their current scope and extend their services to cater to a broader audience. Indian Postal Service is one such government service that has an enormous scope of leveraging existing government initiatives. We conducted a study to identify the key issues and opportunities for India Post and see how they could be translated into developing new services.

2.2.1 Secondary Research Findings

The study commenced with gathering literature findings around India Post and associated service ecosystem. During the study, it was found that digital transformations are happening in the world and the entry of various private players in the market are creating challenges for India Post. The need for creating new services to be offered by Post for its survival in the competitive market while also embarking upon its extreme outreach in every corner of India was evident. While India post has advantages like ubiquitous network and the experience in distributing enormous numbers of products and services, it also enjoys the trust and proximity that it has developed over the years. However, the overall decline in the traditional postal market has always been a threat to India Post. Issues like customer orientation and connectivity, technological reforms have been significant reasons hindering the growth of India Post. In order to identify the strength, weakness, opportunities and threats for India Post, secondary research was conducted [4]. Social media analysis was also done to understand the most prevalent problems that customers are facing by analysing data through Complaint Board Forum and Twitter feed analysis. About 1300 posts were collected for review. The key findings of the secondary research are structured in the form of a SWOT analysis, shown in Table 2.1. Based on the literature review, the initial design brief for the study was to re-imagine Indian Post Services and what role it can play in urban cities.

2.2.2 Primary Research Findings

Purpose of the study. As India Post has a stronghold within the rural areas of the country, it became necessary to understand how it has been able to retain its identity within a rapidly shifting urban landscape. The purpose of the study was to understand multiple viewpoints of different citizens and their perceptions about India Post. Also, to know how citizens foresee India Post to transform in upcoming years, probably integrating with existing government initiatives like Smart City Mission, in India.

Table 2.1 SWOT analysis—India Post

Strengths	Weakness
• Provides communication, banking and insurance services and products • Enjoys extraordinary goodwill among citizens • 85% of the offices are in rural India, with local staff, giving employment opportunity to residents • Has a strong network for national and international services • Cost-Effective • Post office situated in prime situations and some also have historical significance • Government is the backbone • Reach is very high—Last-Mile Delivery • Willingness to restore the human touch during digital transformation	• Unaltered working culture • No effective feedback mechanism • No job rotations & Job stress • Very limited direct recruitments in the past five years • The current workforce is ageing and has limited IT skills • The infrastructure is inadequate in newly developed urban pockets and rural areas [4] • Customers are unable to get full benefits of IT induction such as track and trace, due to limited use of technology [4] • Customers have a perception of poor service [4] • Mode of transport has not evolved • Manual handling and sorting
Opportunities	**Weakness**
• The increasing number of customers • Inflationary market • Technological Advancement • Leverage the last-mile delivery network • There is untapped potential in terms of money transfer for the migrant population and other small businesses • E-commerce • Mobile penetration is an opportunity • E-governance	• Increasing the market share of competitors • An (internal) threat—India Post and its employees may be unwilling or unable to meet the challenges it faces and to seize the opportunities presented to them • Customer Dissatisfaction • Digitization—Shift from postcard to mails • Change in citizen needs • Payment banks, digital currency and online money transfer via Paytm, PayPal, UPI • Private Insurance Companies

Moreover, the study was directed to find the scope of new services that India Post can envision to achieve, leveraging their existing strengths in the marketplace. As the research was directed towards understanding how the citizens perceive India Post, it did not involve any members from the India Post organization. This was taken up as an independent study.

Methodology. The approach followed, in this case, was a cohesive blend of research through design methodology along with the service design process. After gathering the literature findings, the next step was, conducting a primary user research study through mixed-research methods. Derived insights were used during brainstorming and ideation session. An initial user journey and personas were created based on the derived insights and initial ideas from brainstorming sessions. Then complete ideation and conceptualization were done for a few newly identified services to be proposed as new service offerings by India Post.

A key aspect of the service design process for this case study was the adoption of CraftChange—Service Design for Behaviour Change framework [16, 17].

The CraftChange Framework is a guided set of tools and techniques that allows service designers to imagine new service models holistically by considering all the stakeholders that are affected by the service ecosystem. The key components of CraftChange that were used during the study are—Current Intervention Cards to understand the current context, ignite cards that were based on behavioural principles to ignite various ideas, Empathy Square that guides to address all the primary stakeholders' needs through the solution, considering society and environment concerns while ensuring completeness of the designed solution.

A multi-disciplinary team of designers, engineers, researchers and psychologist collaborated to perform this study. The members of the team, being Indian citizens, were also users of the India Post service. The team members belonged to different regions of India (North, North-East, West). They represented respective cultures of India, and hence, the multicultural aspect of the team was leveraged to get diverse insights. Primary research was conducted by a mix of quantitative and qualitative methods based on focus areas defined, explained further. The participant recruitment for the study was a mix of various age groups, gender and locations to get diverse views leading to further data triangulation.

Additionally, the Current Intervention cards of CraftChange framework were used to ensure the coverage of the questions and focus areas covered in the protocols used while conducting the study. The following are the methods as described in Table 2.2.

The design team was divided into two for analysis of the collected data. Affinity diagram and data clustering methods were used to analyse the data, and the findings were converted into insights through a series of team discussions. SWOT analysis data was also updated with the primary user research findings, that led to extracting key areas that must be explored for ideation of new service concepts. The knowledge gained from user research was externalized to reflect analysis insights through various in-process artefacts such as personas, a detailed SWOT and ecosystem maps and perception map as shown in Fig. 2.1.

Findings from the primary research. From the survey, it was found that although people are aware of postal and banking services of India Post, there is very little or no knowledge about other services such as retail—railway tickets, RD, PPF accounts. Majority participants preferred other private courier services over India Post, because of their speed of delivery, proximity to their location and tracking services. The overall satisfaction about India Post is not great, based on factors like interaction with the staff, office infrastructure and time is taken for processing. Most of the survey participants believed that the overall transformation of India Post could be a significant contributor towards the smart city mission of India.

The interviews findings indicated certain conventional notions that a diverse set of citizens held about India Post. There were some variations too in the responses from small town versus city participants and different age groups. India Post delivery services ensure guaranteed delivery; their services are affordable; and the factor of trust in the postman was more evident among the middle-to-old-aged population originally from smaller cities. Urban citizen participants raised certain doubts about the reliability of postman. They fear to send valuable articles and belongings through India Post. Factors, like unawareness of the services, paperwork hassle, lack

Table 2.2 User research—objective, methods and protocol

Data collection methods/tools	Number of participants	Demographics	Research objective	Questions protocol	Location (India)
User survey	42	69% male; 31% female respondents, Working professionals, Age bracket 20–50yrs	Penetration of India Post in terms of offerings awareness, competitors and prior experience	Questions: 20 Time: 10-15 Mins	Mumbai Pune Nagpur Thane
Semi-structured interviews	15	Working professional, Age bracket 25–55yrs, Mixed gender	To find commonality and differences in experiences	Questions: 7–10 Time: 30 Mins	Mumbai Pune Thane
Onsite naturalistic observation and quick feedback	6	India Post visitors, Age bracket 20–60yrs Mixed gender	Understand user experience through user stories and identify service issues	Questions: 5 Time:15 Mins	Mumbai Pune
Focus Group Discussion	5 participants	Migrated working professional, Age bracket 20–50	Opinions, attitude, expectations of citizens and how they foresee smart cities, leveraging India Post	Questions: 8 Time: 1.5 h	Mumbai Pune

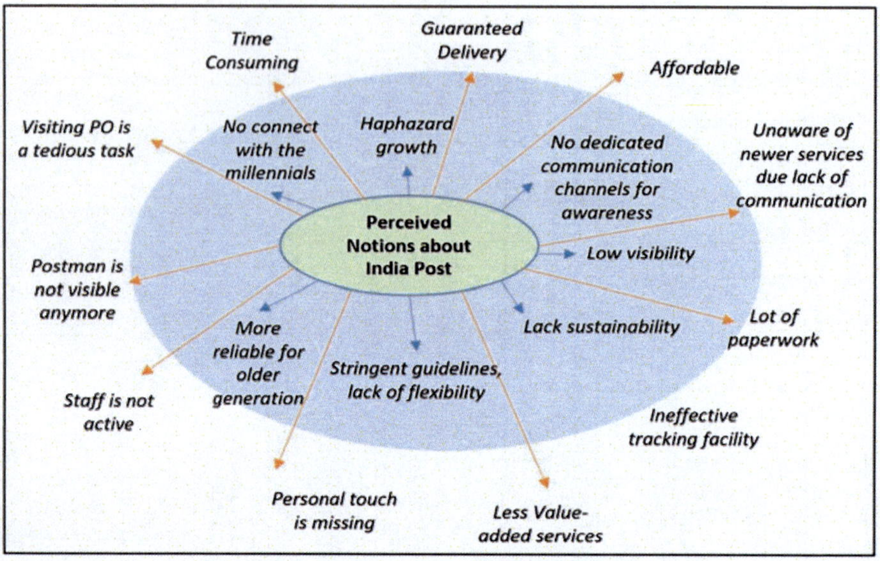

Fig. 2.1 User study findings—Perceived notions about India Post

of personal assistance while facing issues, are the commonly reported concerns. Participants reported the non-modern infrastructure of post offices. The participants also perceived that the Post staff lacks a customer-centric attitude and tech-savviness. During naturalistic observation also, users reported the staff to be irresponsible, disinterested, untrained, not taking charge and not giving out the required information. Users also reported infrastructural issues such as outdated technology, limited seating space, unorganized information boards and restricted office timings.

Through focus group discussion, it was found that citizens foresee that cities must be prepared to adopt measures to increase collective belongingness within communities and improve existing government services through digital initiatives. The discussion highlighted the aspect of mass migration among working professionals, who seek better job opportunities and require a set of facilities for a smooth transfer from one city to another. The participants felt that India Post in all aspects is beginning to fade and that it lags far behind the private players in the market. Ideas for new or improved services of India Post revolved around India Post acting as an information centre, integrating all citizen services, enhancing postal services through pick and drop services and much more.

Collectively, a profound dissatisfaction, among India Post users across multiple service touchpoints, was revealed. This reflects that, though India Post has only been improving to address certain aspects of their service, like introducing more services and bringing new technology in place to attract the users, unfortunately, these individual problems are not looked from a service systems perspective. There is a lack of holistically addressing the shortcomings. To conclude, the scale at which India Post operates is massive as compared to any other private organization in this sector.

Which is why it becomes difficult for the government to keep up with this aspect of scalability and catering to the concerns or issues faced by such a diverse population becomes challenging. India Post also has multiple stakeholders involved that work together within the service ecosystem. Introduction of new interventions involves analysing the challenges within scalability to make the service more acceptable and sustainable in the long run.

2.3 Exploiting opportunities and Re-imagining Service Offerings

Further to understand the India Post context, its services, user expectations and experiences; the speculative design study proceeded with ideation and conceptualization stage. Ideation intended to leverage the strength and opportunities for India Post and design new services to be offered by India Post for its users, stay relevant and useful for a more extended time. The exact parameters for identifying new services were that new service(s) should (1) act as an alternate revenue source for India Post; (2) align with the identified expectations and needs of users through primary research (3) align with the vision, mission, goals of India Post and new government initiatives (4) exploit its impressive outreach in rural and urban areas of the country; (5) build upon its existing core mail service by redesigning it to fit in today's digital and modern world.

The designed services should also guide the users and the human touchpoints to interact with the services at any point in time, in an intended manner. For example, the staff should be motivated enough and be able to deliver the experience to its users during any of the interactions. The users should be motivated, through design interventions, for example, to opt for the environment-friendly options made available by India Post.

Addressing behaviour-driven concerns through CraftChange Framework. The redefined design brief for the study was to identify 'India Post as a first choice and one-stop solution for all citizen needs'. We used 'CraftChange—Service Design for Behaviour Change' framework [16, 17] in order to make Indian citizens aware of the proposed service, engage them with the service, enable them to use the service and sustain the usage over some time such that they can also advocate the service. CraftChange—Service Design Framework canvases enabled the design team to ideate for each stage of the users such as aware, engage, enable to use, use regularly and advocate. Later, ideas where clustered to create detailed service concepts. The cards used were based on behavioural principles and theories to ignite the ideas and guide the design team to think from society and environment perspective, ensuring completeness of the proposed service.

Through the usage of CraftChange framework and toolkit [16, 17], during multiple workshops and brainstorming sessions, many ideas were proposed for new service offerings, that embark upon the identified needs of citizens as opportunities, align with the new government initiatives and can generate revenue for India Post. There were also many incremental ideas to improve existing India Post services. Some design interventions specifically catered to motivating and enabling the expected staff behaviour. A few of the newly proposed services were designed in the form of detailed concepts. CraftChange Empathy square [18] guided us to ideate and created a balanced solution addressing the concerns of all the involved stakeholder; service users, human touchpoint, organization and society as a whole. One of empathy square nodes being 'society and environment', helped the design team to focus on cultural, societal and environmental aspects of the proposed service.

2.4 Reflections

We reflected on the process followed, outcomes and feedback received, based on which, the following are our findings and learnings from this study:

1. *Need for scaled-up qualitative research to design complex citizen services that involve diversity, complexity and large scale.* While service design allows us to indulge in the needs and expectations of citizens, it is a challenge to cover the diversity among citizens that gives richer and deeper insights. The need to evolve qualitative research techniques and methods, to reach out to a broader diverse user base of such large-scale services, was observed and felt. The quantitative research is already scalable and can be used during triangulation.
2. *The service design approach is useful to find alternate revenue generation by introducing new services or service options*: The way current context is studied, presented and utilized during the ideation phase helped to come up with alternate revenue services. Ideas that encourage improved channels of revenue for India Post also highlights that service design can foster business impact. Using service design approach for India Post, we could identify and design new service offerings that can be developed by India Post to generate revenue in today's modern digital era by aligning with new government initiatives and fulfilling citizen needs.
3. *The integrative approach of service design and Behaviour Change seems apt for citizen services*: User research done for India Post brought out the particular perception of users about staff behaviour such as lack of customer-centric attitude, negligence, lethargy, rudely speaking, irresponsible behaviour. For citizens also, there is a need to engage them with the service actively. To build successful services, apart from building the core features of the service; it is required to make the users aware, engage, actively use and advocate the service. Hence, designing the service and changing the behaviour works simultaneously, addressing greater adoption of services that can potentially stay longer.
4. *Service design approach as an in-built mechanism to be a self-sustained organization, in rapidly changing times.* There is a need to have an in-built mechanism within government and its departments like India Post, to regularly evaluate,

update and expand their services in order to be self-sustained. Service design seems to be an apt approach to achieve this aim.

5. *Addressing the need of all, in a balanced manner through designed services is required to make services acceptable for all stakeholders.* Service design's approach suggested by CraftChange Empathy square [18], to design balanced solutions that address the needs and concerns of all the primary stakeholders including society and environment, enables all stakeholders to embrace the new services effortlessly. Since the needs of citizens, concerns of India Post and Government of India, concerns of India Post staff and concerns and constraints of environment and society can all be addressed through the designed services, there is a higher possibility of broader acceptability.

2.5 Conclusion and Further Direction

Citizen services are associated with large scale, complexity and diversity-related challenges, leading to the need for innovative approaches. This study was an attempt to apply service design approach for citizen services like the post office. Overall, service design approach and its methods are useful in improving the current services, identifying and generating new service offerings such as alternate sources of revenues. In order to be self-sustaining in the rapidly changing times, identification of other revenue-generating services needs to be made a part of the government structure and processes to evaluate, update and stay relevant continuously. Service design can go deeper and understand the ground-level reality, needs and expectations of users. At the same time, it has the potential of holistically designing strategies, business models and processes to create favourable conditions for business organizations or government, its users as well as society and environment. Designed services, in such a balanced manner, have the potential to last longer and stay relevant in the changing world.

Moreover, the integrative approach of service design and behaviour change, based on behavioural principles and theories, enables designers to design interventions that achieve the desired behaviour of citizens and service provider staff. At a broader level, it has the potential to achieve desirable human behaviours and societal transformations. Therefore, we believe service design process of understanding the context, finding services for alternate revenue sources, filling the gaps between citizen needs and offered services should be continuously applied to citizen services to be always up-to-date, useful and relevant.

We understand that India Post was a speculative design study with the limited availability of time and resources. However, it was an attempt to verify our belief that service design and behaviour change approaches could benefit citizen services tremendously. Further, there is a need to explore methods of service design in-depth, that can solve the large-scale-related issues and evaluate those through multiple citizen service case studies, to be ready to be applied in the real scenarios.

References

1. India at a Glance—Profile. https://www.india.gov.in/india-glance/profile
2. Census Of India 2011 (2011)
3. Bisht, P.: Social Stratification in India, IOSR J. Human. Soc. Sci. (IOSR-JHSS), 28–29 (2015)
4. India Post, Ministry of Communication & Technology. https://www.indiapost.gov.in
5. Carbaugh, R.: The future of the US postal service. Cato J. **27**, 459 (2007)
6. Purohit, K.J.: Measurement of customers' satisfaction in selected indian postal services (2017)
7. Keating, J.E.: The World's Best Post Offices (2013). Available: https://foreignpolicy.com/2013/02/07/the-worlds-best-post-offices/
8. Buchanan, C., Junginger, S., Terrey, N.: Service design in policy making. London (2017)
9. Council, U.D.: Design for public good. Ann. Rev. Policy Des. **1**(1), 1–50 (2013)
10. Baig, A., Dua, A., Riefberg, V.: McKinsey Center for Government. How to improve citizens' experience and satisfaction with government services, PUTTING CITIZENS FIRST (2014)
11. Mintrom, M., Luetjens, J.: Design thinking in policymaking processes: opportunities and challenges. Aust. J. Public Adm. **75**(3), 391–402 (2016)
12. Carstensen, H.V., Bason, C.: Powering collaborative pol-icy innovation: can innovation labs help. Innov. J.: Public Sect. Innov. J. **17**(1), 1–26 (2012)
13. Mager, B.: Service design as an emerging field. Des. Serv. Innov. Methods **1**, 27–43 (2009)
14. Stickdorn, M., Hormess, M.E., Lawrence, A., Schneider, J.: This is service design doing: applying service design think-ing in the real world, O'Reilly Media, Inc. (2018)
15. Government Trends 2020 (2020)
16. Mahamuni, R.: Crafting a change—service design perspective for sustained behaviour change, Thesis (2020)
17. Mahamuni, R., Khambete, P., Punekar, R.M.: Behaviour Progression Framework for Designing Sustained Behaviour Change. Research into Design for a Connected World, Singapore (2019)
18. Mahamuni, R., Meroni A., Punekar, R.M.: Empathy square: an aid for service design for behaviour change to support sustainability. Designing Sustainability for All-3rd LeNS World Distributed Conference, vol. 3 (2019)

Chapter 3
Design Intervention for Improving Accessibility and Affordability of Ocular Care—A Context-Specific Study of Rural Assam

Abhijit Kakati and Amarendra Kumar Das

Abstract Practices in conducting eye screening camps in rural underserved areas seem to have not changed lately. Majority of the settings organizing eye screening camps, lack the necessary facilities and do not meet the general medical standards. Adverse terrain conditions also do not allow for medical mobile unit vans to be an option in such areas. Such factors result in the exclusion of a significant portion of the population, who are at risk of developing ocular abnormalities due to diabetes and other farm-related injuries. Field studies conducted in eye camps of rural Assam and subsequent interviews with healthcare providers and recipients established the novelty of the information. Considering these constraints as an opportunity, the author engaged the healthcare experts for mining a context-specific design from the perspective of the specialists. The information collected through ethnographic narratives, focus group discussions, interviews, and field studies were shared with a team of healthcare experts. The healthcare team, by means of a design workshop, tried to address user needs in their concept design. This research examines the role of association of designers with the healthcare professionals in bringing about solution-oriented design services and systems, meeting the specific needs of the indigenous people as well as the healthcare facilitators. The paper aims to reflect the versatility of design education in addressing socially significant problems, intending to direct it toward the unmet healthcare needs in low-resource settings. Such context-specific interventions can open the window to valuable insights into the position of design education in other professional domains.

A. Kakati (✉) · A. K. Das
Indian Institute of Technology Guwahati, Guwahati, India
e-mail: abhi.kakati@gmail.com

A. K. Das
e-mail: dasak@iitg.ac.in

A. Chakrabarti et al. (eds.), *Design for Tomorrow–Volume 2*, Smart Innovation, Systems and Technologies 222, https://doi.org/10.1007/978-981-16-0119-4_3

3.1 Introduction

Providing quality eyecare services in rural and remote communities has always been considered daunting. People living in such remote places are often vulnerable to several ocular abnormalities arising from work-related injuries, aging, and complications arising from non-communicable diseases like diabetes. With the growing epidemic of diabetes, diabetic eye diseases are affecting a significant portion of the underserved population of India. This requires consideration for viable alternative services that are contextual.

The current practices for delivering eyecare services to the remote population are through community outreach camps. These camps have remained an integral part of the underserved population for access to eyecare services. Community outreach camps act as a bridge between the eyecare providers and patients at risk in the underserved areas and the population relies on such health camps for their well being. Mobile unit screening vans have somehow tried to make its presence felt in such areas. However, there are many areas in the state of Assam that still lacks proper road connectivity and has adverse terrain conditions. Under such environments, medical mobile unit vans are often not considered as a choice in delivering eyecare services. Rural communities present unique social and physical features and challenge the effective implementation of digital technology [1]. Such factors finally result in the exclusion of a significant portion of the population that are at risk of developing ocular abnormalities due to diabetes.

The implementation of design methodologies can somehow help in optimizing the complex nature of healthcare services. It is imperative to focus on how health care can fit to the complexity of people's lives and what role design can play in providing suitable interventions in the preventative and curative approaches of health care [2]. The role of design is crucial in developing interventions by responding to the needs of the people and the society where they live in [3]. Using a focused sense of design strategy will help in developing better products that will be both functional and pleasant [4].

The context of the proposed study is the rural and remote areas of the northeastern state of Assam that lacks access to quality eyecare services and remains vulnerable to developing ocular abnormalities throughout their lifetime. Since the outreach to the rural population in the context of retinal screening is negligible, the rural community is unaware of the retinal complications that can arise from diabetes and corneal injuries.

More than 80% of the population of Assam depends on farming as a source of income. Agriculture-related corneal injury, retinal vessel damage, and retinal detachment are predominant among the farmers and daily wage workers of rural Assam. A general practitioner having basic knowledge often serves all the patients, and so the persons encountering retinal damage or corneal injury are often left undiagnosed. This gradually results in severe conditions of the eye, and, if left untreated, may lead to irreversible blindness. There is a primitive need to address this gap prevalent in low-resource settings and make quality eyecare services accessible for the population at risk. The research aims to engage the healthcare experts and extract their design sense in designing for rural health camps.

3.2 Barriers Contributing to Poor Conditions of Eye Health Care in Rural Areas

The prevalence of non-communicable diabetic eye diseases is increasing sharply day by day. The majority of the people with diabetes are not seeking ocular care that could prevent them from blindness and visual impairments [5]. Lack of awareness among the people is a significant barrier for undertaking regular retinal screenings of diabetic patients who could develop diabetic eye diseases. People are not aware of the complications of diabetes in their eyes. Although the people had a good grip on the causes of diabetes, the majority were unaware of the association of eye disease with diabetes [6]. Poverty is also considered to be one of the most important barriers in accessing ophthalmic services in both the developing and developed countries [7]. Due to the economic crunch among the rural dwellers, they are afraid to access health care due to the exorbitant cost resulting in deterioration of eye health [8].

3.2.1 Lack of Awareness and Negligence in Diagnosis

The rural population has faith in folk practices and beliefs for any healthcare needs. People in such areas suffering from eye diseases often ignore visiting an eyecare specialist for factors like cultural, social, and psychological like fear of treatment, mobility issues, time of travel, poor quality outcomes, family obstacles, fatalistic beliefs, and reliance on folk medicine. They believe that the prevalence of blindness is due to supernatural forces, and no interventions can help in this regard. In a study by the Public Health Foundation of India (PHFI), Madras Diabetes Research Foundation (MDRF) and Harvard School of Public Health; it was found that about 1 in 2 Indians with diabetes is unaware of their condition [9].

Awareness related to diabetes and its complications is extremely low in rural areas of Western India [10]. Although around half of the patients were aware of diabetes affecting their eye health, they were not aware of retinal screening [11]. This emphasizes the need for increasing awareness related to diabetic eye disease diagnosis and complications.

3.2.2 *Delay in Follow-Ups for Treatment*

Visiting for timely follow-up might not be possible in many instances for the rural dwellers because it involves time, travel cost, and other opportunity costs. In certain circumstances, taking a day off from work or household work for both men and women could make arranging money difficult and treatment delayed.

3.2.3 *Lack of Communication Between General Physician and Ophthalmologists*

A study mentioned that only 1 out of 5 ophthalmologists in India have access to diabetes case records of the patients [12, 13]. The lack of information between the two health domains is a severe concern in the prevention and management of diabetic eye disease. There is a lack of information sharing network between the two healthcare domains [13]. The diabetic experts were not aware of diabetic eye disease screening by ophthalmologists and were also unaware of whether the patient has undertaken any diabetic eye screening [14].

Communication between treating physicians and ophthalmologists as well as a timely referral is essential for diagnosing and preventing irreversible blindness due to diabetic complications.

3.2.4 *Lack of Facilities at the Rural Eye Screening Sites*

People in rural areas often rely on primary health centers and outreach screening camps for their eye health. However, the health centers located in rural areas lack the necessary facilities for proper eye care. Eye screening camps organized in the rural areas lack the essential equipment for retinal screening and conditions under which the testing has to be done is not adequate. A majority of the government-aided hospitals do not have proper facilities for managing complex cases of diabetic retinopathy [15].

3.3 Research Methods

During the research, the author adapted qualitative methods for obtaining the contextual factors of rural health camps. Expert interviews and field visits to rural health campsites helped in unraveling the factors responsible for poor conditions of health camps. At the final stage of the research process, a team of healthcare experts was engaged in a design workshop to design a concept for solving some of the underlying issues extracted through interviews and field visits. The motive in involving the healthcare experts in developing a design concept was their expertise in this field. For obtaining a clinically safe and practical design, the perspectives of the specialist should be considered.

3.3.1 Expert Interviews

At the initial stage of the research, the author visited several eye clinics and hospitals to get an insight into the process of eye screening examinations. Different categories of eye health check-ups undertaken in rural health camps were discussed with the ophthalmologists and optometrists. The researcher interviewed six eyecare experts consisting of both ophthalmologists and optometrists. The formal discussion series with the experts revealed that the rural health camps are focused mainly on cataracts and visual acuity tests. Of the ophthalmologists interviewed, two mentioned that diabetes has a significant impact on the eye health of an individual, and it can lead to damage to the retina. Therefore, it becomes necessary for the eye health camps to also focus on diabetic-related eye diseases. Such practice is seldom performed in rural eye camps mainly due to lack of facilities, awareness, and time. Though certain camps perform the examination of the retina at the rural sites, the results obtained are sometimes erroneous due to interference of the environmental light.

The researcher interviewed five diabetic experts by visiting their clinics to gain their perspective regarding diabetic eye diseases. It was obtained from the discussion that the referral rate of diabetic patients to go for retinal screening is significantly low. Three diabetic experts mentioned that they seldom refer their patients for an eye check-up as most of them never turn back for consultation in the future.

3.3.2 Field Visits to Community Eye Screening Camps

This phase concentrates on the issues faced by the patients as well as the healthcare experts during the process of rural eye camps. Qualitative research methods have been primarily used for the study. The study helped to unravel the different problems faced by the stakeholders during the vision screening process in community groups. As most of the ocular care in the rural areas is provided through eye screening camps, the author found it appropriate to attend eye camps for having an insight into the issues faced by the stakeholders.

The first step involved field visits to five different rural health campsites. Seven experts and two hundred five health seekers were interviewed in relation to eye screening in community camps and other rural clinical settings. The objective was to understand the underlying issues associated with eye screening processes for various eye disorders. Interviews and focus group discussions provided us with valuable insights from the perspective of the experts and as well as the patients.

The experts mainly revealed the issues they face in interacting and imparting care to the patients. On the contrary, the patients focused on the complications they face in attending the eye screening camps. An unexpected finding of the study was that eighty-nine patients who had a background of diabetes never consulted an ocular care specialist for prevention or diagnosis of diabetic eye diseases. They were unaware of the eye complications that can arise from diabetes. Some even thought that the vision abnormalities which they were experiencing was merely normal and will be cured using spectacles. Out of sixty-four persons with age more than seventy years, thirty-five of them felt that the vision impairment is a part of their aging process and was not ready to accept that it can be associated with diabetes.

3.3.3 Problems Encountered by Healthcare Providers in Rural Camps

Unsuitable lighting conditions at campsites. Due to the lack of proper lighting conditions, accurate refraction and visual acuity tests could not be performed. Streak retinoscopy examination and retinal examination must be performed under a dark-room environment, which can hardly be seen in the rural eye camps. Most of the camps are organized in open areas, and there is an illumination problem from the surrounding (Fig. 3.1), which results in an erroneous finding. A low level of ambient light is necessary for efficient screening of the eye. As evident from the figure, ambient light levels are of great concern in the proper diagnosis of retinal abnormalities. The procedure for retinal diagnosis must be performed under proper light levels with very less ambient light interference. All the experts interviewed (seven numbers) mentioned this issue at some point during the interview.

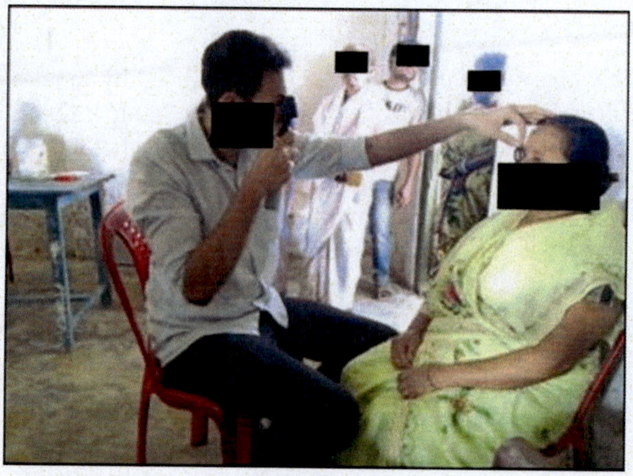

Fig. 3.1 Ambient light interference at the campsite. Source: Author

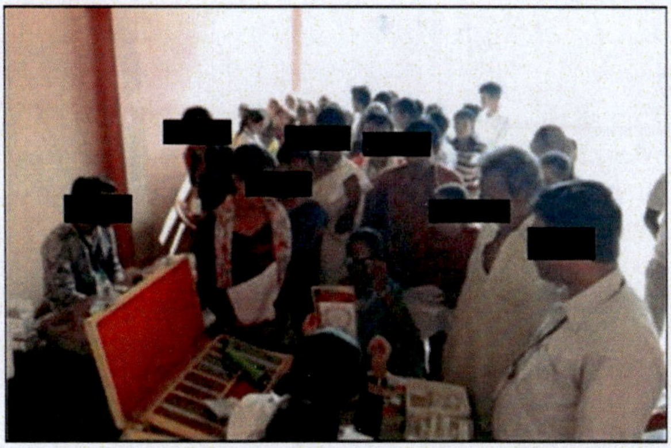

Fig. 3.2 Congestion of patients at the campsite. Source: Author

Congestion at the campsite. Overcrowding of the patients at the campsites sometimes escalates patient–provider conflicts. The healthcare providers are not able to focus on the care process due to such congestion of patients. This ultimately leads to a delay in the screening process and impacts the outcomes of screening. Such practices are seen in eye screening camps at rural locations (Fig. 3.2). Five experts mentioned that lack of a system for managing the stream of the patients affects the screening process during eye camps, whereas two of the experts mentioned that they use token numbers during the camps for managing the patient flow, and most of the time, it gives good results.

Problems encountered by rural health seekers. The open-ended interview and focus group discussion with two hundred and five rural health seekers unfolded several issues. A majority of them revealed that they often hesitate to go to the community hospital for treatment, as ocular specialists are hardly found to be present and lack the necessary facilities for proper eye care. They are mostly dependent on rural eye camps for their check-ups and treatment.

Eighty-three persons interviewed, mentioned that it becomes challenging for the aging and disabled patients to attend the eye screening camps as they need a company to assist them to the campsite, which is not possible at all times.

Secondly, a hundred and forty-two people stated that the distance between the campsite and the patient's home is a significant barrier for them to undergo screening at the health camps. Due to the distance, most of the patients at risk ignore going to the eye screening camp. It becomes tough for the daily wage workers to attend the screening camps if the site is too far from their home. They are not able to leave their work unattended and go for screening even if they want to.

Around eighty-five percent of the participants (a hundred and seventy-four participants) mentioned that, there are privacy issues in most of the health camps they have attended so far. In the health camps, the healthcare providers ask them questions which they only want to answer in private. As there is no privacy during the examination process, they feel shy to open up, and at times provide false information to hide their condition.

3.3.4 Design Workshop

This phase involves actively engaging the eyecare experts in developing a design brief having the potential of solving the underlying issues encountered by both the providers and recipients of eye care in rural health camps. The researcher decided on not engaging any patient in the workshop since issues faced by patients were extracted beforehand during the interview phase. The author shared the extracted information from the patients with the experts before the start of the workshop. Moreover, since healthcare professionals are well qualified and experts in their domain, it was considered pertinent to the researcher to extract design solutions from them.

The workshop took place in a seminar room of a private hospital, and one of the doctors working there helped the researcher in obtaining permission from the hospital authorities. The participants of the workshop included three ophthalmologists (two from the hosting hospital and one from a private eye clinic), four optometrists (one from the hosting hospital and three from different eye clinics), and two nurses (from the hosting hospital). At first, the author provided the healthcare experts a brief about the findings from interviews and field visits. The participants were different from the healthcare experts interviewed earlier. They were given the task of addressing some of the issues faced by the patients and the healthcare experts during eye screening camps in underserved areas. They were informed that their intervention could include a framework, service or system design, medical device design, or any other medical setup. The team was equipped with some blank A4 sheets, chart papers, color pencils, and sticky notes for developing their concepts. The participants first hesitated in sketching their ideas and were conveying their ideas verbally. The researcher helped the participants during the sketching process as they were not used to sketching.

Among the several issues encountered in rural health camps, the participants chose solving the problem of congestion at the health campsites, darkroom facility for a proper eye examination, and privacy concerns raised by the patients. The participants of the design experiment upon mutual consent came with an idea of designing a chamber for use in rural health camps. The chamber will have the functionalities of conversion into a dark room as and when required. As per the view of the participants, the chamber will also be able to solve the problem of congestion and privacy concerns at the rural campsites. During the design concept stage, the researcher never tried to overpower the participants with their thought process. The participants were listened to carefully and helped them in the sketching process and encouraged them to make some rough sketches of their concept for a better understanding of the mechanism. The participants came out with two design concepts.

Concept 1 mainly involved providing a closed chamber with dark screens for limiting the ambient light inside the chamber. This will enable a proper setup for retinal examinations for diabetic-related eye diseases like diabetic retinopathy and other abnormalities requiring retinal screenings. The design parameters as identified by the participants include connectors for connecting the screens, pole design, different types of joints, screen material, the distance between the patient and the plane mirror (a plane mirror is used for forming an image of a Snellen chart at a

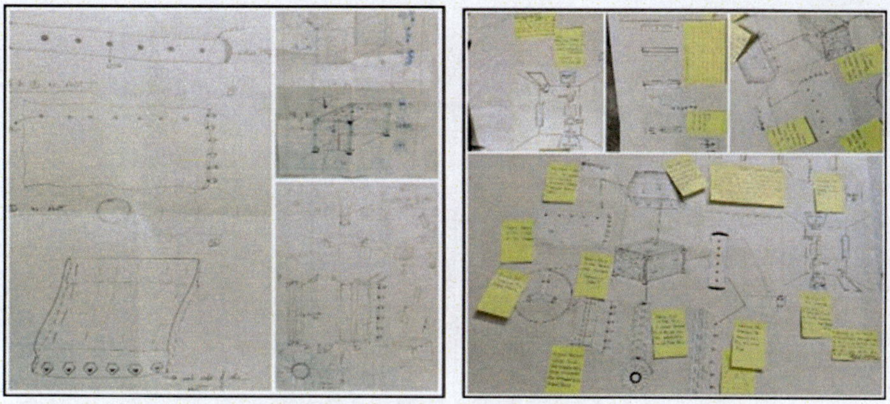

Fig. 3.3 Rough sketches along with refined sketch plan during ideation of Concept 1 by the team of healthcare experts

distance of 6 m from the patient). After identifying the design factors, they made some rough sketches of their ideas. To bring in more clarity to their design, the researcher helped them in refining their sketches. After sketching the concept, the participants explained the mechanism and used the sticky notes for explaining, in brief, the mechanism of the various components of the designed concept (Fig. 3.3).

The participants exhibited a sense of accomplishment after their concept sketch was ready. They were very happy seeing their concept and started discussing among themselves regarding the feasibility of the design. They asked whether the designer can help in developing a prototype of their concept for testing purposes. At this, one of the optometrists mentioned that the concept though having potential, may not be feasible due to mobility issues. As the design was considered mainly for underserved areas with no proper road connectivity, carrying the model would come as a significant challenge. All the participants agreed to this and discussed that if the same concept is made modular, then it may serve the purpose. Due to time constraints, the doctors were not able to discuss it further on that day. They expressed their enthusiasm and asked for another date for developing the modular concept.

The participants were approached on the date of appointment for discussing their second concept. On the day of the appointment, one of the doctors did not turn up due to an emergency. The team this time comprised two ophthalmologists, four optometrist, and two nurses. The same process was adapted in sketching concept 2 as in concept 1. The design parameter this time included modular pole and screen design. In concept 2, the team tried to improvise concept 1 by making it mobile for the ease of carrying the setup. The team distributed each of the poles of concept 1 (Fig. 3.3), into three different parts. It indicates converting a single-pole measuring around 6.5 m in concept 1 to three smaller poles of 2 m, 2.5 m, and 2 m, respectively. Each of the poles will have a screen attached to it in a rotor. The participants in this concept also changed the screen design of the front and rear sides of the chamber. They said that instead of using screens, they would use square blocks for stability.

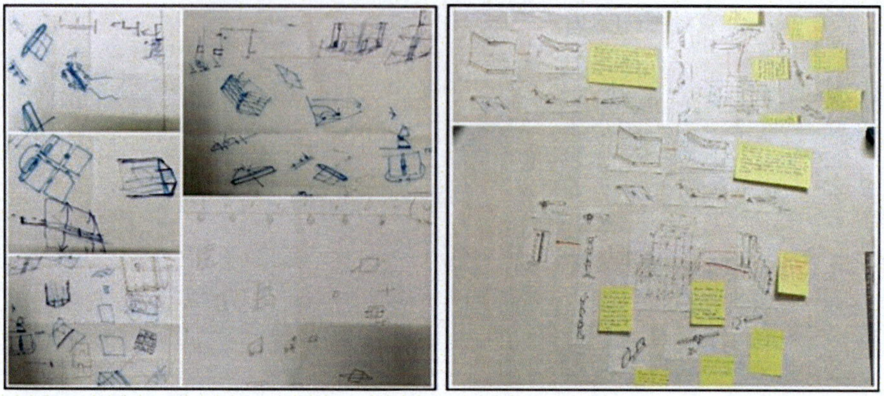

Fig. 3.4 Rough sketches along with refined sketch plan during ideation of Concept 2 by the team of healthcare experts

So, in place of the screens, several square blocks will be integrated into one. The purpose of using several numbers of blocks is mobility. It will be easy to transport smaller chunks than a larger single piece of block.

The participants also mentioned that they would provide markings on where to place the plane mirror and Snellen chart (standard chart used for visual acuity testing). Providing markers by maintaining the required distance will save time in assembling at the campsites. Target fixation is another issue encountered in rural eye campsites. If the target is already marked, then there will be no need to measure the distance at the site. By placing the mirror and the chart at the proper markings will serve the purpose very quickly. Thus, apart from solving issues like darkroom facility, congestion, and privacy, concept 2 (Fig. 3.4) may also solve problems like target fixation, and mobility.

Since concept 2 is designed to be modular, according to the team of experts it will be feasible to carry it in different terrain conditions. Concept 2 will also address the issues faced by the aging and the vulnerable population in attending rural eye camps by making it convenient to arrange the set up in close proximity with the vulnerable and aging population. This will address the distance issues faced by the rural community in attending health camps.

3.4 Conclusion

This design practice reflects on how the versatility of design can help in solving socially significant problems. Here the designer engages the healthcare experts in developing solution-oriented designs for improving the service experience in rural health camps. In this approach, the healthcare experts received a sense of design education from the conceptual stage of their design intervention. The healthcare experts expressed their enthusiasm in learning design from the designers as they have found it very useful in solving real-life problems. The team of healthcare experts successfully designed a chamber addressing the needs of the patients and health carers at rural health camps that, too, without much help from the designer. The designer, during the design stage, has never tried to overpower the health carers with their expertise in design. The concepts designed is solely the contribution of healthcare experts. The designer only helped the team in sketching their ideas. The team of healthcare specialists got a picture of the stages of design thinking through the design process. Encompassing design education among different domains outside the design fraternity will prepare them to think and act like designers. This will finally help in developing various solution-oriented designs based on the perspective of the experts on their respective grounds. A little momentum from the designers will help in bringing out the latent designer among personnel of different spheres. Design education should thus gear up for educating and imbibing ethics and social values not only to the design fraternity but also to the general public for contributing toward the upliftment of the society, which is a social responsibility.

3.5 Limitations and Scope for Future Work

While it is beyond the scope of this paper, practical assessment and analysis of the concepts with the patients, medical professionals, and designers in low-resource settings may well provide us with varied insights. Although a team of well-experienced healthcare professionals developed the concepts, these are to be tested in practice before any assumption can be made regarding its feasibility, viability, and desirability. This continues to be a work in progress, and future work will focus on developing prototypes of the concepts (following a collaborative approach among the healthcare experts and designers) and inclusion of patients in the testing phases.

References

1. Goodridge, D., Marciniuk, D.: Rural and remote care: overcoming the challenges of distance. Chronic Respir. Dis. **13**(2), 192–203 (2016)
2. Chamberlain, P., Craig, C.: Design for health: reflections from the editors (2017)
3. Tosi, F., Rinaldi, A.: Design and usability of the next medical devices for the home care. Des. J. **20**(sup1), S2033–S2043 (2017)
4. Foque, R., Lammineur, M.: Designing for patients: a strategy for introducing human scale in hospital design. Des. Stud. **16**(1), 29–49 (1995)
5. Beaser, R.S., Turell, W.A., Howson, A.: Strategies to improve prevention and management in diabetic retinopathy: qualitative insights from a mixed-methods study. Diab. Spectr. **31**(1), 65–74 (2018)
6. Srinivasan, N.K., John, D., Rebekah, G., Kujur, E. S., Paul, P., John, S.S.: Diabetes and diabetic retinopathy: knowledge, attitude, practice (KAP) among diabetic patients in a tertiary eye care centre. J. Clin. Diagn. Res.: JCDR, **11**(7), NC01 (2017)

7. Robin, A.L., Nirmalan, P.K., Krishnadas, R., Ramakrishnan, R., Katz, J., Tielsch, J., Friedman, D.S.: The utilization of eye care services by persons with glaucoma in rural south India. Trans. Am. Ophthalmol. Soc. **102**, 47 (2004)
8. Oduntan, O.O., Mashige, K.P., Hansraj, R., Ovenseri-Ogbomo, G.: Strategies for reducing visual impairment and blindness in rural and remote areas of Africa. Afr. Vis. Eye Health **74**(1), 5 (2015)
9. Chacko, B.: 1 in 2 Indian diabetics unaware of their condition: study available: https://www.indiaspend.com/1-in-2-indian-diabetics-unaware-of-their-condition-study/. Accessed on 22 Nov 2019
10. Shroff, P.H., Patel, K.J., Kapadia, P.R., Patel, S.D.: Prevalence of diabetic retinopathy (DR) in rural areas of Western India. Natl. J. Community Med. **9**(9), 647–650 (2018)
11. Manu, A.S., Davalgi, S.B., Aithal, S.S., Dilip, B.: Awareness of diabetic retinopathy and barriers for eye screening among adults with type 2 diabetes mellitus attending tertiary care teaching hospital, Davanagere, Karnataka. Int. J. Med. Sci. Public Health **7**(9), 686–691 (2018)
12. Murthy, G.: The emerging epidemic of diabetic retinopathy in India: report of a situational analysis and evaluation of existing programmes for screening and treatment of diabetic retinopathy. IAPB report 2015
13. Poore, S., Foster, A., Zondervan, M., Blanchet, K.: Planning and developing services for diabetic retinopathy in Sub-Saharan Africa. Int. J. Health Policy Manage. **4**(1), 19 (2015)
14. Chakrabarti, R., Harper, C.A., Keeffe, J.E.: Diabetic retinopathy management guidelines. Expert Rev. Ophthalmol. **7**(5), 417–439 (2012)
15. Pandey, S.K., Sharma, V.: World diabetes day 2018: battling the emerging epidemic of diabetic retinopathy. Indian J. Ophthalmol. **66**(11), 1652 (2018)

Chapter 4
Frugal-IDeM: An Integrated Methodology for Designing Frugal Innovations in Low-Resource Settings

Santosh Jagtap

Abstract People living in low-resource settings at the base of the world income pyramid (i.e. base of the pyramid—BOP) face several constraints. To satisfy their unmet needs, integrated frugal innovations are necessary. Several studies, discussed using many names such as 'design for base of the pyramid', 'design for development', 'product service systems', 'frugal innovations', 'humanitarian engineering', 'appropriate technology', have explored the design of such integrated frugal innovations for the BOP. Based on ten key design aspects, gleaned from such studies, this paper aims at developing an Integrated Design Methodology for designing frugal innovations for the BOP (Frugal-IDeM). We have related the key design aspects with the various phases in the design and development process. The Frugal-IDeM also includes recommendations for how to implement the key design aspects.

4.1 Introduction

People living in low-resource settings at the base of the pyramid (BOP) face many problems such as lack of income generation opportunities and poor access to basic facilities such as healthcare, education and clean drinking water (e.g. [17]). Design, with its central idea of changing an existing situation into a desired situation, is essential to alleviate problems of BOP people [5]. Appropriately designed frugal innovations can potentially satisfy their needs [1]. Such frugal innovations include, for example, water and sanitation facilities, systems providing access to healthcare services, etc.

Frugal innovations for satisfying unmet or underserved needs of BOP societies need to address many different constraints such as poor physical infrastructure, shortage of resources, illiteracy of BOP people and their low income [21]. Therefore, activities of designing such innovations must address several requirements and deprivations in these societies, resulting into holistic solutions (e.g. product service

S. Jagtap (✉)
Blekinge Institute of Technology, Karlskrona, Sweden
e-mail: santosh.jagtap@bth.se

A. Chakrabarti et al. (eds.), *Design for Tomorrow–Volume 2*, Smart Innovation, Systems and Technologies 222, https://doi.org/10.1007/978-981-16-0119-4_4

systems—PSS). Such holistic, integrated innovations include, for example, appropriate systems to address the problem of weak distribution channels or they may include suitable systems to repair implemented products [18, 21].

Whilst some authors have proposed design methodologies for developing integrated innovations in the context of Western countries, there is the absence of such design methodologies for the BOP context (e.g. [6]). Since design is context-specific, there is a profound need of developing integrated design methodologies, specifically tailored to the BOP. The objective of this study is to develop an Integrated Design Methodology (IDeM) for designing frugal innovations to satisfy unmet or underserved needs of BOP people. We call this methodology as Frugal-IDeM. The Frugal-IDeM is based on key aspects of designing integrated frugal innovations for the BOP, gleaned from the analysis of several studies undertaken in the BOP context. The research methodology used for searching and analysing these studies is available in our previous study [6]. Following this introduction, the rest of the paper is organized as follows. Whilst Sect. 4.2 presents key design aspects, they are synthesized in the form of an integrated methodology in Sect. 4.3. Finally, Sect. 4.4 presents conclusions.

4.2 Key Aspects of the Integrated Design Methodology

There are several models of design and development processes. In general, the process of designing and developing innovations consists of the phases (e.g. [7]): (1) *project planning and task clarification* (e.g. contextual understanding, need-identification); (2) *conceptual and detail design* (e.g. generation and evaluation of concepts, detail design); and (3) *realization and implementation* (e.g. testing, manufacturing, maintenance, etc.). This section presents the ten keys aspects of designing integrated frugal innovations for the BOP and relates these aspects to above-mentioned phases of the design and development process.

4.2.1 Holistic Contextual Understanding

BOP societies face many different problems. Several scholars have devised taxonomies to categorize these problems. Some scholars have classified them into individual, institutional, infrastructural and technological categories (e.g. [1]), whereas others gave organized them into economic-, psychosocial-, physical- and knowledge-deprivations (e.g. [15]). Therefore, satisfying needs of BOP societies demand identifying and addressing many problems. Innovations designed for these societies must be based on comprehensive contextual understanding of various issues in these societies.

Several scholars have offered methods for developing and establishing comprehensive understanding of the target context. For example, Aranda-Jan et al. [1]

developed a method called 'holistic design framework' to support designers in understanding the target context. Another method is people, objects, environments, messages and services (POEMS) framework [22]. It is essential to develop comprehensive understanding of the context early in the design process—that is—in the project planning and task clarification phase (see Fig. 4.1a).

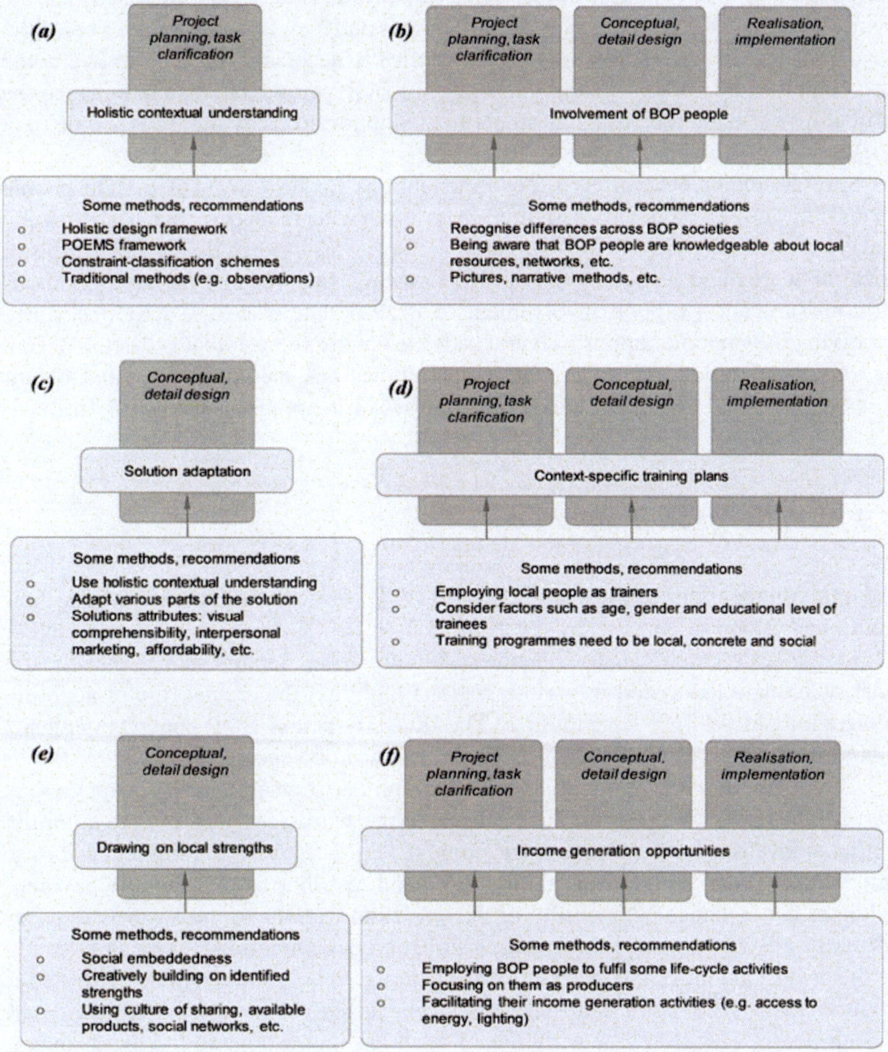

Fig. 4.1 Key activities and recommendations to support activities in the design and development process

4.2.2 User Involvement

Innovations designed without involving BOP people in the design process may not be adopted and used on sustained basis when the teams developing those innovations leave the community or begin working on some other projects (e.g. [16]). Outsiders, living in developed countries or relatively wealthy areas of developing countries, are typically unfamiliar with the lives of BOP people and their circumstances. Therefore, involvement of BOP people in design activities is necessary not only to overcome their unfamiliarity about life circumstances of BOP people, but also to enhance the probability of continued usage of implemented innovations by the BOP people (e.g. [5]).

Several studies have offered recommendations for how to involve BOP people in design and development activities. It is crucial to recognize that BOP people have expertise in living in low-resource settings and have in-depth knowledge of local networks and resources (e.g. [14]). Drawing, pictures and narrative methods of communication support involvement of BOP people in design activities. Since involving BOP people supports several activities such as conceptual and detail design as well as realization and implementation of innovations, the activities related to their involvement need to be undertaken in all phases of the process (see Fig. 4.1b).

4.2.3 Innovation Adaptation

Adapting innovations to the specificities of BOP communities is essential for their continued adoption and use by the BOP people [6]. Ernst et al.'s [3] and UNDP [21] large-scale empirical studies undertaken in various sectors such as healthcare, energy, transportation and in various regions highlight the crucial role of adapting innovations to the BOP specificities. The diversity across BOP contexts demands adapting innovations to the needs and circumstance of a specific context.

Another recommendation found in the literature is to synergistically adapt various parts of the innovation (e.g. products, necessary networks, services, etc.) to tailor the entire innovation for a specific BOP context (e.g. [2]). Visual comprehensibility, interpersonal marketing, affordability, atomized distribution and flexible payment schemes can enhance adaptability of designed innovations to the specific requirements of a BOP context. Implementing holistic contextual understanding, gleaned in the project planning and task clarification phase, to tailor innovations to the target context of the BOP communities needs to be carried out in conceptual and detail design phase. Therefore, the activities of adapting innovations to the local context are mainly undertaken in the conceptual and detail design phase (see Fig. 4.1c).

4.2.4 *Drawing on Local Strengths*

Because BOP societies face many different constraints and deprivations, they exhibit peculiar socio-cultural characteristics. For instance, the values of sharing resources are not uncommon among BOP communities (e.g. [19]). They typically give priority to building trusting relationships over business transactions. Previous studies have found that drawing on the strengths of BOP societies is a key strategy in designing innovations in this field (e.g. [21]). Besides drawing on socio-cultural strengths of BOP societies, many studies suggest building on existing available resources, supporting the attempts of reducing costs and enhancing affordability (e.g. [17]).

Previous studies have offered recommendation for how to build on strengths of BOP communities. Leveraging strengths of BOP societies demands identifying their strengths. Social embeddedness supports design teams in gaining deep understanding of various socio-cultural strengths and resources available in these communities. Some studies also suggest creatively building on the identified strengths of BOP societies. For example, designed innovations can use sharing culture and available products available in BOP societies (e.g. [9]). The activities of building on these strengths are predominantly undertaken in conceptual and detail design phase (see Fig. 4.1e).

4.2.5 *Context-Specific Training Plans*

Designing innovations for the betterment of BOP societies requires many inputs from several actors. These actors may lack all the knowledge and skills required to design, develop, implement the innovation (e.g. [13]). Training these actors to build and enhance their knowledge and skills in some specific areas is therefore essential. For example, training programmes are essential to overcome BOP people's illiteracy and lack of required skills to perform some design activities or to implement and maintain innovations. In a similar fashion, suitable training programmes can be necessary to train other stakeholders. Jagtap et al.'s [9] study, analysing many design cases, found that most of the cases employed training to support involved stakeholders.

Some studies suggest training a few local people in the initial phase and then employing them to train others in the community (e.g. [2]). This strategy builds on the language skills of the local trainers and their knowledge about cultural norms acceptable in the local community. Factors such as age, gender, educational level of trainees ought to be considered [9]. Programmes designed to train BOP people need to be local, concrete and social. The activities of designing and implementing training plans need to be considered throughout the design process (see Fig. 4.1d).

4.2.6 Income Generation Opportunities

Income is a key dimension of resource scarcity in BOP communities. Increasing income is an effective alternative to address their problems. BOP people prefer income generating solutions. For example, a study, examining the design of an irrigation system, found that low-income farmers participated in the implementation of those systems as they were aimed at increasing their agricultural productivity and thus their income [18].

Some studies recommend creating income generation opportunities by employing BOP people to address functions such as manufacturing, maintenance (e.g. [12]). Their income can also be raised by focusing on them as entrepreneurs. The activities of designing and implementing income generation opportunities need to be considered throughout the design process (see Fig. 4.1f).

4.2.7 Suitable Awareness Programmes

BOP people typically face problems such as illiteracy, knowledge-deprivations and short-changed education (e.g. [9]). As a result of these problems, they may face difficulties in reading product labels or selecting a more suitable product option. They may also lack awareness about availability of existing products or benefits of an innovation. As such, innovations for BOP societies must consist of awareness programmes that are appropriate for their life conditions and ways of thinking (e.g. [21]). Previous research suggests that appropriate awareness programmes support the diffusion of innovations in BOP communities (e.g. [19]).

Some studies have offered recommendations for how to design awareness programmes for BOP communities. Such awareness programmes ought to consider pictographic and concrete thinking styles of BOP individuals (e.g. [19]). The awareness programmes can also use demonstration strategies—that is—demonstrating potential advantages of an innovation, how it works, or how it needs to be used. The activities of designing and implementing suitable awareness programmes need to be considered throughout the design process (see Fig. 4.2a).

4.2.8 Implementation of Life Cycle Requirements

A broad range of individual-, institutional- and structural-level deprivations in BOP societies hinder many activities in the life cycle of an innovation such as implementation, repair and recycling [21]. When a product needs corrective maintenance, it can be difficult to repair it because of lack of maintenance management or shortage of

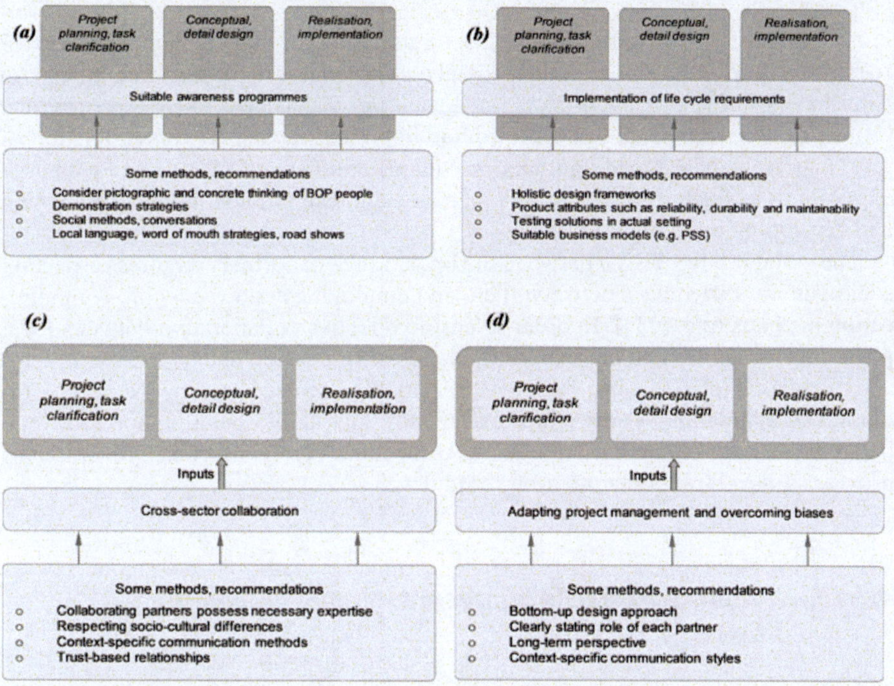

Fig. 4.2 Key activities, recommendations to support the activities, and phases in the design and development process

skilled people to repair a product. Therefore, in designing innovations for BOP societies, it is imperative that design teams devote efforts in identifying and implementing life cycle requirements.

Several scholars have developed recommendations for implementing life cycle requirements in designing innovations for the BOP. Some recommendations centre on enhancing product attributes such as reliability and maintainability (e.g. [1, 9]). It is also crucial to test innovations in actual setting. Life cycle activities can also be implemented by employing suitable business models such as PSS in which a solution developer remains responsible for life cycle activities (e.g. [11]). The activities of identifying and implementing life cycle requirements need to be considered throughout the design process (see Fig. 4.2b).

4.2.9 Cross-Sector Collaboration

Designing innovations to support development of BOP societies requires inputs from a broad range of stakeholders such as companies, NGOs and local governments (e.g. [21]). Collaboration between such stakeholders is useful to bring together their knowledge and skills in order to address a multitude of value chain gaps in BOP societies (e.g. [9]). NGOs, with their social embeddedness in local communities, can contribute towards gaining in-depth understanding of the needs of BOP people (e.g. [20]). Local governments can also provide important inputs, e.g. dissemination of information about designed solutions, providing subsidies, etc. Companies, with their expertise in technology and business management, can provide inputs in designing and developing products [18].

The collaborative design between the broad range of partners requires respecting socio-cultural differences between them and employing context-specific communication methods (e.g. [21]). In order to ensure effective collaboration between local governments, NGOs and companies, it is crucial to develop and support long-term and trust-based relationships. The involved partners also need to recognize that developing and maintaining partnerships can require long-term operating horizons (e.g. [9]). Cross-sector collaboration between companies, NGOS and local governments provides inputs throughout the design and development process (see Fig. 4.2c).

4.2.10 Adapting Project Management and Overcoming Biases

Jagtap et al.'s [10] study comparing processes of designing products for BOP and non-BOP contexts revealed that designers may handle requirements in a biased manner when they design for the BOP context. For example, they tend to give little attention to requirements about aesthetics and ergonomics despite their applicability in BOP context. Biases can also emerge from the ways in which NGOs, local governments and companies view each other. These partners need to view each other positively. In addition, innovation development for the BOP demands adapting project management to the specificities of the local context (e.g. [4, 10]).

In project management, the role of each partners needs to be explicitly stated. The involved partners need to adapt communicating styles to local specificities and conditions. The bottom-up approach can also assist stakeholders throughout the design and development process, supporting them to overcome biases and in adapting project management to local context. The activities of adapting project management and overcoming biases serve as inputs throughout the solution development process (see Fig. 4.2d).

4.3 Integrated Design Methodology for the BOP

Satisfying needs of BOP societies requires integrated innovations. In Sects. 4.1.1–4.1.10, we presented key aspects of designing and developing such innovations. Each of these aspects has been supported by findings of several studies, while providing methods and recommendations for how to implement these aspects. We also related these aspects with the phases in the design and development process. Whilst we presented these aspects in separate subsections in the preceding part of the paper, they are here put together in the form of an integrated design methodology—Frugal-IDeM (see Fig. 4.3). Of the ten aspects, the two aspects—'cross-sector collaboration' and 'adapting project management and overcoming biases'—provide requisite inputs and resources to undertake various tasks in the process. The remaining eight aspects are about undertaking essential activities in the process.

While some aspects focus on addressing value chain gaps in BOP societies, some focus on leveraging the strengths of these societies. For example, developing context-sensitive awareness and training programmes, creating income generation opportunities and designing systems to address life cycle activities are key aspects in addressing issues related to weak physical infrastructure, low literacy level of BOP people, their low and uncertain income, etc. Alternatively, aspects of drawing on local strengths and involving BOP people throughout the process build on the strengths of BOP societies.

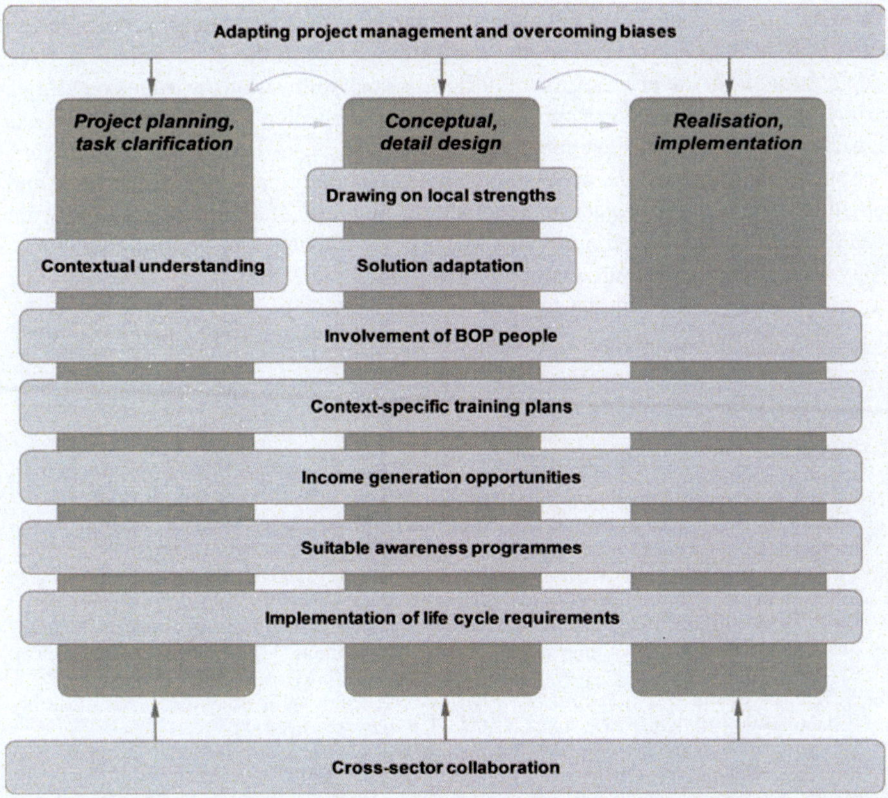

Fig. 4.3 Frugal-IDeM—Integrated methodology for designing frugal innovations

Some design aspects are about addressing weaknesses or building on strengths of outside partners, who typically come from non-BOP contexts, but contribute towards the design and development of innovations for BOP societies. For example, the outside partners bring in their expertise in design or technology. On the other hand, biases (if any) of outside partners or their lack of knowledge about BOP societies are addressed by understanding the local context holistically and by co-designing with BOP people.

4.4 Conclusions

Integrated innovations are necessary to satisfy needs of BOP people. We identified ten key aspects of designing such integrated innovations. These aspects are related to various phases in the design and development process, while offering methods and recommendations for their effective implementation. Innovations for the BOP ought to be adapted to the specificities of the local context by grounding them in the holistic contextual understanding. To enhance acceptance of these innovations, their design needs to be drawn on local strengths and resources, with the involvement of BOP people throughout the process. Since designing such innovations warrants tackling a variety of constraints, their design requires context-specific training and awareness plans, income generation avenues and implementation of life cycle requirements. Designing such innovations necessitates inputs from a broad range of partners, and demands adaptation of project management to local specificities, with a significant need to avoid biases. Based on the ten key aspects, we developed a unified methodology for designing integrated frugal innovations. Future studies can test and refine the methodology, for example, by seeking feedback of relevant stakeholders or by its implementation.

References

1. Aranda-Jan, C., Jagtap, S., Moultrie, J.: Towards a framework for holistic contextual design for low-resource settings. Int. J. Des. **10**(3), 43–63 (2016)
2. Devisscher, T., Mont, O.: An analysis of a product service system in Bolivia: coffee in Yungas. Int. J. Innov. Sustain. Dev. **3**(3–4), 262–284 (2008)
3. Ernst, H., Kahle, H.N., Dubiel, A., Prabhu, J., Subramaniam, M.: The antecedents and consequences of affordable value innovations for emerging markets. J. Prod. Innov. Manag. **32**(1), 65–79 (2015)
4. George, C., Shams, A.N., Dunkel, F.V.: Lessons learned in an international service-learning collaborative: Shea butter case study. NACTA J. **55**(2), 71–77 (2011)
5. Jagtap, S.: Design and poverty: a review of contexts, roles of poor people, and methods. Res. Eng. Des. **30**(1), 41–62 (2019)
6. Jagtap, S.: Key guidelines for designing integrated solutions to support development of marginalised societies. J. Cleaner Prod. **219**, 148–165 (2019)
7. Jagtap, S.: Design creativity: Refined method for novelty assessment. Int. J. Des. Creativity Innov. **7**(1–2), 99–115 (2019)

8. Jagtap, S., Kandachar, P.: Representing interventions from the base of the Pyramid. J. Sustain. Dev. **3**(4), 58–73 (2010)
9. Jagtap, S., Larsson, A., et al.: Design and development of products and services at the Base of the Pyramid: a review of issues and solutions. Int. J. Sustain. Soc. **5**(3), 207–231 (2013)
10. Jagtap, S., Larsson, A., Hiort, V., Olander, E., Warell, A., Khadilkar, P.: How design process for the Base of the Pyramid differs from that for the Top of the Pyramid. Des. Stud. **35**(5), 527–558 (2014)
11. Lemaire, X.: Fee-for-service companies for rural electrification with photovoltaic systems: the case of Zambia. Energy Sustain. Dev. **13**(1), 18–23 (2009)
12. Letaifa, S., Reynoso, J.: Toward a service ecosystem perspective at the base of the pyramid. J. Serv. Manage. **26**(5), 684–705 (2015)
13. López, A.M., Musonda, F., Sakao, T., Kebir, N.: Lessons learnt from designing PSS for Base of Pyramid. Procedia CIRP **61**, 623–628 (2017)
14. Murcott, S.: Co-evolutionary design for development: influences shaping engineering design and implementation in Nepal and the global village. J. Int. Dev. **19**(1), 123–144 (2007)
15. Nakata, C., Weidner, K.: Enhancing new product adoption at the base of the pyramid: a contextualized model. J. Prod. Innov. Manage. **29**(1), 21–32 (2012)
16. Nieusma, D.: Alternative design scholarship: Working toward appropriate design. Des. Issues **20**(3), 13–24 (2004)
17. Prahalad, C.K.: The fortune at the Bottom of the Pyramid: Eradicating Poverty Through Profits. Wharton School Publishing, Upper Saddle River, NJ (2004)
18. Jagtap, S., Larsson, T.: Resource-limited societies, integrated design solutions, and stakeholder input. She Ji: J. Des. Econ. Innov. **5**(4), 285–303 (2019)
19. Jagtap, S., Larsson, A.: Design of product service systems at the base of the pyramid. In: Chakrabarti, A., Prakash, R.V. (eds.) ICoRD'13, pp. 581–592. Lecture Notes in Mechanical Engineering, Springer India (2013)
20. Rivera-Santos, M., Rufín, C.: Global village vs. small town: Understanding networks at the Base of the Pyramid. Int. Bus. Rev. **19**(2), 126–139 (2010)
21. UNDP: Creating Value for All: Strategies for Doing Business with the Poor. United Nations Development Programme (2008)
22. Whitney, P., Kelkar, A.: Designing for the Base of the Pyramid. Des. Manage. Rev. **15**(4), 41–47 (2004)

Chapter 5
Artificial Neural Networks Supporting Cause-and-Effect Studies in Product–Service System Development

Omsri Kumar Aeddula, Johan Wall, and Tobias Larsson

Abstract A data analysis method based on artificial neural networks aiming to support cause-and-effect analysis in design exploration studies is presented. The method clusters and aggregates the effects of multiple design variables based on the structural hierarchy of the evaluated system. The proposed method is exemplified in a case study showing that the predictive capability of the created, clustered, dataset is comparable to the original, unmodified, one. The proposed method is evaluated using coefficient of determination, root mean square error, average relative error, and mean square error. Data analysis approach with artificial neural networks is believed to significantly improve the comprehensibility of the evaluated cause-and-effect relationships studying PSS concepts in a cross-functional team and thereby assisting the difficult and resource-demanding negotiations process at the conceptual stage of the design.

5.1 Introduction

Concept evaluation in product development has historically been overly reliant on experts' ambiguous and subjective judgments and qualitative descriptions [1]. Hence, a major quest in engineering design is the search for information and the subsequent analysis and build of knowledge. An efficient way to generate data is through experimentation. This means running some kind of test procedure for various input data and to study how these data affect the output of the procedure. Experimentation is commonly done using models, simplified representations of the studied system, nowadays predominantly mathematical models. Experimentation with mathematical models is here referred to as simulation. Building and solving mathematical models, considering a diverse set of systems and disciplines, have received significant attention within the research community during the last decades. However, the question regarding how designers should interpret and use these models and solutions in design has been largely neglected [2]. This problem is even more pronounced considering

O. K. Aeddula (✉) · J. Wall · T. Larsson
Blekinge Institute of Technology, Karlskrona, Sweden
e-mail: omsri.kumar.aeddula@bth.se

A. Chakrabarti et al. (eds.), *Design for Tomorrow–Volume 2*, Smart Innovation, Systems and Technologies 222, https://doi.org/10.1007/978-981-16-0119-4_5

that designing industrial products or services is nowadays normally a group effort and rarely an individual act [3].

The transition away from product view, toward providing product–service systems, increasing product complexity, and the long-prevailing concurrent engineering approach, makes inroads for this development [4]. This emphasizes the need for collaborations, internally within the organizations and externally with suppliers, customers, and other stakeholders in the value chain, especially in the early phases of product development, where the information is scarce, vague, and potentially conflicting.

Supporting cross-functional design teams with data analysis and visualization, helping them to interpret and utilize simulation results, hence become a key success factor, while providing a potential setting for information sharing, communication, understanding, and building of knowledge. Data analysis assists in transforming information into knowledge, exploiting the relationships among the data, to drive the product or service design process [5].

Bititci et al. [6] emphasize exploring the cause-and-effect relationships among the multiple variables involved, as a part to measure the quality of design improvement. Wall et al. [7] propose a data analysis method based on partial least square regression supporting cause-and-effect analysis in design exploration studies for linear and continuous variables. The proposed method clusters and aggregates the effects of multiple design variables based on the structural hierarchy of the evaluated system. According to [8], multiple linear regression analysis is a tool to understand the relationship between one dependent variable and several independent variables. Linear regression methods are easy to implement and simpler in design. The limitation of multiple linear regression analysis is that they are modeled for linear data, owing to which they cannot discern any nonlinear relationships in data [9]. However, design problems often contain nonlinear relationships between design variables and attributes. Hence, a method able to cope also with nonlinear relationships is sought [10].

In recent years, artificial neural networks have been widely employed in modeling dynamic systems, as they are found to accurately model continuous functions and shown to have good predictive performance in simulations of nonlinear dynamic systems, due to their adaptability, flexibility, and optimization capabilities [11]. They have the ability to learn the system behavior from samples by inspection. This paper extends and modifies the method proposed by [7].

An artificial neural network-based data analysis method to understand quantified cause-and-effect relationships of the studied system is proposed. More specifically the aim is to develop a method able to support quantification of dependencies between design variables (independent variables) and design attributes (dependent variables), irrespective of the data linearity, on a component or subsystem level in cases where more than one independent variable drives the configuration of the component or subsystem.

The remainder of this paper is organized as follows; In Sect. 5.2, the proposed data analysis method is presented. In Sect. 5.3 the proposed method is exemplified through a case study. The paper concludes with discussion, conclusions, and directions for future work.

5.2 Data Analysis Method

A cause-and-effect relationship may be studied by establishing mathematical equations between design variables (independent variables) and design attributes (dependent variables). Design variables are entities that may be changed in an experiment affecting the shape or properties of the studied system. An attribute can be defined as any aspect of the product itself or its use that can be used to compare product alternatives [12]. As an example, the attributes of an automobile might be acceleration or fuel consumption. Quantification of dependencies between design variables and design attributes is established by analyzing the design variable's contribution toward the design attribute. Linear regression methods analyze the design attributes dependencies on the design variables from the generated linear mathematical equation coefficients [13]. This type of analysis fits for linear observed variables and limits the applications to data with nonlinearities and for clustered variables.

The variables in a design exploration study might be classified as a two-level hierarchy including design variables and design attributes. Typically these design variables are not identifiable as components or subsystems. Rather, a subset of design variables in a design exploration drives the configuration of a particular component or subsystem and also attributes associated with that component or subsystem. As this "intermediate level" of components and subsystems are not directly included in the conducted experiment, data to populate, and in the next step analyze it, does not exist. Rather it is to be inferred from the set of observed design variables by clustering the identified subset of design variables and aggregate their effect on the design attribute. Creating an intermediate level without losing the significant contributions of other non-clustering design variables is a vital task.

Data linearity affects the "goodness" of fit of the intermediate level in the structural hierarchy [7]. This paper proposes a variable clustering method, in combination with an artificial neural network to fit the intermediate level, irrespective of the linearity without altering the significant contributions of other non-clustering design variables. The independent variables are clustered based on the structural decomposition given by the structural hierarchy of the studied system. This inline with the research aim presented in the introduction. Generating new, clustered, variables inferred from the observed design variables is challenging, and adding the element of relationship exploration without losing the significant contributions of other design variables serves to confound the process even further.

Figure 5.1 depicts an overview of the proposed method, schematically also showing its role in the concept evaluation process. In the proposed method, data is initially analyzed to identify the continuous and categorical variables. A subset of design variables, according to the structural hierarchy, is clustered and each design variable forms a node in the input layer of the designed artificial neural network. The

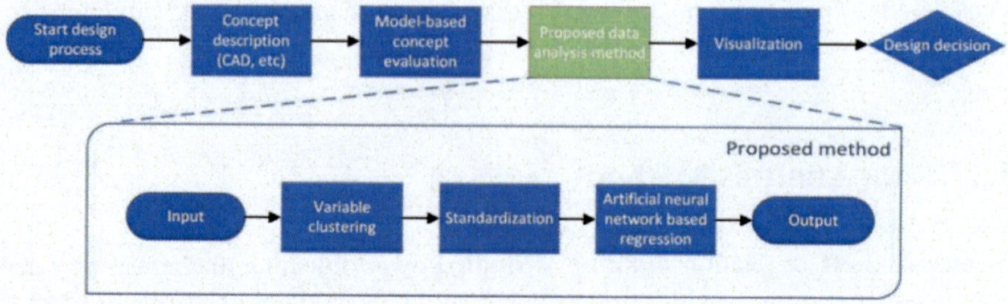

Fig. 5.1 Overview of the proposed method

network has a single output node with clustered variable constructs. Exploration of the neural networks layer weights reveals the contributions of design variables on the design attribute [14]. Design variables and design attributes are normally of different metric units. In this work, they are standardized to scale down all the variables to a common metric scale. Standardization reduces the mean value of the variables to "0" and their standard deviation to "1", aligning all the variables, distributed normally, on the same scale. This helps to understand the contributions of design variables on a design attribute when they have different metric units [15]. An artificial neural network with designated nodes based on the number of design variables and a predetermined number of hidden layers is employed to understand the cause-and-effect relationships for the desired hierarchical levels or between desired design variables and a design attribute.

5.2.1 *Artificial Neural Networks*

An artificial neural network is a mathematical and computational model that simulates the human brain functions of perception, computation, and memory. The ability of neural networks to learn and generalize from the input data makes them a powerful tool to solve numerous real-world applications [16]. An artificial neural network is a system consisting of processing elements namely nodes in each layer with connections (Synapse) between them. Its inherently nonlinear structure is particularly useful in exploring complex relationships of real-world problems. Artificial neural networks learn from the data through iterations without any prior knowledge, capable of handling data with noisy, linear, and nonlinear relationships [17].

In the studied system, the architecture of the artificial neural network is characterized by a single input layer, a single hidden layer with a nonlinear activation function, and a single output layer. A multilayer perceptron, also referred to as a feedforward network, is employed featuring supervised learning. The network maps a set of inputs, design variables, onto a design attribute linking through the hidden layer. The neural network is depicted in Fig. 5.2. The input layer is a set of neurons with design variables and the output layer contains a single neuron with a design

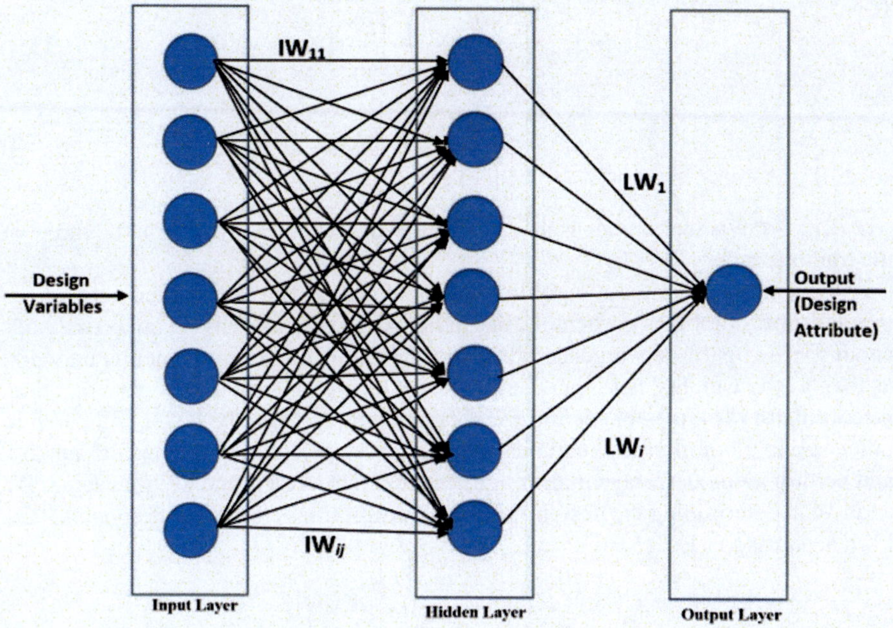

Fig. 5.2 Artificial neural network architecture

attribute. IW and LW are the weights associated with the input-hidden layer and hidden-output layer, respectively.

The adapted neural network is described [18], according to Eq. (5.1):

$$y = (\sum_{j=1}^{h} \mathrm{LW}_j * f_{\mathrm{inner}}\left(\sum_{i=1}^{d} \mathrm{IW}_i * x_i + b_0^{(1)}\right) + b_0^{(2)}) \tag{5.1}$$

where y is the design attribute; IW_i and LW_j indicate the weights connecting the input layer with hidden layer and hidden layer with output layer, respectively; h is the total number of hidden layer neurons; d is the number of input layer neurons; $b_0^{(1)}$ and $b_0^{(2)}$ are the biases for the hidden neuron and the output neuron. f_{inner} is a hyperbolic tangent sigmoid activation function. A hyperbolic tangent sigmoid transfer function is given by Eq. (5.2):

$$f_{\mathrm{inner}}(n) = \frac{2}{(1 + \exp(-2 * n)) - 1} \tag{5.2}$$

Network weights are identified using the following iterative method [19], according to Eq. (5.3) and (5.4):

$$\{w\}_{i+1} = \{w\}_i - \eta \frac{\partial E}{\partial \{w\}}(\{w\}_i) \tag{5.3}$$

$$E = \sum_{n=1}^{N} \{t_n - y_n\}^2 \tag{5.4}$$

where t_n is the actual output value; y_n is the estimated value; and n is the index of the training data.

A Levenberg–Marquardt algorithm is employed to train the adapted network, as the algorithm combines the benefits of gradient descent and Gauss–Newton methods and it speeds up the convergence rate [20]. The algorithm trains the neural network, as long as the weights, net inputs, and transfer functions have a derivative function. According to [21], the connection weights method proved to be better compared to other variable contribution methods such as the stepwise method, profile method, and perturb method. Connection weights method make use of the weight vectors IW and LW in determining the design variables contributions (*RI*) to the design attribute [22], according to Eq. (5.6):

$$Q_{(h,i)} = \frac{\left|(\mathrm{IW}_{(h,i)})\right| * |(\mathrm{LW}_h)|}{\sum_{i=1}^{n}\left|(\mathrm{IW}_{(h,i)})\right| * |(\mathrm{LW}_h)|} \tag{5.5}$$

$$\mathrm{RI}_n = \frac{\sum_{h=1}^{n} Q_{(h,i)}}{\sum_{i=1}^{n}\sum_{h=1}^{n} Q_{(h,i)}} \tag{5.6}$$

where h is the total number of neurons in the hidden layers; and n is the total number of neurons in the input layer.

5.2.2 *Variable Clustering*

Design variables clustering involves constructing latent variables, inferred from the observed design variables. In the studied system, design variables are clustered according to the referenced structural hierarchy. Each design variable to be clustered is expressed as a product of two matrices with weights for each variable value and original data with a rotated coordinated system, according to Eq. (5.7).

$$P = [w] \times Z^T \tag{5.7}$$

where $[w]$ is the weight matrix of the design variable; and Z is the rotated coordinated system matrix.

The rotated matrix is analyzed to estimate the variance (v) of the rotated matrix. Clustered variables are identified as a sum of the product of total variance of the rotated matrix with the rotated matrix of associated design variables to be clustered, according to Eq. (5.8):

$$X_{\text{new}} = \{(V_1 \times Z_1) + (V_2 \times Z_2) + \cdots + (V_n \times Z_n)\} \tag{5.8}$$

where X_{new} is the new clustered variable in the intermediate hierarchy level, according to variable hierarchy; V_i is the total variance of the rotated matrix of the ith design variable; and n is the number of variables to be clustered.

5.3 Case Study

To exemplify the proposed method, it was used to analyze an existing dataset originating from a design exploration study evaluating proposed concepts of road construction equipment. In the study, 700 variants of the vehicle platform were studied. Variants driven by seven design variables and 16 design attributes were assessed. The dataset is divided into two sets with 70% of data for the artificial neural network process and 30% for testing the derived observations from the proposed method. The adapted neural network divides the input data into three sets, training, validation, and testing sets, with 70% of input data going to the training set.

A value-driven design (VDD) approach was applied in the design experiment as systems engineering (SE) research has stressed the importance of the value model to frontload engineering design activities [23]. This model is expressed as a single objective function that aims at measuring the "goodness" of the design. More information on how this dataset is generated and the applied modeling and simulation scheme can be found in [24]. The model-based experimentation relies on a hybrid model environment evaluating both performance and resource space. The value model was fed with the output of these models to render a value score for the design configuration under analysis. All aspects of value were quantified in monetary terms, enabling easy tradeoff with more traditional requirements. This quantification process was based on the implementation of net present value (NPV) from the VDD literature [25]. The dataset also contained information enabling mapping each design variable directly to components or subsystems in the structural hierarchy.

The studied dataset contained three categorical variables rendering nonlinear behavior. These variables were transformed into continuous variables, generating dummy variables equal to the number of classes in a categorical variable [26]. Dummy variables are continuous variables with logical values representing subgroups of the categorical variable. The model was custom trained to distinguish between categorical and continuous variables and transforming all categorical variables to continuous variables.

To exemplify the method, NPV was chosen as a design attribute, and the relative importance of the continuous design variables was assessed. The relative importance values could be used to understand the cause-and-effect proposition on the magnitude scale, where it represents the relative change in the design variables for a unit change in the design attribute value. The relative importance of the design variables for the

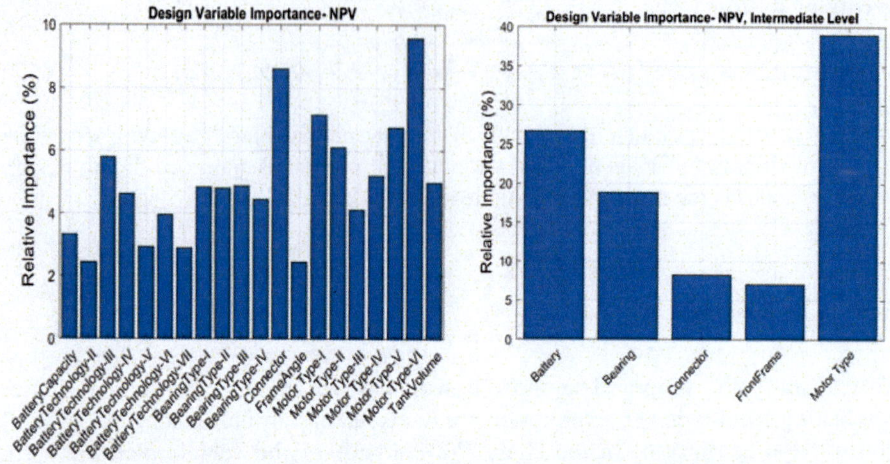

Fig. 5.3 Variable importance

design attribute "NPV" is shown in the left part of Fig. 5.3. The design variables were clustered according to the structural hierarchy, rendering an intermediate level with five clustered design variables. At this stage, these five variables are interpretable as subsystems or components of the studied system. The contributions of the clustered design variables are shown in the right part of Fig. 5.3. The non-clustered variable "connector" has a contribution value of 8.6 both in the left and right part of Fig. 5.3, indicating that the significance of the non-clustered variable remains the same.

The mathematical equation framing design attribute from design variables, according to Eq. (5.1), is a nonlinear function, thereby the proportionality is also a nonlinear function. The cause-and-effect relationship between a specific design variable and the design attribute was analyzed by varying the value of a design variable while keeping all other design variables constant. This method of estimation is cumbersome, as the artificial neural networks popularly referred to as "black box" [16] does not provide a direct method to estimate the variable contribution with magnitude and direction directly.

Statistical analysis technique, coefficient of determination (R^2), was used for methodology validation. It determines how close the predicted data matches the original data, according to Eq. (5.9) [27].

$$R^2 = 1 - \frac{\sum (y - Y)^2}{\sum (y - \bar{y})^2} \tag{5.9}$$

where y represents the actual data values; Y is the predicted (estimated) data values; and $\bar{y}$ is the mean value of y.

Figure 5.4 shows visualization of the goodness of fit for the NPV design attribute based on the proposed neural network architecture.

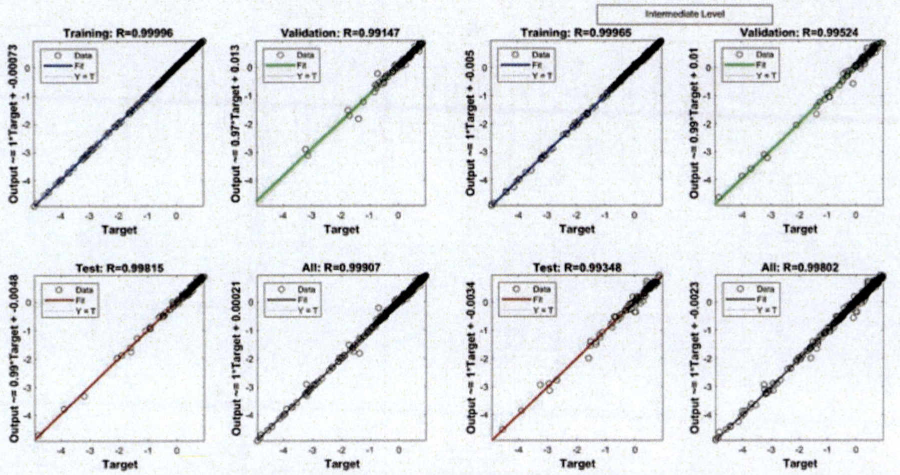

Fig. 5.4 Artificial neural network regression plot

The value of the regression coefficient of determination ranges from [0,1]. As the value of R^2 approaches unity, the predicted values are equal to the actual data values, i.e., the closer the value of R^2 to 1, the greater is the fit of the data, and the closer the value of R^2 to 0, the poorer is the fit of the data. For NPV, R^2 tends out to be 0.9990, in comparison coefficient of determination of assessed attributes are 0.98 and above. Clustered variables were derived from a set of observed design variables, thereby an indirect method was used for validation of intermediate level. Figure 5.5 shows the predictive capability of the derived neural network model using 30% of the unused design attribute (NPV) data both at the intermediate level and component level. Root mean square error (RMSE) and average relative error (RE) measure the deviation of the predicted values from actual observed values. Prediction of unused design attribute values, along with RMSE and RE, according to Eq. (5.10) and (5.11), [15] validates the methodology.

$$\mathrm{RMSE} = \sqrt{\frac{1}{n}\sum_{i=1}^{n}(Y_i - y_i)^2} \tag{5.10}$$

$$\mathrm{RE} = \left|\frac{Y - y}{y}\right| \tag{5.11}$$

where y represents actual data values; and Y represents predicted data values.

At the intermediate level, RMSE for the estimated design attribute tends out to be 0.0026 and average relative error of 0.0033, similarly, at the component level, RMSE for the design attribute tends to be 0.0023 and average relative error of 0.0029.

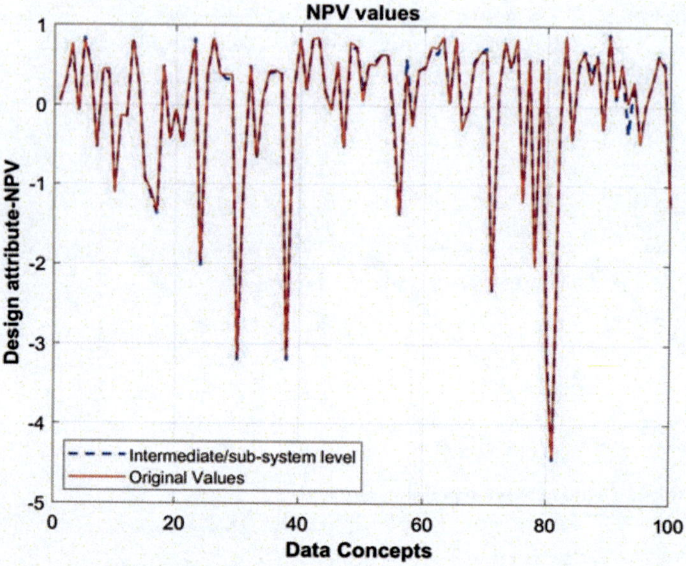

Fig. 5.5 Estimated data versus observed data

5.4 Discussion and Conclusion

An artificial neural network-based data analysis method that clusters and aggregates the effects of multiple design variables in a study is presented. The proposed artificial neural network is trained using the Levenberg–Marquardt algorithm, with 70% input data as training data. The method focuses on the exploration of cause-and-effect analysis providing a structured input to a visualization construct. The intention is to generate an intermediate level and map the analyzed data based on the structural hierarchy of the proposed concept.

The proposed method is based on artificial neural networks and matrix decomposition for variable clustering, generating an intermediate level in the structural hierarchy, thereby reducing the dimensionality of the data and hence decrease in computational complexity compared to the original data. The proposed method ensures that the significances of the variable contribution of the non-clustered design variables are not lost during the experimentation. The goodness-of-fit measure in the intermediate level reveals that the adapted neural network model closely fits the data, with clustered variable preserving the original information of the design variable. The proposed method extends the predictive range of linear relationship variable modeling, and the model is trained to solve complex data, overcoming the assumptions in [7]. The proposed method has no limitation to the number of variables, and the intermediate level analysis supports in reducing the computational resources and time compared to the computation of all design variables.

The method is exemplified through a case study including visualization of cause-and-effect relationships by bar graphs of the studied concept. Applying a validation scheme based on the coefficient of determination, testing of non-trained data, RMSE, and relative error, it is shown that the data structure using a "virtual" intermediate level performs, considering the nonlinearities in the original dataset, comparably to the original dataset when it comes to predicting attributes. Future works aim to extend the proposed method by implementing a cascaded network for clustering based on unsupervised learning and comparing it with other aggregating variable methods such as the Karhunen–Loeve method.

Acknowledgements The research leading to these results has received financial support by the Swedish Knowledge and Competence Development Foundation (Stiftelsen för kunskaps-och kompetensutveckling) through the Model Driven Development and Decision Support research profile at Blekinge Institute of Technology.

References

1. Zhang, Z., Gong, L., Jin, Y., Xie, J., Hao, J.: A quantitative approach to design alternative evaluation based on data-driven performance prediction. Adv. Eng. Inform. **32**, 52–65 (2017). https://doi.org/10.1016/j.aei.2016.12.009
2. Komoto, H., Masui, K.: Classification of design parameters with system modeling and simulation techniques. CIRP Ann. **63**(1), 193–196 (2014). https://doi.org/10.1016/j.cirp.2014.03.098
3. McComb, C., Cagan, J., Kotovsky, K.: Rolling with the punches: an examination of team performance in a design task subject to drastic changes. Des. Stud. **36**, 99–121 (2015). https://doi.org/10.1016/j.destud.2014.10.001
4. Sundin, E., Sandström, G.Ö., Lindahl, M., Rönnbäck, A.Ö., Sakao, T., Larsson, T.C.: 'Challenges for industrial product/service systems: experiences from a learning network of large companies', p. 7
5. Van Horn, D., Olewnik, A., Lewis, K.: Design analytics: capturing, understanding, and meeting customer needs using big data. In: Volume 7: 9th International Conference on Design Education; 24th International Conference on Design Theory and Methodology, Chicago, Illinois, USA, Aug 2012, pp. 863–875. doi: https://doi.org/10.1115/DETC2012-71038
6. Bititci, U., Nudurupati, S.: Driving continuous improvement. Manuf. Eng. **81**(5), 230–235 (2002). https://doi.org/10.1049/me:20020506
7. Wall, J., Aeddula, O.K., Larsson, T.: Data analysis method supporting cause and effect studies in product-service system development. Proc. Des. Soc. Des. Conf. **1**, 461–470 (2020). https://doi.org/10.1017/dsd.2020.123
8. Hair, J.F. (ed.): Multivariate Data Analysis with Readings, 4th edn. Prentice Hall, Englewood Cliffs, N.J (1995)
9. Jobson, J.D.: Multiple Linear Regression. In: Applied Multivariate Data Analysis. Springer, New York, NY, pp. 219–398 (1991)
10. Tenenbaum, J.B.: A global geometric framework for nonlinear dimensionality reduction. Science **290**(5500), 2319–2323 (2000). https://doi.org/10.1126/science.290.5500.2319
11. Haykin, S.S.: Neural Networks: A Comprehensive Foundation, 2nd edn. Prentice Hall, Upper Saddle River, N.J (1999)
12. Grunert, K.G.: Attributes, attribute values and their characteristics: a unifying approach and an example involving a complex household investment. J. Econ. Psychol. **10**(2), 229–251 (1989). https://doi.org/10.1016/0167-4870(89)90021-4

13. Understanding Regression Analysis. Springer US, Boston, MA (1997)
14. Olden, J.D., Jackson, D.A.: Illuminating the "black box": a randomization approach for understanding variable contributions in artificial neural networks. Ecol. Model. **154**(1–2), 135–150 (2002). https://doi.org/10.1016/S0304-3800(02)00064-9
15. Devore, J.L.: Probability and Statistics for Engineering and the Sciences, 8th edn. Brooks/Cole, Cengage Learning, Boston, MA (2012)
16. Hassoun, M.H.: Fundamentals of Artificial Neural Networks. MIT Press, Cambridge, Mass (1995)
17. Baughman, D.R., Liu, Y.A.: Neural Networks in Bioprocessing and Chemical Engineering. Academic Press, San Diego (1995)
18. Bishop, C.M.: Neural Networks for Pattern Recognition. Clarendon Press, Oxford; Oxford University Press, New York (1995)
19. Marwala, T., SpringerLink (Online service): Artificial Intelligence Techniques for Rational Decision Making. Springer International Publishing, Cham, Springer, Imprint (2014)
20. Gavin, H.P.: The Levenberg-Marquardt method for nonlinear least squares curve-fitting problems c © (2013)
21. Olden, J.D., Joy, M.K., Death, R.G.: An accurate comparison of methods for quantifying variable importance in artificial neural networks using simulated data. Ecol. Model. **178**(3–4), 389–397 (2004). https://doi.org/10.1016/j.ecolmodel.2004.03.013
22. Gevrey, M., Dimopoulos, I., Lek, S.: Review and comparison of methods to study the contribution of variables in artificial neural network models. Ecol. Model. **160**(3), 249–264 (2003). https://doi.org/10.1016/S0304-3800(02)00257-0
23. Collopy, P.D., Hollingsworth, P.M.: Value-driven design. J. Aircr. **48**(3), 749–759 (2011). https://doi.org/10.2514/1.C000311
24. Bertoni, M., Pezzotta, G., Scandella, B., Wall, J., Jonsson, P.: Life cycle simulation to support cross-disciplinary decision making in early PSS design. Procedia CIRP **83**, 260–265 (2019). https://doi.org/10.1016/j.procir.2019.03.138
25. Price, M., Soban, D., Mullan, C., Butterfield, J., Murphy, A.: A novel method to enable trade-offs across the whole product life of an aircraft using value driven design. J. Aerosp. Oper. **1**(4), 359–375 (2012). https://doi.org/10.3233/AOP-120028
26. Yip, P.S.L., Tsang, E.W.K.: Interpreting dummy variables and their interaction effects in strategy research. Strateg. Organ. **5**(1), 13–30 (2007). https://doi.org/10.1177/1476127006073512
27. Kurz-Kim, J.-R., Loretan, M.: A note on the coefficient of determination in models with infinite variance variables. SSRN Electron. J. (2007). https://doi.org/10.2139/ssrn.996664

Chapter 6
The Factors Influencing Usage Intention of Urban Poor Population in India Towards Mobile Financial Services (Mobile Payment/Money)

Raman Saxena and Ravi Mokashi Punekar

Abstract Considering the promise mobile telephony offers and the dramatic increase in mobile telephone usage in India, several initiatives have emerged in extending financial services to the marginalised population. However, these mobile financial services have not resulted in the desired acceptance among this targeted population. Following the contextual inquiry methodology, this paper presents the outcome of two empirical studies that aimed to identify factors that underlie this lack of usage of mobile payment options among the urban poor population. While the first study examines mobile phone features and services' usage patterns and services, the second study examines the prerequisites to access and use BHIM and Paytm, the two most popular mobile payment/money in India. The two studies' outcomes help understand the acceptance and adoption levels and their shortcomings in the existing design features. The study suggests desired features that may guide in designing of these financial services.

6.1 Introduction

The development of information and communication technologies (ICT), access to inexpensive mobile phone devices and decreasing mobile service tariff rates have made mobile phones globally affordable to large sections of society. Today, ICT is widely used and accessible among poor communities in developing countries [1]. There is a significant increase in mobile service penetration across the global population. Financial service delivery is no exception. There has been significant worldwide growth and penetration of mobile financial services among developing countries, including India. Kirui et al. [2] reported that there is growing evidence to

R. Saxena (✉)
USID Foundation, Noida, Delhi NCR Region, UP 201301, India
e-mail: raman@usidfoundation.org; raman.saxena@alumni.iitg.ac.in

R. Saxena · R. M. Punekar
Department of Design, Indian Institute of Technology Guwahati, Assam 781039, India
e-mail: mokashi@iitg.ac.in

A. Chakrabarti et al. (eds.), *Design for Tomorrow–Volume 2*, Smart Innovation, Systems and Technologies 222, https://doi.org/10.1007/978-981-16-0119-4_6

show a rapid increase in the access, adoption, and usage of mobile financial services (MFS) [2]. M-Pesa (Kenya) and PayPal mobile [3] are two such successful global case examples.

As per IAMAI and KANTAR-IMRB report published in India Today, March 2018 [4], it is estimated that there are 478 million mobile Internet users in India, including 291 million urban mobile Internet users and 187 million rural. This phenomenal growth is a potential digital and financial inclusion tool, reaching over 2.5 billion unbanked poor in developing countries [5]. However, as per the Gartner group report 2008, mobile-based financial services' overall adoption has relatively been slow. A similar pattern is seen in India [6]. As per the TRAI, in the year 2018, India has reached 1188.99 million mobile subscriptions, an increase of 25% since 2012 [7]. Mobile Internet subscribers have grown from 78.7 million [8] to 472.72 million [9]. However, during the same period in 2018, mobile money/ payment subscription has remained only 73.9 million despite having several offerings.

Such a potentially promising digital financial service offering received a required impetus with the Government of India's decision of demonetisation and subsequently encouraging digital payments in all modes of financial dealings that followed this decision. These drastic decisions initially forced people to switch to digital financial transactions. However, with the Indian currency's cash-to-GDP ratio returning to pre-demonetisation levels by May 2018 [11], such a monetary transaction trend among the poor and the marginalised communities have also drastically dropped down. BHIM and Paytm are currently the two leading mobile payment/money offerings in India. The Government of India has also laid renewed thrust and promoted digital modes of financial transactions. However, it is found that the adoption of mobile financial services is marginal compared to mobile phone ownership among the urban poor.

Given the accepted benefits of digital payments, it is pertinent to seek a potential explanation for the low adoption of mobile financial services among the urban poor. Could the demand side barrier such as affordability, low literacy and an understanding of technology usage behaviour be a critical focus area that needs to be examined?

These form the primary inquiry of this design research study. The outcomes of which subsequently feed as critical factors for consideration in the design of pro-poor mobile financial services' targeted India's urban economically poor population.

6.2 Problem Statement

In this study, we ask the following research questions:

1. Why is there such low adoption of digital payments and the waning interest in access to mobile financial transactions among the urban poor?
2. Is the failure due to technology acceptance and adoption? Could there be a mismatch in the service features not meeting this massive low income and urban poor segment's real-world needs in India?

The study investigates why the lack of acceptance and the factors and barriers limiting mobile payment/money acceptance. Towards meeting the above aim, this study set the following three research objectives:

1. Understand the functional/digital literacy (urban poor-demand side) related to mobile phone usage (skills, comfort level) and analyse if they hold any implication on acceptance and adoption of mobile financial services.
2. Understand the prerequisites (hardware, software, registration/set-up, cost, etc.) for using BHIM and Paytm if these prerequisites/preconditions (service providers-supply end) serve a barrier in acceptance and usage of mobile payment/money amongst the urban poor population.
3. Understand what "useful" means to the urban poor in their context of mobile financial services-mobile payment/money.

6.3 Literature Review

Design researchers follow various theoretical research models in the domain of user research and technology acceptance methods. These broadly include diffusion of innovation theory (DOI) [12], theory of reasoned action (TRA) [13], technology acceptance model (TAM) [14], theory of planned behaviour (TPB) [15], TAM2 [16], unified theory of acceptance and use of technology (UTAUT) [17]. All these aim to explain the acceptance and adoption of the users' behaviour towards technology. Among these, the technology adoption model (TAM) has been most widely used by researchers to study technology acceptance behaviour in various IT contexts.

6.3.1 Socio-economic Background and Cultural Context

Low literacy levels often serve as barriers to MFS's acceptance and usage among the urban poor population. Laforet and Li [18] found a lack of understanding of the concepts and benefits as the main barrier to consumers using mobile banking. A study in China by Trappey and Trappey [19] found that the traditional customer using cash for any financial transactions had no prior exposure and confidence in the new payment culture, which has impacted their acceptance of mobile payments. In a study undertaken in India, Singh [20] reported that more males than females from high-income groups are mobile banking users. Medhi et al. [21], research on the adoption and usage of existing mobile banking services by low-literate, low-income individuals reported a lack of awareness about the availability and features of M-banking services as the cause affecting their adoption. Bashir and Madhavaiah [22], in their study, found that perceived usefulness (PU), ease of use, trust, self-efficacy and social influence had a positive influence while perceived risk exerted a negative effect in the use of Internet banking. Singh [23] identifies "security/privacy, reliability, and responsiveness" essential factors in customer's perception towards mobile banking among 18 to 46+ years users with annual income above Rupees two lakhs.

Potnis [1], outlining a pro-poor perspective, suggests a "bottom-up" approach that considers specific contextual factors that influence access; community factors that influence adoption and cultural factors that influence the continuous usage of mobile financial services by the poor. Kim et al. [24] suggests the need for further research on aspects other than individual differences such as innovativeness and knowledge about mobile payment. They hinted that there could be many more characteristic variables that can affect the intention to use mobile payment.

6.3.2 Technology Acceptance Model (TAM)

Researchers more popularly apply the technology adoption model (TAM) to understand technology acceptance behaviour in a variety of information technology (IT) contexts. TAM examines parameters such as perceived usefulness and perceived ease of use to understand their relationship with external variables and their influence on information system adoption [25, 26]. Perceived usefulness is defined as the degree to which a person perceives that adopting the system will boost their job performance resulting in an immediate effect on one's adoption intention. Kim et al. [24] have defined the perceived ease of use as the degree to which an individual believes that adopting the system will be free of efforts resulting in both an immediate effect and an indirect effect adoption intention. The meta-analysis of these factors reveals that perceived usefulness and ease of use are the most influential factors that impact the behavioural intention towards the adoption of MFS [27].

6.4 Research Studies and Methodology

For gaining insights into the research query, we planned the following two qualitative studies with the target user segment following a contextual enquiry research methodology using a semi-structured interview.

Sample size: Considering budgetary constraints on the study, the participants were identified living in Delhi and NCR region. They included 121 adults (73 male and 48 female), belonging to the urban poor, sub-class—churning and transient poor user segment. All the participants owned a mobile phone. The convenience sampling method was used after confirming ownership of a mobile phone.

6.4.1 *Study 1: Patterns of Mobile Phone Usage Among Urban Poor Users*

Considering the economically weaker section of mobile phone users as our target user group, our study focused on understanding the users' functional literacy and comfort levels in using the various features of their mobile phone, the pattern of use of mobile Internet, social media applications like Facebook, YouTube, WhatsApp and mobile payment/mobile money. It specifically made inquiries if low functional literacy in terms of skills and comfort levels in the use of mobile phones forms a barrier in acceptance and usage of mobile financial services (mobile payments/mobile money).

Questions included demography profile and awareness and usage of phone features. These included:

1. The type of (basic/feature or smart) mobile phone they have?
2. If they have data/Internet connection on their phones?
3. If they used the phone book/contact book for making a phone call?
4. If they use SMS for messaging?
5. If they use social media such as YouTube and Facebook on their mobile?
6. If they have mobile payment/money apps installed on their mobile phone?

After the interview, the participants were asked to perform specific tasks on their mobile phones to assess their comfort using the phone features. The researcher made direct observations and noted down their engagement. These tasks included making a phone call using a phone book, storing a phone number in their phone book, sending an SMS, using mobile internet, watching a YouTube video, posting a video/ photo on their Facebook timeline and sending/forwarding a WhatsApp message.

Results/Findings of study 1

Types of mobile phone: Out of 121, 36 males (49%) and 33 females (69%) own a basic/feature phone, whereas 37 males (51%) and 15 females (31%) owns a smartphones.

UI Language Selection: Hindi is the choice of UI language selection across the users irrespective of their literacy profile. 91% of them have Hindi UI on their phones.

Access to the internet on mobile: 81 of the 121 respondents (65%), 53 males and 28 females could access the Internet on their mobile phones. We noted that schemes that offered free or a low data fee plan by the mobile service providers had considerably increased intranet access on mobile among this population segment. Nearly, all the respondents (except four) have Internet facilities provided by JIO mobile service.

Usage of Mobile Phone Feature including Calling, Phone Book and SMS

Chart 6.1 summarizes the literacy usage patterns of mobile phone features. Chart 6.2 summarizes the profession-based usage of mobile phone features.

Chart 6.1 indicates that respondents with high literacy levels have higher usage of mobile phone features, especially phone book and SMS, compared to those with low

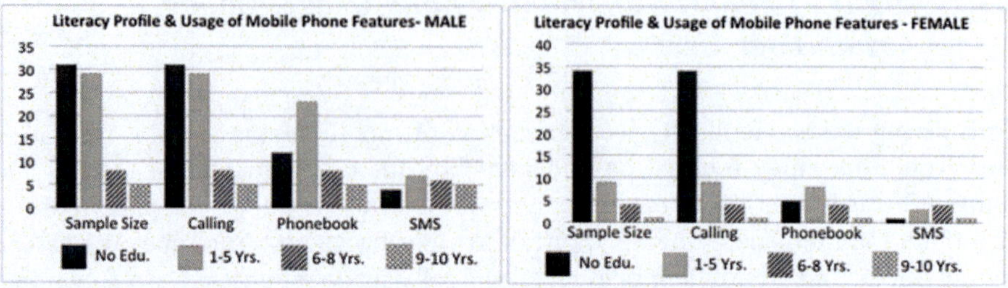

Chart 6.1 Literacy profile-wise usage of mobile phone features (Male & Female)

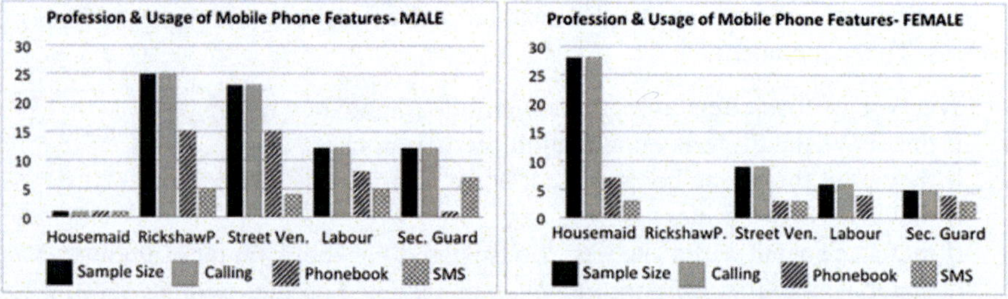

Chart 6.2 Profession-wise usage of mobile phone features (Male & Female)

or zero literacy. Similarly, Chart 6.2 indicates that the security guards have a higher comfort level in using features like phone book and SMS than other vocation.

Out of 121 respondents, only 66 (55%), 48 male, 18 female, reported using a phone book to call, and except one female, all others were able to make a phone call dialling a number from the phone book. However, when asked to store a number in their phone book, only 42 males (88%) and 11 females (61%) could do this. We were told that their family members/friends had helped them store the phone book numbers. Their phone book consists of very few contacts limited to family members, friends, employers, and suppliers/contractors.

Usage of SMS: 31 (25%) respondents (22 males, 9 females) claimed using the SMS, however, when they were asked to send an SMS, only 15 male (69%) and 6 female respondents (67%) could do it. We observed that they had no difficulty accessing SMS, but they had difficulty entering the text using the keypad due to low literacy and not knowing the script and constructing the words.

Usage of Social media platforms including YouTube, Facebook and WhatsApp on Mobile

Chart 6.3 summarizes the literacy-based usage of social media apps, and Chart 6.4 summarizes the profession-wise usage of social media apps on a mobile phone.

YouTube: 48 male (59%) and 18 female (22%) respondents reported using YouTube for watching videos and movies for entertainment purposes.

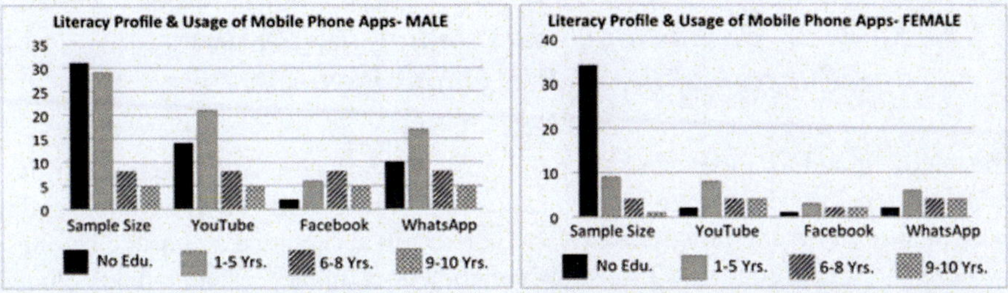

Chart 6.3 Literacy profile-wise usage of social media apps on a mobile phone (Male & Female)

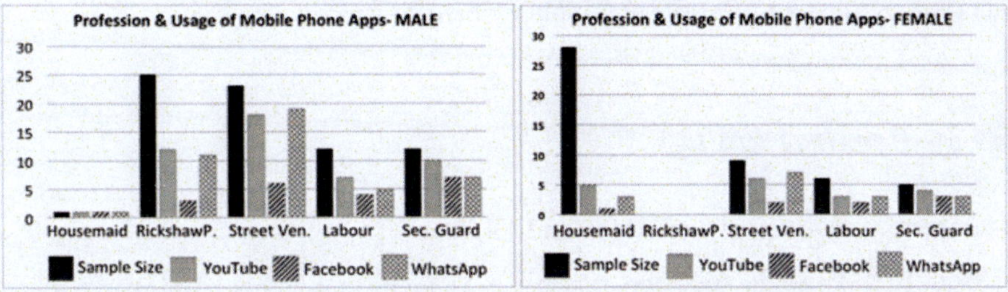

Chart 6.4 Profession-wise usage of social media apps on mobile phones (Male & Female)

Chart 6.3 indicates that respondents with high literacy levels have higher usage of social media apps, including YouTube, Facebook and WhatsApp, compared to those with low literacy. Chart 6.4 indicates that street vendors and guards have a high usage of social media apps on mobile and WhatsApp has high usage among street vendors.

Facebook: 21 male (26%) and 16 female (10%) respondents reported using Facebook and all except two (male and female each) were able to show their FB timelines. Facebook usage seems to be limited to entertainment (watching/sharing videos).

WhatsApp: 40 (49%) male and 18 female (33%) respondents reported using WhatsApp. When asked to send a WhatsApp message/image, all (except 3) could do that successfully. For them, WhatsApp is the primary medium to share images, videos, and messages with their family and friends. The self-employed respondent reported using WhatsApp to communicate /share with their customers.

Usage of Mobile Payments (BHIM/Paytm): Inspecting the respondent's mobile phones, we found that of 27 respondents (19 males, 8 females) have a mobile payment /money apps (Paytm or BHIM) installed on their mobile phone, however, only 14 respondents (10 males and 4 females) reported at least one transaction using mobile payment/money. They mainly use the mobile payments/money app is for receiving rather than making payments. The respondents did not indicate any preference for receiving payments, and they generally go with customer choice of payment mode, fearing they may lose business by not accepting their preferred payment mode.

6.4.2 Study 2: A Study on Understanding the Features and Functionalities of BHIM and Paytm Wallet

This study aimed to understand the prerequisites (hardware, software, regulations, etc.) and steps/process for registering and using "BHIM" and "Paytm Wallet" and if these prerequisites and processes serve as barriers in acceptance and usage amongst the urban poor user segment. This study conducted secondary research that included gathering information sourced from user guides and support information available on BHIM, NPCI, Paytm and other related material, including papers/articles available in the public domain. This was followed by a study of BHIM and Paytm, considering that they are the two most popular mobile financial service providers.

6.4.3 Results/Findings

Prerequisites for using the BHIM and Paytm.

- Both BHIM and Paytm offers services in versions that do not require access and connectivity to the Internet.
- In instances where there is no access to the Internet, the user needs to dial *99# using any mobile phone to initiate the setting up process and access BHIM features over the phone.
- BHIM serves as a gateway for the user to transfer money from bank to bank. This aspect is limiting since the user should mandatorily hold a bank account to use BHIM for financial transactions.
- While having a bank account is not mandatory for transferring money using Paytm wallet, the user needs to access a formal financial service to transfer money into their m-Wallet. Only after loading the money in the m-Wallet can one use the same for sending or paying money using Paytm.
- The precondition of having an active bank account and access to a formal source to load money is a significant barrier in using these by the urban poor.

User Interface (UI) of BHIM App and Paytm Wallet App.

- The multilingual user interface adds to higher usability.
- The UI of the opening screen of BHIM is much straightforward than Paytm.
- Paytm Wallet has links connecting it to other in-house E-commerce portals and shows on the phone screen UI elements for many different services like bill payment, ticket booking and online shopping. This makes it more matched to suit the needs of more diverse socio-economic user groups. It, therefore, is not sharply targeted for the urban poor population. This also makes the UI more complicated and a limiting factor for the population with low literacy.

6.5 Discussion

Our two studies show very high usage of mobile phones among the economically poor in India. This user segments of the urban poor are indeed very comfortable using mobile phone features and apps. Self-entertainment and communication using social media applications such as YouTube, Facebook and WhatsApp are its primary use among this targeted population. The mobile phone facilitates staying in contact with family members [28], enabling social bonding through messaging services, networking for sustaining livelihood by presenting work samples, product images, and taking orders from customers. Meeting these targeted user groups' needs is the prime driver in using the mobile phone as an enabler among this targeted population.

The user perceives mobile phone usage as a change-maker in their economic and social situation, making them feel more empowered and equal [29]. WhatsApp remains their popular choice and is widely used for sharing photos and videos with their family, relatives, friends and peer group. Among street vendors and micro-entrepreneurs, WhatsApp serves as a useful tool for carrying out their vocation or business activities by sharing work samples and receiving delivery orders. Their WhatsApp communication exchanges are limited to the sharing of images and photos. However, many respondents have used WhatsApp's voice message feature, making it easy for those who have limitations in writing using script due to their low literacy.

Considering the frequency of mobile phone features and mobile phones, it is evident that the users' have high functional literacy in using a mobile phone. However, mobile financial services, such as mobile payments/ mobile money, are still minimal among this population and relatively marginal. The push from the customer end to accept digital payment and the fear of losing income opportunities form the critical driver in registering for the mobile payment/mobile money platform. Even in these instances, they use the mobile payment/mobile money platform largely restricted to accepting money. Market demand primarily drives these economic activities.

The monetary loss due to the fee charged for transferring money from mobile wallets to the bank also drives them to use the mobile payment to make the payment. Both these cases highlight a push method or fear factor that seems externally imposed rather than self-motivation as the resulting cause for mobile payment/mobile money.

Addressing our research objective 1, "if the functional literacy of using a mobile phone has any implication on the acceptance of mobile financial services," Our study indicates that the comfort level in using the mobile phone is primarily to fulfil the day-to-day communication, entertainment and business/vocational needs among this targeted urban poor user segment. The phenomenon of the digital divide seems to be non-existence in this case. However, our study also indicates that low illiteracy levels among the user segment result in certain limitations in using SMS features, such as sending messages using SMS, creating a group on Facebook and difficulty navigating the user interface filing long forms. Limitations are reflected in their writing inabilities and understanding of finance—prevalent in this user segment [21]. In contrast, our study also confirms that functional literacy related to other mobile phone communication and entertainment features and services is high among the targeted population.

This ability does not impose a barrier but promises a high potential in acceptance if the design can factor some of these shortcomings in planning suitable modes that make the usage of mobile payment/money acceptable to these users'.

Our second study in seeking answers to research objective 2 and 3 viz.—"if prerequisites for using BHIM and Paytm" and "What Useful means to the urban poor in the context of mobile financial services," respectively indicate that BHIM and Paytm Wallet facilitate cashless transactions for the users. However, both the apps do not offer enough help and motivation for the targeted population who are more often unbanked or do not have access to formal financial services. These user segments have limited access to money due to their low and irregular income. Therefore, the "usefulness" of the mobile financial service (mobile payment/mobile money) is directly linked to the demand side's socio-economic attributes. The interaction involved in making a payment using the "Paytm Wallet" involves only scanning, making it simple to use. However, other monetary transactions involving money transfer into the Paytm Wallet or from Paytm into bank account prove restrictive, as they are required to fulfil any two of the following three prerequisites, namely

1. The person should necessarily have money.
2. The person must have either his bank account or formal financial services.
3. Alternatively, request others to transfer the money into the wallet.

These prerequisites prove to be impeding barriers to the acceptance and usage of mobile payment/mobile money for the economically poor user segment. As stated earlier, Paytm also charges a fee for transferring money from the wallet to the bank account. This aspect further discourages these low-income earners who feel burdened in paying this fee for transferring their hard-earned money into the bank.

The observation and findings indicate that the mobile financial services available are not very useful and usable for the targeted urban poor population, as they do not seem to model the offering from an understanding of the user and their real-world, day-to-day financial needs. This calls for a critical need for a human-centred design-led intervention applying empathy, and bottom-up approach for designing a mobile payment/ money service focused on day-to-day financial transaction needs and barriers of low-income/urban poor population. Designers engaged in outlining a business model aimed at such a large urban poor segment are well-advised to factor in the above strengths and limitations of this real-world consideration while designing financial service offerings that meet the aspiration and real-world needs of this user segment. Micro-financing needs, ease of communication, perhaps with voice-based interfaces and utilitarian patterns of financial features may offer better chances of accepting and adopting these services among the urban poor.

6.6 Limitations and Future Scope

This study forms a part of a self-funded doctoral research study. The financial and human resources were minimal; hence, field studies were limited to qualitative studies within the Delhi and NCR region undertaken with limited sample size. The insights are drawn based on firsthand face-to-face interactions with this identified focused group over four months. The study exemplifies an approach that offers scope for conducting a more extensive study across cities for validation and pan Indian coverage.

6.7 Conclusion

This study examined the low acceptance of mobile financial services, including mobile payment/money being offered by BHIM and Paytm, to the targeted user comprising India's urban economically weaker section. The study considered understanding this from two perspectives. **Study 1** examined it considering the targeted user segment to understand their patterns in the value they attach and the socio-economic drivers in owning a mobile phone. Secondary research indicates the staggering increase and the enormous potential such a segment can offer to businesses should they make some interventions in packaging their services that suitably match this potential segment's aspirations and needs. Usage patterns of mobile phones indicated that their usage is limited to interpersonal communication, keeping them connected to their family, friends and stakeholders related to their vocation and their personal economic growth in the environment of a competitive urban landscape. The other draw toward ownership is the ease of access to content that meets their entertainment aspirations, including downloads of music and films. However, a platform that offers ease of access to financial services and monetary transactions through a digital platform can be a significant economic game-changer and is still seen with suspicion and discomfort. It also pointed towards the economically weaker section being illiterate. Navigating through the different layers of information, data entry required to use the present features proved taxing to the user and resulted in the rejection.

Study 2 primarily shifted focus to address the reasons for this lack of enthusiasm among the targeted user segment to understand those shortcomings that are presently being offered by two popular service providers in India—BHIM and Paytm. These two platforms were examined, referring to literature and conducting direct observations of use case scenarios with participants from the targeted audience. There are pointers that these services are more suitable for the educated and digitally literate segment comprising a large section of the middle-income group.

So, from our studies, we recommend the following:

1. Develop a suitable app that creatively outlines a profitable business model that offers features that match the identified needs of this potentially large section of urban low segments living in our ever-expanding urban cities.
2. Develop appropriate UI design that matches culturally identifiable icons and motifs related to financial transactions, factoring illiteracy among them.
3. Build in a component of human interaction in the form of an avatar into the interface to build effective verbal communication to make the financial service more personalised and suspicion free with the target user segment. This feature of personalisation may enhance its usefulness.

In conclusion, our study indicates a human-centric approach to understand, identify, specify and develop mobile financial services that meet the targeted segment's primary needs for their acceptance and success in terms of use and perceived use.

References

1. Potnis, D.D.: Examining mobile banking in developing nations from pro-poor "context, culture, and community" perspective. Proc. Am. Soc. Inf. Sci. Technol. **51**(1), 1–4 (2014)
2. Kirui, O.K., Okello, J.J., Nyikal, R.A., Njiraini, G.W.: Impact of mobile phone-based money transfer services in agriculture: evidence from Kenya. Q. J. Int. Agric. **52**(892-2016-65177), 141–162 (2013)
3. Wolfe, D.: The tech scene: Paypal, Skype Link play for remittances, American. Banker **172**(37), 1 (2007)
4. India Today website.: Mobile internet users in India seen at 478 million by June: IAMAI (2018, March 29). https://www.indiatoday.in/pti-feed/story/mobile-internet-users-in-india-seen-at-478-million-by-june iamai-1200669-2018-03-29. Accessed on 17 June 2019
5. Mohan, L., Potnis, D.: Mobile banking for the unbanked poor without mobile phones: comparing three innovative mobile banking services in India. In: 2015 48th Hawaii International Conference on System Sciences, IEEE, pp. 2168–2176 (2015, January)
6. Gartner Group: Gartner Group Dataquest Insight: Mobile Payment, 2007–2012. Gartner Group, Stamford, CT (2009)
7. https://trai.gov.in/sites/default/files/TRAI-English-Annual-Report-10032014.pdf. Accessed 23 Sept 2020
8. https://dot.gov.in/sites/default/files/statistical%20Bulletin-2018.pdf. Accessed 24 Sept 2020
9. https://cms.iamai.in/Content/ResearchPapers/664f59c2-8473-48ed-a952-b93660af0e73.pdf. Accessed 24 Sept 2020
10. https://www.trai.gov.in/sites/default/files/Annual_Report_21022019.pdf. Accessed 23 Sept 2020
11. Bloomberg Quint.: BQ Desk. Chart: Cash To GDP Ratio Approaches Pre-Demonetisation Levels (2018, March 3). https://www.bloombergquint.com/business/chart-cash-to-gdp-ratio-approachespre-demonetisation-levels. Accessed on 29 Sept 2020
12. Rogers, E.M.: Diffusion of Innovations, 1st edn. The Free Press, New York (1962)
13. Fishbein, M., Ajzen, I.: Belief, attitude, intention, and behavior: An introduction to theory and research (1977)
14. Davis, F.D.: A technology acceptance model for empirically testing new end-user information system: theory and results. PhD Thesis, Sloan School of Management, Massachusetts Institute of Technology, USA (1986)
15. Ajzen, I.: The theory of planned behavior. Organ. Behav. Hum. Decis. Process. **50**(2), 179–211 (1991)

16. Venkatesh, V., Davis, F.D.: A theoretical extension of the technology acceptance model: four longitudinal field studies. Manage. Sci. **46**(2), 186–204 (2000)
17. Venkatesh, V., Morris, M.G., Davis, F.D., Davis, G.B.: User acceptance of information technology: toward a unified view. MIS Q. **27**(3), 425–478 (2003)
18. Laforet, S., Li, X.: Consumers' attitudes towards online and mobile banking in China. Int. J. Bank Mark. **23**(5), 362–380 (2005)
19. Trappey, C.V., Trappey, A.J.: Electronic commerce in greater China. Ind. Manage. Data Syst. **101**(5), 201–210 (2001)
20. Singh, S.: The impact and adoption of mobile banking in Delhi. Int. Res. J. Bus. Manage. **1**, 22–26 (2014)
21. Medhi, I., Ratan, A., Toyama, K.: July) Mobile-banking adoption and usage by low-literate, low-income users in the developing world. In: International Conference on Internationalization, Design and Global Development, pp. 485–494. Springer, Berlin, Heidelberg (2009)
22. Bashir, I., Madhavaiah, C.: Determinants of young consumers' intention to use Internet banking services in India. Vision **18**(3), 153–163 (2014)
23. Singh Sambhy, G.: Study of Mobile Payment Services in India: Distribution of the Roles, Responsibilities and Attitudes Amongst Actors of the Payment System (2014)
24. Kim, C., Mirusmonov, M., Lee, I.: An empirical examination of factors influencing the intention to use mobile payment. Comput. Hum. Behav. **26**(3), 310–322 (2010)
25. Wu, J.H., Wang, S.C.: What drives mobile commerce? An empirical evaluation of the revised technology acceptance model. Inf. Manage. **42**(5), 719–729 (2005)
26. Yang, S., Lu, Y., Gupta, S., Cao, Y., Zhang, R.: Mobile payment services adoption across time: an empirical study of the effects of behavioral beliefs, social influences, and personal traits. Comput. Hum. Behav. **28**(1), 129–214 (2012)
27. Dass, R., Pal, S.: Exploring the factors affecting the adoption of mobile financial services among the rural underbanked. In: European Conference on Information Systems, ECIS 2011 Proceedings, 246 (2011)
28. Portus, L.M.: World-making: mobile phone discourses among selected urban poor married couples. Plaridel J. **2**(1), 108–109, 112–116 (2005)
29. Ondrus, J., Pigneur, Y.: Cross-industry preferences for the development of mobile payments in Switzerland. Electron. Markets **17**(2), 142–152 (2007)

Chapter 7
Bridging the Gap Between Service Design Specification and Technical Specification

Ravi Mahamuni, Supriya Mantry, and Mayur Jadhav

Abstract Well implemented services can have a transformative impact on any organisation. The implementing team receives service design specifications once the design phase is complete. These specifications include blueprints, storyboards, touchpoint guidelines, among many other custom artefacts. Implementation of technical touchpoints, as envisaged by the designer, will bring the experience alive not only for service users but also for all stakeholders, including those from the service provider. However, after the handover, inconsistencies may often be introduced during the development of touchpoints. The touchpoints may differ from the designers' original vision and affect the stakeholders' unified experience. The place where touchpoints' implementation went wrong is hard to trace. Hence, the long-term involvement of the designers with the development effort is recommended but may not always be feasible. The handover process, and the form and content of the service specifications are critical to reducing the strong dependency on the designers post-handover. In this paper, we propose a framework to map service specification to technical specifications, focusing on traceability and completeness. After scrutinising various technical specification templates to find the essential components, these components were mapped directly or indirectly, by diving deep into multiple artefacts of service specifications. Mapping this extensive technical specification with design specifications early in implementation makes it comprehensive and complete. They are hence reducing the dependency later in the implementation cycle. Such a technical specification document leads to seamless implementation.

R. Mahamuni · S. Mantry (✉) · M. Jadhav
Tata Consultancy Services Ltd, Mumbai, India
e-mail: supriya.mantry@tcs.com

R. Mahamuni
e-mail: ravi.mahamuni@tcs.com

M. Jadhav
e-mail: mayur.shankarjadhav@tcs.com

A. Chakrabarti et al. (eds.), *Design for Tomorrow–Volume 2*, Smart Innovation, Systems and Technologies 222, https://doi.org/10.1007/978-981-16-0119-4_7

7.1 Introduction

Organisations are getting attracted to service design due to its systematic, holistic, user-centred and interdisciplinary nature. One of the ways to define service design is, it is designing experiences that happen over time and across different touchpoints [1]. Holistic service experience is given by an ecosystem of these touchpoints, interacting with the service user or with each other. One touchpoint cannot ensure the complete service experience in silos [2], making it necessary that they have excellent communication and maintain the same standard of experience.

We can broadly divide touchpoints into technical, human and physical touchpoints. Here, technical touchpoints consist of mobile application and web portal, although sometimes the product itself is a technical touchpoint. Furthermore, with new devices and innovations, almost all services have one or many technical touchpoints, providing varied channels and giving more personalisation and owning to the user. The interaction of a customer with these touchpoints is a critical defining stage of the customer's journey with the organisation [3]. From an implementation perspective, this paper is focusing only on the specifications of technical touchpoints.

The methodology used to carry out this research is as follows:

- Identify the gap and issues faced in the handover from a service design team to the implementation team, through practitioner experience and literature.
- Hypothesise a framework to bridge the gap between the service specifications and the technical specifications.
- Use the Reverse Cases Study Mapping method [4] to map the findings from inhouse ongoing service design to the proposed framework to identify the gaps, if any. This method helps to validate the framework during the initial phases of evolution.

7.2 Technical Specification

At the beginning of traditional methodology for implementing technical touchpoints, system or software requirement specification (SRS) documents, also known as technical specification, are created. SRS is a document(s) that describes the features and behaviour of the system or software. With development teams moving towards agile methodology [5], the way documentation and capturing of requirements is done have evolved. An equivalent of SRS in agile methodology is the product backlog. The product backlog is a full list of requirements (called user stories) for a system, ordered by priority.

One of the early definitions of requirement is stated by Ian Sommerville and Pete Sawyer [6]."*Requirements are a specification of what should be implemented. They are descriptions of how the system should behave, or of a system property or attribute. They may be a constraint on the development process of the system*". This definition imbibes all the aspects of requirement from the feature, behaviour to constraints, required for implementation. While implementation doing the right things as well as doing things right depends on the quality of the stated requirements.

Characteristics. The importance of technical specification has already been established by researchers and practitioners alike. Accessing quality has been of significant importance that industry standards from the eminent IEEE 830-1993 [7] to the more recent ISO/IEC/IEEE 29148-2011 [8] have been created.

The following are the essential characteristics of technical specifications:

- **Complete**—The requirement should not need any explanations to describe its capability and features.
- **Unambiguous**—The requirement should be stated in such a way so that it can be interpreted in only one way.
- **Traceable**—The requirement should be upwards traceable to specific documents, highlighting user needs.
- **Verifiable**—The requirement should possess the ability to be checked, proving the system satisfies the requirement.

Content. Every implementation team has one or more templates selected for SRS. Many templates are available from ISO and various sites, with ample flexibility to adapt to the team needs. By analysing many such templates [8–11]. We listed some of the most common components available in all templates and which are needed from a completeness perspective in the traditional framework. Table 7.1 explains the components identified.

In agile software development, vision document and product backlog contain these parts individually or in combination. The roadmap and release plan are also decided at inception.

7.3 Service Design Implementation

A great design can only be fruitful when it is implemented and provides the outcome as designed. However, service designers have been criticised for lack of creative skills to facilitate service implementation [12], leading to more services on the drawing board rather than enhancing the users experience on the ground. The implementation of service design is different from various other design fields. Products and services share similar characteristics [13], but service design is broader, holistic and co-created which differentiate it with one-time focused product design. In terms of complexity, a product's response to any input from the user is predictable. However, in service, it depends on multiple interactions between the service user, service staff and various touchpoints. In product design implementation of the outcomes starts in early phases. Implementation of service can be placed at various stages of service design. However, it can also be said that implementation has to be in the later stage of service design as it requires adaptation of generic service resources that are the outcome of a design process [14].

Table 7.1 Requirement document components

Requirement document components	Explanation
System purpose	It covers the expectation of the system that needs to be built. Key components of this section include definition, vision, system overview and references
Scope/Exclusions	It covers the benefits, objective and goals of the system. It can also specify the exclusions explicitly
Policies/Regulations/Constraints	It covers all the policies from the external environment (laws, standards, physical/natural environment), organisation, business and so on
User characteristics/Actors	These are user classes and characteristics such as, educational level, experience, disability, technical expertise that may affect the working of the system
Functional requirements	These are all the functions and behaviour a system is supposed to perform
Non-functional requirements/Quality attributes	This specifies criteria that are used to judge the complete system, rather than its specific features/behaviours
External interface requirements	It consists of all interfaces to elements which are required by the system to provide a functional/ non-functional requirements

There are very few studies on how the design phase can be systematically connected to implementation, especially for technical touchpoints. However, for human touchpoints, some studies are done for engaging and informing people (human touchpoints) using change management and human-centred principles [15]. Also, in the product-service system, how human-centred approach can help the networked collaboration for development [16]. Thus, showing how change management and networked collaboration can be a starting point for technical touchpoint implementation.

7.4 Problems in Handover

Generally, the 'double diamond design process' is followed in service design [17]. The last diamond 'deliver' where the service is realised to launch, is where the handover occurs for implementation. Many methods and tools have been proposed for discovering insights, defining problems, generating ideas, but there are very few methods available to assist in the implementation of service design. Various service design artefacts, as listed in Table 7.2 are generally handed over to the implementation team.

Table 7.2 Service design artefacts used for handover

Service design artefacts	Sources
Service blueprint	[17–21]
Service roadmap	[18, 19]
System map	[19]
User stories/Design scenarios	[17, 19, 20]
Business model canvas	[17, 19, 20]
Live prototypes (theatrical, desktop walkthrough, staging, roleplay)	[17, 20]
Customer lifecycle map	[20]
Storyboards	[20]

Based on the literature and reflection of our multiple service design projects executed during the last few years, following is the list of most significant problems during the handover of service specifications to the technical team.

Lack of actionable details. Artefacts such as service blueprint were adequate for older services where encounters were straight forward. Most services today are highly complex, which use multiple technical touchpoints, making service blueprints long and complicated. There are attempts to enrich the service blueprint with the various notations to represent such services [21]. It is also well acknowledged that long and complicated services make it difficult to comprehend and analyse [22]. The receivers of service design specifications, that is, the implementation team have also shown their confusion while getting started with these long, complex artefacts [23]. The artefacts such as service blueprint show interactions of all touchpoints involved in an encounter, making it challenging to carve out specifications of a touchpoint and its interactions with others. Although service design is considered as a process [24] rather than just the list of features, it is difficult to absorb and pass it to the implementation team.

Lack of traceability. The design decisions and ideas in service design are based on the knowledge and understanding of insights that were uncovered. This knowledge is built mainly in the early phases, using various user involvement methods [20]. A deeper understanding of this knowledge resides with the designer and is reflected in the decisions taken to reach the final design. The artefacts created at every stage of design are the outcome of knowledge acquired. It is also emphasised on how the insights and its evidence should be backwards traceable to either reaffirm the outcome or to understand when service might have started to fail [25].

When implementing technical touchpoints, traceability become a vital characteristic [8]. Tracing back of any feature in service can give insights related to the objective or concerns of users which lead to the feature. Designers take design decisions for any feature after a deep understanding of the user, which implementers might be unaware of, these design decisions need to be connected with the insight behind it to communicate rationale [26]. Traceability in service is also essential to coordinate and align, how service is designed and how it is implemented, which is crucial [27]. Currently, the deliverables at the handover process are the result of the complete service design process and give a complete view of service, thus leaving the knowledge and user insights with the designer. In agile methodology, where implementation is done in sprints and requirements are kept as a live document which can change if there is a change in user need or market, complete traceability of every requirement to the insight which leads to it becomes critical.

Lack of touchpoint specific quality measure. While designing any service, many quality aspects are identified. In the technical implementation, these quality aspects are termed as non-functional requirements. Addressing non-functional requirements is already tricky in implementation [28]. Quality aspects in service design can be expressed in blueprints by adding quality measure swim-lane [29]. These are the factors that measure the success or value of interaction/encounter. The quality measure is defined for one touchpoint but could be affected by one or many touchpoints or its features involved. Adding to this ambiguity of what quality aspects are stated for which touchpoints, the blueprint also fails to accommodate how vital the aspect is [22]. Consider it with an analogy of a coach and a team of football players. A coach can place an individual player on a specific location by gauging his/her strengths. However, it is the coordination between the strengths of all players which, when used cumulatively leads to an overall victory and not just the individual performance of the said player.

7.5 Current Solutions for Better Handover

Researchers have started looking into this grey area of implementation in the service design process [30, 31]. Execution-based implementation methods for service design requires more focus and study [32]. Following are some of the proposed directions which are being followed to realise the implementation of the design along with its limitations.

Co-creation and visual language. Co-creation of any service involving multidisciplinary team is long-discussed to give a common understanding of the service. This, with visual language, has proved to develop a shared understanding and can help to work in one direction while implementing the service [30]. While co-creating transferring individuals' knowledge into innovative outcomes needs careful selection and of team members based on their domain and motivations [33]. However, little is known about the optimal degree of heterogeneity and the optimal degree of shared understanding [34]. Collaboration, on the other hand, has challenges such as conflict, disagreement in the team and slow decision-making. Unless these challenges are overcome, collaboration is likely to fail [35].

Designer involvement in the implementation. To solve the root cause of handover, the involvement of a designer in the implementation process seems to be suitable for seamless implementation. However, practically it varies from an active participant, helping and evaluating touchpoints at all steps to just a handover of service design artefacts. However, it is debated whether the implementation responsibility of the designer, what are the new skills designer will be needed to learn in order to facilitate successful implementation [14].

Service implementation a part of the service design phase. Timely articulation of how to implement and what resources need to be shaped can be helpful for successful implementation. This also takes into consideration or co-creation that we discussed above. Ignoring the challenges, even if the implementation is made as a concept part of service design, the involving parties still pose a risk of viewing incomplete future service in place and defining the resources which might be needed to shape it [30]. Hence, defeating the purpose of better and fast implementation.

User Experience (UX) team as a bridge. Depending on the organisation, UX designers can highly be involved in a service design project and helping with validating difficult concepts with prototypes making their connection with both strategy and implementation. [36], thus making them an excellent bridge to connect service design to implementation. This makes implementation more dependent on some team rather than the artefacts created, risking the task if UX designer leaves the team.

7.6 Proposed Solution

With all the problems in handover discussed above and the limitations of currently available solutions, we focused on two primary goals. First, to maintain that the technical specification created remains traceable, complete, verifiable and unambiguous. Secondly, to make sure that the designer is not overburdened by learning and conceptualising new artefacts altogether. While coming up with the proposed solution, the main criteria was to how we can come with a new framework which does not replace the current service design artefacts but work seamlessly with them.

To introduce traceability, as shown in Fig. 7.1, it is critical to understand the progression of service design specification into the technical specification. Service design starts with the initial design brief and gets refined into the design brief after understanding the problem space. This refined brief gives the design direction during the service design life cycle. The 'problem space' is where the stated or latent intents, concerns and constraints and the resources available with all the stakeholders, are discovered and explicitly mentioned. CraftChange Empathy Square [37] helps to identify the critical stakeholders. After understanding the problem space, during the 'ideation space', the ideas are generated considering various personas, scenarios and user journeys that emerged by analysing of data collected during the user research phase of the service design life cycle. Initially, ideas may be discrete and need to be converted into service concepts by methods like clustering. These service concepts are individually and holistically specified in the 'Service definition space'. The service specifications contain elements such as policy changes if any, process changes if any, the details of human touchpoints (e.g. service staff), technology touchpoints (e.g. mobile application) and physical touchpoints (e.g. flyer). It also provides details of interactions among various touchpoints and service users,

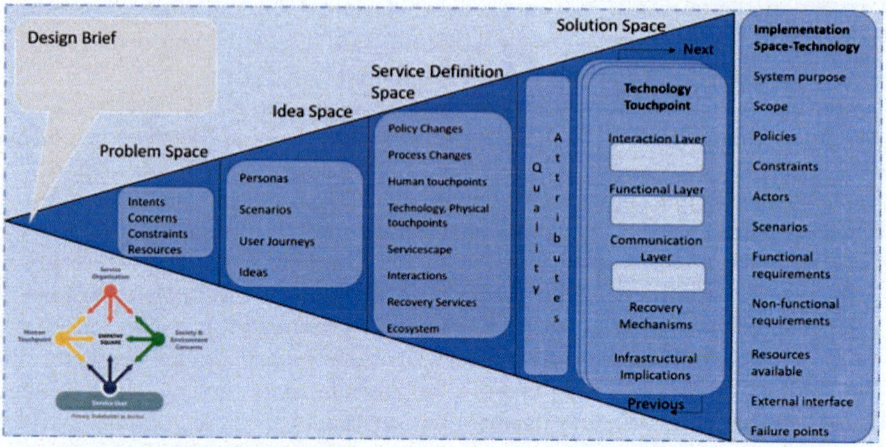

Fig. 7.1 Progression from Service design specification to Technical design specification

servicescape, that is, physical layout if required, the list of failure points and associated recovery services, and the required ecosystem of internal and external elements. The next 'solution space' provides details of various touchpoints along with their interconnections. The expectations, concerns and constraints of various stakeholders help to derive the quality attributes of the designed service system, e.g. turnaround time and scalability. The 'solution space' also provides detailed, layered information about each technology touchpoint which includes the interaction layer, functional layer and communication layer. Here communication layer details how the technical touchpoint will communicate with the other elements of the ecosystem at the backend. These layers help to arrive at the implications for the existing operational infrastructure if any. This space provides the details related to recovery mechanisms required to handle the various service failure points.

As shown in Fig. 7.1, the specifications get detailed and realisable while moving from left to right, i.e. progression of service design life cycle. The next section talks about the validation of our hypothesised framework.

7.7 Case Study—Mobile Application for Lateral Hiring

The Reverse Cases Study Mapping method is adopted from the 'Retroductive' method [38]. As part of the Reverse Cases Study Mapping method [4], there is a need to map the findings from the practice to the proposed framework to identify the gaps, if any. This method helps to get close to the real-life successful or otherwise examples depending upon the context to validate during the initial phases. It is supposed to be a quick and effective method to validate the framework for its applicability in practice. With one of our service design for lateral hiring and on-boarding process, a mobile application was playing a crucial touchpoint, being used in most of the service encounters.

Lateral hiring is one of the ongoing processes in our organisation, involving many different stakeholders. Candidate, Acquisition team, Human resource and Project team were the most significant stakeholder. However, the Security team, Facility team and Administration team also were identified as crucial stakeholders as well as human touchpoints while defining the service. A mobile application allowed multiple human touchpoints to connect across service, becoming too complex to understand from mere service blueprint. Hence, as a proof of concept, we developed a working prototype of the mobile application. As an output of this service, a package of service blueprints, personas, ecosystem and user journeys were created.

As part of the reverse case study mapping process, we tried to map the technical specifications using the given framework to check for the traceability. The following components of the requirement document were extracted from in-process artefacts.

- System purpose—Using design brief/ re-designed brief and Intents
- Scope—Concerns and Resources identified while defining problem space
- Policies—Policy changes and Process changes identified while defining service definition space
- Constraints—Constraints identified while defining problem space
- Actors—Persona identified while defining idea space
- Scenarios (user stories)—Scenarios identified while defining idea space
- Functional requirements—Interaction layer, Functional layer, Communication layer identified while defining solution space
- Non-functional requirements—Quality attributes, Ecosystem and Infrastructure implications identified while defining solution
- External interface—Ecosystem identified while defining service definition space
- Failure points—Recovery mechanisms identified while defining solution

While doing reverse mapping, these components made a clearer picture from an implementation perspective. This also helped to trace the reasons for any business justifications for a specific feature in the mobile application. For example, in the technical specification, there was a change in policy where a candidate wants to apply for a job vacancy through his/her friend working in the same organisation. This framework gave an easy way to trace back to every detail for that policy change. Figure 7.2 explains how policies defined in the requirement document was defined in service definition space, and it was based on personas (candidate's friend), scenarios (applying the job for a friend), concerns (help the friend) and constraints (unavailability of candidates' personal information). Thus, making the change in policy's intent clear.

It helped to validate the traceability using the proposed framework as well as find out the missing elements in the technical specification. Interestingly, it also helped to understand the rationale behind the features mentioned in the technical specification. When this framework was showcased to the technical team, they felt that this framework would help to provide traceability, check the completeness of technical specifications as well as understand the spirit and rationale behind the

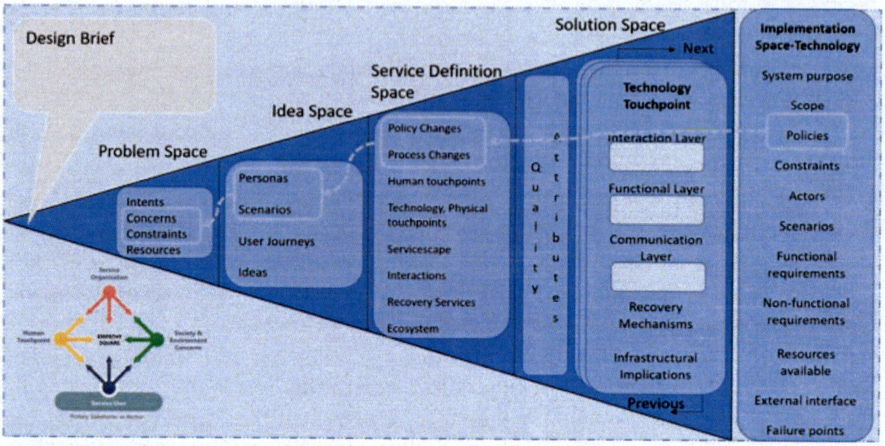

Fig. 7.2 Traceability for Policies in the Technical specification

technical specifications. Generally, the technical team is not so conversant with the whole picture of the proposed solution. However, the team felt that this framework would help them to understand it very quickly and comprehensively. These are early findings and need to be validated in multiple real-life projects and refined accordingly.

7.8 Conclusion and Future Work

With the availability of new devices, more services would be based around technical touchpoints. For the translation of proposed design to implementation such that experiential and effectiveness parameters are achieved as expected; without much involvement of the designer, there is a need for seamless handover. We tried to bridge the gap to form a traceable technical specification. Since the process led to an increase in artefacts, managing versions and traceability becomes very effort-intensive. Moreover, complex projects having an ecosystem of the interconnected touchpoints, manually connecting and tracing each designed requirement to the features in touchpoint is not feasible. Hence, there is a need for a formulated tool easing the process and keeping the traceability intact. Along with a tool, there is a need for a behavioural shift on both the sides; service designers and technical touchpoint implementers, to realise the potential of service design. As a proof of concept, we created the prototype of the mobile application, to show the effectiveness of the service and to give a starting point for implementation, the only handful of features were prioritised and developed. This prioritisation is also needed for the agile framework, helping to give priorities to features. The artefacts, like a roadmap, which are not frequently used in the handover, also need to be discussed for converting service specification to the agile framework's product backlog. This is a significant step in bridging the gap but need further exploration. We are planning to use it in multiple service design projects to check its applicability.

References

1. Clatworthy, S.: Service innovation through touch-points: development of an innovation toolkit for the first stages of new service development. Int. J. Des. **5**(2), 15–28 (2011)
2. Polaine, A., Lovile, L., Reason, B.: Service Design: From Insight to Implementation, Rosenfeld Media, pp. 22–23 (2013)
3. Rawson, A., Duncan, E., Jones, C.: The Truth about Customer Experience, vol. 91 (2013)
4. Mahamuni, R.: Crafting a change—Service Design Perspective for Sustained Behaviour Change. Thesis (2020)
5. Agile Methodology: An Overview. https://zenkit.com/en/blog/agile-methodology-an-overview
6. Sommerville, I., Sawyer, P.: Requirements Engineering: A Good Practice Guide. Wiley, England (1997)
7. IEEE Recommended Practice for Software Requirements Specifications. https://standards.ieee.org/standard/830-1998.html
8. ISO/IEC/IEEE International Standard—Systems and software engineering—Life cycle processes—Requirements engineering. https://standards.ieee.org/standard/29148-2011.html
9. Wiegers, K., Beatty, J.: Software Requirements. Microsoft Press (2013)
10. Software requirements specification. What do you need to know. https://medium.com/globalluxsoft/software-requirements-specification-what-do-you-need-to-know-22a6b8585945
11. Guidelines: Software Requirements Specification. https://sceweb.uhcl.edu/helm/RationalUnifiedProcess/process/modguide/md_srs.htm
12. Mulgan, G.: Design in Public and Social Innovation. What works and what could work better. NESTA (2014)
13. Wagle, P., Mahamuni, R., Mehta, K.: Products and Services—Strangers or Siamese Twins? ICoRD (2019)
14. Overkamp, T., Holmlid, S.: Views on Implementation and How they Could Be Used in Service Design. ServDes.2016. Geographies, Copenhagen,Denmark (2016)
15. Lin, M.C., Hughes, B.L., Katica, M.K., Dining-Zuber, C., Plsek, P.: Service design and change of systems: human-centered approaches to implementing and spreading service design. Int. J. Des. **5**(2), 73–86 (2011)
16. Henze, L., Mulder, I., Stappers, P.J.: Understanding networked collaboration: fields and patches of interactions. In: ICE & IEEE International Technology Management Conference, The Hague (2013)
17. Design methods for developing services. https://www.designcouncil.org.uk/
18. Service Design. https://www.servicedesigntoolkit.org
19. Service Design Tools. https://servicedesigntools.org
20. Stickdorn, M., Schneider , J.: This is Service Design Thinking. Wiley (2010)
21. Lobo, S., Sharma, S., Hirom, U., Mahamuni, R., Khambete, P.: Extending service blueprint for new age services. In: Research into Design for a Connected World. Springer, Singapore, pp. 809–821 (2019)
22. Haugen, M.: Service blueprints Persistent qualities and future potential. https://www.ntnu.no/documents/10401/1264433962/MargretheHArtikkel.pdf
23. Almqvist, F.: Service design in the later project phases: exploring the service design handover and introducing a service design roadmap. ServDes (2018)
24. Mager, B., Sung, T.J.D.: Special issue editorial: designing for services. Int. J. Des. **5**(2), 1–3 (2011)

25. Parkes, P.: Create actionable, traceable user insights. https://medium.com/qualdesk/create-actionable-traceable-user-insights-703c5c161283
26. Lønvik, S., Pettersen, N., Verhulst, E.: Achieving readiness for change for successful implementation of service design projects: learning from. NordDesign (2016)
27. Ponsignon, F., Smart, P., Maull, R.: Service delivery system design: characteristics and contingencies. Int. J. Oper. Prod. Manage. (2011)
28. Prehofer, C.: An adaptive control model for non-functional feature interactions. In: 37th EUROMICRO Conference on Software Engineering and Advanced Applications (2011)
29. Ross, I.: Service blueprints: laying the foundation. https://www.cooper.com/journal/2014/08/service-blueprints-laying-the-foundation
30. Overkamp, T., Ruijs, F.: Involving stakeholders towards service implementation. Co-designing change of practices using a visual language. Des. J. (2017)
31. Sangiorgi, D., Eun,Y.: Service design as an approach to new service development: reflections and future studies. ServDes, Lancaster (2014)
32. Martins, R.: Analysis of available design implementation methods. Des. J. **20**(sup1), S4730–S4733 (2017)
33. Trischler, J., Kristensson, P., Scott, D.: Team diversity and its management in a co-design team. J. Serv. Manage. **29**(1), 120–145 (2018)
34. Bittner, E.A.C., Leimeister, J.M.: Creating shared understanding in heterogeneous work groups—why it matters and how to achieve it. J. Manage. Inf. Syst. (JMIS) **31**, 111–143 (2014)
35. Weiss, J., Hughes, J.: Want collaboration. Harvard Bus. Rev. **83**, 93–101 (2005)
36. Bridging the gap between Service Design and Scrum: Why UX is key. https://uxplanet.org/bridging-the-gap-between-service-design-and-scrum-why-ux-is-key-c81f1e66d203
37. Mahamuni, R., Meroni, A., Khambete, P. Punekar, R.M.: Empathy square: an aid for behaviour change to support sustainability. In: 3rd LeNS World Distributed Conference (2019)
38. Blaikie, A.: The secret world of subcultural ageing: what unites and what divides?. In: Cultural Gerontology. Greenwood Press (2002)

Chapter 8
Service Design for Scale—Overcoming Challenges in Large-Scale Qualitative User Research

Ravi Mahamuni, Shivani Ganwani, Sylvan Lobo, Bhaskarjyoti Das, Radhika Verma, Ulemba Hirom, Supriya Mantry, and Mayur Jadhav

Abstract Services must scale up to meet the varying contexts and demands of diverse consumer groups, especially for global consumers, citizen services and organisational services. To enable services for scale, the activity of service design must scale up concerned processes, tools, techniques and people. Beyond traditional design approaches, rather than depending on a handful of designers and stakeholders, we can reap benefits of a more significant number of people and designers, across regions, and considering a wide variety of issues, users and methods to inform the design of the services. In this paper, we deliberate on this conversation of scaling service design by primarily considering the phase of primary research, specifically qualitative user research. Although qualitative user research is accepted as an academic approach, it is often questioned on the grounds of validity and practicality of logistics. When considering service design for scale across very diverse user groups, data saturation may be challenging to achieve with a small number of samples. We argue that it is necessary to tackle more extensive qualitative studies if we want to scale up designing

R. Mahamuni · S. Ganwani · S. Lobo (✉) · B. Das · R. Verma · U. Hirom · S. Mantry · M. Jadhav
Tata Consultancy Services Ltd, Mumbai, India
e-mail: sylvan.lobo@tcs.com

R. Mahamuni
e-mail: ravi.mahamuni@tcs.com

S. Ganwani
e-mail: shivani8.s@tcs.com

B. Das
e-mail: bhaskarjyoti.das@tcs.com

R. Verma
e-mail: radhika.verma@tcs.com

U. Hirom
e-mail: ulemba.h@tcs.com

S. Mantry
e-mail: supriya.mantry@tcs.com

M. Jadhav
e-mail: mayur.shankarjadhav@tcs.com

A. Chakrabarti et al. (eds.), *Design for Tomorrow–Volume 2*, Smart Innovation, Systems and Technologies 222, https://doi.org/10.1007/978-981-16-0119-4_8

for services, with technology as an enabler. We highlight this need for conducting large-scale qualitative studies while covering theory and critique in literature and propose some of the approaches we have attempted in our service design practice.

8.1 Introduction

Modern complex services are continually innovating to cater to a high volume and variety of consumers. Such services may include public services that need to reach out to all citizens, or consumer services like online shopping and door-to-door delivery platforms. These services need to consider regional variations in culture, laws and prevailing trends in order to provide better services. Further, employee services in large multi-national and multi-cultural organisations also must address issues of scale, plurality and complexity in order to cater to the needs and concerns of the entire diversity of its workforce. Service design for such large heterogeneous target population requires special attention to the scale involved in its approaches, and there is a growing demand for its application on a larger scale [1] to arrive at effective solutions and offerings.

Service design as a discipline is continuously evolving its processes and techniques to increase its application at large scale. In the effort to scale up design activities for the development of large-scale services, the traditional design processes and techniques may face bottlenecks. One challenge is to conduct user research activities for a large sample of the target population in the service design lifecycle.

In the design of large-scale services, connecting and collaborating with users are challenging as the user base is large, geographically widespread and has a wide variety of needs and problems. As service design practitioners, performing qualitative studies for a large number of geographically dispersed users is time consuming, rigorous, and resource-intensive. Service designers need to be enabled to plan, collect and analyse large-scale data. They need to maintain standardisation of process, necessary rigour, comprehensiveness in quality and traceability of research data.

The research question in focus, in this paper, is how to conduct qualitative user research at large-scale effectively, for service design. We discuss the literature on performing qualitative user research at scale and the related issues and opportunities. We followed specific hypothesised approaches and techniques in service design projects in our organisation to conduct qualitative user research at scale. We have outlined our reflections from these case studies and learnings from literature. Our research opens the discussion for developing tools and approaches for successful large-scale qualitative user research meant to enable service design at scale.

8.2 Literature Review

User research, including various qualitative methods like user interviews, observations, focus groups and ethnography, is an essential part of the service design process to arrive at an empathetic understanding of users. The scope of user research in service design remains broader than other related design fields. Besides developing and understanding of service users, it includes arriving at an understanding of the context, physical infrastructure, processes and activities of human touchpoints and other stakeholders. Since the focus of service design lies on addressing the holistic end-to-end journey experience rather than individual touchpoints alone, qualitative user research in service design plays a crucial role to take a deep dive into the journeys of all service stakeholders. It allows us to understand their perceptions, beliefs, attitudes, values, opinions, behaviours, social and cultural aspects. Qualitative research helps challenge preconceptions and reveals newer perspectives, making it widely adopted by researchers and practitioners to address challenges associated with increasingly complex services and wicked problems [2]. Qualitative user research basis is in interpretivism, constructivism and inductivism [3]. It usually does not focus on generalizability but rather on developing a rich and in-depth understanding of people's lives. Hence, it is often acceptable to conduct with small sample size. However, it is often questioned for lack of transparency about the research process including sampling design, recruitment of respondents, and data analysis and interpretation processes; the variable nature of qualitative data; possible cherry-picking of examples and quotations; the lack of opportunity to verify or replicate the research and reliability of data [4]. Qualitative user research faces similar challenges in the management field too such as lack of standardisation in methods and techniques of data collection and analysis, reflexive research design and accepted researcher bias [5].

There are approaches and frameworks to ensure the quality of qualitative user research based on criteria such as credibility, transferability, confirmability and dependability for establishing trustworthiness [6–8]. Mays and Pope [9] discuss how reliability, reproducibility and validity of qualitative user research findings can be achieved through a systematic and documented approach, expert review, and triangulation considering many alternative sources of data.

Some qualitative researchers believe that collecting extensive data does not define good research indicating that rather than the big or small size of data, what matters is how the data is utilised [10]. At the same time, researchers also find value in engaging in large-scale qualitative user research. They have emphasised the importance of both quantity as well as the quality of the data. Funded projects often focus on large-scale qualitative user research considering it to have more rigour and benefits as compared to smaller-scale studies [11]. It is argued that quantity is a crucial aspect to support granular themes from analysis, to strengthen the trustworthiness of the findings and to provide a sense of completeness [12].

The rapid advancement of technologies such as big data is impacting the way qualitative user research is conducted, assessed and shared [13]. The opportunity of using big data, thick data and open data is being explored mainly in the fields of service innovation and open innovation; to collect data from various distributed sources, analyse it in real time and identify underlying patterns [14, 15]. Researchers strive to achieve the same rigour and quality as established approaches through improved processes and technological interventions. Large-scale qualitative user research should focus on aspects such as accelerating the pace of research process, building accuracy and reliability among insights.

Sampling strategies. The time, effort and resources involved in collecting and analysing data can be overwhelming and is a significant consideration in deciding the sample size [16]. Usually, studies are limited to a local population [17] and small sample sizes of 20–30 or 30–40 which may change as the actual study unfold [18–20], primarily if grounded theory and theoretical sampling strategies are used. Here, the initial theory is developed by selecting an initial sample, and the remaining informants are selected iteratively as the theory develops [9]. Purposeful sampling strategies [21] are also common for gaining an in-depth understanding of specific research objectives, rather than generalizability through probability samples. Larger sample sizes are recommended when informants are plenty and easy to recruit, or when the group under study has varied perceptions, roles, statuses and problems with the subject [16] which is often the case in large-scale services. The sample size and sample should be such that it adequately represents the cases or individuals necessary to address the research questions and achieve the intended depth in the study [22].

While for a homogeneous target population, the smaller sample size may be sufficient, and for a heterogeneous target population (that large-scale services cater to), the larger sample size is required [23]. It has been claimed that, while considering large heterogeneous samples across geographies and demographics, it will be relatively removed from real-life settings, and challenging to find cross-case themes during analysis. Careful consideration of homogeneity/heterogeneity trade-off is required to achieve the research aims [24]. Large-scale qualitative research conducted in a EUROCARE study involving participants across 14 countries indicates that with resource availability, large-scale qualitative research consisting heterogeneous sample is entirely possible [25, 26]. A heterogeneous sample is often required for cross-cultural qualitative research, where individuals from different cultures, demographics and geographies are selected in the sample in order to look for any similarities and differences.

Distributing the effort into teams. One way to conduct qualitative user research at a larger scale is to utilise large teams and aggregate separate relevant studies together to form a more extensive synthesis [24]. Team-based projects involve many coordination challenges when a larger number of people perform data collection and analysis, posing risks like insights being lost in translation to the team [27, 28]. It requires carefully addressing: (1) Communication, coordination and ethical issues that appear due to a multi-cultural and multi-site study; (2) Issues about the reliability of data when data passes from one point to the next in the research process and (3) Difficulties in achieving coherence at the end after dividing tasks among multiple players [27]. The pairing of researchers is recommended in order to make sure knowledge is not lost if one researcher leaves.

Another perspective to scaling user research, especially organisation-wide is to enlighten and enable the workforce to conduct their user research programs. Organisations seem to be shifting from a single shared user research team to distributed and trained researchers for each project in order to satisfy the parallel demands of multiple research projects [29]. Pokerface [30] is an internal program in Google to develop user empathy across multidisciplinary teams of engineers, managers, designers and enabling them to conduct user research internalising and advocating the learnings across the organisation. Various organisations adopt such strategies first to develop awareness, adopt a process and eventually to scale by use of tools and various strategies such as having at least one designer in each team to bring in a user-centredness [31].

Technology and computation. Approaches other than face-to-face interviews are commonly adopted, such as telephone interviews, email, instant messaging and video conferencing software. Further, researchers are now discussing how sensors and technologies can be used to gather extensive data, especially in ethnography study of the product when the product is in actual use [32]. For instance, Paco is an experimental sampling methodology tool, which enables researchers to design surveys, add triggers, make use of device sensor data and collect quantitative and qualitative data [33]. Such approaches may be increasingly relevant post the COVID-19 pandemic, where researchers may want to utilise and enrich approaches other than face-to-face interviews.

Dealing with sheer volumes of data that get generated through large-scale qualitative user research and analysing those to derive insights is a crucial challenge. Even small sample sizes in qualitative user research can generate large volumes of data. Therefore, it becomes essential to establish procedures that can handle the data in its multiple forms [34]. It is recommended to plan and focus on qualitative data collection and analysis together via the usage of coding frameworks, training researchers to build expertise in analysing the qualitative data and maintaining an audit log of data collection and analysis methods [35]. Use of specific tools for coding and analysis such as HyperResearch, Nvivo, MaxQDA, and AtlasTi is common in qualitative research. However, these tools are aimed at helping researchers for coding and manual analysis [36]. Deriving insights from extensive data remains a time and effort consuming process [36].

Overall, it is found that there is a need for large-scale qualitative user research, especially for the design and development of effective large-scale services. There are challenges involved such as (1) lack of time and resources, (2) communication and coordination among multiple researchers, (3) maintaining consistency throughout the research process when different researchers are involved in data collection and analysis, (4) handling massive sets and multiple forms of data, (5) maintaining the rigour of the process and quality of data and (6) efforts required to analyse voluminous qualitative data.

8.3 Reflections on Qualitative User Research from Service Design Practice

As service design practitioners, we have conducted multiple service design projects within our organisation. We present some findings here that highlight the need for scaling up qualitative user research activities across the business, citizen and organisational services and discuss approaches and challenges to achieve large-scale research.

8.3.1 Project 1—Designing for an Enhanced Air Travel Experience

In this service design project, we aimed at enhancing the experience of air travellers at airports. They could be travelling to and from any corner of the world under various circumstances—business or leisure, domestic or international trips, short-haul or long-haul trips, travelling alone or with family or groups, with physical challenges. To design a seamless and delightful experience for such air travellers, there was a need to know their current needs and experiences. Since the focus was on experience, it was required to take a deep dive into the various stories that they might have about air travel and hence qualitative studies was aimed. The needs and experiences would vary across age, gender, culture and ethnicity, languages, regions of the world to which the travellers belong, the purpose of their travel, frequency of the travel, their health conditions and so on. However, it seemed practically impossible to get access to such diverse air travellers for conducting qualitative studies in a limited timeframe.

Leveraging large organisation diversity, and their reach to others. We reached out to the global employee population of our large organisation as it has a global presence an employee base of over four lakhs with the right gender and age diversity. Many employees require to travel nationally or internationally, frequently or infrequently for business purpose, as well as leisure with their friends and family. If we could reach to the whole employee base of the organisation, we could not only get to know their first-hand experiences of air travel but also the viewpoints of their friends and families indirectly. We decided to explore the possibility of considering such diverse employee population of an organisation as a representative sample of the air travellers across the world. While the approach will have an implicit bias, skewing towards salaried professionals, including their families and friends, would help mitigate to an extent. Additional external participations and triangulation would also be necessary.

Large-scale surveys with open-ended questions enabled with nudges to elicit rich responses. We decided to roll out a survey which would contain a right blend of questions seeking objective and subjective responses so that respondents could freely share their thoughts and experiences. We also decided not to administer this survey as a form-filling activity but to gamify it such that respondents could feel as if they are planning for air travel and going through the process of air travel at an airport. This could be effective to elicit their past experiences and share those freely with us.

Need for analysis tools. A large number of targeted survey participants and open-ended responses could generate a large volume of qualitative data which would be difficult to analyse manually. There was a need of using new modern technology to analyse the data. Though technology alone may not bring out richer insights, it could help in activities such as data cleansing, identify useful data and perform partial analysis of data.

8.3.2 Project 2—Designing for Citizen Services in India

India Post is the largest postal network in the world and one of the oldest citizen services in India. With its large population, India is known for its diversity across regions, religions, socio-economic strata, cultures and ethnicity. We undertook this as an experimental project to reimagine India Post service offerings by applying a service design sprint approach. This involved various challenges related to diversity, complexity and large scale. As part of primary research, we aimed at reaching out to the diverse user population across regions of India, but with limited time and resources.

Leveraging team's multi-cultural diversity, and their reach to others. The design team was formed such that it involved members that were not only from different disciplines but also belonged to different cultures, ethnicity, and regions in India and speaking different local languages. The regions that they belonged to also varied in terms of urbanisation and availability of various facilities. To design for large-scale complex citizen services to meet the needs of diverse citizen population, we decided to use the formal and casual acquaintances of this multi-cultural multi-located design team and reach out to diverse population for large-scale qualitative user research. The study was conducted with users from different age groups, across gender and some urban cities and small towns in Northern, Western and North-East regions of India.

Need for ease in reaching out to a large, diverse group. Although the coverage of complete diversity during primary user research was not feasible due to time and resource constraints, the study uncovered essential aspects. It developed richer insights that varied across cultures, age groups, genders, urban and rural regions. It was evident that reaching out to users in other different regions of India that are drastically different, could have revealed many more varying needs of citizens, grounded in their cultures, religions, regions, etc. The study acted as evidence that, for large-scale services, it is required to reach out to more extensive and diverse user population in order to meet their varying needs and concerns.

8.3.3 *Project 3—Employee Experience Within Organisational Services*

We engaged in a service design project within our organisation for an employee service to manage employee concerns. It was necessary to gain an in-depth understanding of the system, not just from the employee's perspective but also from the perspective of human touchpoints such as personnel from the Human Resources or administration, as well as organisational leadership. To make informed design decisions while designing the service, as the organisation is large and spread across locations, it was necessary to reach out to a large volume of service users and stakeholders. The chosen sample was spread across multiple regions for regional variations, and with different characteristics such as age, gender, location, roles, and the types of employee concerns faced. We aimed at reaching out to 500 employees and 50 stakeholders through semi-structured interviews rather than quantitative or written surveys, as we wanted to arrive at maximum actionable insights through this effort-intensive activity.

Enabling HR personnel to conduct user interviews in a short span. Since the design team had few members, to conduct 500 interviews, it was decided to take support from a large team of over 70 Human Resource personnel (HRs). The HR was already familiar with speaking to employees and were spread across different locations and were a natural choice. They were trained to conduct these interviews within their business units.

The structured response format for consistency. The HR was provided with a structured response sheet to capture their observations. The response format guided the HRs to conduct the interview, like a protocol and contained nudges and guidance to elicit their observations in detail. The response format allowed us to assemble the collected responses in a streamlined manner for ease of consolidation, tool-based processing and allowed some descriptive quantitative statistics.

Team-based manual analysis with the support of automated tools. Consequently, the data was processed through a blend of manual and technological methods and tools. The tools helped in cleaning up the data and identifying points of interest. Manually, the data was divided among the entire design team to individually derive findings and then consolidated to come up with insights collectively.

Need for tools to enrich and support data collection, coordination and analysis. In comparison to the pilot study initially conducted with a smaller sample, this large-scale activity resulted in more in-depth and richer insights. However, there were various challenges, some similar to the challenges in team-based approaches highlighted in the literature, especially in coordinating the data collection activity when such a large team is involved. The need for technological tools was observed in conducting the interviews, gathering the data in one place, doing data sanity checks and cleansing, as well as analysing the data.

To summarise, the above three projects highlight the need for conducting larger-scale studies, especially for the design of services where heterogeneous user groups and stakeholders are involved. It is possible to conduct these with rigour and quality, but there is a dire need for tools and approaches to make the process manageable. In line with the challenges summarised in Sect. 8.2 regarding time and resource constraints, coordination, consistency, rigour, quality and effort in analysis, the following are some additional challenges that need to be addressed when scaling qualitative user research: (1) Identifying and reaching out to the right user groups, (2) Eliciting rich written responses and (3) Enabling the teams to capture notes and observations in detail.

8.4 Discussions and Recommendations

The previous sections have highlighted various needs, challenges and perspectives of the process of scaling up qualitative user research. This section presents our initial approach and recommendations to scale up qualitative user research activities in service design practice. In order to conduct systematic qualitative user research for a large target population, the research journey that service design researchers and practitioners undergo can be primarily segregated into three crucial stages—(1) Planning stage; (2) Execution stage and (3) Analysis stage. The planning stage includes the preparation for scale before collecting data. It involves mobilising and distributing activities among different teams of researchers, planning inquiry protocols, identifying and recruiting participants with comprehensive coverage of representative demographics. The execution stage involves multiple teams of researchers conducting the studies at different locations. This includes rigorous interactions with research participants for collecting participation consent, conducting the inquiry sessions, and capturing notes and multimedia. The analysis stage involves aggregating participant responses, cleaning of data, triangulating findings, reporting the insights and presenting the research outcomes to different types of audiences.

The different researcher activities involved in conducting qualitative studies at large scale can be categorised into five components in the research journey: (1) Scope of activities: involves deciding the objectives that a research group should focus upon depending on the team size and skills; (2) People management: includes the tasks involving efficient management and training of multiple research teams, and enabling them with essential tools; (3) Processes and methods: includes operational guidance to maintain quality of research data and reproducibility of study process; (4) Documentation: includes notes, audio recordings, video clips and other artefacts generated during the research period for reliability and traceability of data and transferability of knowledge and (5) Aggregation: involves the compilation of various primary and secondary data sources, and studies to conduct extensive analytical synthesis. Table 8.1 highlights some of the approaches to facilitate qualitative user research for a wide target user population.

Performing qualitative user research at scale needs management skills, along with qualitative research skills. The use of technology plays a crucial role in planning and executing user research at scale. In our experience, we found that identification and

Table 8.1 Approaches to facilitate large-scale qualitative user studies

	Planning stage	Execution stage	Analysis stage
Scope	Identify the research objective(s), organise the research team and identify various perspectives to be investigated	Organise response collection from research participants, and compile notes, raw user data files, and data from secondary sources	Aggregate and analyse multiple data sets from one or multiple relevant studies to generate insights and visualise
People management	Facilitate formation of core research team and enablers, guide research design, sampling and participant recruitment strategy and training of researchers through pilot study run	Allocate researchers to different studies, enable them with tools and templates for data collection, and provide regular feedback guidance to ensure consistency and quality of collected data	Collaboratively identify the analysis strategy. Distribute tasks to research groups and enable them with various data cleansing and coding tools
Process and methods	Create a homogenous protocol for research inquiry, distribute the problem scope and conduct multiple separate studies if required	Provide nudges to participants to enrich responses by adding contextual details. Provide ready access to the protocol, and guidance to the researcher	Identify and suggest methods to triangulate insights with other relevant sources and studies, both primary and secondary
Documentation	Use different data collection templates, record meeting notes, multimedia data, artefacts	Enable provisions to use from multiple templates and mediums for participant consent and response collection Enable backup data storage to ensure no loss in participant data due to device errors	Availability of different suggestive formats for reporting and visualising insights based on audience
Aggregation	Enable data organisation, creation and storage of multiple studies and media files	Compile field research notes, raw participant responses and secondary research data files from various researchers. Enable transfer and accessibility of data from multiple sensors, recording devices and computer formats	Compile and share insights from multiple researchers and research groups and enable different formats for visualisation and reporting

usage of the right technology at the right time helps to reduce the manual effort and time. It improves the overall quality of the user research.

The right use of technological tools is crucial in the various phases of user resarch to plan, reach and effectively capture responses of the diverse target. Followings are the design directions which emerged through the case studies and literature, for introducing technology with each stage to enable qualitative user research at scale:

1. Creation, usage and distribution of homogenous research protocol are essential to perform qualitative research activity at scale. Technological tools could be useful to guide researchers throughout the protocol creation and piloting process.
2. Tools can aid designers for appropriate sampling and recruitment strategy for the required research objective.
3. Use of technology for recruitment and training of researchers and interviewers is critical to maintain appropriate rigour, quality and consistency of methods used and activities performed throughout the entire user research.
4. During interviews, a researcher is often laden with multiple activities such as taking verbal or written consent, ensuring proper audio or visual recording, taking regular notes, capturing observations and verify compliance to the research protocol. The tool can be introduced to provide real-time nudges and guidance to the researcher to ease activities.
5. Technological tools are essential to aggregate all the primary research data and ensure traceability about who did what and when. Further, data cleansing activity for vast sets of primary data can be aided with technology to reduce overall human effort and improve the quality of data for analysis.
6. Analysis for large sets of qualitative data is both time and resource-intensive. Technology can ease navigating the vast data and help focus on specific information, stakeholder responses or dimensions, and aid the extraction of insights.

8.5 Conclusion

New-age business, as well as social sector services, are becoming complex while catering to the diverse user base. These services are becoming borderless and catering to a large set of users. Designing of such services is fueling the need to scale the overall service design process, including qualitative user research. This paper establishes the need and benefits of the large-scale qualitative user research and how it can aid while designing scaled-up services. The need for technology is emerging through literature as well as multiple service design projects we executed. This paper further proposes the technological intervention points in qualitative user research to meet the need of reaching out to the required diversity. The proposed approach may not be applicable to all design projects, but to the large-scale design projects catering to large sets of user base. We believe that this is an initial but significant step in achieving service design for scale. This can be an open discussion around service design for scale and how technology can enable to make it happen. Going further, although technology to handle large sets of data and draw insights from those is still evolving, we need to design modern technological interventions and experiment those for large-scale qualitative research.

References

1. Grimes, J., Fineman, A., Kronqvist, J.: Service design at scale. Touchpoint—J. Serv. Des. **9**(3) (2018)
2. Bansal, P., Smith W.K., Vaara, E.: New ways of seeing through qualitative research. Acad. Manage. J. (2018)
3. Jupp, V.: The Sage Dictionary of Social Research Methods. Sage (2006)
4. Baxter, J., Eyles, J.: Evaluating qualitative research in social geography: establishing 'rigour' in interview analysis. Trans. Inst. Brit. Geogr. **22**(4), 505–525 (1997)
5. Bluhm, D.J., Harman, W., Lee, T.W., Mitchell, T.R.: Qualitative research in management: a decade of progress. J. Manage. Stud. **48**(8), 1866–1891 (2011)
6. Shenton, A.K.: Strategies for ensuring trustworthiness in qualitative research projects. Edu. Inf. **22**, 63–75 (2004)
7. Rolfe, G.: Validity, trustworthiness and rigour: quality and the idea of qualitative research. J. Adv. Nurs. **53**(6), 304–310 (2006)
8. Carcary, M.: The research audit trial—enhancing trustworthiness in qualitative inquiry. Electron. J. Bus. Res. Methods **7**(1), 11–24 (2009)
9. Mays, N., Pope, C.: Rigour and qualitative research. BMJ **312**, 109–112 (1995)
10. Borgman, C.L.: Big Data, Little Data, No Data: Scholarship in the Networked World. MIT press (2015)
11. Camfield, L.: Rigor and ethics in the world of big-team qualitative data: experiences from research in international development. Am. Behav. Sci. **63**(5), 604–621 (2019)
12. Williams, E.N., Morrow, S.L.: Achieving trustworthiness in qualitative research: a pan-paradigmatic perspective. Psychother. Res. **19**(4–5), 576–582 (2009)
13. Mills, K.A.: Big Data for Qualitative Research. Routledge (2019)
14. Kimbell, L.: The Service Innovation Handbook: Action-Oriented Creative Thinking Toolkit for Service Organizations. BIS publishers (2014)
15. Lehrer, C., Wieneke, A., Vom Brocke, J., Jung R., Seidel, S.: How big data analytics enables service innovation: materiality, affordance, and the individualization of service. J. Manage. Inf. Syst. **35**(2), 424–460 (2018)
16. Baker, S.E., Edwards, R: How Many Qualitative Interviews is Enough (2012)
17. Mack, N., Woodsong, C., MacQueen, K.M., Guest, G., Namey, E.: Qualitative Research Methods: A Data Collector's Field Guide, North Carolina. Family Health International, USA (2005)
18. Creswell, J.W., Poth, C.N.: Qualitative Inquiry and Research Design: Choosing Among Five Approaches. Sage publications (2016)
19. Marshall, B., Cardon, P., Poddar, A., Fontenot, R.: Does sample size matter in qualitative research? a review of qualitative interviews in IS research. J. Comput. Inf. Syst. **54**(1), 11–22 (2013)
20. Morse, J.M.: Designing funded qualitative research: In: Handbook of Qualitative Research. Sage Publications, Inc., pp. 220–235 (1994)
21. Patton, M.Q.: Qualitative Research & Evaluation Methods: Integrating Theory and Practice. Sage Publications (2014)
22. Guetterman, T.: Descriptions of sampling practices within five approaches to qualitative research in education and the health sciences (2015)
23. Padgett, D.K.: Qualitative methods in social work research, vol. 36. Sage Publications (2016)

24. Robinson, O.C.: Sampling in interview-based qualitative research: a theoretical and practical guide. Qual. Res. Psychol. **11**(1), 25–41 (2014)
25. Schneider, J., Murray, J., Banerjee, S., Mann, A.: EUROCARE: a cross-national study of co-resident spouse carers for people with Alzheimer's disease: I—factors associated with carer burden. Int. J. Geriatr. Psychiatry **14**(8), 65 (1999)
26. Murray, J., Schneider, J., Banerjee, S., Mann, A.: EUROCARE: a cross-national study of co-resident spouse carers for people with Alzheimer's disease: II—a qualitative analysis of the experience of caregiving. Int. J. Geriatr. Psychiatry **14**(8), 662–667 (1999)
27. Guest, G., MacQueen K.M.: Handbook for Team-Based Qualitative Research. Rowman Altamira (2008)
28. Hunt, G., Moloney, M., Fazio, A.: Embarking on Large-Scale Qualitative Research: Reaping the Benefits of Mixed Methods in Studying Youth, Clubs and Drugs. Nordisk Alkohol Nark (2011)
29. Scaling User Research in an Agile R&D Organization. https://medium.com/athenahealth-design/scaling-user-research-in-an-agile-r-d-organization-71f9be1097e2 (visited on 12/06/2020)
30. Liu, A., Sosik, V.S., Singh, K.: Building empathy: scaling user research for organizational impact. In: CHI '18 Extended Abstracts on Human Factors in Computing Systems, Montréal, Canada (2018)
31. Strategies in adopting and scaling user research within 1.5 year. https://uxdesign.cc/strategies-in-adopting-and-scaling-user-research-within-1-5-year-ad7ca28b3490 (visited on 12/06/2020)
32. Gorkovenko, K., Burnett, D., Thorp, J., Richards, D., Murray-Rust, D.: Supporting real-time contextual in-quiry through sensor data. In: Ethnographic Praxis in Industry Conference Proceedings (2019)
33. Evans, B.: Paco—applying computational methods to scale qualitative methods. In: Ethnographic Praxis in Industry Conference (2016)
34. Miles, M.B., Huberman, A.M.: Qualitative Data Analysis: An Expanded Sourcebook. Sage Publications (1994)
35. White, D.E., Oelke, N.D., Friesen, S.: Management of a large qualitative data set: establishing trustworthiness of the data. Int. J. Qual. Methods **11**(3), 244–258 (2012)
36. Paredes, P., Ferreira, A.R., Schillaci, C., Yoo, G., Karashchuk, P., Xing, D., Cheshire C, Canny, J.: Inquire: large-scale early insight discovery for qualitative research. In: 2017 ACM Conference on Computer Supported Computer Supported Cooperative Work and Social Computing, pp 1562–1575 (2017)

Chapter 9
Design for Small Businesses in India—Helping the Real Entrepreneurs

Vipul Vinzuda, Jitendra Singh Rajput, and Amresh Panigrahi

Abstract Sustainable development goals (SDGs) set in 2015 by the United Nations aims to achieve a better and sustainable future for all, intended to accomplish by 2030. This paper presents the innovative and creative efforts taken up by a multidisciplinary team of designers, researchers, engineers, and architects worked on a pilot project, which was presented to the Ministry of Micro-Small Medium Enterprises (MSME), India. The focus of the project was to understand the challenges faced by a non-organized business or small business trades ranging from tea vendors, vegetable vendors, etc. How to design interventions can bring an improved 'value perception' in the society while providing ease of work, more earnings, cleanliness, and overall well-being to these hardworking people in India. Design is considered an essential strategy for innovation to create differentiation in the market. During this project, the designing is explored as a catalyst to bring positive social change. The team studied selected popular trades, which are generally present in every town of India, by conducting field trips, photography, and observational studies to collect data. The map of problems versus opportunities was generated. The specific requirements derived from this study offered rich insights and established the need for design interventions at the product-service-system level. Further, a co-design workshop was conducted with design students focusing on future of vending, improving employment opportunities in small businesses through mobility. The paper highlights the design process, creative explorations, and design developments of solutions. This pilot project provides future directions and lists ample opportunities for designers to play an active role in bringing social transformation.

All the authors have contributed equally to this paper.

V. Vinzuda (✉) · J. S. Rajput · A. Panigrahi
National Institute of Design, Gandhinagar 382 007, India
e-mail: vipul_v@nid.edu

J. S. Rajput
e-mail: jitendra_s@nid.edu

A. Panigrahi
e-mail: amresh@nid.edu

A. Chakrabarti et al. (eds.), *Design for Tomorrow–Volume 2*, Smart Innovation, Systems and Technologies 222, https://doi.org/10.1007/978-981-16-0119-4_9

9.1 Introduction

9.1.1 Premise of the Project

Based on the Ministry of Housing and Urban Poverty Alleviation, India, there are more than 10 million street vendors [1] running small businesses on streets by selling fruits/vegetables, tea-shop, and snacks on the wheels, or sitting/standing/moving around footpaths/streets. These vendors belong to the vast pool of unorganized enterprises. The challenges faced by these enterprises are majorly in the form of financial crunch, lack of skills, poor/no infrastructure, etc., [2] However, we realize their challenges to simplify processes, optimizing related products/services, utilizing simple tools, and excellent branding.

The project aimed to study the underlying design issues related to entrepreneurship, behavior, ergonomics, products, and services. As an initiative of the Ministry of MSME to design and implement products for better usability and perception, the project was assigned to the National Institute of Design (NID). NID took up the research initiative under the leadership of faculty members and delivered the initial concepts to the ministry for further implementation and feedback. As part of the extended effort, the faculty members continued the research activity and workshops to acquire more insights in the small businesses. The paper is the outcome of these insights generated from the research activity.

The project's primary goal was to identify design intervention possibilities and create the shift, sustenance, and growth in earnings of individual small businesses. Further to provide core and cosmetics improvements, while encouraging more women participation [3]. Ahmedabad-based Self Employment for Women's Association (SEWA) provides 60,000 street vendors into union. [4] Street vending provides 'affordable' and 'convenient' array of services to majority of citizens and play an important role in contributing to nation's economy [5].

Following clear objectives were defined at the onset of this project:

- To improve the value perception of the trade, individuals associated
- Provide ease of working
- Generate more business and increase earnings
- Bring more happiness in life

9.1.2 Overview of Small Business in India

In times of rapid urbanization, many people migrate from rural parts of the country to seek better work opportunities, living, and hoping to provide a good future to their younger generation. Lack of education and a competitive environment makes earning livelihood and survival in Indian cities difficult, and several other factors add to the overall challenge of working and living.

Many street vendors make their living by selling on footpaths. Congested roads have more probability of attracting a broader customer base at the cost of affecting traffic flow or smooth pedestrian movements. Street vendors are generally blamed for piling up garbage and plastic paper waste thrown on roads. However, National Association of Street Vendors of India (NASVI) plays an instrumental role in advocacy and ensured provisions for street vending at the stage of city planning by various state governments [6]. There is no formal social security system meant for these people who earn their living by such means [7–9]. The practice-based design research approach was considered while working on this project.

9.2 Methods

The project comprises of two research studies done to contribute by design for small businesses in India. A practice-based design research approach was selected; the design process and outcome from an ethnographic research followed by a co-design workshop are shared in this paper.

9.2.1 Study-1: Ethnography Study on Selected Small Businesses in Ahmedabad and Gandhinagar, India

It is imperative to understand all the stakeholders' concerns in this context, especially the central user. [10, 11] Conducting ethnographic research to study user needs, challenges, and opportunities were considered the first step in the right direction. The clear intent of making contributions to a large number of people in this sector and transform their livelihood and family life positively by design. Field visits and immersion study experience of select trades were documented through diaries, photographs, and illustrations. A total of six trades were selected for the research and design studies in the first phase. As shown in Fig. 9.1, each occupation was classified based on parameters, which can be crucial for design and business development.

Is category such as service, trading, or product making is the key to a sale? What kinds of spaces are used to conduct the trade? Do low investment and low skill make it easy to start earning and toward self-sustenance? What importance duration, time, and season makes to the success of the shop? These questions were pertinent to understand and classify the diverse trades into a common lens for study purposes. Rapid sketching, illustrations to capture nuances of the actual context provided an apparent reference of scenarios. It gives details on posture, movements, real space, and setup, interactions, and exchanges, as highlighted in Fig. 9.2.

The research diary notes were later translated into a mind map diagram understanding the intangible aspects of user stories. Figure 9.3 shows a typical work routine of one of the trade— a tea vendor that depicts work routine and motives

Fig. 9.1 Categorization of trade based on parameters crucial to design

Fig. 9.2 Fieldwork: Immersion study—representation of vegetable vendors in the Indian context

during the day. It is an efficient method to capture important information into a visual format and provide designers a clear reference picture of the user's needs for further development.

Most of them use customized pushcart/cycles as per their trade requirement and budget. There are practical issues of usability and safety; however, beyond that, the 'perception' associated with their profession(s) and hygiene concern is pivotal one to be addressed soon. With competition from the latest mall, a chain of restaurant and advancement in infrastructure all around, these set of vendors with lower socio-economic background and migrant from small town/villages face a lot of challenges

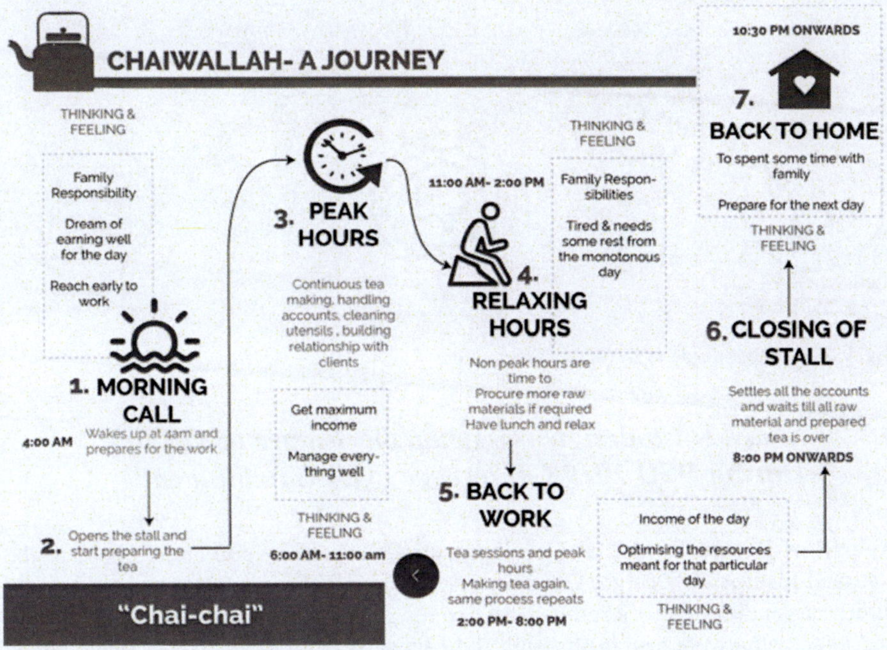

Fig. 9.3 Typical work routine of a tea vendor: mapping thinking and feeling aspects

and issues which can be resolved with innovative solutions and design thinking. One of the main research questions was formulated as to how we might design to enhance customer perception and service experience from street vendors? The question served as an anchor point for the research.

Specific challenges were identified for each trades, and major issues associated with food purchased on consumed from vendors are following:

- Hygienic and price
- Fresh food
- Sanitation and cleanliness
- Value perception

Figure 9.4 represents the potential factors affecting for tea vendors; these are common problems across the trade, which are associated with value generation. Many small businesses, street vendors consider owning a shop as their aspirations, primarily to gain visibility, upgrade and get more organized in running a business. This provides profitability, an increase in business, and a stable income—the pride and ownership associated with having their own shop avoid all legal problems and hassles. However, the key issues in owning a fixed shop are a higher investment, over-designed, adaptability to change, less modularity, optimized use of space, and lack of awareness of branding. Presently, there are no standards that are developed by government/ municipalities to regulate or maintain such fixed shops.

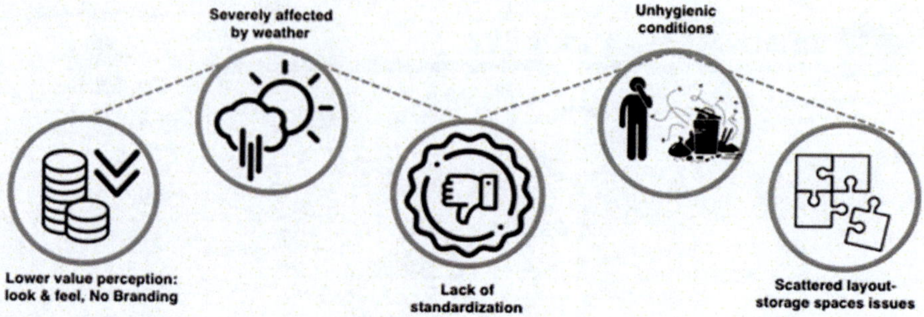

Fig. 9.4 Potential negative factors for tea vendors

9.2.2 *Study-2: Co-design Workshop on 'Future of Vending' During DSD 2019 Conference, NID Gandhinagar*

As part of Design for Social Development (DSD), 2019, conference held at the National Institute of Design, Gandhinagar, India, in February 2019, a workshop on future of vending was offered to participants. A total of 12 participants signed up and joined who were design students and educators from India. The critical agenda of the workshop was to engage participants in design sprint activity in 3-h duration. Authors shared the findings from study-1, highlighting significant issues, a list of potentially detrimental factors, challenges faced by small businesses for three kinds of trades (Fig. 9.5).

The central focus of the discussion for the workshop revolved around tea vendor, vegetable vendors, and pani-puri stall. How small business opportunities can be enhanced on a vehicular platform? Participants in a team of two were required to select one of the trades, select three Sustainable Development Goals (SDGs) [12] from the given list, and consider core design elements as depicted in Fig. 9.6.

Fig. 9.5 Focus on select trades: Tea vendor, vegetable vendor, pani-puri stall

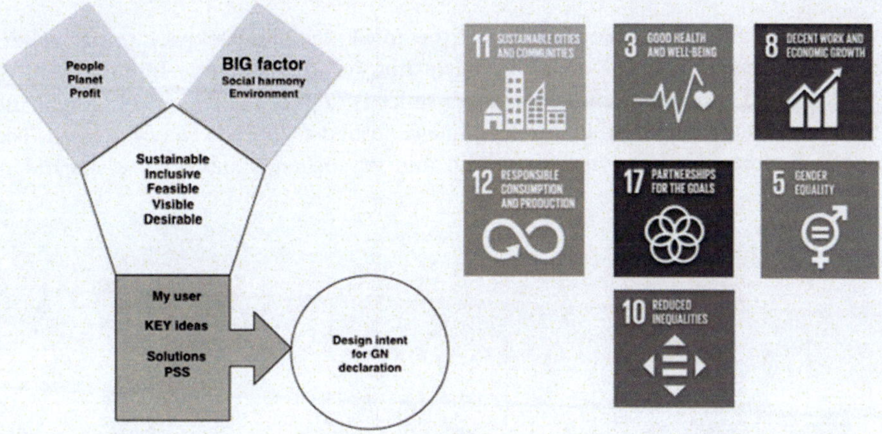

Fig. 9.6 Core design elements & a select list of SDGs provided to the participants

9.3 Results

9.3.1 Study-1: Outcome: Design Ideation and Development for Selected 6-Small Businesses

Design team comprising of designers, researchers, architects, and engineers identified the core and cosmetic areas for design interventions, as compiled in Table 9.1. Examining literature findings, specific areas of improvement based on field research, and current trends. The scope of design became more evident for short-term and long-term goals. Conceptual ideas explored mainly focused on improving the existing products and working while creating newer impressions of street vendors.

Creative exploration of visualizing product, system, and service design was prepared using manual and digital mediums. *Tea on the go*, a concept as showed

Table 9.1 Design intervention approaches for small businesses

Sr.no	CORE	COSMETIC
1	Structure and layout: Cart, shops, shelves, counter, storage design	Visual communication/branding of small business
2	Objective-based design: Design for service, hygiene, product quality, logistics	Improving product packaging and display
3	Technology adaptation for improving efficiency	Mobile wallet-based payments and offers
4	Behavioral changes through promotion, education, training, feedback, appraisals	Following guidelines and display at the stall

in Fig. 9.7 highlights the key features of the novel idea of a backpack for tea sellers increasing reach, business, and serving of hot tea to customers. Improvement of the perceived value and comfort of a tea seller is kept at the center. Photorealistic renderings to indicate how an alternate image around each trade can be created. The proposals created a keen interest within ministry officials and other stakeholders from MSMEs (Fig. 9.8).

Fig. 9.7 Concept exploration: Tea on the Go

Fig. 9.8 Creative visualization: CAD and photorealistic renderings

9.3.2 *Study-2: Outcome: Connecting SDGs with Selected 3-Trades- Co-Design Workshop on 'Future of Vending'*

Participants created concept sheets for the chosen trades, the ideas generated in response to Sustainable Development Goals (SDGs) specifically for small businesses in India. The ideas ranged from a product, communication to service-based innovative solutions. As indicated in Fig. 9.9, the pani-puri stall can emphasizes on hygienic and cleanliness. Effective communication and the implicit message of following the sustainable way of operating build image. The use of natural material for packaging, digital payment, and offers to attract customers were explored. Certifications from government, health, and food authorities renew trust. A convergence of such ideas to uplift overall service experience and regain trust amongst customers was a central motive. The results of the workshop helped in articulating challenges and exploring design opportunities to enhance their performance and perception amongst customers. The workshop streamlined the issues faced in this unorganized sector and uncovered priority areas for future developments. Proposed design ideas and strategies indicated a valid direction toward feasible and low-cost solutions thereby improving service experience and enhancing earnings.

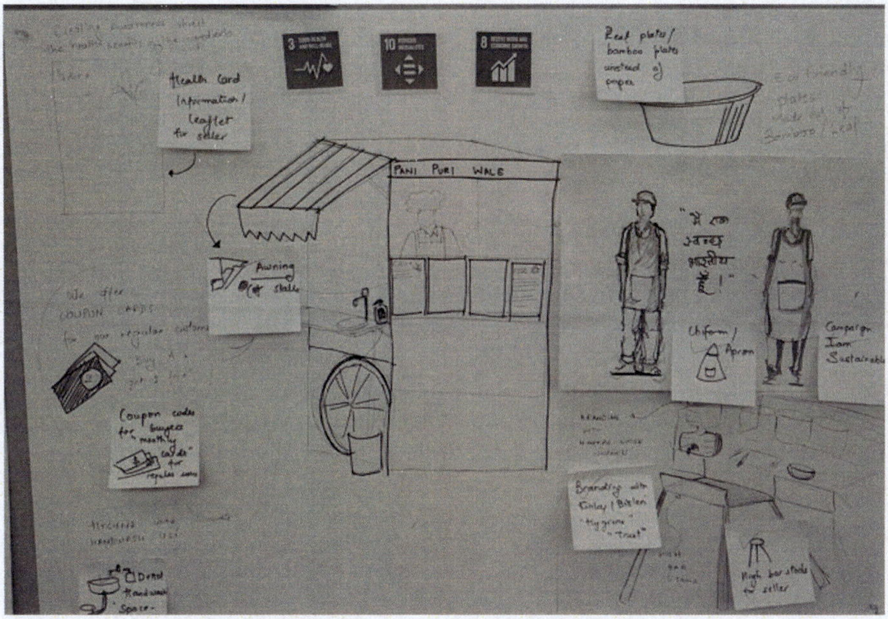

Fig. 9.9 Concept sheet for pani-puri stall by a team of participants

9.4 Discussions

The government support and initiatives can be enhanced by offering specific incentives and overall support models for people who earn their livelihood from a small business with limited or no initial investment.

The basic roadster cycle and pushcart become the main platform to the function of the business as its advantage of being low cost, storage space, and mobile in reaching out to more customers. In the present context, while door-to-door delivery is prevalent and popular with smartphone-based apps, it is essential to focus on the specific requirements of Indian context as India is a really unique country, vast, and diverse with its scale and cultural practices, livelihood, and our range of mobility solutions varies from human-powered carts to latest personal mobility innovation made by a startup.

The orientation, training support, and periodic up-gradation for every kind of street vendors are necessary to keep them competitive in changing times and making them earn more. Complex problems with overlapping needs would require more holistic and systemic design-driven change, especially while addressing the needs of socially and economically weaker populations. The core and cosmetic changes should be implemented in stages with close and active coordination amongst street vendors, government, and manufacturing sectors. A strategic communication design approach can build a new perception of society toward these trades.

9.5 Conclusion

The research and workshop activities generated exciting insights in the form of present-day challenges and opportunities for street-based small businesses. Difficulties encountered by the small entrepreneurs are as per below have been in the areas of non-standardized products, issues of workstation/workplace, display, arrangement, transportation, etc. The research activity tried to understand the problems related to the small businesses. It generated multiple models for product, service, and business enhancements. The other vital possibilities of research work are based on the broader systemic domain. The areas where the design would help in creating more long term and positive impact for small businesses and entrepreneurs.

Future scope of work

(1) Spreading awareness among small businesses for better design and sophistication possibilities. Small businesses are, most of the time, established out of the urgent need for earning a livelihood. Therefore, the setup of these businesses is mostly makeshift and through available resources. The challenges of efficiency and productivity are most common in small businesses due to lack of sophistication and order in the processes and equipment. The design plays a vital role in creating products and services to enhance the outcome and profitability of

the business through contextual solutions. There is a need for such design-led solutions for small businesses, which is missing due to the lack of awareness, and defined need in the sector. The research has identified it as one of the main challenges in the way of design intervention in the sector.

(2) Government policies to handhold and facilitate small businesses beyond financial support. The majority of government initiatives in the small business and MSME sector has been focused on providing financial assistance to enterprises. Beyond financial support, there is also a policy framework required for supporting the small enterprises to train, educate, and handhold for best practices. The innovation in products, and services, best hygiene practices, better packaging, and communications also need to be part of policy initiatives. The report suggested the ministry and other stakeholders have reviewed the inclusion of such initiatives.

(3) Decentralization and democratization of design models for small businesses. The ease and access of products for fulfilling the needs of small businesses are must for reviving and establishing them. The products being used for supporting small businesses like handcarts, tea preparation and distribution products, vegetable carts and storage systems, etc., should be manufactured locally, across the country to make them cheaper and accessible. The designs of these products should be made available to local manufacturers and the local authorities/industry department should monitor support for these activities.

(4) Recognizing small businesses and bringing pride in entrepreneurs for sustenance and growth. The pride associated with small businesses should be recognized and encouraged at various stages. Small businesses need to be brought at an aspirational level and young generations should be trained to adapt to small businesses. The research highlighted the lack of positive recognition with small businesses leads to loss of pride and innovation in the sector.

Acknowledgements DC, Micro Small Medium Enterprise (MSME), New Delhi

Director, National Institute of Design, Ahmedabad, India

DSD 2019 conference Workshop participants and vendors who voluntarily supported in the study.

References

1. Report on conditions of work and promotion of livelihoods in the unorganized sector. https://msme.gov.in/sites/default/files/Condition_of_workers_sep_2007.pdf, last accessed 2020/06/01
2. Bhowmik, S.K., Saha, D.: Financial Inclusion of the Marginalised, 1st edn. Springer, India (2013)
3. WIGO Homepage. https://www.wiego.org/, last accessed 2020/06/02
4. SEWA Homepage. https://www.sewa.org/, last accessed 2020/06/02
5. National Policy for Urban Street Vendors. https://muepa.nic.in/policies/index2.htm, last accessed 2020/06/01

6. Mahadevia, D., Vyas, S., Brown, A., Lyons, M.: Law, Rights and Regulation for Street Vending in Globalising Ahmedabad (2012)
7. Shah, D.: Informal Markets, Livelihood and Politics: Street vendors in urban India, 1st edn. Routledge, India (2016)
8. Bhan, G., Srinivas, S., Watson, V.: The Routledge Companion to Planning in the Global South. Routledge, UK (2018)
9. Bhowmik, S.K.: Street vending in urban India: The struggle for recognition (Book Chapter), Street Entrepreneurs: People, Place and Politics in Local and Global Perspective. Routledge, UK (2007)
10. Bhowmik, S.K.: Street Vendors in the Global Urban Economy, 1 st edn. Routledge, India (2010)
11. Mahadevia, D., Vyas, S., Brown, A., Lyons, M., Jajoo, K., Mishra, A.: Street vendors in Ahmedabad: Status, Contribution and Challenges. CUE, CEPT working paper series, India (2013)
12. Le Blanc, D.: Towards integration at last? The sustainable development goals as a network of targets. Sust. Dev. **23**, 176–187 (2fwrr015). doi: https://doi.org/10.1002/sd.1582

Chapter 10
A Tool to Design a User-Centred Town Plan

Abhishek Singh and Pratul Chandra Kalita

Abstract The race of rapid urbanization has ended up developing numerous smart cities around the globe. A large amount of funds and resources are exhausted in this sector but the returns are not satisfactory. The infrastructure development is as per the vision and brainstorming of the experts and policymakers of the society. But the same is not accepted by the end user of the recently developed smart cities. So, it is very important to understand the requirement of the end-user, to make these projects successful. The conventional methods of data collection for town planning like questionnaires, personal interviews, group discussions, public hearings, etc. are having their limitations. The major issue with these data collection tools is that they all are very time-consuming and effort-intensive. The end users lose their patience or interest and end up giving biased data which fails the entire purpose of end user data collection. Hence, in this paper, we have designed a tool to collect user-centred data from a large number of people, in a short duration of time. We have used card sorting techniques for making it interesting and easy to understand for the users. The zoning of the city can be easily done, and the same can be used as a framework for detailed planning. So, this paper is on designing a user-centred town planning tool. This tool will ensure that the requirement of majority of the population is taken into consideration. It can be converted into a mobile app-based tool to have the vote of the entire population for designing their own town.

10.1 Introduction

The trend of smart cities has fascinated the ambitions of the citizens and governments around the globe. In spite of the growing interest and developments of smart cities in the current years, there is no established definition of smart cities. And framing of strategic plans for development of smart cities is unexplored in multiple facets [1–4].

A. Singh (✉) · P. C. Kalita
Department of Design, IIT Guwahati, Assam, India
e-mail: deeabhi@gmail.com

A. Chakrabarti et al. (eds.), *Design for Tomorrow–Volume 2*, Smart Innovation, Systems and Technologies 222, https://doi.org/10.1007/978-981-16-0119-4_10

Their study [5] has revealed that there is no exclusive definition of smart city; the trends of all the evolving smart cities are largely local context specific. It may be completely different in the technology or any other aspect of implementation.

Smart cities signify a framework of urban development with human requirements and technology as the foundation for the development of urban set-up [6].

In the case study of Songdo in South korea, Masdar city in Abu Dhabi, numerous ghost towns of China and many upcoming smart cities, it was observed that they have adopted state of the art technologies and majority of the parameters of smart cities identified in the literature implemented with proper planning and policy framework but still failed to attract occupant as planned. It raises a question whether to consider such hi-tech project successful, where technology and environmental realms have attained a level of sanctity but failed to have a human scale.

So, it was important to understand the requirement of the end users to make it a user-centred town, which is missed in most of the cases. A user-centred town planning means to plan a city by keeping the priorities and preferences of the end users as the main focus in the planning. Their opinion should be given more importance than the policy makers or the urban planners.

There are multiple methods available for collection of data from the end users like questionnaire, public hearing, personal interview, group discussions, suggestion box and discussion forums. These data collection tools have their own drawbacks, which compromises the significance of the data collected.

10.1.1 The Problem Statement

Generally, the politicians and policy makers focus more on the census data for proposal of any town. Their main focus is on the strategic or the commercial viability of the project, and how they can utilize the available fund so as to influence the maximum population. The policy makers pass down their budget sanctions for specific target population/segments of the society for which it is to be utilized; it is passed on to the field experts and urban planners, whereas the urban planners use the census and GIS data to understand the geographical, metrological, population density, etc. for checking technical feasibility of the project. Finally, the project is sketched on the paper after checking the commercial, technical and environmental feasibility of the project. In the entire process, the opinion of the end users was never taken into consideration.

It was clear from the literature review of smart cities that the formulation of smart city policy should be city and people specific. To make a user-centred town plan, it is important to have a robust data collection tool.

While studying the data collection tools, it was found that questionnaire is the most popular choice among all researchers for collecting data, since it is well-structured tool for quantitative analysis. But this method is linked with multiple drawbacks also. Questionnaire filling is time taking activity, and people lose their interest for multiple reasons like shortage of time, or have questions beyond their field of interest, or if they have difficulty in understanding the questions, if the subject is not comfortable with the content of the question, and in most of the cases, subjects consider it as a liability to complete the task as early as possible and may answer in waste without understanding the intent. The data collected may be skewed.

Similarly, in other data collection tools, like in personal interviews, it is very time-consuming and not suitable for large sample sizes. In case suggestion boxes are installed, it is observed that very few people participate actively in it. While conducting focus group discussion, there are chances that any outspoken speaker may skew the opinion of the entire group skewing the actual desired data. So, all the existing tools were having multiple limitations.

Hence, there is no optimum tool for data collection to collect the opinion of mass population related to town planning. Another issue was that all the people are not an expert in this field of town planning; hence, they may not be able to actually plan a city or even give relevant suggestions. Hence, the tool should be designed in such a way that their requirement can be extracted irrespective of their expertise in town planning and can be used to frame a zoning of town, which can be used by town planners to convert into the blue prints of town plan.

10.1.2 Card Sorting

'Card sorting was originally developed by psychologists as a method to the study of how people organize and categorize their knowledge. As the name implies, the method originally consisted of researchers writing labels representing concepts (either abstract or concrete) on cards, and then asking participants to sort (categorize) the cards into piles that were similar in some way. After sorting the cards into piles, the participants were then asked to give the piles a name or phrase that would indicate what the concepts in a particular pile had in common'.

There are two types of sorting technique, open sort and closed sort. In open sort, the categories in which the cards are to be sorted are not identified, and the users are asked to identify the categories and then sort, and in closed sort, the categories are predefined, and the users are asked to sort the cards in the mentioned categories.

10.2 Methodology

Since the city dwellers or the target population may not be an expert in town planning, but they are expert in their own daily requirement and day to day activities. The people are interacting with the city 24 × 7 and 365 days a year, throughout their life. So, on the basis of their experience, everyone has list of the services and utilities in their mind which when compiled together form a mind map. The mind map of an individual with respect to town planning means that people are already living in the cities and are satisfied with some of the services of the cities, and on the other hand, they are struggling to access some of the services. Hence, on the basis of their experience, they can suggest that which services to be prioritized and where to be placed in a city. So the image in their minds about the positioning of the services for an ideal city is their individual mind map. The challenge is to compile all the mind maps of residents of the city into a single-mind map and then convert it in to blue prints of town planning.

The sample size of our research was 50. The age variation of the people was from 18 to 60 years. The demographic of the population was that 15 samples were students, 20 people were working professional at various levels, and remaining 15 were businessman and from unorganized sectors. The monthly family income of the participants varied from Rs. 10,000/- per month to Rs. 1.5 Lakhs Per Month. The samples were present in Guwahati city but are residents of ten different states of India. Nearly 30% of the population are from north-eastern states, and remaining samples are from across India.

The study started with listing of all the services of a town. The list was made while interviewing multiple people from different localities and backgrounds. After the list was prepared, we need to plot these services in a town.

We used card sorting technique for zoning of the listed services. We have used hybrid card sorting technique in which we have decided some categories on the basis of distance. 0–1 km range is considered as comfortable walking distance if the quality of the route is good. This is the apt distance considered in majority of the cities by the architect. Majority of cities have their city centre in the radius of 1 km to ensure comfortable walking (e.g. Zurich, Brisbane, Pittsburgh, Copenhagen, etc.). This range is considered for utilities and services required on daily basis by majority of the citizens for which no extra transport is required. In larger cities, the comfortable walking distance is the same but they might have multiple city centres [7]. Second range was considered for cyclist, rickshaw and E-rickshaws. It was considered on the basis of comfortable cycle able distance. This range was discussed with all the participants and considered as 2–5 km in Indian context. 'This distance is considered as the average suited to cycling in urban areas'. This range was linked with services of daily use but need not be close to a residential area. The third range was 6–10 km. Other than six cities in India, all other cities average radius is below 10 km. The fourth range was 11–20 km; this area is the outskirt of cities or the neighbourhood of the cities. The fifth range is the range beyond 20 km from the starting point. The starting point of every range is not the city centre but the residence of the subject. Since the subjects may or may not be able to quantify the importance of distance in perspective of city as a whole but they can easily asses the distance on the basis of their own daily requirements.

A chart was prepared with clear demarcations of the ranges and was placed in front of the subject. All the subjects were given cards with names of the services and utilities of any town. Blank cards were also provided to the subject to add any additional service of their choice.

The subjects were asked to arrange the cards with services and categories mentioned on them in the chart with different ranges of distance as per their requirement. The data was collected from 42 samples and was analysed. The data collected were in-line with one-on-one interview with another group of eight people. The data was used to establish relations between the services and the distance at which they were placed.

10.3 Results

A table of preference by the users was calculated from the card sorting data. The calculation of preference is done by calculating the percentage of the total number of times the service was sorted in a particular range divided by the total number of subjects participating in the sorting. The percentage of a particular service across all ranges was calculated. This gives a clear picture about the preference of the service in a particular range with exact percentage to identify its significance in the range.

The values with the highest percentage of preference were the most demanded services in the said ranges of the distance. Each range has their own preferred services. The top preferred services of each range are indicated in their respective charts.

In Fig. 10.1, the percentage of preference of services was plotted in the graph. This range belongs to the area accessible by walk or activity of frequent nature. This range has special preference since the user has to decide what are the services which will hamper the sanctity of any residential area and need to be discarded from the range. At the same time, they should not be dependent on any transport facility,

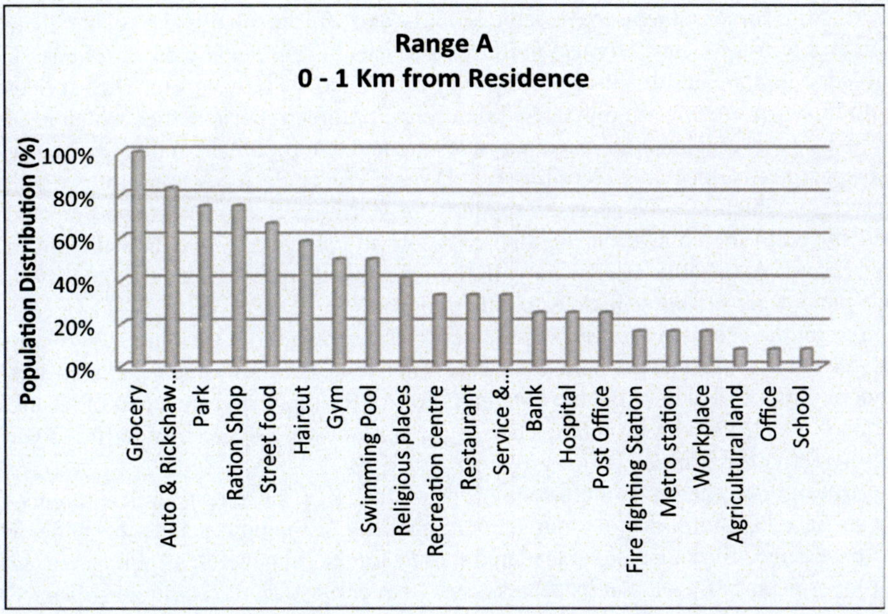

Fig. 10.1 Range A services ranking

public or private to access this range. The comfortable range of walk is considered as 1 km radius as found in the literature study. Out of 36 services, users have opted to keep 15 services strictly out of the range A with common preference.

Remaining 21 were ranked as per their preference in range A. In the chart, most preferred are starting from left and the level of preference decreases as we move towards right. The top scorer of the entire experiment was grocery with 100% preference in range A. It shows that all the subjects want that grocery to be available within walking distance from their residence.

Similarly, other ranges were also plotted. Range B was focused as the range for bicycle, rickshaw, autorickshaw or any private vehicle mobility area. The services in this range are also considered as the frequent use services but which should be kept at some distance from the residential area. There may be various concerns linked with this decision. As per the discussion with the subjects, some of them stated that these services are important but increase in the density of traffic, when kept near residential area, generates risk of accident and makes the residential area unsafe place for children and elderly people while they use the road side for leisure. This also increases the disturbance due to use of horns, traffic noise. It will generate more dust and pollution in the residential area.

The service considered in this range caters service to large number population. Hence, these are to be positioned at some common points, where people from different residential zones can access it easily; at the same time; any residential zones are not disturbed.

Range C was considered by the subjects for services which are not very frequently used. Still some subjects have kept schools and offices in this range but there percentage is very low. Highest preferred services in this range were insurance or investment office, malls followed corporate offices, court, movies, etc. The services with higher preference in this range broadly are of three types of office which need to be accessed at fixed intervals of days, leisure and entertainment and long distance transport network. Two exceptions are there of corporate office and college and university which may be accessed on daily basis. Majority of subjects participated had stayed in the college during their exposure to college and may have biased in this regard. As per the discussion with the subjects, the preferences of mobility in this range were private vehicle or public transport.

As per the standard size of Indian cities, range D will be in the out skirts of any city. So the service sorted here-were used after long duration of time. Places like zoo, museum and picnic spot are the top scorers of the range and are a part of leisure activity which may be preferred once in a month or year depending on the other leisure activities available in the city.

Airport is another most preferred service of the range, since its technical requirements have kept airport in the out skirts in majority of the city; hence, the subjects have perceived the same and considered it in the range. In general, airports are used for long-distance travel, and its access is not frequent; hence, the additional distance is not a major concern. Similarly, in range E, the services which were opted beyond the radius of 20 km from the residence of the user were opted. In Indian context, majority of city limits are within the radius of 20 km from the centre, (If we assume that the cities are in a circle or some symmetrical shape). As discussed with the subjects, these are the services that they want to keep out of the bounds of their city.

The top preferred services of the range are industries, picnic spots, agricultural lands and court. Industries are kept out of the city bounds to keep the pollution and heavy vehicle movement away from their residential areas. Whereas picnic spots are kept in this range so that when they are out for picnic, then they get a peaceful environment away from the busy life of city. Agricultural land is kept in this range due it its huge land requirements, and the agricultural set-up does not jell very well with the city life. Since the crop will be affected by the dust and pollution of the city and for city dwellers, it will be awkwardly unsafe area with huge lonely lands. Considering court in this range is because of their perception of staying away from any disputes and unlawful act. This notion is debatable and seems biased from user perspective since their reasoning is not in-line with the logical thinking.

10.4 Discussion

Figure 10.2 is a scatter chart of range A, in this top preferred services as per the collected data are placed. If the service has received highest preference in this range, then only they are place in the scatter chart. It signifies that these services will be considered an integral part of this range only. These services will not be considered in

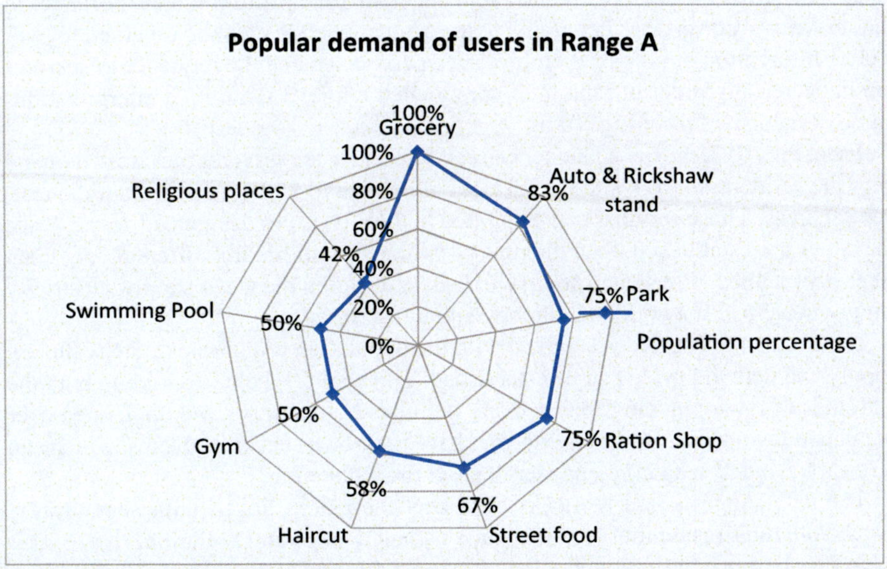

Fig. 10.2 Range A popular demands

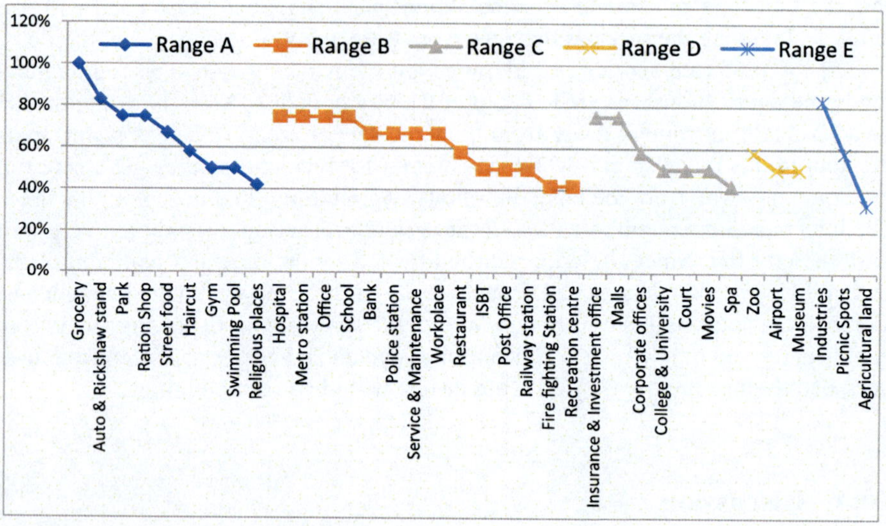

Fig. 10.3 Range-wise popular demands of users

other zones unlike the range service ranking charts discussed before. Similar plotting was done for all other ranges.

Figure 10.3 gives the exhaustive list of all the services with their associated ranges established from the data. Different colour codes are used to differentiate the ranges of the services. And services belonging to the same range are clubbed together. It gives a clear bifurcation of ranges; their preferred percentage and the sequence of services are in decreasing order of rank in the designated range. So, multiple interpretations can be established from Fig. 10.3.

From Fig. 10.3, the town planner can easily identify the services that are in demand near the residential areas and the services which are to be kept away from the residential areas. These services can be plotted in the city as per designated zones. Once the zones are established, then the transportation system linking different zones can be easily planned. The same zoning will help in deciding the water supply, electricity supply, sewerage network and other essential networks.

If the density of services is linked with the population requirement, then efficient supply and demand cycle can be established. This tool can help in avoiding both the extremes of town planning like scarcely populated city sprawl and unplanned over populated town ships. Both the extremes have their own sets of problems, which can be easily avoided with efficient user-centred town planning.

The data suggests that the users have kept the places linked with heavy traffic away from their residential area to avoid unwanted noise and pollution.

In Fig. 10.1, we have plotted all the services preferred by the users in range A, i.e. 0–1 km; similarly, all the ranges (A–E) were analysed. Then, we have identified the most preferred range for each service identified by the user, and the same was plotted in Fig. 10.3. In Fig. 10.4, we have plotted the data of Fig. 10.3 with respect to the

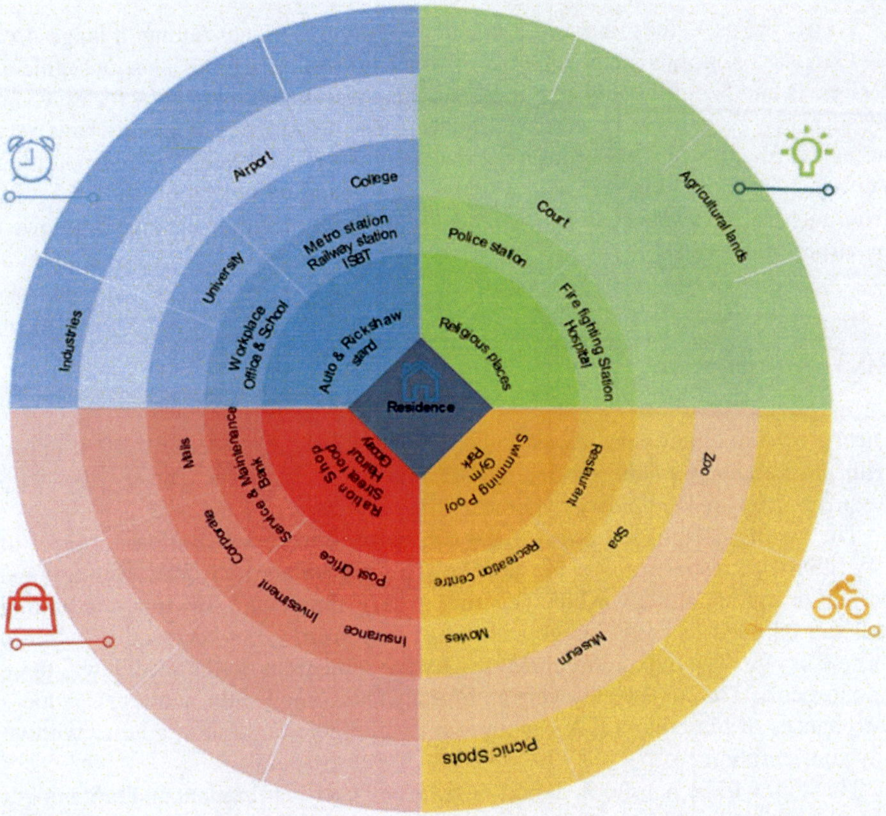

Fig. 10.4 Zoning of utilities

user perspective by keeping their residence as point of reference for every service in the city. Since for an individual, their residence is point of reference from where they view their city and accesses each and every service of the city. So, it should be the most logical point to start a town planning, if we want to design a town from users perspective. Hence, town planners can use this perceived town plan of an individual to build a more user centric city.

Figure 10.4 is having 5 cores, each core represent a range starting from A to E. The services of range A are plotted in first core, and that of range B is plotted in second core and so on.

We have further grouped the services with the help of colour code such as: The orange colour was for the services related to leisure activities; the red zone is for services related to commercial activities on regular intervals; the blue colour is for services which are of prime importance on everyday basis, whereas green colour services are for need-based important services. These were interpreted on the basis of interview data gathered from the subject.

As per the user study, it was found that majority of services falling in range A of 0–1 km was of unorganized sector, out of which unorganized retail is major segment. Religious and health-related activities were also prioritized in range A for easy and frequent access. Auto and rickshaw stand were also prioritize in range A to have easy access to other ranges. It was observed that there is a link between the transportation networks depending on the distance of destination and frequency of visits. Starting from auto rickshaw stand in range A to metros and railway station in range B followed by airport in range D.

10.5 Conclusion

During personal interview, it was found that the priority of distance is often linked with the frequency of the service requirement by the subjects. The services with frequent visits are kept closer to home by the subjects.

The zoning established in Fig. 10.4 can be used to establish a basic model of township with respect to user perspective. This can be a switch from conventional way of designing any township. This tool is very simple and less time-consuming and can be used to collect large amount of data from users perspective within a minute or two. The maximum time taken by the subjects to do the sorting was three minutes only. The subjects felt very comfortable and enjoyed the activity; it reduces the chances of biasedness linked with user to complete the task in a hurry without any consideration.

This tool will be helpful in identifying the end user requirement and preferences but cannot predict the population size, city size, geographical conditions, etc. The benefit of this tool is that it can be easily converted into a mobile app, and the entire city can vote for their preferred town within no time. The most significant part of this tool is that\ even if the user is not very well verse with the town planning method, still they can contribute efficiently for suggesting a good framework for town plan.

References

1. Hollands, R.G.: Will the real smart city please stand up? Intelligent, progressive or entrepreneurial? City **12**(3), 303–320 (2008)
2. Nam, T., & Pardo, T.A.: Conceptualizing smart city with dimensions of technology, people, and institutions. In Proceedings of the 12th Annual International Digital Government Research Conference: Digital Government Innovation in Challenging Times, pp. 282–291 (2011, June)
3. Komninos, N.: Intelligent cities: variable geometries of spatial intelligence. Intell. Build. Int. **3**(3), 172–188 (2011)
4. Wolfram, M.: Deconstructing smart cities: an intertextual reading of concepts and practices for integrated urban and ICT development, pp. 171–181. na
5. Neirotti, P., De Marco, A., Cagliano, A.C., Mangano, G., Scorrano, F.: Current trends in smart city initiatives: some stylised facts. Cities **38**, 25–36 (2014)
6. Angelidou, M.: Smart city policies: A spatial approach. Cities **41**, S3–S11 (2014)
7. Spencer, D., Warfel, T.: Card sorting: a definitive guide. Boxes Arrows **2**, 1–23 (2004)

Chapter 11
The Cancer Positive Journey: A System Design Thinking Perspective

Alishka Shah

Abstract *Background* Breast cancer, one of the most common invasive cancers for women both in the developed and developing world, poses a threat as a multi-dimensional malignancy branching out into an array of medical, physical, financial, social, emotional and sexual turmoil. This paper reports research that has been carried out in an academic pursuit for answers to queries encapsulating the social perception, impact and aftermath of breast cancer—affiliated healthcare systems, effective caregiving, healthy coping and holistic healing mechanisms. *Objective* The study aims at presenting the illness and its negative imprints as a cumulative concern, instead of singularly scrutinizing it through a clinical lens. It urges practitioners and caregivers to innovate and intervene at three identified and overlapping target phases of the journey: (i) awareness and diagnosis, (ii) short-term healing and (iii) long-term recovery. *Methodology* With underlying system design practices, this qualitative study was conducted as a part of a visual communication project under the graphic design department at National Institute of Design, Ahmedabad. The deployed methodology made use of empathy mapping, opportunity mapping, gigamaping, interviews and questionnaires as tools to engage the two stakeholder groups, one including health care providers, patients and their support systems personifying direct stakeholders of the journey and the second being a group of general participants embodying indirect stakeholders. *Conclusion* This approach devised cancer positive, a collective and curative movement, which provides a systemic solution to downsize the trauma of the illness, foster interpersonal relationships and eradicate the perceived and actual stigma attached to breast cancer. The proffered system provides strategies for accessible and responsive caregiving, remote monitoring, telemedicine and behavioural modification by proposing an allocated breast cancer data unit. This paper primarily elaborates on the design rationale that suggests the imperative need of an all-inclusive recovery-centric approach instead of a mere survival outlook towards the life-threatening phenomenon.

A. Shah (✉)
National Institute of Design, Paldi, Ahmedabad 380007, India
e-mail: alishkashah@gmail.com

A. Chakrabarti et al. (eds.), *Design for Tomorrow–Volume 2*, Smart Innovation, Systems and Technologies 222, https://doi.org/10.1007/978-981-16-0119-4_11

11.1 Introduction

Cancer is a fatal disease of the cells. Breast cancer is contracted as a result of an uncontrolled growth of abnormal cells and genetic mutations that otherwise regulate the development of healthy cells in the breast. The subsequent tumour when diagnosed as malignant is declared cancerous and can be spotted by symptoms such as lumps in the breast, skin dimpling, nipple pain, discharge and changes in nipple shape or texture. Incessant urbanization and the escalating adoption of western lifestyles have been an augmenting factor in breast cancer incidences worldwide [1]. Non-communicable diseases account for a lion's share of global deaths in this century with cancer dominating the list as the greatest barrier to expanding life expectancy worldwide [2].

11.1.1 Breast Cancer

Breast cancer is the most common and terminal cancer among women, universally, with an approximate of 2.1 million facing the brunt of this disease every year [3]. As per the 2018 WHO reports, causing an estimate of 9.6 million casualties globally, cancer is known to be the second most prevalent cause of death worldwide [4]. In addition to that breast cancer contributed to approximately 15% of all women deaths from above-mentioned statistics, summing up to a figure of around 627,000 demises annually [5]. In India, approximately, 165,000 new breast cancer cases are diagnosed every year, contributing to a substantial 14%, the greatest of all new cancer cases in both sexes [5]. Breast cancer has embedded itself as a systemic health issue, woven into the complex framework of the Indian society, as the most commonplace cancer among urban Indian females and second most rampant amongst the rural women [6].

Multi-Dimensional Health Concern. Keeping aside the daunting death statistics, the term 'cancer' itself carries a hefty dosage of attached stigma and trauma for all stakeholders. Although cancer deaths may be preventable with early detection and adequate treatment, it inflicts a whirlwind of not only medical and physical but also emotional, financial, social and cultural anguish [7]. Over the years, while numerous strategies have been implemented by the government towards enhancing breast cancer awareness, availability of detection units, subsidized treatment and supportive services [8], an evident gap emerges in the system when we fail to tackle breast cancer as a multi-faceted global health crisis, addressing each tier of concern. It is significant to acknowledge that the manifold strains exerted as effects and aftermaths of breast cancer are interlinked, associated, each arousing and overstating the other [9].

Insufficiency and Imperfection of Statistics and Soft Data. When examined solely from the perspective of its medical profile, these cancer journeys have been reduced to digits that do injustice to the intricate and nuanced realities outlining the full picture.

Giorgia Lupi, data humanist and partner at Pentagram, states, '*Data is always just a tool that we use to represent reality. They are always used as a placeholder for something else, but they are never the real thing*'. [10]. Numbers are never the point, more like means to an end, incomplete, insufficient and lacking context. The real story lies in the experience, behaviour, knowledge and humans themselves, and only from there is it that we can draw our solutions. Hence, this report has been constructed around breast cancer journeys from the point of view of examining soft data in the light of the immense potential of qualitative data studies. In an attempt to frail the diminishing of these journeys into empty graphs and figures, it also advocates shifting the dialogue, wholly, from 'survivorship' to 'recovery', promoting the need for a pleasant quality of life post diagnosis and treatment [11].

Awareness, Perception and Social Stigma. Prior research conducted with the aim of testing waters with breast cancer awareness revealed the lack of general awareness about the disease, irrespective of gender or socio-economic status [12]. In comparison with the several studies that were undertaken to document awareness towards preventive oncology, cancer detection and care, close to negligible evaluate the overshadowing belief systems and social stigma that elevate the challenge in an Indian context [9]. The controversial affected region of the body in tandem with the sensitive nature of the disease further build onto the issue, comprising two tabooed subjects—'breast' and 'cancer'. Perceived or personal stigmatization refers to the negative connotations associated with having a condition, internalizing the prejudices and developing a fear of discrimination and social exclusion. With reference to breast cancer, female patients have faced physical impairment, body image and sexuality concerns and downtrodden self-esteem [13], whereas male breast cancer patients encounter stigmatization in terms of suffering from what is typically considered a 'woman's disease' [14]. Actual or social stigmatization, on the other hand, refers to stereotypes, discrimination and prejudices projected towards those who were socially declared ill [15]. In all, the cultural myths, stigmas and taboos have a silencing impact on cancer care continuum, setting back the efforts made towards cancer prevention, treatment services and perception [16].

Cancer Caregiving and Social Support. Social networks and support systems occupy a crucial yet highly underrated and underutilized role in the cancer caregiving journey. Ideal and effective caregiving services place the diagnosed individual at the centre of an extensive support system instead of exclusively focusing on the patient-clinic relationship [17]. It is pivotal to investigate the psychosocial implications of breast cancer on interpersonal relationships as comfort and communication levels determine the potency of support extended and requested [18]. Informal caregiving, a strenous activity surrounded by critical decision making, gap of knowledge, emotional processing, uncertainty of the future and fear of loss, has been reported for causing an alarming number of anxiety and depression cases amongst those adopting the role of a caregiver [19, 20].

11.2 Materials and Methods

This paper briefly reports on research conducted as collateral to a system design project undertaken at National Institute of Design, Ahmedabad in an attempt to use visual communication strategies at a system level to erase adversities and aftermaths of breast cancer. Systems thinking refers to a philosophy of addressing a complex problem by synthesising and analysing its context, the structure and order in which it dwells, acknowledging the plurality of the problem, inquiring into the multiple layers in its ecosystem, their interactions with one another while identifying relationships, patterns and emerging opportunities throughout its vast landscape. Design thinking, the broader umbrella that houses the systems approach, presents an approach that thrives on alternating between macroscopic and microscopic, even nanoscopic views of inspecting the issue and reframing and refocusing the lens as requisite. The study undertaken as a part of this design process employed a combination of tools such as empathy mapping, participatory design methods and iceberg model for systems thinking, opportunity mapping, gigamaping, interviews and questionnaires. The two design groups were discreetly chosen for this exercise based on their contact with and the nature of their stakeholderships in cancer journeys.

Design Group 1: Direct Stakeholders. The first group involved participants termed as direct stakeholders of the process, i.e. people that have already been or are actively involved in a breast cancer journey. Given the sensitivity of the subject, convenience sampling was used to gather inputs from a mixed bunch ($n = 19$) of oncologists, support group staff, breast cancer survivors, patients and supporters playing the roles of children, spouses and siblings of the patient. For the purpose of qualitative material and authentic contribution, data has been partially collected by visits to one of the nation's premier cancer hospitals, and participants anonymity has been maintained keeping in mind the vulnerability associated with the situation. Duty and role specific assessment helped in classifying commonalities and patterns, while demarcating shortfalls and expectations for individual needs and demands. The samples drawn from the first group were responses from an urban demographic with ages ranging from 21 to 65. Table 11.1 gives a gist of the individualized questionnaires and surveys that were drawn up to derive intimate reflections of their personal journeys, drawbacks in the existing cancer care framework, stunted participation in seeking counsel and potent coping mechanisms.

Design Group 2: Indirect Stakeholders. For the latter half of the experiment, a breast cancer awareness and scenario workshop was conducted with the second design group consisting of general participants, urban males and females, young adults ($n = 20$), ranging from 20 to 24 years. The general participants selected for this study introduced volunteers that had not necessarily been in contact with breast cancer patients or caregivers but were involved to measure the knowledge, perception and reaction towards the condition. The exercise probed an internal dialogue about comfort in one's own skin and breasts as fundamental aspects of femininity.

Table 11.1 Range of topics covered in the questionnaires drawn up for design group 1

Medical/support staff	• How frequently do you get patients diagnosed with breast cancer? Did you notice any patterns among the diagnosed—age, background or lifestyle? • The ideal manner in which one is to disclose the results of the diagnosis? How do the patients and their supporters react when you break the news? • As per your observations and interactions, how did the disease affect them medically, emotionally and financially? • What do you recommend as effective and healthy coping strategies to patients and their supporters? Does complementary medication serve to be beneficial in the process?
Supporters	• Elaborate on your relationship with the diagnosed and your age at the time • How was the result of the diagnosis revealed to you? • How did you react and cope with the diagnosis? • How did your behaviour around the diagnosed changes, and what efforts did you make to extend your support? • Share the emotions you experienced whilst watching a loved one go through the cancer treatment and journey. How did it affect your relationship with the diagnosed, during and post treatment? • As per your observations, what do you think helped the patient most during this tough journey? • Looking back, what is one aspect of cancer treatment and care that is not addressed enough? • Any personal regrets where you feel you could have contributed more effectively as a caregiver but failed to do?
Patients (former and current)	• Personal details, medical history and background • What was your immediate reaction to the diagnosis? • How did you reveal the same to your family and friends? • How did the cancer journey take a toll on you—medically, financially, socially and emotionally? • How did you cope with this, personally and what engagements and activities that you indulged in, helped you during this period? • In your personal experience, what role did your support system play during these trying times? How did it affect your interpersonal relationships, social behaviour, self-perception and self-esteem? • How has surviving cancer changed you as a person? Any reflections that you would like to share or advice that you would offer to people that get diagnosed with breast cancer and their supporters?

While the data collection method, revolving around hypothetical situations and visualizing 'what-if' scenarios and metaphor-centred drawing exercises, might serve as a limitation to the study, it suggested recurring key themes, transcended any linguistic barriers and helped convey complex experiences and emotions [21]. Empathy maps were adopted to cluster the visualizations and observations, to yield well-rounded interpretations, remarks and critiques. In short, the workshop tried to capture reflections on the shortcomings of caregiving, psychosocial effects of physical deformities and human tendencies towards acceptance, grief and lack of control in the context of cancer.

11.3 Results

The patients, survivors and social support systems from design group 1 were at the minimum graduates holding bachelor degrees while the healthcare providers and staff were highly trained professionals in their respective fields of practice with a minimum of 22 years of practice. The participants from the design group 2 were young adults, actively enrolled into courses, receiving education and experience in the domain of design. The inputs and expressions from both groups led to the formulation of the below mentioned inferences.

11.3.1 Breast Cancer Awareness, Social Perception and Occurrence Patterns

While breast cancer was predominantly perceived as an elderly women disease, oncologists in this study strongly pointed out a rapid rise in the number of cases amongst younger women, under the age of 40 with contemporary routines of gluttonous food intake, nutrition deficient diets, physical stagnation and the normalization of excessive alcohol consumption serving as common risk factors. One participating oncologist confirmed the diagnosis of 5–10 new cases per month, with a male to female ratio of 1:100. Survivors and patients agreed to have perceived cancer as unconquerable and synonymous with death, prior to treatment and survivorship; however, their viewpoints and learnings towards cancer changed post a successful journey. Survivors and supporters attributed their prior lack of awareness, caution and curiosity to the notion and belief that something as mammoth as cancer would never affect them.

Responses from the workshop conducted with design group 2 suggested that their immediate word associations to the term cancer were ‘terminal’, ‘death’, ‘chemotherapy’ and ‘hair loss’. Figure 11.1 displays the participants visual outcomes for the assigned task of picturing *what cancer looks like*, exhibiting a prominence of mesh-like patterns flooding the picture area with intense strokes. Repetition of bold imagery and abstract expressions as opposed to portrayals of the medical or tumour-like diagrams indicated that for majority of the partakers, the psychological and emotional impact linked to cancer outweighs the clinical and scientific perception of the condition [21].

Fig. 11.1 Visual perception of the term 'cancer' as sketched out by participants of the workshop

11.3.2 Diagnosis, Reflection and Consultancy

While medical professionals agreed upon the fact that it was vital to communicate the diagnostic results as matter of fact, laying down the specifics of the situation honestly, they also shared a need to act based on observed reactions. *It would be ideal to have a counsellor when one is breaking the news to the patient and their caregivers,* as stated by one of the oncologists; *however, there is a severe dearth of addressing the emotional baggage of the illness, especially in a developing country like India.* They also stressed on the need to clarify that the ailment is certainly non-contagious to eliminate any self-guilt and foreboding or stigma within the support system itself.

The most frequent immediate reaction to the diagnosis, as stated by the patients, survivors and supporters was a '*Why me*'? or '*Why them*' response while undergoing an array of emotions—shock, fear and disbelief. Patients and supporters discussed the challenges that they dealt with when conveying the gravity of the situation to close family, friends or colleagues. Additional resources and explanatory materials such as books, brochures, seminars, online articles, recommended by specialists or acquired from support groups, served as supportive descriptors when having to communicate the acute medical condition. However, this method proved to be fallible when applied to children that were incapable of entirely fathoming the phenomena. The invasive cancer diagnosis induces an essential discussion as to whether to even share the news with certain individuals, based on their age, maturity and outlook-linked inhibitions, familial or financial circumstances and social awkwardness. The non-disclosure, hesitation and barriers in unrestrained communication, further, lead to the tabooed tagging of cancer as 'the c word' and other such epithets, only accumulating more mystery, terror and suspicion around the remediable disease.

11.3.3 Value of Social Support Systems and Personal Accounts of Supporters

Participants stressed on the merits of an actively involved and informed social support network when proceeding to clinical stages of surgery, treatments and therapies. Survivors recalled that the endearment, strength and resilience of supporters prevailed as a source to replenish positivity, purpose and inspiration and vice versa. Frequent visits planned group activities and contribution or proxy in menial chores posed as a constructive distraction in times of distress. Dialogues suggest that remote caregiving, where the supporters are unable to maintain consistent physical proximity with the patient, turn to telephonic and digital interactions to keep a check on the status. Survivors and patients both mentioned attending support group sessions, targeted at physical and psychological aid, as a boon to the initial process. However, they also admitted to having struggled and failed to keep up with the schedules due to the eventual loss of physical ability as a side effect of intense medicine dosage.

The empathy mapping exercise revealed supporter's constant initiatives to indulge in more qualitative time with the patient and participate in everyday errands with more enthusiasm, in an attempt to have the patients experience a more 'fulfilled life'. While simultaneously, attending to their physical suffering and bed ridden needs without trying to disrupt any sense of normalcy or express preferential behaviour. A common resonating theme surfaced as the need to exude an unwavering brave aura, disguising their personal emotional battles of frequent and familiar encounters with—uncertainty, devastation, fear of loss, anxiety, hopelessness, helplessness, agitation, bitterness, empathy, pain with salvaging spurs of positivity, optimism and numbness (all recurring terms that were gathered from the research). Supporters reflected that accompanying the patient to all their medical appointments, actively seeking further expertise, reading and researching the condition assiduously helped them feel more 'in control' of the helpless situation. When prodded with similar scenario-specific questions, participants from group 2 unanimously admitted to offering the patients consolation by fortifying the momentariness of the storm, expressing affection and re-instilling their valiance. However, supporters readily confessed to the hardships of caregiving, inferred anxiety and depression that followed the dread of losing a loved one, a sense of helplessness and inadequacy as a caregiver, trying to prepare oneself for all possible outcomes while masking any form of panic or negativity when in contact with the patient and simultaneously having to oversee medical and financial decisions. They expressed the guilt that they experienced while watching the patient suffer mixed with their inability to be present incessantly and provide sufficient care.

11.3.4 Self-image and Stigmatization as a Consequence of Physical Deformities, Fear of Relapse and Long-Term Recovery

It was affirmed that aggressive treatment regimens overlook the preservation of quality of life and the palliation of side effects and symptoms, solely targeting survival of the patient. Chemotherapy drugs have subjected patients to severities of hair loss, fatigue, nausea, hormone and fertility complications and so on, while mastectomy surgeries can lead to changes in appearance of the breast, loss of one or both breasts, reconstructive surgeries and use of external breast prosthesis. The participants reported physical displacement, discomfort and pain, emotional stress, embarrassment and wavering self-esteem, apart from a psychological blow to their self-image and faltering sense of femininity. Survivors stated that dwelling upon the radical psychophysical modifications in their self-perception and physical appearances resulted in self-imposed isolation and a decline in their sexual relations. The unwillingness to participate in social activities, however, was derived from a combination of perceived and actual stigmatization, with shame, sympathy and alienation directed onto them by society. Doctors determined cancer to be a 'controllable' not a 'curable' disease with a lingering likelihood of relapse and hereditary occurrences that haunt patients post survival. Deep-seated psychological effects of cancer like wounded self-image issues, strained interpersonal relationships and relapse-related anxieties could persist lifelong.

11.3.5 Post Cancer Perspectives and Suggested Measures from Multiple Stakeholders

Oncologists emphasized on encouraging conversation around breast cancer awareness and prevention to normalize the disease and combat internal and social stigmatization. Participants indicated the catalyzing role of a solid support system in holistic recovery and strongly recommended enrolling into counselling when interpersonal relations flounder. They also underlined the advantages of complementary medication techniques and unconventional therapies but not as an alternative to formal treatment. Participants affirmed that the cancer journey invoked a sense of gratitude, enhanced their appreciation towards life, urging them to reconsider their own capacity, strength and tolerance, push away trivialities and work towards things that actually matter. *It made me a better mother, a better wife, a better daughter and a better person. Time is short, uncertain and being aware on that makes you want to live a fuller life.* Survivors endorsed sticking to one's daily routines, keeping oneself busy, maintaining private journals and interacting with other individuals that share similar experiences. Supporters advised the need to accept the situation with the powerlessness it entails, while being transparent, patient and present for the diagnosed throughout the journey.

11.4 Discussion

This study examined the present-day landscape of breast cancer in India with respect to social awareness strategies, effective communication material, accessible coping mechanisms, caregiving guidance, quality of life post treatment and comprehensive recovery. It is essential to address the cloud of ambiguity and inflicted anxiety that floats around the conundrum, multiplying myths, misconceptions and following societal inhibitions towards the disease. In fact, the growing interest in the semantics of medical discourse questions the grave connotations and inadequacy of the term 'cancer' with several researchers, pathologists and cancer institute's altogether recommending dropping the title [22]. The study suggests that the language ideologies and psychosocial constructs manifest an instant correlation of 'cancer' to death, myopic to the inflating chances of survival. This paper aims to derail the 'cancer curse' by replacing the daunting terminologies of battling cancer—being a warrior, survivor, a competent caregiver to being diagnosed *cancer positive* or being a *contributor of the cancer positive journey*. Positive affirmation embedded in disease nomenclature and corresponding clinical discourse can prove beneficial in the demystification of the disease, decision-making stages and adoption of coping strategies. The landmark 'Asian Disease' problem research conducted by A. Tversky and D. Kahneman not only verified the cognitive biases in the psychology of choices, now tagged as the 'Framing Effect' [23] but also formed the basis of the medical implications of the same with respect to disease prevention, detection and course of treatment [24]. There is enough evidence pointing to the potential of medical language to influence a patient's response to coping mechanisms and recovery [25]. Reviewing the currently circulating cancer care communication material, physical and virtual, supplied by medical institutions, NGO's and support groups demonstrated the imperative need for a similar shift from non-emotive medical jargon to humane, reflective and comprehendible data.

The participants' feedbacks were assimilated and allocated to identify gaps that could possibly make use of design intervention by conducting an opportunity mapping exercise. Macro-level analysis of current facilities with the needs indicated as a result of the research, unfolded a three-phase target model with the primary diagnosed targets as awareness, short-term healing and long-term recovery. This structure was developed to better tackle breast cancer journeys as a multifarious threat, addressing respective tiers and onslaught via customized tactics. Figure 11.2 serves as visualization to the target model with an additional and consequential phase of diagnosis. Apart from being resolved as isolated targets, addressing them in the following combinations could be increasingly profitable. Gender-specific strategies when applied to challenge the two interdependent attributes of awareness and recovery can proliferate knowledge pertaining to risks and prevention, encourage dialogue and understanding and suppress stigmatization. This, in turn, regulates the number of cases, accentuates the probability of early detection and addresses social concerns of current or former breast cancer patients. Utilizing role-specific strategies, when dealing with phases of diagnosis and short-term healing, encapsulates the

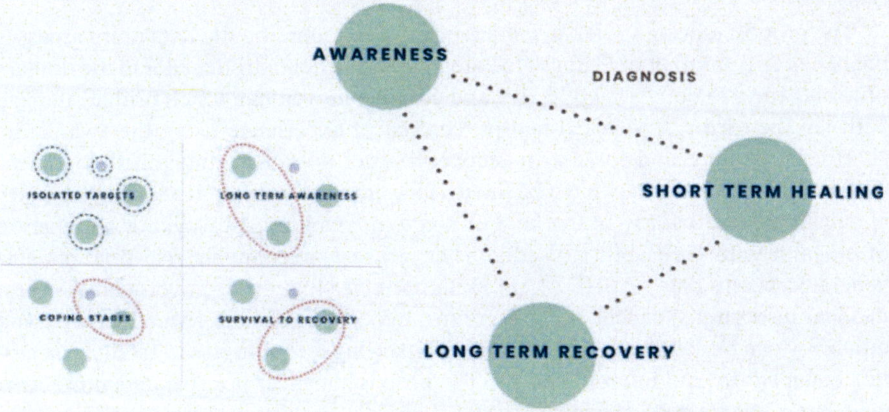

Fig. 11.2 Three-phase target model (awareness, healing and recovery)

counsel of coping mechanisms for the patient and caregiver during various stages of treatment, assistance in acceptance of the situation and communication strategies to maximize awareness and sound decision-making. Likewise, the phases of short-term healing and long-term recovery, when collectively targeted, facilitate the shift of overall outlook from survival to recovery rather than stunting efforts post termination of medical treatments.

The study put forward the requirement of a one-stop role-specific platform with the ability to record, disseminate, connect and simplify the access of health information, medical consultancy, government health plans and subsidies, complementary therapy, status trackers, cancer related resources, enable reminders and schedules, request proxy involvement or care services, palliate remote caregiving and connect support networks, cancer communities and help groups. The ongoing inquiry into health tech has made provisions for public health record services like Google health [26], efforts of the national telemedicine task force set-up by the ministry of health and family welfare enabled commendable telemedicine services like mammography at SGRH, Delhi, Regional Cancer Center, Trivandrum and Rajiv Gandhi Cancer Institute and Research Centre, New Delhi [27], caregiving service platforms like nightingales [28] and so on. Pervasive technology can to a great extent eradicate the inhibitions that function as barriers to seeking care, ensuring that the information or assistance demanded is fulfilled, and caregiving is more empowered, less challenging and more efficient [29]. However, existing enterprises are largely disjointed, singular in utility, inaccessible or unheard of. An ideal all-inclusive system would be one that puts to use unobtrusive digital developments, guarantees medical relevancy, privacy, hassle-free tracking with coherent data outputs and provide personalized training and coping strategies, adhering to the dynamic roles and relationships warped into this collaborative journey of the diagnosed, the practitioner, the caregiver and the bystander.

The proposed design system, cancer positive, comprising the data unit harnesses the benefits of Internet of Things (IoT) technology when introduced into the domain of health care, to deliver a collective and curative movement which bridges the gap between the medical, physical and psychological experience using the awareness, short-term healing and long-term recovery target approach. Integrated IoT technology and smart devices have been sparking interest with their multidisciplinary applications, particularly in the field of medicine, paving pathways for automation of e-health with their ability to administer, process, accumulate real-time patient-generated health data (PGHD) [30]. With the participation of government bodies, medical institutions, cancer support groups, the constraints of privacy, security and reliability can be avoided while leveraging emerging technologies to facilitate seamless connectivity and interoperability. The proposed cancer positive data unit forms the heart of the framework gathering real-time, role-specific and age-specific health-related responses via the cancer positive application. The assessment, allocation and interpretation of the real-time database are hence capable of enabling telemedicine, remote monitoring and behavioural modification to make medical consultancy immediate, personalized and accurate. It aims to manifest a close-knit ecosystem of all stakeholders to foster interpersonal relationships by providing customized guidelines and recommending personalized strategies to make the overall caregiving experience more responsive and accessible. The acquired and constantly updated database can be further studied by researchers and experts to formulate efficient modules for training professional and informal caregivers. Individuals register to cancer positive with the name of the diagnosed positive and by stating their contributory role in the journey, to receive an automated and customized dashboard with all relevant medical records, schedules, services and facilities. The application is designed to incorporate wellness and counselling modules, task and schedule manager, emotional tracker, complementary medication services, request external contributors, track medicinal and dietary intake, with the reflection of data collected from the integrated smart devices such that it can be shared and administered by the other contributors of the journey. A gamified iteration of the system earns you 'redeemable positives' on completion of tasks and probes the members to indulge in qualitative activities. The earned positives can be exchanged for mementoes with the cancer positive collective which undertakes the sale of life-post-treatment products like wigs, breast prosthetics, lingerie, etc., organizes seminars, exhibitions events and campaigns towards awareness, sensitization and normalization of cancer throughout society. It provides resources and platforms for community groups, offers external caretaking services and encourages and connects former patients from all walks of life to embrace, share and exhibit their journeys.

11.5 Conclusion

At this very moment, more people living with cancer than dying from it [29]. In conclusion, this paper aims to address the missing links in breast cancer-related research, development and design in the light of the plurality of affiliated challenges and the persons affected. It briefly investigates factors of breast cancer treatment and care continuum in a contemporary Indian scenario from manifestation of stigma, faith in medication techniques, patient's internalization of guilt and other associated psychological turmoil, roadblocks to acquiring social assistance and external services and influence of caregiving expectations on interpersonal relationships. The study proposes a three-phase target model in tandem with smart-integrated system design solution based on emerging technologies in the domain of health care to bridge the gaps between medical and psychological, physical and emotional, social and personal. The limitation of the method lies in the small sample size volunteer views, while the proffered system is a futuristic solution with financial constraints, data privacy concerns and banks on government cooperation and endorsement. Urging future research to take into consideration the all-inclusive role-specific approach and conduct rigorous user-testing and implementation of progressive solutions and technologies to make cancer journeys, endurable, recoverable and perhaps, even unifiable.

Acknowledgements Thanking all the participants for their time, cooperation and valuable suggestions. I would like to express my gratitude to Tanishka Kachru from National Institute of Design for her mentorship and guidance with the project.

Reference

1. World Health Organization: Cancer: Detection Breast Cancer. International Agency for Research on Cancer. Available: https://www.who.int/cancer/detection/breastcancer/en/. Accessed 18 Apr 2020
2. Bray, F., Ferlay, J., Soerjomataram, I., Siegel, R.L., Torre, L.A., Jamal, A.: Global cancer statistics 2018: GLOBOCAN estimates of incidence and mortality worldwide for 36 cancers in 185 countries. CA: Cancer J. Clinic. **68**(6) (2018).
3. World Health Organization: Cancer: Early Diagnosis and Screening. International Agency for Research on Cancer (2018). Available: https://www.who.int/cancer/prevention/diagnosis-screening/breast-cancer/en/
4. World Health Organization: Health Topics: Cancer (2018). Available: https://www.who.int/health-topics/cancer#tab=tab_1. Accessed 14 Apr 2020
5. Globocan: Breast Factsheet. International Agency for Research on Cancer (2018). Available: https://gco.iarc.fr/today/data/factsheets/cancers/20-Breast-fact-sheet.pdf. Accessed 14 Apr 2020
6. Agarwal, G., Ramakant, P.: Breast cancer care in India: The current scenario and the challenges for the future. Breast Care **3**(1), 2008 (2008)
7. Alexander, A., Kaluve, R., Prabhu, J. S., Korlimarla, A., Srinath, B., Manjunath, S., Patil, S., Gopinath, K., Sridhar, T.: The impact of breast cancer on the patient and the family in Indian perspective. Indian J. Palliative Care **25**(1) (2019)
8. India Against Cancer: Cancer treatment centers. National Institute of Cancer Prevention and Research (NICPR). Available: https://cancerindia.org.in/cancer-treatment-centers/. Accessed 19 Apr 2020
9. Gupta, A., Dhillon, P.K., Govil, J., Bumb, D., Dey, S., Krishnan, S.: Multiple stakeholder perspectives on cancer stigma in North India. Asian Pacific J. Cancer Prevent. **16** (2015)

10. Lupi, G.: Data humanism: my manifesto for a new data world. Available: https://giorgialupi.com/data-humanism-my-manifesto-for-a-new-data-wold
11. Gangane, N., Khairkar, P., Hurtig, A.-K., Sebastián, M.S.: Quality of life determinants in breast cancer patients in Central Rural India. Asian Pacific J. Cancer Prevent. **18**(12) (2017)
12. Gupta, A., Shridhar, K., Dhillon, P.K.: A review of breast cancer awareness among women in India: cancer literate or awareness deficit? European J. Cancer **51**(14) (2015)
13. Wang, Q.-X., Bai, Y., Lu, G.-F., Zhang, C.-Y.: Perceived health-related stigma among patients with breast cancer. Chin. Nurs. Res. **4** (2017)
14. Midding, E., Halbach, S.M., Kowalski, C., Würstlein, R., Weber, R., Ernstmann, N.: Men with a "Woman's Disease": stigmatization of male breast cancer patients—a mixed methods analysis. Am. J. Men's Health **12**(6) (2018)
15. Corrigan, P.: How stigma interferes with mental health care. Am. Psychol. **59**(7) (2004)
16. Daher, M.: Cultural beliefs and values in cancer patients. Ann. Oncol. **23**(3) (2012)
17. Weiss, J., Lorenzi, N.: Online communication and support for cancer patients: a relationship-centric design framework. In: Proceedings from American Medical Informatics Association Annual Symposium (2005)
18. Skeels, M., Unruh, K., Powell, C., Pratt, W.: Catalyzing social support for breast cancer patients. In: Proceedings of the SIGCHI Conference on Human Factors in Computing Systems, Chicago (2010)
19. Sahadevan, S., Namboodiri, V.: Depression in caregivers of patients with breast cancer: a cross-sectional study from a cancer research center in South India. Indian J. Psychiatry **61**(3) (2019)
20. Devi, A., Siddik, S., Ariaratnam, S.: Depression and anxiety among family caregivers of cancer patients in an oncology clinic. Malaysian J. Psychiatry **13**(1) (2005)
21. Gameiro, S., Guevara, B., Refaie, E., Payson, A.: DrawingOut—an innovative drawing workshop method to support the generation and dissemination of research findings. PLOS ONE **13**(9), e0203197, 20 Sept 2018
22. Kolata, G.: Cancer' or 'Weird Cells': which sounds deadlier? The New York Times, vol. Section D, p. 1, 22 Nov 2011
23. Tversky, A., Kahneman, D.: The framing of decisions and the psychology of choice. Science, AAAS **211**(4481) (1981)
24. Mazzocco, K., Cherubini, P., Rumiato, R.: Effetto framing: implicazioni in ambito medico. Annali Italiani di Medicina Interna **20**(1) (2005)
25. Vranceanu, A.M., Elbon, M., Adams, M., Ring, D.: The emotive impact of medical language. HAND (N.Y.) J. **7**(3) (2012)
26. Google Health. Available: https://health.google
27. Chellaiyan, V.G., Nirupama, A., Taneja, N.: Telemedicine in India: where do we stand? J. Family Med. Primary Care **8**(6) (2019)
28. Nightingales: Services: caregivers. Available: https://www.nightingales.in/services/caregivers. Accessed 8 May 2020
29. Hayes, G.R., Abowd, G.D., Davis, J.S., Blount, M.L., Ebling, M., Mynatt, E.D.: Opportunities for pervasive computing in chronic cancer care. In: International Conference on Pervasive Computing, vol. 5013 (2008)
30. Sunanda, D., Sharma, C.: Survey on smart healthcare: an application of IoT. Int. J. Emerg. Technol. **8**(1) (2017)

Chapter 12
A QFD Approach for Selection of Design for Logistics Strategies

Mahadharsan Ravichandran, K. E. K. Vimal, K. Jayakrishna, and Asela K. Kulatunga

Abstract Designing products considering logistic costs and improved customer service in the field of supply chain management is termed as design for logistics (DfL). By improving the design of the products for logistics, organizations can reduce the costs spend on logistic and delivery. Some of the critical factors of adopting DfL are ease of transport, ease of packaging, ease of loading/unloading, minimize logistic cost, and so on. To improve the DfL characteristics at the product design stage, five strategies, namely flat packaging strategy, design for non-circular subparts, modular design principles, and design for ease of fabrication, have been identified. Quality function deployment (QFD) approach, a successful method often used for new product development, was used in the selection of strategies for designing products from a logistics viewpoint. The results of QFD show that flat packaging strategy and ease of handling are the critical DfL strategies for the improvement of logistics characteristics of the product at the design stage. The methodology has been tested through a real-case application in a packaged drinking water manufacturing organization.

12.1 Introduction

Logistics refers to the overall process of managing and controlling the flow of goods, energy, information, and other resources, e.g., products, services, and people, from the source to the marketplace [1]. Logistics include transportation, inventory maintenance, order processing, purchasing, warehousing, materials handling, packaging,

M. Ravichandran · K. Jayakrishna (✉)
School of Mechanical Engineering, VIT University, Vellore, Tamil Nadu, India
e-mail: mail2jaikrish@gmail.com

K. E. K. Vimal
Department of Mechanical Engineering, NIT Patna, Patna, Bihar, India

A. K. Kulatunga
Department of Manufacturing and Industrial Engineering, University of Peradeniya, Peradeniya 20400, Sri Lanka

A. Chakrabarti et al. (eds.), *Design for Tomorrow–Volume 2*, Smart Innovation, Systems and Technologies 222, https://doi.org/10.1007/978-981-16-0119-4_12

customer service standards, and product scheduling, which must be continuously developed to meet the challenges of the market [2]. The expenses correlated with the logistics activities are significant. Therefore, integration of logistics in product design is an important and intricate problem that must be addressed. Design for logistics (DfL) is defined as a series of concepts involving product and design policies that aid in regulating logistics expenses and increase customer service quality in the discipline of supply chain management [3]. The charge of creating design modifications is little in the initial stages of the product development course compared to constructing changes at last [4]. Therefore, forthcoming manufacturing and logistics activities have a considerable impact when the product development decisions were made in the early phases of the product development process [5].

Organizations use quality function deployment (QFD) to make decisions on design strategies to effectively translate customer requirements to comprehensive manufacturing specifications that help to produce the products that full fill the customer needs [6]. QFD can be a powerful tool because it can reduce time to market, improved quality, and increased customer satisfaction [7]. Therefore, organizations adopt QFD in the process of product development since there is a constant need to improve time to market, quality of the manufactured product, and customer satisfaction in the prevailing competitive environment.

The packaging is the means of guaranteeing the safe delivery of the products to the consumer in good condition at the least possible cost [8]. Paine states that packaging is a synchronized system of preparing goods for transport, distribution, storage, retailing, and end use [9]. The packaging is considered a fundamental element in logistics and is regarded as having a significant effect on logistics cost and performance [10]. It is observed that the prospects for DfL are remarkable, whereas the literature on the concept of DfL has been found to be limited. The primary objective of this case reported in this article deals with providing an alternate design for the case organization, considering their DfL requirements captured via QFD. Further, sections of this article include a review of existing literature on DfL, and QFD, methodology adopted in this study, HoQ development, providing alternate design based on the key expectations from the case product in the viewpoint of customers and the delivery team, followed by discussions on the implications and other benefits of this design changeover using the DfL strategies and QFD to the case organizations.

12.2 Literature Review

12.2.1 Review on Design for Logistics

DfL is a sequence of concepts in the field of supply chain management that focuses on three key components: economic packaging and transportation, concurrent and parallel processing, and standardization [10]. Researchers suggest that reducing manufacturing time and cost are essential for an organization to be competitive. It

Table 12.1 Review on DfL strategies for product development

DfL strategy	Description	References
Ease of assembly, services, reuse, recycling, and other product life cycle	Integrated modular design	[12]
The required customer delivery lead time, with the production capacity needed to meet the customer orders	A customer order-driven production planning is applied to determine the WIP cap, the work-ahead-window of a CONWIP controlled production and used to implement a new market-driven production planning	[13]
Lower the total cost from the supplier to the customer	A new packaging solution requires product redesign along with modern manufacturing and packaging equipment. Redesign the packaging for an existing product to minimize space	[14]

can be achieved by standardization, i.e., the use of standard parts and design strategy, which may also shorten the design portion of the new product development cycle [11] (Table 12.1).

12.2.2 Review on Quality Function Deployment

QFD provides a means of translating customer requirements into engineering terms for each stage of product development and production [15]. Quality function deployment (QFD) is a systematic means of advancement of functions and operations of company efforts through the utilization of consumer requirements and desires, which result in the development of product quality parameters compatible with consumer requirements at the quality assurance of the produced product [16]. It offers a comprehensive understanding of categorization and prioritization of product or service elements, which identified this model as a simple and effective tool, particularly for executives to focus on attributes of greater importance [17–20].

It is observed that there is a minimal number of studies addressing the concept of DfL. Hence, this paper attempts to bridge the gap by providing a framework model for DfL using a case study by adopting QFD.

12.3 Methodology

The methodology adopted in this study is shown in Fig. 12.1. Extensive interviews and brainstorming sessions were carried out. The development of a new design of an existing product, for the betterment of the organization, has been set as the goal for the brainstorming session. Department heads of the organization participated in

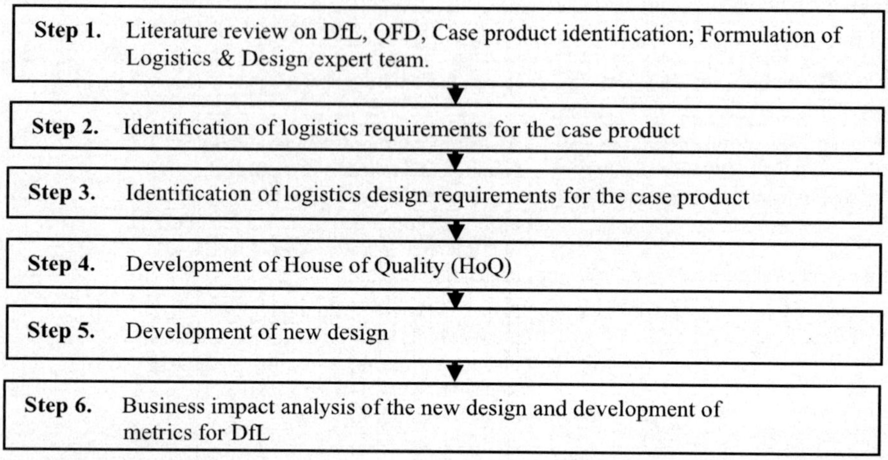

Fig. 12.1 Methodology adopted

the brainstorming session, and their viewpoints presented during the discussion were taken into account. Various constraints and challenges that will be encountered were also foreseen. The durability of the newly developed product must not be compromised, and the profit should be increased by the new product, and the degree of comfort during usage of the redesigned product should be increased, and tested are some of the notable inputs from the brainstorming session. The design team of the case organization analyzed their need for design change of the case product. The study starts with the identification of the design requirements, i.e., what is actually expected from the product (*WHAT*?), and how the design requirements can be achieved by dealing with design parameters (*HOW?*), followed by the development of HOQ between WHAT and HOW. This study adopts the action research approach by working in close liaison with a cross-functional team formed. The members of the cross-functional team were from all departments of the case organization and their logistics providers. The critical design requirement and strategies of DfL can be identified, and the corrective design changes can be made to arrive at the new design. Thus, the business impact of the newly designed product can be analyzed. The results achieved by this study can be used as a toolbox for design experts in designing products for logistics.

12.4 Case Study

12.4.1 Case Organization

The case organization located in the southern part of Tamil Nadu, India, is involved in the processing of packaged drinking water with a total capacity of 5000 jars per day. It

Fig. 12.2 CAD model of the existing water jar

makes a profit of 20 million USD per year. The case organization holds all necessary facilities to maintain their quality of water as per Bureau of Indian Standards (BIS) and Food Safety and Standards Authority of India (FSSAI).

12.4.2 Case Product

12.4.2.1 Requirements of the Case Product for Logistics

The case product considered in this study is the jar used for packaging processed drinking water. Often the jars are difficult to handle during the transportation and use phase. Currently, they are producing jars made out of polycarbonate and are planning to change their design for easy transportation and handling. Figure 12.2 shows the CAD model of the existing jars.

12.4.2.2 DfL Strategies Identified for the Case Product for Logistics

The DfL strategies identified for the current design of the jar are flat packing, easy, quick, and safe to handle and deliver, durable, and resistive to any physical damage, apt size, and fit.

12.4.2.3 Development of House of Quality

In general, QFD analyses the relative importance of the customer requirements and the organization will act in accordance with the results. In DfL, the prominent use of applying QFD in design for logistics is to confirm that the changes in design of the product corresponding to logistical cost and efficiency also meet customer requirements. This is indeed a crucial factor in DfL, as sheer inclusion of ideas

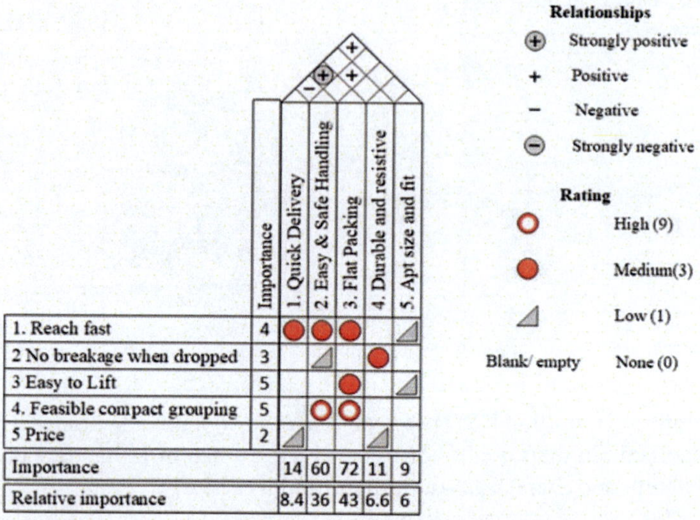

Fig. 12.3 House of quality

related to DfL neglecting the fact that the ultimate profit depends on the likings and comfort of the consumer is substandard as well as unworthy. This QFD will be used as reference by the design as well as production team while designing and production.

Figure 12.3 shows the HoQ developed based on the DfL strategies identified for the existing design. The HoQ shows us the design strategies that are relatively important as compared with DfL strategies identified. The significant DfL strategies that should be given more focus during the early stages of product design are flat packing and secure handling as far as design of logistics is considered. It is significant to note that the importance of a flat packing strategy is much higher. The key results obtained from HoQ have been used by the case organization in the redesigning their water jars. An expert panel was formulated, including executives from several departments of the organization with an experience of above seven years. Brainstorming sessions were conducted with the logistics provider, the delivery teams, and customers to enlist their logistics requirements and their respective importance. The importance weighing was scaled in the range of 1–5 with 5 given the highest significance. The importance rating scores were obtained from the logistics and design expert team, and eventually, the relative importance of the DfL strategies was achieved, as shown in Fig. 12.3. It is found that the relative importance of the strategies such as flat packing and ease of handling is exceptionally higher than the other strategies [14]. Flat packing strategy makes the shipment of jar easier and allows compact transport of products within the available area. From the scores, it can be observed that the customers and the delivery team were keen on modifying the design for easy handling, to lift, carry, and to transfer without much effort and support.

12.5 Results and Discussion

From the results of the QfD, the design team of the case organization developed a new alternate design for the water jars as shown in Fig. 12.4. Three handles were added to the new design. Two handles on one side and one on the lower bottom of the opposite side. The provision of handles will enable anyone to lift and carry the jars at ease. The head of the jar was made flat, with a knob to pour the water, precisely like a beak of a bird. The flat head helps to transport more units in a single trip by occupying the space available. The bottom of the jar was designed with bumps to add rigidity and durability of the jar, and this bumps also helps to lift the jar even from the lower position.

The new design is easily manufacturable. Though the cost of production has been lowered only to a smaller extent as compared to the existing model, the logistic costs have been reduced drastically. In regard with the business impact and the packing efficiency are greater since the design is made for logistics. The jars, when transported, can be arranged one above the other and occupies most of the truck volume available. The number of existing water jars that could fit inside a usual mini-truck is 20 per trip. The transportation cost from the production plant to the delivery end is approximately Rs. 100 per trip, which is Rs. 5 per water jar. After the usage of newly developed water jars, the truck could accommodate 40 of them per trip, which is Rs. 2.50 per water jar. Again, depending on the number of trips and total quantity of water jars transported may add up to a huge gain for the organization on a larger scale. Moreover, reduced production time coupled with faster delivery using the DfL characteristics decreased the lead time considerably gaining good customer satisfaction rate and reduced carbon foot print. The jars are majorly handled by women domestically in India have to depend upon others for lifting and changing them. But now, they find it extremely easy and comfortable to lift and changeover the jars. Thus, the reputation of the product manufactured by the company has also created an impact on its customers.

Figure 12.4 shows the CAD model of the new jar developed based on the key inferences derived through QFD based on the HoQ.

Fig. 12.4 CAD model of the new water jar

12.6 Conclusions

The case organization has determined to manufacture a water jar that should adopt the concept of design for logistics to increase the profit and efficiency of delivery. Therefore, this case study is brought to the spotlight for examination. There exists numerous literature that used QFD to address the redesigning of the product. But the research which considered logistical aspects is relatively few. Moreover, inclusion of design for logistics concepts in development of the product, coupled with the help of computer-aided design model and customer requirements is rather undiscussed. Therefore, this study primarily focuses on considering customers viewpoint in developing new product using the application of QFD combined with the concept of DfL. It is aspired to design a product such that it is inclusive of concepts related to logistics. Furthermore, this study showed that how QFD would yield the design concepts that can be developed from the voice of customers themselves. HoQ has been adopted to the WHATs of the customer, and HOWs of the design parameter and strategies requirements and the significant design strategies are derived from the HoQ. Thereby, the design of the existing model was reformed and manufactured and when commercialized, received positive feedback from its customers and its logistics providers.

Future directions include conducting a case study of other products from similar organizations with the application of design of logistics concepts. Conducting similar studies will help us to arrive at developing a standard guideline for designing products considering logistics aspects. Further, the study can be extended on using alternate sustainable materials in the manufacturing of water jars for developing an eco-friendly product.

References

1. Bramklev, C., Hansen, C.T.: On the logistics effects of integrated product and package design. Published in the Proceedings of the International Conference on Engineering Design, ICED'07, 28–31 Aug 2007, Paris, France
2. Andersson, A.: A framework supporting the collaboration between the logistics-and the product development process. Doctoral dissertation, Institutionen för innovation, design och produktutveckling (2007)
3. Kao, H.P.: Design for logistics in set-based concurrent engineering environment. J. Chin. Inst. of Ind. Eng. **23**(1), 34–47 (2006)
4. Hallstedt, S.I., Isaksson, O.: Material criticality assessment in early phases of sustainable product development. J. Clean. Prod. **161**, 40–52 (2017)
5. Steimer, C., Cadet, M., Aurich, J.C., Stephan, N.: Approach for an integrated planning of manufacturing systems based on early phases of product development. Procedia CIRP **57**, 467–472 (2016)
6. Tambunan, M.M., Napitupulu, H.L., Rizkya, I., Syahputri, K.: Design of work facilities using quality function deployment (QFD) and macro ergonomic analysis design (MEAD). In: IOP Conference Series: Materials Science and Engineering, vol. 801, No. 1, p. 012119. IOP Publishing (2020)

7. Vatthanakul, S., Jangchud, A., Jangchud, K., Therdthai, N., Wilkinson, B.: Gold kiwifruit leather product development using quality function deployment approach. Food Qual. Pref. **21**(3), 339–345 (2010)
8. Müller, P., Schmid, M.: Intelligent packaging in the food sector: A brief overview. Foods **8**(1), 16 (2019)
9. Paine, F.A.: Fundamentals of packaging. Brookside Press, Leicester (1981)
10. Sohrabpour, V., Hellström, D., Jahre, M.: Packaging in developing countries: identifying supply chain needs. J. Human. Logist. Supply Chain Manag. (2012)
11. McDermott, C.M., Stock, G.N.: The use of common parts and designs in high-tech industries: a strategic approach. Prod. Invent. Manag. J. **35**(3), 65 (1994)
12. Gu, P., Hashemian, M., Sosale, S., Rivin, E.: An integrated modular design methodology for life-cycle engineering. CIRP Ann. **46**(1), 71–74 (1997)
13. Jodlbauer, H.: Customer driven production planning. Int. J. Prod. Econ. **111**(2), 793–801 (2008)
14. Klevas, J., Johnsson, M., Jonson, G.: A packaging redesign project at IKEA (2006)
15. Govers, C.P.: What and how about quality function deployment (QFD). Int. J. Prod. Econ. **46**, 575–585 (1996)
16. Klochkov, Y., Klochkova, E., Volgina, A., Dementiev, S.: Human factor in quality function deployment. In 2016 Second International Symposium on Stochastic Models in Reliability Engineering, Life Science and Operations Management (SMRLO), pp. 466–468. IEEE, Feb 2016
17. Sivasamy, K., Arumugam, C., Devadasan, S.R., Murugesh, R., Thilak, V.M.M.: Advanced models of quality function deployment: a literature review. Qual. Quant. **50**(3), 1399–1414 (2016)
18. Chen, K.J., Yeh, T.M., Pai, F.Y., Chen, D.F.: Integrating refined Kano model and QFD for service quality improvement in healthy fast-food chain restaurants. Int. J. Environ. Res. Publ. Health **15**(7), 1310 (2018)
19. Tan, K.C., Shen, X.X.: Integrating Kano's model in the planning matrix of quality function deployment. Total Qual. Manag. **11**(8), 1141–1151 (2000)
20. Kathiravan, N., Devadasan, S.R., Michael, T.B., Goyal, S.K.: Total quality function deployment in a rubber processing company: a sample application study. Prod. Plann. Control **19**(1), 53–66 (2008)

Chapter 13
Design Interventions for Improvement of Adoption Rate of Micro-irrigation in Assam

Manoj Kumar Verma and Amarendra Kumar Das

Abstract Agriculture contributes to about 16.5% of India's Gross Value Added (GVA) and provides direct and indirect employment opportunities to a large proportion of India population, nearly 70%, which more than any other sector in the country. Increasing demand for food has put stress on the sector to increase its production. This necessitates increased dependence on irrigation to meet crop water needs. However, due to decreasing water table, it is imperative to popularize water efficient technologies for irrigation. Despite progressive schemes of the government, the highly efficient micro-irrigation system has not been well adopted in Assam. This paper studies the present system in the state and correlates with the feedback received from 225 beneficiaries to develop a framework for design of an effective system. Design intervention can play a major role in designing such system.

13.1 Introduction

In India, agriculture contributes 16.5% to the Gross Value Added (GVA) while providing livelihood to the 70% of the rural households. It is also contributing to the food security for the increasing population. Proportion of Indian population depending directly or indirectly on agriculture for employment opportunities is more than that of any other sector in India [1].

Primarily, the Indian agriculture is rainfed and depends on the monsoon every year; however, during the past few years due to insufficient rainfall, irrigation is playing a major role in the agricultural production. According to the report published by International Water Management Institute (IWMI) [2], there is a growth forecast of 17% in the potential irrigation water demand. This implies lesser rainfall dependence and more reliance on irrigation systems.

M. K. Verma (✉)
Center for Rural Technology, Indian Institute of Technology, Guwahati, India
e-mail: mk.verma@iitg.ac.in

A. K. Das
Department of Design, Indian Institute of Technology, Guwahati, India
e-mail: dasak@iitg.ac.in

A. Chakrabarti et al. (eds.), *Design for Tomorrow–Volume 2*, Smart Innovation, Systems and Technologies 222, https://doi.org/10.1007/978-981-16-0119-4_13

Central water commission of India estimates that the annual water availability in terms of natural runoff in the rivers of India is 1869 Billion Cubic Meter (BCM) per year; however, due to uneven distribution and physical constraints, only about 1123 BCM/year water is available. Out of this, 690 BCM/year is the surface water, and 433 BCM/year is the groundwater. In this same report, the CWC reveals that extracting water from the groundwater for irrigation purposes is the major component of groundwater use in India [3]. As per the Agricultural Statistics of the Ministry of Agriculture and Farmer's Welfare, Government of India [4, 5], the use of tube wells for irrigation is increasing at a sharp rate, and it is putting pressure on the groundwater table.

In such a scenario, using an irrigation system with higher irrigation efficiency is the most direct way to address sectoral water shortage [6]. A separate study done in reference to the smallholders in developing countries [7, 8] reviewed the various factors influencing the installation of these systems and found that the higher value crops are more suitable for this type of irrigation system owing to the higher investment costs. However, to overcome this barrier, the author argues that the government must come out with a subsidy-based system to bear a part of the installation costs and to some extent should also provide technical support. Studies carried out by numerous other authors [9, 10] established that micro-irrigation technologies are the best bet technology considering the decreasing water resources and limited input costs in the agriculture sector. Studies highlight that micro-irrigation technologies if used properly can yield water savings to the tune of 50–80% [6, 10–12]. A recent study [13] established that unsustainable water use, over drafted aquifers, seasonally dry rivers, and disappearing freshwater lakes are a worldwide phenomenon. The author [13], asserts that using latest irrigation technologies like micro-irrigation, saves water as evidenced from various studies over the years.

The authors of this research paper have presented the findings based upon a study done in the State of Assam. Economy of Assam is primarily based upon rural sector as about 70% of its population is directly dependent upon agriculture as the source of livelihood [14]. As per the latest available economic survey of Assam [14], the total geographical area of the state is 78.44 lakh ha, out of which the land use pattern in Assam indicates that the gross cropped areas are 40.82 lakh ha. The net sown area comes out to 28.19 lakh ha in 2014–15. There is vast scope of irrigation in temporal and spatial considerations. From October to March, i.e., the period of post-monsoon to pre-monsoon, there is lack of irrigation water in the State of Assam [15]. To meet the irrigation demand, farmers using conventional irrigation methods of flow and furrow irrigation, waste around 50–60% of water, and this in turn wastes energy. Micro-irrigation on the other hand minimizes this loss and increases the productivity, thereby improves overall economy of farmers even though the initial investment is higher.

During the preliminary phase of this research, discussions were held with officials of the department of agriculture, research stations, and staff who are working in the field; it was gathered that farmers were less inclined to adopt this technology due to higher cost of investments and lack of awareness. Though the Government of India has launched many schemes for adoption of micro-irrigation systems by the farmers, since 2006, including latest initiatives like Pradhan Mantri Krishi Sinchayee Yojana (PMKSY) and micro-irrigation fund to increase the adoption of micro-irrigation in the country, the adoption rate in Assam is very less [11, 16]. Despite government giving subsidies, there are not many takers of this efficient technology of irrigation. As such the groundwater level in Assam is taking additional load apart from the surface water sources owing to the smaller reach of existing irrigation facilities and lack of coordination among the various departments.

Most of the earlier studies done in the field of micro-irrigation take up the subject of crop productivity and yield in context to other parts of the country, i.e, excluding Assam. Considering this backdrop, this work which is part of Doctoral research is being carried out in the State of Assam for understanding the irrigation requirements of farmers and analyzes the various factors involved which influence the selection of type of irrigation and study the constraints associated with their adoption.

13.2 Aim and Objective

The research aims to understand whether design interventions can help in facilitating adoption of micro-irrigation among farmers of the State of Assam. The objective of this study is to develop a representative model of variables affecting the adoption of micro-irrigation system in the State which would help the designers in developing effective design intervention strategies. Design is ubiquitous in application and is capable of understanding challenges while devising strategic and holistic systems.

13.3 Methodology

13.3.1 Data Collection

The research study is done in the State of Assam covering seven districts, viz. Dhemaji, Dibrugarh, Golaghat, Jorhat, Nagaon, Sivasagar, and Sonitpur. A total of 225 nos. of farmers who have received subsidy or are enrolled for getting subsidy from the state government for implementing micro-irrigation in their farm were contacted and interviewed using a structured questionnaire. The questionnaire was framed after discussions with various experts on the subject and government department officials. The questionnaire included the following details from beneficiaries:

i. Age.
ii. Educational status (Illiterate, pre-primary, primary, secondary, highersecondary, graduate, graduate and above).
iii. Farming experience (in years).
iv. Landholding size (in hectare).
v. Non-farm income no. of crops.

vi. Area under micro-irrigation (MI) (in hectare).
vii. No. of years of adoption.
viii. Type of MI adopted (drip/sprinkler/micro-sprinkler/others).
ix. Continual of adoption (yes/no).
x. Fertigation along with MI (yes/no).
xi. Availing government subsidy (yes/no).
xii. Source of water (own/hired).
xiii. Type of source (pump/lake/river/canal).
xiv. Water issues (iron impurities/clear).
xv. Technical support (own/hired/expert from supplier/government).

Apart from this, it was revealed during preliminary discussions that the following factors pose hindrance to the government efforts to increase adoption of micro-irrigation system:

i. Abundance of Rainfall in the region.
ii. High cost of investment.
iii. Inadequate subsidy from government.
iv. Difficulties in getting subsidy from government.
v. Lack of awareness about micro-irrigation.
vi. Non-availability of water source/pump.
vii. Maintenance issues after/before installation (e.g., clogging of drippers).

Accordingly, while collecting responses, the farmers were also asked to rank these constraints in their order of preference (from 1 to 7).

13.3.2 *Data Analysis*

All the data received from 225 nos. of farmers were analyzed using statistical software tools like Jamovi [17], MS-Excel for their frequency distribution and variation of responses. The ranks given by farmers were analyzed using **Garrett ranking method** [18, 19]. The **Garrett score conversion formula** is

$$\text{Percent position} = \frac{100(R_{ij} - 0.5)}{N_{ij}}$$

where

R_{ij}—Rank given for the *i*th variable by the *j*th respondents.

N_{ij}—Number of variables ranked by *j*th respondents.

With the help of Garrett's table, the estimated percent position is converted into scores, and then, for each factor, the scores of every respondent are added. Then, we compute the total value of scores and calculate the mean score. The factors having highest mean value is the most important factor [18]. The main advantage of using Garrett's method is that one can arrange the constraints in the order of their severity from the viewpoint of respondents. For any constraint, it can be ranked as no. 1 by one respondent, while another respondent may rank it as no. 2 or no. 3 as per their choice of preference. Therefore, when we are to rank many constraints together at time for many respondents, the Garrett formula converts them into percent position. So, we would have mean score for all constraints which can be accordingly studied.

13.3.3 Design Intervention

One of the objectives of this research is to understand how design intervention can be used to significantly help the farmers in adopting this technology of micro-irrigation. A detailed strategic analysis shall help to develop an alternative design of system which will help in increasing the adoption rate of micro-irrigation among farmers of Assam. Through this end, the Modular System Design for Sustainability (MSDS) has been used in this research. The MSDS method aims to understand the entire system and support the process of system innovation development. It is equally suitable for public institutions and Non-government organizations (NGOs) and hence can help design a system which is sustainable in nature. The method is organized in four main stages, viz. strategic analysis; exploring opportunities; designing system concepts; designing system [20]. Since the method employs a modular structure, it can also be adapted for different context and applications for developing alternative solutions.

13.4 Results and Discussion

The statistical analysis of basic data collected during the study reveals the following pattern:

(a) The study was done for a sample size of 225 farmers/beneficiaries who have received subsidy from the government or are in the process of getting such subsidy for installation of micro-irrigation system in their plot of land. This is facilitated under the PMKSY—per drop more crop scheme of Government of India.
(b) About 56% of the respondents do multiple cropping in a year (i.e., 3 crops in a year) followed by 37% of beneficiaries doing 2 crops a year (Table 13.1). It may

Table 13.1 Nos. of crops cultivated in a year

Levels	Counts	% of total	Cumulative %
1 crop	4	1.8	1.8
2 crops	84	37.3	39.1
3 crops	127	56.4	95.6
4 crops	10	4.4	100.0

Table 13.2 Type of micro-irrigation system installed

Levels	Counts	% of total	Cumulative %
Sprinkler	197	87.6	87.6
Drip	17	7.6	95.1
Micro-sprinkler	6	2.7	97.8
Rain gun	5	2.2	100.0

Table 13.3 Responses of beneficiaries for continuity of micro-irrigation

Levels	Counts	% of total	Cumulative %
Yes	180	80.0	80.0
No	45	20.0	100.0

be noted that as per government records, the state is dependent on irrigation for cropping during post-monsoon period (i.e., October to March) [15].

(c) About 87% of the respondents have opted for sprinkler irrigation, 7.6% for drip irrigation, and the balance have installed micro-sprinklers, etc., (Table 13.2). It may be noted that this is due to the lower cost of installation in case of sprinkler irrigation for a similar size of plot. Since this entire exercise is subsidy driven, the farmers have chosen the least costing option.

(d) 80% of beneficiaries positively responded for continuing with installation of micro-irrigation system as it was convenient (Table 13.3).

(e) 87% of beneficiaries have availed government subsidy for installation of micro-irrigation system, while the balance was yet to receive the subsidy and carry out installation in their farm (Table 13.4).

(f) Nearly, 63% of respondents have their own source of water for using the micro-irrigation system while 37% of respondents use water source from somebody else, on payment of fee (Table 13.5).

(g) Iron and other impurities are present in the water source for nearly 63% of respondents, and they have to do active filtration for separation of these impurities (Table 13.6). This is an additional cost incurred by them.

Table 13.4 Respondents availing government subsidy

Levels	Counts	% of total	Cumulative %
Yes	196	87.1	87.1
No	29	12.9	100.0

Table 13.5 Water source ownership

Levels	Counts	% of total	Cumulative %
Owned	141	62.7	62.7
Hired	84	37.3	100.0

Table 13.6 Presence of impurities and iron in water source

Levels	Counts	% of total	Cumulative %
Iron and impurities	141	62.7	62.7
No or less impurities	84	37.3	100.0

Garrett analysis: The respondents were also asked to rank in order of preference, the various constraints which prevent them from taking up micro-irrigation. The responses were analyzed using Garrett technique, and it reveals the following pattern:

(a) High cost of investment was ranked as no. 1 deterrent in adoption of micro-irrigation.
(b) Maintenance issues were ranked as no. 7 deterrent in adoption of micro-irrigation.

The results of Garrett analysis are presented as in Tables 13.7, 13.8, 13.9, 13.10, and 13.11.

From Garrett analysis, it can be inferred that apart from higher cost of investments, difficulties in getting government subsidy is the second most ranked deterrent. This can be attributed to government systems and procedures. Lack of awareness about the benefits of this system of irrigation, how to use it and maintain in long run is the third ranked constraint among beneficiaries. Since the area receives good amount of rainfall in general, people are not much inclined to take up this method of irrigation.

Table 13.7 Garrett ranking variables identified for this research

Sl. no.	Variables for ranking
1	Abundance of rainfall (C1)
2	High cost of investment (C2)
3	Inadequate subsidy from government (C3)
4	Difficulties in getting subsidy from government (C4)
5	Lack of awareness about micro-irrigation (C5)
6	Non-availability of water source/pump (C6)
7	Maintenance issues after/before installation (e.g., clogging of drippers) (C7)

Table 13.8 Percent position calculated as per Garrett technique

Sl. no.	$100\,(R_{ij} - 0.5)/N_{ij}$	Garrett value
1	7.14	78
2	21.43	66
3	35.71	58
4	50.00	50
5	64.29	43
6	78.57	35
7	92.86	21

Table 13.9 Rank given by respondents to the variables

Sl. no.	Description	Rank given by respondents						
		1st	2nd	3rd	4th	5th	6th	7th
1	Abundance of rainfall	90	45	42	38	10	0	0
2	High cost of investment	206	19	0	0	0	0	0
3	Inadequate subsidy from government	61	68	84	13	0	0	0
4	Difficulties in getting subsidy from government	145	74	6	0	0	0	0
5	Lack of awareness about MI	109	84	32	0	0	0	0
6	Non-availability of water source/pump	29	35	74	77	10	0	0
7	Maintenance issues after/before installation (e.g., clogging of drippers)	10	3	80	41	46	16	29

However, as attributed above in this paper, the groundwater level is very much stressed, and it is imperative upon the beneficiaries to take up such highly efficient irrigation system. Among other ranked constraints, we have inadequate subsidy from the government, non-availability of own pumping system and maintenance issues of drippers and sprinklers. Based upon the field study and analysis, it is imperative to make design interventions in areas of system and procedures to develop a system which is sustainable and has components of knowledge and finance inbuilt in it, so that it can attract farmers.

13.4.1 Understanding the Present System

A system map (Fig. 13.1) was prepared to link all the actors involved in the process of installing a micro-irrigation system using government subsidy. It was decided to study this system only, as at present most of the installation has been subsidy and policy driven. There are four major actors in this system. First is the farmer, who applies for getting the subsidy. Second, Government of India who allocates financial resources to each state and has developed the scheme of PMKSY—Per Drop More Crop. The entire system follows the norms and guidelines set forth in this scheme document [21]. The state government (receiver of funds) which has designated a department (usually department of agriculture) to disburse the funds and oversee the project implementation is the third actor. Entire monitoring and evaluation are being facilitated by the state implementing agency. The suppliers of micro-irrigation system are the fourth and last actor in this system, who get themselves registered for supplying and installing micro-irrigation systems at the farmer's site.

For availing the subsidy, a user (farmer/beneficiary) must fill up an application (offline/online mode) and submit it to the concerned authorities of state implementing agency. The department of agriculture has field officers, staff in all the districts of the state, and they receive these applications from the farmers. A portal has also

Table 13.10 Garrett values for the responses received

Sl. no.	Description	1st	2nd	3rd	4th	5th	6th	7th	Total	Garrett score	rank
1	Abundance of rainfall	7020	2970	2436	1900	430	0	0	14,756	65.58	4
2	High cost of investment	16,046	1273	0	0	0	0	0	17,319	76.97	1
3	Inadequate subsidy from government	4764	4455	4847	643	0	0	0	14,709	65.37	5
4	Difficulties in getting subsidy from government	11,282	4879	373	0	0	0	0	16,534	73.49	2
5	Lack of awareness about MI	8524	5516	1864	0	0	0	0	15,904	70.69	3
6	Non-availability of water source/pump	2256	2334	4288	3857	415	0	0	13,150	58.44	6
7	Maintenance issues after/before installation (e.g., clogging of drippers)	752	212	4661	2025	1990	563	608	10,810	48.05	7

Table 13.11 Final ranking of constraints after Garrett analysis

Constraint code	Garrett score	Rank
C2	76.97	1
C4	73.49	2
C5	70.69	3
C1	65.58	4
C3	65.37	5
C6	58.44	6
C7	48.05	7

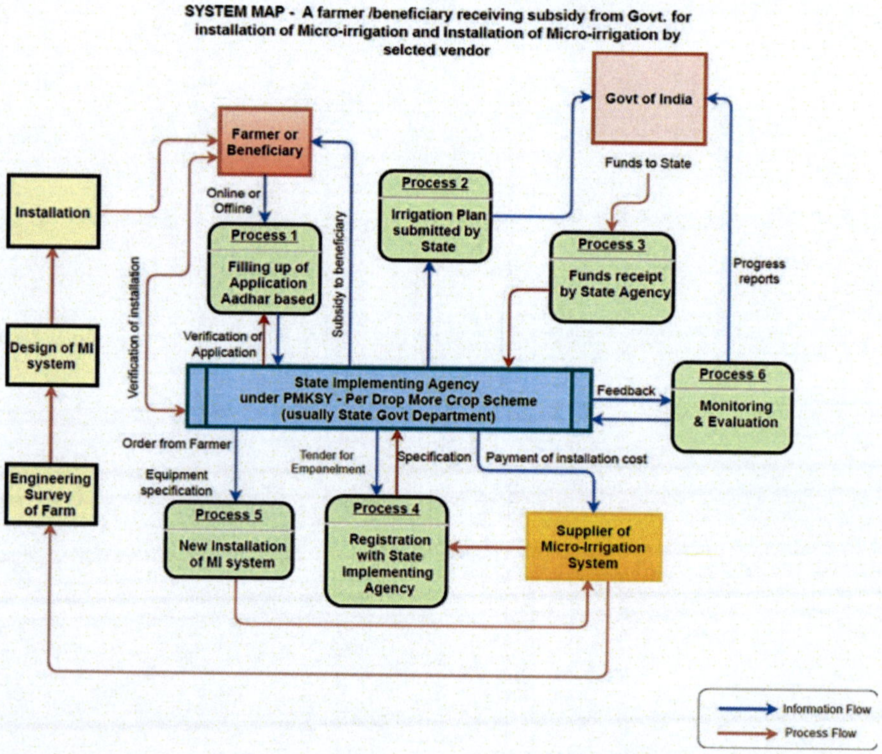

Fig. 13.1 System map showing the interaction between farmers, government, and suppliers

been developed by the state government [15] to receive online applications and disseminate information about the scheme. The application is necessarily linked with Aadhar (as in case of all Government Direct Benefit Transfer schemes) so that once the approval and installation is done, the subsidy amount can be remitted directly to his/her account.

The user has option to choose among a list of suppliers, as the government guidelines has mandated a fixed cost of items which can be given to the beneficiaries.

Each supplier can however add-on additional benefits to the scheme for a premium which is to be borne by the farmers. The state implementing agency forwards these applications to the selected supplier, and an agreement is signed between the user, supplier, and state government before the installation is carried out [21].

13.4.2 Understanding the Present Implementation Strategy

As per the present system, the government has taken an approach to promote the use of micro-irrigation system among farmers of the state through the Pradhan Mantri Krishi Sinchayee Yojana (PMKSY) scheme. However, there are gaps in the implementation strategy adopted by the state implementing agency which is reflected in the ranking obtained in Table 13.11. Higher cost of investment is the top ranked constraint which hinders the planning and approach of the government in promotion of micro-irrigation system. Due to higher cost of investment, people are attracted toward government subsidies, but this has its own drawbacks, and owing to a complex system interaction, people face difficulties. Another issue is that because of lack of knowledge about installation of micro-irrigation system, farmers are not much inclined to use it. Micro-irrigation is supposed to be more effective when used with fertigation, i.e., application of fertilizers through the irrigation water, and this has been established in research studies [6, 9, 10].

Therefore, one of the proposed strategies by government under the scheme of PMKSY is to make effective use of fertigation through micro-irrigation. States being the implementing agency are supposed to encourage the use of liquid fertilizer and make necessary awareness among farmers on the benefits of fertigation. However, no such information or feedback was given by the beneficiaries. As per the PMKSY guidelines, focus and priority are to be given to the water scarce, water stressed, and critical groundwater blocks/districts to conserve water. But there seems to be unavailability of such data as per the feedback of respondents.

13.5 Inference

The various constraints identified as per Garrett analysis were mapped with the present strategy and system of implementation to explore opportunities and design system concepts. It is reflected in Table 13.12. As per the adopted MSDS approach [20] for system design, the background information was analyzed, and it was gathered that various factors like knowledge, system, and finance, pose hindrance toward adoption of micro-irrigation among farmers of Assam. It is important to understand these factors so that a revamped system can be designed.

At this stage of research, it is noticeably clear that if the cost of installation is less and there are attractive financing options other than government subsidy, farmers

Table 13.12 Inference system design parameters

Constraint	Rank	Comments on present system	Factor
High cost of investment	1	As per the study, farmers are not much inclined to take up this system without government. subsidy. It can be also due to lack of knowledge and awareness about micro-irrigation system	• Finance • Knowledge
Difficulties in getting subsidy from government	2	This can be attributed to nos. of issues in the complex system map and procedures involved in getting subsidy. The respondent must go through number of channels as presented in the system map	• Finance • System
Lack of awareness about micro-irrigation	3	As evidenced from many studies referred in this paper, the micro-irrigation systems have many advantages over the conventional system of irrigation; efficiency of system is one such parameter. Fertigation (application of liquid fertilizers through the system) is an especially important feature, and the respondents were unaware of this possibility. The farmers are unaware of this mostly in the present system	• Knowledge • System
Abundance of Rainfall	4	As established from several studies, abundance of rainfall cannot be taken for granted for a geographical region, owing to many natural changes	• System
Inadequate subsidy from government	5	The scheme necessitates subsidized grant of finance for only 5 ha of land for a period of 7 years and the same beneficiary cannot avail it again. Here, again, there is a gap of alternative financial support	• System • Finance
Non-availability of water source/pump	6	Since micro-irrigation system is a pressurized irrigation system, the presence of a pump of appropriate capacity is of utmost importance. Purchasing a set of pumps for the system may put additional financial burden on the farmer, and there must be a system to support this element	• Finance • System

(continued)

Table 13.12 (continued)

Constraint	Rank	Comments on present system	Factor
Maintenance issues after/before installation (e.g., clogging of drippers)	7	The after sales service provided by suppliers must be upgraded or a different system must be there to take up the usual maintenance issues. Awareness about micro-irrigation can save the day for any beneficiary	• Knowledge • System

may be interested. But this statement needs validation which will be part of next course of action in this research.

13.6 Future Course of Action

The present work is a part of Doctoral research and future course of action includes development of representative model of variables affecting the adoption of micro-irrigation system in the region. This will also include working on a design of system which is sustainable in nature and tailor made to suit the geographic area. The design of system will have to invariably include all the parameters identified in this study to make it an attractive option for the farmers to adopt.

References

1. Economic Survey: Min. of finance, Govt of India, vol. 2 (Chap. 7), pp. 193–194 (2019–20). https://www.indiabudget.gov.in/economicsurvey/doc/vol2chapter/echap07_vol2.pdf
2. Rosegrant, M.W., Cai, X., Cline, S. A.: Global water outlook to 2025: averting an impending crisis (No. 572–2016–39087), p. 7 (2002). https://agris.fao.org/agris-search/search.do?recordID=GB2013200328
3. Suhag, R.: Overview of ground water in India. PRS (2019). https://dspace.bmsit.ac.in/xmlui/bitstream/handle/123456789/1785/Case%20study-IA1-GWH.pdf?sequence=1&isAllowed=y
4. Government of India: Water Related Statistics (2015). https://www.indiaenvironmentportal.org.in/content/412064/water-and-related-statistics-2015/
5. Government of India: Agricultural Statistics at a Glance 2015, Directorate of Economics and Statistics (2016). Ministry of Agriculture and Farmers' welfare. Google Scholar]
6. Dinesh, M., Sharma, B., Singh, O.: water saving and yield enhancing micro-irrigation technologies: how far can they contribute to water productivity in Indian agriculture? (2006). https://www.researchgate.net/publication/237425984
7. Cornish, G.A.: Pressurised irrigation technologies for smallholders in developing countries–a review. Irrigat. Drain. Syst. **12**(3), 185–201 (1998). https://doi.org/10.1023/A:1006037817381
8. Polak, P., Nanes, B., Adhikari, D.: A low-cost drip irrigation system for small farmers in developing countries 1. JAWRA J. Am. Water Resour. Assoc. **33**(1), 119–124 (1997). https://doi.org/10.1111/j.1752-1688.1997.tb04088.x
9. Ali, M.H., Talukder, M.S.U.: Increasing water productivity in crop production—a synthesis. Agric. Water Manag. **95**(11), 1201–1213 (2008). https://doi.org/10.1016/j.agwat.2008.06.008

10. Dinesh Kumar, M., et al.: Water saving and yield enhancing micro-irrigation technologies in India: when and where can they become best bet technologies? https://publications.iwmi.org/pdf/H041700.pdf
11. Jha, R.: Introduction and the state of Indian agriculture. In: Facets of India's https://doi.org/10.1057/978-1-349-95342-4_1Economy and Her Society, vol. II, pp. 3–33. Palgrave Macmillan, London (2018).
12. Palanisami, K., Mohan, K., Kakumanu, K., Raman, S. Spread and economics of micro-irrigation in India: evidence from nine states. Econ. Pol. Wkly. Supplement EPW xlvi (2011).
13. Perry, C., Steduto, P.: Does improved irrigation technology save water? A review of the evidence (2017). https://doi.org/10.13140/RG.2.2.35540.81280
14. Economic Survey of Assam: Govt. of Assam (2017–18). https://des.assam.gov.in/information-services/economic-survey-assam
15. Website of Government of Assam–Department of Agriculture & Horticulture. https://dirhorti.assam.gov.in/schemes/micro-irrigation-under-prime-minister-krishi-sinchayee-yojana-pmksy-per-drop-more-crop-pdmc
16. Government of India—Report of Task Force on Micro-irrigation. Ministry of Agriculture, Government of India (2013)
17. The Jamovi Project: *jamovi*. (Version 1.1) [Computer Software] (2019). Retrieved from https://www.jamovi.org
18. Aleeswari, A., Merline, W. L., Martin, N.: Study on industrial problems using Garrett ranking technique. BULMIM J. Manag. Res. **4**(1), 1–5 (2019). https://www.icmis.net/icmis18/ICMIS18CD/pdf/S177-final.pdf
19. Garrett, H.E.: Statistics in Psychology and Education (book). https://archive.org/details/statisticsinpsyc00henr
20. Vezzoli, C., Kohtala, C., Srinivasan, A., Xin, L., Fusakul, M., Sateesh, D., Diehl, J.C.: Product-service system design for sustainability. Routledge (2017). https://www.lens.polimi.it/index.php?M=0
21. Website of Pradhan Mantri Krishi Sinchayee Yojana: https://pmksy.gov.in/ Website of Department of Agriculture & Horticulture, Govt. of Assam https://agri-horti.assam.gov.in/

Chapter 14
Inclusive Design in Higher Educational Institutes for People with Disability

Priyanka Yadav, D. Udaya Kumar, and Sougata Karmakar

Abstract Concepts of universal design are gradually making its roads in the design and development of public spaces (airports, shopping malls, metro-railway stations, etc.) and public utilitarian facilities (ATM, ticket vending machines, telephone booths, etc.). Although the educational institutions are an essential category of the public space, it is still deprived of harnessing the benefits of universal design in the majority of the countries. When infrastructures are created for educational institutions' buildings, due considerations are given for architectural space layout, strength and stability of the architectural structure, aesthetics and functional aspects, sustainability parameters, etc. Barrier-free environments are created mainly for wheelchair users to make the institutional space inclusive to accommodate physically challenged/specially-abled stakeholders (students, teachers, supportive staffs, and visitors). Still, the requirements of different types of specially-abled people are not aptly addressed to make the built-space inclusive. Here, it is worth mentioning that the difficulties due to their various limitations among specially-abled people are diverse. Thus, all these necessities are to be addressed collectively in the infrastructure of an educational institute. Besides infrastructures, various assistive aids are crucial to facilitate quality education among specially-abled stakeholders. Through a systematic review, the current research aimed at determining the extent to which the issues/problems of the specially-abled stakeholders have been addressed in the design and development of infrastructures and assistive aids in educational institutes.

14.1 Introduction

More than one billion of world populations have some form of disability, estimated to be 15% of the whole population [1]. According to the Census of India 2011, the population of India was 1.2 billion, of which 26 million people are disabled/ specially-abled, which makes up to 2.21% of the Indian population. Many families are reluctant to share data about their specially-abled family members; this results in

P. Yadav · D. Udaya Kumar · S. Karmakar (✉)
Indian Institute of Technology Guwahati, Guwahati, Assam 781039, India
e-mail: karmakar.sougata@gmail.com

A. Chakrabarti et al. (eds.), *Design for Tomorrow–Volume 2*, Smart Innovation, Systems and Technologies 222, https://doi.org/10.1007/978-981-16-0119-4_14

lesser reported data than originally existing [2]. This data does not include children, pregnant women, people with strollers, and the elderly who are also required to be included to make a building inclusive [3].

United Nations Convention on the Rights of Person with Disability (CRPD) article 1 states that "disability results from the interaction between persons with impairments and attitudinal and environmental barriers that hinders their full and effective participation in society on an equal basis with others". Lin et al. [4] acknowledged that equal accessibility for persons with disabilities has become an important area that has led to increased awareness globally. This awareness gave way to the need to improve current knowledge on the interaction of disability and built environment [5].

Evcil [6] concluded that moving through a city safely and independently is every citizen's right, so is accessing built environment. If the built environment is not inclusive, opportunities for social integration are lost. However, accessibility is an issue that is not going to vanish, but the conditions can be improved by evolving designers' awareness about it [6]. Heylighen et al. [7] explained that universal design, inclusive design, and design for all are design approaches that aspire to include a range of users. In this article, inclusive design will be used as an umbrella term, as it reflects collective design approaches directly. By keeping diverse abilities into account, the built environment can be made to provide opportunities and spatial qualities to people through inclusive design [7].

Facilities are created mainly for wheel-chair users to make the institutional space inclusive to accommodate physically challenged/ specially-abled stakeholders (students, teachers, supportive staffs, and visitors). Still, the requirements of different types of specially-abled people are not aptly addressed to make the built-space really inclusive. Through a systematic review, it is aimed at current research to determine the extent to which the issues/problems of the specially-abled stakeholders have been addressed in the design and development of infrastructures and assistive aids in educational institutes all over the globe.

14.2 Methodology

A detailed literature search was conducted using the online Scopus and Web of Science database to shortlist papers using suitable keywords and their combinations. These keywords were associated with different aspects of universal design, physically challenged/specially-abled and, infrastructures for educational institutes. Apart from research papers, various Govt. reports and guidelines/ standards related to universal design have also been explored. Available information was segregated and presented under different headings and sub-headings to show the state-of-the-art review.

14.3 Various Challenges Faced by People with Disability

Besides residential spaces, there are types of public buildings that are accessed by different types of disabled population. Considering the built environment, population with disabilities can be categorized as wheelchair-bound, sensory disabled, ambulant and temporarily disabled [3]. The description of the categories is provided in Table 14.1.

14.3.1 Public Buildings and Environments

Kadir and Jamaludin [3] highlighted that standards and codes are majorly used for measurement of a building's accessibility for wheelchair-bound users. The other types of disabled users are hardly considered for defining space allocations in a building. It is suggested that a study from a diverse user's point of view will be beneficial for refining the existing standards. The following procedure was carried out:

Each participant was chosen from each category like sensory disabled, wheelchair-bound, ambulant and temporarily disabled who were from related institutions of disabilities. Five public buildings that get high footfall from diverse users were marked for study. They were asked to assess 15 facilities which were ("1. PWD Parking Space, 2. Pedestrian Pathways, 3. Guiding Blocks, 4. Ramp, 5. Main Entrance, 6. Door and Doorways, 7. Interior Pathways, 8. Information Counter, 9. Stairways, 10. Elevators, 11. Escalators, 12. Signage, 13. Praying Room and Ablution Area, 14. Public Restroom, 15. PWD Restroom") and emergency escape of the building. Then they were asked to give their feedback on building's comprehensive accessibility.

The result of the audit was that the level of ease of accessibility varied based on the type of disability of the participant. Visually impaired users gave the least rating for satisfaction; the result shows that the design considerations were concentrated on the needs of wheelchair-bound users, while users with sensory disabilities were neglected. In conclusion, Kadir and Jamaludin [3] insisted on the need to make

Table 14.1 Categories of disabled people [3]

Categories	Description
Sensory disabled	Person with partial or total hearing or sight impairment
Wheelchair-bound	Person dependent on mobility assistive devices for movement
Ambulant	Person able to walk but requires handrails for support
Temporary disabled	Person who is pregnant/sick/recovering from an accident

inclusive buildings so that people with varied disabilities can be catered to. It also suggests to warrant an inclusive environment in built spaces. A thorough study in the application of inclusive design in architecture is required, as it will also be able to include users with unseen abilities such as children, pregnant mothers, people with strollers, library staff carrying a stack of books, and the elderly [3].

Commercial buildings. Accessibility study of shopping complexes in Malaysia done by Hashim et al. [8] indicated that the accessibility in buildings is required to be improved to provide a better quality of life for the disabled population [8]. Bromley et al. [9] assessed commercial complexes in Malaysia involving people with different disabilities. It gave satisfactory result from the perspective of accessibility. However to improve the quality of life of disabled people, improvements were recommended. In a review done by Calder et al. [10] to assess public indoor fitness centres on universal accessibility parameters, in which 10 out of 14 studies were from USA. It was identified that accessibility of fitness center was poor for people with disabilities [10].

Service sector. In many countries, large restaurants have to be abided by the laws to have accessibility but small and medium restaurants are not required by law to provide accessible environment. This ends up creating a service gap between possible person with disability customers and small and medium restaurants [4]. Wazzan [11] did a study to evaluate, establish and document disabled friendliness of accessible hotel rooms, surveyed the complaints regarding accessible hotel rooms. They concluded that the accommodations needed to suit different types of disabilities but because of lack of consideration in standards made it hard to achieve. These issues arise because of neglecting principles of human factors accommodations [11].

A study by Tutuncu [12] brought it to notice that hotel satisfaction of person with physical disabilities is connected to forms of assistive devices, disability types and accessibility factor. Different physical disabilities cause different range of issues, characterization and identification of all possible impediments to accessibility is challenging. Hotel satisfaction is suggested to be predictable by accessibility of rooms, bath in rooms, public areas, recreation areas, areas of food and beverages. As every physical disabled person experiences disability differently, it can affect their hotel satisfaction as well. Researchers concluded that people with acquired disabilities, wheelchair-bound users and powerchair users are the most disadvantaged in terms of accessibility and it may become the reason of poor hotel satisfaction [12].

Travel and transportation. Community integration can be achieved through good accessible design practices that are safe, accessible and efficient and that caters to the need for visually impaired, hearing impaired and people with reduced mobility [13, 14]. A study done on a bus and metro barrier-free system, highlighted the importance of maintaining the barrier-free facilities and administering them so that these can be fully utilized by the disabled and elderly population [15]. Besides the maintenance of resources, the Aarhaug and Elvebakk [16] had the opinion that the accessibility should be available “from door-to-door, not only from bus stop to bus stop” [16].

Darcy [17] explains that for wheelchair-bound travellers, their impaired body and their management for it; assistive aid they use; an accessible built space and accessible customer service attitudes, these four components make up their embodiment for themselves. While air travel, if any of these components are disrupted because of air travel chain's inaccessible services, travelers find themselves as "(DIS)embodied 'nature of being'", where their citizenship is removed by damaging their independence and dignity [17]. These studies represent the ignorance that person with disability endure because of inaccessible built environments and facilities.

Recreation. Darcy and Dickson [18] presented the case of accessible tourism by examining the concept of whole-of-life approach. They analyzed the relationship between tourism, access, aging, and disability. Followed by reviewing accessible tourism and places from the perspective of universal design and explored the legislations that shape the future of accessible tourism in Australia [18]. Mullick [19] discussed the accessibility in National Park and points out the aesthetic, economic, and ecological implications of providing greater access in natural environments, concluding that there is a need to define policies to prevent misuse of natural resources [19].

According so the survey by Lovelock [20], which was done on 400 tourists and residents with mobility disabilities and able bodied. It was carried out to study the desire for accessing wild spaces. The survey concluded that the people with disability showed similar enthusiasm for tourism in wild spaces, but the inaccessibility keeps such spaces out of their reach [20].

14.3.2 Institutional Buildings and Environments

An audit was done by Osman et al. [21] at the University of Malaya, Kuala Lumpur, by surveying the campus for facilities provided for people with disabilities. It aimed to check the campus area based on standards and opinions of people with disabilities on the facilities provided in the campus area and provide recommendations to improve the accessibility of the campus using the users' perceptions. It was found that the campus was not able to follow the inclusive design in some areas as pathways were inaccessible, and since most of the respondents were visually impaired, they were found having issues with the signage system because it was making wayfinding hard for them [21].

Stetieh [22] in a similar study, done at the University of Jordan, used interviews with students, staff members, and other stakeholders. The analysis unveiled that users faced problems in class and campus. Peer and lecturer related issues, e.g., hearing-impaired student was not able to concentrate as they were bothered by the background noises and were unable to follow the lecture if the lecturer was not facing them and was not able to use the induction loop out of fear of standing out from peers. Some issues were based on unfavorable weather conditions such as water clogging; it caused problem for cane using blind students and ramps became slippery because of snowfall. Other issues were emergency alarms not being favourable for hearing impaired, as a faculty suggested incorporating lights in emergency alarms [22].

Ambati and Ambati [23] found inaccessible built spaces as a hindrance in educational experiences of students with disabilities, though the focus was on the change that is required in policies to improve management of colleges. They suggest that by providing supporting disability coordinator who can assist and help students with disability in getting assistive devices or interpreters campus can become accommodating for students with disability. They also recommended that there should be funding provided for technical training of staff to help in getting academic success [23].

Taking a new approach by interviewing 119 faculty members who follow inclusive pedagogy, from 10 Spanish universities, Morina and Orozco [24] based their qualitative study on barriers and aids in learning and participation. Bureaucracy and social stigma were emphasized as hindrances in accessibility while it was suggested that the rigidity of the curriculum should be relaxed to make education inclusive. This approach might help in supporting the students with different abilities because they have a different pace from abled-bodied students [24].

Lopez-Gavira et al. [25] in a four-year-long multidisciplinary study included experts from different fields (educational sciences, economics, health sciences, experimental sciences and humanities). The study was done on 44 students with disability, for the purpose of knowing the perspective of university students toward the required improvements in classrooms. The key requirements for improvements were accessible spaces, change in teaching methods to make them more accessible, and the necessity of trained teachers, so that they are able to provide help to students with different needs [25].

A quality of life assessment was done on a sample size of 203 handicapped students, integrated into higher educational institutions. Through the survey, it was revealed that with the physical barriers, students were also facing issues in context of course and the delivery of it. These issues made the students to report their quality of lives to be dissatisfactory. Nandjui et al. [26] suggest that the change can be brought by providing financial measures to these students and by improving the legislations in their favor [26].

14.3.3 Institutional Architectural Spaces and Stakeholders

Study by Raheja and Suryawanshi [27] done at Jawaharlal Nehru University in India, focusing on the informal spaces, concluded that these spaces play a major role in creating a cohesive social structure for students [27].

Raheja and Suryawanshi [27] defined educational campuses as a complex, city-like structure composed of people from varied cultures, genders, castes, sizes, and abilities. User mapping was used before the assessment to define stakeholders from the perspective of accessibility in an educational campus setting (Fig. 14.1) [28].

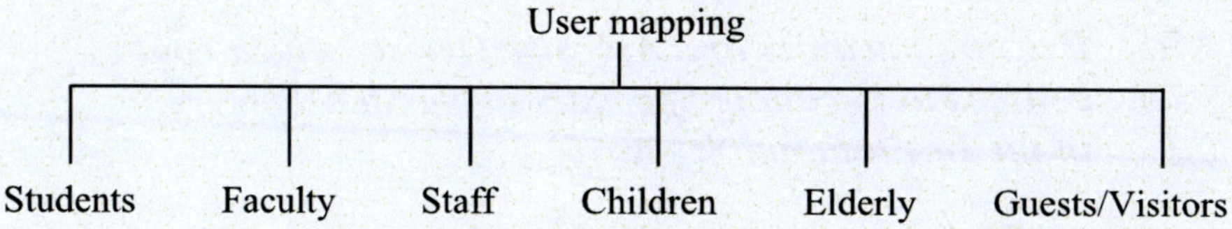

Fig. 14.1 User mapping framework for the educational campus. *Source* Inclusive strategies for universal access in an educational campus environment

Ceresnova [29] suggests that introducing inclusive design in the course of students can sensitize them for people with varying needs; it could improve the future of inclusive design as students are the main stakeholders of the future [29].

14.3.4 Existing Design Guidelines for Architectural Spaces and Buildings for Specially-abled People

European Committee for Standardisation compiled a report for analysing accessibility in the built environment. Klenovec and Isabela [30] inspected the feasibility of existing European and international standards, building codes, technical regulations, and guidance documents for inclusive buildings. The study highlighted that there are little data and guidance available for the issues faced by people with sensory disabilities and learning difficulties than what is available for people with motor disabilities.

It pulls the attention to the shortfalls of efficient and adequate conformity assessment and enforcement processes currently in force. These shortfalls make introducing new and existing regulations, guidance documents, mandates, and directive redundant and as a result the situation of accessibility in-built environment stays stagnant.

Klenovec and Isabela [30] call for change in the education system by introducing courses on Universal Design or Accessibility to sensitize the stakeholders in building industry, i.e., construction engineers and architects. It was concluded on the basis of a survey done on 336 European universities that offer architectural courses in which only 5% were providing obligatory lectures on Universal Design or Design for All [30].

14.3.5 Existing Facilities and Assistive Devices which could be Used to Facilitate Specially-abled Stakeholders in the Institutional Scenario

To complete course requirements through improving their abilities, people with disabilities need to use assistive technologies in the educational campus [23]. Assistive technologies for accessibility and learning used by people with different abilities are discussed below:

Partially sighted or visually impaired: Visually impaired people need the following aides for learning such as magnifying lenses, large print, Braille books, Braille embossers, audio recorder, information in electronic format, Information and Communication Technology (ICT), Optical Character Recognition software, electronic readers, Video magnifiers and magnification software, speech output systems (Job Access With Speech (JAWS)) and electronic Braille devices [31]. Assistive devices required for accessibility in the built-environment are white cane, elevators with tactile buttons, and emergency alarm with sound notifier [22, 32].

Hard of hearing or hearing impaired: Devices required for learning for hearing impaired people are hearing-aids and induction loop. Emergency alarm with visual indicators is required for an inclusive built environment [22].

Speech Impaired: Assistive device required while learning is augmentative or alternative communication devices [33].

Physically impaired: Assistive devices for accessibility are crutches or four-point cane, walkers, manual and powered wheel-chairs. For upper extremity reacher and grabber, mechanized page-turners for reading and add-on enlarged handles for writing [34] are available.

14.4 Research Gap and Way Forward

The paper is an attempt to investigate the existing relevant studies on the condition of accessibility in higher educational institutes. It is an effort to provide exclusive data on disability issues in institutes on the level of policies for disabled stakeholders, social inclusivity, and guidelines for the inclusive built environment.

The existing studies fail to discuss the requirements of assistive technology for people with different disabilities. They also lag to identify exclusive requirements of people with visual, hearing, speech, and physical impairments for learning and accessibility in educational institutes.

Further research in the area of inclusiveness of architectural space and assistive facilities for people with disability in higher educational institutions can examine following research questions:

1. What are the issues that a person with disability faces in their everyday life?

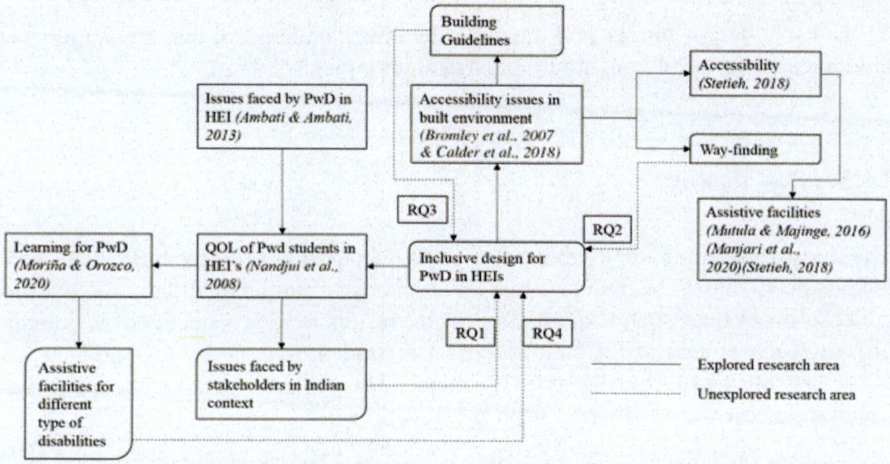

Fig. 14.2 User mapping framework for the educational campus. *Source* Inclusive strategies for universal access in an educational campus environment

2. What are the issues that are specific to built environment?
3. What are the existing guidelines that are followed to make a built environment accessible?
4. What are the assistive devices that are being used in the context of higher education by people with disabilities?

Identification of the challenges faced by person with disability in higher education institutes and a systematic research on assistive facilities and guidelines that are available in other domains can be the aim of further study. The aim could be achieved through following objectives (Fig. 14.2).

1. To identify the research gap and frame appropriate research questions to understand people with disability.
2. To identify the challenges faced by people with disability in higher educational institutes.
3. To study exiting design interventions and assistive facilities available.
4. To propose and develop design interventions with favourable guidelines.
5. To validate the effectiveness of developed design interventions and appropriateness of guidelines for providing assistive facilities.

Scott et al. [35] suggested that while systematic research on universal design in higher education institutes is encouraged, it is also recommended to collaborate and utilize the experience of professionals across boundaries.

It will be beneficial in the Indian scenario to emulate the strategies of including professionals from different fields (i.e., architects, engineers, designers, and building managers, etc.) with disability advocates and users from institutes while the design is in primary stages. These collaborations and discussions will direct the discussions on inclusive design further and untangle the issues underneath that are acting as a hindrance to the development of inclusive environments.

14.5 Discussion

The study draws attention to different aspects that are important to attain inclusivity in higher educational institutes. While the studies used similar qualitative methods to collect primary data, such as questionnaire, the results were dependent on the domain of study such as analysis of built space, universal learning, pedagogy and analysis of factors causing inaccessibility in a campus. The general issues mentioned are as follows:

1. Issues in signage.
2. Inaccessible parts of campus.
3. Inaccessible pathways.
4. Peer and faculty behaviour and support.
5. Role of disability coordinator and management of the institution.
6. The renovations being done failed to facilitate PWD adequately.
7. Students feeling disabled in absence of required facilities.
8. Building guidelines being focused on wheelchair-bound, ignoring sensory disabilities.
9. Social stigma and anxiety of being seen as taking advantage of the system through their disability.
10. Students reported to have low quality of life, and there was no significant difference if students with physical or sensory disabilities.

These studies helped in defining the issues that stakeholders with disability face, but they are focused on one institute and its built environment. One study examines ten public colleges, but it is focused on pedagogy and faculties' view on the condition of disabled students [24]. The literature is missing regarding the combined perspective of different stakeholders that play a major role in designing, managing and maintaining the institute, i.e., architects, policymakers, disability coordinators, and faculties, as they all play a major part in making an institute inclusive.

14.6 Conclusion

There are different factors affecting inclusivity in higher educational institutions. These factors depend upon policies at the level of campus management, stakeholders in campus (i.e., disability coordinators, faculty, support staff, students), availability of assistive devices (hearing aids, cane) or technology (digital study material, Information Communication Technology), built structure (i.e., elevators, ramps, stairs, classrooms) and built environment (i.e., signage, wayfinding, ease of accessibility), the social environment around the person with disabilities, financial factors and universal design in learning. From the perspective of a designer and architect, it limits the focus on built structure, built environment, and assistive technology.

In the context of Indian educational institutes, there is a need to build guidelines for bringing inclusivity in the built environment, as the current guidelines are defined focusing on the requirements of people with physical disability while the needs of people with other disabilities are ignored.

Designers, engineers, and architects should focus their research to address this relatively ignored domain of making the public space like academic institutions universal in nature to overcome the limitations of specially-abled toward enabling them as an equal independent part of the society without the feeling of separation/segregation.

By providing inclusive design in a built environment, the barriers between disability and education can be broken. This will clear the path for an inclusive and diverse society and workforce.

References

1. who.int: https://www.who.int/news-room/fact-sheets/detail/disability-and-health. Last accessed 2020/05/29
2. livemint.com: https://www.livemint.com/Opinion/1rx8tSYGwHB0ZRsvdNFiBP/Indias-missing-disabled-population.html. Last accessed 2020/05/29
3. Kadir, S.A., Jamaludin, M.: Users' satisfaction and perception on accessibility of public buildings in Putrajaya: access audit study. Procedia Soc. Behav. Sci. **50**(July), 429–441 (2012)
4. Lin, P.M.C., Peng, K.L., Ren, L., Lin, C.W. : Hospitality co-creation with mobility-impaired people. Int. J. Hospital. Manag. **77**, 492–503 (2019)
5. Lid, I.M., Solvang, P.K.: (Dis)ability and the experience of accessibility in the urban environment. Alter **10**(2), 181–194 (2016)
6. Evcil, A.N.: Raising awareness about accessibility. Procedia Soc. Behav. Sci. **47**, 490–494 (2012)
7. Heylighen, A., Van der Linden, V., Van Steenwinkel, I.: Ten questions concerning inclusive design of the built environment. Build. Environ. **114**, 507–517 (2017)
8. Hashim, A.E., Samikon, S.A., Ismail, F., Kamarudin, H., Jalil, M.N.M., Arrif, N.M.: Access and accessibility audit in commercial complex: Effectiveness in respect to people with disabilities (PWDs). Procedia Soc. Behav. Sci. **50**(July), 452–461 (2012)
9. Bromley, R.D., Matthews, D.L., Thomas, C.J.: City centre accessibility for wheelchair users: the consumer perspective and the planning implications. Cities **24**(3), 229–241 (2007)
10. Calder, A., Sole, G., Mulligan, H.: The accessibility of fitness centers for people with disabilities: a systematic review. Disabil. Health J. **11**(4), 525–536 (2018)
11. Wazzan, W.: My accessible room is not accessible, applying human factors: principals to enhance the accessibility of hotel rooms. Procedia Manuf. **3**(AHFE), 5405–5410 (2015)
12. Tutuncu, O.: Investigating the accessibility factors affecting hotel satisfaction of people with physical disabilities. Int. J. Hosp. Manag. **65**, 29–36 (2017)
13. Soltani, S. H. K., Sham, M., Awang, M., Yaman, R.: Accessibility for disabled in public transportation terminal. Procedia Soc. Behav. Sci. **35**, 89–96 (2012)
14. Zajac, A.P.: City accessible for everyone - improving accessibility of public transport using the universal design concept. Transp. Res. Procedia **14**, 1270–1276 (2016)

15. Wu, C., Li, X.: Using community planning method to improve effect of urban barrier-free transportation system. Transp. Res. Procedia **25**, 4330–4337 (2017)
16. Aarhaug, J., Elvebakk, B.: The impact of universally accessible public transport-a before and after study. Transp. Policy **44**, 143–150 (2015)
17. Darcy, S.: (Dis)embodied air travel experiences: disability, discrimination and the affect of a discontinuous air travel chain. J. Hosp. Tourism Manag. **19**(1), 91–101 (2012)
18. Darcy, S., Dickson, T.J.: A whole-of-life approach to tourism: the case for accessible tourism experiences. J. Hospit. Tourism Manag. **16**(1), 32–44 (2009)
19. Mullick, A.: Accessibility issues in park design: the National Parks. Landsc. Urban Plann. **26**, 25–33 (1993)
20. Lovelock, B.A.: Planes, trains and wheelchairs in the bush: attitudes of people with mobility-disabilities to enhanced motorised access in remote natural settings. Tour. Manage. **31**(3), 357–366 (2010)
21. Osman, M.M., Radzi, F.H.M., Bakri, N.I.M., Ibrahim, M.: Barrier-free campus: University Malaya, Kuala Lumpur. Procedia Soc. Behav. Sci. **168**, 134–144 (2015)
22. Stetieh, H.: An evaluation of the School of Engineering buildings at the University of Jordan with regard to accessibility. Sustain. Cities Soc. **42**, 240–251 (2018)
23. Ambati, N., Ambati, H.: Paradigm shift in German disability policy and its impact on students with disabilities in higher education. Int. J. Soc. Sci. Interdis. Res. **2**(4), 22–42 (2013)
24. Morina, A., Orozco, I.: Spanish faculty members speak out: Barriers and aids for students with disabilities at university. Disabil. Soc. 1–20 (2020)
25. Lopez-Gavira, R., Morina, A., Melero-Aguilar, N., Perera-Rodríguez, V.H.: Proposals for the improvement of university classrooms: the perspective of students with disabilities. Procedia Soc. Behav. Sci. **228**(June), 175–182 (2016)
26. Nandjui, B.M., Alloh, D.A., Manou, B.K., Bombo, J., Twoolys, A., Pillah, A.: Quality of life assessment of handicapped students integrated into the ordinary higher education system. Annales De Readaptation Et De Medecine Physique **51**(2), 109–113 (2008)
27. Raheja, G., Suryawanshi, S.: Inclusive informal campus spaces through universal design India principles. Assist. Technol. Res. Ser. **35**, 195–203 (2014)
28. Raheja, G., Suryawanshi, S.: Inclusive strategies for universal access in educational campus environments. Incl. Des. (2014)
29. Ceresnova, Z.: Student engagement in assessment of universal access of university buildings. Incl. Des. 143–151 (2014)
30. Klenovec, M., Isabela, N.: Accessibility in built environment. Comité Européen de Normalisation, Europe (2011)
31. Mutula, S., Majinge, R.M.: Information behaviour of students living with visual impairments in University libraries: a review of related literature. J. Acad. Librarianship **42**(5), 522–528 (2016)
32. Manjari, K., Verma, M., Singal, G.: A survey on assistive technology for visually impaired. Internet of Things **11**, 100188 (2020)
33. Simion, E.: Augmentative and alternative communication—support for people with severe speech disorders. Procedia Soc. Behav. Sci. **128**, 77–81 (2014)
34. Johnson, K.L., Dudgeon, B., Amtmann, D.: Assistive technology in rehabilitation. Phys. Med. Rehabil. Clin. North America **8**(2), 389–403 (1997)
35. Scott, S.S., Loewen, G., Funckes, C., Kroeger, S.: Implementing universal design in higher education: moving beyond the built environment. J. Postsecondary Educ. Disabil. **16**(2), 78–89 (2003)

Chapter 15
Modular Communication as a Structure for Sustainability Within Social Enterprises

Mehak Gupta

Abstract How can communication designers think systemically, within the context of modular communications, to design for social enterprises aiming to create socio-environmental impact? This research explores modular communications as a structure to empower sustainability within social enterprises and to create social impact through messaging. It contextualizes the working of *more with less* over less with more. By using a multiple case study approach (five social enterprises mainly from India, South Africa, and Kenya), the findings of which communication strategies work (education, participatory action, positive messaging, action-oriented engagement) and which do not (negative messaging, blind reliance on social media, no social proof), become core themes. Interviews with the enterprises' key leaders address the intention and outcome of their messaging. Thematic analysis was used as a systematic yet flexible approach to achieve the core themes. The study also led to the creation of a diagram to visually analyze the communication touchpoints of the enterprises, in modules. The core themes (as foundation) and modular communication (as structure) together suggest the achievement of sustainability within social enterprises and on-going effectiveness of their communication systems. This paper can act as a guide for designers passionate about creating impactful communication. It sprouts conversations about *processes* that are required in order to achieve the desired end-product.

15.1 Introduction

Ground-level work that deals with real-world problems is predominantly associated as a volunteering activity or a selfless deed. The resources for such work are majorly

The original version of this chapter was revised: Deleted words in the abstract now has been inserted. The correction to this chapter is available at https://doi.org/10.1007/978-981-16-0119-4_81

M. Gupta (✉)
ISDI School of Design and Innovation, One India Bulls Centre, Tower 2A 6th Floor, Parel, Mumbai, Maharashtra 400013, India
e-mail: mehak.g123@gmail.com

A. Chakrabarti et al. (eds.), *Design for Tomorrow–Volume 2*, Smart Innovation, Systems and Technologies 222, https://doi.org/10.1007/978-981-16-0119-4_15

acquired in the form of charity or in-kind. Donating clothes to the poor or being conscious about switching off the lights before leaving a room has often been termed as *selfless* or *good* behavior [1]. "The norm of giving implies that one should want to give, not because of any anticipated returns but for its own value" [1]. The social, environmental, or even cultural centric work is considered to promise no conventional incentivizing return on investment (be it time, energy, or resources) [2]. It is because of this reason that social work is mainly considered an act of selflessness. Even though altruistic work is needed for effective survival of our society, our global problems need consistent focus and a sustaining approach for resilient change [3–5]. Social enterprises bring with them a sustainable approach to tackle real-world problems [5].

Young designers today are sensitive to the changes in society. Political unrest, climate change, disease and pandemics, world peace, etc. are some phenomena they critically think about [6–8]. Global super-systems are layered with complex sub-systems, and hence to move one node, one must shift several others [9]. Designers need to *unfocus* [10], that is take a step back from directly coming to the design solution and spend more time in the questioning and iteration phase to *focus* on the processes that come before [11].

The objective of this paper is to conduct case studies of five social enterprises working for socio-environmental impact. The study aims to illustrate the communication strategies (core themes) that worked and did not work to fulfill their social and business motives. The research has resulted in the creation of a *communication analysis diagram (CAD)*, designed by the author as a novel approach to visualize communication touchpoints of existing enterprises, in modules. It facilitates this analysis by allowing the user to see one module in context of the whole system. The user can then point out the constraints in the communication system up to the minutest *node.*[1]

This paper suggests that modular communication can serve as an effective tool for sustainability within social enterprises. Through the course of this research, an understanding of the terms *modular communication* and *systems thinking* has evolved. These become *key tools for sustainability within a social enterprise.*

Modular communication is an incipient term. Communication that uses modules to flexibly communicate information is modular communication. Modules are almost like building blocks. Coupling the definitions of "modular" and "communication," "modular communication" can be defined as: "exchange of information constructed with standardized units for variety in use, ex: modular furniture" [12, 13].

Sustainability is defined as "being a method of using a resource so that the resource is not depleted" [14]. Within a social enterprise, these resources are tangible and intangible (i.e., time and energy).

This research aims to aid a link between modular communication (as a *structure* of communication systems) [15] and the core themes (as the foundation) that together lead to the sustenance of resources within the social enterprise. The findings and discussions can act as a guide for passionate designers wanting to create social impact through their work.

---

[1] See Fig. 15.2, nodes represent details about each communication channel.

"Design can and must become a way in which young people can participate in changing society"—Victor Papanek [16].

15.2 Methodology

15.2.1 Multiple Case Study Approach

The five social enterprises chosen are based in different geographical locations, mainly India, Kenya, and South Africa. Each of them is at varying stages of growth. The research methods include:

1. Observing the enterprises' visuals/messaging digitally.
2. Conducting interviews with the key leaders of the enterprise.
3. Studying secondary resources to understand existing workable and unworkable communication strategies for social change.

15.2.2 Conducting Research

Research Statement. Modular communication as a structure for sustainability within social enterprises.

Primary Research.

Interviews. Purpose: understanding business strategies and the enterprises' intentions, successes, and measurement of messaging. An interview with one key member, founder, or communication strategist, aided in bringing the creator's perspective to all the communication, which otherwise would only be seen from the receiver's end. Table 15.1 shows the two types of interviews conducted: (1) semi-structured and (2) structured. To maintain anonymity, the interviewees are coded as I-1, I-2, I-3, I-4 and I-5.

Secondary Research.

Observation. Purpose: identifying visual touchpoints to analyze the link between the designer's intention and the actual output. This was done by observing all social media platforms, assessing the website and testimonials, media mentions, and the communication material provided by the interviewees (if any).

The key study areas were type of business, brief of the work they do, brand core, strategy management, target audiences/stakeholders, visual communication channels, most innovative thing, and business analysis. Figure 15.1 shows the *brand core* of each social enterprise.

Studying secondary resources to understand existing workable and unworkable communication strategies for social change. Purpose: identifying and correlating

Table 15.1 Interview methods table based on Bleich and Pekkamen [17]

Specifications	Ecobrick Exchange	Kabaadey.com	Sampurn(e)arth	UnSchool	The FlipFlopi
Location	Cape Town, South Africa	Haryana, India	India	Portugal and Web-based	Republic of Kenya
Interviewee	Founder	Founder	Founder	Communication strategist	Communication strategist
Date	17.12.2019	07.01.2020	07.01.2020	29.01.2020	29.01.2020
Status	Video based	Telephonic	In person	Email	Email
Format	Semi-structured	Semi-structured	Semi-structured	Structured	Structured
Length	1 h 5 min	1 h 6 min	1 h 15 min	NA	NA
Recording	Audio recording and concurrent notes w/i 30 min	Audio recording and concurrent notes w/i 30 min	Audio recording and concurrent notes w/i 45 min	Email transcript	Email transcript
Language	English	Hindi	English	English	English
Interviewee code	I-1	I-2	I-3	I-4	I-5

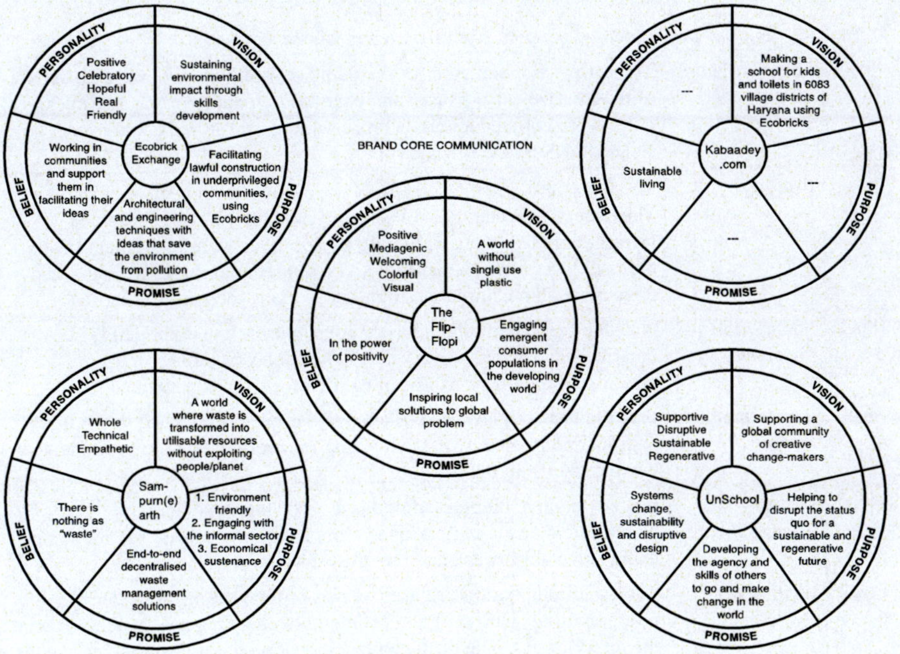

Fig. 15.1 Brand core of the five social enterprises. The content is adapted from the interviews and/or the brand websites [18–22], as per year 2020

existing frameworks. All the social enterprises are picked because they have innovative business models making socio-environmental impact. This secondary study provided key insights directly applicable toward communication design outcomes.

15.2.3 Data Analysis

Thematic analysis, as a six-phased method documented by Braun and Clarke, is used flexibly [23, 24]. The process has been more iterative and reflective, rather than linear [25] due to the nature of the study. The qualitative data includes observations of branding elements and communication channels, along with five interviews. The data was in the form of texts, documents, multimedia, public domain sources, and photographs [26] (Table 15.2).

Along with the findings of the thematic analysis, an outcome of this study was a communication analysis diagram as a structure for modular communication.

Table 15.2 The six-phase method adapted from Braun and Clarke [24]

Phase 1	*Familiarization*	Transcripts were created for the audio-recorded interviews with details of the interviewee's pauses and changes in tonality of voice. Reflexive notes were collated about the observations and stored on an excel sheet alongside the raw data
Phase 2	*Coding*	The interviews are the "creators" point of view for every enterprise. The observations are the "observers" point of view, i.e., received information. Data sets from both perspectives were triangulated with the raw data itself. The sorting and organizing of these data sets was done manually. Coding was done in as many themes as seen fit
Phase 3	*"Searching" for themes*	A deductive, latent, and interpretative approach was applied [23]. A creative way (maps) was used as a data analysis tool [25], to understand the relationships of the diverse themes generated
Phase 4	*Reviewing themes*	Five themes are generated. Each one has been referenced to see that it is grounded in the raw data. A scope for every theme has been created; what it entails and why
Phase 5	*Defining and naming themes*	Each of the five themes defined is a workable or unworkable communication strategy derived from the data analysis. All are named using some keywords stated by the interviewees
Phase 6	*Writing the report*	Short and long quotes are used as part of the writing alongside interpretations, in Sect. 15.3. No member checking was conducted with the interviewees. The diffusion of innovation theory created by Everett M. Rogers, 1962, as cited by Maloney [27] was used to examine the practical application of the themes, in Sect. 15.4.2

15.2.4 Limitations

1. The interviews with key leaders demonstrate their sole opinion and reasoning which can differ from the other designers within.
2. The social enterprises studied are on a growth curve and are being recognized in the markets for their work; their communication strategies will keep evolving.
3. This study is conducted with the lens of communication design and does not directly take into account exclusive product or service outreach.
4. The communication analysis diagram (CAD) is *suggestive* and facilitates the visual analysis of communication channels, as modules. It is created after several rounds of feedback from design experts guiding this study. It is not however evaluated by communication designers. Its function beyond this research paper has future scope of study.

15.3 Results

Communication Strategies as Core Themes. This paper suggests that modular communication is a structure for efficient communication systems. However, the use of such a structure was not a significant in the case studies. What emerged from the qualitative data analysis were workable and unworkable communication strategies. The most prevalent ones are derived as the core themes. However, these themes are comprehensive only to the extent of their function within the studied enterprises and their current communication systems.

Core Themes.

Education as a participatory action tool is empowering. Social enterprises that educated their target audience and stakeholders as part of their communication saw the possibility of lasting impact. For third world countries like South Africa and India: "environmental impact is not very high on the majority's agenda. If people have to hustle and work very hard in order to feed their family, protecting the environment from plastic waste isn't something on their list of priorities." (I-1) "The crowd here isn't very rich. The main goal of people here is to make money. They don't know the havoc that plastic has created everywhere. They have nothing to do with nature." (I-2). Initially, Ecobrick Exchange propagated incentives for the making of ecobricks[2] within communities. Slowly, they changed their approach: "if we can teach people to have unique skills, but also to use those skills to start their own micro businesses, then we can see that we will have more sustaining environmental impact." (I-1).

Kabaadey.com's founder demonstrates the struggles of not having an audience who gets his work. For students after him, he wants to bring an "entrepreneurship curriculum" that will foster an environment of growth. (I-2).

The FlipFlopi educates its audience about a global problem with a local solution. This contextualizes a sense of citizenship and serves as a message for the rest of the world, to ask questions about.

Understanding cognitive skills inspires action. To create messaging for global problems, three studied enterprises explicitly stated that *positive messaging* ensures sustaining impact. "Fear is generally an ineffective tool for motivating consistent engagement" [28].

> The FlipFlopi is a hugely positive, visual and mediagenic project - the concept of a boat sailing around the world is a communications and a campaign tool in itself: because she is in motion and multi-colored, she attracts the eye of the media as well as brings people together when she arrives at community destinations. (I-5)

Their message "celebrates the amazing work being done by FlipFlopi recyclers, sailors, dhow-builders as well as all the other artists, women, community members, activists—who are the real unsung heroes of the plastic revolution." (I-5).

Ecobrick Exchange was also named around messages of hope and not despair. "We wanted to celebrate people taking practical steps to protecting the environment" (I-1).

Actions beyond awareness give results. The enterprises saw that their work and messaging needed to inspire action. Social change occurs when shifts in human thinking and behavior are translated into action. "We deliberately design ourselves

[2] "Ecobricks are thermally insulating bricks that are made by simple compressing unrecyclable plastic into 2L bottles [18]."

out of being an addictive system, focusing on developing the agency and skills of others to go and make change in the world." (I-4) "We want people to start habits that are good for the environment and we want them to have reasons to continue those habits so that if we are no longer around to support or remind them, they will be motivated to continue those habits." (I-1).

Kabaadey.com and UnSchool involve their end users as part of their iterative processes. "Plastic is harmful: if I explain it like that, nobody is going to understand." But if "an object is created out of recycled plastic; a child will understand because he is seeing and using it." (I-2) "Check out participatory action research ideas… include the population you are designing for in the conversation - you are not designing FOR them; you are designing WITH them so as to not impose your ideas about what is best." (I-4).

Blind reliance on social media (SM) can have social, economic and environmental drawbacks. Not every time are seemingly "popular" communication channels the best suited for an enterprise's purposes. Sometimes, the beneficiaries of social enterprises do not have access to SM. Ecobrick Exchange experiences that SM's efficiency lasts only till "updating" the public and not as a measurement for change. SM platforms also go wrong to update any "failures or removal of campaigns," limiting the audience's real-time knowledge.

In 2020, UnSchool is "dramatically reducing" their extensive SM activity based on an in-house report they did on the "social and environmental impact of the internet" (I-4). Their "rate of success" is "best measured in the testimonials and amazing projects their alumni go on to complete, and the… successes of the organizations they design interventions and workshops for" (I-4), not social media metrics.

No social proof of an enterprise's innovations can lead to fallouts. If communication design is not planned as part of a business system, it could lead to enterprise fallouts. Example—when Sampurn(e)arth was founded, it gained media appeal and was listed as one among the top businesses. However, failing to display credible media mentions on communication channels, lost them their visibility. This emphasizes the need for social proof to keep the enterprise known in the population. This proof can take many forms, media mentions, testimonials, word of mouth, etc.

15.4 Discussion

15.4.1 Communication Analysis Diagram (CAD) as an Outcome

The core themes create the *foundation* for the enterprises' communication. However, these core themes are one of two key elements of the whole communication process, for which the second element is the *structure*, i.e., modular communication.

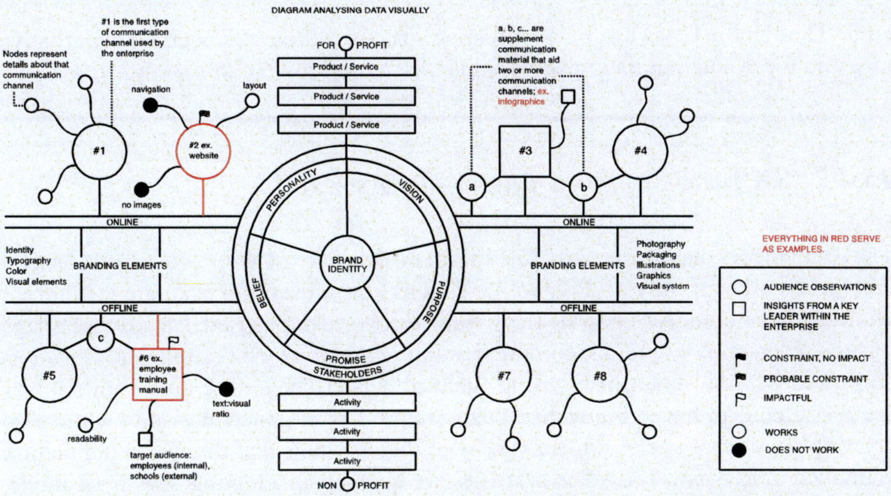

Fig. 15.2 Communication analysis diagram created as an outcome of this research process—aimed to facilitate visual analysis of the communication touchpoints within every social enterprise

The observation and analysis of the communication touchpoints led to the creation of the CAD (Fig. 15.2), which presents digital, print, and media platforms. CAD presents these touchpoints as modules. It becomes clear to see one module in context of the whole system and easily identify constraints right up to the minutest detail. Each module has a distinct intent and can be improved without affecting the whole—lacunas in one will not affect performances of the others [15]. CAD gives a visual structure to facilitate an analysis, as otherwise, it could get overwhelming. It is intended for use by communication designers and other key leaders within an enterprise to distinguish how the system can be restructured to make it modular. A periodic review of the diagram will make these systems all the more efficient in making the desired impact.

The several modules of the CAD include—brand core, stakeholders, branding elements, #number represents communication channels: ex. #2 is website and #6 is a training manual. The nodes attached to these channels enable detailing. "a, b, c …" represent supplement communication material to every channel. The vertical axis (top and bottom) divides the for-profit and non-profit work into modules.

Modular communication is a structure ensuring the working of more with less over less with more [15]. Messages can be organized and disseminated to the intended target audiences through each module. This will ensure lesser wastage of tangible and intangible resources. Additionally, consistent review of these systems will ensure that the communication is making the intended socio-environmental impact.

One of the main reasons for the fallout of social enterprises is when the business models are not sustainable [29], i.e., the profit-making motives and the non-profit motives are not parallel (one takes precedence over the other). The CAD helps put a structure for every communication module which is important for both of these ends. Though the CAD can be an effective structure, there has been no quantitative measure for evaluating its impact. That is the future scope of this study.

15.4.2 *Diffusion of Innovations Theory (DOI)*

The DOI theory has been used to explain which part of their journey the social enterprises are currently at (Fig. 15.3). This has been useful to examine the current working of communication strategies, within the world population. It also emphasizes whether early use of modular communication will help reach desired goals faster as compared to those enterprises using them at a later stage. The social enterprises are considered to have "innovative business models to generate profits while also creating significant social impact" [5]. The theory states that the world population is divided in five parts [30]. The first 16% are quickest in adopting new innovations, as they are introduced [27]. The chasm is a break after that—which is the hardest for any innovation to cross over as the 84% population after that depends on "social proof" to buy a new idea [27].

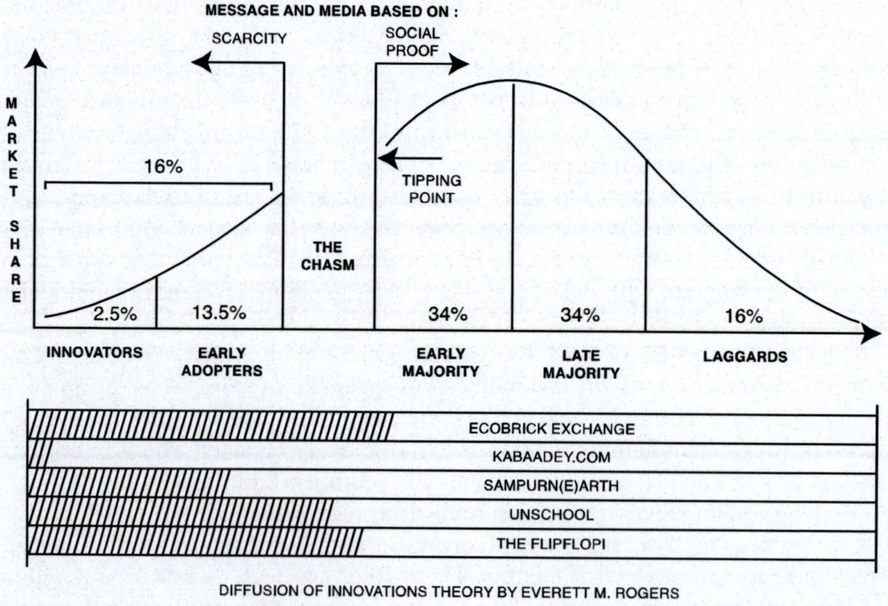

Fig. 15.3 Analyzing the population outreach of the social enterprises using the DOI theory adapted from Everett M. Rogers by Maloney et al. [27]

15.5 Conclusion

The research analyzes case studies that are rooted in socio-environmental impact. It presents strategies (core themes) as findings from the thematic analysis for effective communication. These strategies will help reach people for whom it is otherwise difficult to realize the value of the work being done. When designing for social causes, measurement of impact will convey if communication is facilitating change. The DOI theory is one way of measuring impact.

The core themes act as foundation to the communication systems. Modular communication is the structure. One way to visually analyze the communication systems of social enterprises is the CAD. The analysis will present *how* the system can then be restructured to be made modular. Modularity can be achieved by any enterprise designing its communication in flexible standardized units.

Sustainability is the mindful use of tangible and intangible resources. Every part of the system must be modular enough so as to enable such sustainability. From "modular brochures" (I-3) and "modular presentations" (I-1) to an overall system thinking approach, all of these ultimately lead to systemic change. Systems thinking is "seeing the whole beyond the parts," and "on the other hand, it means seeing the parts in the context of the whole" [31]. Systems thinking enables the design and implementation of social impact projects from a holistic perspective.

Leaving the audience with a "sense of agency" (I-4) and giving them "reasons to continue" (I-1) without the physical presence of the social enterprise ensures self-sustenance—which in the longer run allows for scaling sustainably. Communication design can facilitate such autonomy with action-oriented modular messaging.

Acknowledgements I wish to thank Prof. Laskarina Yiannakaris and Prof. Anando Dutta of the ISDI School of Design and Innovation for supporting the research process. I thank Harshil Vora from Proinnovate [32] for his feedback throughout. I also thank Rukmini Banerjee, Anne Panicker, and Mookesh Patel for proof-reading this paper.

References

1. Leeds, R.: Altruism and the Norm of Giving. Merrill-Palmer Q. Behav. Dev. **9**(3), 229–240 (1963)
2. Schweitzer, D., Chianello, T., Kothari, B.: Compensation in social work: critical for satisfaction and a sustainable profession. Administration in Social Work **37**(2), 147–157 (2013)
3. Kropotkin, P.: Mutual Aid: A Factor of Evolution. Sargent, Boston (1955)
4. Sorokin, P.: Altruistic Love. Beacon Press, Boston (1950)
5. Rangan, K.: Business and Social Impact, s.l.: s.n (2014). Available at: https://www.hbs.edu/socialenterprise/Documents/BUSIForum2014Summary.pdf
6. Mohieldin, M., Petkoski, D.: How Youth can help solve the world's toughest problems [Interview] (30 March 2018) (2018)
7. Affairs, U. N. D. o. E. a. S.: World Youth Report 2018. New York: United Nations

8. House, C.: The Biggest Global and National Issues for Young People (2019). Available at: https://medium.com/chatham-house/the-biggest-global-and-national-issues-for-young-people-c1e3b810fed1. Accessed 14 Oct 2020
9. Draper, S.: Systems thinking: unlocking the sustainable development goals (2018). Available at: https://www.forumforthefuture.org/blog/systems-thinking-unlocking-the-sustainable-development-goals. Accessed 30 Mar 2020
10. Pillay, S.: Your brain can only take so much focus (2017). https://hbr.org/2017/05/your-brain-can-only-take-so-much-focus. Accessed Oct 2020
11. Chaudhuri, I.: Design Talk with Itu Chaudhuri [Interview] (16 January 2020) (2020)
12. Merriam-Webster: Modular, Merriam-Webster Dictionary AND Thesaurus (n.d.). [Online]. Available from: https://www.merriam-webster.com/dictionary/modular/ Accessed Oct 2020
13. Merriam-Webster: Merriam-Webster Dictionary AND Thesaurus (n.d. Communication). [Online]. Available from: https://www.merriam-webster.com/dictionary/communication/. Accessed Oct 2020
14. Merriam-Webster: Sustainability, Merriam-Webster Dictionary AND Thesaurus (n.d.). Available from: https://www.merriam-webster.com/dictionary/sustainability/. Accessed Oct 2020
15. Bradley, S.: Vanseo Design 2013. https://vanseodesign.com/web-design/modular-benefits/. Accessed 14 Oct 2020
16. Papanek, V.: Design for the Real World: Human Ecology and Social Change, 2nd edition, revised ed. Academy Chicago Publishers, Chicago (1971)
17. Bleich, E., Pekkanem, R.: How to report interview data. In: Interview Research in Political Science. Cornell University Press, s.l., pp 95–101 (2013)
18. EcoBrickExchange: Ecobrick Exchange (2020). https://www.ecobrickexchange.org/views/home.php. Accessed 7 June 2020
19. Kabaadey: kabaadey|Instagram 2020. [Online] https://www.instagram.com/kabaadey/?igshid=nt4b3nxjfr5t. Accessed 28 Mar 2020
20. Sampurn(e)arth Environment Solutions Pvt. Ltd.: Sampurnearth Homepage (2018). https://www.sampurnearth.com/. Accessed 8 June 2020
21. DISRUPT DESIGN LLC: UNSCHOOL Homepage (2020). https://online.unschools.co/. Accessed 30 Apr 2020
22. The FlipFlopi Project Foundation Limited: The FlipFlopi Expedition (2020). [Online] https://www.theflipflopi.com. Accessed 9 June 2020
23. Clarke, V., Braun, V., Hayfield, N.: Thematic Analysis. In: Smith, J.A. (ed.) Qualitative Psychology, A Practical Guide to Research Methods, pp. 222–248. SAGE Publications Ltd., London (2015)
24. Braun, V., Clarke, V.: Using thematic analysis in psychology. Qual. Res. Psychol. **3**, 77–101 (2006). https://doi.org/10.1191/1478088706QP063OA
25. Nowell, L.S., et al.: Thematic analysis: striving to meet the trustworthiness criteria. Int. J. Qual. Methods (2017). https://doi.org/10.1177/1609406917733847
26. Thorne, S.: Data analysis in qualitative research. Evidence Based Nurs. **3**, 68–70 (2000)
27. Maloney, C., 2010. The secret to accelerating diffusion of innovation: The 16% rule explained [Online]. https://innovateordie.com.au/2010/05/10/the-secret-to-accelerating-diffusion-of-innovation-the-16-rule-explained/. Accessed 7 June 2020
28. O'Neill, S., Nicholson-Cole, S.: Fear Won't Do It **30**(3). SAGE J. 7 Jan 2009
29. Cain, M., 2014. Five reasons social enterprises fail. [Online]. Available at: https://www.theguardian.com/social-enterprise-network/2013/jun/28/five-reasons-social-enterprises-fail-business. Accessed 21 Oct 2020
30. Rogers, E. M.: Diffusion of Innovations. 5th Edition ed. Free Press, A Division of Simon & Schuster, Inc., New York (2003)
31. Shaked, H., Schechter, C.: Systems thinking for principals of learning—focused schools. J. School Admin. Res. Dev. **4**(1), 18–23 (2019)
32. Proinnovate. https://www.proinnovate.co

Part II
Design Training and Education

Chapter 16
Introduction to Self-help Books/Texts for Design Students

Shefali Yadav

Abstract Time spent at college is a happy experience and a fond memory for most of the students but for some, it has many rough patches. Everyone's situation is unique, but there are a few problems that almost all college students deal with at least once during the time they spend in college. 72% of students in India are unaware of how to deal with stress and its ill-effects as studied by Dr. S.G. George. A study by NIMHANS has also found that 11% of college students have attempted suicide because of stress. The purpose of this study is to analyze the kind of stress design students go through and how taking help from a self-help books/texts can guide students to deal with the feelings, thoughts or behavior that come with stress and anxiety and can act as bibliotherapy for them. Self-help books take insights from psychological science and draw in particular on the newly developing "positive psychology". There is evidence that bibliotherapy is effective in the treatment of psychological disorders. To know how self-help books can help the design students, eight in-depth interviews with the current and former design students were taken who have experienced the stress and triumphed over that with the help of self-help books. A survey amongst the students was conducted to know about the stress they go through and how they deal with it, what they most need at that moment and their awareness about self-help books.

16.1 Introduction

College is a time of significant transition. Many students stay away from home for the very first time and have less access to support from family and friends. Though they have more freedom at that time, some students go through stress because of pressure of assignments, increase in competition, new environment, homesickness, language issues, sometimes even lack of support from family because of choosing field which family does not prefers, etc.

S. Yadav (✉)
PLC State University of Performing and Visual Arts, Rohtak, Haryana, India
e-mail: shefali2y@gmail.com

A. Chakrabarti et al. (eds.), *Design for Tomorrow–Volume 2*, Smart Innovation, Systems and Technologies 222, https://doi.org/10.1007/978-981-16-0119-4_16

Center for the Study of Developing Societies conducted a survey in which about 4 in 10 students in India have experienced bouts of depression in the last few years. Prime Minister Narendra Modi also addressed the issue of mental health among students and urged for "expression of depression instead of its suppression" in his radio address. As per Lancet 2012 report, India has world's highest suicide rates for youth aged 15–29. It also illustrated the need for urgent interventions for this demographic. About 4 in 10 students in India have experienced bouts of depression in the last few years as per a survey done by the Center for the Study of Developing Societies [9].

Students cannot always share these problems with their parents because in some cases, they stay in different city, their tricky age or even sometimes they have fought with their parents to join creative courses like design. They cannot share with their school friends as they are also going through same kind of feelings or are enjoying one of the most memorable time of their life. Mentors are available in some colleges but they come on particular days and some students feel uncomfortable sharing their personal stuff with unknown person at that age.

At that time, students can be introduced to self-help books which can guide them and be of great help to them and can act as bibliotherapy. In bibliotherapy, literature is used in any form that can help in finding a way through psychological, emotional and social problems.

16.2 Literature Review

16.2.1 Stress in Students

Due to complicated emotional eco-systems, young adults are easily afflicted by emotional, behavioral, learning or mental disorders which also include anxiety, depression, bi-polar disorder, disruptive behavior disorders, intellectual disability and ADHD. Mental concerns, racial, sexual or religious discrimination, emotional issues, body shaming, low self-esteem or insecurities, family or financial problems and even hormonal issues can also occur due to environmental challenges. Mental concerns often manifest themselves as lack of motivation, poor academic performance, social interaction issues and sometimes even self-harm. They also have to contend with the beast of stigma attached to mental issues. Students, parents and teachers—all 3 play an active part in ensuring mental well-being. In India, parents often go into denial mode when it comes to their child's mental health, stating, "My child has no such problems!" [2].

Counselors revealed that youth finds it difficult to cope with failure in careers and in that case neither families nor the institutions offer adequate support. India endures an 87% shortage of mental-health professionals due to which it is difficult to find professional help. India does not spend enough on mental health. India spends 0.06% of its health budget on mental health only currently. According to 2011 WHO report, many developed nations spend above 4% of their budgets on mental-health research, infrastructure, frameworks and talent pool. India, faces an 87% shortage of mental-health professionals. There are 3800 psychiatrists, 898 clinical psychologists, 850 psychiatric social workers and 1500 psychiatric nurses nationwide, according to a reply by the Ministry of Health and Family Welfare in the Lok Sabha in December 2015. According to data from WHO, there were three psychiatrists per million people, 95% fewer than the Commonwealth norm of 5.6 psychiatrists per 100,000 people. India is short of around 66,200 psychiatrists [10].

16.2.2 Introduction to Self-help Books

Dictionary description of self-help is "the acts of helping or bettering oneself without the aid of others". The books with which way one copes with his/her personal or emotional issues without professional help are self-help books [1]. Self-help books are utilized as a cost-effective way of reducing psychological or emotional difficulties [13].

Categories of Self-Help Books. The common themes of self-help books are self-discovery, finding inner strength and living happier and fulfilling life [12]. Main themes of books are growth, relationships, coping and identity. The common message shared by all the self-help books is that you can improve yourself [1].

Nature of Writing. Majority of authors want to keep the writing style accessible. Readers are able to recognize that addressing readers explicitly using personal pronouns is a linguistic feature of self-improvement books. Keeping the chapters shorter and concise messages keeps the readers engaged. It is unnecessary for authors to explain an idea extensively. One of the ways to be concise is to organize the message like a series of bullet-points. Readers do not read the full book instead they read chapters as per their needs [8].

16.2.3 Effectiveness of Self-help Books

When examined effects of self-help for unipolar depression and anxiety with 272 participants by comparing the difference in effectiveness between treatments like bibliotherapy, individual therapy, group therapy or waiting list control group it was found that "Bibliotherapy is an effective treatment modality, which is no less effective than individual or group therapy" [3].

Advantages of self-help books are they are helpful for those who are literature-oriented, they may expedite treatment gains when used in conjunction with therapy, they are cost effective, they may serve as a—if not the only—viable alternative to those who are resistant to, or are restricted from, traditional mental-health treatments [5].

Considering self-help books optimistic approach, they propose "at least some answers" to readers' issues or problems [6]. Bibliotherapy for clinically significant emotional disorders is more effective than waiting list or no treatment conditions. No difference was found between bibliotherapy and psychiatric treatment of relatively short duration [4]. Self-help books can be effective if the books are chosen by professionals. She pointed out that self-help book lifts the spirit, engenders and supports hope and keeps people striving toward their goals. It helps them to fend off feeling of hopelessness, helplessness, despair and depression [1].

When one just reads 20 pages of self-help in the morning, he or she can see the difference around them. They start interpreting themselves more positively. It does not improve one's life but enhances it. A daily diet of inspiring words spurs you to explore what one is capable of, and what are his/her limitations so that they can be smashed. While they may not necessarily tie into one's career, they address the part that yearns to feel vibrantly alive and fulfilled in every moment. When there is awareness of the thinking behind actions, one can erase negative thought tapes and replace them with words of love, possibility, power and magic [7].

Four pragmatic factors that explain the success of the self-help books are: *Cost.* As price of them is low compared to a consultation with a psychologist. *Accessibility.* They are easily available and can be read over a lunch or a sleepless night. *Privacy.* Written solution for problems offers the opportunity to work on problems without "going public" or having to speak to a doctor or psychologist. *Excitement.* Self-help books quite often become best sellers and buying and reading such a book gives one the opportunity to become part of an in-group [11].

16.3 Methodology

16.3.1 Qualitative Research

Qualitative method employed in the research is **In-depth Interview**. With eight current and former design students, who have felt stressed at some point during their college life. Five of them have taken help of self-help books at that time and three have taken major decisions because of the stress. Age group of the interviewees was from 18 to 30. Snowball sampling was done for conducting interviews as stress is a sensitive subject to talk about and not everyone is comfortable to talk about it.

The interviews were semi-structured and was between 30 and 40 min long. The interviews were conducted on phone and through video calls because of Covid-19 pandemic. The entire interview period took about one month (April 2020–May 2020) and the interviews were spread more or less evenly across that period. Questions were added to the list throughout the interview period as per the issues raised from text analysis.

16.3.2 *Quantitative Research*

Quantitative methods employed in the research are: **Survey**. The questionnaire had ten general questions designed to capture the reasons behind student's non-willingness to attend college, availability of psychologist and their interest in reading. The questionnaire had multiple choice questions, predefined answers offering respondents the possibility to choose among several options and for some questions, optional space was provided to elaborate on the answer. *Sampling*. First-year students of Bachelor of Designing course of NIFT, Delhi, and PLCSUPVA, Rohtak, were selected for the research. Total number of students are 240; with simple random sampling, keeping confidence level 95% and confidence interval 5, sample size of 148 students was taken. Simple random sampling is done for the conduction of survey.

16.4 Findings

16.4.1 *Interviews*

Reasons of stress. All eight design students interviewed have felt stressed at some point in their college life. Jigyasa (former design student) has experienced stress because of low attendance. She missed her classes of module-based subject as she suffered from typhoid. There was no problem of re-semester but still due to already existing anxiety issues, she used to cry a lot. One day, she broke down in front of her best friend and said that "*I cannot pay attention to any other subject and I am feeling uncomfortable all the time.*" (telephonic interview, May 01, 2020).

Shiv (design student) stated, "*Except my mother no one was supportive with me doing designing course. My uncle once taunted that I'll never be able to earn as much I am spending and after that I used to be constantly worried about my future. I even started doubting myself and stopped going to classes*". (telephonic interview, April 28, 2020).

One out of eight interviewees, Bhawna (former design student) is still uncomfortable sharing the reason of stress. She just said that it was the darkest phase of her life (telephonic interview, May 4, 2020).

Ishita (design student) said, "*I used to feel left out and inferior and to ignore that I started eating lot of fast food and gained weight. After discussing with my parents, I have taken a therapy for 8 to 9 sessions that cost around 1500 to 2000 per session. Though my parents are always supportive but I always feel guilty that already the course is so expensive and my parents have to spend more because of me*". (video interview, May 15, 2020).

Amit (design student) also use to feel left out due to which he got homesick and because of it, he left the course mid-semester (video interview, May 15, 2020).

Deepshika (design student) has anxiety issues because she could not complete her assignments at home as she has to help with house chores and she is not allowed to stay in hostel. She mentioned "*My paternal family has even started forcing me to get married as soon as I turned 18 and this idea haunts me.*" (telephonic interview, May 19, 2020).

Reaching out to first self-help book. The books were either gifted to the interviewees by someone or recommended by a friend and in case of the students who were reader before found them while browsing the book store.

Jigyasa (former design student) mentioned that she was going through difficult time in college and while during a book store visit, she bought her first book because of its catchy name and it was also one of the bestsellers and first row of the self-help section was filled with it (telephonic interview, May 01, 2020).

Similarly, Pratibha (former design student) stated, "*I bought my first book while browsing a book store but it never helped me.*" (telephonic interview, May 17, 2020).

Shiv (design student), said, "*after seeing me struggling through stress my friend recommended me a book 'The Power of Now' but because my college was very far from the book store and I never liked online shopping that's why my friend gifted me my first self-help book.*" (telephonic interview, April 28, 2020).

Aakriti (design student) mentioned "*I have always struggled with asking for help, no matter how difficult things get. As I loved reading books since school, I always knew about self-help books but stayed away from those because my brain subconsciously thought that reading one is admitting that I need help and it will imply that I am broken.*" (telephonic interview, May 30, 2020).

Others received their first book as a gift from family and close friends who saw them in stress and tried to help them.

Impact of content of self-help book. All the interviewees mentioned that content is the main factor to make them able to complete it. When asked about what they like most in book Aakriti said, "*I really like the content of books like 'Good Vibes, Good Life'. The best part is it's practical and it doesn't tell you to think positive and ends like other self-help books but it teaches you about self-love, humility, failure, meditation, empowering thoughts and living a greater life which I find the best in it.*" (telephonic interview, May 30, 2020).

Pratibha (former design student) commented, "*I don't like content of all the self-help books because of repetitive messages that the books give and I sometimes find authors contradicting themselves. I loved the way some books like 'Calm' are written. They are also perfect for just picking up and having a flick through whenever I need it. After reading the book 'Calm', I found that I am more oriented towards different design stream and that's why after foundation year I opted and changed it as well so for me it's life changing.*" (telephonic interview, May 17, 2020).

Bhawna (former design student) stated that she has never completed a single self-help book till date because she feels sleepy or wants to do something else just after reading few lines *(telephonic interview, May 4, 2020).*

Jigyasa (former design student) mentioned that she found books which are in storytelling format or the books which give you tasks are the best as they never bore you *(telephonic interview, May 01, 2020).*

Change or improvement in behavior after reading. Jigyasa (former design student) said "*we are unaware of where to start or which path is the right one for our individual journeys. For me, that book has steered me into the clear and onto a path of hard work, success, and growth*". After that book, she has read so many books of this genre and she is proud of herself and says that she owes it a lot to the authors of those books and their incredible insight and opinion of direction (telephonic interview, May 01, 2020).

Shiv (design student) that book, "The Power of Now" taught him to live in present as it runs on the concept that we live only in the "Now" and it takes time to understand, intellectually and we may seem to understand but to put it in practice is very important. He says, "*It taught me how to deal with painful emotions, how to awaken from the preconditioned self, what I used to think I am but in reality, they were lot of ideas and beliefs coming from friends, parents,* etc. *I recommend this particular book to all the people as it offers helping hand to everyone weather, they are going through stress or not. The book frees you from that old self and helps to live in the now.*" (telephonic interview, April 28, 2020).

Pratibha (former design student) stated reading the book after some days, "*I used to forget about the learnings but now I try to remember them and apply them in the situation I am in. I have found peace and safe haven in many of these amazingly written pages because of which I have become more optimistic.*" (telephonic interview, May 17, 2020).

Aakriti (design student) said, she does not gets completely healed after reading books but it is always a new start (telephonic interview, May 30, 2020).

Summary. The main issues of stress in students are homesickness, low attendance, feeling left out and unsupportive family. The reaction to the stress was different from student to student like; started eating more, left the course, crying a lot, taking a therapy, etc. The first self- help books they read were either gifted to them or they found them in book store and bought because of catchy, finding it as bestseller name and in that case, it could or could not help. An uneasy feeling of asking for help if they choose self-help book is also there. Content is the most important part of the book; it should be practical and students like it to be in a story telling, giving tasks format to make it more interesting and help them in discovering themselves. Though it is difficult to get completely healed but some have taken it as a start, some have found peace and now see their life path better than before.

16.4.2 Findings from Survey

Table 16.1 depicts age, gender, student's current stay and their willingness to attend college. The chart (Table 16.2) was answered by only those who have "no" or "sometimes" in previous question. Above chart shows that 38% respondents miss their families. 34% feel lack of companionship. No respondent has been bullied.

Table 16.1 Age, gender, currently staying and willingness to attend college

Age	
16–20	67%
21–25	33%
Gender	
Female	67%
Male	28%
Other	00%
Prefer not to answer	05%
Respondents currently staying	
At home	15%
Hostel	53%
As paying guest	22%
Local guardian's place	10%
Willingness to attend college	
Like attending college	56%
Do not like attending college	24%
Sometimes like attending college	20%

Table 16.2 Reason for not attending college

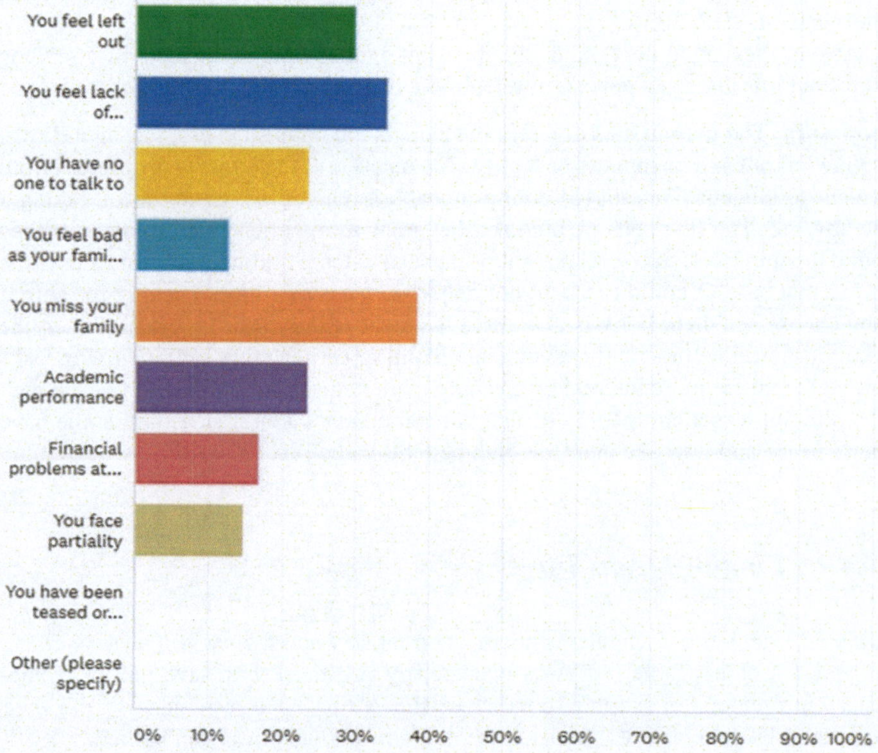

Table 16.3 Respondents on being teased

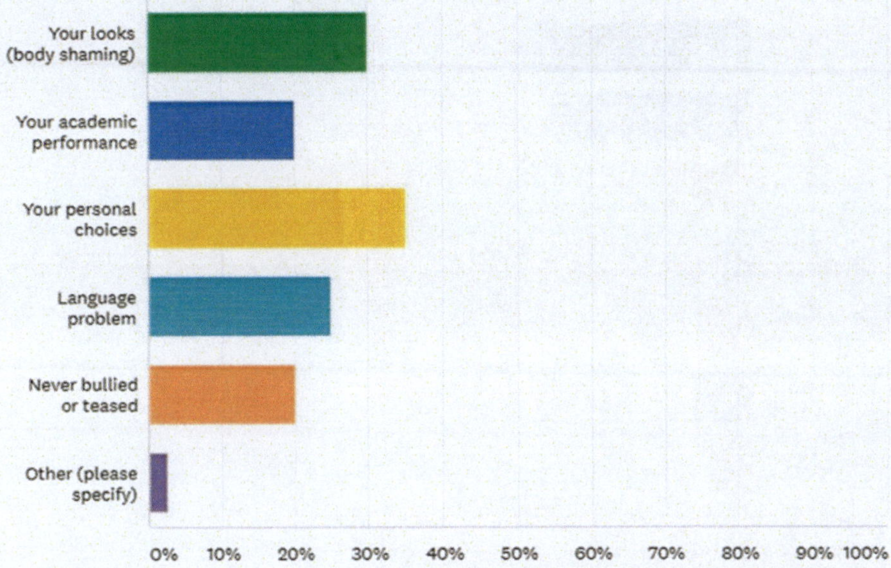

The findings from chart (Table 16.3) indicates 35% of respondents have been made fun of because of their personal choices, 30% have faced body shaming, 25% have been teased due to language problem, 20% for academic performance and 20% have never been teased or made fun of. One of the female respondents chose other option and wrote that her friends ask her in a fun way, "Why are you so flat?".

Chart (Table 16.4) shows what respondents do when they face above-mentioned issues. 30% of them skip college, 28% talk to friends, 12% to family, rest of them either eat or sleep, panic, follow a hobby. Charts (Tables 16.5, 16.6 and 16.7) depict psychologists visit in college, students who like to read books and students knowledge about self-help books.

Table 16.4 Reaction of respondents

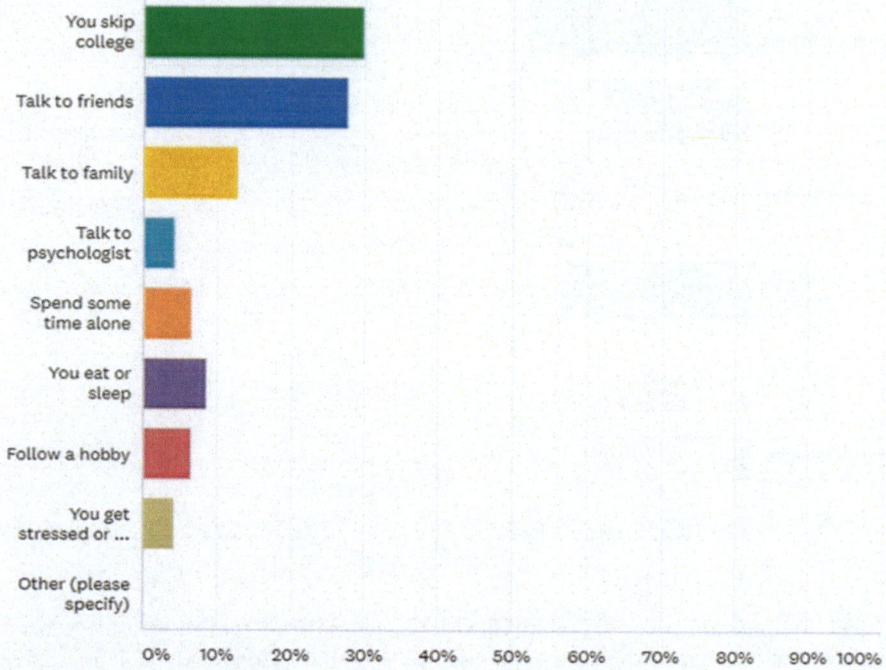

Table 16.5 Psychologist visit in college

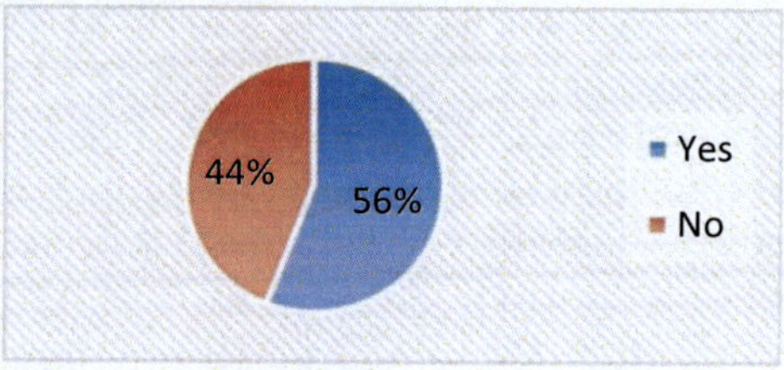

Table 16.6 Like to read books

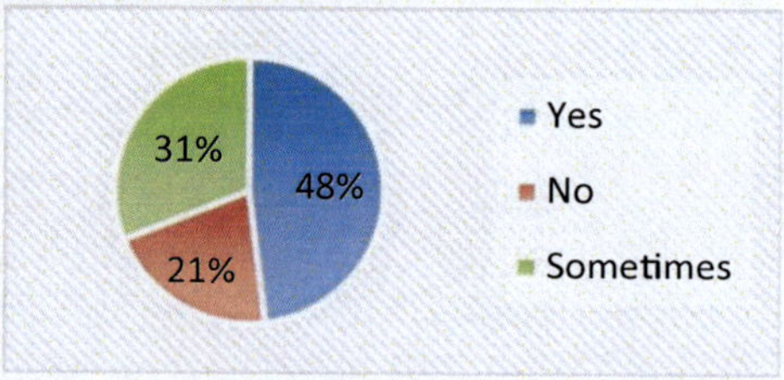

Table 16.7 Knowledge about self-help book

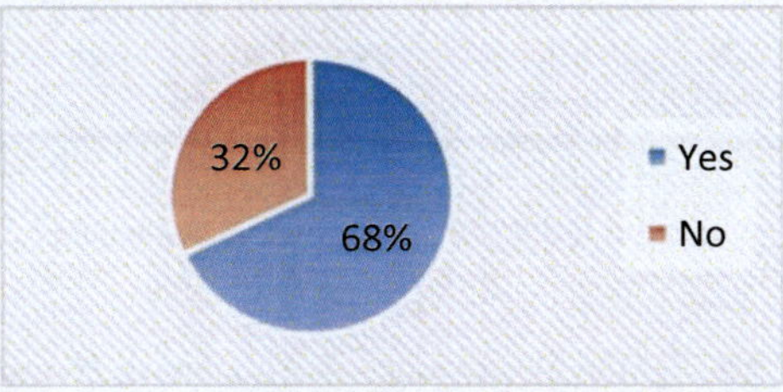

16.5 Conclusion

After primary and secondary data analysis, it can be concluded that for many students, it gets hard to adapt in new environment due to which they feel stressed and have anxiety issues. Self-help books can be used as a way to help them come out of it. Students are in their critical age and with the help of self-help books, they can avoid the fear of evaluation, judgment and consequent embarrassment. It need not to be in form of complete books and can also be in form of short stories or of two–three pages especially for the students who have less attention span because of young age or cognitive impairment. Self-help books are also useful as they offer a non-threatening forum. In addition, less commitment is required compared to traditional psychotherapy as it is certainly much easier to make a commitment to read a book than to enter therapy. The validity of the results obtained is dependent on the cognitive abilities of the individual and this may vary.

The selection of the books is critical thing as students can misdiagnose their own problems. It should be done with the guidance of professionals as they know the best and can select the books which could be useful for the students and can teach them how to deal their feelings. Students will also understand that these kind of experiences and feelings are not unique so that they will be motivated to come out of them.

References

1. Bergsma, A.: Do self-help books help? J. Happiness Stud. **9**, 341–360 (2007)
2. Birla, N.: A student commits suicide in India every hour; how can our educational system prevent this? The Economic Times (2018)
3. Cujipers: Effects of self-help for Unipolar Depression and Anxiety (1997).
4. Den Boer, P.C., Wiersma, D., van den Bosch, R.J.: Why is self-help neglected in the treatment of emotional disorders? A meta-analysis. Pub Med **34**(6):959–971
5. Ellis, A.: The rational-emotive therapy (RET) approach to marriage and family therapy. Family J. (1993)
6. Forest, J.J.: Self-help books. Am. Psychol. **43**(7), 599 (1988). https://doi.org/10.1037/0003-066X.43.7.599.a

7. Jasmine, O.: 5 Reasons to Read Self-Help Books (Even If You Think You Don't Need To) (n.d.). Retrieved May 24, 2020, from https://www.mindbodygreen.com/0-15734/5-reasons-to-read-selfhelp-books-even-if-you-think-you-dont-need-to.html
8. Liang, K.D.: Self-improvement books: a genre analysis. Victoria University of Wellington, Wellington (2015)
9. Mukhopadhyay, A.: When will India address its student suicide crisis? Fair Observer (2019)
10. Saha, D.: Every hour, one student commits suicide in India. Hindustan Times, New Delhi (2017)
11. Starker, S. (1989). Oracle at the Supermarket: The American Preoccupation with Self-Help Books. New Brunswick: Transaction, ©1989
12. Susan McKay, F.B.: Evaluating illness in women's magazines. J. Lang. Soc. Psychol. (2002)
13. Wild, M.: Preferences of self-help books and their presumed credibility in the college population. Murray State University, Murray (2018)

Chapter 17
Mapping the Research Thread of PhDs in Design: A PhD Citation Analysis of the Portuguese Doctorates

Nina Costa, Rui Costa, Afonso Borges, Vasco Branco, Raul Cunca, António Modesto, and Ana Catarina Silva

Abstract The present study undertaken within the DesignOBS project is based on 172 PhD theses in Design submitted to the Portuguese National Design Schools between 2005 and 2019. It focuses in particular on the extraction and analysis of 522 PhD citations appended to design doctoral work. The analysis is used to observe school impact, explore the weight of previous design focused and non-design doctoral work to develop PhD research in design in the country. The results reveal few connections between doctorates and few overlaps in-between as well as outside design schools; thus, indicating poor continuity and reproducibility of domestic doctoral work, little tradition of PhD citation, and an important weight of non-design schools. A network-based visualization of the connections between PhDs in design within PhD thesis, by use of a citation analysis method, enabled to draw reflections on the status of domestic doctoral research in Portugal and provides an empirical approach to explore the reproducibility of this type of research which may be used in other countries.

N. Costa (✉) · R. Costa · A. Borges · V. Branco
Aveiro University, Campus Universitário de Santiago, 3810-193 Aveiro, Portugal
e-mail: ndc@ua.pt; ninacostandc@gmail.com

R. Costa
e-mail: ruicosta@ua.pt

V. Branco
e-mail: vasco.branco@ua.pt

R. Cunca
Faculty of Fine Arts, Lisbon University, Largo da Academia Nacional de Belas Artes 4, 1249-058 Lisbon, Portugal

A. Modesto
Faculty of Fine Arts, Porto University, Av. Rodrigues de Freitas 265, 4049-021 Porto, Portugal

A. C. Silva
Polytecnique Institute of Cávado E Do Ave, Lugar Do Aldão, 4750-810 Barcelos, Portugal

A. Chakrabarti et al. (eds.), *Design for Tomorrow–Volume 2*, Smart Innovation, Systems and Technologies 222, https://doi.org/10.1007/978-981-16-0119-4_17

17.1 Introduction

Research at universities, regardless of the area of knowledge, is intimately linked to the existence of post-graduate programs [1]. These programs need to be anchored upon a strategic line of thought that provide defined objectives for research and development [2]. Moreover, a discipline is about theories and methods that accumulate through academic research and reflection [3]. Thus, it should establish its grounds in robust and systematic research efforts that expand its knowledge base. In the particular case of design research, this maturation process has evolved quite haphazardly [4]. Characterized as "internally scattered and confused", the nature and culture of design research is yet an open and evolving debate [4, 5]. Despite the diverse challenges of working within an integrative and inherently multidisciplinary discipline, finding a common corpus of knowledge is important [6, 7]. In the particular case of academic design research, written theses are the most commonly known ways of capturing the whole research process of the study in a single, endurable and searchable medium [8]. Referencing in doctoral—or any—work is central to the growing literature of design research as it helps to support arguments, access common evidence, build the field, and improve the intellectual and practical quality of the discipline [9].

Design education and research is one of the key objects of observation of the "DesignOBS: Towards a design observatory in Portugal" project [11]. It aims to collect, map, and interpret data about the Portuguese Design Ecosystem that may support the creation strategies and policies for its promotion and evolution [10, 11]. Previous studies in DesignOBS looked at the design doctorates undertaken in Portugal and identified a problem regarding their communicability, hampering the quick access to the content, and connections through the most common search cues, i.e., title, keywords, and abstracts [12]. Although useful to provide a first scan of doctoral work in the country, the analysis did not enable the establishment of connections between doctorates in a coherent and meaningful way.

Based on Durling's [8] argument and the recent criticisms of Margolin and Dorst, this study evolves the previous analysis by exploring the research thread of the PhDs undertaken in Portugal via citation analysis [13]. We focused in particular on exploring only the citations of PhD work (design and non-design focused). The in-degree and modularity scores—typical statistical methods used within network analysis, calculated within Gephi [14]—were included for the creation of the citation network to explore the weight of PhD work across design schools, and identify the connections amongst different communities. By using this method, we were able to gain important insight on some foundational aspects of doctoral research in design, namely its reproducibility and usefulness within academia. Based on these insights, we propose additional guidelines for future design research.

17.2 Background

17.2.1 PhDs in Design: The Portuguese Case

Portugal has six national schools that provide doctorate-level degree in design namely, Aveiro University, Faculty of Fine Arts of Porto and Lisbon Universities, Faculty of Architecture of Lisbon, European University, and Beira Interior University. The first thesis in design produced in the country was in 2001 [12]; since early 2010, the number of PhDs has been growing steadily, with new PhD programs being developed (e.g., Fashion Design in University of Beira Interior). Previous work developed within DesignOBS [10–12] focused on making a numerical analysis of design doctorates undertaken in Portugal based on title, key words, and abstracts to understand the scope of design research in the country [12]. The same study concluded, however, that some of the body of knowledge categorized in this area is not always design-focused, but rather more aligned with fine arts, architecture, or sociology. Moreover, some of these doctorates do not mention design in either title, key words, or abstracts [12].

This first effort aimed to depict the PhD design research landscape in the country which was useful to partially corroborate Dorst [4] view of design research via empirical data. However, additional research is needed to trace, visualize, and interpret PhD research in design via empirical data. To address this challenge, this paper combines network analysis with bibliometric studies.

17.2.2 Networks and Citation Analysis

A network can be defined as a system of interrelated nodes and edges. In the context of bibliometrics, nodes can be constituted by journals, authors, or other, while edges are the relationship between the nodes. There are mainly two types of relationship: author co-citation, that may reveal the intellectual structure of a field, or collaboration network (or co-authorship) that show the social network of a research field [15]. Networks can be used to identify the intellectual structure of a discipline; the most influential authors, articles or other objects/subjects; explore the influence of specific journals, the core canons within a discipline, consensus and disparities; the flow and knowledge transfer between researchers and institutions and progression of thoughts [16, 17]. The exploration of the properties of a network can be used to extract insights based on empirical data, and drive decision making.

Citation analysis is a specific method within bibliometric studies that can be used to create networks amongst researchers that cite the same article, book, or other document/objects. Each citation originates an edge/line between the source (e.g., author of the PhD work) and the "target" (e.g., authors/co-authors of the items in the bibliography of the PhD); thus, establishing a connection between nodes/objects/subjects. The links created in co-authorship and co-citation networks are different because of the nature of the network. Whereas the first is characterized as an undirected network because the relationship of co-authors is mutual (they have all participated in the creation of a specific object); the latter is characterized as a directed network since author A can cite another author B, but author B may not cite author A back. In both networks, the link established between nodes is as strong as the number of times an author cites/co-authors a work with another author. In directed networks, the in-degree (number of incoming links) increases whenever an author is cited by numerous other authors.

Studies in multiple disciplines have used academic journals to create citation networks of specific subjects or knowledge areas (e.g., marketing studies, [16]; sociology studies, [18]; Technology and other, Lewis [17]). In the design area, these methods are also starting to emerge with datasets based on journals (e.g., [15]). Although the number of academic journals in design research has expanded considerably in the last decades [19, 3], citation analysis made through those means can be partial because the existing indexed design journals in the Thomson Reuters Science Citation Index, Thomson Reuters Social Science Citation Index or in Scopus is yet limited [19].

This study aims to map PhD research in design adopting a PhD citation analysis approach. It uses Portuguese PhD research as a case of observation to explore the impact and reproducibility of PhD research in design.

17.3 Methodology

To map the trace and visualize the PhD research in design in Portugal, the PhD citations network was extracted in doctoral work undertaken in Portugal. The collection and selection of PhDs were based on a previous study developed by Costa et al. [12, 20] which already identify the design schools that provide the degree in the country, namely Aveiro university (UA), Faculty of Fine Arts of Porto University (FBAUP), Faculty of Fine Arts of Lisbon University (FBAUL); Faculty of Architecture of Lisbon (FAUL), Faculty of Architecture of Lisbon Technical University (UTL), European University (UEU), Creative University (IADE), and—more recently—Beira Interior University (UBI). The authors undertook a preliminary scan of the content of the doctorates collected, distinguishing between those that were more aligned with sociology, arts, history or others, from design. The present analysis takes into account the method used in Costa et al. [12] and considers design-focused PhDs as source material. PhDs which were more aligned with arts or other were not included as source material (see [12] for more details). National databases were consulted to check if any other PhDs were undertaken until the end of the PhD.

A total of 172 PhDs were included in the sample, going from 2005 to late 2019. Some doctorates were completed between 2001 and 2005 but were not considered in this study as they were classified as non-design focused by Costa et al. (2020a). Figures 17.1 and 17.2 provide the overall characterization of the source. Several schools have changed or been combined into others, i.e., UTL became FAUL in 2012;

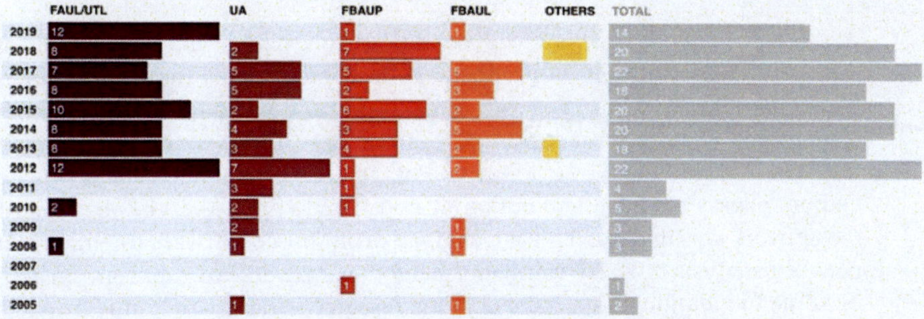

Fig. 17.1 Characterization of the sample per year and school (172 PhDsource)

Fig. 17.2 Percentage of PhDsource per school

IADE and UEU were also combined. Our analysis, thus, identifies these schools as one (FAUL/UTL; UEU/IADE). In addition to the database provided by Costa et al. [20], the analysis was complemented with the latest information updates in the national governmental database RENATES until December 2019. The school, author, year of completion, and supervisor of the source (which we will name further on as "PhDsource") were extracted to characterize the sample.

Two of the schools (UBI and IADE/UEU) were combined in the "other" category since they represent a small portion of the overall source (~2%). Most of the PhDsource is from FAUL/UTL (44%). The year 2019 also mainly presents input from this school. University repositories were consulted to check if any other design thesis was completed until the end of the year, but no other documents were found.

17.3.1 PhD Citation Extraction and Analysis

The author, year, country/school of the PhDs cited in the bibliography of the source dataset were extracted.

(1) To explore the weight/influence of Portuguese doctorates, we distinguish the cited PhDs in two categories: "undertaken in Portugal" or "undertaken abroad". Portuguese authors which undertook their PhDs abroad were categorized as "undertaken abroad";

(2) To understand the influence of Portuguese schools within the doctoral system—and thus, the roots/ground of the domestic design PhD research—we also extracted the name of all the Portuguese schools from which the PhDs cited were undertaken (design school or other);

(3) A citation network analysis was conducted to examine the distribution of domestic PhDs focusing in particular in in-degree scores (number of times a PhD by other PhDs). In network analysis, the nodes and links are important to map and explore the connections. In our case, these connections constitute the overall domain of intellectual PhD references that domestic doctorate authors express. After the curation of nodes and edges, the visualization of the network was undertaken via Gephi: an open source software for graph and network analysis [14]. By using this method, we are able to gain some insight on some foundational aspects of doctoral research in design in Portugal and its reproducibility in the country.

17.4 Results

We counted 522 PhD citations appended in 172 doctoral theses in design submitted to the Portuguese National Design Schools which provide the degree (i.e., Aveiro University, Faculty of Fine Arts of Porto and Lisbon Universities, Faculty of Architecture of Lisbon, European University and Beira Interior University) between the period of 2005–2019. As seen in Fig. 17.3, the distribution of citations is proportional to the number of PhD theses considered as the source (PhDsource), with the exception of FAUL/UTL which presents a slightly inferior number of citations (−10%) in comparison to the number of PhDsource.

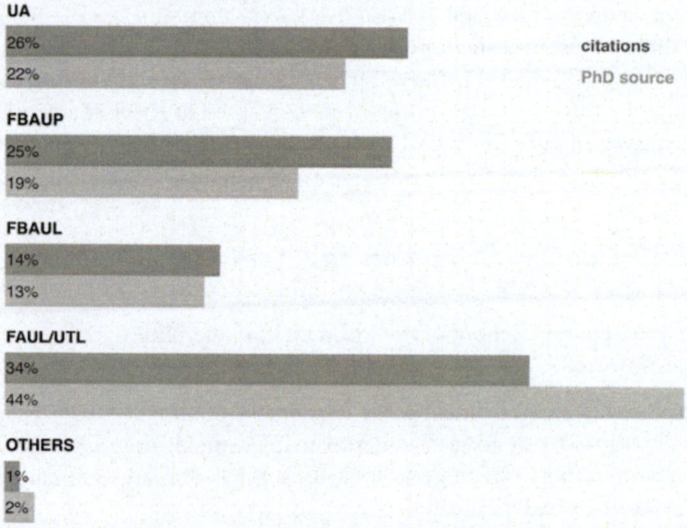

Fig. 17.3 Percentage of PhD citations versus percentage of PhDsource per design school

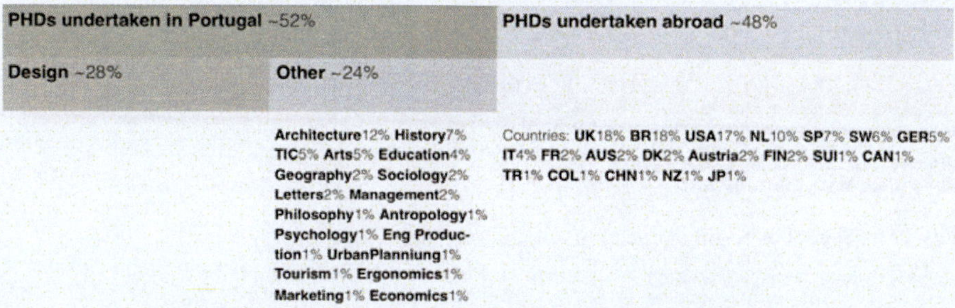

Fig. 17.4 Cited PhDs undertaken abroad and cited PhDs undertaken in Portugal

A total of 350 PhDs (or single authors) constitutes the universe of the doctorates cited, from which about half is undertaken in Portugal (~52%). The next countries with a higher citation index are the United Kingdom and Brazil (18%), the USA (17%) and the Netherlands (10%). Spain, Sweden, and Germany constitute 20% of the cited PhDs followed by Italy (4%). Other 13 countries are mentioned in the cited PhDs (Fig. 17.4). European countries constitute approximately 27% of the PhDs cited (besides Portugal). Design theses constitute at least 28% of all the citations and the remaining PhDs undertaken in Portugal are mostly related to architecture (12%), history (7%), communication and information technologies (5%), arts (5%), and education (4%). Other PhDs are focused on subjects such as marketing, economics, ergonomics, tourism, psychology, engineering production, letters, and urban planning (Fig. 17.4).

Results show that 32% (117) of the PhDsource do not cite any doctoral thesis in the bibliography; a quarter of the sample cites one to two PhDs; 19% cites 2–4 doctorates, 13% mention 5–6; and 11% cites 7 or more doctorates (Fig. 17.5). Only one thesis cites 23 PhD documents.

When looking at the distribution of PhDs cited "undertaken in Portugal", an important percentage of the results points at multiple universities and polytechnic institutes mostly concentrated in the city of Lisbon ("other PT"). UL (Lisbon University), UTL (Technical University of Lisbon), and UNL (New University of Lisbon) are also represented in "others (PT)" since the PhDs were not undertaken in the Fine Arts nor the Faculty of Architecture (FBAUL and FAUL/UTL). Most of "other PT" doctorates are from Lisbon University. Minho University and Coimbra University

Fig. 17.5 PhDs cited in bibliography

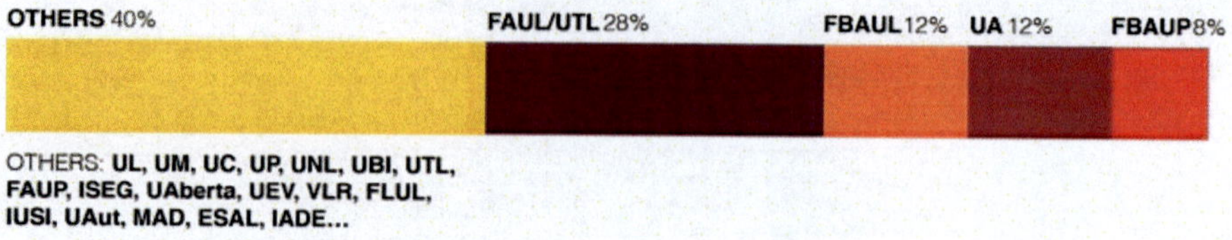

Fig. 17.6 School of origin of cited PhDs undertaken in Portugal (n182)

Table 17.1 PhD citations per school of PT schools including self-citations (total citations = 278)

Mentions>	UA (%)	FAUL/UTL (%)	FBAUL (%)	FBAUP (%)	OtherPTschool (%)
UA	34	13	19	4	30
FAUL/UTL	8	40	15	3	33
FBAUL	17	10	57	0	17
FBAUP	24	9	11	30	26

Table 17.2 PhD citations per school of PT schools excluding self-citations (total citations = 171)

Mentions>	UA (%)	FAUL/UTL (%)	FBAUL (%)	FBAUP (%)	OtherPTschool (%)
UA	0	19	29	6	45
FAUL/UTL	14	0	25	6	55
FBAUL	38	23	0	0	38
FBAUP	34	13	16	0	37

do also have an important weight (6 and 3%). The remaining schools (score: ~1%) are mentioned in one thesis (Fig. 17.6).

We analyzed the percentage of corresponding design-focused PhD cited "undertaken in Portugal" by schools to understand their global impact in doctoral domestic research. Table 17.1 shows domestic citation per school (source) of PT schools (including self-citations). Table 17.2 shows domestic citation excluding self-citations. UEU/IADE is not represented in these tables since the PhDsource do not cite PhDs undertaken in Portugal. UBI was also not represented since only one PhD in design was identified, and thus it is not representative.

The analysis shows that on average and considering only domestic research, the design schools mostly cite themselves, especially FAUL/UTL and FBAUL which account for more than 40% of self-citations. When looking at the same patterns, excluding self-citations (Table 17.2), the doctoral research impact changes abruptly indicating that most cited doctoral work is made in other non-design universities. FBAUL and UBI are an exception. FBAUL in particular shows a stronger connection to both UA and other non-design schools. When looking at citations of doctorates undertaken abroad, the main influencing references are European for all the schools (Table 17.3), especially the United Kingdom (Table 17.4).

Table 17.3 PhD citations per school of foreign schools/continent (total citations = 204)

Mentions>	North America (%)	South America (%)	Europe (%)	Asia (%)	Australia (%)
UA	24	5	71	0	0
FAUL/UTL	15	20	45	0	20
FBAUL	18	9	73	0	0
FBAUP	4	7	78	4	7

17.4.1 Network of the PhD Research in Design in Portugal

The map in Fig. 17.7 shows the connections established via cited PhDs in the bibliography of PhDsource. The nodes represent both cited authors as well as PhDsources. Directed links are drawn when an author/source mentions another PhD thesis. The size of the node is related to their in-degree value, that is, the number of incoming links or number of citations in PhDsource. The bigger nodes have received at least two citations from different sources in the dataset; thus, indicating some degree of semantic connection [15]. Figures 17.7a, b are colored according to the school in which the doctorate was undertaken. The layouts were created with Force Atlas.

The analysis shows that an important part of the PhDsource and PhD citations are isolated as most of citations are only mentioned once, in one thesis (Fig. 17.7b, one node, one/two links). Thus, few overlaps between the sources can be established via PhD citation. When analyzing the overlaps per design school, we conclude that the results are similar. There are only a few cases of cross-pollinating PhD works that connect the PhDsource (overlaps in-degree ≥2, Fig. 17.7b). Table 17.5 indicates that the only school with some degree of cross-citations is FAUL/UTL. The case of FBAUL is also curious: despite the results presented in Table 17.1 (self-citation ~ 55%), the overlaps of the PhD citations are only a few (~2% in Table 17.5).

Second, there are numerous small clusters of one node pointing to other nodes. This indicates that a source mentions other PhDs (one or more) but is not continued nor connected to other doctorates via PhD-based bibliography. Since this is a directed network (authors mention other authors, but may not be mentioned back), this data shows that many research topics are not continued, evolved, nor connected to existing PhD research. When applying the in-degree filter of "more than 1 link" (mentioned in/mentioning more than one other node), only 56 nodes of the entire universe of PhDs (~11%) remain and 15 links remain visible (~4%). In Fig. 17.7b, we can visually identify the cross-pollinating nodes. From this network, most of the authors are still connected with a lot of isolated nodes with few links. The nodes with higher in-degree are mainly from the two national design schools. The rest of the map is abundant with "other PT" schools.

Table 17.4 PhD citations per school of European schools (total citations = 115)

Mentions>	SP (%)	FIN (5)	UK (%)	SW (%)	IT (%)	NL (%)	SUI (%)	GER (%)	DK (%)	AST (%)	TRK (%)
UA	11	0	33	22	11	15	4	0	0	4	0
FAUL/UTL	9	0	47	5	5	18	0	13	0	2%	0
FBAUL	11	0	44	0	0	22	11	0	0	11	0
FBAUP	13	0	25	17	0	13	0	13	17	0	4

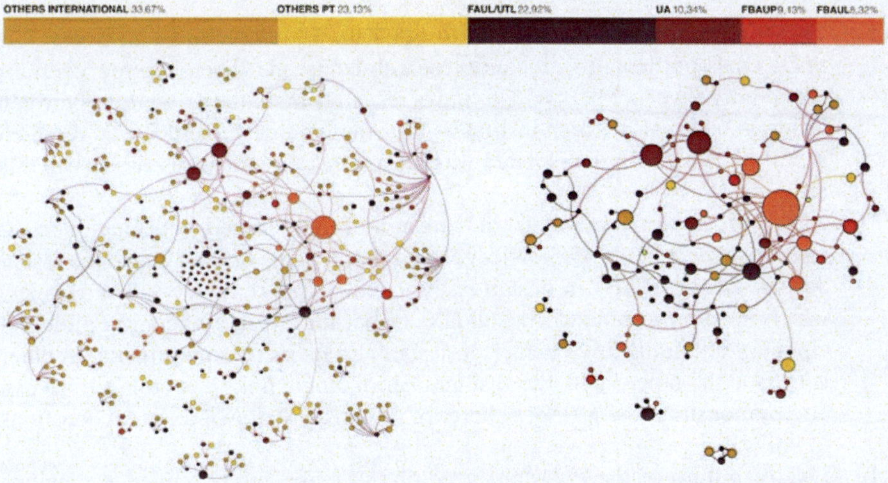

Fig. 17.7 **a** PhD cited in PhDsource (color according to school), **b** PhD cited in PhD source with degree ≥ 2 (with at least two links)

Table 17.5 PhD citations overlaps per school

School	# PhDsource	# nodes	# edges/links	Overlaps (in-degree ≥2)
FAUL/UTL	76	257	246	25 (33%)
UA	37	95	84	7 (5%)
FBAUP	32	94	84	9 (4%)
FBAUL	23	49	43	5 (2%)

17.5 Conclusions and Future Research

This study aims to trace, visualize, and interpret PhD research in design adopting a PhD citation analysis of domestic academic research. It applies a network-based approach in an empirical case to explore the common core connections between PhDs in design. From this process and delimitation of our case, we observe several aspects, namely:

(1) PhD theses constitute a very small portion of the bibliographic work, with approximately 2% of representation in the entire dataset. PhDs in design constitute, more specifically, about half of this percentage of the source dataset. So far, we found no studies to compare this PhD percentage with other countries or other disciplines that could help us assess if it is common case. PhDs in design being such a small part of the bibliography of PhD research in design is also quite indicative that this type of output may not be relevant in academic design research—which is quite contradictory.

(2) The connections are not evident through the PhD citations since very few PhDs cross-pollinate the doctoral research landscape. There are few overlaps observed between PhDs of the same research institute as well as between different national research institutes. This may indicate that doctoral research is still undertaken in an isolated way, despite having more researchers in this area.
(3) An important part of the cited PhDs are non-design focused, with citations scattered in diverse fields of knowledge and schools. This may be due to the recent development of PhDs in design in the country (approximately two decades) which need more time to consolidate, or because of the inherently multidisciplinary and integrative nature of design and its increasing infusion in other areas. On the other hand, the vast number of single links and connections can also demonstrate the lack of coherence or narrowly drawn research questions in PhDs.
(4) Finally, a third of the PhDs analyzed do not make any reference to previous PhDs (national or foreign). This observation is critical and raises important questions regarding the usefulness of PhD work—even within academia. Are there truly no connections which may be established with previous PhD work? Which type(s) of output(s)/production(s) are mostly used to support the development of a PhD in design? Moreover, given that Portugal has two decades of PhD research in the area—should aspiring design PhD students have access to a more structured research training? Do current research methods used and imported from other disciplines produce interesting and useful results for those who practice and investigate Design? Existing studies point at different directions when it comes to undertaking academic research at the PhD level (e.g., [21–25]). The questions raised, however, fall out of the scope of this research and require further inquiry. We encourage to further explore this issue in other countries, for comparison purposes.

This study has also some limitations which indicate directions for future research: First, (1) the paper only takes into account the PhD citations of PhD theses undertaken in Portugal. A more holistic view of the domestic doctorate system should account for all the research production cited by Portuguese authors to further understand the existing domestic PhD research system in design, including books, chapters, conference proceedings, or other. Second, (2) a more in-depth analysis of the cited research production of foreign doctorates could be useful to trace the core/origins and fundamental ideas of domestic doctoral research. Third, (3) although this is not the aim of this paper, the analysis of the collaboration networks is a core part of the ecosystem. Connecting the dots via doctorate citations—both inter and transnationally—could provide an overall map of the research collaborations between Portugal and the rest of the world; thus, providing a more robust and comprehensive view of the status quo of PhD design research in the country.

Doctoral research in Portugal is still evolving. This study, undertaken within DesignOBS project, contributes to advanced design research by presenting an application of a network-based observation approach to map the doctorate landscape, based on empirical data, and draw some reflections regarding the status quo of domestic PhDs in design. At the international level, it provides important insights and raises questions regarding the traceability and usefulness of doctoral work within academia.

Acknowledgments This article is a result of the project Design Obs. Para um Observatório de Design em Portugal: Modelos, Instrumentos, Representação e Estratégias, (Towards a design Observatory in Portugal: models, instruments representation and strategies) supported by Lisbon Regional Operational Programme (LISBOA 2020) and the Competitiveness and Internationalisation Operational Programme (POCI-01-0145-FEDER-032445), under the PORTUGAL 2020 Partnership Agreement, through the European Regional Development Fund (ERDF) and FCT – Fundação para a Ciência e a Tecnologia (Foundation for Science and Technology).

References

1. Durling, D., Friedman, K.: Foundation for the future of doctoral education in design. La Clusaz (8–12 July), France (2000)
2. Branco, V., Branco, J., Aguiar, C., & Providência, F.: Universities and design research. In Durling, D., Friedman, K. (eds.) La Clusaz—Foundations for the future of doctoral design education, pp. 345–351. France (2000)
3. Gemser, G., de Bont, C.: Design-related and design-focused research: a study of publication patterns in design journals. She Ji **2**(1), 46–58 (2016). https://doi.org/10.1016/j.sheji.2016.05.002
4. Dorst, K.: Design practice and design research: finally together? In: DRS2016: Future-Focused Thinking, vol. 7, pp. 1–10 (2016). https://doi.org/10.21606/drs.2016.212
5. Margolin, V. (2000). Building a design research community. In: Pizzocaro, S., Arruda, A., De Moraes, D. (eds.) Design Plus Research: Proceedings of the Politecnico di Milano Conference, pp. 1–7. Retrieved from https://design.osu.edu/carlson/id785/designcommunity.pdf
6. Buchanan, R.: Design research and the new learning. Design Issues **17**(4), 3–23 (2001)
7. Margolin, V.: Design research: what is it? What is it for? In: DRS2016: Future-Focused Thinking, vol. 1, pp. 1–11. https://doi.org/10.21606/drs.2016.9(2016)
8. Durling, D.: Discourses on research and the PhD in design. Qual. Assur. Educ. **10**(2), 79–85 (2002). https://doi.org/10.1108/09684880210423564
9. Friedman, K.: Reference, argument, and evidence. how good referencing and citation serve design research and professional design practice. Centre for Design Innovation, Swinburne University of Technology, Melbourne (Revised Reprint) (2015).
10. Borges, A., Silva, A.C., Modesto, A., Cunca, R., Costa, R.C., & Branco, V.: REDE#01: Reunião de Escolas de Design (2017). Retrieved from https://hdl.handle.net/10773/24199
11. Branco, V., Costa, R., Borges, A., Silva, A.C., Modesto, A., Cunca, R., & Costa, N. (2019). Towards a Portuguese design observatory: models, instruments, representation and strategies. Retrieved 7 Nov 2019, from https://designobs.pt/
12. Costa, N., Branco, V., Costa, R., Borges, A., Modesto, A., Silva, C., Cunca, R.: Towards a design observatory: the case of scholarly design research in Portugal. In: Proceedings of the Design Society: DESIGN Conference, vol. 1, pp. 827–836 (2020) https://doi.org/https://doi.org/10.1017/dsd.2020.327
13. Van Dalen, H.P., Henkens, K.: What makes a scientific article influential? The case of demographers. Scientometrics **50**(3), 455–482 (2001). https://doi.org/10.1023/A:1010510831718

14. Bastian, M., Heymann, S., Jacomy, M.: Gephi: An open source software for exploring and manipulating networks. In: BT—International AAAI Conference on Weblogs and Social. International AAAI Conference on Weblogs and Social Media, pp. 361–362 (2009)
15. Perna, S.: Design ResearchScape. a visual exploration of design research publications. Design J. **20**, S952–S963 (2017). https://doi.org/10.1080/14606925.2017.1353040
16. Baumgartner, H., Pieters, R.: The influence of marketing journals: a citation analysis of the discipline and its sub-areas. Discussion Paper, Dec 2014 (2000)
17. Lewis, J.: How transdisciplinary is design? An analysis using citation networks. Design Issues **36**(1), 30–44 (2020). https://doi.org/10.1162/desi_a_00573
18. Zafrunnisha, N.: Citations in the sociology doctoral dissertations: a quantitative analysis. Int. J. Inf. Disseminat. Technol. **2**(3), 212–218 (2012)
19. Gemser, G., De Bont, C., Hekkert, P., Friedman, K.: Quality perceptions of design journals: The design scholars' perspective. Des. Stud. **33**(1), 4–23 (2012). https://doi.org/10.1016/j.destud.2011.09.001
20. Costa, N., Branco, V., Costa, R., Borges, A., Cunca, R., Modesto, A., Silva, A.C.: Design doctorates undertaken in Portugal (database) (2020). https://doi.org/10.5281/zenodo.3701614.
21. Hilton, K.: The future of doctorates in design. In: DS 43: Proceedings of E and PDE 2007, the 9th International Conference on Engineering and Product Design Education, (September), pp. 197–202 (2007)
22. Melles, G.: Global perspectives on structured research training in doctorates of design–what do we value? Des. Stud. **30**(3), 255–271 (2009)
23. Melles, G., Wölfel, C.: Postgraduate design education in Germany: motivations, understandings and experiences of graduates and enrolled students in master's and doctoral programmes. Design J. **17**(1), 115–135 (2014). https://doi.org/10.2752/175630614X13787503070079
24. Vaughan, L., Morrison, A.: Unpacking models, approaches and materialisations of the design PhD. Stud. Mater. Think. 11 (2014)
25. Vaughan, L., Morrison, A.: Form, fit and flair: considering the design doctorate. In: Complexity and diversity on the design PhD (May), pp. 14–17 (2013)
26. Curado, R.: Centro Português de Design vai ser extinto. 1–6. Retrieved from https://www.dn.pt/artes/interior/centro-portugues-de-design-vai-ser-extinto-3248523.html (2013)
27. Margolin, V.: Doctoral education in design: problems and prospects. Design Issues **26**(3), 70–78 (2010). https://doi.org/10.1162/DESI_a_00031

Chapter 18
Pedagogic Influences of Art and Design Schools on Architecture Education in India

Harshitha G. Raju

Abstract This study investigates the pedagogic relevance of art and design schools that have had major influences on schools of architecture in India. The data presented is a culmination of literature review of published articles and papers and analysis of curriculum. The curriculum of top ten architecture schools as ranked by NIRF in 2020 are considered here as representative of what is largely being taught in schools of India. The conclusion of the study establishes the prevalent influences of pedagogy of art and design schools, subjects that have gained importance and vice versa, and their relevance to the future.

18.1 Introduction

Arts, crafts, and design have been allied fields to architecture. The ideologies and thoughts in each of these have surpassed boundaries through varied forms of expression creating paradigm shifts within every movement and era. The pedagogic methods followed in art and design schools have been formative grounds of creativity and inspiration in architecture domain. Rather, innovations and techniques borrowed from mathematics, science, and social studies have found place in curriculums of architecture schools. This paper investigates the pedagogic influences of different 'schools of thoughts' both local and overseas, on formal architecture education in India. The aim of the study is to provide a comprehensive literature on possible linkages of 'design subjects' taught in architecture schools with other established institutions of arts and design. The objectives are (1) to identify and elaborate pedagogic content followed in different art and design schools, and (2) establish their relation with curriculum and design methodology followed in renowned schools of architecture in India. The methodology of data collection is based on secondary sources such as published articles in renowned journals, websites, research papers, conference proceedings, and books. The analysis is by correlating the different 'models of art and design education' with that of pedagogic methods followed in architecture

H. G. Raju (✉)
NID, Ahmedabad, India
e-mail: Harshitha_r@nid.edu

A. Chakrabarti et al. (eds.), *Design for Tomorrow–Volume 2*, Smart Innovation, Systems and Technologies 222, https://doi.org/10.1007/978-981-16-0119-4_18

schools that were established in early and mid-twentieth century. The study does not include content stipulated by council or governing body and details of credit system.

18.2 Art and Design Schools

The art and design schools listed here are based on its acquaintance with architecture schools in the methods and subjects followed in schools in present day. The description of 'schools of thought' with an understanding of course of their establishment in Indian context helps in understanding the correlations and influences on pedagogy.

18.2.1 South Kensington System

The art and design education in South Asia was established in the latter half of nineteenth century—i.e., in Madras-1850, in Calcutta-1854, in Bombay-1857, in Lahore-1875, and in Jaipur-1884 [7]. In Bombay, Sir Jamshetjee Jeejeebhoy who was a Parsi-Indian merchant with interest in arts and crafts, opened Sir J. J. School of Art. His initial idea to set up School of Art came after he attended the Great Exhibition held at London, in 1851 [13]. As said by Mitter [20], he envisioned an institution 'for the improvement of arts and manufacturers of the middle and lower classes.' In addition, India being one of the colonies of British during that time with their appreciation toward Indian art-wares saw education as one of the several means to reinforce and endure their interests [1]. Both these aspects, preserving India's crafts and to meet the growing needs of manufactured goods of British market, resulted in formalizing 'South Kensington system' as a method of education [13]. The method supported an industrially based system for teaching 'drawing' that worked as a tool to imbibe social discipline [Ibid].

In the following years within the arts department, the Sir JJ School of Architecture, started in 1917, marked the beginning of formal architectural education in India [20]. When the course of architecture took off, the principal objective was essentially to equip students with technical knowledge and basic drafting skills, such that students later in their career would render skilled assistance to foreign or native employers [28]. Architect John Begg, consultant to government of Bombay who started classes, trained pupils a technical draftsmanship course focusing on drafting and reading drawings. It can be noted that the technicality required for architecture course has relevance with Cole's South Kensington system that reinforced correct observation, accurate expression, and monochrome drawings. The drawings created were mechanical in nature, tedious, and involved reproduction of drawings such as machines, screws, and other details [15].

18.2.2 Ecole Beaux Arts

While the pedagogy at JJ School of Architecture saw several faces of development and underwent revival with time and people who administered it, Westerners mostly influenced the curriculum. Among them, the first appointed Professor and the Head of the Department was Robert W. Cable who was trained at Ecole Beaux Arts.

Historically, Ecole des Beaux Arts was started in 1819 during the period of French revolution, and it had courses on painting, sculpture, music along with architecture. The architecture course taught there was integrated with other artistic fields, rather than to construction and engineering [8]. As mentioned by Cunningham [4], the pedagogies developed in several arts, architecture, and design schools in present day have taken cues in relation to structure, contents, and methods from the Ecole des Beaux Arts. To elaborate, the characteristic methods of teaching at Ecole were:

- To consider drawing and painting as skills required for representing and defending one's projects.
- Cartesian pedagogical program which stressed understanding of buildings as solid objects that could be analyzed graphically as plan, section, and elevation (Hewitt, 1985)
- Systematic resolution of design problems with sketch designs called as *esquisse* [4] with time and importance given mostly in making plans followed with sections and elevations.
- *Ateliers* (design studios) that trained pupils for designs, with character and theory developed for the century ahead future generations.
- Visit to significant buildings guided by tutors to observe; followed by exercises where one would criticize and suggest corrections.
- In addition, inspection of building construction under the guidance of practicing architect for a year to obtain the degree.

While, not all these were implemented at JJ School of Architecture, the dean was able to shift the course importance to drawing, composition, history [29] ornamentation, practicality, and architectural detailing [Ibid] from the previously mundane Kensington system. The subjects thought during the first years included materials, geometry, applied mechanics, construction, specifications, surveying, and leveling [25]. History included introduction to Greek architecture, and there were courses on town planning and zoning and brief understanding of professional practice [28].

18.2.3 Santiniketan

The British in India established and governed the art schools in an Imperial manner and thereby imposed Western techniques producing 'copyists' and ruining the local talent and creativity [26]. On the contrary, Santiniketan was started in 1901, by the Nobel laureate Rabindranath Tagore as an experimental university for arts, painting, music, crafts, and design [Ibid]. The pedagogy at the arts department was unique and rooted in Indian rural socio-cultural aspects. The studios were conducted in open air, emphasizing one to live in harmony with nature. As envisioned by the poet, 'life and learning were to be pursued through a direct engagement with mental and mundane realities' that required empathetic recognition toward others [24]. The study of 'rhythms of everyday life, traditions, and natural phenomena' took form as art of various expressions [26].

The program made students not only aware and conscious of local surroundings but also globally sensitive, as it was open for cultures of West and East. The other programs involved 'community activities' such as architectural projects that acted as contribution to development of public spaces like Kalo Bari [5]. The curriculum was also designed to incorporate events and festivals like Vrikshopan (tree planting festival) which made them collaborate and participate with the community [Ibid]. The pedagogical methods of Santiniketan devoid of influences of industrialization and mass production, with importance given to crafts and reverence to learnings from rural culture, and incorporating Gandhiji's 'learning by doing' as a principle of education have been incorporated in curriculums of several architecture and design schools such as CEPT, NID, and SPA.

18.2.4 Bauhaus

In 1919, Bauhaus school was created by uniting 'Fine art' and 'Art and Crafts' school [10] and lead by Walter Gropius with a vision of 'the fundamental unity underlying all branches of design' [9]. Minahan [19] claims that Bauhaus school was influenced by not alone the English arts and crafts movement, but also Dutch De Stijl group, Architect Le Corbusier, Dada movement of Switzerland, Russian revolution of 1917 along with World War-I and Industrial revolution. During Post-World War-I, while industrialization and mass production were seen as methods of future development, Bauhaus adhered to 'arts and crafts' as language of design and in its pedagogy of teaching architecture design [4]. It was the first to create a curriculum for formal design education [Ibid] and the reason behind developing 'modern design' [27] present today.

As described by Cunningham [4], the pedagogy at Bauhaus included creative intuitive process that involved integration of both 'thinking and handling,' i.e., 'mental and practical' understanding of materials for design explorations. The materials explored included textile-looms, clay-ceramic, metal, stone, glass, pigments, and the likewise for which experts 'masters' from the field would also conduct workshops to students. Among the masters who taught at Bauhaus, Albers lectured on 'optical illusion and its relation to form and material appearance,' Klee was interested in 'ways of Nature study,' Kandinsky explored on 'colors and fundamental elements of form,' Maholy-Nagy shared on 'modern typography,' and 'furniture design' was dealt by Breuer [Ibid].

The school advocated 'functionality' as a principle of design, where the role of design was to 'temper and direct,' which worked as 'antidote to over-mechanization' as stated by Gropius and thereby addressed the shortcomings of industrial world [11]. It was the first school to incorporate classes on mass housing, city planning, district planning, and environmental relationships with a 'rational' understanding as opposed to Beaux Arts classical formulation [4]. Whereas, Beaux Arts' 'Decorative Composition Study' that valued intuition was adapted in architectural education [*Ibid*]. Standardization was an important aspect designed at Bauhaus to suit the industry requirements and measures were taken to improvise production and sales of products designed [23].

The school functioned in three different cities—Weimer, Dessau, and Berlin for a brief period of 14 years, but due to lack of internal legitimacy in the organizational structure of Bauhaus [19] and the growing powers of Nazi regime, it lead to the closure of school in 1933.

18.2.5 Hochschule für Gestaltung, Ulm

In Germany, about eight years after World war-II, in 1953 Hochschule für Gestaltung, Ulm was founded to commemorate the Scholl siblings who were executed for their resistance movement against Nazi party [23]. The school was founded by Inge Scholl, with Max Bill as the first dean who had his education from Bauhaus.

The school had specialized departments of industrial design, visual communication, film-making, and industrialized building which students choose after finishing one year of foundation program [23]. The school grew progressively as it was concerned with the curriculum and revised it annually where tutors, student representatives, and technicians were involved in the discussions [12]. Thereby, design methodology was dealt in-depth to incorporate scientific and technological knowledge and identify strong grounding for design as a core domain [2].

The philosophy of the school was different from Bauhaus as it did not strive to integrate 'art and technology' but rather drew a line between art and Industrial production and establish design as an independent field [2]. The notion was that industrial production was not to be seen as a subsidiary resultant of art or other cultural influences [Ibid]. Rather, with the functioning over years, the school emphasized the importance of function and market needs and ignored aesthetics [23].

In addition, this thought led to revision of curriculum, what came to be called as Ulm model in which theory based on scientific innovations and practical training from industry was provided to pupils. As described by Leopold [16], the pedagogic development phases at HfG Ulm happened in three phases, firstly foundation based on Bauhaus course, secondly emphasizing scientific understanding of design process, and lastly incorporating social sciences in design practice.

The foundation course under this model included understanding of scientific theories such as perception, gestalt theory, topology and basic geometry [16], and explorations of materials such as plaster, plastic, wood, metal [14]. As mentioned by

Leopold [16], the experiments in teaching were such open ended that 'they shifted between precise and imprecise, order and chaos, rule and random that such rules help even today in parametric and digital designs.' Ulm school was advanced to study forecast of design field to identify the importance of communication and information technology in a post-industrial era. Hence, their contribution in explorations of digital design that involved conceptualizing and coding as design process is relevant to this day [11].

18.3 Pedagogy Followed in Architecture Schools in India

The pedagogy followed in several architecture schools in India has a direct or indirect bearing from different art and design schools as discussed in the paper. Even though, it is challenging to find an exact trajectory of how the existing curriculums have rooted in them, by the facts of timeline and the people associated with architecture schools, it is evident that the subjects and design processes experimented are associated with either of the above schools. The curriculums of top ten architecture schools as identified by NIRF rankings 2020 are studied here; among which brief curriculum details of two renowned institutes, i.e., School of Planning and Architecture (SPA) and Center for Environmental Planning and Technology (CEPT) are stated here.

1. The curriculum followed at School of Planning and Architecture at New Delhi in 1959 and later revised in 1967 of each academic year is as follows [18]:
   - First year foundation studios involve design problems of two-dimensional nature.
   - Second year has rural planning that involves design of simple structures for village scenarios by conducting site visits and by understanding materials and methods of construction.
   - Third year addresses issues of urban environment and designing public buildings.
   - Fourth year focuses on complex problems such as urban housing, multistoried buildings, and the design of industrial buildings.
   - The final year, one works on redevelopment of large-scale urban areas involving complex sociological, economic, and technical considerations and last semester meant for thesis submission; that reflects ones complete understanding of academic career.
2. The curriculum at School of Architecture SA, CEPT [2012] published as booklet explains the pedagogy prepared during the years 1976, 1988, and 2001 and describes goals, objectives, methodology, and credit distribution system. Few of the goals mentioned are as follows [3]:
   - *To develop an individual rooted in his/ her context of society and place in India, capable of the disciplines and skills through which to interpret them into a qualitative physical environment,*

- *Sensitivity to environmental surroundings to create balance between manmade and natural systems,*
- *Appropriate choice of material and techniques with appropriate use of resources, and*
- *To continue with Indian traditions yet suitably make modern day transformations.*

The course at CEPT includes mainstream subjects such as basic drawing, design, history, structures, environmental science, skills and tools, and humanities. Subjects of drawing, painting, sculpture making, photography, ceramics, poetry, drama, appreciation of dance, literature, and others are incorporated under arts and crafts/design as workshops, electives or as exchange programs. The related study program (RSP) in the fifth semester involves visit to rural areas where students and professors document, measure draw a village staying on-site for weeks together. Internship is during the sixth semester, and in ninth semester, students individually propose their studio program, followed by thesis in the tenth semester [6].

18.4 Analysis

Architecture is called as 'mother of all arts,' but in terms of pedagogy, 'all other arts have been guiding force for architecture.' To elucidate this claim, (Table 18.1) provides a list of characteristics of each art school described in this paper that assist in identifying design methods and ideologies used in Architecture schools.

The discipline and accurateness, bare pencil drawings and orthographic projections that were imbibed in Cole's South Kensington system are practiced to this date in subjects of basic design, architectural graphics, and measure drawing subjects. The emphasis on sketching as a tool of thinking as in Ecole Beaux Arts is an evident method of design process even for several famous architects. As a method of teaching in Ecole, where the importance was given to plans than compared to elevations and sections is a practice in schools even today, wherein few professors have a keen eye toward a linear process and focus mostly over plans than compared to whole. Santiniketan as a colloquial art school that provided the unique ideology of finding inspiration in the rural atmosphere has been an important tool of gaining 'sense of place' and understanding 'Indianess' through subjects such as 'related study programs' now adapted in several architecture schools. Bauhaus as an education model of 'modern movement' has implicitly got engrained in teaching of architectural design through its contemporary notion of ascertaining intuition and exploration of materials in building industry. In addition, as mentioned by Ranjan [21], the influence of HfG Ulm can be observed in 'basic design assignments and other early projects that was adopted to meet local needs and challenges.' Quijano who has worked with HfG archive has remarked that 'Ulm School of Design made an essential contribution to the consolidation of a methodology which has had a lasting effect on design education worldwide' [22]. Lastly, Ulm and its explorations of incorporating

Table 18.1 Characteristics of art schools adapted in architecture pedagogy

South Kensington	• Correct observation • Accurate expression • Monochrome drawings • Tedious reproduction
Ecole des Beau Arts	• Drawing and painting skills • Cartesian pedagogical program • Sketch designs—Esquisse • Ateliers (design studios) • Guidance under practicing architect for a year
Santiniketan	• Indian rural socio-cultural aspects • Community activities devoid of influences of industrialization • Importance to crafts • Gandhiji's 'learning by doing'
Bauhaus	• Arts and crafts • Creative intuitive process • Thinking and handling, material exploration • Nature study • Fundamental elements of form • Classes on mass housing, city planning • Environmental relationships
HfG Ulm	• Scientific and technological knowledge • Basic geometry, gestalt theory • Importance of function and market needs • Practical training from industry • Social sciences in design practice • Material exploration • Coding and digital design

mathematics, science, and humanities have been means of analyzing qualitatively and thereby keeping human and ecology as the center of 'need of design' in architecture.

Table 18.2 helps in understanding the origin of a certain 'subject' and further its adaptation in other art schools that have further developed in similar lines. From this, one realizes that South Kensington system has a minimal role in terms of number of subjects in architecture curriculum, compared to Bauhaus and Ulm that have been key influencers. Yet, few significant ideologies and thought processes such as conducting design studios and professional training are contributions of Ecole des Beau Arts. Likewise, in establishing the importance of local study and cultural awareness, Santiniketan takes a vital position.

In Table 18.3, the top ten architecture schools against subjects taught in all the five years that possibly have roots from arts school are listed. The analysis explains 'most commonly' adapted subjects and vice versa. From, this one deciphers that subjects such as Basis for design from Bauhaus, housing and community from Santiniketan, design studio and professional training from Ecole Beau Arts are widely accepted in all the architecture schools, whereas ornamentation and transformation of Ecole is least taught. Subjects such as digital fabrication, programming,

Table 18.2 Relationship chart of art and design schools with subjects of architecture

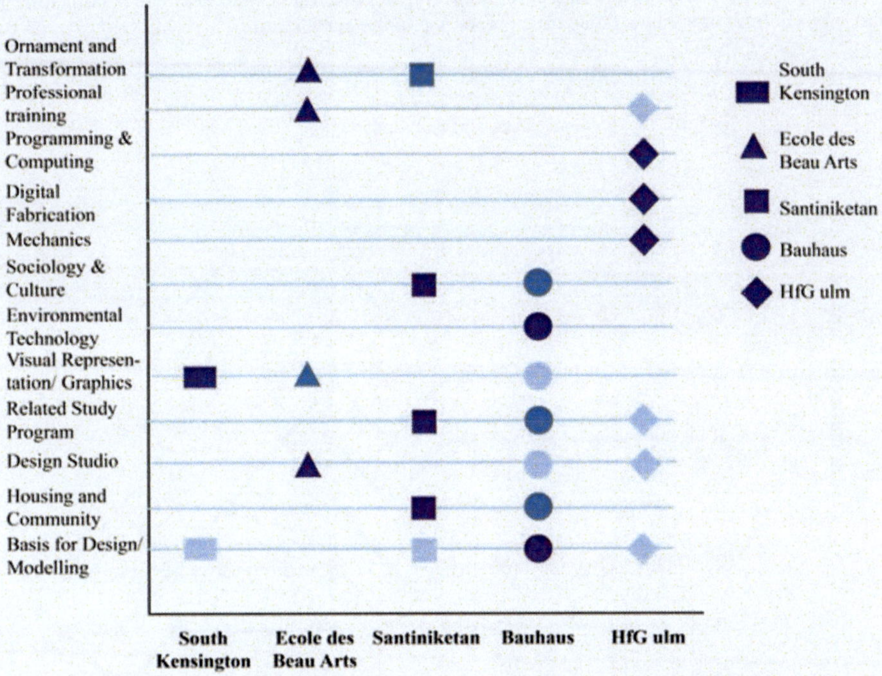

Note: The darkest tone of colour in each column indicates it to be the initial Art school to have introduced the subject and gradual decrease in tone indicates active role in developing the subject domain but not to be the first school to have introduced the same.

and computing from HfG Ulm, have not yet found their place in most of the architecture schools. While related study program credited to Santiniketan, is familiar in 9 out of 10 schools which incorporates student visits to rural areas for documentation and community activities.

18.5 Conclusion

The pedagogy at architecture schools has not taken a phenomenal leap in content and methods but has been minor deviations and experimentations within the scope of 'tried and tested' age-old techniques. As asserted by Mehta, J [17] 'most schools have not come to grips with the pedagogical structure that arrived here from the West at the advent of the modern college level and education has more or less remained.' Desai [6] is of the opinion that 'curriculum model is almost the same all over the country with minor differences and the course has been of five years duration with the office or practical training shifted one semester here or there.' Thus, the influences on architecture curriculum previously when there was minimal distinction between

Table 18.3 Top ten architecture schools as according to NIRF ranking 2020 (not in ranking order), and subjects from arts schools that are incorporated in the curriculum

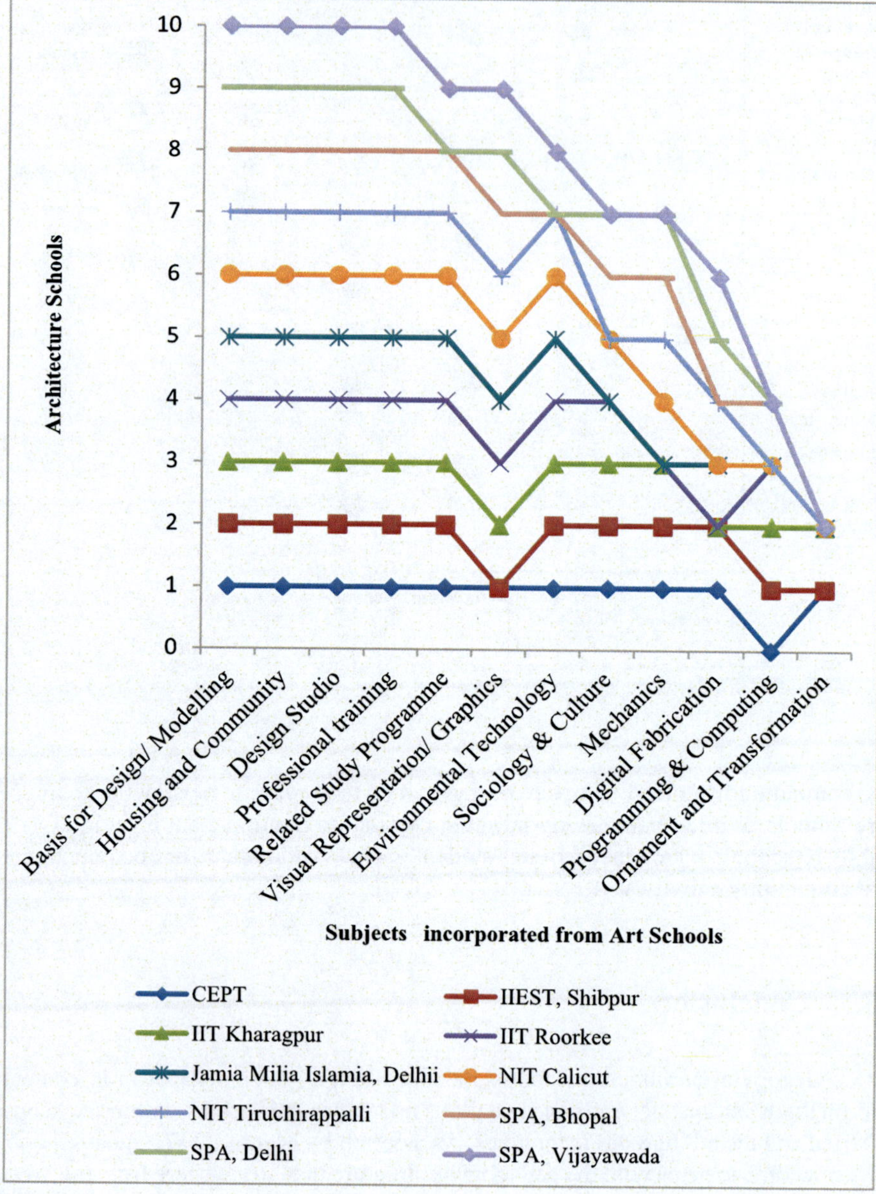

courses of art, design, and architecture are evidently still persistent in the present context, even now while architecture is distinguished and has a defined niche. In the future, to identify what lies in India as a means of inspiration in whole, of people and geography, of politics and culture, of real and mundane, than just aesthetic or an appeal of perfection, than just scientific or technology to please the West, is where architecture pedagogy needs to carve its path.

References

1. Altbach, P.G., Kelly, G.P. (eds): Education and Colonialism, p. 367. New York: Longman (1978)
2. Bonsiepe, G.: Cullars, J.: The Invisible Facets of the Hfg Ulm. Design Issues, Summer, 1995, vol. 11, No. 2, pp. 11–20. The MIT Press (1995)
3. Chhaya, N., Shankar, P., Kashikar, V. (eds): Curriculum at SA, CEPT, 1988. In: Course Curriculum at the Centre for Environmental Planning and Technology. FA–CEPT University Booklet Series. Faculty of Architecture, CEPT University, Ahmedabad (2011)
4. Cunningham, A.: Notes on education and research around architecture. J. Archit. **10**(4), 415–441 (2005)
5. Dasgupta, A.: Rules of metaphor and other pedagogic tools (2019). https://www.bauhausimaginista.org/articles/4343/santiniketan?0bbf55ceffc3073699d40c945ada9faf=7e11bb37766d77a6a5358c8ac695df8c. Last accessed 2020/05/12
6. Desai, M.: Aspects of curriculum in the architectural education of India: a view towards the philosophy of it.In: Workshop on Professional Program in Architectural Education, Bandung, Indonesia, Institut Teknologi Bandung (2006)
7. Dewan, D.: Crafting Knowledge and Knowledge of Crafts: Art Education, Colonialism and the Madras School of Arts in 19th Century South Asia. University of Minnesota (2001)
8. Garric, J.-P.: The French Beaux-arts. In: Bressani, M., Contandriopoulos, C. (eds) The Companions to the History of Architecture, vol. III, Nineteenth-Century Architecture. Wiley, New York (2017)
9. Gropius, W.: The New Architecture and the Bauhaus, English edition, London, p. 57 (1935)
10. Hughes, Q.: Before the Bauhaus: the experiment at the Liverpool School of Architecture and Applied Arts. Archit. Hist. **25**, 102–113 (1982)
11. Holt, M., Holt, M.: Is the Bauhaus Relevant Today?: Design Theory and Pedagogy at the Hochschule für Gestaltung, Ulm (1953–1968). University of Technology Sydney (Insearch), p. 15 (2011)
12. Jacob, H.: Ulm: a personal view of an experiment in democracy and design education. J. Des. History. **1**(3/4):221–234 (1988)
13. Kantawala, A.: Art education in colonial India: Implementation and imposition. Stud. Art Educ. **53**(3), 208–223 (2012)
14. Krippendorff, K.: Designing In Ulm and Off Ulm, pp. 55–72. In: Czemper, K.-A. (ed.) HfG, Ulm; Die Abteilung Produktgestaltung. 39 Rückblicke. Verlag Dorothea Rohn, Dortmund, Germany (2008). https://www.rohn-verlag.de/04.html. Last accessed 2020/05/12
15. Krut, R.: South Kensington To South Africa: art education in government elementary schools and schools of art in South Africa 1800–1910. A Dissertation, Submitted to the Faculty of Education, University of the Witwatersrand, Johannesburg (1983).
16. Leopold, C.: Precise Experiments: Relations Between Mathematics, Philosophy and Design at Ulm School of Design, Nexus 2012: Relationships Between Architecture and Mathematics, Milan (2012)
17. Mehta, J.: Architectural Education in India, An Overview. Vadodara, India (2006)
18. Mehta, M.: Architectural education in India. Archit. Sci. Rev. **21**(1–2), 35–37 (1978)
19. Minahan, S.: The organizational legitimacy of the Bauhaus. J. Arts Manag. Law Soc. **35**(2), 133–145 (2005). https://www.tandfonline.com/loi/vjam20. Last accessed 2020/05/21
20. Mitter, P.: Art and Nationalism in Colonial India (1850–1922): Occidental Orientations. Cambridge University Press, Cambridge (1994)

21. Ranjan, M.P.: Web of connections: Indian design education with influences from the HfG Ulm. www.bauhaus-online.de. Last accessed 2019/ 03/14
22. Rinker, D., Quijano, M., Reinhardt, B. (eds.): Ulmer modelte-modelte nach ulm-ulm school of design, from the Ulmer Museum and HfG-Archiv. Hatje Cantz, Ostfildern Ruit (2003)
23. Rhi, J.: Two different ways of HfG Ulm: design for industry and design for humans. Manchester School of Art, Manchester Metropolitan University (2019)
24. Sarker, S.C.: Rabindra Nath Tagore: Educational Philosophy and Experiment. Visva Bharati (1961)
25. Scriver, P.: India: Modern Architectures in History. Reaktion Books, pp. 97–99 (2015)
26. Singanapalli, B.: Design Pedagogy in India: A Perspective Design Issues, vol. 21, No. 4. MIT Press, Santiniketan (2005)
27. Wingler, H.M.: The Bauhaus: Weimar, Dessau, Berlin. Chicago. MIT Press, Cambridge (1969)
28. Woods Hill, S., Solomon, W.E.G. (ed): I. The School of Architecture of the Bombay School of Art. In: The Bombay Revival of Indian Art—A Descriptive Account of the Indian room constructed and decorated by the Staff and Students of the School of Art, pp. 125–133. Sir J. J. School of Art, Bombay (1924)
29. Woods, M.N.: Women Architects in India: Histories of Practice in Mumbai and Delhi. Routledge (2016)

Chapter 19
Product Design with Form, Strength, and Function for Undergraduate Product Design Students—A Case Study

Saurabh Kumar Mukherjee

Abstract At the undergraduate level, many students studying product design often focus on the visual and tactile aspects of the design process. Those students inclined toward a wider perspective with respect to a desire to also focus on the functional, and strength aspects of products, may need a framework to understand how a design moves from an idea to a product. This study, through the example of designing a bicycle (for road use), provides a study on how undergraduate students could be engaged to better understand certain elements in the design of a tangible product. (1) Form, function and strength are key elements of any design which needs to be linked with the user. (2) Trade-offs of various types are needed at every stage of a design process. (3) The tools (including computing software) used are easily available to students at no charge. Product needs to be safe, efficient, attractive, and comfortable. Its complexity, in terms of mechanics and mathematics, is provided at levels where a 'high-school student' who has studied mathematics and physics with some training in mechanics and kinematics can understand; while also allowing application of other aspects of knowledge learnt by them, such as 'color theory'. This, and other studies of similar nature, focused on the incorporation of 'applied sciences' more intimately with product design courses, at undergraduate levels, might help graduating students to be better prepared to serve in industry.

19.1 Introduction

The curricula in many undergraduate programs in product design focus on the aesthetics of the product without, perhaps, devoting enough attention into the functioning and usage of the end product. However, there is evidence that incorporation of this aspect enhances product design. For example, Bates et al. [1] have documented their experience at the 'Internal Design Program' of San Jose State University that served as a model which provided better outcomes. Moreover, Self et al. [2] have demonstrated that while students taught by a single instructor were more appreciative

S. K. Mukherjee (✉)
Independent Consultant, Pune, MH 411014, Republic of India
e-mail: learningdesign@yahoo.com

A. Chakrabarti et al. (eds.), *Design for Tomorrow–Volume 2*, Smart Innovation, Systems and Technologies 222, https://doi.org/10.1007/978-981-16-0119-4_19

about the quality of instruction, a team-teaching approach was perceived to provide greater opportunities to understand the relevance of different disciplines to the course subject. Despite these documented advances in teaching methodology, only a third (or less) of the courses in India emphasize the functional and strength aspects of product design. See 'Heading 4'.

The 'Vitruvian Trinity'[1] has been used by Kepler [3] to explain relevance of strength, function, and form in designing products. All tangible products are invariably made of some material, have certain geometry, and have colors. This results in a form conveying a particular visual aesthetic. Any material of certain geometry (form) inherently has certain strength, and geometry drives the stiffness. Wenham [4] has described how the dimensions and geometry of the structure and the method of force application, impact the magnitude of the force required to bend it. Incidentally, a product needs to perform a function. These characteristics have a significant impact on the approach to designing products.

Today, leveraging accessible technology might permit a student, without formal and broad studies in applied sciences (like completing an engineering course), to effectively consider these aspects when proposing a design. Linear finite element (FE)[2] modeling and solving have become more user-friendly and accessible for design focused practitioners vis-à-vis engineering focused ones.

Studies have already been done in verifying strength and stiffness of a design once an architype proposal for a design exists, or an iteration prepared [5]. Kim et al. [6] have presented a course which covers the discussed aspects, in an undergraduate course context, starting from sketches and right until testing. Optimization studies related to lateral handling, for specific parameters, have been reported by Moore et al. [7]. In this study, the example of designing a bicycle to illustrate steps a student could incorporate into the design workflow to factor in functionality while designing is shared. However, while the bicycle is used as a 'muse,' the steps in the process described may be adapted to other design projects.

19.1.1 Context—A Bicycle

Three unique elements in a bicycle are: 'it's a single-track vehicle', 'it's non-polluting in use' and for its basic version 'it's not a complicated machine with aggregates like suspension, gearbox, and clutch which are needed in other vehicles'.

In a bicycle, the frame is one aggregate which decides the overall look of the product significantly and connects most other aggregates together. The ubiquitous

---

[1]Named after Markus Vitruvius Polliio, author of a treatise of Architecture. See: https://en.wikipedia.org/wiki/Vitruvius.

[2]As explained by Prof. Bathe[17]: "This method is used for linear analysis of solids and structures that starts with idealization of the system, discretization of the system into a set of springs which follows the Hooke's law, formulation of the equilibrium equation and then finding the solution to the equation." From 'Lecture-1' of an online course available under the Creative Commons License BY-BC-SA. Link: https://youtu.be/oNqSzzycRhw.

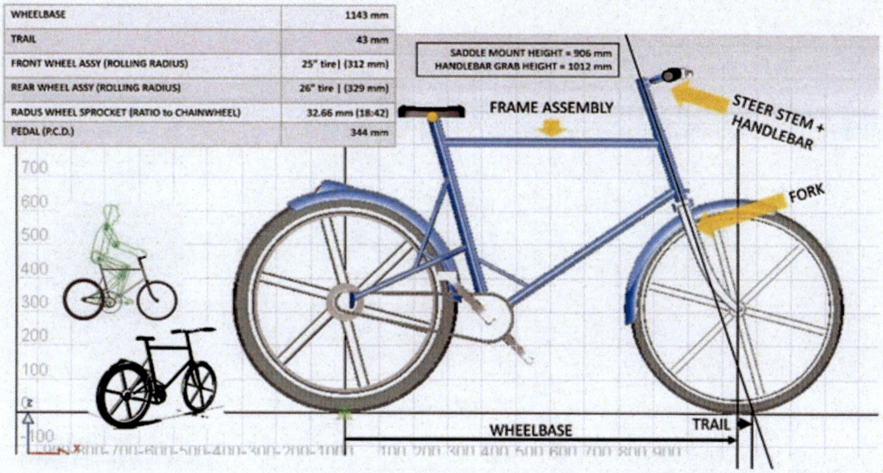

Fig. 19.1 Layout of a bicycle

design of a bicycle (based on a design from the 1880s[3]) has chain-driven traction on the rear-wheel-sprocket, and both wheels are of a similar rolling radius.

The layout of the bicycle (see Fig. 19.1) involves: starting with a particular size of wheels; relative positioning of the saddle, the pedal and handlebar; that are dependent on anthropometrics as well as desired handling characteristics. Patterson et al. [8] have reported that trail and the 'moment of inertia' are important factors to determine the handling qualities of a bicycle, when designed.

Within the presented framework, the focus is to develop a design proposal for the frame; as other aggregates such the sprockets, pedals, and pedal cranks can be selected off-the-shelf.

As 'material (and its processing)' and 'geometry' have a significant impact on the strength and stiffness of an object, this study focuses on these two aspects. The software—i.e., tools—chosen for this project, available at no cost, includes:

'FreeCAD 0.18' [9]; 'Z88 Arion V2' [10]; 'Netgen—as embedded within FreeCAD' [11]; 'CalculiX—as embedded within FreeCAD' [12].

19.2 The Process—for the Framework

The process described is relevant to any physical product that is subject to application of forces, though all steps are not necessarily needed to propose a design.

The process starts with 'estimation of loads and constraints,' followed by 'creating a packaging space,' and 'running an optimization to evolve a topology.' It is then

---

[3]As described in the page regarding 'John Kemp Starley' at Wikipedia. See: https://en.wikipedia.org/wiki/John_Kemp_Starley.

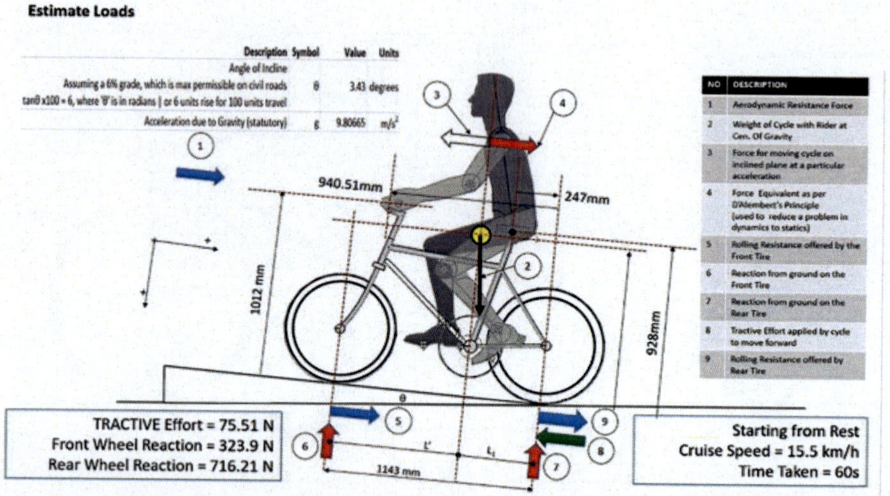

Fig. 19.2 Estimation of loads

followed by 'interpretation of the topology into a manufacturable part' and ends with 'specifying the proposed design in the form of a drawing and specification sheet.' The final purpose of a 'product design' is practical usage. Manufacturing a product requires sharing a drawing that the manufacturer's team can use. For this study to be effectively used, it is assumed that a student has an understanding of introductory courses in 'Applied Mechanics' and 'Strength of Materials.'

19.2.1 Estimation of Loads and Constraints

From a 'system level' standpoint, tractive effort is an important parameter needed for designing any vehicle (see Fig. 19.2). This effort is an input to estimate the pedaling force (see Fig. 19.3). As the handlebar controls the steering and supports the inertia of the rider, estimating the load on the handlebar is important (see Fig. 19.4). The saddle supports some rider inertia (see Fig. 19.5).

To simplify the study, loads along the lateral direction of the bicycle are not included in this case study. For finding the tractive effort, a process explained by Krishnakumar[4] has been adapted.

Using mechanics, one can estimate the forces required on the pedal, once a value for the tractive effort on the rear wheel is known. The force on the chain is more than the tractive effort on the rear wheel since the 'rear wheel sprocket' is of a smaller

[4]See Lecture-02, by Krishnakumar [18] titled: "Mod-01 Le-02 Longitudinal Dynamics", which is part of a course provided courtesy IIT Madras through NPTEL. Link: https://youtu.be/T7el7XL1aK8.

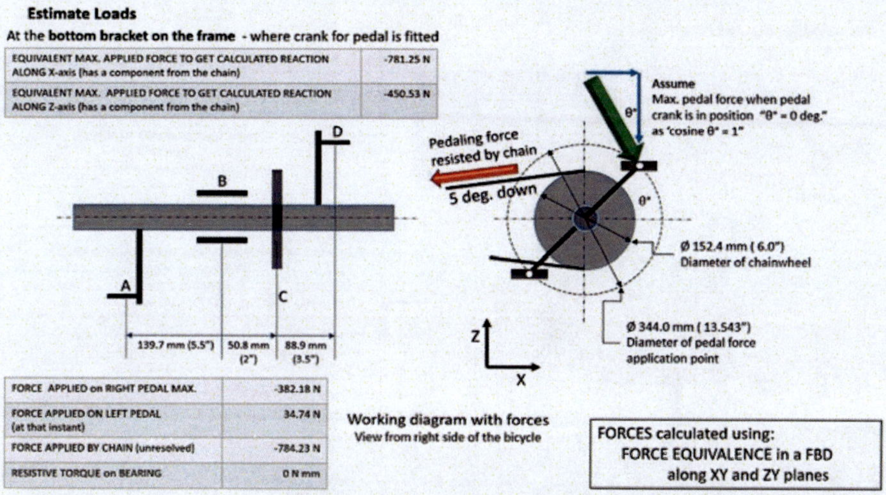

Fig. 19.3 Forces on the 'bottom bracket'

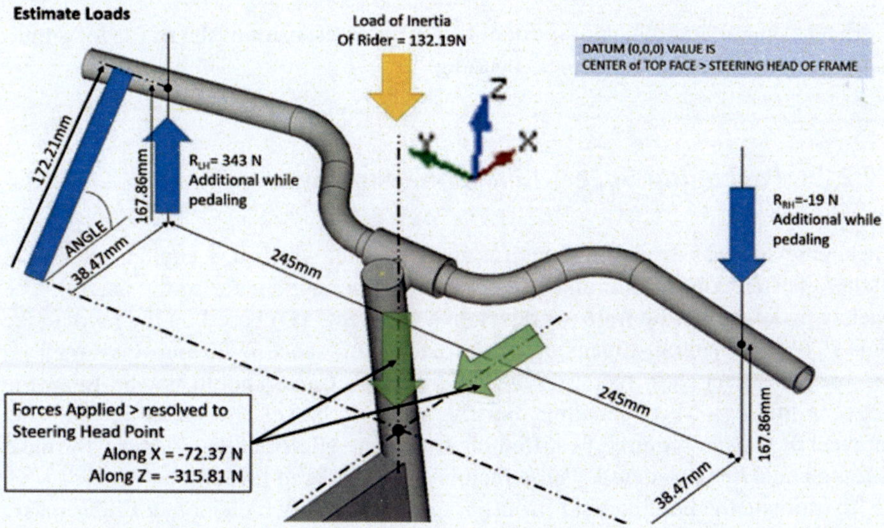

Fig. 19.4 Load on frame due to handlebar forces (ignoring lateral)

radius vis-à-vis the rear wheel rolling radius. Based on the pedal (force required) and the 'resistive force on the chain,' forces on the bottom bracket of the frame can be estimated.

Resolving forces applied on the handlebar allows estimating forces on the top of the frame (the steering stem top).

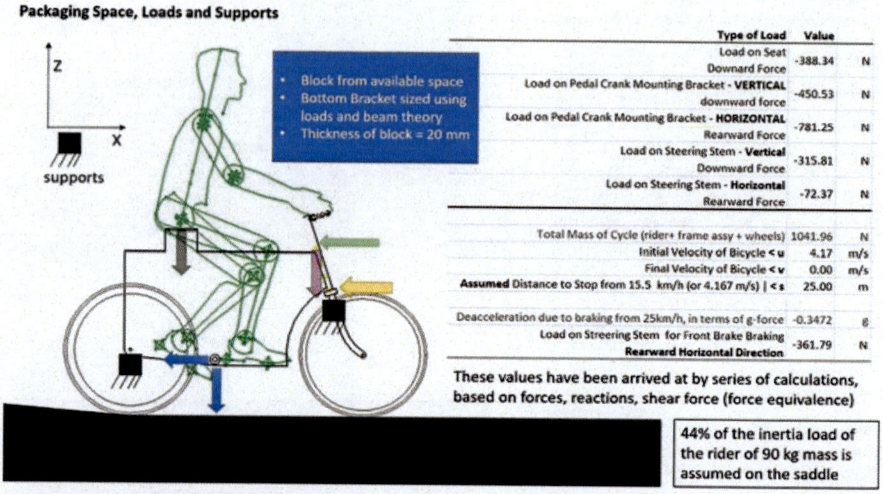

Fig. 19.5 Space for packaging, loads and supports

Forces due to braking, an important factor for assessment of the frame longitudinally, are discussed under the next heading.

19.2.2 Packaging Space, Loads and Supports

Packaging space is available space for an aggregate, which if occupied, would not hinder the function, positioning and control of other aggregates or the user(s). The packaging space for the frame is developed (in FreeCAD) based on the layout (see Fig. 19.5). The lateral dimension of the packaging space is 20 mm (assumed), a reasonable aspect ratio to represent the frame. A frame should not be too wide between the legs while pedaling. For the speeds a bicycle is ridden at, the 'area moment of inertia' required to stiffen the frame enough for lateral loading, for most riders, should be considered. These factors are considered for reasonableness.

To estimate the braking load (front), a certain inertia (which includes rider mass) is assumed. Deceleration is calculated as a 'g' value and converted to a force. It is applied at the bottom part of the steering stem, in a rearward direction.

Reaction forces always develop at a support. Support at points 'where the rear axle fits' and where the 'frame is held on the fork' are assumed as rigid points. While at the rear end of the frame (near the stays) the true geometry of a frame splits to provide for space to accommodate the 'rear wheel assembly' and the chain, this slotting is ignored at this stage (as the impact on longitudinal studies such as this would be minimal).

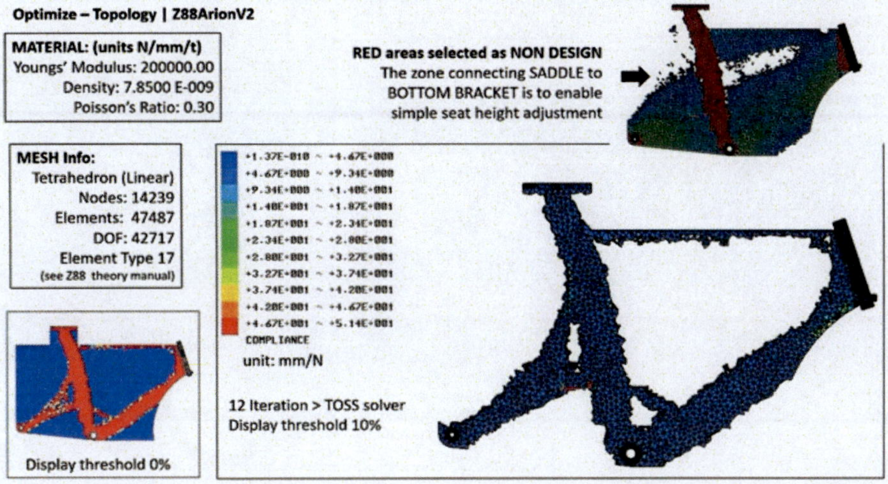

Fig. 19.6 Topology optimization

19.2.3 Topology Optimization

Topology optimization enables the development of the most efficient path (topology) for loads to transfer from the 'location of application' to the 'location of supports' provided in a structure. The efficiency in this context is the 'minimum volume' of the material. The specified objective may be to 'maximize the strength' or 'maximize the stiffness' and/or other custom requirements.

Bendsøe et al. [13] while proposing a 'homogenization method,' highlight the difficulty encountered to change the structural topology during the design process. Specifically, structurally significant members often drive the correct position, function, and specification of other aggregates in a product. Any late changes impact these other elements (dependencies) increasing the complexity of execution.

The 'Z88 Arion' tool is used for optimization of the design concept (see Fig. 19.6). The packaging space is pre-processed (meshing, boundary conditions, materials, loads, and selection of non-design spaces), solved and post-processed (showing the results in a way that is easier to interpret). Material selection (metals in this context) is a trade-off between price, properties of the material (such as: density, Young's modulus, modulus of rigidity, Poisson's ration, yield strength) and processing complexity. For this study, a type of steel is selected.

There are several solver options within this tool. 'TOSS' is the solver option used as its optimal here for its breadth.[5] The results are reasonable.

[5]See 'Page 14′ of the 'Theory User Manual' of the software. Kindly translate online since this manual is written in German. See: https://download.z88.de/z88arion/V2/benutzerhandbuch.pdf.

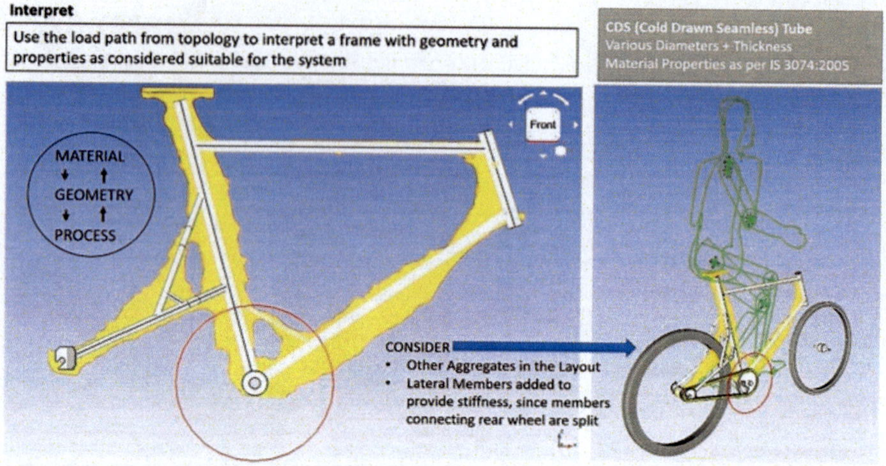

Fig. 19.7 Interpretation of the topology result

19.2.4 Interpretation of Results—Creating a Design Proposal

A topology optimization result provides the guidance for developing a frame. A frame representing 'material, geometry, and manufacturing' feasibility needs to be created (see Fig. 19.7). The material, for example, is a standard material normally used on bicycles ('IS 3074:2005'[6]). Certain additional members are added to connect members laterally. These are positioned in a manner that they do not interfere with other aggregates in the product.

19.2.5 Verification of the Design Proposal

For virtual verification, see Figs. 19.8, 19.9, and 19.10.

The FE workbench within 'FreeCAD' is used. For meshing 'Netgen' and for solving 'CalculiX' are used (embedded inside the workbench). Two aspects Viz. 'stiffness' and 'propensity to yield' are assessed. The results are reasonable for the given conditions.

The 'Von Mises' stresses provide a 'better correlation with experimental behavior than 'Tresca' yield criterion'[7] and has been used to assess the developed stresses.

---

[6]Bureau of Indian Standards. See: https://bis.gov.in.

[7]Teaching and learning packages of the 'University of Cambridge'. See: https://www.doitpoms.ac.uk/tlplib/metal-forming-1/yield_criteria.php.

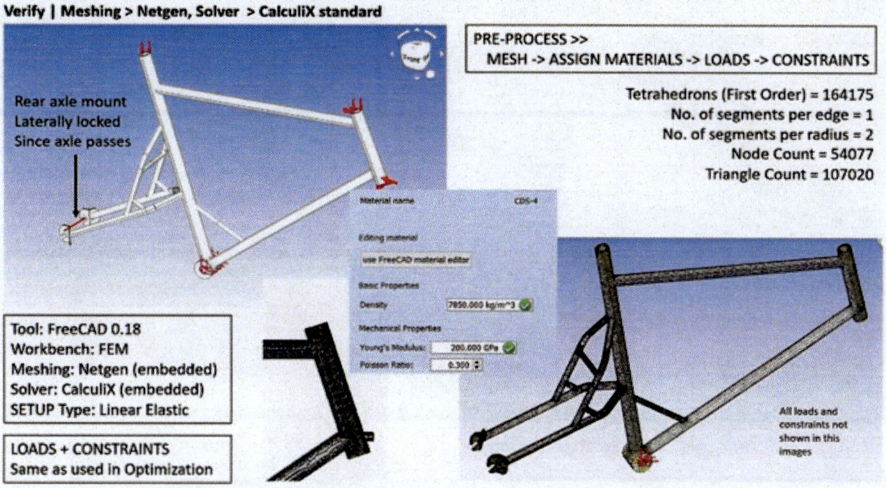

Fig. 19.8 Verification of interpreted design

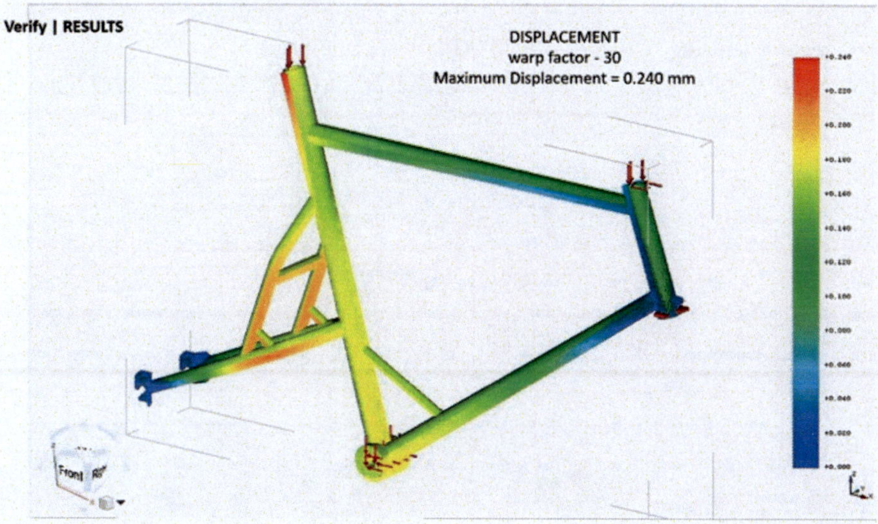

Fig. 19.9 Stiffness behavior

19.2.6 Specification of the Design for Manufacturing

A manufacturing team needs a drawing and specifications communicating the design (to the extent it impacts—functional performance) (see Figs. 19.11 and 19.12).

The framework described above leverages technology to drive efficient design.

Fig. 19.10 Stresses due to loading

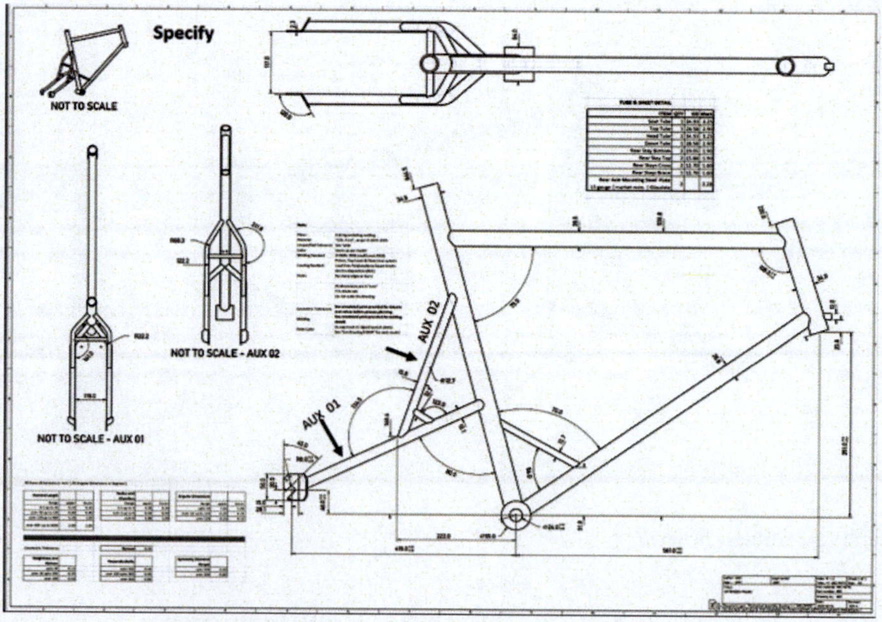

Fig. 19.11 Frame drawing

Specify – Details of NOTES, TUBE, UNSPECIFIED tolerances

Note:	
Mass :	6.63 kg excluding welding
Material:	"CDS -4 (c)" , as per IS 3074: 2005
Unspecified Tolerance	See Table
Color	Varies as per model
Welding Standard	IS 9595: 1996 (reaffirmed 2003)
Coating Standard	Primer, Top Coast & Clear Coat system: Lead Free Polybutadiene-based Anodic Electro-Deposition (AED)
Finish	No sharp edges or burrs
	All dimensions are in "mm"
	If in doubt ask
	Do not scale the drawing
Geometric Tolerances	See annotated print provided for datums and values before process planning
Welding	See annotated print provided for welding details
Paint Color	As approved on signed swatch plates See "List Drawing # 000P" for each model

TUBE & SHEET DETAIL (in mm)

ITEM	QTY	OD	thick
Seat Tube	1	34.93	3.25
Top Tube	1	28.58	2.90
Head Tube	1	34.93	3.25
Down Tube	1	28.58	2.90
Rear Stay Bottom	2	19.05	1.90
Rear Stay Top	2	15.88	1.90
Rear Stays Cross	2	12.70	1.60
Rear Stays Brace	1	12.70	1.60
Rear Axle Mounting Steel Plate 13 gauge | market nom. | Obsolete	2	-	2.28

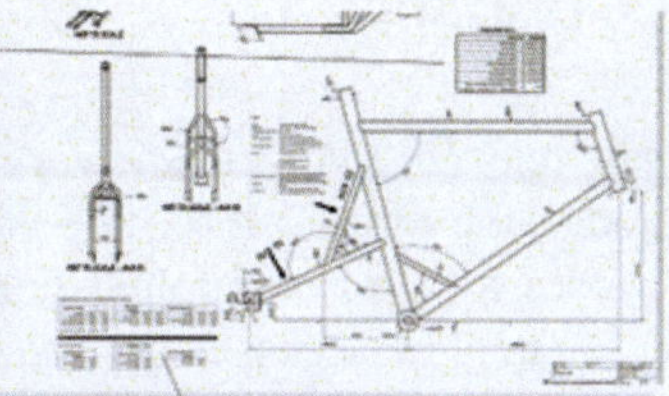

All linear dimensions in 'mm' | All Angular Dimensions in 'degrees')

Nominal Length			Radius and Chamfers			Angular Dimensions		
Range	upper	lower	Range	upper	lower	Range	upper	lower
0.5 up to 30	0.20	0.20	0.5 up to 3	0.20	0.20	upto 10	1.00	1.00
over 30 up to 120	0.80	0.80	over 3 up to 6	0.50	0.50	over 10 up to 120	0.25	0.25
over 120 up to 400	1.20	1.20	over 6	1.00	1.00	over 120	0.13	0.13
over 400 up to 1000	2.00	2.00						
Geometric Tolerances			Runout	0.20				
Staightness and Flatness			Perpendicularity			Symmetry (Nominal Range)		
upto 30	0.10		upto 100	0.40		upto 300	0.60	
over 30 upto 300	0.30		over 100 upto 300	0.60		over 300	0.80	
over 300	0.50		over 300	0.80				

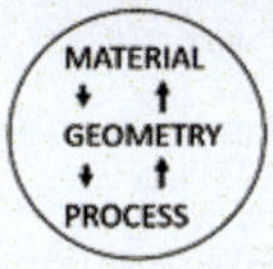

Fig. 19.12 Details in drawing

It provides an example of proposing concepts that are a suitable option from strength and stiffness standpoint, while meeting form and functional requirements.

19.3 Potential of This Framework for Students

Students love to design products which consider human factors and are pleasant to use. The framework embodied in this example could be applied to other products and/or aggregates as an integral attempt to propose a design that is 'desired by customer' and which is appreciated by engineers and manufacturers. For those interested in designing bicycles for design competitions, this case study could be adapted to comply with the 'approval protocol' of UCI.[8]

Optimization as a process has been used in industry for over a decade [14]. Virtual testing has been used in automotive industry to estimate 'fatigue life' [15]. The hypothesis that 'drawings are important in mechanical design process' has been evaluated, and support for the hypothesis has been reported [16].

There may be other case studies that already provide frameworks which leverage technology to incubate aspects of the 'applied sciences' in design of products. Considering the existing focus on inter-disciplinary studies in product design among researchers, this case study is an addition to these efforts.

[8]UCI is an abbreviation of 'Union Cycliste Internationale'—Document: 'Approval Protocol for Frames and Forks'. See: https://www.uci.org.

19.4 Data

Data relevant to this framework including: calculations for estimating loads, the 3D model, the optimization, the verification, and a study to assess the 'proportion of courses focused on applied sciences in undergraduate product design programs in India' may be accessed at:

https://drive.google.com/file/d/1A27WF_3wpTckmdtm8kpO30odewMDrXQ1/view?usp=sharing.

Password: 'XL1632ptq#!acd4'

Acknowledgements I thank the teams who have developed and provided access to the modeling and simulation tools used in this case study.

References

1. Bates, S., Backer, P.: Introduction to product design and innovation: A cross disciplinary mini curriculum. In: 2005 Annual Conference Proceedings . ASEE Conferences, Portland, pp. 10.835.1–10.835.10 (2005)
2. Self, J.A., Baek, J.S.: Interdisciplinarity in design education: understanding the undergraduate student experience. Int. J. Technol. Des. Educ. **27**, 459–480 (2017). https://doi.org/10.1007/s10798-016-9355-2
3. Kepler, J.A.: Structural Optimization as a Design and Styling Tool. Aalborg (2002)
4. Wenham, M.: Mechanical Properties of Materials and Objects. SAGE Publications Ltd. (2001)
5. Lin, C.C., Huang, S.J., Liu, C.C.: Structural analysis and optimization of bicycle frame designs. Adv. Mech. Eng. **9** (2017). https://doi.org/10.1177/1687814017739513
6. Kim, I.Y., De Week, O., Nadir, W., Young, P., Wallace, D.: Innovative modern engineering design and rapid prototyping course: a rewarding CAD/CAE/CAM experience for undergraduates. In: ASEE Annual Conference Proceedings. pp 7405–7416 (2004)
7. Jason, K.M., Mont, H.: Expanded optimization for discovering optimal lateral handling bicycles. In: Proceedings, Bicycle and Motorcycle Dynamics, p. 2 (2019)
8. Patterson, W.B., Leone, G.L.: The application of handling qualities to bicycle design. In: Proceedings, Bicycle and Motorcycle Dynamics 2010 Symposium on the Dynamics and Control of Single Track Vehicles, pp. 20–22 (2010)
9. Riegel, J., Mayer, W., et al.: "FreeCAD" Software. Version 0.18. https//wiki.freecadweb.org/About_FreeCAD (2002)
10. Rieg, F., Alber-Laukant, B., Lange, C., et al.: "Z88 Arion" Software (1985). Version 2 https//en.z88.de/z88arion/
11. Egger, H., Gopalakrishnan, J., Hochsteger, M., et al.: "Netgen" software. Interg. ver. Free. V0.18 (2017). https//ngsolve.org/
12. Mtu, A., Wittig, K., Dhondt, G., et al.: "CalculiX" software. Integr. Insid. Free. v0.18 (1998). http//www.calculix.de/
13. Bendsøe, M.P., Kikuchi, N.: Generating optimal topologies in structural design using a homogenization method. Comput. Methods Appl. Mech. Eng. **71**, 197–224 (1988). https://doi.org/10.1016/0045-7825(88)90086-2
14. Schramm, U., Zhou, M.: Recent developments in the commercial implementation of topology optimization. In: Solid Mechanics and its Applications, pp. 239–248. Springer, Berlin, (2006)
15. Huizinga, A.T.M.J.M., Van Ostaijen, M.A.A., Van Oosten Slingeland, G.L.: A practical approach to virtual testing in automotive engineering. J. Eng. Des. **13**, 33–47 (2002). https://doi.org/10.1080/09544820110090304

16. Ullman, D.G., Wood, S., Craig, D.: The importance of drawing in the mechanical design process. Comput. Graph **14**, 263–274 (1990). https://doi.org/10.1016/0097-8493(90)90037-X
17. Bathe, K.-J.: RES.2-002 Finite element procedures for solids & structures. In: Open Course Ware (2010). https://ocw.mit.edu
18. Krishnakumar, R.: Mod-01 Lec-02 longitudinal dynamics. In: Veh. Dyn. (2014). https://nptel.ac.in/courses/107/106/107106080/

Chapter 20
Non-technical Competencies for Succeeding as a Designer in a Digitally Transforming Workplace

Farrokh Mistree and Janet K. Allen

Abstract Digitization is disrupting our world and shaping the challenges that newly minted graduates will need to address in their professional careers. The solution to manage such disruptions is anchored in the principles of sustainability and values that are foundational to mitigating inequities by managing the tensions between the pulls of people, planet, and profit. To succeed in the digitized world, we suggest that we provide the opportunity for our soon to be designers to develop five non-technical, career sustaining competencies:

1. To continue learning through reflection and the associated creation and articulation of knowledge.
2. To speculate and identify gaps that foster innovation.
3. To ask questions, actively listen, reflect, and identify gaps and opportunities worthy of further investigation.
4. To make decisions using incomplete information.
5. To think critically (deductive reasoning and inductive speculation) and identify a way forward.

Our intent in this paper is to promote reflection, dialog, and action on modifying curricula to provide our soon to be designers the opportunity to internalize non-technical career sustaining competencies and values that empower them to foster societal and technological innovations that promote sustainable development and mitigate societal inequities.

This paper is based on course notes for AME5303—Designing for Open Innovation and a talk given at the Future Steel Forum in Budapest in 2019.

F. Mistree · J. K. Allen (✉)
The Systems Realization Laboratory, University of Oklahoma, Norman, OK, USA
e-mail: janet.allen@ou.edu

F. Mistree
e-mail: farrokh.mistree@ou.edu

A. Chakrabarti et al. (eds.), *Design for Tomorrow–Volume 2*, Smart Innovation, Systems and Technologies 222, https://doi.org/10.1007/978-981-16-0119-4_20

20.1 Frame of Reference

Our aim in this communication is to spur dialog amongst the academic community. Much has been written about Industry 4.0 and its attendant ramifications on industry, society, and the workforce. Without question, there has been "hype" foreshadowing significant changes and prospects for social and economic development. Therefore, we could wonder whether the current writings are perhaps just new wine being poured into old bottles. Be that as it may, Darwin's statement rings as true today for industry and society as it does for evolution: "It is not the strongest of the species that survive, nor the most intelligent, but ones most responsive to change." We recognize that students being educated for the workforce need both domain dependent competencies (e.g., technical, business, and management) and non-technical, career-sustaining competencies. In this paper, we suggest five non-technical, career-sustaining competencies for success in a digitally transforming enterprise. It is in this context that we pose the question: For success of people in enterprises characterized by the digitally transforming forces released by the adoption of Industry 4.0 what are the non-technical competencies required?

The quality of life has been dramatically improved by industries and enterprises around them. Companies have dramatically contributed to the increase in GDP. This has mainly been the result of design and deployment of *cyber-physical systems,* and the result has been the inequity in living standards. We suggest that the design of *cyber-physical-social* systems is one way to mitigate inequities in the quality of life which we see around us. To design cyber-physical-social systems, it behooves us to embrace the principles of sustainability that are foundational to mitigating inequities in managing the tensions between the pulls of people, planet, and profit.

To provide context for the design of *cyber-physical-social-systems,* we use India by way of example. Although India is one of the fastest growing developing economies in the world, more than 800 million people live in rural areas and about one third of those live in poverty [6]. Weerawardena and Mort [10] suggest that one answer to this problem is that social enterprises be developed. However, for their long-term sustainability and growth, these social enterprises must respect the sustainability drivers. These authors contend, "the role of the social mission goes hand in hand with the sustainability of the organization. Sustainability resulting from a balance of the entrepreneurial drivers of innovativeness, proactiveness, and risk management is not seen as an end, but sustainability is focused on ensuring the continuation of the organization because of its social mission."

There are many challenges to improving the lot of the rural poor, for example, the lack of proper education, health care, housing, sanitation, electricity, droughts, and floods. Further, the replication of strategies embodied in industrialization and globalization for rural development are not appropriate. For a critical review of the literature and suggestions on addressing, the issue of minimizing inequities in rural development, see Kamala [4].

A key role in uplifting conditions of the poor can be played by social entrepreneurs [10]. They also provide leadership toward achieving sustainable advantages, to

Table 20.1 A Comparison of social and business entrepreneurs

	Business entrepreneurs	Social entrepreneurs
Goal	Make a profit, capture the market	Improve quality of life by filling a market gap
Objective	Build the business, make a profit	Create social change by developing sustainable solutions
Profit motive	Maximize shareholder value	Profit as a means to advance social aims
Risk	Basic business risks	Both basic business risks and social risks
Link to social problems	Indirect	Direct
Feedback	Market and communication sources have been established	No standard methods of feedback available
Competition	Surpass others in the market	Social enterprises exist because no one else is adequately addressing the problem
Growth	Competitive	May rely on collaboration to enhance social impact
Capital	Generally robust and reliable sources of funding	Financing unpredictable and fragmented

accomplish their social missions [10]. In Table 20.1, developed by Tschang et al. [9], the differences between a business and social entrepreneur are shown.

In order to address societal needs, social entrepreneurs create non-profit solutions which have social value [1, 2]. In comparison with business entrepreneurs, a social entrepreneur seeks a market gap which, when filled, will bring about social change. The business entrepreneur seeks a profit coupled with a substantial market share; the social entrepreneur works to create social change and social value. If the business entrepreneur happens to create social value while making profit, that is an added advantage, but is not the primary motivation of the business entrepreneur. These differences in goals have a considerable impact on the growth path chosen. In the next section, we address the following question: Given the paradigm of Industry 4.0, what is a cyber-physical-social system for promoting sustainability through the mitigation of tensions between people, plant, and profit?

20.2 A Cyber-Physical-Social System to Promote Sustainability

To provide context for the ensuing dialog, we offer the following definitions of some keywords and phrases which we have used in this communication:

Autonomy is a state in which a robot or piece of equipment operates independently, without explicit instructions from a human. https://www.oqton.com/zh/from-automation-to-autonomy/.

Automation is a set of human-defined functions performed by a robot or piece of equipment. https://www.oqton.com/zh/from-automation-to-autonomy/.

Competence is the ability to perform a specific task, action, or function successfully. Incompetence is its opposite. https://en.wikipedia.org/wiki/Competency.

Competencies are the result of integrative learning experiences in which skills, abilities, and knowledge interact to form bundles that have currency in relation to the task for which they are assembled. https://nces.ed.gov/pubs2002/2002159.pdf.

Digitization is the transformation from analog to digital or digital representation of a physical item with the goal to digitize and automate processes or workflows.

Digital business is the creation of new business designs by blurring the digital and physical worlds. It promises to usher in an unprecedented convergence of people, business, and things that disrupts existing business models—even those born of the Internet and e-business eras. https://www.forbes.com/sites/gartnergroup/2014/05/07/digital-business-is-everyones-business/#660921477f82.

Digital transformation is the novel use of digital technology to solve traditional problems. These digital solutions enable—other than efficiency via automation—new types of innovation and creativity, rather than simply enhance and support traditional methods. Success is anchored in both digital business and digitization. https://cn.bing.com/search?q=https://en.wikipedia.org/wiki/Digital_transformation&PC=MENEPB.

Internet of Things is the interconnection via the Internet of computing devices embedded in everyday objects, enabling them to send and receive data.

Internet of People (IoP) is associated digitally with a cyber-physical system. This network is responsible for resolving exceptions, interferences, or conflicting issues in a cyber-physical system that is vested with autonomy.

20.2.1 Digital Transformation in a Cyber-Physical-Social System

A digital *cyber-physical-social* system is a digital model of an end-to-end network which enables the of transfer autonomy from the physical realm to the cyber-physical realm. The human is critical for the cost-effective and safe operation of the cyber-physical-human (social) system. Digitization has made it possible:

- For an IoP to communicate with cyber-physical systems (machines, devices, sensors) and each other via the IoT and innovate.
- For *cyber-physical* systems to perform tasks near—autonomously, a human in the IoP is needed for resolving exceptions, interferences, or conflicting goals.
- For an IoP to be supported by assistance systems anchored in a virtual copy of the physical world (digital facility models enriched with sensor data). Assistance

systems are designed to provide appropriate information for humans in the IoP to make informed decisions to resolve critical problems on short notice in an IoP, where decisions are decentralized.

The leadership of any *cyber-physical-social* enterprise must not only worry about productivity, efficiency, and cost-effectiveness, but it must also focus on innovation. It must respond to new and different societal pressures, usually while internalizing new environmental disruptions and addressing social justice concerns. To survive in such a difficult business climate, a contemporary enterprise must embrace digital transformation for it to innovate and survive.

20.2.2 What Does Digital Transformation Mean for the Future Workforce?

Clearly, with more automation, there is going to be a shift in manufacturing from operational/reactive problems (firefighting) to proactive, system-focused approaches, where potential problems are predicted and addressed before they become actual problems (policing). Undoubtedly, some firefighters will always be required but being able to solve a problem will not be as valuable of a skill as being able to identify a systemic flaw at a stage before it becomes an actual problem.

Not only do members of the IoP have to have the skills to do firefighting, but they also must be trained to do policing. Members need to become experts at sandboxing, piloting, trialing, debugging code, etc. If one takes the Boeing 737 Max disaster, for instance, the failure was not just with the plane crashes themselves. It had more to do with the fact that the code had not been properly tested with a representative set of human pilots; it did not help that Boeing executives initially refused to acknowledge that they had a systemic problem. Boeing has spent a lot to fix the code over the past few months (firefighting), but that cost will pale in comparison to the resources that the company will spend in the coming years and decades to make its software testing processes more robust (policing). There will be organizational resistance to change as well, which means that in a policing world conditioned by increasing automation, non-technical (soft skills) competencies (e.g., ability to continue learning coupled with systems thinking) will be even more essential for members of the IoP.

Peter Senge believes that real learning is central to being human. Through learning, we re-create ourselves; see [8]. Both individuals and organizations can and must learn. For a learning organization, survival learning is necessary. This is often called adaptive learning and essential for survival. However, a learning organization must have both adaptive learning and generative learning, learning which can enhance our capacity to create. Equally important is that only those in the workforce who can recognize and respond to change brought about by digitization will succeed. We suggest that generative learning is foundational to being adaptive.

Enhancing talent involves improving the skills of a person, usually through training, so that they will be better at firefighting and policing. Typically, in a manufacturing enterprise of yore, **talent** could be expressed as follows:

$$\text{Talent} = \text{Experience} + \text{Skills} + \text{Training}$$

However, for a digitally transforming enterprise, we suggest that the preceding expression for talent is inadequate. A person in a digitally transforming enterprise, in addition to being able to adapt to advances in technology, needs to be able to communicate and relate to people (from different disciplines, cultures, values) who may not be co-located. We suggest that talent in a digitally transforming enterprise (**Talent$_{dt}$**) can loosely be expressed as follows:

$$\text{Talent}_{dt} = \text{Talent} + \text{Generative Learning}$$

Although people can be trained to use new technologies and how to communicate and relate to other people, we cannot teach people how to learn, unlearn what is no longer relevant and relearn that which is needed. We can, however, provide an opportunity for people to learn by reflecting on doing (see https://en.wikipedia.org/wiki/Experiential_learning). Through learning, unlearning, and relearning people can recreate themselves. Generative learning enhances our capacity to innovate (see https://www.businessdictionary.com/definition/generative-learning.html).

20.3 Non-technical, Career-Sustaining Competencies for Generative Learning

Engineering designers are engaged in building the future, and, the future is increasingly being determined by a digital transformation, and the students we educate must also be prepared for this changed environment and its further evolution. In the manufacturing industry, this takes the form of Industry 4.0—but other industries are being affected similarly. This will require a substantial modification in our approach to education, both in the specific content which is to be taught and in the competencies which our students will need to develop in order to be successful in future. Here, our focus on in the competencies to be taught rather than the specific engineering content. We propose that the following competencies will be required in future workforce:

1. *The ability to learn, unlearn, and relearn through reflection on doing and the associated creation and articulation of knowledge.*
2. *The ability to speculate about the future and to identify gaps in state-of-the-art or state of practice in ways that may lead to innovation.*
3. *The ability to ask questions, actively listen, reflect, and identify opportunities worthy of further investigation.*
4. *The ability to make decisions to move forward using incomplete information.* Innovation involves risk, and in order to move forward, it is necessary to accept this risk, make a decision and proceed.

5. *The ability to assess and think critically and identify a way forward.* Critical thinking is necessary to move forward on a technical project and in identifying what one needs to learn next. Two perspectives are important—a perspective which is external to the individual, and one which is internal.

20.4 The Way Forward: Let the Dialog Begin

In this paper, we offer some thoughts on enhancing the non-technical competencies that are essential for generative learning of individuals in an IoP in a digitally transforming learning organization. In keeping with our desire to foster discussion, we offer the following questions to critically evaluate what is presented in this paper:

1. How should the description of a digitally transforming enterprise in this paper be enhanced?
2. How should the proffered non-technical, career-sustaining competencies to foster generative learning in people constituting the IoP be modified to cover what is missing in our description of the digitally transforming enterprise?
3. How should we modify our curricula to facilitate newly minted engineers to gain these non-technical career sustaining competencies? For an example of an attempt, at fostering the internalization of the career sustaining non-technical skills see [3, 5, 7].
4. How can we ensure that people constituting the IoP, through generative learning, continue to improve productivity, efficiency, and cost-effectiveness in a learning organization through innovation and under societal pressure to integrate sustainability and social justice with technical efficiency?
5. What is needed to motivate the people that constitute the IoP to join/form a generative learning organization?
6. How should a digitally transforming enterprise facilitate people constituting the IoP develop the non-technical, career-sustaining competencies?

To repeat Charles Darwin's maxim "It is not the strongest of the species that survive, nor the most intelligent but ones most responsive to change." A transformation is taking place in the workplace and this creates a need to adapt, so let us consider the question *what non-technical, career-sustaining competencies are needed for the success of newly minted engineers in digitally transforming social enterprises that mitigate the tensions between people, planet, and profit?* An answer to this question is particularly relevant to a rapidly industrializing country such as India with private and government academic institutions graduating the world's second largest number of engineers.

Let the dialog begin.

Acknowledgments Janet K. Allen and Farrokh Mistree acknowledge the financial support from the John and Mary Moore Chair and the L. A. Comp Chair accounts at the University of Oklahoma, Norman.

References

1. Austin, J., Stevenson, H., Wei-Skillern, J.: Social and commercial entrepreneurship: same, different, or both? Entrep. Theory Pract. **30**(1), 1–22 (2006)
2. Dees, J. G.: The meaning of social entrepreneurship. In: Case Studies in Social Entrepreneurship and Sustainability, pp. 34–42. Routledge (2017)
3. Ifenthaler, D, Siddique, Z., Mistree, F.: Change of attitudes, self-concept, and team dynamics in engineering education. In: Ge, X., Ifenthaler, D., Spector, M.J. (eds) Emerging technologies for STEAM education, pp. 201–215. Springer (2015)
4. Kamala, V.V.R.: A Computational Framework to Foster Sustainable Rural Development in Indian Off-Grid Villages, Masters of Science, University of Oklahoma, Norman Campus (2020)
5. Mistree, F., Panchal, J.H., Schaefer, D., Allen, J.K, Haroon, S., Siddique, Z.: Personalized engineering education for the 21st century. In: Gosspar, M., Ifenthaler, D. (eds.) Curriculum Models for the 21st Century, pp. 91–112. Springer, Berlin (2013)
6. OECD: Policy Brief on Social Impact Measurement for Social Enterprises. Organisation for Economic Co-operation and Development (2015)
7. Peng, S., Ming, Z., Allen, J.K., Siddique, Z., Mistree, F.: Quantification of students' learning through reflection on doing based on text similarity. In: ASME Design Education Conference, online, virtual. Paper DETC2020–19439 (2020)
8. Senge, P.: The Fifth Discipline: The Art and Practice of a Learning Organization. Century Business, p. 14 (1990)
9. Tschang, M., Hajj, A., Lang, A.: The Business of Social Entrepreneurship. Cisco Internet Solutions Group, Slide 17 (2011). https://www.cisco.com/c/dam/en_us/about/ac79/docs/ps/BoSE_IBSG.pdf
10. Weerawardena, J., Mort, G.S.: Investigating social entrepreneurship: a multidimensional model. J. World Bus. **41**(1), 21–35 (2006)

Chapter 21
Application of ASPECTSS™ to the Design of Vocational Training Spaces for Individuals with Autism Spectrum Disorder

Nithya Venkataraman and Kudrat Kashyap

Abstract Autism spectrum disorder (ASD) is a developmental disorder that leads to a characteristic pattern of perceiving, thinking and learning. ASD, due to its complex nature, has not been given enough attention in design guidelines or building codes. Designing spaces in which an individual with ASD could be sensorily comfortable can potentially help in the management of as many of the symptoms as could be manifestations of their sensory discomfort. The personal interpretation of a physical environment affects the meaning attached to it. Thus, it becomes essential to design the built environment for individuals with ASD in order to maximise their independence and efficiency. Vocational training in craft-based skills involves a specific sequence of process requirements. The working environment can embody these requirements to be more conducive to learning for individuals with ASD. The study explored the application of ASPECTSS™, a design index, for the design and layout of the training institute. Based on the preliminary observations, the study mainly focused on three areas, acoustics, compartmentalisation and escape spaces. Improvement of acoustics through installing high noise reduction coefficient materials, provision of an escape space and collapsable structures for compartmentalisation were the interventions that were suggested to the institute.

21.1 Introduction

Autism spectrum disorder (ASD) is a neurodevelopmental condition characterised by deficits in reciprocal social interactions and communication skills, accompanied by restrictive and repetitive behaviours [8]. Stereotyped movements and engagement in repetitive activities, resistance to environmental change or change in daily routines and unusual responses to sensory experiences are some behaviours that characterise

---

N. Venkataraman · K. Kashyap (✉)
National Institute of Fashion Technology, Bengaluru, India
e-mail: kudrat.kashyap@nift.ac.in

N. Venkataraman
e-mail: nithya.venkataraman@nift.ac.in

A. Chakrabarti et al. (eds.), *Design for Tomorrow–Volume 2*, Smart Innovation, Systems and Technologies 222, https://doi.org/10.1007/978-981-16-0119-4_21

autism. Autism research often describes some of the behavioural factors as a relative strength of the cognitive profile. As early as 1944, the cognitive skills of individuals with autism have been observed and penned down. A study described children with autism as unquestionably endowed with good cognitive potentialities and possessing intelligent physiognomies [2].

Autism spectrum disorder, as the name suggests, is called a spectrum disorder due to the complexity of the types and the degree of disorder. The three main areas of difficulty faced by individuals with autism are known as the "triad of impairments". They are difficulty with social communication (difficulty with verbal and non-verbal communication), difficulty with social interaction (difficulty with social relationships) and difficulty with social imagination (difficulty in the development of interpersonal play and imagination) [7]. The best predictors of social outcome in children with an ASD diagnosis are cognitive skills and language skills. Both are currently listed as specifiers for ASD in DSM-5 [1].

Since a few decades ago, architecture practice considers that there are individuals with different types and degrees of disabilities (mainly visual, hearing and motor impairments). Some disabilities are not so "visible", and that is, usually, ignored in the task of making the environment accessible [9]. Autism spectrum disorder is one of these disabilities. It is often said in the autism community, "If you have met one person with Autism, you have only met one person with Autism", thus elucidating the idiosyncrasies of ASD.

Individuals with ASD often experience difficulty in processing the information they receive through their senses. Several elements from the environment become a barrier which somehow impedes the understanding of the environment, and, in turn, leads to the so-called socially unacceptable behaviour (gestures, verbal utterances and body movements). In the eyes of a lay observer being at that exact time in that exact space, the whole scenario would be perceived as inappropriate behaviour in a given situation [11]. Whereas it has been the imbalance between the environment and an individual's ability to adapt to it, that has triggered the seemingly peculiar behaviour. The built environment, therefore, has to be considered vital as it greatly influences, directly and indirectly, individuals with "less visible" impairments.

According to the existing literature on environment and behaviour relations, how a person behaves in a particular situation reflects the interaction between the two, not the person alone or that person's environment [4]. Performance of pupils with autism is enhanced in an appropriate physical environment [3]. To work towards building such an environment, architecture or space design plays a vital role. The physical architectural surroundings of colour, texture, ventilation, sense of closure, orientation and acoustics must be designed before, rather than after the sensory malfunction occurs [6]. Perhaps by altering this sensory input in a manner designed to accommodate specific needs of individuals with autism, the behaviour may be improved, or at least a more conducive environment created, for more efficient skill development.

Research in this field has shown this to be a successful approach, particularly in the area of acoustical design [6]. However, much of the literature on accessibility has focussed on physical disability, with less attention paid to "unseen" disabilities or differences. Despite the growing incidence rate, autism is excluded from design guidelines, even building codes which address the requirements of people with special needs [6].

Craft provides a viable medium for individuals with autism to gain vocational training. It also provides engagement with fine motor skills that may be beneficial for their management. Vocational training, especially in skill-based crafts, involves a specific sequence of activities and process requirements which must be a part of the working environment of individuals with ASD, in order to make it more conducive to learning. Workspaces for people with ASD in India are not inclusive, especially concerning skill-based vocational training. There are limited studies which focus on the layout design of a vocational training institute for individuals with learning disabilities. Despite the importance of vocational training being recognised as a bridge to an actual working place in the life of an individual with autism, this area has not been studied yet.

The gap in the existing literature suggests that designing sensorily favourable training spaces for individuals with autism may help individuals in experiencing the world with relatively lesser sensory overloads. Thus, this study aims at understanding the interaction of individuals with ASD, within the spaces of a vocational training institute. It also explores the application of ASPECTSS™ [5] for suggesting feasible alterations to the space in a skill-based vocational training institute. ASPECTSS™ is a group of design principles consisting of acoustics, spatial sequencing, escape space, compartmentalisation, transition zones, sensory zoning and safety.

21.1.1 Research Question

How can we design a space for individuals with ASD who train in skill-based activities, in order to enhance their learning experience and augment their behavioural outcomes?

21.1.2 Objectives

The study aims at the following objectives:

- To study the existing workspaces for skill-based activities designed for teenagers and adults with autism spectrum disorder.
- To understand the workspace requirements for skill-based activities for individuals with autism spectrum disorder.
- To explore the possibility of using ASPECTSS™ as a design framework for re-designing the workspace.

To the best of our knowledge, this was the first study that explored the application of ASPECTSS™ to the design of a vocational training institute in India.

21.2 Methodology

21.2.1 Participants

Participants for the study consisted of the trainees at a vocational training institute located in Bengaluru. The institute has prior permission from the parents for observation. The total number of trainees observed at the institute is 13, ten boys and three girls. The ages of the trainees range from 14 to 33 years. The objectives of the study were explained to the authorities at the institute before commencing the study. The observation took place for a period of 20 days from January 2020 to March 2020. Interviews were conducted for teachers at the institute. Teachers were asked for their views on the influence of space on the behaviour of trainees. Apart from the teachers, an architect with prior experience in the field of special needs was interviewed. Parents who have children with autism also work at the vocational training institute to better understand the needs and demands of their children. A therapist undergoing training in special needs was also interviewed. She too has a son with autism.

21.2.2 Procedure

The research followed an ethnographic approach. The study follows an exploratory research design as it aims to understand the interaction of individuals with autism and the spaces around them. Both participation and non-participation observation were employed to observe the trainees. The observation was carried out from 11 a.m. till 3 p.m. on most days. The trainees participate in many activities during the day, each scheduled according to the abilities and interests of the trainee. The institute provides training in three broad categories of skills, namely craft-based skills (e.g. block printing, the printing of table cloth, cushions, spreads and other home decor items, jewellery making and sewing) hospitality skills (e.g. back end kitchen jobs, baking cakes and cookies, maintenance of workspaces, table setting and catering) and digital skills (e.g. training in graphic designing and multimedia). The institute works towards creating valued social roles for individuals with autism. A class on pragmatic language is also conducted at the institute.

The researchers devised a procedure to develop suggestions on three areas from the ASPECTSS™ framework, as they were found to be most relevant for the study. These are—acoustics, compartmentalisation and escape spaces. The trainees were observed for 20 days, and their behavioural practices mapped concerning these three selected areas. Based on the preliminary observations, a structured interview was conducted with an architect. It was followed up by detailed observation and interviews of teachers at the institute. Suggestions were made based on feasibility and cost-effectiveness (Table 21.1).

Table 21.1 Plan of the study

Methodology	Sample characteristic	Number of observation days	Principles selected under ASPECTSS™, based on observation
Ethnographic approach; Exploratory research design	$N = 13$ 10 boys 3 girls	20	Acoustics Escape space Compartmentalisation

21.3 Analysis

Post the preliminary observations, the researchers prepared a report for the institute. The observations were collated and studied along with the content from the interviews. Suggested areas of intervention were derived from content analysis and observations (Table 21.2).

For ease of understanding, a simplified version of the floor plan was drawn using basic shapes. The floor plan is illustrated in Fig. 21.1. It represents the area where the craft-based and hospitality training is conducted. This area is multifunctional. The suggested changes in this area are compartmentalisation of work stations in block printing area and installation of high noise reduction coefficient material (e.g. cork, hemp husk, acoustic panels) on the shutters to reduce outdoor noise. The institute lacks an unoccupied space which can be converted into an independent escape space. However, an escape space can be created in the classroom where pragmatic language is taught. The class is conducted only once a week. A collapsable structure that can be drawn when the class is not in session will prove useful as an escape space. Partition sheets made of Cane are in use at the centre. Thus, Cane is the suggested material to construct the partition for the escape space.

21.4 Results

A user-centric approach is necessary when designing a physical environment for individuals with autism. We aim to augment the insights received through this approach with design principles. The observations and the corresponding suggestions are presented in a tabular form in Table 21.3. Some behaviours in the trainees occurred repetitively throughout the observation. According to the three-dimensional model (basic level, integrative level and logical level) [10], repetitive activities may relate to different deficits and compensatory processes within and between dimensions. Ben Shalom's model of frontal integration is non-deterministic, allowing different aetiologies (causes) to similar symptoms [12]. Such is the case with behaviours observed in individuals with autism. We may not be sure of the cause; however, through trial and error, the environment around these individuals can be altered to fit their needs best. Behaviours presented in the table can be modified for the benefit of the individual through design intervention.

Table 21.2 Observed behaviour and suggestions for the space design at the training institute

Day of Observation	Observed behaviour	The applied principle under ASPECTSS™	Reason for application of principle under ASPECTSS™
Day 1	Noise of a motor running outside. Stimming (self-stimulatory behaviour) observed in one of the trainees. Slapping and hitting his head with force		To improve the acoustics of a space, the hypersensitive profile must be given priority. It is relatively easier to add elements to space rather than to subtract [6]
Day 2	A trainee sat next to speakers while they were on. His ears were closed. Although the trainee has a hypersensitive profile, he continued to sit in the class expressing discomfort	Acoustics	The training institute works out of a shop cum office complex. There are shutters in the front wall, which allow a significant amount of noise from the road to enter the activity area. Since some of the trainees are sensitive to loud noises, this disrupts their behaviour
Day 3	A trainee hit himself on his head, screaming loudly & making sounds that expressed discomfort		
Day 1	A trainee lied down next to my feet, on the floor amid a session		
Day 2	Loud shouting by a trainee may suggest the need for a space to recalibrate senses	Escape Space	Trainees are often in discomfort or are overwhelmed. An observer cannot be confident of the cause of discomfort. Probable causes are: • Unfiltered noise • Visual stimulus in the space allocated for craft-based activities
Day 3	Another trainee sat in a corner all day long. He did not participate in any class. The teacher on duty said he had fought with his mother in the morning		

(continued)

Table 21.2 (continued)

Day of Observation	Observed behaviour	The applied principle under ASPECTSS™	Reason for application of principle under ASPECTSS™
Day 1	Trainee continuously explored the objects kept right behind him during block printing. Visual stimulus is in excess	Compartmentalisation	Space can be customised based on function to achieve maximum productivity. Multifunctional spaces should be avoided to reduce sensory confusion [6]
Day 2	Teachers conducted a meditation session post lunch, in the lunch area itself. Most of the trainees were distracted		Visual stimulation leads to sensory overload, resulting in behaviour change. The area allocated to block printing is vast in comparison with areas allocated to other activities. Compartmentalisation in this area will prove beneficial
Day 3	A trainee was exploring a pair of cloth scissors, which were lying on the block printing table. The trainee could injure himself and others with a large pair of scissors made of metal		

The researchers selected three design principles, namely acoustics, compartmentalisation and escape space for application in this study. The training institute runs in a rented space. Hence, the construction of a new space or addition of elements was not economically and practically feasible. The suggestions are limited to the use of a collapsable structure for creating an escape space, rigorous use of cupboards and use of colour coding for compartmentalising the work stations and the installation of high noise reduction coefficient material on the shutters to improve the acoustics.

21.5 Discussion

The study explored the application of ASPECTSS™, a set of design principles to the design of a vocational training institute. These guidelines have been developed to design efficient classrooms for children with autism [6]. However, the design index

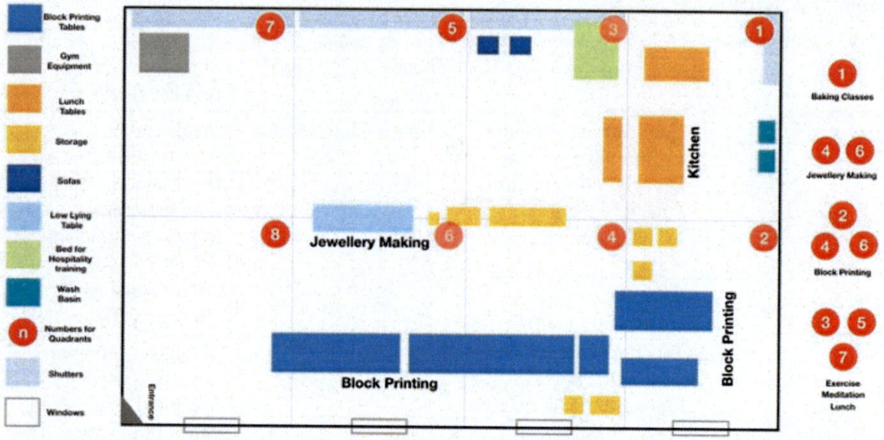

Fig. 21.1 A simplified version of the floor plan where craft-based activities such as block printing, jewellery making, hospitality and baking are conducted

itself has a vast scope of application. Out of the suggestions made, improvement in acoustics is of great consequence for the training institute. Acoustics are the most influential architectural factor in the behaviour of individuals with autism [6]. Dr Temple Grandin, professor, inventor and a spokesperson for autism, affirms that reactive behaviours may result through a mere expectation of being subjected to some form of objectionable noise. Natural materials such as cork and hemp husk function as sound absorbers. We suggest the installation of panels made with such materials to improve the acoustics of the activity area.

Compartmentalisation of work stations can prove useful, especially in the block printing area as it would limit the visual stimulus for the trainees. The physical compartmentalisation of activities will decrease visual distractions and limit the field of peripheral vision. These issues typically limit the attention span of an individual. If the visual stimuli are limited, the child will learn to focus on the educational task at hand [6]. Here, we speculate that an adult or teenager with autism will also be benefitted by decreased stimuli in his/her environment. The provision of an escape space is of importance as well. The proposed escape space can be envisioned as a quiet corner. The use of accessories such as textured-rugs and cushions is suitable.

Calming toys for individuals with autism can be used as well. Such a space can help in managing over-stimulation and its manifest behaviours. Of particular mentioned here is the stimming behaviour. We speculate that stimming, which is a self-stimulatory behaviour, can be effectively managed through time spent in an escape space. Each individual is different; however, stimming usually occurs due to a lack or overload of sensory information from the environment. Such behaviours are comforting to the individual. Stimming behaviours can be controlled in a person who engages in them. However, the opinion of experts is conflicted on whether or not it should be controlled.

Table 21.3 Keywords from interviews and the corresponding suggested ASPECTSS™ design principles

Interviews with teachers & architect	Identified keywords from interviews	Principles under ASPECTSS™	Suggestions made to the training institute
Architect, Teacher 2	Calibration, Respite, Sensitive students	Acoustics	Material with high noise reduction coefficient to cover the shutters in the craft-based activity area, as shown in Fig. 21.1
Architect, Teacher 2	Place to rest between sessions, respite, Texture,	Escape space	An escape space is particularly useful for the trainees with a hypersensitive profile. It can help them to recalibrate their senses as and when they feel the need A teacher suggested the need to create a space that can be used by the teachers as well to take rest between sessions. As ASPECTSS™ suggests, an escape space can have benefits not only for the individuals with ASD but also for their teachers
Architect, Teacher 1 and 2	Colour-coded spaces, Activity-based spaces, Classrooms for activities, Partitions in classrooms	Compartmentalisation	Compartmentalisation of work stations. It will prove beneficial in the block printing area. Under compartmentalisation the following is suggested based on feasibility: • Strict use of cupboards for tools by teachers • Colour coding of the floor and walls according to type of activity

The intention of designing a training institute based on a design index such as ASPECTSS™ is to maximise learning and productivity. Trial and error will occur with an intervention thus made. Learnings from observation and interviews of those occupying such spaces can substantially increase our understanding of the idiosyncrasies which would otherwise go amiss. A designed configuration in space can help develop skills to handle the chaos in a real world, uncontrolled situation. Creating an environment where the individual with autism is more focused and less prone to sensory overloads will provide a platform for more efficient skill development in a shorter period.

21.5.1 Scope and Limitations

The primary research of the study is limited to skill-based vocational training. The intervention aimed to follow the suggestions enumerated by a pre-test and post-test design of experimentation. The researchers envisioned a quasi-experiment which would reveal the change in behaviour if any, after the design intervention. The lockdown imposed by the government to curb COVID-19 limited the intervention. The study reveals, in its limited progress, the idiosyncrasies of individuals with ASD in a vocational training centre. The aim is to follow up with the intervention once the lockdown restrictions are lifted. The study faced some methodological limitations. The sample of the study is limited to one vocational training centre in Bengaluru, wherein the number of individuals with ASD is 13. Although the study is in the area of special needs and results are not generalisable by this very nature of the study, the findings may not be verifiable independently. The method of data collection also involves self-reported data in the form of interviews. This technique in itself suffers from some shortcomings like selective memory and attribution. The nature of questions in the interviews attempted to eliminate any biased retrieval. The researchers did not have permission to talk to the teachers and the trainees without prior consent which hampered the rigour of data collection. Collection of data is limited to a defined time frame when the researcher was visiting the facility. Video recording was not permissible due to the number of activities which take place at the centre. The government enforced lockdown also impeded the progress of primary research.

21.6 Conclusion

The study echoes the relevant literature through its suggestions. A space for individuals with autism spectrum disorder must be designed through a user-centric approach. The ASPECTSS™ design index is one such tool that enables designers to create inclusive and sensory-friendly spaces for individuals with autism. The researchers have applied three most relevant principles, acoustics, compartmentalisation and escape space in the study. Despite the limitations, the study adds to the pool of knowledge in fields such as special needs, human-centred design, autism spectrum disorder and space design. The proposed ideas are economical and feasible according to the space available at the institute. Based on data analysis, they afford serviceability and convenience to the trainees and the teachers at the institute.

References

1. Harris, J.: Leo Kanner and autism: a 75-year perspective. Int. Revi. Psychiat. **30**, 1–15 (2018)
2. Kanner, L.: Early infantile autism. J. Pediat. **25**, 211–217 (1944)
3. Khare, R., Mullick, A.: Educational spaces for children with autism: design development process In: CIB W 084 Proceedings, Building Comfortable and Liveable Environment for All, Atlanta, USA, pp. 66–75 (2008).
4. Mead, G.: Mind, Self and Society. University of Chicago Press, Chicago (1934)
5. Mostafa M.: Architecture for autism: autism ASPECTSS TM in school design. Int. J. Archit. Res. **8**, 143–158 (2014)
6. Mostafa, M.: An architecture for autism: concepts of design intervention for the autistic user. Int. J. Archit. Res **2**(1), 189–211 (2008)
7. National Autistic Society, United Kingdom. https://www.nas.org.uk. Last accessed 13 April 2020 (2016)
8. Salazar, F., Baird, G., Chandler, S., Tseng, E., O'sullivan, T., Howlin, P., Pickles, A., Simonoff, E.: Co-occurring psychiatric disorders in preschool and elementary school-aged children with Autism spectrum disorder. J. Autism Dev. Disord. **45**, 2283–2294.
9. Sánchez, P.A., Vázquez, F.S., Serrano L.A.: Autism and the Built Environment. INTECH Open Access Publisher (2011).
10. Shalom, D.B.: The medial prefrontal cortex and integration in Autism. Neuroscientist **15**, 589–598 (2009)
11. Smith, D.: Spatial design as a facilitator for people with less visible impairments. Aust. Med. J. **1**(13), 220–227 (2009)
12. Tzur, G.: Thinking in three dimensions: a different point of view for understanding autism. Front. Hum. Neurosci. **9** (2015)

Chapter 22
Ways of Teaching—Simulating Real-Life Scenarios into Twenty-First Century Interior Design Education

Amrita Ravimohan

Abstract The dichotomy that exists between the sanitised atmosphere of campuses versus the messy reality of 'sites' often strikes interior design educators and practitioners alike as anomalous. It is imperative for students to better understand real-life scenarios, but the very nature of interior design sites means that it is difficult to recreate those conditions on campuses, and hence, the experience is incomplete. All design schools christen the core courses in which projects are undertaken by their students as 'studios' (similar to what design practices call themselves), but in their present form these do not allow for several crucial issues to be addressed. While certain other design disciplines find it easier to simulate these conditions as the nature of their creations is of a standalone variety, the proposed design of an interior space is a conception that is rooted and hence dependent on many external factors. Secondly, if an interior design proposal (or even a mock up) is to be created in three dimensions for the purposes of demonstration, it requires such a high investment of resources (time, labour and money) that it becomes unfeasible. This paper aims to investigate the ways in which students, through their undergraduate interior design education journey, can learn in a studio environment that accurately reflects the potential which an ongoing site holds, along with its intrinsic constraints and challenges. While the wind beneath the students' wings is the educator who encourages them to dream, experiment, make mistakes and be free of the commercial shackles that often dictate design decisions on sites, in order to have a complete education and become well-rounded designers, today's design studio must better equip them with a holistic outlook of the profession, which includes the ability to recognise and acknowledge practical considerations. This, in the long run, will only improve the employability and placement prospects of the graduates of tomorrow.

A. Ravimohan (✉)
School of Design & Innovation, ISDI, Mumbai, India
e-mail: amritar0208@gmail.com

A. Chakrabarti et al. (eds.), *Design for Tomorrow–Volume 2*, Smart Innovation, Systems and Technologies 222, https://doi.org/10.1007/978-981-16-0119-4_22

22.1 Introduction

> Students are supposedly being prepared for practice, but they are usually led to think of themselves as independent designers, which is a rare condition in real life. The larger, team-oriented form of practice is often poorly understood and sometimes denigrated, although it is now the dominant and most vital structure in the profession - Caesar Pelli

This paper aims to deal with the transformation of a student into a professional, using the tools available in Design Education, through the author's particular lens of interior design to demonstrate certain concepts and innovations. Design itself is subjective, so how can one successfully teach it? Therefore, Design Education itself has many varied approaches, but most educators are in agreement that the current models leave the student under-equipped as interns, and then further as working professionals. This paper aims to study the lacunae that exist in the current systems and then investigate and explore what altered methods can be applied by educators and imparted to young minds.

22.2 Research Methodology

The methodology undertaken for the purposes of this research paper is a combination of primary and secondary research sources. These include

- A student survey (Appendix 22.1).
- A customised questionnaire regarding design education methods (Appendix 22.2).
- Some observations about the functioning of institutes and offices that are gleaned through empirical data from the author, as a practitioner of interior design and design educator since the past 15 years.
- Secondary research that was conducted through an online literature review.

22.3 Academia versus Practice: Understanding How Interior Design Projects Work

22.3.1 The Design Studio

A typical design studio project given to individual students lasts for a duration of 15–19 weeks. Usually, a group of students is assigned to a faculty mentor/guide, who gives them feedback on their individual progress on a weekly basis. While research on the typology is conducted by the students, there is a nominal teaching/instructional component from the faculty side. Characteristically, the following stages are followed by the students during the design studio:

- Research on the given typology/stakeholders/users

- Site study
- Conceptualisation
- Design development
- Design resolution

Final outcomes usually include drawings, sketches, renders, mood boards, physical models, etc.

22.3.2 *The Office*

Interior design projects in the office (also known as the practice or the firm) typically run for a period of 4–6 months. Here, the main impetus is towards getting work started on the site. The major chunk of the project time goes towards the 'execution' of the design. After the final drawing sets are explained and handed over to the team of contractors, there are periodic site visits by the design team to ensure that the designs are carried out as per the desired specifications and quality control with respect to materials and workmanship is maintained. This forms a crucial part of the project, close supervision ensures that timelines and budgetary constraints are met, and the project is completed to the highest standards.

Therefore, as demonstrated above, because of the stark difference of the working styles of the academic studio vs. the office studio, when a student or a fresher comes into the office, it becomes very difficult for her/him to quickly assimilate and change gears. In addition, every office has its own unique work culture and subjective ways of handling the project, thus minimising the newcomer's role to an extent that the early years of their career are often spent unlearning the theoretical knowledge gained in the institute.

22.4 Gaps in Interior Design Education Today

22.4.1 *A Bird's Eye View*

Design is a process. Though the steps that a designer takes may differ from project to project, it is not merely landing on a solution but building and developing it through multiple iterations, discussions, feedback, resolution and more. Conversely, the industry is totally outcome driven, due to budgetary constraints and tight deadlines, among other factors, and the students have to make this switch immediately when they transform into professionals.

In the questionnaire, when quizzing design educators about the gaps in today's design education, the question that came up was that what is the role of the institute towards preparing the students for the industry? Many educators believe that the years spent by the design students at the undergraduate level should be spent honing and developing their higher-order thinking skills, such as conceptualising, ideating and building a bank of history and theories. While successful courses do impart these, the lack of practical knowledge is sorely felt when these same students go out and become job seekers. While the institute needs to clarify that the education the students receive is related to the field of design, and not practice or job related, this difference is not clearly illustrated, leading outgoing batches of students to believe they are equipped as designers as well as practitioners. If that does not turn out to be the case, they believe that the institute has failed to give them a well-rounded education.

Further, many practitioners believe that instead of getting an education, if an aspiring designer was to spend an equivalent amount of time working in the capacity of an apprentice, then the learnings would be the same, if not more. What then is the problem with such a system? To quote Gaurish Chandawarkar, senior academician and mentor for the Centre for Interior Environment and Space Design (IES Institute for Skill Development)—'The river which is the profession, has two banks, practice and academics. One participates with the other. However large the river shall become, the banks have to bridge this gap and communicate. There should be a healthy discussion between the practice and academics. Both have to work towards a vision for the profession. The day we shall bridge the gap, we shall affect the profession in a positive manner. Till then, challenges such as gaps in theory and practice will have to borne by the students and young architects'. Academics and practice are not interchangeable entities, each has its specified role, and current academics is not in a position to replicate the modes of practice.

22.4.2 Gaps that Come to Light During Internship

A student, who leaves the cocoon of the institute to intern at an office, is often surprised to learn within a few days that most academic learnings are not helpful when it comes to practice. The offices on their part throw varied tasks into the intern's lap, and what invariably happens is learning on the job, i.e. trial and error, and in an optimistic scenario, over time one develops a mutually acceptable way of working that is suitable to the individual, the project and the firm.

However, this lacuna becomes grossly apparent when an interior designer is expected to act as a manager of contractors, vendors, agencies and liaise with them as well as clients on a number of issues. Even though an overwhelming 83% of undergraduate students surveyed believe that developing innovative or path-breaking designs should be prioritised, as they grapple with these site-related coordination issues in the real world, they also realise that sticking to the budget and timelines is given more importance.

The primary issue currently is that firstly, the 2–4 month period of internship does not accommodate the longer project timelines. Secondly, an intern might join the office when the assigned project is already underway, resulting in the intern being put onto sundry tasks without getting an opportunity to engage with many aspects of the project. Among the students surveyed, during internships, 89.7% were put onto drafting or rendering jobs, 79.3% were given an opportunity to go on accompanied site visits and 48.3% were engaged in follow-up or coordination tasks. The dissonant or out of sync nature of the internship experience points towards a disengagement of practice from academia and vice versa.

The concern also lies within the internship structure that most institutes follow. The internship occurs in a single block somewhere nearing the final year of the undergraduate course, which does not allow the students to apply their internship learnings or experience different working cultures of studios. In Design after all, 'doing is learning'. It is only by constant practice and exposure can students get better at their chosen vocation. So instead, if academia and practice were more closely aligned via regular Summer Schools, Design Cells or Design Charrettes that fostered interaction right from the beginning, not only would the students be better prepared when they finally went in for internships, but the offices also would be better receivers of these interns.

22.5 Holistic, Reflective and Designer-Centric: Transforming the Current Models of Learning for the Twenty-First Century

From the previous sections, the various hurdles that the students face when they step out into the real world have been established. No person can transform magically into a consummate professional within a singular stint of working, i.e. the internship. The difference in the aims of the academic learning outcomes taken up by the institute and the more pragmatic ones learned out on the field while working must be distinguished right from the start. Arjun Sharma, academician with KRVIA goes so far to say that institutes should 'stay away from this false pretence of preparing students for Day 00001 of entering an industry which is so entrenched in history, theory, international and regional culture, politics, technology and the vicissitudes of daily life.'

There must be opportunities open for the student to be '*shishyas*' or 'apprentices' working with an experienced designer at multiple junctures during the undergraduate journey. This can be achieved in multiple ways.

Among the students surveyed, only 15.5% were unsure of simultaneously being involved in course work as well as office projects throughout the academic year. Thus, it is imperative that academia and industry are not disengaged from each other with the student being the only link between them. Just as how institutes prefer getting faculty on board who have on field experience, practices also must encourage their staff to participate in academia if not through regular teaching, then through the introduction of collaborative projects on which students can actively participate. Multiple practices can bring their projects at various stages, and students can be involved in these stages of design depending on which office has which project at which stage. According to Rajratna Jadhav, senior academician with the Academy of

Architecture, Mumbai—'For this to work, over the years, a large network of offices will have to be built that are willing to work with various institutes in an integrated way for the long term. The extent of the network depends on the number of students, but the network can keep growing accordingly'.

> A building has at least two lives – the one imagined by its maker and the life it lives afterward – and they are never lived the same.—Rem Koolhas

In design schools, students often take up 'case studies' where particular buildings and spaces are studied and analysed. But these case studies are looked at only from the designer's point of view leading to a myopic understanding that the spaces are created as if frozen in time—the singular point just before the keys of the completed project are handed over to the client, when everything is perfect. Sangeetha Solanki, senior academician with the Pillai HOC College of Architecture, Navi Mumbai, tried a small alternate experiment with a group of students in order to acquaint them with a project in a real-life office scenario. The students documented the entire design process from the beginning to the end, which allowed them to understand its various aspects from the design team and all other stakeholders. Of course, for this one needs cooperation from design practices but it must be accepted that it is mutually beneficial as most of the time, office deadlines do not allow the staff time to document processes and analyse processes.

Institutes, traditionally, believe that in order to instil the right design values and be able to fully explore the potential each given site offers, students must be shielded from the costing or budgeting aspects that their proposals would incur. However, there are ways around this. It should be clarified that project cost is not merely a number; it encompasses specifications (which material is used where and why), estimation (forecasting material as well as labour costs) and time frames (certain processes may be more time consuming than others and consequently more expensive). Two innovative suggestions for specific courses received from faculties of design must be mentioned here. Dnyanesh Madgavkar, senior faculty at KRVIA and ISDI, suggests a plug-in to the primary studio project wherein students would be asked to design a small project for two budgets, helping them explore materials, and finishes using the same design concept. Along the same lines, Shreyas More, faculty of interior design and researcher at ISDI, suggests an elective course that enables the students to account for all statistical data regarding resources, labour, services, finishes, construction and maintenance of project materials in one semester followed by a design challenge in the subsequent semester to apply this knowledge.

When the industry players are constant, active participants within the academic system, there could be a weekly engagement where the students participate in the daily proceedings of the office or the design team visits the classroom studio and both parties equally contribute to the stage of the design process that is ongoing at the time. In the survey conducted, an overwhelming 84.5% of students believed in the value of visits to ongoing as well as completed sites. If physical exchanges are not possible, one can use technology to virtually broadcast 'events' such as site visits, client meetings and discussions with contractors, etc. For some reason, if practices are unwilling to participate in this process, this can be simulated within the campus by the method of role play. Groups of students can assume the role of stakeholders such as the client, project lead, contractor, vendor, designer and intern, work with a given situation and ideate on possible mutually acceptable solutions. This can help in fostering the value of how to work as a team and inculcate the importance of a professional working environment.

22.6 At the Threshold of New Technology

We are today at the cusp of a technological revolution when it comes to visualising spatial design. Conventionally, interior design has relied on physical models that exhibit the spatial qualities of their designs along with the material palette. But certain key aspects such as natural and artificial lighting, tactility and durability of materials along with design details are invariably lost. A physical model of 1:50 scale can be effective in this regard but is often not practical as the model becomes too large in size and expends valuable resources. Currently, one of the most standard ways of representing a design proposal especially when pitching it to a client is a series of three-dimensional representations, commonly known as a '3D views' or 'renders'. While this enables the client to better understand scale, extent of space, material palette, among other things, at a later stage it also becomes an unachievable quest to convert that 3D image into reality, thus causing several problems with the executing team on site as well as clients.

On the other hand, in the last 4–5 years, there have been experiments within the realm of Augmented Reality (AR) that have demonstrated that it has the potential to open up new possibilities, and as the technology becomes more readily available and commonly used, it can be harnessed by academia to encourage students to not only showcase their ultimate designs but as an intermediate tool for design development, testing and gauging if their designs have fulfilled the intended studio goals/client expectations. Among the students surveyed, 94.6% of them believe AR maybe the way forward to realistically simulate site conditions within the design studio on campus. Shreyas More, faculty of interior design and researcher at ISDI, offers that methods such as projections of real-life designs in space, virtually manipulating models with one touch, modelling in space using hands and sensors and interactive simulations can help budding designers to better understand their own designs and gives them an opportunity to upgrade accordingly. Further, Dnyanesh Madgavkar, senior faculty at KRVIA and ISDI, is more in favour of an industry-academia tie-up in this aspect too. 'Practicing faculties/designers could make videos of their projects for pre-construction, during construction and post-construction and, this basis can be used for making AR material which could even be interactive. This will enable students depending on their interest to go in as much detail as they want. However, I recognise this will involve some major investment. Realistically, over time, if we can build a resource/library for projects of various typologies (like offices, residential, restaurants, spas, hospitals, etc.) of simulations for real-life conditions of construction phase of projects, it could be an effective tool in the integrated studios'. So perhaps, this is the way forward with respect to technology—not as a tool for merely impressing clients and making showy demonstrations of the final outcome, but as a learning and development method for the student designers themselves at various crucial stages.

22.7 Conclusion: Integration and Collaboration Leading to Innovation

This paper began by examining the distinct working methods and timelines within the institutes' design studio vs. a design office. Thereafter, gaps which exist within the interior design education process at a macro- and micro-level were identified, and their impact on a student's professional preparedness was analysed. Next, the student's experience during the internship period was broken down, and some possible reasons for certain drawbacks in the system were discussed. The last section of the paper puts forth certain alternate modes of learning for students using an integrative approach that encourages the introduction of a collaborative model between the industry (the offices) and academia (the institutes).

The potential of positive influence that these transformed methods can have on graduating students being more well-rounded professionals, aware of the possibilities and the limitations that each site has to offer can be demonstrated using the analogy of medical colleges, according to Divya Vijaychandran, senior faculty of interior design at ISDI. While studying medicine, the period where the student is a resident, attached to a hospital interning while under the guidance a senior doctor, is built into the programme. Hence, there is also a clear understanding that the theoretical education that the institute offers is not a substitute, but a necessary supplement to the practical knowledge one gains on field, for example during surgery. Both complement each other and along with the students, even the experienced doctors recognise the value of this system and hence are ready to invest time and effort preparing the interns for a full-fledged dive into the profession. This exchange also opens the possibilities for innovation such as using path-breaking simulative methodologies to understand how to perform better which also become learning aids in institutes. If a similar model is followed by interior design schools, by setting up a system of integration and collaboration with the industry, which benefits both parties, then we can finally look towards achieving holistic learning.

Appendices

Appendix 22.1: Responses to the Survey Circulated Among Undergraduate Interior Design Students and Fresh Graduates (58 Responses)

1. You are a:
   - Incoming 3rd year student 27.6%
   - Incoming 4th year student 46.6%
   - Graduating student 20.7%
   - Working Professional 5.2%
2. As an intern, what were the tasks assigned to you (tick all that are applicable)
   - Drafting 89.7%
   - 3D Rendering 43.1%
   - Site visits (only with senior designer) 79.3%
   - Site supervision (independently) 43.1%
   - Follow up/Coordination (on phone or in person) with contractors, vendors, agencies 48.3%
   - Negotiations of budget, bargaining, tendering etc. 20.7%
   - Preparing timelines, BOQs, other documentation 36.2%
   - Meeting clients 29.3%
   - Allied tasks (Photography, Photocopying, Taking Messages) 46.6%
   - Sourcing materials 1.7%
   - Preparing mood boards 1.7%
3. Which of these tasks did you feel most unequipped for? (Tick all that are applicable)
   - Drafting & 3D Rendering 29.3%
   - Site visits & Site supervision 15.5%
   - Client Meetings & Follow up with team 15.5%
   - Costing, Negotiating, Tendering, Preparing Timelines and BOQs 38%
   - Allied Tasks (Plotting, Photocopying, Photography) 6.9%
   - Designing Related 3.4%
   - No particular problems 13.8%
4. Do you feel that a part time office work and part time course work (college studios, lectures) would work for you simultaneously?
   - Yes 44.8%
   - No 15.5%
   - Maybe 39.7%

5. What do you believe is the most important criteria for a project (you can select multiple criteria)
   - Completed within the pre-decided timeline 62.1%
   - Completed within the pre-decided budget 74.1%
   - Experimental/Innovative design 82.8%
   - Bespoke details (hand crafted, customised) 55.2%
   - Winning Awards/Getting Published 15.5%
   - Fees being paid on time by client to Designer/Contractors/Vendors 51.7%
   - Pleasing the user 1.7%
   - Keep creative juices flowing 1.7%
6. During your student years, which according to you enables you to learn more?
   - Visits to ongoing sites 15.5%
   - Visits to completed sites 0%
   - Both have Value 84.5%
7. Do you believe a 3D render is the most appropriate way of representing your design to the client/faculty?
   - Yes 43.1%
   - No 13.8%
   - Maybe 43.1%
8. In the near future, do you believe it would be feasible to use AR/VR to realistically simulate site conditions within the design studio on campus?
   - Yes 50%
   - No 3.4%
   - Maybe 46.6%

Appendix 22.2: Questionnaire Circulated Among Design Faculties (18 Respondents)

1. In the field of Interior Design, the gap between theory and practice is something which is often cited as an issue. This is currently resolved during the early working years, but how can institutes better prepare graduating students for the challenges and demands of the workplace?
2. In the present scenario, an intern rarely gets to fully participate in the design projects due to the discrepancies between the project timelines and internship period. Do you believe this leads to an incomplete experience? If yes, how can this be addressed?

3. Considering the way VR & AR technologies are advancing, how do you think they can play a role in simulating real life conditions for students in their design studios?
4. Studios are often kept free from commercial constraints, i.e. we do not ask students to design within a budget. Is this in your opinion a right approach or does this leave them under-equipped?
5. Within our design studios do you feel the need to simulate a more realistic working environment for the students? If yes, what are the methods can we use/should we use to bring about this change?

Names and Specialisations of Design Faculties from various institutes who responded to the Questionnaire

1. Mr. Gaurish Chandawarkar—Interior Design and Architecture
2. Ms. Divya Vijaychandran—Interior Design
3. Ms. Elizabeth Jerome—Interior Design
4. Ms. Neha Shah—Interior Design
5. Mr. Shreyas More—Interior Design
6. Mr. Dnyanesh Madgavkar—Interior Design and Architecture
7. Ms. Kimaya Keluskar—Architecture
8. Ms. Sangeetha Solanki—Architecture
9. Mr. Rajratna Jadhav—Architecture
10. Ms. Isha Patel—Communication Design
11. Ms. Prarthana Patil—Design Studies
12. Mr. Mohan Neelakanthan—Fashion Communication and Styling
13. Ms. Solange Suri—Fashion Design
14. Ms. Ritika Karnani—Product Design
15. Mr. Shamit Shrivastav—Product Design
16. Ms. Prachee Velankar—Landscape Design
17. Mr. Arjun Sharma—Landscape Design
18. Mr. Sandeep Menon—Landscape Design

References

1. Maii, E., Dina, T., Zeyad, E.S.: Collaborative pedagogy in architectural design studio: a case study in applying collaborative design. Alexandria Eng. J. **58**(1), 163–170. Accessed https://www.sciencedirect.com/science/article/pii/S1110016818301704 (2019) (Elsevier)
2. Shaheen, Z.R.: A qualitative investigation to develop interior design pedagogy that fosters students' creativity. In: Doctor of Philosophy. Faculty of Arts, Law and Social Sciences, Cambridge School of Art, a.r.u., Anglia Ruskin University. Submitted February 2018, Accessed https://arro.anglia.ac.uk/703783/ (2018)
3. Dr. Choi, S.M.: The potential of augmented reality and virtual reality technologies as interior design visualization and presentation tools. In: Interior Design Educators Council, U.S.A., Accessed https://www.idec.org/i4a/pages/index.cfm?pageID=4544 (2017)

4. Martinez, C.: Bridging the gap between design education and practice. In: Gensler on Work. Published 2016, Gensler. Accessed https://www.gensleron.com/work/2017/3/10/bridging-the-gap-between-design-education-and-practice.html (2016)
5. Thomas, K.: Between making and action—Ideas for a relational design pedagogy. In: Emerging practices: professions, values and approaches. Ma, J., Lou, Y. (eds.) College of Design and Innovation, Tongji University. Shanghai, China: China Architecture and Building Press, Accessed https://www.studiochronotope.com/between-making-and-action--ideas-for-a-relational-design-pedagogy.html (2015)
6. Is there a gap between design education and the creative industries? https://www.itsnicethat.com/articles/opinion-piece-lcc-debate

Chapter 23
Ethnography as a Tool to Study Indigenous Craft Clusters to Build Cultural Sensitivity and Inclusivity Amongst Design Students

Lavina Nimba Bhaskar

Abstract The narrative of traditional Indian craft and traditional textile is woven tightly with sub-continent's culture. These indigenous crafts and traditional textiles in India are clustered in small weaver communities or tucked away in rural artisan pockets far from the urban settlements. These craft cluster communities make artisanal handmade products for unchanged rural, massy urban and niche luxury markets. Spradley (1979) defines, ethnography as the work of describing a culture from the native point of view. Ethnography in that sense, in this study, bridges the rural and urban divide to understand indigenous art, craft and the people in them [20]. This paper describes ethnographic research methods that are used as a way of studying different indigenous Indian art and craft forms, and the craft communities who make these artisanal products. It enlarges how design students co-create and innovate design solutions in the textile and craft sector. And within these inter-cultural exchanges and interactions, students learn the life lessons of being culturally sensitive. Through an extensive review of existing literature, decade long ethnographic studies in various craft clusters through out India (conducted in past by the author), and semi-structured interviews with the students, the study aims to answer the core questions of the research with an objective of perhaps finding more harmonious and inclusive societies.

23.1 Introduction

This paper is an ongoing learning of shared experiences with students. It is 10 years of experiencing ethnographic studies with design students. Over these years, field studies have been conducted through the length and breadth of the country. From Honavar to Channapatna, Bhuj, Udaipur, Nathdwara, Molela, Jaipur, Ajmer, Bhilwara, Dharamshala, Barabanki, Peepli, Bhubaneshwar, Norbulingka and many more. And in all the silent exchanges and loud interactions with new people, places

L. N. Bhaskar (✉)
National Institute of Fashion Technology, New Delhi, India
e-mail: melavina@gmail.com

A. Chakrabarti et al. (eds.), *Design for Tomorrow–Volume 2*, Smart Innovation, Systems and Technologies 222, https://doi.org/10.1007/978-981-16-0119-4_23

and art; we as instructors or mentors have tried to observe, analyse and take notes of how our students engage in the ethnographic studies from a native point of view. And how we can tune our ways to construct a learning experience for our student researchers which is culturally sensitive, creative and inclusive. These are the research questions this paper attempts to answer. Research methods adopted to write this paper include an extensive review of existing literature, decade long ethnographic field studies in various craft clusters through out India (conducted in past by the author as mentioned above) and semi-structured interviews and informal conversations with the students; before, during and post the craft cluster visits over decade long period. The study aims to answer the core questions of the research with an objective of perhaps finding more harmonious and inclusive societies.

23.2 Ethnography Over the Years

Gerhard Friedrich Müller, a professor of history and geography constructed the concept of *Völker-Beschreibung* or ethnography, while he was participating in the Second Kamchatka Expedition (from 1733 to 1743). August Ludwig von Schlözer and Christoph Wilhelm Jacob Gatterer of the University of Göttingen introduced the term "ethnography" into the academic discourse in an attempt to reform the contemporary understanding of world history.

Though Herodotus an Ancient Greek historian who lived in Persia had significant works on the cultures of various peoples beyond the Hellenic realm, which earned him the title of philobarbarian, and he possibly created the first ever works of ethnography.

Spradley (1979) defines, ethnography as the work of describing a culture from the native point of view. Ethnographic research seeks to learn from those they are studying and to understand their perspectives, customs, behaviours and meanings within the context of their culture. And at the same time suspending judgements, preconceived notions or stereotypes, and interpretation of behaviour based on their personal cultures and construct an image of the world view of new culture. Spradely's definition finds its roots in Malinowski's definition of ethnography, "from the native's point of view". He emphasizes knowing the nature of "local knowledge" and viewing social life as consisting of negotiated meanings [18].

Communication scholars like Harris also Agar, widely used ethnographic research and elaborated that the purpose of ethnography is to describe and interpret the shared and learned patterns of values, behaviours, beliefs and language of a culture-sharing group. Ethnography is defined as both, a process and an outcome of the research.

Ethnography as a method is a storied, careful and systematic examination of the reality-generating mechanisms of everyday life. Ethnographic work in communication studies explain how ordinary methods, practices, performances construct the ordinary actions used by ordinary people in the accomplishments of creating and maintaining their identities [15].

Like anthropology scholars, communication scholars often immerse themselves and participate in and/or directly observe the particular social group being studied [1].

Clifford and Marcus [5]. in their book *Writing Culture: The Poetics and Politics of Ethnography* described ethnography as postmodern, reflexive, literary, deconstructive or poststructural in nature [5].

Ethnographic research can range from a realist perspective, in which behaviour is observed, to a constructivist perspective where understanding is socially constructed by the researcher and subjects. Research can range from an objectivist account of fixed, observable behaviours to an interpretive narrative describing "the interplay of individual agency and social structure". Critical theorists address "issues of power within the researcher-researched relationships and the links between knowledge and power" [5]

An interesting insight into ethnography is when a native carries out the research.

Traditionally, ethnography was a western gaze towards the far "exotic" east, but now eastern researchers are undertaking ethnographic studies in their own social environment. According to Dewan [7], a researcher may be a native or not; a native researcher may find commonalities that connect him or her to the people and other characteristics that perhaps highlight differences [7].

Ethnography in past has most captured the imagination of researchers in the field of education. Although ethnographies of schooling have been done by a small group of anthropologists for some time, the ethnography "movement" began in the field of education during the late 1960s and early 1970s. The works of various scholars provided examples of the genre that later educational ethnographers would emulate [12].

Collaborations between ethnography and conceptual art have evolved in past decades, drawing on older convergences between anthropology and art practice (Gell 1998). Conceptual artists entered into broadly ethnographic terrains and found inspiration in ethnographic methods, while social and cultural anthropologists and archaeologists pursued experimental methods beyond social scientific realism, gesturing towards or learning from conceptual art—variously contriving social situations and observing their unfolding, studying the social by way of material objects and forms, or emphasizing performative and playful dimensions of fieldwork, seeking poesis and surprise rather than "data" [20].

Ethnography today is studying from the native point. Understanding their perspectives, customs, behaviours, values, beliefs, language, negotiated meanings, local knowledge, artefacts, etc. *Ethnography is both, a process and an outcome of the research. It is postmodern, reflexive, literary, deconstructive and poststructural in nature. Used widely in educating students about cross-cultural phenomena and has a positive relation to conceptual art and design.*

This study further discusses how ethnography can be an important tool to look at traditional art-craft from a native point. It enlarges how design students co-create and innovate design solutions. And within these inter-cultural exchanges and interactions, students learn the life lessons of inclusivity and cultural sensitivity.

23.3 Ethnographic Methods

Traditionally ethnographers to study a cluster or ethnicity selected the most knowledgeable informant who knows the activities of the community well. And that person would identify other relevant subjects to gather information from just like snowball sampling. This process is often effective in revealing common cultural denominators connected to the topic being studied (Fig. 23.1).

Ethnographers mainly use qualitative methods, though they may also employ quantitative data. Dewan [7], the researcher is not looking for generalizing the findings; involves investigation of very few cases, maybe just one case, in detail and involves working with primarily unconstructed data. Emphasis is on exploring social phenomena rather than testing hypotheses [7].

Van Maanen (1988) describes ethnography as an objective study of the situation towards the individual being studied. It is third person's perspective by getting the data from the members on the site. Study is uncontaminated by individual predisposition, political objectives and judgement. Ethnographers will give a detailed report of the everyday life of the individuals under study. He/she also uses standard categories for cultural description (e.g. family life, communication network). And produces the participant's views through closely edited quotations and has the final word on how the culture is to be interpreted and presented (Qualitative Inquiry and Research Design, 93).

In ethnography, the researcher gathers what is available, what is normal, what it is that people do, what they say and how they work. Aim is to observe a situation without imposing any deductive structure or framework upon it and to view everything as strange or unique.

Various methods to study ethnography are personalized, multifactorial, inductive, dialogic, holistic and reflexive [13].

Fig. 23.1 Depicts a model to describe ethnographic study

23.3.1 *Ethnographic Coursework*

Learning research methods should be incorporated in the curriculum of design studies or taught as a special workshop/module to design students. Students in class to prepare for field study must perform ethnographic assignment as a part of the subject—be it interviewing classmates or observing a chosen subject. Craft cluster studies should not be limited to research or documentation of traditional craft, but expose students to societal constructs of livelihood, family and how marginalized sections (women or low caste communities) negotiate their identities in various craft clusters. Evolution of the traditional art-crafts and how it relates to history and extends to contemporary world should be incorporated in the subject.

23.3.2 *Literature Review*

Ethnographers turn to search for others to understand different cultures. Patton and Welsby (1992) and Green (1982) suggest that students need to learn critically when consulting written material watching for cultural bias, and checking to see how well this research aligns their own observations. The writing of others can provide important insights and hypothesis on different culture. Students must learn to judge the veracity of written records by comparing them to their own interviews with the cluster they will take in the field. Before going to the field students or in this case design students must be encouraged to gather already present literature on the craft, cluster and people in them [18].

23.3.3 *Pre-field Visit Talk*

A pre-field visit talk orients students how craft cluster studies sits in the course of Design. From studying research methods as subject to going on craft cluster visit, co-creating new designs, writing a research paper to likely graduation project. Next, students are advised to plan their entire visit. Meeting the artisans/family, documenting the craft, market visits, visiting local museums, scheduling interviews, capturing film footage and photographs and the importance of keeping a free day in hand.

Students are briefed on the importance of maintaining a continuous dialogue with the artisan. From greeting the artisans on the first phone conversation to first meeting them in person, maintaining an organic relationship throughout the visit, building relationship with artisans through craft bazar visits in student's native city, to colloquium paper visit (second stage of cluster studies), craft-based graduation project and beyond maybe in entrepreneurial venture or social research. In this entire engagement its indispensable that the artisan gets fairly paid. When students are at the field, they are encouraged to listen to artisan's concerns, experiencing their culture with local cuisine, talking to locals, meeting master artists, visiting the local museum, shopping at local market to understanding native point of view, looking beyond their own urban life but still gazing at the commonalities in human life.

Students are advised to meet their mentors or instructors everyday in the evening to discuss their progress and experiences with artisans, any problems they are facing in the new town, and overall planning of documentation. Though it is important to note that deliverable planning in terms of co-creating and cluster experience builds together and perhaps cannot to be planned before hand.

Safety of students especially in small town is a major concern as an instructor / mentor (specially when we do travel with young girls). Prior to the visit students are asked to submit details on travel and lodging. In the light of both student's safety and being sensitive to cultural norms of the cluster being studied, students or student researchers are asked not to wear clothing which may be offensive to the town's norm, keep a humble conduct, respect the environment they will adopt for few weeks, and not to travel with expensive things.

23.3.4 Interviews

Ethnographic interviews are to understand artisan's point of view. Leigh and Green (1989) and then Roswell (1987) suggest there is no hurry to begin defining the problem; the idea of student designers as a helper is suspended temporarily in favour of student designers as learner. Listening and understanding artisan's life and unique experiences becomes prime. Ethnographic researchers ask global open questions and are flexible in following conversational interest and direction of the interviewee. These interviews are semi-structured fluid yet alert on the level where student designers are taking part in active observation [18].

Students ask questions about—Traditional Craft itself? How artisans got introduced to the craft? How has the craft changed over the years? What is their daily routine? How is their family structured? Are their children interested in carrying out the craft traditions? As a community do they let the women of the household practice the craft? And a possible request to demonstration of craft?

The interview moves between global, open-ended questions and follow up questions that may represent key cultural issues, conventions, assumptions and colloquial (Spradley 1979). Many times artisan belong to a marginalized caste structure of Indian society. When students are interviewing artisans they are explicitly asked to address artisans by their profession not by caste [18].

23.3.5 Observation

Students spend time in the field interviewing artisans at their homes, neighbourhoods, place of work; actively observing them in their native environment. Food, activities, communication, patterns, interactions (especially in a shared culture group), relationships, dress, jewellery, songs and dance (e.g. weaving songs) are observed and described by students. Ethnographers learn much they wish to know by carefully watching the members' movements, facial expressions and posture [4, 18]

Ethnographers often become participants in casual, friendly conversations suggest Spradley (1979). This observation method is participant observation. But at this point its important to unbiased learners. Students often have difficulty differentiating observations from inference points Carkhuff [18]. Ethnographers must learn to observe impartially without assuming the meanings or causes of behaviour [18].

In past, we as instructors have observed that design students not only adopt the role of a participant observer but sometimes because of the liking to the region, craft, people, or holistic cultural experience students become accultured and emulate how natives dress, talk and enjoy the experience of being native. We see a reflection of this behaviour and extending of self-identity on Insta posts and stories/social media.

23.3.6 Video Ethnography

One of the methods that offers novel opportunities is video ethnography.

Video ethnography is a cinematic approach to recording ethnographic expressions of lived experiences. This approach provides a new angle on qualitative analysis in general and enriches conventional ethnographic analysis in particular. The possibility facilitates for continuous replay provided by video ethnography facilitates the detailed analysis needed to gain a better understanding of the use of tools and technologies or the display of emotions. It also enables multi-modal analysis, in which verbal communication can be linked with visual and aural elements. Video ethnography is particularly useful if research requires exceptional details on conversational analysis [14].

The method may also be appropriate for situations where direct participation of the researcher is forbidden as a cultural norm. Nevertheless, video recording may also bring its challenges, including information overload in the sense that the researcher may focus too much attention on details instead of analysing more important patterns, practices or processes [11].

Video ethnography is widely used in women's studies and documentary making in past. We as instructors have observed it to be an interesting method to study craft clusters. Craftsman and woman sometimes are not comfortable with cameras recording them and may feel shy or apprehensive. Students are advised if it does not work in first go, then student researchers should try to make the subject comfortable first. In one instance students handed over their camera to the craftswoman and asked her to take student researcher's photographs and later students did not push to do the interview the same day but casually spoke to the craftswoman and took the interview next day, while the camera was on.

Video ethnography as a tool allows for long take or extended sequence, which is an unbroken shot, recorded over a prolonged duration, which evokes varied aesthetic experiences comprising of whatever the camera shoots, which creates a submersive experience for audience. And by crafting this the lived experience into a shareable form of media, video ethnography creates a form of "vibrant knowledge" [14].

Video ethnography may be a fairly a new method to carry out ethnographic studies, but students still chose to document the craft, region and people in photographs.

23.4 Craft Within the Socio-Cultural Fabric of India

Furthering the discussion on craft it is important look at the social–cultural fabric in which the narrative of traditional Indian craft and traditional textile is woven tightly. These indigenous crafts and traditional textiles in India are clustered in small weaver communities or tucked away in rural artisan pockets far from the urban settlements. These craft cluster communities make artisanal handmade products for unchanged rural, massy urban and niche luxury markets.

Indian crafts are traditional occupations of a particular community but also plays a vital role in the socio-religious celebration of the region. Traditional crafts are the knowledge systems, creations, innovations and cultural expressions of people in a given geographical cluster. They are transmitted to enhance, safeguard and perpetuate the identity, well-being and rights of the indigenous peoples [16] (Figs. 23.2 and 23.3).

Traditional craft education is transmitted from one generations to other, through the skill learning—technical substance which is the knowledge of materiality and the value transmission which is enculturation. From the perspective of the artists who had learnt the skills from direct ancestors. This then forms a model of art education construction within society [8].

Crafts are not simply a particular way of making objects, but are inextricably bound up with the structures, values, history and identity of the communities in which they are practised.

Indian handicrafts are, historically, part of a generation to generation learning, passed down along a line of inheritance and the transfer of tacit knowledge. Craft has multiple meanings, and functions, ranging from personal consumption, to ritualized, and religious significance, and, or economic activity, which has a commercial outcome. It plays a role in defining regional, national or ethnic identity and culture [19].

While craft already incorporates a relation to nature, its ability to claim something durable from mere resource is yet to be properly exploited. Craft also incorporates a particular mode of production, and a relation to technology that fosters the value of living labour, as well as itself producing technological instruments to increase

Fig. 23.2 Depicts design students with the artisans of Norbulingka. Photograph taken on an ethnographic field trip to Norbulingka

Fig. 23.3 Depicts Thangka artisan with his art. Photograph taken on an ethnographic field trip to Norbulingka

Fig. 23.4 Depicts A silver filigree artefacts. Photograph taken on an ethnographic trip to Odisha to study the craft

the sustainability of practices. In order to be fully realized as sustainment, craft also must be constituted as a new cultural concept and supported by the (re)formation of social institutions to more formally support and legitimize newly formed sustainable practices of fabrication [10]. (Figs. 23.4 and 23.5).

Traditional crafts are evolving and in many cases perishing because of new modern materials. And newer generation of artisan is leaving the profession all together sometimes because of decreasing economical viability of the craft and sometimes the charms of working in big city [16].

Modernization and globalization had a major impact on the craft practice, which involves many stakeholders. Craft interventions act as great initiative in facing commercialization by developing the new contemporary design based on market needs [3].

Craft is evolving. Through the past and in present times researchers, designers, local governments and NGOs alike have redefined features of crafts which helps to create a possible innovation through the collaboration and exchange between "craftsman" and "designer" [17].

USTAD project is one such example of co-creation.

Fig. 23.5 Depicts a family of Oriya filigree artisan. Photograph taken on an ethnographic trip to Odisha to study the craft

23.5 Co-creating Design Solutions

Clifford and Marcus [5] in their book *Writing Culture: The Poetics and Politics of Ethnography,* helped to encourage the development of "collaborative ethnography". This exploration of the relationship between writer, audience and subject has become a central tenet of contemporary ethnographic practice. In certain instances, active collaboration between the researchers and subjects has helped blend the practice of collaboration in ethnographic fieldwork with the process of creating the ethnographic product resulting from the research [5].

Collaborations between ethnography and conceptual art also evolved in past decades, drawing on older convergences between anthropology and art practice (Gell 1998). Conceptual artists entered into broadly ethnographic terrains and found inspiration in ethnographic methods, while social and cultural anthropologists and archaeologists pursued experimental methods beyond social scientific realism, gesturing towards or learning from conceptual art—variously contriving social situations and observing their unfolding, studying the social by way of material objects and forms, or emphasizing performative and playful dimensions of fieldwork, seeking poesis and surprise rather than "data" (Marcus 2010; Pearson 2004; Schneider and Wright 2013; Ssorin-Chaikov 2013) [20].

Before discussing on co-creating design solutions with artisans, its important to note that indigenous crafts and traditional textiles in India are clustered in small weaver communities or tucked away in rural artisan pockets far from the urban settlements. Craft cluster micro-scale sectors make artisanal handmade products for unchanged rural, massy urban and niche luxury markets.

Native craftsman and craft-woman are well aware of the likings of their community and mostly create artefacts that are religious in nature. Massy urban and niche luxury market on the other hand is a foreign market with changing needs and preferences. Mostly to cater to the design aesthetics and urban needs students collaborate and co-create with the craft clusters to innovate new designs and products keeping it as close to traditional craft as possible. To site an examples in many cluster artisans kept making pen stands, which is redundant as products. To them depending on what craft they do new products are suggested. This kind of design was limited to material play for making sure livelihood of artisans keeps going, but failed to look at the native point (Figs. 23.6 and 23.7).

While co-creating with craftsman and craftswoman it is crucial to maintain livelihood of artisans, incorporate urban needs, understand native point of view, understanding local issues, local design sensibility, local source resources and sustainability of a design.

Fig. 23.6 Depicts English learning material for primary schools in illustrative craft of Phad paintings

Fig. 23.7 Depicts English learning material for primary schools in illustrative craft of Phad paintings

An ethnographic research to create English learning material while reviving oral traditional stories through the illustrative craft of Phad paintings was co-created by Ms. Nehal Singla, under NIFT Craft Cluster grant of 2019 and Mr. Kalyan Joshi a Phad Craftsman from Bhilwara. This research looked at a social problems which local primary schools were facing of in learning English language. The challenge was to localize English and make it approachable by using visuals that were native, stories that were of the soil and with easy hindi language translations. This was well received by primary schools. In a mentor capacity, it was important that the above-mentioned methods were taught/discussed with her, she was connected to right artisan and local. Since the perspective is not how urban design students can cash in from indigenous craft but how we can add value to their cluster and learn to be culturally sensitive.

23.6 Building Cultural Sensitivity and Inclusivity

Ethnography relies greatly on up-close, personal experience. Participation, rather than just observation, is one of the keys to this process [9].

According to Brewer (2000), a leading social scientist, data collection methods are meant to capture the "social meanings and ordinary activities". The goal is to collect data in such a way that the researcher imposes a minimal amount of personal bias in the data.

As much as design students are instructed and conditioned to be unbiased researchers, a problem that ethnographers face is that of reflexivity. Reflexivity is not a problem of craft cluster being studied rather, it refers to deepening our understanding of ourselves and our position in the world to understand the social reality of others. When we are able to connect that as researchers we are a part of the bigger picture we are able to understand reflexive nature of the study of our society.

Historically, scholars like William-Dorothy Thomas, Merton, Popper, Nagel have looked at reflexivity as mere predictions. Later viewed as an issue and a solution to the problem of structure and agency, by Anthony Giddens and Pierre Bourdieu. Michael Foucault in his book Order of Things, while defining reflexivity says man is both knowing the subject and the object of his own study. And in that context, it includes both a subjective process of self-consciousness inquiry and the study of social behaviour with reference to theories about social relationships.

Within cluster studies and during the pre-field visit talk, students are asked to maintain continuous dialogue with the artisan. From greeting, the artisans on the first phone conversation to first meeting them in person, maintaining an organic relationship throughout the visit, building that relationship with the craftsman or craft-woman through craft bazar visits in student's native city, to colloquium paper visit to artisan's town (second stage of cluster studies), craft-based graduation project and beyond maybe in entrepreneurial venture or social research. In this entire engagement its indispensable that the artisan gets fairly paid. A pre-field visit or pre-cluster study talk is always a good idea before students head for field study.

Many a times students are in the habit of texting, and they text the artisan that too in English language. The artisan may or may not be tech savvy to text back and mostly is unlikely to know read English so even if text is written in roman hindi its ineffective communication and fails. Students are asked to think from the subjects' point of view, calling may be the best medium to reach out. Even if it means calling multiple times before connecting.

When students are at the field they are encouraged to listen to artisan's concerns, experiencing their culture with local cuisine, talking to locals, meeting master artists, visiting the local museum, shopping at local market to understanding native point of view, looking beyond their own urban life but still gazing at the commonalities in human life. In past, we as instructors have observed that design students not only adopt the role of a participant observer but sometimes because of the liking to the region, craft, people, or holistic cultural experience students become acculturated and emulate how natives dress, talk and enjoy the experience of being native. We see a reflection of this behaviour and extending of self identity on Insta posts and stories/social media.

During the visit, artists invite students to their homes for a meal or tea. This activity has proven a success to build inclusivity. When a student eats with the subject he or she is studying, it strikes a human connection. Many students report back to their mentors or instructors, saying the they were humble, homely and food was very delicious. This activity extends to women students to engage in talking to wives or daughters of the artisans and get the chance of observing the subject closely in the basic unit of community or extension of self which is their family. Students can also understand how women negotiate their identities in small towns and villages.

Many students reported that they find it easy to talk to artisan in field study but when the artist meets them in their social environment they may hesitate to talk to them at first or may find themselves bound to purchase the products that they are selling (perhaps in market place), to which mentors responded by encouraging students to make a warm greeting and expressing shared memories.

Saussure and Pierce in their study describe how language form a semiotic meaning. Charles S. Peirce's viewed language as a vehicle for scientific inquiry, giving equal emphasis to signification and communication (Charles Peirce 1931–58). The process of understanding meaning of a word and resultant can vary in different cultures [6].

Many times artisan belongs to a marginalized caste structure of Indian society. When students are introducing artisans they are explicitly asked to address artisans by their profession or as artists not by the caste they belong to.

To build cultural sensitivity amongst design students or young researchers, it is crucial that they drop their bias and preconceived notions and observe a situation without imposing any deductive structure or framework upon it and to view everything as "strange or unique". Design students here are both knowing the subject and the object of their own study. To understand craftsman or craftswoman students need to view life from their perspective. Like any other human relationship or emotion to develop cultural sensitivity continuity of relationship is required. Typical social activities like eating together, or finding commonalities in shared culture are starters for conversations. And if these conversations are effective and form continuous dialogue over a period of time, students build sensitivity towards a culture and artisan also no longer feels marginalized but included.

23.7 Conclusion

Ethnography is not an oriental gaze. (Edward Said 1978) It is to be seen or studied from the native point. Ethnography is understanding native's perspectives, customs, behaviours, values, beliefs, language, negotiated meanings, local knowledge, artefacts, etc. Ethnography is both, a process and an outcome of the research. It is postmodern, reflexive, literary, deconstructive and poststructural in nature. Used widely in educating students about cross-cultural phenomena and has a positive relation to conceptual art and design. Through the immersive experiences of instructors and research mentors with design students an effective way undertaking ethnographic is suggested which includes ethnographic coursework, literature review, pre-field visit talk, field visit (observation, interviews and video ethnography), on-visit everyday experience sharing and co-creating. This study sees co-creation of designs with designer and artisan collaboration as a way forward to meet the challenges of modern world.

To build cultural sensitivity among young researchers, it is crucial that student researchers drop their bias and preconceived notions and observe a situation without imposing any deductive structure or framework upon it and to view everything as "strange or unique". To understand craftsman or craftswoman students need to view life from native perspective and engage continuously in a dialogue with the artisan which cannot only bring the cultural sensitivity but bridge the othering between rural and urban.

References

1. Bentz, V.M., Shapiro, J.J.: Mindful Inquiry in Social Research. Thousand Oaks, California (1998)
2. Bhaskar, L.: Indian textiles with narratives. In: Presented at WASET Conference at Sydney (2018)
3. Bintan, T., Rigout, M., Cassidy, T, Dallabona, A.: Designing an Appropriate Technology for Revitalising Traditional Craft Practice. Case study: Indonesian Stitch Resist Dyeing, Design Journal. Routledge Publication (2019)
4. Carkhuff, R.: The Art of Helping, 6th edn. Human Resource Development Press, Amherst, MA (1987)
5. Clifford, J., Marcus, G.E.: Writing Culture: The Poetics and Politics of Ethnography. University of California Press (1986)
6. Crow, D.: Visible Signs, 3rd edn. Bloomsbury Publishing (2016)
7. Dewan, M.: Understanding Ethnography: An 'Exotic' Ethnographer's Perspective, Asian Qualitative Research in Tourism. Springer, Singapore (2018)
8. Eko, S., Rohidi, T.R., Florentinus, T.S., Kartika, D.S.: The art education construction of woven craft society in Kudus Regency, HARMONIA. Journal of Arts Research and Education, CrossMark Publication (2017)
9. Genzuk, M.: Tapping Into Community Funds of Knowledge, Effective Strategies for English Language Acquisition: A Curriculum Guide for the Development of Teachers (1999)
10. Matthew, K.: Theorising a transformative agenda for craft. In: Craft + Design Inquiry, ANU E-Press Publication, Originally published by Craft Australia 2011 (2011)
11. Mikko, V., Vaara, E.: Strategic ethnography 2.0: Four methods for advancing strategy process and practice research. In: Strategic Organization, vol. 12, Sage Publications (2014)
12. Ogbu, J.U.: School Ethnography: A Multilevel Approach, Anthropology & Education Quarterly, vol. 12, no. 1, Issues in School Ethnography, Wiley (1981)
13. Preece, J., Sharp, H., Rogers, Y.: Interaction Design: Beyond Human-Computer Interaction, 4th edn. Wiley (2015)
14. Rodman, D.: Video Ethnography: Theory, Methods and Ethics. Routledge Publication (2019)
15. Rubin, R.B., Rubin, A.M., Piele, L.J.: Communication Research: Strategies and Sources. Thomson Wadworth, California, Belmont (2005)
16. Simon John, S.: Traditional Knowledge of Folk Crafts in Tamil Nadu. Indian Journal of Traditional Knowledge, vol. 9. Published by Anthropological Survey of India Southern Regional Centre, Mysore, Karnataka (2010)
17. Temeltaş, H.: Collaboration and exchange between "Craftsman" and "Designer". In; Symbiosis Towards Product Innovation, The Design Journal. Published by Routledge Taylor & Francis (2017)
18. Thornton, S., Garrett Kendra J.: Ethnography as a Bridge to Multicultural Practice. Journal of Social Work Education, Taylor & Francis, Ltd. on behalf of Council on Social Work Education (1995)

19. Vermeulen, H.: Early History of Ethnography and Ethnology in the German Enlightenment: Anthropological Discourse in Europe and Asia. Privately Published Leiden (2008)
20. Wenzel, G., Kelly, A.H.: Science, Ethnography, Art, Social Studies of Science, vol. 46. Published by Sage Publications (2016)

Chapter 24
Game-Based Learning for the Awareness of Culture & Tradition: An Exploratory Case Study on the Indigenous Naga Tribe

Susmita Roy, Pankaj Pratap Singh, and Abhijit Padun

Abstract Game has always been an enjoyable part of human life. Nowadays, gamification technique efficiently relates learners from various domains to their concerned area of interest. This paper attempts to integrate the ancient culture of the Naga tribe of North-east India with the digital gamification realm. The Naga culture survived centuries through oral tradition of tales, myth, customs and beliefs. This traditions mainly flourished in their dormitories called "Morung".The advent of modernization replaced the traditional system with western education, which led to the constriction of sharing cultural values. Gamification, if designed with indigenous insight, serves its own outlook and can go far beyond gameplay session. For understanding the acceptance of the content of the conceptual game, a survey was carried out within a group of Naga people. Their affinity towards the gaming world enhances the further possibilities of development of the conceptual game, which may serve as an archive of their cultural heritage.

24.1 Introduction

Tribal Culture is perceived to be a significant part of human society. Nagaland, located in the North-east of India, is an abode of fascinating tribal culture and rich history. The ancient Naga tribes attained the status of self-sufficient and self-reliance as the community had plenty of reserve forest and land for cultivation [1]. Game as a learning tool is an age-old concept used by the Nagas for teaching children

S. Roy (✉) · A. Padun
Department of Multimedia Communication and Design, CIT Kokrajhar, BTAD, Kokrajhar, Assam, India
e-mail: s.roy@cit.ac.in

A. Padun
e-mail: a.padun@cit.ac.in

P. P. Singh
Department of Computer Science and Engineering, CIT Kokrajhar, BTAD, Kokrajhar, Assam, India
e-mail: pankajp.singh@cit.ac.in

A. Chakrabarti et al. (eds.), *Design for Tomorrow–Volume 2*, Smart Innovation, Systems and Technologies 222, https://doi.org/10.1007/978-981-16-0119-4_24

safety, strength, agility, and sportsmanship. The elders of this tribal community were powerful storytellers of their creation and history. These were the effective means to carry forward their cultural values to their younger generation. All these customs were shared in the dormitory called "Morung". However, with the rise of globalization and modernization, this traditional youth-centric dormitory system was replaced with the western schools and colleges [1]. This shift brought drastic changes in their lifestyle, which is gradually distancing them from their tradition and culture. The present education system too lacks the concern for cultural conservation. This paper attempts to put forward a novel approach of spreading awareness among the new generation towards the necessity of acknowledgement, restoration and revival of their age-old culture and belief. The proposed conceptual gamification tries to create an environment where young generation can practice the real-life situations and challenges in the form of role-play, thereby making decision to safeguard and restore the lost customs and traditions in a fictional settings.

24.2 Background Study

Nagaland, one among the sister states of north-east India, covers an area of 16,579 km^2 located between 25° 6′ and 27° 4′ N latitude and 93° 20′ and 95° 215′ E longitude. Nagaland is the abode of 16 major tribes situated amongst the hills and high mountains with beautiful landscapes and exquisitely rich flora fauna. The tribes carry folklore depicting the brave acts of their ancestors and costumes displaying their ancestral lineage. Before modernization, culture was an indispensable inherent asset of the Naga tribe. However, with the arrival of western culture and literacy, most of the natives got inclined towards the modern lifestyle. The historical facts and contemporary literature are evident that in the process of chasing modernization, the tribe may have gradually shifted apart from their traditional culture. Some researchers also argued that such a drift may contribute to the shunning of traditional customs and belief. The researcher also believed that presently, only remote areas carry the authenticity of their culture [2]. However, another researcher Horam [3] claims that very recently, Naga people are making serious efforts to revive their vanishing traditions and half-forgotten cultural heritage. Celebration of cultural events like the Hornbill Festival is one such example, but these festivals generally ends up being superficial and simulated, completely not resembling the earlier times [4].

24.3 Literature Review

A study has been conducted on gamification techniques and gamification framework based on the following criteria:

Game Versus Gamification According to Marczewski [5], there is a precise distinction between game and gamification design and its features. Games are created to be dynamic and engaging mainly through "action, feedback, and emotion" [6]. Gamification is the process of using such elements and mechanics in non-game contexts [7].

Gamification in Education Research studies have widely acknowledged the impact of gamification elements in stimulating high levels of engagement [8]. Gamification has explored to increase engagement in the context of several domains. One of the contexts in which gamification has recently begun to emerge is education. The main goals of using gamification in education are to increase student motivation, participation and community building [9] and promote student engagement within the course [10]. By determining learner's characteristics, defining learning objectives, creating education content and activities and finally adding game elements and mechanism to it, can be an effective strategy for implementing gamification for education.

Game Elements Game elements are the building blocks that craft the game experience. Different authors have different viewpoints on how a game should be defined and what elements game should have [11]. Adams et al. [12] categorize game elements based on the aspects of game definition: challenging goals (challenges and goals); play (interactive activities, feedback, competition and collaboration); rules (core mechanics, levels, balance, luck and four risk); and pretended reality (game world, characters, game aesthetics and story). Werbach and Hunter [13] divided game elements into three levels: components—lower-level elements (achievements, avatars, badges, big boss, combat, leader board, levels, points, quests, teams and more); mechanics—rules that use lower-level elements (challenges, chance, competition, collaboration, feedback, resources, reward and turns); and dynamics—concepts that shape the game (constrains, emotions, narrative, progression and relationships). Deterding and associate [7] categorized game elements by game design levels: interface patterns (e.g. badges, leader board and levels); game mechanics (e.g. time constraints, limited resources and turns); game heuristics (e.g. constant play, clear goals and game stiles); game models (e.g. challenges, fantasy and curiosity); and game design methods (e.g. testing, play centric design and participatory design). Different authors list the same game elements, but they categorize them differently.

Connecting Motivation and Engagement in Gamification Many studies has revealed that learner's level of engagement increases when game elements are introduced to any domain of interest. Motivation is another key factor to take into consideration when designing a gamification. The Self Determination or the Theory of Motivation Technique proposed by Edward Deci and Richard Ryan presents three basic needs of humans: *Relatedness*—the desire to socially and emotionally connect to others, *Autonomy*—the need for freedom, and *Competence*—desire to be good at something.

Study of Various Gamification Design Frameworks Di Tomasso's framework for successful gamification is based on the self-determination theory which involves the following steps based on player differences and social influences: the reason to gamify, identify players' profiles and motivational drivers, set up goals and objectives, describe skills, track and measure, define interest, desired outcomes (feedback and the epic win) play-test, and polish. RAMP is another framework used in gamification design based on the self-determination theory. Furthermore, Yu-kai Choua proposed a Complete Gamification Framework called Octalysis. His framework places the most emphasis on human motivation resulting in a human-focused design approach to get the job done quickly. The approach is based on an octagon shape with eight-core drives, viz. epic meaning and calling, development and accomplishment, creativity and feedback, ownership and possession, social influence and relatedness, scarcity and impatience, unpredictability and curiosity and loss and avoidance. All these frameworks served their purpose of integrating non-gaming context with gaming elements, explored to increase engagement and motivation of learners in the context of several domains.

24.4 Methodology

A content-based gamification technique with a well-structured approach is proposed to demonstrate the intended player-centric gamification design framework. The methodology is an integration of non-gaming context of Naga culture with game thinking approach. The diagram given below describes the framework in five phases, viz. *Define, Analyse, Plan, Design* and *Validate*. The sub-category under each section is stated below with a brief description of each (Fig. 24.1).

24.4.1 Define

Based on the background study, it can be assumed that it is high time for the new generation of the Naga tribe to understand, acknowledge and preserve their cultural heritage before it goes to oblivion. To bridge the gap, a probable solution in the form of a gamification design framework can be conceptualized which would render real-world challenge of restoring the culture in a fictional setting through gameplay. The target users of the game are mainly the students aged between 13 and 16 years of Nagaland and other parts of North-East.

24.4.2 Analyse

The literature review unlocks the possibility of exploring the gamification technique with a new aspect of non-gaming context of tribal cultural heritage. The case studies

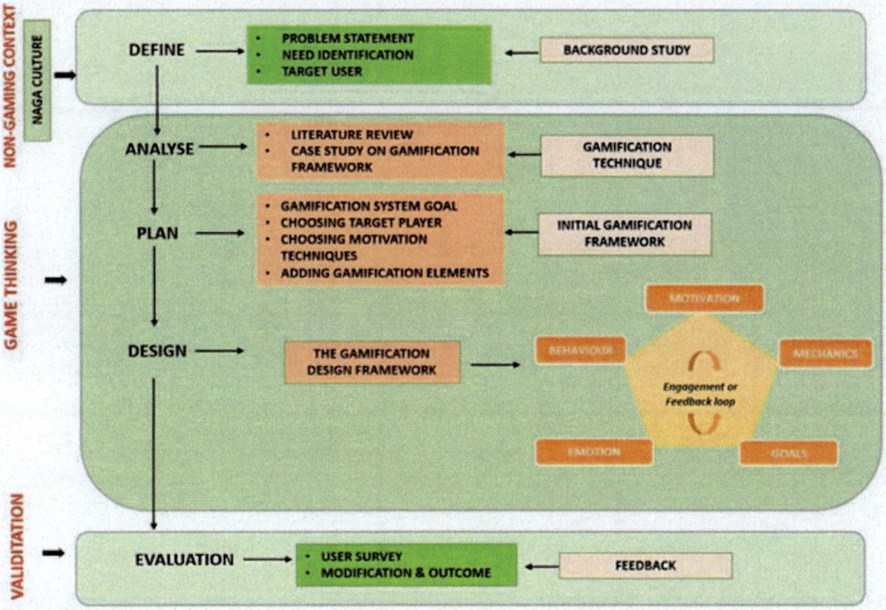

Fig. 24.1 Flowchart representation of the methodology

of different gamification framework further open up the prospect of proposing new methodology to conceptualize a gamification framework to address the integration of this possibility.

24.4.3 Plan

The overall planning of the gamification system is summarized with the following objectives: (1) Designing a gamification system that attract and educates young generation of Naga tribe, (2) Target player type as "*Receiver and Explorer*", (3) Use of motivation technique both *intrinsic and extrinsic* and (4) Use of game elements like *Motivation, Behaviour, Emotion, Goal and Mechanics.*

24.4.4 Design

The gamification framework design emphasizes upon the player's journey depicting its interaction with the five core drive game elements, viz. Motivation, mechanics, behaviour, emotion and goal (Fig. 24.2).

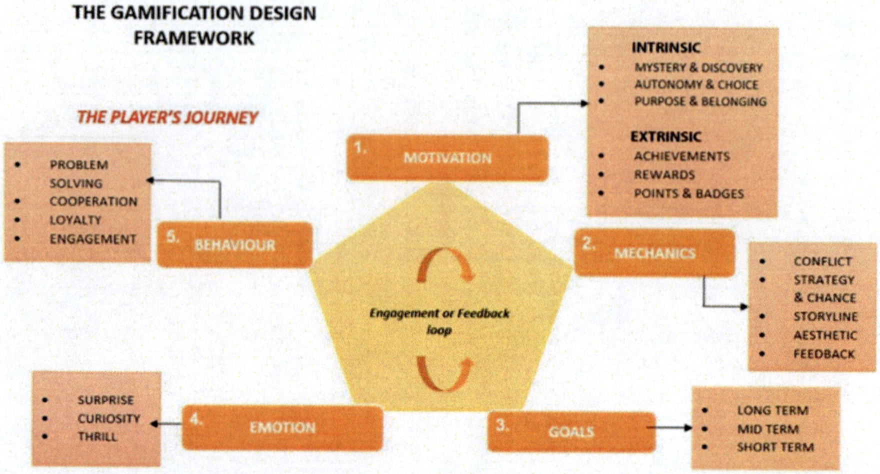

Fig. 24.2 Gamification design framework

(1) Motivation The motivation technique in the gamification framework reflects a well-balanced approach of intrinsic and extrinsic motivation techniques which are as follows:

- **Mystery and Discovery**: The gamification puts the player in the role-play of a saviour of "The Lost Legend Valley". The player must find clues to reveal the story and solve the puzzles to restore the lost valley.
- **Autonomy and Choice**: The player has the full control over choosing the avatars with customized tribal attires and accessories. The player can also navigate using the geographical map of the region.
- **Purpose and Belonging**: While embarking on quest and accepting challenges, players get engaged in the storyline. The quests and challenge gives the player a sense of direction or a purpose in the gamified environment.
- **Achievement, Rewards and Badges**: The progress of the player as learner is represented by earning titles, progress marker, tattoo inking on body, hall of fame, mission trees, collections of skulls, points and unlocking of higher levels, etc.

(2) Mechanics The game mechanics is structured in the following way:

- **Conflict**: Antagonist, the demons, in the gameplay challenges and creates every possible obstacle and tension for the player to overcome which makes the gamification very interesting and challenging.
- **Strategy and Chance**: The gradual difficulty in progressing the levels and obstacle course that the players encounter within the game can be centred on chance while the solution that they come up with involves the element of strategy.
- **Storyline**: The Story basically revolves around a fictional place called "The lost legend Valley" which is under the spell and control of demonic power. The player serves as the only hope of the people for their rescue and for restoration of their

(a) (b)

Fig. 24.3 Visuals of the "The Legend Valley" before and after the demonic attack

cultural assets. The story deliberately tries to portray the glimpse of the real-world concern to redeem the cultural heritage of Naga tribe (Fig. 24.3).

- **Aesthetics**: The aesthetic appeal of the gamification environment and the characters tries to capture the tribal essence of the Naga culture. The background of the gamification also tries to depict the scenic beauty of the bio-diverse region (Figs. 24.4, 24.5, 24.6 and 24.7).
- **Clear Feedback**: The instant feedback system in the gamification generates the gradual flow of the events and gives information to the player how far the player is from achieving the goal.

(3) Goals The goals includes long term, midterm and short term. Restoration of the cultural heritage and reconstruction of the traditional culture centric dormitory "Morung" and earning "The title of Legend" is the long-term goal. Rescuing the hallucinated local tribe hostages from the trap of demons is the midterm goal. Quest for Wisdom, Quest for Potion, Demon Headhunting, Earning Tattoo, Points and Badges are the short-term goals.

(4) Emotion Surprises during the quests, curiosity to solve the puzzles and thrill while unfolding the backstory serve the good factors for motivation and develop various positive emotions.

(5) Behaviour The problem-solving ability during the quest or the combat between the player and demons improves the cognitive skill and decision-making ability and keeps the player engaged in the gameplay. Multiplayer feature of the game allows collaboration which enhances loyalty and develops the team work spirit.

24.5 Evaluation

To evaluate the acceptance of the conceptual gamification, a survey was carried out within a group of 40 Naga people which included both youngsters and adults. They

Fig. 24.4 Game icon

Fig. 24.5 Welcome screen

Fig. 24.6 Home screen

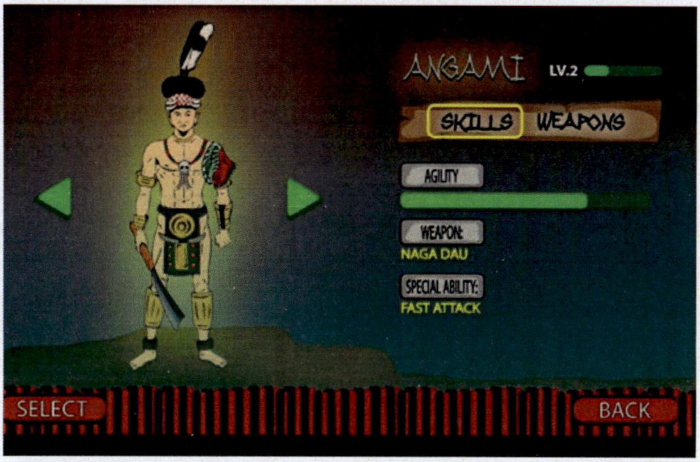

Fig. 24.7 Avatar selection

were asked to share their experience of playing the demo version of the game and render their respective views on the overall impact of the gamification.

24.5.1 User Survey

The survey was carried out in two phases—questionnaire followed by one-to-one interview session. The questionnaire was based on three responses, viz. "Like", "Dislike" and "Not sure" with keen focus on the following five criteria—(1) The concept idea of the gamification (2) Region-specific information sharing capability (3) Motivational factor 4) Authenticity of the content of the game with Naga culture (4) Visual appeal and Fun factor. The Naga people who had actively participated in the questionnaire session were called for the one-to-one interview session to furnish their valuable comments and suggestions.

24.5.2 Result and Analysis

In the given table (refer Table 24.1), column 2 represents the description of evaluation criteria, column 3, 4 and 5 exhibit the three parameters on which the responses of people were furnished. Some of the valuable comments and suggestion of the interview session of the Naga people were included in column 6 (refer Table 24.1).With reference to Table 24.1, it can be observed that majority of the Naga people gave positive response to the criteria they have been given for judgement. The positive reaction (refer Fig. 24.8) on each parameter lays the foundation for justification

Table 24.1 Evaluation of gamification based on responses and feedback of 40 Naga people

Sl. No.	Description of the criteria	Like	Dislike	Not sure	Comments and suggestions
1	The concept of the gamification	28	10	2	Innovative concept Very interesting Amazing idea
2	Information sharing capability	24	8	8	Well executed but needs more focus on facts and information
3	Motivational factor	28	6	6	Appreciable Effective and engaging
4	Authenticity of the content	24	8	8	Interesting storyline Requires a little more exploration of content
5	Visual appeal and Fun factor	34	3	3	Interesting with lots of amazing fun elements

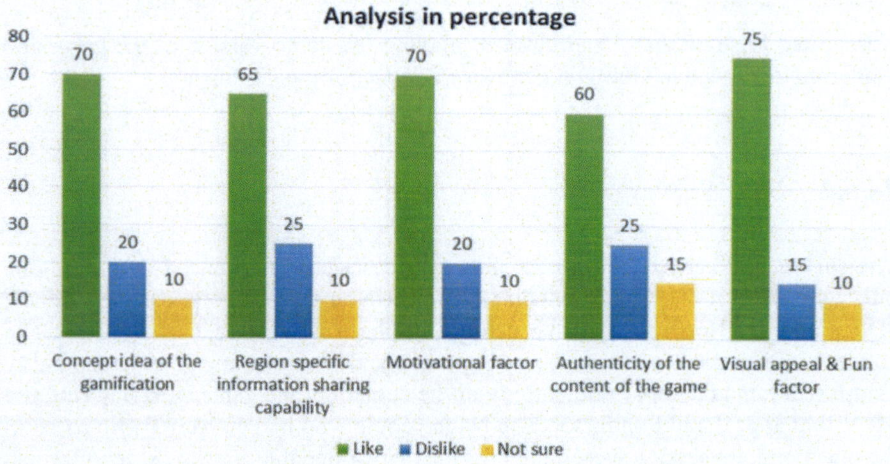

Fig. 24.8 Graphical representation of the responses of the survey in percentage

that the gamification technique was implemented in the right direction, and has the potential to embrace boundless benefit for the learners.

Hence, it can be perceived that the gamification technique gives a positive implication on creating cultural awareness among children with a high possibility in uplifting their motivation and inclination towards their cultural heritage.

24.5.3 Further Modification and the Final Outcome

Based on the interaction session and feedback, rectifications were made in some key areas for more clear and prompt vision. The positive response also created the possibilities of further development of the game in near future.

24.6 Discussion

This paper describes the gamification design methodology and its implementation for creating awareness among young generations of Naga tribe with their ancient culture. Gamification if used effectively can be a powerful educational tool with promising futuristic trend capable of communicating complex subject matters in a very efficient way. Furthermore, using gamification as a cultural awareness tool can be a promising approach in motivating the least interested present generation towards their ancient cultural heritage.

24.7 Conclusion

Based on the positive response, it can be concluded that the experimental gamification has achieved its foremost goal of connecting learners with their cultural roots. The appreciation and acceptance of the concept have opened a new dimension for gamification, as a new learning tool of culture heritage. The experiment may also open up a new approach for digital preservation of tribal cultural heritage. Culture and traditions are an integral part of human life, and while moving forward in the journey of acquiring development and modern pleasure, it is very essential to acknowledge, value and preserve the age-old customs and belief.

Reference

1. Nshoga, A.: Traditional Naga Village System and Its Transformation. Published by Shipra Publications (2009)
2. Aglaja, S., Van Ham, P.: The Hidden World of the Naga: Living Traditions in Northeast India and Burma. Published by Prestel Publishing (2003)
3. Horam, M. (1992). Social and Cultural Life of Nagas. Low Price Publications, 425 Nimri Ashok Vihar Phase-IV, Delhi 110052 (2003)
4. Oppitz, M., Kaiser, T., von Stockhausen, A., Wettstein, M.: Naga Identities: Changing Local Cultures in the Northeast of India. : Snoeck Publishers, Gent (2008)
5. Marczewski, A.: What's the difference between Gamification and Serious Games? https://www.gamasutra.com/blogs/AndrzejMarczewski/20130311/188218/Whats_the_difference_between_Gamification_and_Serious_Games.php. Accessed 10 Jan 2020 (2013)

6. Werbach, K., Hunter, D.: For the win: How game thinking can revolutionize your business. https://fliphtml5.com/ndhs/wtqf/basic. Accessed 05 Jan 2020 (2012)
7. Deterding, S., Dixon, D., Khaled, R., Nacke, L.: From game design elements to gamefulness: defining gamification. In: Proceedings of the 15th International Academic Mind Trek Conference on Envisioning Future Media Environments (2011)
8. J. Hamari, J., Koivisto, Sarsa, H.: Does gamification work?—A literature review of empirical studies on gamification. In: International Conference on System Sciences. IEEE (2014)
9. Moccozet, L., Tardy, C., Opprecht, W., Léonard, M.: Gamification-based assessment of group work. In Interactive collaborative learning (ICL). In: 2013 International Conference on IEEE (2013)
10. Denny, P.: The effect of virtual achievements on student engagement. In: Proceedings of the SIGCHI Conference on Human Factors in Computing Systems. ACM (2013)
11. Salen, K.,. Zimmerman, E.: Rules of Play: Fundamentals of Game Design. MIT Press Cambridge (2003)
12. Adams, W.K., Reid, S., LeMaster, R., McKagan, S.B., Perkins, K.K., Dubson, M., Wieman, C.E.: A study of educational simulations part I—Engagement and learning. J. Interact. Learn. Res. (2008)
13. Werbach, K., Hunter, D.: For the win: How game thinking can revolutionize your business (2012)

Chapter 25
THINK Lab: An Initiative to Foster Creative and Critical Thinking Amongst First Year Design Students

Prachi Mittal

Abstract Synonymous with imagination and reasoning, creative and critical thinking used together increases the possibility of desirable outcomes in situations requiring problem-solving which is a prime function of design. Teaching creativity and criticality appears to be an inevitable task in design education. However, it is often assumed that learners will gain these thinking skills tacitly as a by-product of certain course content delivery that focuses on conceptual or factual knowledge and technical skill development, for the creative disciplines. This paper examines the effectiveness of "Think Lab", a domain-specific, experimental learning module, that aims to develop the creative and critical thinking potential of foundation (first) year undergraduate students of design. Drawing from the work of creativity experts Edward De Bono and Michael Michalko, amongst a few others, Think Lab equips students with formal tools, techniques and strategies for creative and critical thinking. It takes a constructivist approach to learning-teaching, strategically combining the tenets of experiential learning theory and collaborative learning. Primary data has been collected from three batches of students who attended Think Lab, to investigate the appropriateness of the content (what is taught) as well as the effectiveness of the learning-teaching methodology (how it is taught), in context of their future design learning and practice.

25.1 Introduction

The P21 model lists creativity and critical thinking amongst the twelve "twenty-first century skills" that students need to succeed during today's information age [1]. **Creativity** can be defined as the ability to transcend traditional ways of thinking or doing, to generate ideas, methods or possibilities that may be useful in solving problems [2]. **Critical thinking** is the logical, sequential and disciplined process of rationalising, analysing, evaluating and interpreting information to make informed judgements and/or decisions. Synonymous with imagination and reasoning, creative and

P. Mittal (✉)
Indian Institute of Art and Design (IIAD), New Delhi, India
e-mail: nvisage.p@gmail.com

A. Chakrabarti et al. (eds.), *Design for Tomorrow–Volume 2*, Smart Innovation, Systems and Technologies 222, https://doi.org/10.1007/978-981-16-0119-4_25

critical thinking together increases the possibility of desirable outcomes in situations requiring problem-solving.

One of the unique abilities of design professionals is their capability to seamlessly navigate between creative and critical thinking, to create innovative solutions for problems not encountered earlier, and respond effectively to the demands of multiple stakeholders. Teaching creativity and criticality appears to be an inevitable task in design education. The 2007 conference "The student experience in art, design and media education: the drivers for change" posed the uneasy question of whether the tried and tested twentieth century curricular structures (of creative disciplines) are still effective in the twenty-first century?" [3] The world has undergone drastic restructuring from the industrial and post-industrial service economy of the twentieth century, to the age of knowledge economy in the twenty-first century, that is marked by upheavals in technological innovations, and new products and processes coming out of research labs [4]. These shifting paradigms are pressing us to relook at curricula and teaching practices, to forge the development of creative and critical thinking abilities of our students, for "innovation, adaptability and survival" in an increasingly unpredictable and rapidly changing world [5].

This paper is an outcome of an academic initiative titled "Think Lab". The main objective was to develop and test a domain-specific learning module, that uses evidence-informed approaches from research in pedagogy, to bring a change to the creative potential of design students. This initiative operates on two premises–The first is that within design curricula the development of creative and critical thinking is usually implicit to other learnings [5]. The second is that "thinking skills are teachable and learnable", and creativity and criticality is a skill that can be developed and a process that can be managed [6].

25.2 Background

For decades, design curricula have been including several opportunities for creative and critical thinking skills to be developed through practical learning [7]. However, it is often assumed that learners will gain creative and critical skills tacitly, as a by-product of certain course content delivery that focuses on conceptual or factual knowledge and technical skill development for the respective creative discipline [5]. Like most design educators, I have often encountered students struggling with creative idea generation. Some believe that they have maxed their potential, some are satisfied with their current efforts, and some others feel "stuck"–wanting to do more but not knowing how to. While teaching first year students, their lack of any prior knowledge or experience of the creative process, adds to this challenge. There is a need to find ways of helping students realise their creative potential to the full and get back their *creative confidence* [8].

To meet this need, Think Lab was introduced as an academic experiment to the students in their first year of undergraduate design education, to equip them with formal tools, techniques and strategies for creative and critical thinking, that would benefit their future design learning and practice.

25.3 Objective of the Study

This study examines the effectiveness of "Think Lab", taught in the first (Foundation) year of study, in fostering creative and critical thinking skills amongst design students, and the impact of the same on their future design learning and practice.

The study focuses on two dimensions–the appropriateness of the content (what is taught) as well as the effectiveness of the learning–teaching methodology (how it is taught).

25.4 Research Methodology

A deductive research approach was applied to explore the effectiveness of the taught content as well as the learning–teaching methodology of Think Lab, through a survey.

The experiment started with a review of work done by creativity experts such as Edward De Bono, Michael Michalko, Sir John Hegarty, Tom and David Kelly, to name a few, to understand the mechanics of creative and critical thinking, as well as discover ways by which students could develop their potential for creative and critical thinking. After some initial research and exploration, teaching plans were developed to test some of the emerging ideas. These were introduced as part of the academic delivery of first year curriculum of undergraduate courses in design, as a module titled "Think Lab".

Think Lab was first implemented in 2017/18 with six weekly sessions of three hours each. The first year was essentially exploratory; and the format and structure gradually evolved. In the following Academic Year (2018/19), a more methodical approach was taken to plan the structure and content for Think Lab, drawing from evidence-informed approaches to learning-teaching drawing from the existing research in pedagogical practices and learning theories. This year, Think Lab was conducted as a series of nine full-day weekly sessions of six hours each. By the third year, the model was viable for scaling-up; in 2019/20 Think Lab was also implemented independently by colleagues using the plans developed previously.

Instrument for Data Collection

Online survey using both open-ended, descriptive questions and close-ended, scaled questions has been used to collect primary data.

25.4.1 *Sampling*

The sample unit includes students in second, third and fourth year of study of B.A. (Hons.) Fashion Design, Communication Design, and Interior Architecture & Design, at the Indian Institute of Art & Design (IIAD), New Delhi, who attended Think Lab in their first year of study (See Table 25.1).

Methods of Data Analysis

- The quantitative data collected has been analysed using the percentage method.
- To analyse and interpret the qualitative data, recurring keywords were coded and categorised based on their inferred meaning.
- The findings of qualitative and qualitative data were then correlated to arrive at insights.

Limitations of the Study

- The results published here are limited to the views of the sample unit used for the study, i.e. undergraduate students of Indian Institute of Art & Design (IIAD).
- This study does not take into consideration the emotive and psychological aspects, or the dimensions of the learning environment, that may have contributed to the success, or the lack of it, of Think Lab, or as part of the outcome resulting from Think Lab.
- Data collection, interpretation and analysis are limited by the variance in interpretation of the questions and terminology used by the respondents, and the level of details covered in their responses as possible via online survey.

Table 25.1 Details of Sampling Unit for the Study

Batch	Year of study at the time of responding to survey	Total sample size (No.)		Total No. of responses received	Viable responses	
		Students who attended Think Lab in first year	Students enrolled at time of survey[a]		No.	% of total students who attended (%)
2017–21	Fourth Year	104	69	17	16	15
2018–22	Third Year	115	102	33	33	29
2019–23	Second Year	113	113	54	53	47
	Total	**332**	**284**	**104**	**102**	**31**

[a]Some students have subsequently withdrawn enrolment from IIAD

25.5 Think Lab

25.5.1 Overview

In 2017/18, Think Lab was introduced in the course delivery of the Foundation (first) year of a four-year undergraduate design program at IIAD, as an academic experiment to scaffold the learning of creative design development process. Over the past three academic years, it has evolved to become integral to the Foundation curriculum, equipping students with thinking tools and techniques, and developing their creative and critical thinking potential. The prime objective of Think Lab is to make students more fluent and flexible in their thinking and help them get over their inhibition of thinking to increase their *creative confidence* [8]. Additionally, Think Lab also aims to build contextual professional skills of team-working, presentation and communication.

Think Lab takes a **constructivist approach to learning-teaching**, strategically combining the tenets of *experiential learning theory* [9] and *collaborative learning* [10]. As per constructivism, knowledge cannot be simply transmitted; in fact, learners actively construct knowledge in their own minds through the transformation of experience. Therefore, teaching plans must provide learners with experiences that aid acquisition of abstract concepts and encourage active construction of knowledge [11].

25.5.2 Schedule

Think Lab is conducted as a series of 09 to 12, full day (6 h), weekly sessions, starting from the first week of foundation (first) year. While maintaining a 20:1-25:1 student–teacher ratio, sessions are scheduled for a learning group of 40-50 students, to allow for sufficient variety of ideas and approaches to be explored during the session. Students work in teams of 4-5 members, on a well-defined problem-based task, which introduces them to a specific thinking strategy, technique or tool. A strict time schedule is followed to ensure that all teams work efficiently and reach the intended conclusion.

25.5.3 Content

Existing research and work by creativity experts on strategies, tools and techniques for creative and critical thinking has informed the content for Think Lab sessions. I have taken reference from the seminal work of Edward De Bono, a brain training pioneer and creativity thinking expert who gave birth to the term "lateral thinking", and the ingenious thinking techniques put together by creativity expert Michael

Michalko in his bestseller book ThinkerToys. By assimilating the review of existing research in this field, I identified some aspects of the creative and critical thinking process; these would form the intended learning objectives for Think Lab [12–15]:

- Develop an ability to shift one's point of view and look at a situation using different frames of references, to find alternative approaches and meanings.
- Polish powers of observation and ability to make connections and draw inferences.
- Learn to break down an idea into smaller pieces and discover interconnections.
- Enhance ability to recognise opportunities and challenge assumptions.
- Practice connecting and combining ideas, including those that seem unconnected.
- Make generating lots of ideas a second nature and inculcate the habit of continuing to dig—not be content with the first few ideas or solutions.
- Train oneself to defer judgements.
- Identify parameters and criteria to base decisions and judgments.

Next, I devised problem-based tasks, with each of them focusing on one of the above listed aspect of creative and critical thinking. Some tasks introduce students to formal thinking tools, while others familiarise them with an approach or strategy. Tasks are based on diverse themes such as visual, social and cultural, so that students can easily connect them to their own everyday experiences. While crafting these problem-based tasks I kept in mind their relevance to other aspects of design learning, students' proficiency level, session schedule and time available, and that the tasks were conducive for collaborative working and the studio environment. The tasks were kept open-ended, making room for multiple ways of approaching them, and multiple outcomes.

The list of tools and techniques that have been introduced through Think Lab include: Altering Perception, Altering Perspectives, Challenging Assumptions, Joining the Dots, Slice & Dice, Six Thinking Hats, Thinking Bubbles, Impossible Hybrid, Brute Thinking, Idea Box, Circle of Opportunities, Cross-Breeding, Free Association and SCAMPER [12–15]. The idea was to not provide these tools and techniques as point solutions, but to help students understand the underlying principles of creative and critical thinking, thereby giving them ways to flex their brains, allow ideas to flow more fluently and provide a framework that could direct their thinking during design projects.

25.5.4 Structure and Learning–Teaching Methodology

Learning–teaching in Think Lab rests on the bedrock of Experiential Learning theory, proposed by Kolb, wherein experiences are transformed into *meaningful learning*, that can be applied flexibly in a range of situations in the future [16]. Through systematically planned progressive activities, every Think Lab session takes the students through the four stages of **experiential learning cycle** (see Fig. 25.1)—Immediate or *concrete experiences* (working on a task/doing) form the basis for observations and reflections. These *reflective observations* are assimilated and distilled into *abstract*

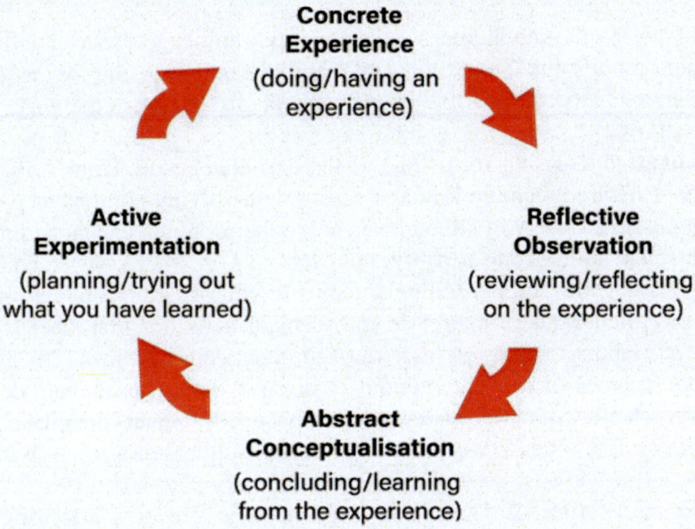

Fig. 25.1 Experiential learning cycle by David Kolb [9]

concepts from which new implications can be drawn. These implications serve as guides for future action, i.e. to create new experiences, wherein the acquired concepts are actively tested through *active experimentation* [17].

Every Think Lab session starts with a simple warm-up exercise that broadly uses the same approach or technique to be introduced through that day's session. This culminates with a quick, guided group discussion to reflect on the current experience and use the acquired understanding as a base for working on the main activity. As students work together in teams to tackle the assigned problem, tutors intermingle with them unobtrusively, supporting them through brief interactions and helpful cues. Students learn by observing how others work, by discussing, explaining and defending their ideas to others, and by analysing different points of view [18]. To facilitate this, session plans incorporate opportunities for students to review each other's work-in-progress as well as the finished outcome.

Think Lab builds-in several opportunities for deliberate **reflection**. Students explore what happened, search for turning points, make connections, analyse actions and their implications, as well as identify their strengths and areas for improvement. At the end of every session, students look back on the day's journey; they reflect in small groups and share their observations with the larger group. Finally, they assimilate their personal reflections in a journal using cues such as: What happened? What succeeded and failed? Why did it happen? What are the implications? What could have been done differently? Now what? This helps in theorising about the experience, making it available for future application [9]. To strengthen the learning from each session, students get a follow-up Home Task to give them a chance for testing the concepts they may have acquired [9].

In this way, Think Lab includes multiple opportunities to experientially try, test, practice, and modify the learnt tools and techniques, thereby improving the chance of storing mental structures of the learnt concepts in long-term memory, for future application [16].

Collaborative learning forms one of the cornerstones of Think Lab. Research shows that structured teamwork increases problem-solving abilities as participants learn to generate a variety of strategies to solve the problem, exchange information and insights, and build strong cognitive connections [19, 20]. However, collaborative learning goes beyond "simply telling students to get into groups and work". Think Lab has an "intentional structure" in the form of activities that steer students to equitably "co-labor" towards attainment of the intended outcomes [20]. It involves formal opportunities of peer interactions to discuss, debate, exchange ideas and to review each other's work. To broaden the perspectives, students are pushed to break out of their comfort zones to work with a diverse group including those with varying abilities.

Another cornerstone of Think Lab is the tactical use of **scaffolding**. Bruner, who proposed the concept of scaffolding in education, describes it as a form of active support that is provided by the "more knowledgeable other", i.e. tutor in this case, while students learn new skills and concepts [21]. Think Lab makes use of worksheets for analytical reasoning and information organisation, visual examples, pictorial representation of concepts, warm-up exercises, as well as cue cards for discussions and reflections, to support students as they grapple with new approaches and ideas. They work much like the scaffolding used in architecture to support the ongoing construction of a building. They also help to manage the cognitive load of the student, and primes them to succeed in the task, thereby keeping up their motivation and engagement.

The strategic combination of **collaborative leaning** where peers help each other to learn, and **scaffolding** where tutor provides active support for learning, allow students "to solve a task or achieve a goal that would be beyond his unassisted efforts" [22]. Working through **experiential learning** cycle helps students in recognising links between the pieces of a concept, enabling them to manipulate and apply the learning to a variety of situations and contexts, in future [9].

25.6 Data Presentation and Analysis

Survey results show that 72.5% of respondents have found Think Lab sessions "very helpful" and 25.5% find it "somewhat helpful" for their future learning and practice (see Table 25.2). This is further substantiated by the number of respondents who report to have continued to the tools, techniques and approaches learnt during Think Lab, in project work during their subsequent years of study (See Table 25.3).

As the data in Table 25.3 suggests, a larger group of respondents reported to using the underlying principle/s of the learnt tools and techniques in subsequent projects, in comparison with those who reported direct use of the taught tools and

Table 25.2 Role of Think Lab in future design learning and practice

Rating (Batch —>)	2017–21 (%)	2018–22 (%)	2019–23 (%)	Total
Very helpful	62.5	69.7	77.4	**72.5**
Somewhat helpful	31.3	27.3	22.6	**25.5**
Not helpful at all	6.3	3.0	0.0	**2.0**

Table 25.3 Continuing use of techniques and approaches learnt in Think Lab, in subsequent years

Frequency (Batch —>)	2017–21 (%)	2018–22 (%)	2019–23 (%)	Total
Often (directly)	25.0	33.3	32.1	**31.4**
Sometimes (directly)	25.0	15.2	24.5	**21.6**
Often use their basic principle/s	50.0	45.5	39.6	**43.1**
Sometimes use their basic principle/s	0.0	6.1	3.8	**3.9**

techniques. As per SOLO Taxonomy by Biggs, which describes levels of student's learning, this implies that learning has advanced to the *relational level* (integrated different aspects of the concept into a coherent whole) and/or the *extended abstract level* (understanding extends beyond what has been taught to generalise the whole to as yet untaught application) [23].

25.6.1 Learnings Acquired Through Think Lab

A detailed inquiry, correlation quantitative data (see Table 25.4) and qualitative data collected provides an insight into the significant learnings experienced by students.

Note: From this point onwards, quotes and key phrases used by students in their responses to the survey are indicated as italicised text.

25.6.2 Creative and Critical Thinking Skills

Data presented in Table 25.4 (see Row 1–6) evidences that respondents have generally found Think Lab sessions to be effective in enhancing their creative and critical thinking. Further, 65 students cite at least one of the following aspects of thinking as a significant learning from Think Lab:

- Ability to explore and experiment with many ideas using strategies, tools and techniques, and "*realising the importance of not being restricted by one idea*".
- Ability to "*organise information and thoughts*".
- Ability to assess idea, approaches and outcomes, and use this assessment to "*work out the way forward*" (critical thinking).

Table 25.4 Effectiveness of Think Lab in imparting specific learnings

Learning gained via Think Lab	Very effective (%)	Somewhat effective (%)	Not really effective (%)	[a]
1. Creative thinking skills	65.7	34.3	0.0	0.0
2. Ways for rapid ideation	52.9	41.2	5.9	0.0
3. Tools and techniques to support creative design development	60.8	38.2	1.0	0.0
4. Critical thinking skills	59.8	37.3	2.9	0.0
5. Ways to make sound judgement	48.0	47.1	3.9	1.0
6. Critically evaluation	49.0	46.1	4.9	0.0
7. Work collaboratively in a team	68.7	30.4	1.0	0.0
8. Presentation and communication skills	68.7	28.4	2.0	1.0
9. Understand own creative process	54.9	41.2	3.9	0.0
10. Ability to analyze an experience in order to draw learning from it	52.9	43.1	3.9	0.0

[a]Don't understand question
Row 1–6: Creative and critical thinking skills
Row 7–8: Professional skills
Row 9–10: Life skills

- "*Enhancement of thinking ability*" including imaginative thinking and the ability to think beyond the obvious (creative thinking).

Professional Skills The effectiveness of Think Lab in enhancing team-working skills, as well as presentation and communication skills is evident in Table 25.4, Row 7–8: 68.7% of respondents find Think Lab to be "Very Effective" in building these professional skills. Furthermore, 26 students have listed development of at least one of the professional skills as a significant learning.

Life Skills Think Lab has also played a role in making students aware of themselves and in developing their capability for learning itself (Row 9–10, Table 25.4)–only 3.9% respondents found it to be "not really effective" in these aspects. Further, 14 students have listed one of the following as significant learning from Think Lab: developing ability to "*manage own thoughts*", building "*creative confidence*", and learning to "*be more objective*", "*break comfort zones*" and "*accept critical feedback and use it for self-improvement*". While this was not the prime objective of Think Lab, their role in maturing students into well-rounded responsible professionals is undeniable.

25.6.3 Effectiveness of Learning–Teaching Methodology of Think Lab

Effectiveness of Collaborative Learning Data in rows 3–8 in Table 25.5 shows how helpful respondents have found the use of collaborative learning in facilitating their learning. An unexpectedly high rating is seen in favour of "Viewing each other's Work" (Row-5) and "Interactions & Discussions" (Row-2): 80.4% and 78.4% respondents find these techniques "very helpful", respectively. Analysis of qualitative data yields the following:

- 34 students highlighted that it enabled them to exchange and value multiple perspectives, as well as "*different ways of thinking and approaching tasks*".
- 26 students say that this increased efficiency and productivity by "*generating larger variety of ideas*", shared responsibility and "*distribution of work*".
- 18 students say that it provides "*insights for improving*" competencies and outcomes as they get opportunities to "*give and receive critical feedback*" from peers.
- 33 students have highlighted its role in developing at least one of the professional skills including teamwork, leadership, time management, communication and listening, and interpersonal skills of "*people management*", consensus building and conflict resolution.
- 38 students say that this helped them to learn essential life skills such as empathy, patience, acceptance (of diversity) and "*to be sensitive to other's views*", as well as "*built confidence*", a "*sense of responsibility*" and "*self-awareness*".

Even though respondents have been generous in their support of collaborative learning, 46 students, while listing benefits, also point out the challenges of working in a team; these include: losing individuality, "*lack of cooperation from group members*" and unequal work distribution, grappling with "*conflicting ideas*", different styles and pace of working, and, dealing with "*authoritarian behaviour of certain members*". This probably explains the data in row-1 of Table 25.6 when compared to data in other rows; 65.7% have found teamwork to be "very effective", a lower rating of effectiveness when compared to that of other collaborative learning tools.

Effectiveness of Experiential Learning Data in Rows 6 and 9–12 in Table 25.5 indicates how helpful respondents have found some of the activities linked to the four stages of experiential learning cycle. The qualitative data collected clearly indicates that students value the experiential learning format of Think Lab as they term

Table 25.5 Effectiveness of various teaching strategies and tools used in Think Lab

Teaching strategy/tools used in Think Lab	Very helpful (%)	Somewhat helpful (%)	Not really helpful (%)	Did not help (%)
1. Working in a team (CL)	65.7	32.4	2.0	0.0
2. Interactions and discussions (CL)	78.4	21.6	0.0	0.0
3. Peer review and feedback (CL)	76.5	18.6	3.9	1.0
4. Presenting work to peers (CL)	77.5	20.6	2.0	0.0
5. Viewing each other's work (CL)	80.4	18.6	1.0	0.0
6. Sharing of learning (CL, EL-RF)	75.5	21.6	2.9	0.0
7. Scaffolds: cues, formats, etc. (SF)	77.5	22.5	0.0	0.0
8. Sharing of examples by tutor (SF)	78.4	20.6	1.0	0.0
9. Warm-up exercises (SF, EL)	63.7	32.4	3.9	0.0
10. Reflections (EL-RF)	73.5	22.5	3.9	0.0
11. Home Task (EL)	47.1	40.2	6.9	5.9
12. Recording in the Journal (EL-RF)	73.5	19.6	4.9	2.0

Key: *CL* Collaborative Learning
SF Scaffolding
EL Application of Experiential Learning Cycle
RF Reflection

"Learning by applying" or "Learning concepts by application" as one of the key teaching approaches that made Think Lab effective for their learning. In this, the role of deliberate **reflection** is one of the key approaches as it helps convert the experience (of working on the task) into a lesson for future (acquisition of abstract concept) [21]. 73.5% respondents find exercises of reflection to be "very helpful" and 22.5% find it "somewhat helpful" for their learning process (Row-10, Table 25.5). Furthermore,

- 36 students highlighted that reflection enabled them to recall, summarise, culminate and record learning in their memory for a longer period, available for future reference. It has helped them to "*become aware of nuances and details that may have been missed (while doing)*" and "*extract meaningful conclusions*"
- 44 students mentioned that reflection helped them to "*analyse what worked and what did not*", identify strengths as well as areas of improvement, and enabling them to make an action plan to improve future performance.
- 14 students indicate the role reflection has played in helping them "*know own self*" by "*understanding their own personal process*", style of working and capabilities.

Finally, data in Row 7–9, Table 25.5 suggests that students recognise the use of **scaffolds** such as cue cards, worksheets, examples, and warm-up exercises as they "*prepare the ground for success*". Critical to the experiential learning process, these provides an elegant way of involving the tutor, while mitigating the risk of spoon-feeding, to extend focused and active support as students face the challenge of learning complex material.

25.7 Conclusion and Recommendation

The findings in this study indicate that systematic and structured learning plans that use evidence-informed, teaching strategies including tools of collaborative learning and experiential learning, to introduce students to formal tools and techniques of thinking, can build the creative and critical thinking potential of first year students of design. The study presents evidence that a number of students continue to use the learnt strategies, tools and techniques of thinking in subsequent creative work–they are able to recall and discuss benefits of the learning gained, after a gap of 8–32 months of attending Think Lab. Whether it is direct use of the tools or an indirect application using the underlying principle, it has resulted in a more fluent creative and critical thinking process. Introducing students to these tried and tested tools and techniques, therefore, provides a well-grounded approach to enhance creative and critical thinking abilities of students.

Furthermore, the use of collaborative learning has been particularly effective in developing diverse professional skills such as team-working, communication, presentation and time management, building interpersonal skills like consensus building and conflict resolutions, and imparting life skills such as empathy, self-awareness, and ability to use feedback for self-improvement. Even as students point out some challenges of team-working, the findings of the study clearly indicate that its role in developing higher-quality problem-solving abilities and autonomous, articulate, thinking individuals make collaborative learning viable for creativity training [19, 24].

The data from the study evidences that a structured learning–teaching plan which leads students to methodically engage with all stages of the experiential learning cycle is an effective way to prepare students for innovative problem-solving by developing their overall creative and critical thinking ability. The benefits of reflection listed by students in converting the experience into lessons for future, closely match those laid out in existing research in experiential learning. If teaching plans stay persistently committed to the use of deliberate reflection, then not only does it take students towards enduring learning for future application, but also helps them to gradually evolve to be self-aware, self-directed and self-regulated reflective practitioners.

The study shows that the various activities and scaffolds used to direct and support students to observe, interact, share, analyse, interpret and reflect, have been considered helpful for the learning. While it may seem that including these activities and scaffolds reduces the time available for working on the main problem task, and/or increases the tutor's effort of planning, their role in getting students to touch base with the stages of the learning cycle is significant. It is therefore important to consider the role these teaching tools play in constructing meaningful learning, before succumbing to the temptation of foregoing or reducing their use in teaching plans.

While Think Lab has been successful beyond our initial anticipation as students cite it as one of the essential fundamental inputs that has formed the foundation for their future creative work, it is humbly recognised that one cannot overstate these claims—it is not implied that the only reason for the development of students' creativity and criticality is the content or teaching–learning methodology of Think Lab. Yet the findings of this study provide an impetus for design curricula to seriously consider specific learning modules that are structured on the basis of research informed pedagogical practices and designed to stimulate creative and critical thinking separately and dedicatedly, in an attempt to better prepare students for the unique challenges set by the twenty-first century.

References

1. Dede, C.: Comparing frameworks for 21st century skills. Bellanca, J., Brandt, R. (eds.), 21st Century Skills: Rethinking How Students Learn, pp. 51–76. Solution Tree Press (2010)
2. Morr, K.: What is creativity? The ultimate guide to understand today's most important ability. https://99designs.com/blog/creative-thinking/what-is-creativity/ (2018)
3. Blair, B., Cummins, A., Dunbar, T., Hayward, D., Weedman, J.: Bau-wow! A model for creative practice, thinking, learning, research and innovation in the 21st century. In: Drew, L., Last, J., Lewis, S., Wade, S. (eds.) The Student Experience in Art and Design Higher Education: Drivers for Change, pp 65–97. Jill Rogers Associates Limited (2008)
4. Powell, W., Snellman, K.: The knowledge economy. Ann. Rev. Sociol. **30**, 199–220 (2004). https://doi.org/10.1146/annurev.soc.29.010202.100037

5. Tudor, R.: The pedagogy of creativity: understanding higher order capability development in design and arts education. In: Proceedings of the 4th International Barcelona Conference on Higher Education, vol. 4. https://www.guni-rmies.net (2008)
6. Cotton, K.: Teaching thinking skills. In: Northwest Regional Educational Laboratory's School Improvement Research Series. https://www.nwrel.org/scpd/sirs/6/cu11.html (1991)
7. Shreeve, A., Wareing, S., Drew, L.: Key aspects of teaching and learning in the visual arts. Fry, H., Ketteridge, S., Marshall, S. (eds.) A Handbook for Teaching and Learning in Higher Education, 3rd edn., pp. 345–362. Routledge (2009)
8. Kelly, T., Kelly, D.: Creative Confidence: Unleashing the Creative Potential Within Us All. William Collins (2013)
9. McLeod, S.A.: Kolb–Learning Styles. https://www.simplypsychology.org/learning-kolb.html (2017)
10. Barkley, E.F., Cross, K.P., Major, C.H.: Collaborative Learning Techniques: A Handbook for College Faculty. Jossey-Bass (2005)
11. Honebein, P.C.: Seven goals for the design of constructivist learning environments. Wilson, B. (ed). Constructivist Learning Environments: Case Studies in Instructional Design, pp 11–24. Educational Technology Publications (1996)
12. De Bono, E.: Lateral Thinking : A Textbook of Creativity. Penguin (2009)
13. De Bono, E.: Six Thinking Hats. Penguin (2016)
14. Michalko, M.: Thinkertoys: A Handbook of Creative Thinking Techniques, 2nd edn. Ten Speed Press (2006).
15. Seeling, T.: Ingenius: A Crash Course on Creativity. Harper Collins (2015)
16. Novak, J.: Meaningful learning: the essential factor for conceptual change in limited or inappropriate propositional hierarchies leading to empowerment of learners. Science Education **86**(4), 548–571 (2002). https://doi.org/10.1002/sce.10032
17. Passrrelli, A.M., Kolb, D.A.: The learning way—Learning from experience as the path to lifelong learning and development. In: The Oxford Handbook of Lifelong Learning (2011). https://doi.org/10.1093/oxfordhb/9780195390483.013.0028
18. Boud, D.: What is Peer Learning and Why is it Important? https://tomprof.stanford.edu/posting/418 (2002)
19. Johnson, D.W., Johnson, R.T., Smith, K.A.: Cooperative learning returns to college: What evidence is there that it works? Change: Mag. Higher Learn **30**(4), 27–35 (1998). https://doi.org/10.1080/00091389809602629
20. Bruffee, K.A.: Sharing our toys: cooperative learning versus collaborative learning. Change: Mag. Higher Learn. **27**(1), 12–18 (1995). https://doi.org/10.1080/00091383.1995.9937722.
21. McLeod, S.A.: What Is the Zone of Proximal Development? https://www.simplypsychology.org/Zone-of-Proximal-Development.html (2019)
22. Wood, D., Bruner, J., Ross, G.: The role of tutoring in problem solving. J. Child Psychol. Child Psychiat. **17**, 89–100 (1976)
23. Biggs, J.: SOLO Taxonomy. https://www.johnbiggs.com.au/academic/solo-taxonomy
24. Qin, Z., Johnson, D.W., Johnson, R.T.: Cooperative versus competitive efforts and problem solving. Rev. Educ. Res. **65**(2), 129–143. https://rer.aera.n

Chapter 26
Active Learning Strategies for Teaching Research Skills to First-Year Design Students

Anshoo Rajvanshi and Gauravi Mittal

Abstract The studies have shown that research is fundamental to design education and to equip students with the nuances of research, research methodology and academic paper writing are included in the curriculum. The purpose of this study is to establish that teaching research skills, rather than research methods, is a more engaged learning process for foundation-level design students at the undergraduate level. Teaching strategies applied in a controlled environment built on Bloom's taxonomy and Vygotsky's zone of proximal development theory using Kolb's experiential learning theory along with motivational strategies have proven to be engaging and enabling for the students. This is mapped through a cultural context-specific research project. To analyze and evaluate the effectiveness of active teaching and learning strategies applied in the entire module, a student survey was conducted on two consecutive batches. According to our findings along with achieving all the intended learning outcomes, the strategies have been instrumental in increasing students' engagement and minimize their inhibitions and have provided opportunities to keep them intrinsically and extrinsically motivated toward practicing research skills.

26.1 Introduction

Research has always been regarded as one of the fundamental skills in design practice as research facilitates sound design solutions. Theoretical approach used for teaching research makes the research concepts abstract, difficult to comprehend and making it into a complex skill. A need was felt to ease out this approach and engage design students more actively by using academic material and employing a range of methods for research analysis and interpretation along with ethical practices to stay more

A. Rajvanshi · G. Mittal (✉)
Indian Institute of Art & Design, New Delhi, India
e-mail: gauravi.mittal@iiad.edu.in

A. Rajvanshi
e-mail: anshoo.rajvanshi@iiad.edu.in

A. Chakrabarti et al. (eds.), *Design for Tomorrow–Volume 2*, Smart Innovation, Systems and Technologies 222, https://doi.org/10.1007/978-981-16-0119-4_26

informed and indulge in more meaningful design practices [1]. With Indian design industry coming of age, the professional boundaries have now extended from design practitioners to independent researchers in design. Considering the pivotal role of research in design both academia and industry, it is considered inevitable and rather mandatory discipline for a strong and holistic development-oriented teaching and training for the future design professionals. The curriculums offering design degrees have included research methodology and academic paper writing for students to learn the nuances of pure research.

Literature has revealed that the students entering higher education possess very weak foundation for conducting empirical research and get overwhelmed when introduced to research methodology, and dread research methods courses as they find research concepts abstract and difficult to comprehend. Many have false assumptions about their research abilities, either they feel intimidated, discouraged and demotivated by extensive details and associated jargons on research methods or they feel contended at collating information directly answered by web sources to successfully submit research-based assignments in school [2–4]. Didactic teaching format for teaching a research methods course to social science students at introductory academic level in higher education has not given encouraging results, whereas use of active teaching strategies using flipped classroom methods and live research projects has been found highly effective in encouraging and enabling students to conduct original research work as well as learn research methods principles [5]. However, in active learning environments, learning is the consequences of one's actions on behalf of the student through constructive engagement in learning activities. Though the responsibility of learning lies on the student, creation of learning environment and learning opportunities lies on the tutor. Active learning strategies using Kolb's experiential learning cycle are a plausible framework to work with as it provides a rationale when a new concept is to be introduced for it to be learned and imbibed as a cognitive skill. Both the theories being learner centric are instrumental in engaging the heterogeneous cohort of learners with varied socio-cultural backgrounds and yet unidentified learning styles [6–10].

Research in design includes observation in the field to gain empathy and identify different factors impacting the user, discovering insights which eventually leads to generating new and contextualized ideas. The process of research is much nuanced than it sounds. Moreover, for a design student, it is important to understand the relevance of context in design. Context in most scenarios is never explicit, and it is to be interpreted from the indicators. Hence, it is important for students to comprehend the perceptual link between learning research and design practice to stay motivated to gain requisite knowledge and skills to be able to conduct research independently in the future. There is a need to map and analyze learning and teaching strategies that are effective in building their confidence, bring down intimidation levels, and raise nuanced awareness toward the merits of conducting research and its application in design process.

This paper is an outcome of learning and teaching strategies applied in one of the four modules to introduce research skills to foundation year design students at undergraduate level. In this module titled 'Design Environment' focus is to introduce historical, cultural and contemporary design practices within the context. The (pedagogy) focus is self-exploration through research skills. The study analyzes and evaluates the effectiveness of active learning and teaching strategies based on the hypothesis that if in an active teaching learning environment a student is exposed to the relevance and use of research skills and processes, it can lead to students imbibing them in their own practice. The strategy like group tasks, pinup on display boards, peer review and discussions, can also be instrumental in reducing (if not eliminating) their intimidation levels and build their confidence with reference to conducting research. This paper discusses the approach applied on heterogenous cohort students of two consecutive batches [2018–19 and 2019–20]. The best practices are presented after reflections and refinements, confirmed by a survey conducted on the students from both the years.

26.2 Experiment

At our institute, a constructivist curriculum [11] with definitive learning outcomes using Bloom's taxonomy [12] provided the allowance to experiment and put students on the path of self-exploration using active learning and teaching strategies. However, active learning focuses on 'how' students learn rather than 'what' students learn which is an important component of any curriculum. Hence, the activities and tasks were designed for students where they could explore the contextual references to design and design practice (what) by applying basic research skills (how).

To address the 'What' discipline-specific contextual relevant themes were chosen ensuring that themes were not restricting students to the classrooms. Students were encouraged to discover immediate surroundings and their community, to think about themselves and their own environment, to develop compassion and empathy toward people and cultures. With engaging and meaningful themes, learners had something to explore, to investigate about, to describe, to compare, and to explain by using the 'How' which facilitated the intended path of exploration using the newly acquired research skills. During the academic year (August–May), the teaching in Term-1 (August-December) focus on students exploring the 'What' of the theme explicitly and 'How' they explored remained implicit. In Term-2 (January–May), had focused on the 'How', the intention was to explicitly explore research as a component of design process through application in design projects. The focus was deliberately kept on 'research skills' rather than research methodology, as a gateway to the discipline of research, inspired by Piaget's theory of Schemas [13, 14]. More complex-scientific principles and theoretical concepts of research along with its wider implications are left to get revealed at advanced levels [12].

For this module, eight weeks of dedicated teaching is spread equally over two terms, all tasks and sessions across the module were planned is a series to facilitate integration and internalization of experiences and learning as a whole. Module learning objectives were broken down into sub-learning-objectives, which were further planned to delivered through smaller sequential tasks to lower the intrinsic cognitive load for the students to facilitate learnings of subtle nuances of research [15, 16]. The tasks were sequenced based on their current zone of proximal development (ZPD) that is their current cognitive level of the learner and the attainable level in the guidance of the tutor. Suitably scaffolded the level of difficulty of these tasks was gradually raised to always keep the learners in their ZPDs [17]. The learning and teaching activities were planned applying Kolb experiential learning theory (KELT) to translate experience into learning through the process of deriving concept by reflecting on the experience and further apply in a new context. Many such Kolb experiential learning cycles (KELC) were applied as a set of spirals for learning and teaching to be more effective. Using the motivational theory, to keep the motivation of the student high, many milestones were marked for students to have a sense of achievement during the process of learning [18, 19].

In term-1, one-week-long briefs were designed supported with a guided investigation to explore, observe, identify connections between different ideas, interpret, provide explanations and evidences. While prime focus was kept on exploration of the theme, selected research tools and skills were introduced. The opportunities were created in a strictly controlled environment for students to learn and apply these tools and acquire the requisite skills. In each consecutive week, students learned a new research skill or refined it. Using the newly acquired research skills, they could explore the theme to unravel hidden aspects which otherwise may not have been possible.

The application of learning and teaching strategies applied in one of the four weeks is mapped in below Table 26.1. The sequence of activities, stages of scaffolds, types of scaffolds, introduction of research tools and skills using Kolb's experiential learning cycle (KELC) is presented along with the intended learning outcomes. In the table given below, the bold cells elaborate on approaches used in white cells. The objective of this table is to present the different approaches applied; with each step a student goes into the next level of ZPD.

Term-2 is oriented toward application of tools and skills acquired in Term-1 in design projects. The KELC in Term-1 was completely guided in terms of 'What' and 'How'. In Term-2, students were guided to apply these learning independently. This phase being 'How' as explicit, students were given the 'What' and had to themselves identify the 'How'. In one of the design projects, students did guided-research and they were provided with helping questions like 'what nature of information and why do you need it? Where will you find it? How do you plan to find it?'. To facilitate, they were grouped with one peer to work together but to ensure their involvement they were to be assessed individually. One major component in the final deliverables was a detailed reflective note on 'Evidencing the use and appropriateness of primary and secondary research skills to collect information as part of the design process'. This becomes an evidence that the students are able to identify and articulate the application and relevance of learned research skills, along with the merits of conducting research. Following this, to showcase their educational accomplishments with reference to the entire learning from all modules in one culminating Capstone project, they submitted a research plan to evident the application of learned research skills. For students, it was part of assessments but for us, it was a way to assess them whether they have entered the next level of ZPD or not.

To analyze and evaluate the effectiveness of the applied active learning and teaching strategies in the module, it is important that we take perspectives of the students.

Table 26.1 Mapping of learning and teaching strategies applied in one of the four weeks in Term1

DO (1)—Pre-work (Individual): Students listed objects from immediate surroundings based on activities done in 24 h REFLECT (2)—During in-class discussion: Students reflected on the role of these objects in their lives THINK (3)—Tutors shared objects from their lives, handmade ceramic mug, spectacles and mangalsutra, discussed material and non-material aspects APPLY (4)—Students found similar pattern in their objects. They differentiated material and non-material aspects, using a visual organizer (VO)
(1) Observation (2) Inquiry (3) Tutors sharing worked as a modelling behavior for understanding the meaning of the concept (4) VO here worked as a scaffold (4) segregate information to make meaning
DO—On display boards, students presented the categorization done using VO and discussed in peer groups REFLECT-Non-material component could be self-created or inherited. Either material or non-material can take priority in perception. Students wrote a 300 words reflection note THINK—During guided discussion, students used terms like culture and lifestyle while articulating the patterns and similarities observed in VO. Tutors introduced academic definitions of 'Culture' with its relevance to design context APPLY—In groups of 5, students were assigned an ethnic community to map their culture with the help of 5W1H[1] using VO and discussed each other's work
(1) Experience a sense of achievement (1) Peer learning (2) Writing reflection notes intensified the experience. (3) The activity was to guide students to analyze and reach on such terms on their own. (3) Academic definitions didn't intimidate students, instead gave them the confidence to reflect on their discussion in the definitions presented (3) Importance of understanding the relevance of a concept in context of their academic learning and in the scope of its application. (4) Group work for more engagement and learning (4) 5W1H worked as scaffold to kickstart the research process
DO-Group presentation and discussions: they made connections within different ethnic communities and also with their own cultures by pointing at similarities and differences REFLECT-Students wrote a reflection note on the similarities (300 words) on- "What and how did you do it? What was the outcome? What was your learning? When can it be used in the future?" THINK-During guided discussion, academic definitions of material and non-material were given. Tutors introduced the terms primary and secondary research and how students have used both data to reach this conclusion APPLY-To conclude the discussion, students wrote 500 words essay on: "Discuss how material and non-material culture informs each other, with one object from the ethnic community as an example". They were encouraged to refer to the VO and reflection notes

(continued)

[1] 5W1H is an abbreviation for What, Why, Where, Who, When & How?
For further information, refer to: https://creatingminds.org/tools/kipling.htm.

Table 26.1 (continued)

(1) There is a sense of achievement with level of categorization, analysis and presentation of work. (2) Writing reflection note (3) Introduction to research skills along with practice was less intimidating. They understood the difference and relevance of both kinds of data. (4) VO and reflection notes as pre-writing tasks facilitated essay writing without feeling the need of using other source of information. At this juncture, students with lesser critical abilities struggled with logical construct
DO-Students who submit their work in the given timeframe received a feedback before submission for final assessment REFLECT—Recognized the relevance of feedback as it gave a direction of improvement based on: width of topic and sources, critical analysis and logical construct, awareness of design practice, language and adherence to the format THINK-They were introduced to the role of feedback in self-improvement and self-regularization. The progress for the final draft is now planned APPLY-Wrote and submitted the final essay along with the journal consisting all pre-writing tasks, VOs and reflective writings
(1) Extrinsic motivation on completion of the task and final submission in time. (2) Feedback based on criteria are more helpful in analyzing the work (3) Student is motivated with an opportunity to improve further. (4) Assessment of the journal along with final essay establishes the value of processes in research

26.3 Methodology

This research is majorly quantitative, with some qualitative aspects accommodated through the design and nature of questions asked to the respondents. This study discusses the approach applied consecutively over two years on 6 batches of an average of 40 students each. This means that this cycle was applied 6 times in 2 years, for teaching research skills to foundation year design students at under-gradation level.

Respondent's Profile It is important to discuss on the L3 and L4 students' skillset to better comprehend the results of this study.

L3 Students (Total 106 Students, 76 Respondents) They operate in predictable, well-defined contexts and guidelines, using given techniques and information sources, under direct supervision. They can collect information to inform a choice of solutions to standard problems in familiar contexts. Yet, their analysis is limited to information of pre-defined frameworks.

L4 Students (Total 104 Students, 45 Respondents) They operate in huge variety of predictable contexts requiring use of specific techniques and information sources, working in limited autonomy as well as supervision though they are responsible for the quality of their output. They have a well-defined focus for enquiry, plan and undertake investigative strategies using a limited and defined range of methods, collect data from a variety of sources, and communicate results effectively in an appropriate format.

Data Collection-Methods, Tools and Sources Conceptualizing the questionnaire: The study uses a questionnaire survey to collect and analyze both quantitative and qualitative data. Two separate questionnaires were developed for two sets of students. While the composition and rationale for these separate student samples will be discussed in further sub-sections, it is important here to discuss the nature and purpose of the designed questionnaires. During the academic year, students currently bi-annually fill feedback forms on module delivery and teaching methods. But to get a clearer picture of their pre-assumptions about research, for better course delivery and to gauge scope of improvement, we have designed a fresh and separate questionnaire. This will hopefully bring us more about their attitude toward practicing research skills. There was also a self-assessment survey, to self-reflect on their research skills and self-confidence, which works as an intrinsic motivation for them to take initiatives for practicing research in the future.

At the time of the survey, the first group (Level 3 students) recently completed their foundation academic year (Batch 2019–2023) and received their results as well. The purpose is to gauge their perspective, personal understanding and grasp on the subject matter over the year. It was designed to trace a timeline of the students' introduction and understanding of research, particularly design research. The questions had four key themes, as following: (1) their impressions of research in school, the content and style of research input they received, if any. (2) Their perception of the course module after introduction of LOs (Learning Outcomes), if they started looking at research skills and its processes as a space for inquiry. (3) The change in perception (if any) after the completion of one-year inputs: What was their experience with research skills, class environment, teaching and understanding of the content after the first few weeks of input? What are their capabilities, how did they navigate through the process of input, what aspects of research skills did they apply during their foundation academic year? Do they see the relevance of research in design education? Do they perceive this module research limited to reading, writing and a space where long lectures happen? What is there understanding of the concept-material and non-material? How many tools like 5W1H have been helpful for them? Lastly, they self-assessed their research abilities, reflected in the culmination project of their academic year.

The second group (Level 4) are seniors who just culminated their second year of design school (Batch 2018–2022). This questionnaire was to access if, research skills learned in L3, were applicable and beneficial in L4. And also, to assess their perception and perspective toward research and research practices along with its role in design practices after two years of input. A set of questions was based on their self-assessment of their capabilities for practicing research-gathering-collection of information from primary, and secondary sources and segregating and analyzing information, and presenting in prescribed formats.

Data Analysis The purpose of this survey was to map student's perspective in terms of their level of understanding along with the changes in their perception toward research skills. Also, we intended to map their perception of relevance of research in design before and after completion of the course, as well as the relevance of this learning at advanced levels. The results from the survey were analyzed against our expectations on the impact of learning and teaching strategies based on applied theories implemented throughout the academic year on the students, acquiring the intended learning and its impact on their attitude toward research (if any).

26.4 Findings

To bring perspective to the discussion of our results, an important clarification is in order. The respondents for the study come from all possible grade categories, thereby ensuring there is no bias in favor of any one category.

School Input on Research The data is significant as this reveals that the students joining for higher education at our institute is a heterogeneous cohort (Set 1, 2, 3 and 4). During the survey of the L3 students, it was revealed that during the school, 43.42% (Set 1) of the total students had practiced research with material that was taken from web sources and added the content in their assignments. These students neither had theoretical input, nor an input on what's a reliable source of information. 15.78% (Set-2) of the total students were given theoretical input on the difference between primary and secondary data, survey and interview. Yet these students never practiced, so their research was limited to theoretical understanding. Only 10.53% (Set 3) of students practiced doing an interview, made a survey, and collected and analyzed the data. This percentage also includes students who received no theoretical input on research. There are students that conducted surveys and interviews without receiving adequate theoretical input, point to a highly inadequate and flawed introduction to research at the school level. This is significant because this is what they start higher education with. Approximately 30.26% (Set 4) students have neither received nor practiced any aspect of research. They never referred to web sources or gathered information for their in-class activities and assignments.

Intimidation Towards Research As mentioned in Fig. 26.1, we found that 94.74% of the L3 students who received input on research in some way (as mentioned in the above section, set 1, 2 and 3) thought research was an intellectual activity beyond their capabilities, before starting this module. A similar pattern was observed in the 81.58% of students who received no input (Set 4), who also dreaded research. This clearly indicates how students have irrelevant and inadequate introduction to research at school level, and they go into higher education feeling intimidated by it. 71.05% students at the introduction of the course and 88.16% after the first two weeks of the input were intimidated by research. However, we must acknowledge that this could include the anxiety someone has when starting a different style of learning than one they were used in the school. This could also be the stress of starting higher education as tutors; one of the main observations we made was that new students were a bit skeptical about working in groups, getting involved in a dialogue with peers and tutors, and concentrated more on the content of their journals. This could be a result of

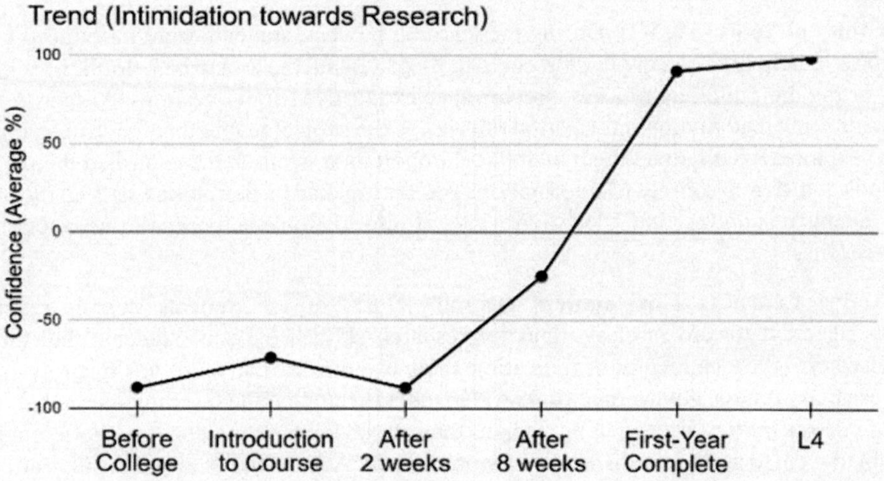

Fig. 26.1 Trend of confidence toward research

the flawed idea of research—if they are introduced to research either through theory or only through practice without theory, resulting in too much focus on content rather than the processes and the meanings. It took some time to show them the benefits of getting involved in the active learning processes. However, their confidence in performing independent research before their capstone project improved by 90.79%. After the completion of their foundation year, 92.11% of the students felt capable of conducting research on their own. It is worth discussing how these numbers increase for L4 students. 92.11% expressed that they could conduct research independently based on the application of learnt research skills from L3. Research is a doable and engaging process for 95.45% of L4 students. A total 97.73% of L4 students could conduct research independently. Another promising finding was that all the students of L4 feel the need for conducting research in their projects, even without instruction.

Understanding the Relevance of Research in Design One of the most important aspects, we wanted our students to understand, is the relevance of doing research in design. Once a student starts seeing the relevance in a subject matter, they intrinsically start getting involved in the process of learning. According to the findings, 98.6% of the total respondents of L3 and 93.2% of the L4 students do think that research is relevant in design.

Conceptual Understanding of Material and Non-material A majority of the students (85.5%) were confident and capable of brainstorming, searching, exploring, making questions, gathering information and segregating. 79% felt capable of drawing insights and doing the analysis. This is something we expect them to achieve at L3. Surprisingly, 40.79% seem to understand the concept in a holistic manner and able to apply in different contexts as well as the capstone project. This group of students has high critical thinking skills.

Utility of Tools (5W1H) During the learning process, students were introduced to tools to help brainstorm the right context. 81.5% found the tool to be helpful in kick-starting their thinking process, preparing an exhaustive list of questions to identify a wide topic, and finding information relevant to the subject matter that they are trying to explore. It has helped them include all important aspects and also pushed them to look at different sources of information. Yet this tool had its limitations, as 39.47% of students responded, that it led to overload of information when used during in-class sessions.

Active Learning Environment Overall, 82.8% of L3 students demonstrated engagement toward in-class input for research skills. 76.3% of students find the research to be challenging, indicating their awareness that there are nuances in research process. An average of 95.61% students find that the teaching method and in-class activities are engaging, helped them to understand the concepts better and abled them to participate in the discussions as well. And 81.14% showed positive attitude toward conducting research (gathering information, segregating and analyzing) in-dependently in the future, as mentioned in the Fig. 26.2.

Students Self-Reflection on Their Research Abilities The results showed that 73.66% of L3 students have scored their capabilities as 'very good' or 'good' for exploring, gathering and segregating information. 62.43% were confident of their analytical skills, and 78.94% were confident about applying the acquired knowledge of research skills in different contexts, and in projects under different modules as well. It was pleasantly surprising to see 78.97% L3 students marked themselves 'very good' or 'good' for knowing and understanding research ethics.

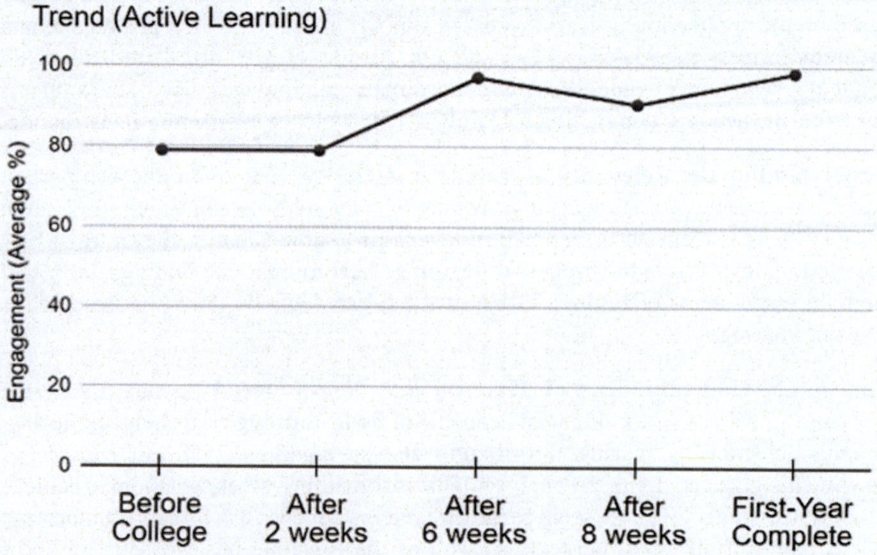

Fig. 26.2 Feedback on active-learning

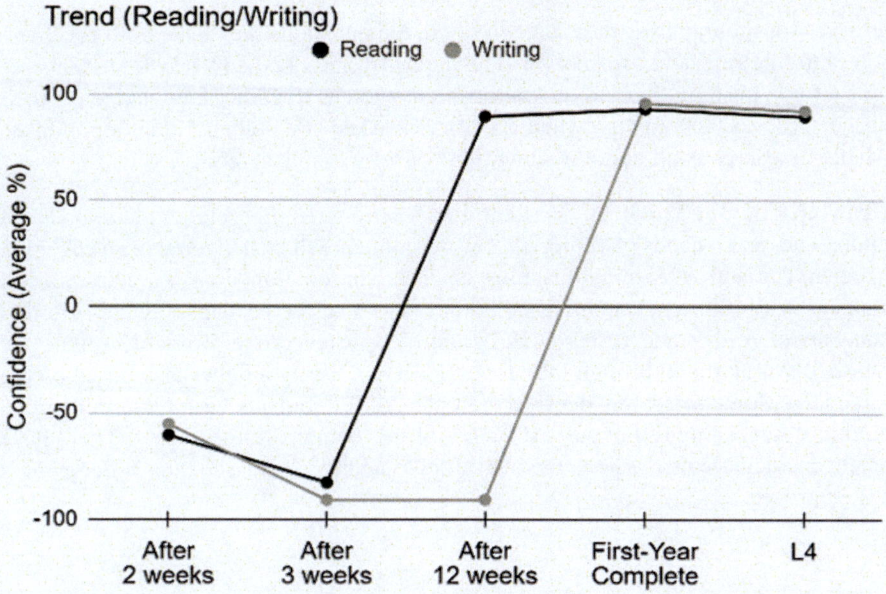

Fig. 26.3 Perspectives towards reading/writing

Perspectives Towards Reflective and Academic Writing in Research As indicated in Fig. 26.3, at the time of introduction to the module, students were strongly inhibited toward research, 55.26% of L3 students had a misconception that's it all about writing (making notes and constructing essays). After giving input for 3 months, 90.79% of the total students were able to develop an argument to write an essay on their own. They started using academic essays as a way of expressing themselves. Overall, 97.37% students have self-assessed their writing-reflection note skills as 'very good' or 'good'. And 94.74% have scored their essay and report writing abilities as 'very good' or 'good'. Another promising finding is that 94.74% students seeing benefit in writing reflection notes, even when they are not asked to; they even said that they will be practicing in future as well. On similar grounds, the findings for L4 students also confirm that 95.45% are still writing reflection notes for their assignments and 88.64% still write reflection notes for themselves, even without instruction.

Perspectives Toward Academic Reading to Gather Information for Research As indicated in Fig. 26.3, students originally had strong inhibitions against research. 60.53% L3 students thought research involves a lot of reading. 89.47% thought that research is about gathering exhaustive information on one topic. Similarly, when they were introduced to the module, 82.89% thought this module will include a lot of reading. This perspective toward reading gradually changed during the course of the delivery, where 96.05% started taking reading material as a source of obtaining answers for the inquiries they have in mind, the questions they have built for themselves to find relevant information. The inquisitiveness helped 97.37% students see it as a force pushing them to look at different sources to gather knowledge for their design process. Eventually when the survey asked the students their opinion on whether research is all about reading, 90.79% of them disagreed.

Drawback of Our Work Active Learning Teaching requires a lot of planning at the tutors' end, and it needs planning on how the content will be delivered in an engaging manner. The active learning–teaching environment combined with a classroom of students with different cognitive thinking meant that we needed to make sure that concepts of design and research skills are clear to everyone. To achieve that, we repeatedly took them through the process of KELC using another task based on the same idea. Sometimes, this became monotonous for students with higher cognitive thinking. According to our survey, 42.1% of the students felt the same. There have been considerable discussions among tutors to address this issue, and refinement is still in process.

26.5 Conclusion

At an undergraduate level, it is crucial to equip first-year design students with research as a tool to explore the unchartered paths and create new avenues in the demanding and dynamic industry, which otherwise is not possible through traditional teaching methods. From the foundation level itself, if cognitive research skills rather than cognitive research ability are introduced it helped them to understand its relevance as well as imbibe it as an inherent part of design process. Demystifying the term 'research' helps lowering their intimidation level and improves confidence.

Inquisitive and challenging (ZPD) active learning environments are instrumental in increasing students' engagement and minimize their inhibitions in terms of responding to their own queries and confusions toward research skills and its applications in design process.

Incorporating inquiry-based exploration of thematic content which is relevant to design domain helps students to comprehend the perceptual link between learning research and design practice, also benefiting their overall academic learning.

It has been realized during the planning and application stage that if the leaning objectives are broken down in smaller components and for each smaller learning outcome, smaller KELCs are applied; then it is easier to scaffold students' learnings. It also provides opportunities to keep them intrinsically and extrinsically motivated and also reduce their intimidation levels.

It is recommended that the activities designed during the inquiry phase of the research process should enable students in generating right questions to arrive at the right keywords to address the inquiry objectively. It has been observed that if the introductory tasks are achieved well by a student, then it helps in retaining the motivational levels and keep engagement high in forthcoming tasks or else it results in disengagement and non-achievement of higher levels of learning. To ensure this academic and planning, the delivery based on Bloom's taxonomy has proven to be beneficial.

It is important for the tutor to stay mindful of expected level of learning objectives from a student's performance, as it has been experienced that when high-caliber students achieve the expected learning outcome while peers are still in process, we felt motivated to extend the task to the next level for them.

However, even in an active learning and teaching environment with a heterogeneous cohort, it has stayed as a challenge to keep a high-caliber student engaged and challenged and a low-caliber student engaged and motivated.

References

1. Stolterman, E.: The nature of design practice and implications for interaction design research. Int. J. Des. **2**(1), 55–65 (2008)
2. Wisecup, A.K.: Take it or leave it. Teach. Sociol. **45**(1), 73–79 (2017). https://doi.org/10.1177/0092055x16658087
3. Ahuado, A.: Teaching research methods: learning by doing. J. Publ. Affairs Educ. **15**(2), 251–260. Retrieved from https://www.jstor.org/stable/40215853?seq=1 (2009)
4. Pfeffer, C.A., Rogalin, C.L.: Three strategies for teaching research methods. Teach. Sociol. **40**(4), 368–376 (2012). https://doi.org/10.1177/0092055x12446783
5. Dorodchi, M., Powell, L., Dehbozorgi, N., Benedict, A.: Strategies to Incorporate Active Learning Practice in Introductory Courses. Faculty Experiences in Active Learning, pp. 20–37. Journal of Murrey Atkins Library at UNC Charlotte
6. Ramsden, P.: Learning to Teach in Higher Education. Routledge, London (2002)
7. Light, G., Cox, R.: Learning and Teaching in Higher Education: The Reflective Professional. Paul Chapman, London (2004)
8. Gibbs, G.: Learning by Doing: A Guide to Teaching and Learning Methods. Retrieved from https://thoughtsmostlyaboutlearning.files.wordpress.com/2015/12/learning-by-doing-graham-gibbs.pdf (1998)
9. Kolb, D.A.: Experiential Learning: Experience as the Source of Learning and Development. Pearson Education, Upper Saddle River, NJ (2015)
10. Kolb, D.A.: Learning Style Inventory, Revised edition. McBer, Boston, Massachusetts (1985)
11. Biggs, J.B.: Teaching for Quality Learning at University, 2nd edn. Open University Press/Society for Research into Higher Education, Buckingham (2003)
12. Bloom, B. S. (1956). Taxonomy of Educational Objectives: The Classification of Educational Goals. Longman, Cognitive Domain.
13. Piaget, J.: Retrospective and prospective analysis in child psychology. Br. J. Educ. Psychol. **1**(2), 130–139 (1931)
14. Wadsworth, B.J.: Piaget's theory of Cognitive and Affective Development: Foundations of Constructivism. Longman, New York (2004)
15. Frank, A.R.: Breaking down Learning Tasks: A Sequence Approach. Teaching Exceptional Children, vol. 6, issue 1, pp. 16–19 (1973). https://doi.org/10.1177/004005997300600104
16. Sweller, J., Ayres, P., Kalyuga, S.: Cognitive Load Theory. Springer, New York (2011)
17. Roosevelt F.D.: Zone of Proximal Development. Encyclopedia of Educational Psychology SAGE Publication (2008)
18. Maslow, A.H.: A theory of human motivation. Psychol. Rev. **50**, 370–396 (1943)
19. Stipek, D.J.: Motivation to Learn: From Theory to Practice. Allyn & Bacon, Boston, MA (1998)

Chapter 27
Design Thinking Instructions and Cognitive Processes

Apoorv Naresh Bhatt, Lavannya Suressh, and Amaresh Chakrabarti

Abstract In the era of the 4th industrial revolution, new jobs demand that the workforce should have several essential skills such as problem-solving, creativity, critical thinking and innovation skill. School education is the right time to impart these skills in young children who are members of the future workforce. Literature shows that several cognitive processes are at the core of these skills. These cognitive processes are also mentioned to be at the heart of the design thinking process, justifying the need for teaching design thinking process at the school education level. In the pedagogic context, the revised Bloom's taxonomy defines 19 specific cognitive processes and classifies these into six major categories. With the help of this taxonomy framework, an attempt has been made to find the association between the instruction for activities within the 'IISC design thinking' process (a specific process that aims to optimise design thinking) and the cognitive processes. Results indicate that following the above instructions while performing IISC design thinking activities enable most of the cognitive processes recommended by Bloom, covering all his six categories. This has the potential to support the development of higher-level cognitive skills that are required for the twenty-first-century workforce.

27.1 Introduction

According to Merriam-Webster dictionary, skill is 'a learned power of doing something competently: a developed aptitude or ability.' Unlike specific skills (e.g., factoring polynomials and solving square-root problems), skills like problem-solving

A. N. Bhatt (✉) · A. Chakrabarti
Centre for Product Design and Manufacturing (CPDM), Indian Institute of Science (IISc), Bangalore, India
e-mail: apoorvbhatt@iisc.ac.in

A. Chakrabarti
e-mail: ac123@iisc.ac.in

L. Suressh
Department of Electronics and Communication (ECE), PES University, Bangalore, India
e-mail: lavannya.suressh@gmail.com

A. Chakrabarti et al. (eds.), *Design for Tomorrow–Volume 2*, Smart Innovation, Systems and Technologies 222, https://doi.org/10.1007/978-981-16-0119-4_27

and critical thinking are general skills because they are useful in acquiring a range of cognitive, motor, and social skills [1]. Authors have reviewed various research papers, international reports, as well as articles and identified the 21st-century skills, which are essential for the workforce to have and be developed by students during the education [2–6]. All the above sources are primarily emphasising creativity (ability to produce a novel outcome), critical thinking (ability to analyze and evaluate information), problem-solving (ability to choose means to reach toward a goal), adaptability (ability to adjust to new conditions) and collaboration (ability to work together) skills as essential skills. In addition, there is a coherence between the skills required from a workforce and the skills needed to be developed by students. [7] noted that in order to develop creative, critical, and innovative thinking, the required core skills are not limited to remembering skills but also focusing, information gathering, organizing, analyzing, integrating and evaluating skills [7]. Thus, school education is the right time to inculcate those skills in young children who are the future workforce.

27.1.1 The Current Indian Education System

Current teaching and assessment techniques in the Indian education system emphasize rote learning over meaningful learning. During teaching and assessment, more importance is given to retention skills. Moreover, repetitions of questions throughout exams indirectly hinder the transfer of knowledge to students. This leads to a lack of inculcation of higher-level cognitive processes in children. According to the National Education Policy 2016 (draft), Ministry of Human Resource Development, Government of India, quality of the education imparted is a critical challenge in the education sector [8]. In most cases, the assessment of learning achievement focuses on memory-based learning and testing the students' ability to reproduce content knowledge [8] where more emphasis is given to the retention of explicit written instructions and leaves no scope for development of essential skills (i.e., creativity, critical thinking and problem-solving skills). Moreover, in the classrooms, learning is receptive (where teacher demonstrates, describes or writes the teaching content and information passes one way), and the assignments given ask students to work individually. This inhibits the development of collaboration and communication skills in students. The above discussion leaves the scope of assessing the current education system and its effects on the development of different skills.

27.1.2 Design Thinking Process, Activities, and Instructions

Design thinking (DT) is the cognitive process from which design concepts emerge [9]. It is an iterative process which involves identifying goals (needs), generating proposals to satisfy the goals, and improving both the goals and proposals [10]. In the previous work, the authors have described a design thinking model known as

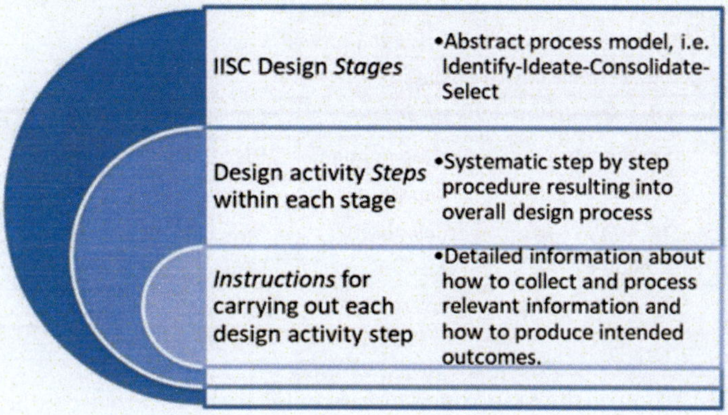

Fig. 27.1 IISC design thinking model: stages, activity steps and instructions

IISC by analysing and combining various activities from existing models [11]. The IISC design thinking model (developed at the Indian Institute of Science, Bangalore) consists of four broad generic stages: Identify, Ideate, Consolidate and Select. Each stage is further divided into several activity steps. IISC design activities direct students and that leads to the generation of solution. Each activity of IISC design thinking has specific instructions which help students or learners to generate different outcomes like a list of requirements, ideas, concepts and prototypes in an effective manner (see Fig. 27.1). In previous work, the outcome produced by the students was assessed by the experts, and mentors' and students' feedback were taken. The results revealed the effectiveness of IISC design thinking as a useful tool for problem-finding and problem-solving [12].

27.2 Elementary Cognitive Operations in Design

Various attempts were made to link steps/activities of DT phases with other theoretical constructs. For example, by conducting a review of the literature [13], shown how specific design thinking processes and tools help decision-makers to reduce their individual-level cognitive biases, and to increase the potential for improving innovation outcomes. [11] established the link between the DTP steps and their potential impact on learning objectives and subsequently derived their favorability to the learning approaches. Similarly, [14] established the link between the activities of the design thinking process with the various constructivist learning principles.

On the one hand, design approaches help us produce a specific outcome in the form of a product (artefact), process or service, which can be novel and have a social value. On the other hand, performing design activities helps designers or learners to acquire certain kinds of skills and instil a school of thought which may contribute to the development of the twenty first-century skillset. From the literature, authors

Table 27.1 Elementary operations in the design process

Elementary operations in the design process				
See	Observe	Read	Listen	Interpret
Measure	Remember	Keep in mind	Recall	Speak
Explain	Write	Report	Sketch	Draw
Dimension	Set up parts list	Calculate	Note, annotate	Order, classify
Compare	Combine	Analyze	Synthesise	Abstract
Concretize	Establish analogy	Invert	Induct	Deduce

have tried to identify the skills/operations/mental processes that can be acquired by performing design activities. Hubka and Eder [15] explained the structure of possible activities in the design process, in which they argued that various elementary operations are used in the design process (As shown in Table 27.1) [15].

Taylor et al. [16] developed a design education program to enable individuals to acquire behaviours related to the following five 'mental processes,' i.e., (1) problem solving, (2) creative thinking, (3) visual thinking, (4) group interaction, and (5) communication skills [16]. Based on a review of top technologists, Halfin identified a total of 17 mental processes used by design practitioners. These mental processes consist of the following: defining the problem or opportunity operationally, observing, analysing, visualising, computing, communicating, measuring, predicting, questioning and hypothesising, interpreting data, constructing model and prototypes, experimenting, testing, designing, modeling, creating and managing [17]. Further, Hill developed and tested a technique for assessing these mental processes as used by students who participated in instructional learning activities in technology education [18]. Similar kinds of work were found in the literature where the effectiveness of technology education in increasing students' cognitive abilities with respect to problem-solving was evaluated by using Halfin's code of mental process [19].

Authors, however, found the following issues in the Halfin's code and definitions of mental processes, as explained below:

- Many of the above that are currently described as mental processes (e.g., communicating) are skills and not mental processes.
- Many mental processes (e.g., defining a problem, constructing models and prototypes, testing) are design activity steps rather than mental processes, and may contain physical and mental processes to be carried out.
- Only a few of the processes (i.e., predicting and hypothesising) are non-compound in nature. For instance, 'observing processes' not only stimulate the senses but also 'differentiate' between relevant parts of a piece of information, and 'retrieve' and 'compare' knowledge from long-term memory that is consistent with external information.

The above confusions created by the current use of overlapping terminology makes it difficult to distinguish between basic and compound cognitive/mental processes (which may or may not involve physical operations), and to investigate as to how these processes relate to and are supported by two systems of interest: the current system of education and the broad activities of design thinking.

In order to resolve this, we propose a common approach for assessment of the current education system as well as design thinking instructions, both using the lens of cognitive processes proposed in the revised Bloom's taxonomy. From a pedagogic point of view, the revised Bloom's taxonomy is a systematic classification for cognitive processes. The revised taxonomy consists of six major cognitive categories (viz. remember, understand, apply, analyze, evaluate, and create) under which it defines a total nineteen cognitive processes (viz. recognizing recalling, interpreting, exemplifying, classifying, summarizing, comparing, inferring, explaining, executing, implementing, differentiating, organizing, attributing, checking, critiquing, generating, planning, producing) [20]. The revised taxonomy is used heavily across the education domain, and it is useful not only for aligning education goals, instructions, and assessment but also for assessing student's abilities [21].

The focus of this work, therefore, is to seek the relationships between these cognitive processes and the two systems of interest: (1) the current teaching activities (in the Indian school education context); and (2) the design thinking activities. The first goal is to understand the extent to which the current school education promotes the cognitive processes. The second goal is to understand the extent to which design thinking has the potential to promote these cognitive processes. This study has been conducted with the following objectives: (1) to understand as to how the questions used in the test papers of Indian school-leaving examinations are associated with (testing of) the revised Bloom's cognitive processes; and (2) to understand as to how the instructions used for carrying out design thinking activities are associated with (promotion and practice of) these cognitive processes.

Based on these two research objectives, the following are taken as the research questions for this study.

Research Questions

1. Do the questions asked in school-leaving examinations have the potential to assess higher-level cognitive processes?
2. Do the instructions for carrying out design thinking activities have the potential to promote higher-level cognitive processes?

In order to answer these questions, the definitions of the cognitive processes, and test questions, and the instructions for IISC DT were analyzed in detail. Then, each question/instruction was mapped with the cognitive processes of the revised Bloom's taxonomy framework. As a rudimentary evaluation of the mapping procedure, an inter-coder reliability test was conducted. The results and future work have been discussed in the subsequent sections.

27.3 Analysis of Question Papers

CBSE is the national board of education followed by a majority of the public and private schools in India. There are approximately 21,000 schools in India and 220 schools in 28 foreign countries affiliated to the CBSE. The board exams conducted annually for class 10 and 12 is a uniform mode of testing throughout the country. As CBSE 10th board exams are conducted over a large number of schools, it is appropriate to investigate the current status of school education by taking CBSE board as a benchmark and analyzing the 10th class CBSE board science question papers assuming that assessment questions are aligned with the curriculum and teaching activities. A detailed analysis of these question papers and their curriculum has been done to evaluate how these exams test a student's brain. The questions in the CBSE class 10, science papers from 2015 to 2019, were classified based on which cognitive category it tests the brain. During the analysis, wherever required, the questions have been divided into sub-questions for a more precise classification. For simplification, association of marks/points with the questions has not been taken into consideration during the analysis.

27.3.1 Key Findings and Inferences from the Analysis

The results of analyzing year-wise data have been shown below in Table 27.2. Whereas number of questions in exams and associated cognitive categories and processes have been shown in Fig. 27.2.

Following a comprehensive analysis of the question papers over the past five years using the Bloom's taxonomy on the cognitive dimension, it was found that the CBSE questions covered mainly three categories: 'Remember,' 'Understand,' and 'Apply.' There were very few questions that tested 'Analyze,' and hardly any that tried 'Evaluate' or 'Create' categories. Based on the nature of questions asked in CBSE board exams, the material provided by CBSE was looked into and the questions asked

Table 27.2 Results of analyzing year-wise data: Percentage of questions to test different cognitive categories over five years

Category	Academic year					Average (Roundoff)
	2015	2016	2017	2018	2019	
Remember	32	35.7	32.6	37.5	31.3	34
Understand	46	40.5	48.8	47.9	47.9	46
Apply	14	19	11.6	8.3	10.4	13
Analyze	6	4.8	7	6.3	2.1	5
Evaluate	2	0	0	0	8.3	2
Create	0	0	0	0	0	0

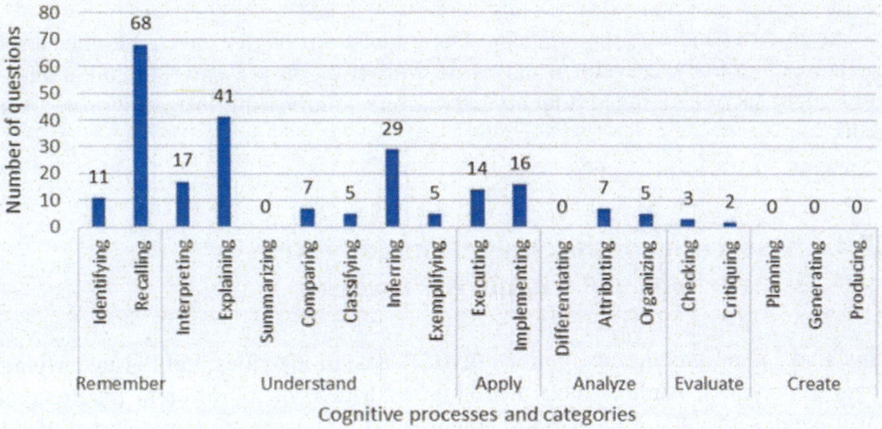

Fig. 27.2 Number of questions to test different cognitive categories and processes over the period of five years

as exercises in the textbook, and the NCERT exemplar book (containing practice questions offered by CBSE) were analyzed, so as to check the extent to which the questions in these textbooks matched the questions asked in the papers. Based on the nature of the questions asked in the materials provided by CBSE, three main categories were created:

'Exactly the same questions from study material', 'Similar to questions asked in the study material (change in values, specifics)' and 'Unfamiliar questions'. Each category of questions was analyzed separately and categorized to see the extent to which the questions match with the CBSE provided material.

From the findings, it is clear that, among the five-year questions, the percentage of precisely the same questions and similar questions from the textbook as well as exemplar book was about 73% of the total questions. This shows that majority of the questions asked were familiar to the student in some way before. According to [15] if the assessment task is identical to a task or example used during instruction, one is probably assessing remembering, despite one's efforts to the contrary. Unless more unfamiliar questions are asked, one cannot ensure that a higher level of cognitive processes rather than remembering is being assessed.

Authors have also observed that some questions have a proxy/auxiliary (optional) question, and it allows students to choose one question to attend out of two. Many of these questions test different categories, e.g., even if CBSE asks questions that test the application part of the brain, the students still have an option not to answer it and answer the proxy question that is purely memory-based. So, between the students who answered the application testing question and the students who answered the remembering type question, both will be evaluated similarly for the given question.

The findings indicate that the assessment questions hardly covered higher-level categories like analyze, evaluate and create. Moreover, the presence of familiar questions might have encouraged students to focus on retention skills instead of on transfer skills.

27.4 Association of the Design Thinking Activity Instructions and Cognitive Processes

This study identifies the instructions of IISC design activities and the underlying cognitive process. Authors have distinguished a total of 52 different instructions covering all steps of IISC design thinking process which are used by a learner while performing design activities. For the consistency and reliability, two authors have comprehended the definitions and examples of the cognitive process dimensions of revised Bloom's taxonomy, and based on their understanding, they have tagged each design instruction with the cognitive process and category independently. The similarity report has been made with the percentage agreement of 81% as a result.

The document was made where the definition of each cognitive process is given along with the example of design activity instruction. The example of each process is shown in Table 27.3.

27.4.1 Results and Inferences

Association of instructions of IISC design thinking activities and cognitive processes shows that out of 19, a total of 17 cognitive processes are mentioned in the Table 27.3 which covers all the six cognitive categories (Fig. 27.3). A significant number of instructions cover the higher level of cognitive categories like analyze, evaluate, and apply (Table 27.4).

Moreover, while mapping the instructions with cognitive processes, authors have identified that:

Some instructions may associate with more than one cognitive process, and one process usually occurs in conjunction with other processes, e.g., making an analogy from any event/object is associated with comparing and inferring along with implementing.

Though exemplifying and inferring processes are not tabulated, they are inherently present in some of the instructions. For example, while observing any process/person/object in a given habitat, one may anticipate the possibility of occurrence of a problem. The process of predicting a problem is nothing but inferring. Thus, those instructions are implicitly associated with the inferring process.

As there are specific cognitive processes underlying the instructions which help in generating design outcomes, the design thinking steps do not only help to produce

Table 27.3 Association of instructions with the cognitive process

Activity step	Instruction	Cognitive process	Definition
Observe habitat	What activities and processes are happening in the habitat?	Identifying/Recognizing (Remember)	Locating knowledge in long-term memory that is consistent with the presented material [20]
	Do you observe any problems that are happening in the habitat?	Differentiating/Discriminating/selecting (Analyze)	Distinguishing relevant from irrelevant parts or important from unimportant parts of the presented material [20]
	If so, why are these problems occurring?	Explaining (Understand)	Constructing a cause-and-effect model of a system [20]
Group problems	Group the problems based on their similarities	Organising/structuring (Analyze)	Determining how elements fit or function within a structure [20]
Benchmark against objects	Check whether there is any existing solution to your problem	Recalling (Remember)	Retrieving relevant knowledge from long-term memory [20]
	Is the existing solution good enough to solve the problems in the situation you identified?	Critiquing/Judging (Evaluate)	Detecting inconsistencies between a product and external criteria, determining whether a product has external consistency; detecting the appropriateness of a procedure for a given problem [20]
Enlist the requirements	Based on the requirements make solution neutral problem statement	Summarising/Abstracting (Understand)	Abstracting a general theme or central point [20]
Enlist process steps and problems	Imagine and enlist the steps that are needed in the desired process to fulfil the requirements you have identified	Planning/Designing (Create)	Devising a procedure for accomplishing some task [20]

(continued)

Table 27.3 (continued)

Activity step	Instruction	Cognitive process	Definition
Generate	Generate as many alternate ideas as possible for carrying out each process step while avoiding the problems if any	Generating (Create)	Corning up with alternative hypotheses based on criteria [20]
Ideas from nature	Explore how does nature addresses similar problems occurring in habitats	Comparing (Understand)	Detecting correspondences between two ideas, objects, and the like [20]
	Can you learn from the principles behind nature's solutions to think of novel solutions to the problem you have identified?	Implementing/Using (Apply)	Applying a procedure to an unfamiliar task [20]
Remove infeasible ideas	Check all ideas for their feasibility and remove all ideas that are clearly not feasible	Checking/Coordinating/detecting (Evaluate)	Detecting inconsistencies or fallacies within a process or product; determining whether a process or product has, internal consistency [20]
	Group the similar ideas; the similarity in terms of the principles used behind the idea, and the process step/problem to be solved using the idea	Classifying/Categorizing (Understand)	Determining that something belongs to a category [20]
Combine Ideas	Combine individual ideas for each process step into a solution	Executing/Carrying out (Apply)	Applying a procedure to a familiar task [20]
Identify conflicts by acting as user	Imagine yourself to be the user and think how good the solution in the sketch/model (that you have made) would satisfy each requirement	Attributing (Analyze)	Determine a point of view, bias, values, or intent underlying presented material [20]
Identify conflicts by asking users	Resolve the conflicts and improve the solution and the model, so that each solution can satisfy each requirement without conflict among these	Producing (Create)	Inventing a product [20]
Make the final prototype	After dealing with the conflicts/discrepancies in the solutions, the changes should be reflected in the models/prototypes you had made earlier	Interpreting/representing (Understand)	Changing from one form of representation to another [20]

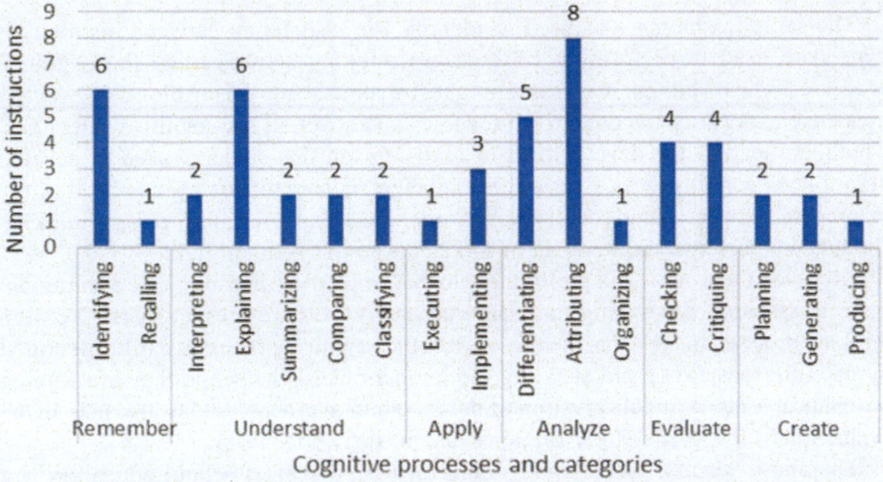

Fig. 27.3 Design thinking instructions and associated cognitive processes and categories

Table 27.4 Percentage of instructions of design thinking activities that cover different cognitive categories

Category	Remember	Understand	Apply	Analyze	Evaluate	Create
Percentage of instructions	13	27	8	27	15	10

outcomes that lead to a solution of the given problem but also can be used as a tool for fostering different cognitive processes. Also, from the above discussion, we can say that IISC design thinking process allows students to get engaged in different activities that contain instructions of a higher level of cognitive processes. Performing which learners may improve their higher-order cognitive processes and these processes can lead to the development of skills which are requisite for the twenty-first-century workforce.

27.5 Conclusion and Future Work

An attempt has been made to examine educational opportunities in India. This has been done by analyzing the last five years of the CBSE board science question papers and classifying them into cognitive categories defined by the revised Bloom's taxonomy. The results suggest that the test questions put a heavy emphasis on the retention abilities and do not demand higher levels of cognitive categories which may lead to the lack of development of skills like critical thinking, creativity and complex problem-solving.

The same taxonomy was used to identify the association between instructions which are used in performing IISC design thinking activities and the cognitive processes. The results show that attaining instructions while performing design activities enables most of the cognitive processes and cover all six cognitive categories, which can lead to the development of skills. Given that students need to develop higher-level cognitive processes to be competent, design thinking seems well suited to prepare a twenty-first-century workforce. Therefore, teaching design thinking should be an essential component of K-12 education system in India.

However, the work is still preliminary in that while overall results are encouraging, there is a scope of covering more instructions which are present in IISC design thinking process. In addition, the presence of the cognitive processes can be identified by carefully observing and studying the learners' activities while they are solving design problems. Also, observation of these cognitive processes may also help in the evaluation of the learners' performance and corrective feedback.

The above results, even though found in the context of school education, are likely to be similar to those in regular engineering and other graduate-level education in India, where rote learning and descriptive content dominate. These will have a consequent impact on the workforce.

References

1. Schunk, D.H.: Learning Theories an Educational Perspective, 6th edn., p. 280. Pearson (2012)
2. Finegold, D., Notabartolo, A S.: 21st century competencies and their impact: An interdisciplinary literature review. In: Transforming the US Workforce Development System, pp. 19–56 (2010)
3. Pacific Policy Research Center: 21st century skills for students and teachers. Res. Eval. 1–25 (2010)
4. The 10 skills you need to thrive in the Fourth Industrial Revolution. https://www.weforum.org/agenda/2016/01/the-10-skills-you-need-to-thrive-in-the-fourth-industrial-revolution/
5. Euroepan Commission: Horizon 2020, Work Programme 2016–2017, 13. Europe in a changing world–inclusive, innovative and reflective Societies. https://ec.europa.eu/research/participants/data/ref/h2020/wp/2016_2017/main/h2020-wp1617-societies_en.pdf (2017)
6. National Education Policy: (Draft) https://mhrd.gov.in/sites/upload_files/mhrd/files/Draft_NEP_2019_EN_Revised.pdf (2019)
7. Adams, D., Hamm, M.: Demystify Math, Science, and Technology: Creativity, Innovation, and Problem-Solving. R&L Education (2010)
8. National Education Policy: https://mhrd.gov.in/sites/upload_files/mhrd/files/nep/Inputs_Draft_NEP_2016.pdf (2016)
9. Cross, N. (2011). Design thinking: Understanding how designers think and work. Berg.
10. Chakrabarti, A. Creative Engineering Design, Presentation Slides. Lecture Notes for the National Programme on Technology Enhanced Learning (NPTEL), India. https://nptel.ac.in/courses/107108010 (2013)
11. Bhaumik, R., Bhatt, A., Kumari, M.C., Menon, S.R., & Chakrabarti, A. (2019). A gamified model of design thinking for fostering learning in children. In: Research into Design for a Connected World, pp 1023–1036. Springer, Singapore. https://doi.org/10.1007/978-981-13-5977-4_85
12. Bhatt, A.N., Bhaumik, R., Moothedath Chandran, K., Chakrabarti, A., IISC DBox: A game for children to learn design thinking. In: International Design Engineering Technical Conferences

and Computers and Information in Engineering Conference, vol. 59216, American Society of Mechanical Engineers (2019). https://doi.org/10.1115/DETC2019-98066
13. Liedtka, J.: Perspective: linking design thinking with innovation outcomes through cognitive bias reduction. J. Prod. Innov. Manag. **32**(6), 925–938 (2015). https://doi.org/10.1111/jpim.12163
14. Pande, M., Bharathi, S.V.: Theoretical foundations of design thinking—A constructivism learning approach to design thinking. Think. Skills Creativity **36**, 100637 (2020). https://doi.org/10.1016/j.tsc.2020.100637
15. Hubka, V., Eder, W.E.: Theory of technical systems and engineering design synthesis. In: Engineering Design Synthesis, pp. 49–66. Springer, London (2002)
16. Taylor, A.P., Vlastos, G., Marshall, A.: Architecture and Children: Teachers Guide Interdisciplinary Learning Activities of the Architecture and Children Curriculum, pp. iv–v. Architecture and Children Institute, Washington (1991)
17. Halfin, H.H.: Technology: a process approach. Doctoral dissertation, West Virgina University. Dissertation Abstracts International, vol. 11(1), p. 1111A (1973)
18. Hill, R.B.: The design of an instrument to assess problem solving activities in technology education. J. Technol. Educ. **9**(1), 31–46 (1997). https://doi.org/10.21061/jte.v9i1.a.3
19. Kelley, T.R.: Cognitive processes of students participating in engineering-focused design instruction. J. Technol. Educ. **19**(2), 50–64 (2008)
20. Anderson, L.W., Bloom, B.S.: A taxonomy for learning, teaching, and assessing: a revision of Bloom's taxonomy of educational objectives. Longman (2001)
21. Forehand, M.: Bloom's taxonomy. Emerging Perspect. Learn. Teach. Technol. **41**(4), 47–56 (2010)

Chapter 28
Problem-Based Learning (PBL) in Undergraduate Education: Design Thinking to Redesign Courses

Shakuntala Acharya, Apoorv Naresh Bhatt, Amaresh Chakrabarti, Venkata S. K. Delhi, Jan Carel Diehl, Ellen van Andel, Andrius Jurelionis, Laura Stasiuliene, Luis De Jussilainen Costa, and Riina Subra

Abstract Problem-based learning (PBL) has profound implications on the motivations of the student to learn and is known to help develop critical thinking, complex problem-solving, self-learning, collaboration and communication skills, thereby enabling fresh graduates to be industry-ready. However, most institutes of higher education in South Asia offering undergraduate programmes have instructional and didactic pedagogical systems. The Erasmus + project, '***Strengthening***

S. Acharya (✉) · A. N. Bhatt · A. Chakrabarti
Indian Institute of Science, Bangalore, India
e-mail: shakuntalaa@iisc.ac.in

A. N. Bhatt
e-mail: apoorvbhatt@iisc.ac.in

A. Chakrabarti
e-mail: ac123@iisc.ac.in

V. S. K. Delhi
Indian Institute of Technology, Bombay, India
e-mail: venkatad@iitb.ac.in

J. C. Diehl · E. van Andel
Delft Institute of Technology, Delft, Netherlands
e-mail: j.c.diehl@tudelft.nl

E. van Andel
e-mail: e.vanandel@tudelft.nl

A. Jurelionis · L. Stasiuliene
Kaunas Institute of Technology, Kaunas, Lithuania
e-mail: andrius.jurelionis@ktu.lt

L. Stasiuliene
e-mail: laura.stasiuliene@ktu.lt

L. De Jussilainen Costa · R. Subra
Aalto University, Aalto, Finland
e-mail: luis.dejussilainencosta@aalto.fi

R. Subra
e-mail: riina.subra@aalto.fi

A. Chakrabarti et al. (eds.), *Design for Tomorrow–Volume 2*, Smart Innovation, Systems and Technologies 222, https://doi.org/10.1007/978-981-16-0119-4_28

Problem-based Learning in South Asian Universities' (PBL South Asia) aims to build capacity of the South Asian partner institutes by collaboratively developing best practices in PBL for undergraduate education, bringing expertise and experience of peers from across Europe and India. Therefore, to gain benefits of the PBL approach, the redesign of existing courses was undertaken and the novel strategy of conducting a Design Thinking workshop to do so, was engaged. During the five-day workshop, faculties from the institutes in Nepal and Bhutan, who are most well aware of the challenges, shortcomings and strengths of their curriculum, were mentored step-by-step, by their Indian and European peers, who have more experience in delivering PBL courses. Backed by the strategy of Design Thinking, the complex problem-solving activity of course design was addressed systematically, and the five institutes proposed redesigned courses which are currently in the process of implementation.

28.1 Introduction

India, Nepal and Bhutan share a common history with respect to modern education practices and have a similar institutional and curricular construct for undergraduate, technical programmes. Graduate programmes are offered either on university campuses or in affiliated colleges, and a national-level body usually oversees curriculum development on technical education, such as All India Council for Technical Education (AICTE) in India and Nepal Engineering Council (NEC) in Nepal. Though the evaluation schemes may vary, the overall assessment is conducted as a common examination across all colleges affiliated to a university. A survey across South Asian universities revealed that the undergraduate curricula are predominantly instructional and not adequately hands-on due to several constraints, such as

- University-directed lesson plans with heavy syllabi to cover and restricted time for practical activities;
- A fewer number of co-instructors to guide in practical, real-world issues that can be addressed in courses;
- Dearth of motivation in students to self-learn and innovate during the stipulated practical hours within a course;
- Poor critical-thinking ability due to a general lack of awareness on sustainable development goals and their local implications in the students;
- Fewer collaborations in these courses and
- Poor communication skills.

Literature corroborates that in traditional engineering education [1]:

- Programs are content-driven instead of need-driven and do not provide sufficient design experiences to students.
- Students lack communication skills and teamwork experience, as well as awareness about social, environmental, economic and legal issues.

– Faculties lack practical experience and are not able to adequately relate theory to practice or provide design experiences, having outdated teaching and learning strategies.

Therefore, to address the above shortcomings and constraints, the introduction of problem-based learning (PBL) as a pedagogical approach is proposed as it is known to support the development of specific skills, such as critical thinking; complex problem solving, self-learning, due to increased motivation and engagement; collaboration and people management and communication [2] and can be implemented without altering the existing course content that is approved by the governing body or university.

This paper presents the unique proposition of using a 'Design Thinking' process to guide the redesign of existing courses into PBL courses.

28.2 Literature Review

28.2.1 Design Thinking

'Design Thinking', as a descriptive model of creative problem-solving, has its roots in architecture and industrial design practice [3]. It is defined as a cognitive process [4] of identifying and resolving 'wicked' or ill-defined problems through iterative stages and activities. Literature notes that this approach is viewed through various lenses, from being a process of reflective practice to being a systematic problem-solving methodology.

Stanford's d.school Design Thinking process [5] and IDEO human-centred design model [6] are among the most renowned, entailing five nonlinear activities, namely Empathize—Define—Ideate—Prototype—Test. These two models stem from human-centred design principles, whereas several other systematic design approaches, such as Cross [4], Pahl and Bietz [7], Hubka and Eder [8], Dieter and Schmidt [9], Eppinger and Ulrich [10] are prevalent in engineering practice. The latter models propose stage-wise progression from problem identification to concept selection, supported by several methods and tools. While each model has its relevance depending on the problem at hand, they may be broadly described by the following four steps [11]:

- ***Step 1: Identification of requirements*** generated and clarified against needs, through observations, interviews, role-play, stakeholder analysis and checklists;
- ***Step 2: Ideation of solutions*** through creative methods such as brainstorming and SCAMPER;
- ***Step 3: Consolidation of solutions into feasible solutions*** through for example TRIZ and morphological chart method and

- ***Step 4: Selection of the most promising solution*** as concept from among all other alternatives, upon evaluation by methods such as weighted objectives and concept selection methods.

28.2.2 Problem-Based Learning (PBL)

Problem-based learning (PBL) is an 'instructional (and curricular) learner-centred approach that empowers learners to conduct research, integrate theory and practice and apply knowledge and skills to develop a viable solution to a defined problem' [12]. This method was developed and implemented in McMaster University around 1965 and is based on the theory that learning is a process in which the learner actively constructs knowledge [13]. It is defined as a 'focused, experiential learning organized around the investigation and resolution of messy, real-world problems' [14] in which students learn through 'facilitated problem-solving that centres on a complex problem that does not have a single correct answer' [15].

Six core characteristics of PBL [16] are identified as follows:

1. Learning is student-centred: Students take responsibility for their own learning, identify the knowledge that is required to learn and determine the way/resources to get information by themselves.
2. Learning occurs in small student groups: A group generally consists of five to nine students who work together along with a tutor. Students share their knowledge and learn from others, and learning happens in collaboration.
3. Teachers are facilitators or guides: Tutor asks students typically questions to better understand and manage the problem.
4. Problems form the organizing focus and stimulus for learning: Problems represent the challenges students will face in real life and provide the relevance and motivation for learning. Students realize what they will need to learn in order to solve the problem.
5. Problems are a vehicle for the development of problem-solving skills: The problem format is in the same way that it occurs in the real world (ill-structured, complex) which allow students to inquire the problem deeper. The students do not restrict to a single subject; instead, they focus on integrating information from many disciplines.
6. New information is acquired through self-directed learning: The students are expected to learn from the world's knowledge and accumulated expertise by virtue of their own study and research.

There are many variants of PBL as it can be modified according to domain or subject, individual course requirements or institute traditions, and can be implemented at a chapter level, course level or even curriculum level. Broadly, every variant has two phases, namely a collaborative-learning phase and a self-directed learning phase. A single phase alone has insufficient impact on learning in PBL [17]. An exemplary PBL structure is given below:

- The PBL process starts with an ill-defined, real-life problem formulated by tutor/teacher.
- Students in small groups start analysing the given problem systematically and try to reach a consensus on the meaning or implication of the problem based on the terms and concepts of the domain, subject or topic.
- Next, they construct a tentative theory explaining the phenomena or events described in the problem-at-hand in terms of its underlying principles or mechanisms and identify the facts that they already know and what they require to know in order to solve the problem. Thus, learning issues for individual study are formulated. These learning issues usually consist of questions arising from the discussions.
- Students search and evaluate resources which can be useful to learn problem domain and pursue these issues through individual, self-directed learning usually using a variety of resources: books, articles, movies and Internet sites, where tutor scaffolding takes place.
- Students return to their tutorial group, review and share what they have learned, propose the solution and elaborate on different aspects of it. Together they discuss and explore the extent to which the students understanding of the problem have developed and whether misconceptions remain that need to be addressed.
- Students self-evaluate and evaluate others in the group (peer evaluation).

Curriculum-wide PBL implementation and single-course PBL implementation show similar findings according to earlier studies [17, 18]. PBL has been found to have profound implications on the motivations of the student to learn, stating that 'the freedom to select their (students) own resources to answer the learning issues gives them ownership over their learning', [19] and the onus of ensuring retained motivation falls on the shoulders of the students as peer–teacher [20]. The role of the mentor is to assure the students and allow constructive discussion while not interfering with the process. The teacher or tutor may also be the mentor. However, the key responsibilities of the tutor are formulating a contextually appropriate problem and facilitating the learning by helping students manage metacognitive activities by providing 'triggers'.

The key element driving PBL is the 'problem'. It guides self-learning and problem-solving and, in turn, develops the other critical skills. Subject or topical problems and learning objectives for a domain maybe considered universal and are well-structured, less complex and domain-specific. However, 'real-world' problems are ill-structured, more complex and cover knowledge of multiple domains [21], and its context is subjectively unique to South Asia. Consequently, borrowing of existing courses and their defined problems from European curricula would not be appropriate. The revised literature also states that different cognitive skills are required to solve well-structured or classroom problems versus ill-structured or real-life problems [22], thereby reflecting that it is inadequate for students to merely be able to solve classroom problems as it has little to no bearing on their abilities for post-undergraduate studies.

The two critical attributes of a 'problem' are as follows [22]:

1. A problem must be an unknown entity in some situation (the difference between a goal state and a current state) varying from algorithmic math problems to complex societal problems, such as violence in the schools.
2. Finding or solving for the unknown must have some social, cultural or intellectual value, i.e. someone believes that it is worth finding the unknown.

Problem complexity is defined by the number of issues, functions or variables involved in the problem; the degree of connectivity among those properties; the type of functional relationships among those properties and the stability among the properties of the problem over time [23]. Problem statement may be formulated with respect to general guidelines in the form of a checklist [24] or criteria identified for constructing problem [25]. Therefore, formulating contextually appropriate problems for South Asian undergraduate students is important.

28.3 Descriptive Study—Redesigning Courses with Design Thinking

A five-day workshop was conducted at the Indian Institute of Science (IISc), Bangalore, with the intent to redesign existing courses into PBL courses, to be implemented in the partnering universities of Nepal and Bhutan for the final year undergraduate students.

28.3.1 Methodology

The workshop had 24 participants comprising 12 faculties from the five South Asian partner universities in Nepal and Bhutan as the key course designers, supported by 12 faculty and research associates from the Indian and European Universities.

Four teams were devised (Table 28.1), based on two factors—(1) status of institute: (S1) autonomous or (S2) affiliated; and (i2) intervention sought: (i1) process and PBL methodology focus; (i2) domain and technical focus and (i3) Soft skill focus. While the status of the institute reflects the ability of the institute to implement the proposed redesigned PBL course, the three broad areas of focus for course development were clarified from previous surveys and need assessments. The four teams were as follows.

Table 28.1 Team composition

	(i1) Process and Methodology focus	(i2) Domain and Technical focus	(i3) Soft skill focus
(S1) Autonomous	Team 1	Team 3	
(S2) Affiliated	Team 2		Team 4

The workshop was planned, such that each day would emulate a stage of the design process, as elucidated by the Design Thinking steps discussed above and, in turn, had (1) tutoring sessions, where the specific design stage and its methods were taught to support the course redesigning task, as well as for further dissemination during the course; (2) collaborative-learning sessions, where each team consisting of the 'course designers' and mentors co-created the solution. Apart from this, expert practitioners of PBL presented cases and examples of PBL course or curricula implementation in India and Europe. At the same time, self-learning sessions were encouraged prior to or during the off hours of the workshop.

The program for the workshop was as follows:

Day 1—Team building

Day 2—Identification/exploration

Day 3—Conceptualisation/ideation

Day 4—Consolidation/discussion

Day 5—Selection/reflection and presentation.

The teams were provided with a template for 'proposal of a new course/PBL course adaptation'. It provided an overall guideline for development and implementation of each proposed course and highlighted the essential elements that need to be addressed, apart from the individual requirement of each course and syllabus.

28.3.2 Results

On the final day of the workshop, the teams presented their proposed 'solutions' for each of the focus areas and drew discussions, reflections and insights to conceptualize institute-specific, redesign of their course with adoption of the PBL approach. An example of such a proposal, detailed with respect to the provided template, is given below.

28.3.2.1 Current Course Description and Justification for Change

This course will teach additional practical skills related to integrated circuit building (including the prototyping of the printed circuit board and integrated circuit) and knowledge of scale integration that are currently missing in the existing course. These missing skills, along with the recent course, will be taught by using PBL methods.

28.3.2.2 Problem Identification

Overview of the Intervention needed is as Follows

Aim: To change the conventional passive learning into active, problem-based learning.

1. Course to be redesigned: Integrated Digital Electronics (Credits: 3) Level: B.Eng., 3rd Year, 1st Sem
2. Course Objective: To impart knowledge different types of Logic Gates, Memory and Switching Systems and apply the same through PBL approach.
3. Duration: One Semester, 15 weeks
4. Learning Outcomes: *On course completion, students should be able to:*

   (a) Develop different digital logic gates using semi-conductor components.
   (b) Analyse, design, simulate and implement digital logic circuits.
   (c) Classify and compare different gates in terms of operation and performance.
   (d) Classify different semiconductor memories.
   (e) Acquire the knowledge to address real-life applications of digital logic gates.

5. Learning Objectives: *Students must be capable of:*

   (a) Independently managing a project;
   (b) Solving real-life problems using digital logic gates/electronics;
   (c) Critically thinking to identify and assess complex problems;
   (d) Working in teams collaboratively, manage projects and people, show leadership; and
   (e) Communicating one's ideas and concepts with clarity.

The Formulated List of Requirements Are Below

1. Course must have the following PBL course elements and ensure that the time is adequately planned:

   (a) Lecture (*L*) delivery time
   (b) Tutorial (*T*) time for mentoring/facilitating time
   (c) Students' group/self-learning time
   (d) Students' collaboration time
   (e) Communication time—presentations (Pr)

2. Course must imbibe PBL through several 'triggers' and 'methods' that aid the process.
3. Course Plan must have the stipulated minimum number of hours per week, as per University:

   (a) 3 h/week—Lectures (*L*) or Tutorials (*T*)

(b) 1 h/week—Presentation (Pr)
(c) 1 h/week—Lab for prototyping (*P*), or Field visit (*F*)

4. Internal Evaluation Scheme is required, with the consultation of the department, as final exam will be conducted as per University.
5. Availability and access to dedicated Team workspace/prototyping space.

28.3.2.3 Ideation and Solution Consolidation

Teams used several ideation techniques to generate various sub-solutions with respect to the requirements identified earlier and consolidated the same into solution variants (Tables 28.2 and 28.3).

Table 28.2 Activities and skills for each PBL course elements

Course elements				
L: Lecture	*T*: Tutorial	*P*: Practical	*F*: Fieldwork	Pr: Presentation
Lecture delivery	Assignment mentorship	Simulation	Industry visit	Presentation/communication
Question answer session	Analytical thinking and self-learning	Testing	Survey Data collection	Report writing/collaboration
Group discussion/ collaboration	Problem finding/ identification	Prototyping	Problem reformulation	Evaluation (by instructor)
	problem-solving/ideation		Solution validation	Feedback (from instructor, mentor, peer)
			Feedback	

Table 28.3 Mapping of PBL course elements to each chapter/unit of course

Chapter/Unit	Topic/Course details	*L*	*T*	*P*	*F*	Pr
1	Review of BJT and MOS	✔	✔	✔		
2	Resistor–transistor logic (RTL) and integrated–injection logic (IIL)	✔	✔	✔	✔	✔
3	Diode–transistor logic (DTL)	✔	✔	✔		✔
4	Transistor–transistor logic (TTL)	✔	✔		✔	✔
5	Emitter–couple logic (ECL)	✔	✔	✔	✔	✔
6	NMOS and CMOS logic	✔				✔
7	Comparison of logic families		✔			✔
8	Memories	✔			✔	✔
9	Switches					

L stands for 'lecture', for *T* stands for 'tutorial', *P* stands for 'practical', *F* stands for 'fieldwork', and Pr stands for 'presentation'

Table 28.4 Proposal for internal evaluation scheme

Attendance	Scheduled test	Laboratory test	Presentation	Report	Prototyping/	Total
5	10	5	10	5 + 5	10	50

28.3.2.4 Concept: Selection of the Most Promising Solution

Teams evaluated the solution variants and selected the most promising as 'concept' to further detail, as described in Tables 28.4 and 28.5.

Table 28.5 Week-wise course plan

Week	PBL tasks	Roles and responsibility	Notes
1	Course introduction, orientation of teaching methodology, timeline, evaluation criteria	Instructor	Introducing PBL
2–4	Lecture delivery, laboratory work, problem identification and analysis	Instructor and supporting laboratory staff	Classroom and laboratory activities
5–6	Case preparation/field visit followed by presentation and preliminary report submission	Instructor, mentor and supporting staff	Group formation, literature review, domain identification, field visit
7–9	Problem-solving assignments, group discussion, lectures, laboratory works and mentoring	Instructor, mentor and supporting laboratory staff	Brainstorming, classroom and laboratory activities
9	Mid-term presentation /assignment evaluation	Instructor, mentor	Group discussion, feedback collection
10	Incorporating the feedback and generating final outcome	Students	Modification, prototyping
11	Deliverables	Students	Prototyping, assignment submission
12	Deliverables	Students	Final presentation and report submission
13–14	Preparation week	Student	
15	Final assessment, university examination	Student	

28.3.3 Discussions

The proposed courses were conceptualized through a systematic approach and mitigated the conflicts between current practice, university demands and the unorthodox approach of PBL. The resulting course plans and evaluation schemes mapped onto the PBL methodology elements were selected upon extensive discussion and evaluation with the mentors/ course co-creators from the other partner universities, from India and Europe, with expertise and experience in PBL.

28.4 Summary, Conclusions and Discussions

Problem-based learning is a pedagogical approach where students pursue self-learning of a subject or domain, driven by real-world problems. This approach is reported to be effective in inculcating hard and soft skills needed to be industry-ready. However, present undergraduate programs in South Asia are instructional and content-heavy, thereby requiring redesign to incorporate PBL methodology. Design Thinking, a creative approach towards problem finding and solving, is employed during a five-day workshop to guide the redesign process and develop a course proposal for each of the participating institutes of Nepal and Bhutan, mentored by Indian and European partner universities. The use of Design Thinking allowed the faculty course designers to identify several issues from different perspectives, ideate large number of solutions, consolidate them into viable solutions and select the most promising one to further detail. At present, these proposals are being implemented at the home institutes and gathering feedback is in progress.

Acknowledgements and Disclaimer This publication is part of the '***Strengthening Problem-based Learning in South Asian Universities***' (PBL South Asia) project, co-funded by the Erasmus + programme of the European Union*. Acknowledgements to the faculty participants from across five countries and ten partner universities, namely Aalto University, Finland; Indian Institute of Science (IISc) and Indian Institute of Technology Bombay (IITB), India; Kaunas University of Technology (KTU), Lithuania; TU Delft, The Netherlands; JNEC, Royal University of Bhutan, Bhutan; and Kathmandu University, Asian Institute of Technology and Management (AITM), Sagarmatha Engineering College and Nepal Engineering College, Nepal.

*The European Commission's support for the production of this publication does not constitute an endorsement of the contents, which reflect the views only of the authors, and the commission cannot be held responsible for any use which may be made of the information contained therein.

References

1. Mills, J.E., Treagust, D.F.: Engineering education—Is problem-based or project-based learning the answer. Aust. J. Eng. Educ **3**(2), 2–16 (2003)

2. Duch, B.J., Groh, S.E., Allen, D.E.: The power of problem-based learning: a practical "how to" for teaching undergraduate courses in any discipline. Stylus Publishing, LLC (2001)
3. Lawson, B. How Designers Think: The Design Process Demystified. Routledge (2006).
4. Cross, N.: Designerly ways of knowing: design discipline versus design science. Design Issues **17**(3), 49–55 (2001)
5. Plattner, H., Meinel, C., Weinberg, U.: Design-Thinking. Mi-Fachverlag, Landsberg am Lech (2009)
6. Brown, T., Wyatt, J.: Design thinking for social innovation. Dev. Outreach **12**(1), 29–43 (2010)
7. Beitz, W., Pahl, G., Grote, K.: Engineering design: a systematic approach (1996)
8. Hubka, V., Eder, W.E.: Theory of Technical Systems: A Total Concept Theory for Engineering Design. Springer (2012).
9. Dieter, G.E., Schmidt, L.C.: Engineering Design. McGraw-Hill Higher Education, Boston (2009)
10. Eppinger, S., Ulrich, K.: Product Design and Development. McGraw-Hill Higher Education (2015)
11. Bhaumik, R., Bhatt, A., Kumari, M.C., Menon, S.R., Chakrabarti, A.: A gamified model of design thinking for fostering learning in children. In: Research into Design for a Connected World, pp. 1023–1036. Springer, Singapore (2019)
12. Savery, J.R.: Overview of problem-based learning: Definitions and distinctions. Essential readings in problem-based learning: Exploring and extending the legacy of Howard S. Barrows **9**, 5–15 (2015)
13. Gijselaers, W.H.: Connecting problem-based practices with educational theory. In New Directions for Teaching and Learning, pp. 13–22 (1996)
14. Torp, L., Sage, S.: Problems as Possibilities: Problem-Based Learning for K-16 Education (2002)
15. Hmelo-Silver, C.E.: Problem-based learning: what and how do students learn? Educ. Psychol. Rev. **16**(3), 235–266 (2004)
16. Barrows, H.S.:What Your Tutor May Never Tell You: A Medical Student's Guide to Problem-based Learning (PBL). Southern Illinois University School of Medicine (1996)
17. Schmidt, H.G., Van der Molen, H.T., Te Winkel, W.W.R., Wijnen, W.H.F.W.: Constructivist, problem-based learning does work: A meta-analysis of curricular comparisons involving a single medical school. Educ. Psychol. **44**(4), 227–249 (2009)
18. Birgegård, G., Lindquist, U.: Change in student attitudes to medical school after the introduction of problem-based learning in spite of low ratings. Med. Educ. **32**(1), 46–49 (1998)
19. Dolmans, D.H., Loyens, S.M., Marcq, H., Gijbels, D.: Deep and surface learning in problem-based learning: a review of the literature. Adv. Health Sci. Educ. **21**(5), 1087–1112 (2016)
20. Caswell, C.A.: Design and facilitation of problem-based learning in graduate teacher education: An MA TESOL case. Interdiscip. J. Problem-Based Learn. **11**(1) (2017)
21. Jonassen, D.H.: Instructional design models for well-structured and III-structured problem-solving learning outcomes. Educ. Tech. Res. Dev. **45**(1), 65–94 (1997)
22. Jonassen, D.H.: Toward a design theory of problem solving. Educ. Tech. Res. Dev. **48**(4), 63–85 (2000)
23. Funke, J.: Solving complex problems: Exploration and control of complex systems. In: Complex Problem Solving: Principles and Mechanisms, pp. 185–222 (1991)
24. Delisle, R. How to use problem-based learning in the classroom. Ascd (1997)
25. Marchais, J.E.D.: A Delphi technique to identify and evaluate criteria for construction of PBL problems. Med. Educ. **33**(7), 504–508 (1999)

Chapter 29
Reinventing Design Education for Twenty-First Century

Madhura Yadav

Abstract Design is everywhere and an integral part of our lives. It is a passion for our 'feel-good factor' adding excellence and efficiency in our activities. Increasing complexity has made design education more challenging. In the process of development, design extended into various disciplines like architecture, planning, fashion, textiles, graphics, visual communications, fine arts, interiors, multimedia, and jewelry. All these design disciplines have design principles common in them, but attributes change as per the discipline. These disciplines mark their presence independently and have been consistently synergizing each other to create a newer vision. There has been a need for reinventing design education. An interdisciplinary approach in design is vital for excellence and furtherance of know-how in allied fields. To bridge the distance between the disciplines, we have been developing a design education model at Manipal University Jaipur (MUJ). The paper will evaluate the model of design education and give recommendations.

29.1 Introduction

Design education in India has developed over the years and is booming. The number of design aspirants in India is increasing every year. As per the UK-India British council report, the potential market for design in India is expected to be INR 188.32 billion (Charles and Ray Eames, n.d.) (GBP 1.43 billion). Only a fifth of the design market has been tapped [8]. As per the report by EY-FICCI and India higher education report, this structural shift in employment will increase demand for designers and thinkers who can thrive in a globally connected and dynamic economy [11]. Design institutions must take these challenges to prepare such professionals.

Design education today is far more complex and challenging. Design is everywhere around us and designs have a significant impact on our lives and vice versa. Design is the creation of an object with a holistic approach that fit societal needs. A holistic approach **is** putting all the attributes together in harmony. Design is a

M. Yadav (✉)
School of Architecture and Design, Manipal University Jaipur, Jaipur, India
e-mail: ymadhura@gmail.com

A. Chakrabarti et al. (eds.), *Design for Tomorrow–Volume 2*, Smart Innovation, Systems and Technologies 222, https://doi.org/10.1007/978-981-16-0119-4_29

multidisciplinary field; it has arts, humanities, social, physical, biological, behavioral sciences, engineering, and business [3]. Design can lead because it cuts across all disciplines.

Design education is both subjective and objective. The objective requirements are easier to comprehend. The technical and business requirements allow for measurement and direct comparison. Design education is complex because of its subjectivity, creative side of design that is hardest to explain and most challenging for most people to understand. The problem is given to the class; each student will come up with different solutions. Each student requires individual attention to develop his/her design. The process of imparting education is based on data, graphics/images/visuals. To boost the student's thinking, various points of view about his/her design are given by different teachers.

In traditional design pedagogy, designers were trained in form, function, materials, and aesthetics. However, new societal challenges, cultural values, and technological advancements require new skills. Design today is more human-centered, assuring that people and technology work harmoniously as collaborative players. In today's context, understanding global issues such as economic, cultural, political, environmental, and technological is essential. To meet the challenge requirement of the twenty-first century, education in design needs to change. It needs better tools, methods, theory, data, analytical techniques, and understanding of how art and science, technology, people, theory, and practice can intermingle effectively and productively.

The traditional teaching method is teacher-centric; it is focused primarily on making a course plan, lesson plan and then delivering as per the lesson plan. It needs to focus on student centric.

In response to this problem, the paper focuses on identifying challenges and solving them through design curriculum, pedagogy, etc., for the all-inclusive growth of a student, which will help to become versatile professionals who can thrive in today's market with a strong sense of leadership, creativity, curiosity, and entrepreneurship.

29.2 Methodology

This paper is based on secondary data through literature review, interviews with experts in design education for last three decades and authors experiences.

29.3 Overview of Design Education

The concept of design education evolved from an exhibition of goods made worldwide at Crystal Palace London in 1851. The first School of Art, Sir J.J. School of Art Mumbai, India, was founded in March 1857. The school started decorative painting, modeling, ornamental wrought ironwork, art, and craft. In the later period, architecture and applied arts were added. The Central School of Arts and Crafts was established in London in 1896. Arthur Wesley Developed an institutional approach to art education in 1906, whereas William Morris and Georgia O'Keeffe Denman Ross developed an academic approach to design education in the 1910s. Architect Walter Gropius opened Bauhaus in 1919 in Germany [4]. This school combined crafts and fine arts and was famous for designing based on the most fundamental principle of the Bauhaus being Form Follows Function. The school emphasized intellectual, theoretical pursuits, combined with practical skills, crafts, and techniques. Craft and Fine Art were brought together to problem-solve for a modern industrial society.

Design institutions were opened in the USA, the UK, and Italy. A brief overview of these renowned institutions is summarized as below.

29.3.1 United States of America

Parson School of design is the number one design school in the USA and among the top three globally. For more than a century, Parsons has been inspired by the transformative potential of design. The school emphasizes interdisciplinary that affords designers the broad design perspective, transdisciplinary design, the business design, the business of design, the core questions of the environment and sustainability, or design itself, advancing art and design and renewing their relevance in the world to creating reflective, resourceful, and responsible designers to address city life's challenges through art and design.

29.3.2 United Kingdom

Design and innovation are critical to the UK, reaching its economic goals. Reigniting the enterprise economy, commercializing science, and technology, and embedding innovation in the public sector. Design education is focused on multidisciplinary courses, working with business schools to develop business skills, science, technology faculties to broaden the knowledge and engineering courses to understand manufacturing and engineering to better prepare tomorrow's design graduates for working in the industry.

29.3.3 Italy

The artistic spirit and outstanding design education in Italy is well known throughout the world. Design education integrates design disciplines art, architecture with technological (engineering and information technology-based), and management (economics linked) disciplines. It focuses on innovative, experimental development that responds to contemporary society's real market needs. Research constitutes a parallel path that is formed by cooperation and alliances with the industrial system.

All these renowned schools emphasized an interdisciplinary, multidisciplinary and transdisciplinary approach to design education.

29.4 Design Education in India

India did not have formal, systematic, or specialized education in the field of design, from ancient time, it was indeed practiced as a trade and a craft passing down the Knowledge and the tricks of the trade from master craftsman to the apprentice, fostering learning by doing approach but not as a profession [4]. The first School of Art, Sir J.J. School of Art Mumbai, India, was founded in March 1857; the School started its journey from drawing, decorative painting, modeling, ornamental wrought ironwork, art and craft and in a later period, architecture and applied arts was also added. After examining many great design pedagogies worldwide, especially that of the Bauhaus, the National Institute of Design was established in India. NID's ethos, philosophy, teaching methodology, and curriculum were developed. An essential addition was a science and liberal arts program to integrate humanities into the inculcation of professional design skills, thus forming a robust human-centered basis for design. After NID, many institutions have started offering design education in India. Education has become the greatest commercial industry. Institutions have become a factory to produce the bulk of professionals and facing many challenges.

29.5 Model of Design Education at MUJ

Design education is a fast-developing area. MUJ intends to nurture it as one of its core strengths and established 'Faculty of Design' with a mission to become the most preferred global destination in design education and research for students, researchers, faculty, collaborators, promoters, investors and developers. The Faculty of Design (FoD) is moving fast in research studies and is expanding its horizon to explore, accommodate, and accelerate the prospects of experimental as well as theoretical research in design studies. All programs are devoted to education and research in architecture, planning, fashion, and fine arts.

These four schools are devoted to education and research in architecture, planning, fashion, interior design, jewelry design and fine arts. All these design disciplines have design principles common in them, but attributes change as per the discipline. These four schools are synergizing each other in education and research. Faculty members are taking classes as per their expertise, e.g., fashion design faculty members take classes in architecture and other programs to teach color theory. Faculty from fine arts is taking classes for sketching and rendering. Faculty from jewelry design is taking classes for three-dimensional modeling. We are exchanging faculty across all programs and promoting interdisciplinary research. Study tours are joint. We promote vertical studios across all programs of FoD.

This model is working fine, and our students are becoming versatile design professionals. Sharing the faculty members has resulted in the economy also for the institution.

29.6 Challenges in Design Education

Universities are currently educating designers in similar ways as they have done in the past decades. There are many challenges in design education.

1. There are two similar streams, one with professional graduated from institutions and others with skills that came as a tradition. However, these two worlds continued to run parallel and could not help each other. There is a gap between these two streams.
2. As per the survey conducted from Industry, it is observed that students are not prepared to handle the task of handling clients. Design schools do not train students about complex issues such as behavioral sciences including business.
3. All students might not practice as designers in the industry and may work as academicians, researchers, journalists, etc. The curriculum is the same for all and not addressing the versatility of students.
4. The computer is becoming a misused tool. Copy paste culture has reduced analytical thinking.
5. Inadequate and poorly designed learning environments with only quantitative aspects of the area but no norms for quality of space hamper effective teaching and learning.
6. Currently, very few institutions are collaborating with national and international design institutions. Indian degrees are not at par with other degrees with global standards. Many Indian students have to appear for exams to get the license to practice in other countries.
7. Presently, research is emerging in design institutions, but interdisciplinary and integrated approach is missing [6].
8. COVID 19 is affecting design education as well. Thus, there has to be a paradigm shift in the thought process of business. The requirements of the design industry are to respond.
9. Faculty members are not fully equipped and trained for conducting online design education. This is to be addressed.

29.7 Discussion

To solve these challenges, the following factors play a crucial role in shaping design education.

- Curriculum
- Pedagogy
- Built environment
- Teachers
- Research

29.7.1 Curriculum

The curriculum is pivotal for any educational institution. How much to teach? How long to teach? What parts of teaching are to be emphasized? What should be the correlation contribution of design and related subjects to the 'other' issues? These are the questions that belong to the sphere of the curriculum. Considering sociocultural and developmental factors in India, innovations should be made in the curriculum. It is not the curriculum, but space it provides the student for learning and broadening their horizons.

29.7.2 Pedagogy

Pedagogy is significant to any education because it is what makes the action take place. The teacher and student get simultaneously engaged in the business of learning. How to teach? How much practical compared with the theoretical inputs/exercises, how to motivate, how to evaluate, etc., are the issues that pertain to pedagogy. Education aims to create teaching and learning environments that would bring about desired changes in learners, be more knowledgeable, better skilled, or influence their attitudes and values positively. The essence of learning and teaching is to plan teaching events (contents, strategies, etc.) and ascertain the learners have acquired the intended competences. Thus, the curriculum is more content-based, whereas the pedagogy is process-based.

29.7.3 Built Environment

Built environment plays an important role in enabling the teaching and learning process. The environment is part of the learning experience, and buildings need to be silent teachers.

29.7.4 Teachers

Teachers are the backbone of any education system. They should become role models for the students. They should not only be teaching from books but lead the students on all fronts. The interest in the subject depends upon the teacher who teaches the subject. Every student has the potential; faculties should help them to explore.

29.7.5 Research

The design profession is demanding interdisciplinary skills and relationships in research. We must logically, organizationally, and strategically meet this challenge as design continues to evolve into an integral part of our lives and society.

29.8 Recommendations

After discussing the challenges, following recommendations are given for design education at different levels.

1. The curriculum should emphasize on newer areas. The old drawing and sketching skills, forming, and molding must be supplemented and, in many cases, replaced by skills in programming, interaction, and human cognition [2]. Rapid prototyping is required, which also means some knowledge of the social and behavioral sciences of statistics, economics, and experimental design so that designers can perform valid, legitimate tests of their ideas before deploying them [7].
2. Open elective courses and choice-based credit system should be introduced for transdisciplinary learning and to address versatility of students. Integrated degrees should be proposed.
3. Design institutions require spaces which fosters creativity and clusters of facilities offering a variety of options for students to work in different ways, increasing the ability for teachers to adopt different instructional approaches. Changes need efforts to overcome the separation of 'formal' learning environments such as classrooms, libraries, and laboratories, on the one hand, and, on the other, 'informal' social spaces such as cafeterias and student lounges and transition spaces [5]. The attention should be on the quality of spaces with physical, visual, and thermal comfort and not only on the number of spaces with prescribed areas.
4. Human resource management including evolving long-term human resource plan, programs of skill upgradation of faculty members, professionalizing R&D management, and support functions through training is to be undertaken.

5. Educational institutions should focus on critical thinking as well as skills to fill the gap between professional and skill-based fields. If this gap is filled up, then our country can be the manufacturing sector in the world and designers can become entrepreneurs
6. Organizational transformation is to be undertaken with respect to: Shared visioning and formulation of detailed action plan is needed for developing strong identification with the vision and mission of institute. Ownership building and initiation of improvement projects for institutionalization of culture of continuous improvement and employee involvement. Periodic review to monitor the change for a motivated workforce.
7. Creating and enabling infrastructure state-of-the-art research and development infrastructure, research and development management structures, Benchmarking as per international levels, generation of database and networking, effective communication—external and internal channels, making design institute a better place to work and live.
8. Strengthening and nurturing linkages with industry, sister institutions, academic institutions and centers of excellence, financial institutions, government organizations [10], user organizations, and individuals and other institutions
9. Skill development of faculty for diverse aspects of business development and marketing activities with the help of professional organizations, Creation of marketing outlets, synergies, the business development of design institutions with professional organizations Institution should act as a resource laboratory for other organisations to evaluate, select, transfer and value addition of R&D and technology inputs in the field of Design
10. External cash flow utilizing existing strengths, promoting yearly advisory contracts, promoting partnership with service industries, working out niche areas, Steps should be taken for self-sustainability Employment focus should be on life skills, analytical skills, technical skills (crisp and clear-unambiguous communication) and development of value systems.
11. Based on emerging societal/industrial needs and markets, an **integrated research** in design approach amalgamates multidisciplinary stakeholders from the beginning and guides them through the entire process. In doing so, it encourages organizations to own the research into their customers and builds customer empathy. It gives organizations the tools to make sense of the research and generate insights. It facilitates turning these insights into solutions that can be tested with customers. Research is co-owned by a multidisciplinary core stakeholder team. The research brief and plan should be created with multidisciplinary stakeholders. Inter-research tools should be co-created and capture insights. As the pace in a global economy increases, clients move away from multiple-source providers and seek out single-source, turnkey providers. To bridge the distance between the disciplines, it is time to review its instructional pedagogy and the inherent value of an understanding and exploring related disciplines. **Interdisciplinary research** in the field of Design is vital for excellence and furtherance of know-how in their fields. World-class database

should be developed as a part of the institutional database. All the work will be available on networked computers to foster interpersonal and inter-groups communication. The concept of **INTERDEPENDENCE** rather than **INDEPENDENCE** should be pursued.

12. Development of virtual reality laboratory for simulation studies particularly in the field of design research is needed. Aesthetics is one of the unquantified aspects of design. Neural networks research is to be undertaken for quantification of aesthetics in different design disciplines. IT (Information technology) is to be utilized as a means of convergence of research and its implementation. Interdisciplinary, interlaboratory, collaboration, and industrial participation.

While solving the challenges and addressing the needs of industry, we must consider:

Speed Speed has become a way of life. We are asymptotically piling data. Speed can be achieved with new design processes, techniques, and technology.

Technology Innovations AI and VR are just some of the many forms of technologies that will play a key role in shaping the design of tomorrow, making it future-ready and ushering in a new age of innovation in this.

Quality Everything is global—'quality' must be world-class—Design detailing should be worked out efficiently. All significant branded designs and infrastructure projects are given to western designers because of their quality and detailing.

Cost Cost is a very important factor. It should be affordable to all.

29.8.1 Conclusion

Education needs to be relevant—We must be self-reliant to boost our economy. Constant experimentation and exposure to experimental work should be mandatory. The curriculum should emphasize on versatility of students. All students might not practice as designers in the industry, but some will work as academicians, researchers, journalists, etc. The design should address physically, mentally, and socio-economically challenged people. Understanding human behavior is fundamental. People in human–computer interaction, cognitive engineering, and human factors or ergonomics are usually ignorant of Design. **A human-centered approach** is the key to design education and can be achieved by understanding the **context**, **monitoring, and mapping trends** that have a far-reaching **impact on the larger society** and the resulting implications on **preparing design professionals**. The future of design education lies in adapting to the population's changing needs and promoting an innovative outlook. We need to focus on self-reliance-Atmanirbhar as a nation. **Design it**, **buy it, use it, appreciate it**, and **make others buy it** and make it sustainable.

Acknowledgements I would like to thank Dr. N. K. Garg, Founder Dean, FoD, Manipal University Jaipur, for giving his valuable inputs through an interview about setting up Faculty of Design at Manipal University Jaipur and other experts for their views on design education.

References

Research Papers

1. Balaram, S.: Bauhaus and the Origin of Design Education in India. https://www.bauhaus-imaginista.org. Retrieved 20 March 2020 (2018)
2. Berno T.: Dangerous Ideas for the Future of Design Education. https://www.dmi.org/blogpost. Retrieved 27 March 2020 (2014)
3. Collina, L., Galluzzo, L., Maffei, S., Monna, V.: Designing Design Education. An articulated programme of collective open design activities. Des. J. https://www.researchgate.net/publication. Retrieved 18 March 2020 (2017)
4. Forgács, É., Bátki, J.: The Bauhaus idea and Bauhaus politics. https://www.worldcat.org/. Retrieved 13 March 2020 (1995)
5. Friedman, K.: Models of design: envisioning a future design education. https://www.academia.edu. Retrieved 18 March 2020 (2012)
6. Hall, P.A.: Re-integrating Design Education: Lessons from History Design Research. https://static1.squarespace.com/. Retrieved 15 March 2020 (2016)
7. Ozkaynak a*, G., Ust, S.: New Forms of Design Education. https://pdf.sciencedirectassets.com/. Mine retrieved 24 March 2020 (2012)

Reports

8. 54th ANNUAL REPORT of NID 2014–2015. Retrieved from https://www.nid.edu/
9. Design-Education The India Report by Charles and Ray Eames., https://www.nid.edu/. Retrieved 1 April 2020 (1958)
10. Report and recommendations from the Multi-Disciplinary Design Network. https://www.designcouncil.org.uk/National. Retrieved 2 April 2020. Design Policy 2007. Approved by National Design Policy. https://dipp.gov.in/. Retrieved 3 April 2020
11. The future of design education in India, November 2016. Retrieved https://www.britishcouncil.in. 5 April 2020

Chapter 30
Do Design Entrance Exams in India Really Test Creative Aptitude? An Analytical Study of Design Tests Conforming Creativity Benchmarks

Nandita Bhanja Chaudhuri, Debayan Dhar, and Pradeep G. Yammiyavar

Abstract Design entrance exams are common nowadays for securing admission to design schools in India, be it private or public institutes. Students are often examined during these competitive exams to test their creative aptitude. Over the years, format of design entrance exams in India has gone through innumerable changes. In some formats, objective questions are preferred, in some subjective tests are conducted, while few prefer a combination of both lack of a standardized format of testing creativity across design schools in India, unlike engineering and medical colleges, raises questions like (1) Do these design entrance tests really capture creativity to its fullest? (2) Do they conform to any standardized format of creativity testing? The study reported in this paper intends to investigate these questions by conducting a detailed comparative study of existing entrance exam formats of Design Schools of India. A detailed analytical approach by using affinity mapping and generating open codes was used to classify and categorize question formats of different Indian Design entrance exams. The open codes were then mapped with the extracted factors of the literature review with an objective to identify the factors confirming to creativity evaluation. The results highlight that existing entrance exam, while capturing major factors for creativity evaluation, lacks in capturing a few essential factors that provide greater insight into the creative instincts of an individual assessed. A model for standardization of creativity assessment has also been suggested based on the study results. This study would provide new light into the nature of design-based entrance examinations in India and might lead the way toward a standardized Design entrance exam in the future.

N. B. Chaudhuri · D. Dhar (✉) · P. G. Yammiyavar
Department of Design, Indian Institute of Technology Guwahati, Assam, India
e-mail: debayan@iitg.ac.in

N. B. Chaudhuri
e-mail: nandi176105102@iitg.ac.in

P. G. Yammiyavar
e-mail: pradeep@iitg.ac.in

A. Chakrabarti et al. (eds.), *Design for Tomorrow–Volume 2*, Smart Innovation, Systems and Technologies 222, https://doi.org/10.1007/978-981-16-0119-4_30

30.1 Introduction

Creativity is essential for design, engineering, science, and other domains, which often require interchanging of ideas among function, structure, and behavior of embodiments [1–3]. It enriches the quality, value, and behavior of an idea or product making them novel and useful [4]. Creativity acts as a backbone in the education system and plays a significant role of catalyst in any innovation process [5]. Design education is highly influenced by natural phenomena and diverse forms of life and possesses a rationale to exhibit an embodiment that is novel in nature [6]. Finding creativity through an examination is a challenge that requires understanding innovative thoughts of students within a stipulated time and a limited number of questions. Creativity in education is defined as " imaginative activity fashioned so as to produce outcomes that are original and of value" [7].

Like other domains, solutions of descriptive aptitude of design education exam require showcasing student's creativity through ideas. The questions in aptitude try to capture creativity in students. Therefore, an effective format of question and standardization of question paper is essential in capturing creativity. However, pattern analysis reveals evidence of inconsistent format of creativity testing in design education. Creativity in education is composed of several indicators, which also reveal the fact of the intensity of creativity extracted from students.

The contribution of this paper is to validate the potential to capture creativity of design tests in India. Further, a creativity assessment model is proposed with the rationale to standardize question paper format. The highlights of this paper are as follows:

- To investigate whether creative aptitudes in design education captures creativity.
- To identify the types of questions in evaluating creativity in design education.
- To confirm the standardization of format in question papers for creativity testing.
- To propose a model for standardized creativity assessment.

30.2 Background Study

India is rich in art, culture, and design, but structured and systematic design education was imparted to only a few of the schools until 2004. After this period, design schools in India increased at an exponential rate, and reports highlight more than seventy design institutes in the year 2016 [8, 9]. Design schools attempt to test creative skills of students; however, each design school has its own strategy of testing creativity. Now, a broader question emerges, i.e., is the test capable of capturing creativity at the fullest? Creativity has various forms, such as big-creativity associated with immense novel inventions like the invention of a steam engine; pro-creativity tends to possess a relatively lesser degree of creativity than the big-creativity. Pro-creativity measures the novelty of outcomes of professionals in their domain of expertise. Little-creativity involves a lesser degree of creativity than pro-creativity and is associated with contributions in day-to-day life activities, whereas mini-creativity possesses the least degree of creativity, and it is evaluated within a particular community, or any person considers an embodiment as creative [10]. Creativity is also classified as H-creativity or Historical creativity and P-creativity or psychological creativity or personal creativity. H-creativity is involved in predominant creativity from the perspective of the entire human race, whereas an idea or a solution is P-creative if it is creative with respect to the mind of the person or the community concerned [11]. Creativity in an examination is mostly associated with mini-creativity or personal or psychological creativity, where creativity is evaluated relative to other solutions [12].

Extensive methods of creativity tests are reported in the literature. Many creativity tests lack reliability and validity in recognizing creativity, while others are not meant to be used in the context of examination. But in examination context, experts may decide to choose a combination of these techniques to extract creativity from different perspectives. Literature highlights some of the creativity tests like Consensual Assessment Technique (CAT) that evaluate products, art, theory, or artifacts based on expert's opinion [13]. Remote Associates Test (RAT) tests creativity based on divergent thinking. It examines the degree of unrelated ideas combined to form a coherent whole of an idea. One of the significant tests reported in literature is the Torrance Tests of Creative Thinking (TTCT), which is similar to the tests that are conducted for selecting students in design schools. It scrutinizes creativity from verbal and figural perspectives. The verbal part is checked for one's creativity by analyzing words with which they frame narrations, whereas the figural part is tested based on the usage of visual elements, completeness of the art, and degree of modifiability of visual elements [14].

Some tests are related to examining creativity of children like Wallach and Kogan's method of creativity testing, which desires to test creativity and intelligence of fifth standard students [15]. Similarly, Getzel and Jackson's study reveals the fact of testing sixth-grade gifted creative students [16]. These categories of tests are associated with testing intelligence and divergent thinking of children. However, creativity of adults is different from creativity of children. An artifact created by children may seem creative, but the same developed by an adult might not appear creative.

Majority of the literature focused on methods of testing creativity are mostly for products and ideas. There is dearth of literature on testing criteria for recognizing creativity of students. However, Park et al. proposed that creativity assessment is composed of two categories of assessment methods—subjective assessment and objective assessment. Experiments reveal the fact that subjective assessment possesses higher mean and significant lesser variance than objective assessment in capturing creativity [17]. Selection process in entrance exams in India depends on cut-off marks. Students exceeding the cut-off marks clears the exam. Objective evaluation depends on the truth or falsity of the solution, whereas subjective evaluation is based on evaluation criteria such as originality, and usefulness. Majority of findings in literature are from the perspective of a general assessment of creativity of individuals or students in day-to-day life or in classwork. However, the rationale of

Table 30.1 Figures demonstrating appearance of students in national level exam

	Undergraduate level	Postgraduate level
Year	Exam 1 (Students appeared)	Exam 2 (Students appeared)
2018	11,567	5507
2019	12,414	5491
2020	12,084	5564

this article is to analyze creative tests of design entrance exams where students aspire selection in design schools.

Many articles have highlighted evaluation of creativity in products, processes, and people. Literature suggests that minimal investigation is performed on examining design entrance tests in India. There are numerous national level and private design entrance tests in India. National level exams are regularized by institutes of government bodies, whereas private tests are associated at an organizational level [18]. Reports reveal the fact that India witnesses large-scale nationalized exams for design education, as shown in Table 30.1. There are other national level and organizational level design entrance exams where the average student ranges between 6000 and 22,000 for each exam [19, 20]. However, these figures associated with the appearance of students varies each year.

A highly consistent selection process is essential in order to qualify students in these large-scale exams. Due to the large population of students participating in exams, most of the tests associated with design education have objective and subjective question structure. The objective part consists of questions related to numerical answer type, multiple-choice, and multiple select questions, whereas the subjective part contains questions associated with sketching, form sensitivity, visual sensitivity, and problem identification that attempts to capture creativity of students [18]. The solutions to objective questions are straightforward and based on a strict set of options, whereas creative solutions are marked by the usefulness of ideas and comparison of novelty relative to other solutions [1]. Subjectivity evaluation is based on individual persuasion and is relatively complex than the objective evaluation.

Literature highlights multiple forms of creativity, creativity tests of young and adults, design tests in India, and its question patterns. A lacuna has been investigated from the evidence of literature that there is less focus on the procedures of capturing creativity in the design entrance exams. There is also a lack of assurance in literature, whether optimized creativity is captured based on all its factors and requirements of the design schools. Further, hardly authors directed toward proposing a standardized format of testing creativity for entrance exams in design schools. This leads to the following research questions:

RQ1 *Do design entrance tests in India really captures creativity to its fullest?*
RQ2 *Do design entrance tests in India conform to any standardized format of creativity testing?*

30.3 Materials and Methods

The first experiment is designed with an intent to identify whether existing design schools of India capture creativity to its fullest in design entrance exam. Creativity has a wide spectrum and is composed of several parameters like novelty, usefulness, divergent thinking, feasibility, and elegance. An extensive literature review is performed to extract features of creativity, and based on that, a survey is conducted to identify acceptance and preferences of factors in capturing creativity by *71* design professionals ($n = 71$) like design educators and design managers. A questionnaire is framed to recognize the preferences of these factors using a Likert-type scale ranging from 1 to 5 *(1—not at all important, 2—slightly important, 3—important, 4—slightly more important, and 5—very important).* The internal consistency is reported using Cronbach's alpha, i.e., *0.726*. The descriptive statistic of the factors is shown in Table 30.2. The percentage of acceptances of features to capture creativity is shown in Table 30.3.

The statistic ensures acceptability and preferences of features to capture creativity. Further, affinity mapping is conducted to verify whether preferred features chosen by experts during the survey to evaluate creativity is present in entrance test question papers of design institutes. This procedure involves exposure of the question paper and brainstorming with a group of *5* ($n = 5$) people possessing expertise in design education. Based on outcomes of the brainstorming session, the rationale behind subjective questions are drafted on the cards. Initially, cards are placed in a category that makes any logical sense without overthinking. In the later stage, cards are re-organized as well as added based on further logical deep thinking but without any discussion within group. Header names are assigned based on the logical relationship within a category. Finally, the categories were analyzed, and gaps were found out. The outcomes of the survey highlighting the acceptance and preferences of factors are mapped with the gaps in affinity maps. The mapping highlights points of expectations of experts to extract creativity and gaps found from the question papers, which further decrypts whether expected factors are predominant in the question papers to capture creativity to the fullest or not.

A second experiment is designed to conform whether creativity testing follows any standardized format. Initially, *five* different exams associated with design education is considered with *two* consecutive years of question formats. Types of questions are identified, and further frequency of types of questions are accounted for all exams. This evaluation exhibits the fact whether any standardized pattern is followed for exams or not. Further, a model for creativity assessment is proposed based on the difference in question formats in exams followed by design institutes.

Table 30.2 Descriptive statistic of the features to evaluate creativity

Statistics		Novelty	Usefulness	Surprising	Genesis	Relevancy	Elaboration	Feasibility	Divergent thinking	Elaboration
N	Valid	71	71	71	71	71	71	71	71	71
	Missing	0	0	0	0	0	0	0	0	0
Mean		4.52	4.21	3.58	3.31	4.32	3.87	4.06	4.24	3.20
Median		5.00	4.00	4.00	3.00	5.00	4.00	4.00	4.00	3.00
Mode		5	5	4	4	5	5	5	5	3
S.D		0.67	0.86	0.98	0.91	0.93	0.98	0.92	0.90	1.14

Table 30.3 Percent of acceptance of factors for capturing creativity

Percent	Novelty	Usefulness	Surprising	Genesis	Relevance	Elaboration	Feasibility	Divergent thinking	Elegance
Not at all important	0	0	0	4.2	0	0	1.4	1.4	5.6
Slightly important	28.2	1.4	15.5	12.7	7.0	8.5	2.8	1.4	23.9
Important	9.9	23.9	31.0	36.6	11.3	29.6	22.5	18.3	31.0
Slightly more important	0	26.8	33.8	40.8	23.9	28.2	35.2	29.6	23.9
Very important	62.0	47.9	19.7	5.6	57.7	33.8	38.0	49.3	15.5

30.4 Results and Discussions

Initially, an experiment is conducted to investigate acceptances and preferences of factors capturing creativity. This study reveals the fact that factors like novelty, usefulness, surprising, genesis, relevance, feasibility, divergent thinking, and elegance are accepted by experts for capturing creativity. In this context, the features are defined as follows—novelty is something that is new and original. A solution having a practical usage is termed as useful. Surprise may be described as the degree of the unexpectedness as compared to state of the art. Genesis provides a new path of conceptualization for novel and non-existing solutions. Relevance is the degree of which a solution maps to its corresponding problem. Elaboration is the degree to which a solution is well-defined. Feasibility clarifies whether a solution is applicable in the real world. Divergent thinking is the ability to combine multiple ideas and form a coherent whole. Elegance is defined as the degree to which an idea is well-formed and proportionate [1, 21, 22]. Novelty shows less variability of data followed by relevance, divergent thinking, usefulness, feasibility, elaboration, surprising, elegance, and genesis. Next, affinity mapping is conducted to identify gaps, if any, in the entrance test question papers of design institutes.

The first phase of the affinity mapping involves categorization based on different exams that are considered in design education. The elements of categorization include the features of creativity involved in question papers corresponding to an exam. In the second phase, factors are further categorized based on their logical relationship. Codes are generated from categories that are found to be similar to features that are extracted from the literature. Next, a significant gap is identified by comparing outcome of the survey and affinity mapping exercise (see Fig. 30.1). The findings of the survey demonstrated the fact that there are significant factors, to be precise nine factors that are responsible for capturing creativity with corresponding high mean and minimum standard deviation. However, an affinity mapping exercise brings out the fact that each of the exams possesses only a few factors that are essential to capture creativity, and a majority of the factors are not considered. Moreover, few of the question paper of entrance tests of design institutes also focuses only on objective questions. It can be solved by referring to existing knowledge-base of an individual and recalling, thereby less associated with creative outcome [17]. Therefore, it can be concluded that exams framed with objective questions and lacking features of creativity in subjective questions, capture creativity up to an extent, but further, there is a chance of optimization in this process. The missing features in question papers for capturing creativity would enable identifying additional aspects of capturing creativity. The result of this experiment confirms the first research question that present question pattern does not capture creativity to the fullest.

The study 1 described in previous paragraph is conducted by triangulation of literature review, survey, statistical analysis, and affinity mapping threw light to the extent of capturing creativity in design-based exams. The second research question attempts to clarify whether there exists any standardization of question paper format maintained in present examination system of design institutes. Any process, product,

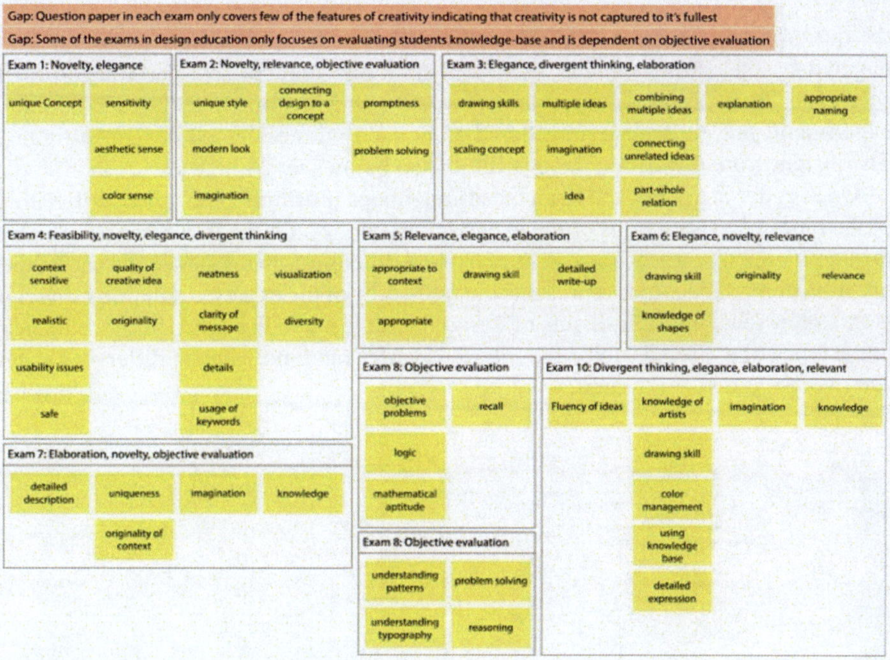

Fig. 30.1 Categorization and finding gaps

or question paper can be termed as standard if it is capable of capturing what it is supposed to do and maintain consistency of its elements depending on an outcome. In this study, the main highlight is to check the consistency of the types of questions in the question paper to confirm whether standardization is maintained in creativity testing. Literature reported the fact of the format of question paper in exams [23] (see Fig. 30.2). This pattern is also identified in design-based exams [18], where the main focus is on objective and subjective questions. This study is concerned with subjective questions and specifically targeting long questions involved in capturing creativity.

The long questions are further analyzed by affinity mapping to perceive the type of expectations of the examiners in the solution. With a thorough scrutinizing of

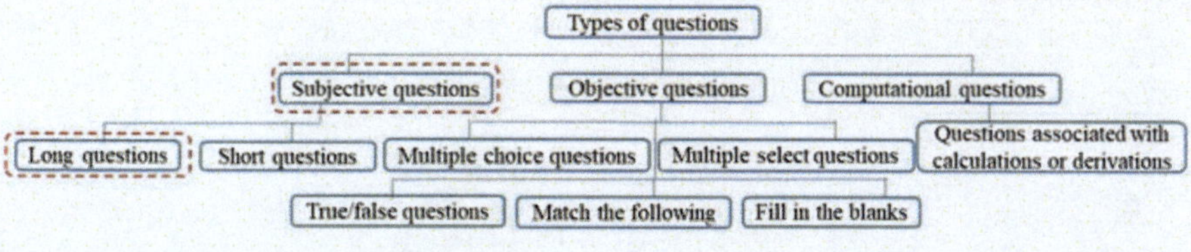

Fig. 30.2 Classification of question patterns

the questions papers by *5 (n = 5)* experts, *three* types of solutions are identified–(1) Text-based solutions, (2) Image-based solutions, and (3) Text-and-image-based solutions. The types of solutions are classified corresponding to their exams. In this experiment, *five* exams are considered associated with esteemed design institutes. The insight from the affinity map is illustrated below (see Fig. 30.3).

Open codes are generated from the affinity maps illustrating the random priorities of the types of expectations of the experts. Frequency analysis is conducted in order to quantify the frequency of patterns of expected solutions in the subjective part of the question paper. The x-axis represents the frequency of the expected solutions, and the y-axis represents multiple exams of design institutes (see Fig. 30.4). The frequency of the types of expected subjective solutions varies randomly among different exams and over M and $M + 1$ years.

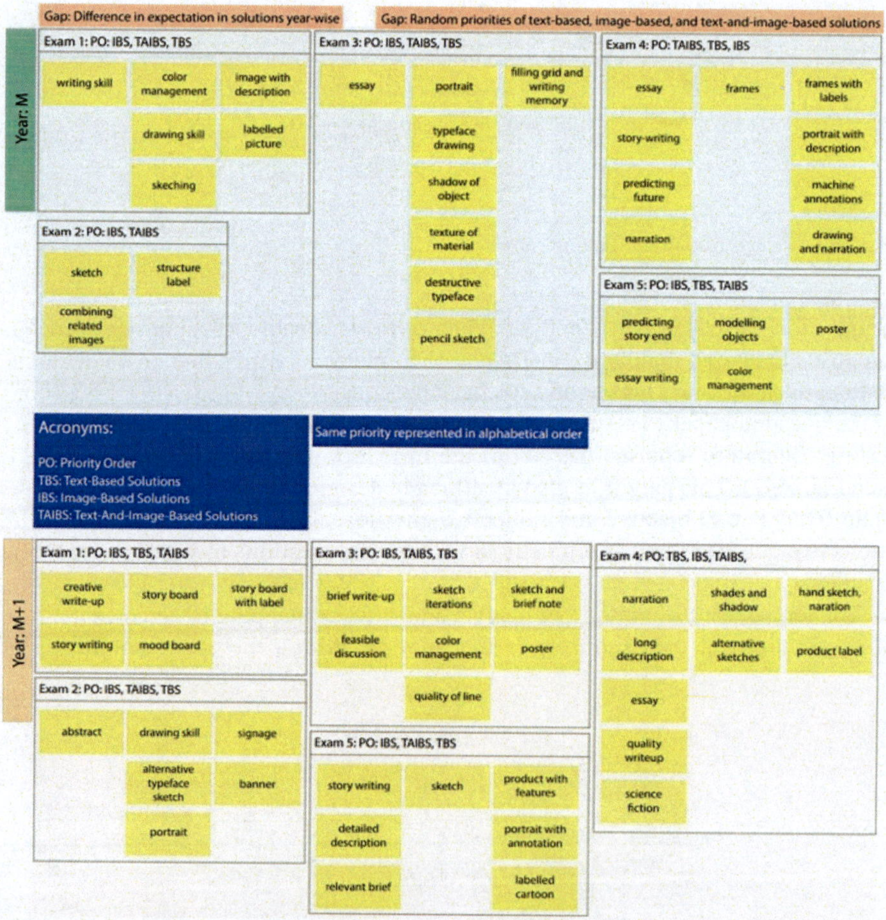

Fig. 30.3 Classification of expectation of solutions by experts

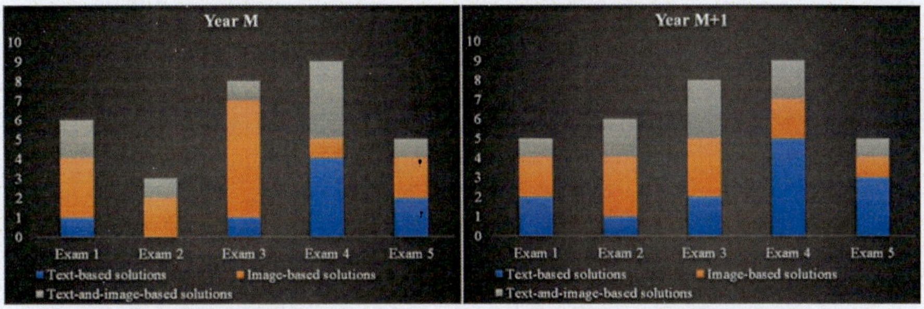

Fig. 30.4 Frequency analysis of patterns in question paper of M and $M + 1$ years

The findings from the above analysis reveal the fact that unlike engineering and medical exams, there is less consistency in the patterns in the question paper of design education where the expectation of the types of the subjective solution varies each year. However, expectation may vary over the years, but critical analysis is essential on the percentage of the types of problems to be given in the question paper. Therefore, a model is proposed to assess creativity to its fullest in a standardized format (see Fig. 30.5).

CAM—creativity assessment model is proposed based on the lacuna found out from the multiple analytical studies reported in this paper. It is essential to identify factors that capture creativity in entrance exam of design education. Next, the proportion of objective and/or subjective questions needs to be estimated to maintain consistency in exam pattern. Objective questions cater to different questions patterns such as mathematical and logical reasoning and depending on examiner's expectations, specific category/categories of pattern is chosen. Percentage of the categories of the questions is determined, and this process is repeated until the question format is ready. Similarly, in subjective questions, also, there are a variety of expected solutions. While a category or multiple categories are selected, the percentage of the questions belong to those categories needs to be determined. This process repeats until the question format is ready. The percentage of the type of questions can be decided by experts based on the type and level of test. Type of test indicates the tests for a specific institute or an exam conducted nationally, whereas the level of test shows the degree of difficulty.

30.5 Conclusion

Design entrance exams aim at capturing creativity of students. Findings from literature and survey reveal the fact that creativity comprises various factors that define multiple aspects of creativity. However, investigations based on analytical studies report that there exists a gap in capturing creativity from students at its fullest. It is also observed that there is hardly any standardization of question papers across

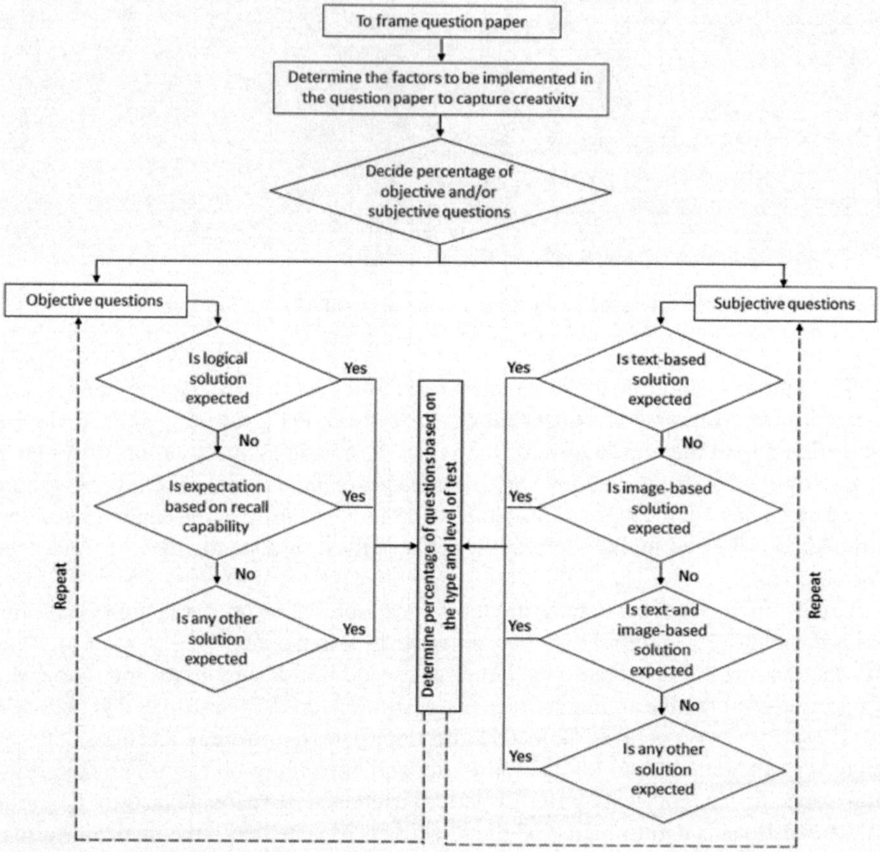

Fig. 30.5 CAM: Creativity assessment model

design institutes. Therefore, a model is proposed that would be able to standardize the question paper format. This study is not only applicable for Indian context but widely applicable for universities of other countries.

Technology adoption might assist in automating the proposed framework which in turn could check the standardization in question papers.

References

1. Sarkar, P., Chakrabarti, A.: Assessing design creativity. Des. Stud. **32**(4), 348–383 (2011)
2. Srinivasan, V., Chakrabarti, A.: Design for novelty–a framework? In DS 48: Proceedings DESIGN 2008, The 10th International Design Conference, Dubrovnik, Croatia (2008)
3. Lobert, B.M., & Dologite, D.G.: Measuring creativity of information system ideas: an exploratory investigation. In: 1994 Proceedings of the Twenty-Seventh Hawaii International Conference on System Sciences (1994)

4. Amabile, T.M.: Social psychology of creativity: a consensual assessment technique. J. Pers. Soc. Psychol. **43**(5), 997 (1982)
5. Shaheen, R.: Creativity and education. Creativity Educ. **1**(03), 166 (2010)
6. Cropley, A.J.: Creativity in Education & Learning: A Guide for Teachers and Educators. Psychology Press (2001)
7. Craft, A., Jeffrey, B., Leibling, M. (eds.): Creativity in Education. A&C Black (2001)
8. Gemmell, B., Vyas, P.: The Future of Design Education in India, pp. 1–36. British Council. https://www.britishcouncil.in/sites/default/files/the_future_of_design_education_in_india.pdf (2016)
9. Sharma, A., Kirloskar, V.: India Design Report. India Design Council. CII National Committee on Design, pp. 1–120. https://www.cii.in/webcms/Upload/a2.pdf (2008)
10. Kaufman, J.C., Sternberg, R.J.: (eds.). The Cambridge Handbook of Creativity. Cambridge University Press (2010)
11. Gigerenzer, G.: Where do new ideas come from. In: Dimensions of Creativity, pp. 53–74 (1994)
12. Csikszentmihalyi, M., Wolfe, R.: New conceptions and research approaches to creativity: implications of a systems perspective for creativity in education. In: The Systems Model of Creativity, pp. 161–184. Springer, Dordrecht (2014)
13. Pritzker, S.R., Runco, M.A. (ed.): Encyclopedia of Creativity. Academic Press/Elsevier (2011)
14. Kaufman, J.C., Plucker, J.A., Russell, C.M.: Identifying and assessing creativity as a component of giftedness. J. Psychoeduc. Assess. **30**(1), 60–73 (2012)
15. Silvia, P.J.: Creativity and intelligence revisited: a latent variable analysis of Wallach and Kogan. Creativity Res. J. **20**(1), 34–39 (2008)
16. Getzels, J.W., Jackson, P.W.: Creativity and Intelligence: Explorations with Gifted Students. Wiley (1962)
17. Park, N.K., Chun, M.Y., Lee, J.: Revisiting individual creativity assessment: triangulation in subjective and objective assessment methods. Creativity Res. J. **28**(1), 1–10 (2016)
18. https://www.ceed.iitb.ac.in/2021/. Accessed 24 June 2020
19. https://www.shiksha.com/. Accessed 07 Oct 2020
20. https://www.ndtv.com/. Accessed 07 Oct 2020
21. Grace, K., Maher, M.L., Fisher, D., Brady, K.: Data-intensive evaluation of design creativity using novelty, value, and surprise. Int. J. Des. Creativity Innov. **3**(3–4), 125–147 (2015)
22. Mohanani, R., Turhan, B., Ralph, P.: Requirements framing affects design creativity. IEEE Trans. Softw. Eng. (2019)
23. Liu, Z., Jansen, B.J.: Subjective versus objective questions: Perception of question subjectivity in social Q&A. In: International Conference on Social Computing, Behavioral-Cultural Modeling, and Prediction, pp. 131–140. Springer, Cham (2015)

Chapter 31
Innovation by Design—A New Post-Graduate Program at SUTD

Arlindo Silva and Lucienne Blessing

Abstract Design thinking has been gaining importance in training and education worldwide, but mostly in the form of short courses and executive education initiatives. Although there is enormous value in short courses and executive education, they often lack the depth required to effectively practice the tools and methods learned and thus to realize design as a strategic investment for both companies and countries. The particular focus of this paper is Singapore. At the Singapore University of Technology and Design, a new Master of Engineering (MEng) program has been set up to address this perceived gap in education. The MEng program in Innovation by Design (MIbD) is a research-based program that takes design thinking and design innovation to the level of other post-graduate programs in other areas worldwide. The organization allows practitioners to participate part-time. Three terms into the program, the balance is extremely positive. The program has been very well received in several presentations to companies. It is expected that these students will either start their own business or easily find jobs in a context that is craving for people with this formal education: a broad view of design and the ability to implement it.

31.1 Introduction

Design has become a strategic investment for companies [1] and countries [2] alike. The investment of Singapore in design has been tremendous in recent years. This investment has led, among other achievements, to the recognition of SUTD as an emergent leader in engineering education [3]. However, there is a perceived gap in post-graduate education, which the Singapore University of Technology and Design (SUTD) has tried to bridge with a new Master of Engineering (MEng). The MEng

A. Silva (✉)
MEng in Innovation By Design, Singapore University of Technology and Design, Engineering Product Development Pillar, Singapore, Singapore
e-mail: arlindo_silva@sutd.edu.sg

L. Blessing
SUTD-MIT International Design Center, Singapore University of Technology and Design, Engineering Product Development Pillar, Singapore, Singapore

A. Chakrabarti et al. (eds.), *Design for Tomorrow–Volume 2*, Smart Innovation, Systems and Technologies 222, https://doi.org/10.1007/978-981-16-0119-4_31

program in Innovation by Design (MIbD) is a research-based program that takes design thinking and design innovation to the level of other post-graduate programs in other areas worldwide. It further develops the SUTD design ethos [4, 5] into the post-graduate level. It contains a comparatively reduced coursework load and instead focuses on research and development projects where the tools and methods delivered in the courses are actively used. The structure of the program is such that there are only three compulsory full-credit courses (green and red in Fig. 31.1) and a significant number of electives (dark blue in Fig. 31.1) that students can take to scaffold their research work. These are complemented by three experiences/accelerators (short, no more than one-week long workshops and seminars, spread around the first year, in light blue in Fig. 31.1, see Sect. 31.5 for details).

The first intake of this program was in September 2019. A total of 30 students were selected (20 male and 10 female), 25 being full-time and five part-time. Of the 30, a total of 15 have on-the-job experience ranging from 1 to 15 years. 24 scholarships were given, and one student is being supported by his company. The students come from seven countries (Singapore, Sri Lanka, India, China, Tanzania, Colombia and Indonesia) and have backgrounds in, e.g., robotics, IoT, chemistry, materials, aging/health care, drones/UAV, food science, design, sports, mechanics,

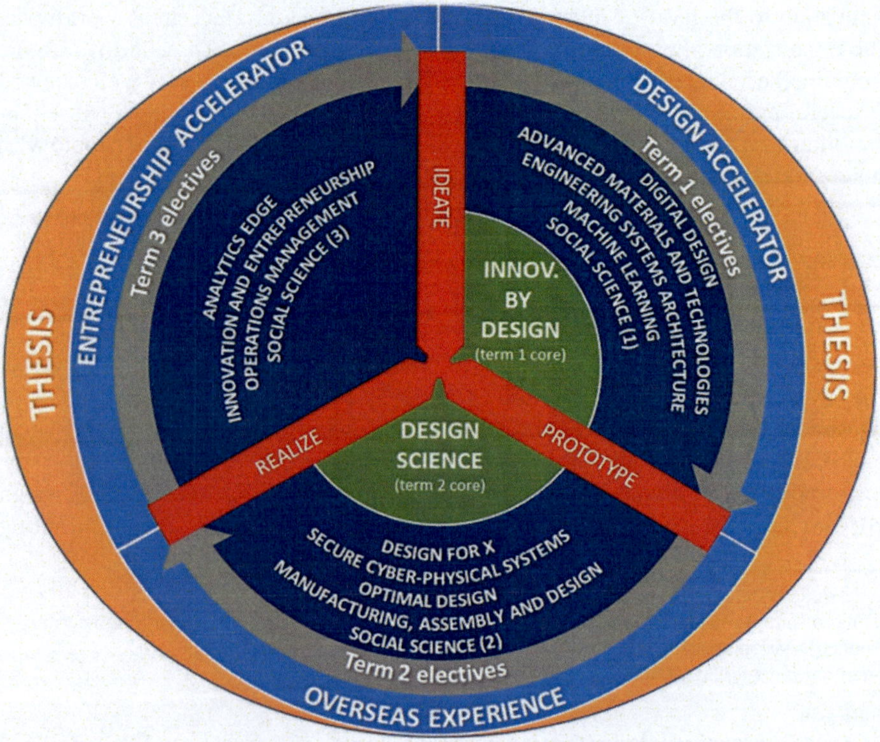

Fig. 31.1 Overall structure of the MEng in Innovation by Design

electronics and education. This intended diversity in background and culture enriches the learning and design process. By the end of the Master, each student will have completed a minimum of seven full-length projects, from discover to deliver. Those taking electives may have done even more. The sections below will describe various compulsory courses and experiences, student feedback on multiple aspects of the program and our plans for improvement.

31.2 The Compulsory Course on Innovation by Design

The first term compulsory course is Innovation by Design. In this course, the students work in teams to develop a product/service/system. The classes are about the tools and methods of product design and development [6] covering the 4Ds (discover, define, develop and deliver) with a mix of presentations, discussions and studio work for the 12 weeks of class. Students have to find a problem within a broad theme and solve it. For the first intake, the theme was "Play." Students have to find a problem that is meaningful to them and then solve it through an engaging, playful product/service/system. Each year, the problem space will change. Four lessons were devoted to invited speakers from industry to talk about their experience in developing new products, or their work in their respective organizations in fostering creativity and innovation.

31.3 The Compulsory Course on Design Science

The second term compulsory course is Design Science. This course aims at making the participants better qualified and equipped for research in Design Science, i.e., research that is focused on obtaining a better understanding of design and innovation, or on developing novel technologies, products, services or systems. The course provides insight into the existing design theories and models; understanding of the DRM design research methodology [7] to support planning and execution of research; an overview of Design Science methods; and the application of these methods to the participant's own research topic in the form of assignment. The course offers an active, project-based learning environment involving lectures, discussions, exercises, presentations, assignments and homework.

31.4 The Compulsory Thread on Ideate-Prototype-Realize

On the first week of the program, the students are enquired about their research interests, and a match is made between their interests and the appropriate faculty advisor(s). Hence, within three weeks of the program, the students can start right away with their research work. The Ideate-Prototype-Realize thread (I-P-R) runs on the first three terms of the program, lasting a full year. The vision behind I-P-R is to scaffold the students' research with their chosen faculty advisor(s) expertise and embed the student in one of the SUTD's research centers, potentially having him/her join an ongoing research project. In this way, the student will be part of a team that is already doing research, instead of having to start from scratch. The student-advisor pair will have to craft a research program that takes advantage of the on-going projects and is also meaningful for the student to learn skills relevant to his/her research interests. I-P-R will link directly to each students' research topic leading to the final thesis.

Each student will have his/her own topic, so a complete and detailed set of guidelines for I-P-R is not possible. The focus for each term is the following, but can vary depending on the topic:

- **Term 1: Ideate**: is discovering the topic and proposing potentially innovative ideas, through literature reviews, benchmarking, ideation techniques, etc.;
- **Term 2: Prototype**: develop prototypes (or experiments) that embody the ideas or gaps found in the literature;
- **Term 3: Realize**: conclude the study with a proof-of-concept of the ideas developed and prototyped earlier.

So far, students are hosted in four SUTD research centers showing the diversity of students and topic: SUTD-MIT International Design Center, iTrust Center for Research in Cyber Security, Digital Manufacturing and Design Center and Lee Kuan Yew Center for Innovative Cities. Some topics of the research projects which students have developed can be seen in Fig. 31.2.

31.5 The Experiences and Accelerators

As seen in Fig. 31.1, the program contains two accelerators and one overseas experience.

The first one is the 2–3 days Design Accelerator, specifically designed as a pre-term learning journey for those which may not be very familiar with design language and processes. The students work in teams under a theme that changes every year. This accelerator mimics directly the executive workshops that SUTD conducts with companies through the SUTD Academy, and it serves to get the student up to speed with a common basic knowledge of design principles, tools and methods, as well as the 4D design model (discover, define, develop, deliver).

The second is the overseas experience. This experience takes place between terms 1 and 2, after the students have gone through the first term Innovation by Design course. Students travel to another country to experience co-design (again within a theme that varies for each year) with students from a different country and background. The co-design experience can last between one and two weeks and is co-funded by the program. For the first intake, students were hosted by the School of

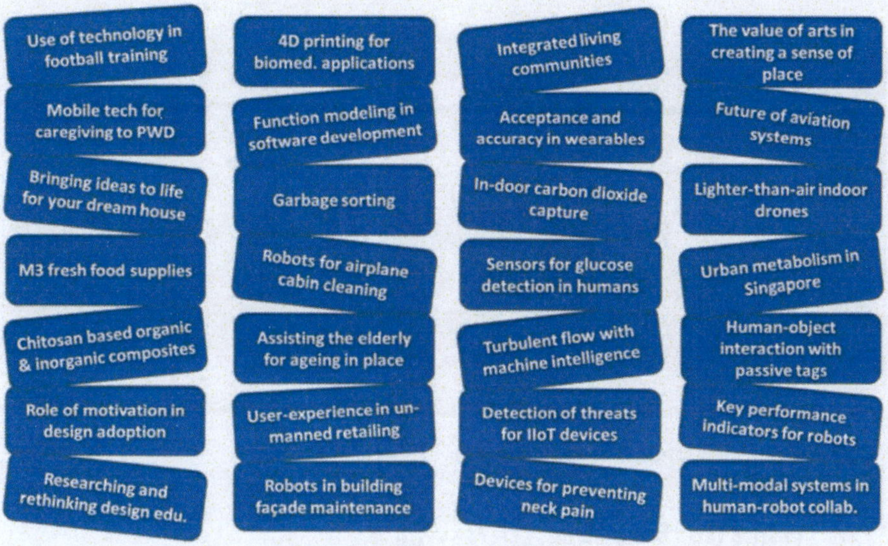

Fig. 31.2 Research areas of the first year students

Design at Jiangnan University, Wuxi, China, for one week and were paired with local students to solve a given problem. By the end of the week, all teams had to present a prototype of their solution. The problem for the first run of the experience was:

Design a personal mobility device for the last mile to cater for users in CITY X with the following profile:

- *Young professional in a growing business;*
- *Married, 25–30 years old;*
- *One child.*

The assignment came in five variants of CITY X: Tokyo, Boston, Lisbon, Sydney and Cambridge. With 10 teams, two teams were working independently on the same city. Figure 31.3 shows some of the prototypes.

The third experience was the three-day Entrepreneurship Accelerator, at the end of term 2. This accelerator caters for those who want to start their own companies or just want to know more about entrepreneurship. Students again work in teams to develop a pitch for a new product/service/system of their own choice, within a given theme. The instructor team will take the teams through the steps of a good entrepreneurship practice, from idea to investor pitch, and touch on a variety of topics such as Singapore law and incentives for starting a company, etc.

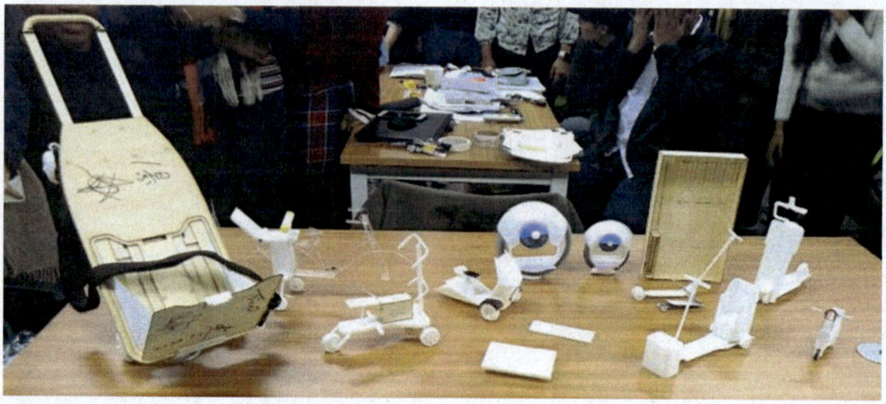

Fig. 31.3 Prototypes resulting from the overseas experience in 2020

31.6 What the Students Are Saying

The Innovation by Design course in term 1 and Design Science in term 2 received an overwhelmingly positive reaction from the students. Students also provided suggestions to improve the course. The survey questions and answers can be seen in Tables 31.1 and 31.2. The answers were given on a Likert scale from 1 (strongly disagree; poor) to 5 (strongly agree; very good) with 3 being neutral/average. The percentages in the tables refer to the positive answers, i.e., agree/good (4) and strongly agree/very good (5).

Suggestions for improvement were also sought and well noted. Some comments were on the lack of time in IbD devoted to classes on more technical content, and the fact that there is only one class per week. Other comments are more related to specific content of classes (examples, exercises, etc.) which the students feel are not directly related to their respective projects. The Design Science course was strongly affected by the sudden COVID-19 measures, for which the very interactive format was ill-prepared. Suggestions for improvement are mainly related to the workload.

Table 31.1 Survey results for the term 1 course on Innovation by Design

Survey question	% positive answers
The course has stimulated my interest to learn more about the subject	81
The course has improved my knowledge on the subject	81
The course is well organized and structured	76
The course workload is manageable	95
The course involved me in active learning experiences	96
After going through all the classes and assessments, I will be able to do what is prescribed in the learning objectives	91
Overall, I would rate this course as	81

Table 31.2 Survey results for the term 2 course on Design Science

Survey question	% positive answers
The course has stimulated my interest to learn more about the subject	85
The course has improved my knowledge on the subject	95
The course is well organized and structured	73
The course work load is manageable	52
After going through all the classes and assessments, I will be able to do what is prescribed in the learning objectives	84
Overall, I would rate this course as	84

For the overseas experience and the Entrepreneurship Accelerator, the survey results can be seen in Tables 31.3 and 31.4, respectively.

The most obvious outcome for the overseas experience was the mixed feelings about the duration. Half of the students felt that it was very good, whereas the other half felt that it should have taken longer, some of them suggesting up to two weeks. There was no significant difference across the two batches of students (from SUTD and from Jiangnan University) on this.

The overall sentiment on the entrepreneurship accelerator was that it was very good and relevant, although not all the students are planning on starting their own companies. They felt that the knowledge they acquired in this course nicely complemented the knowledge acquired in the compulsory courses. Some students expressed

Table 31.3 Survey results for the overseas experience (SUTD and Jiangnan University students)

Survey question	% positive answers
How would you rate the entire experience	86
How would you rate the interaction between team mates during the entire week?	86
How would you rate the delivery of the week long experience?	86
Were the materials/facilities provided adequate?	55
Was the duration of the experience adequate?	50

Table 31.4 Survey results for the entrepreneurship accelerator

Survey question	% positive answers
The program content was relevant and practical	94
The atmosphere and interaction with the other participants were good and contributed to the sessions	85
You are more confident in your ability to start your own company after completing the program	79
The program duration (3 days) was efficient and practical	84
I would recommend this program to fellow students	79

that they are still not confident in applying the entrepreneurship knowledge to their own work, which is an issue we will look into in future years.

31.7 Conclusions and Future Work

The program's first year has just been completed. The number of applicants and the overall results have surpassed our expectation. The structure and content were new and very different from the existing programs, and it was not clear when we developed MIbD how varied the backgrounds of the students would be (see the research areas in Fig. 31.2).

There are multiple elements of the program that require improvement. At the time of writing, the 2020/21 program has just started with a total of 25 students, but with an increase of self-paid and company-sponsored students and a reduced number of foreign students, as compared to the September 2019 intake.

The continuing COVID-19 measures have an impact, but fortunately, in class teaching is allowed, albeit with a large number of constraints (masks, tracing, social distancing, etc.) that reduces the social elements of studying. We hope that early 2021 will see a return to normal. The January overseas experience is likely to remain affected. Traveling restrictions may still be in place. This requires a rethinking of this important component of the program. Alternatives are a national experience with another local university, or a virtual overseas experience conducted online.

Our own experiences in this first year and the students' suggestions have already resulted in some improvements; others require further analysis and preparation. For the term 1 course (IbD), it was suggested to have more technical content and more time devoted to classes instead of external speakers and studio work: A possible improvement will be to shift all invited speakers from IbD to I-P-R, thus freeing up space for more classes and more technical content. I-P-R did not have formal weekly classes, so a one-hour slot will be scheduled every week for external speakers to come and impart their knowledge to our students. This will not significantly affect I-P-R, as this is a research-based course, with work done in the laboratories, and a one-hour slot every week taken out of the laboratory will not be a problem. The number of assignments in Design Science will be reviewed, and the taught content better adapted to the variety of research topics.

The program coordination is considering student exchanges with other overseas universities, with the aim of making this program even more international and vibrant and potentially leading to a joint Master program. The first exchange program with a European university is expected to start in 2021.

References

1. Dore, F., Kouyoumjian, G., Sarrazin, H., Sheppard, B.: The business value of design. McKinsey Q. (4), 58 (2018)
2. Design Singapore Council Design 2025. National Design Center (2016). ISBN 978-981-09-8875-3
3. Graham, R.: The global state of the art in engineering education. Massachusetts Institute of Technology (2018)
4. Telenko, C., Camburn, B., Hölttä-Otto, K., Wood, K., Otto, K.: Designettes: New approaches to multidisciplinary engineering design education. In: International Design Engineering Technical Conferences and Computers and Information in Engineering Conference, IDETC/CIE, 17–20 August 2014, Buffalo, New York, USA
5. Otto, K., Camburn, B.A., Wood, K.L., Nannicini, G., Bouffanais, R., Kyoseva, E., Yong, J.W.H.: Integrated 2D design in the curriculum: effectiveness of early cross-subject engineering challenges. In: 121st ASEE Annual Conference, 15–18 June 2014, Indianapolis, Indiana (2014)
6. Ulrich, K.T., Eppinger, S.D., Yang, M.C.: Product Design and Development, 7th edn. McGraw-Hill (2020)
7. Blessing, L.T.M., Chakrabarti, A.: DRM, A Design Research Methodology. Springer, Berlin (2009)

Chapter 32
Role of Authoritative Elements in Design

Sunalini Esther and Sheeba Chander

Abstract The importance of learning theories is that they expose the learner to a wide spectrum of ideas, thereby expanding the knowledge base while allowing the learners to experiment, analyze and internalize their observations. In all the study of design, the principles, factors and elements that comprise, modify and establish design can be summed into three categories, namely physical elements, modifying elements and authoritative elements. There are factors that physically comprise design, factors of ephemeral and intangible quality that modify design, and authoritative factors that root and establish design at a particular place in time. In most cases of theoretical analysis, there is a sense of looking backward—to the study of established facts and the principles that may be gleaned from the study. The authoritative elements, in particular, comprise the primary establishing elements which form the base for study. The word authority has its roots in the word "author" which means someone who has the power to bring something into existence and sustain it. With respect to design, it is these elements that bring richness to design or add the dimension of depth. The authoritative elements are history, typology and context. At the time of undertaking this research, there is no literature that comprehensively brings the elements of design together to enable place-making. However, such an endeavor would be hugely beneficial to the design community in terms of providing a framework, within which their design may be generated while also proving relevant to its time. The purpose of the research is to establish the impact of authoritative elements on form and space. The methodology of inquiry and research is through the literature study and survey questionnaire.

S. Esther · S. Chander (✉)
Hindustan Institute of Technology and Science, Padur 603103, Tamil Nadu, India
e-mail: sheebac@hindustanuniv.ac.in

S. Esther
e-mail: suna.fisher88@gmail.com

A. Chakrabarti et al. (eds.), *Design for Tomorrow–Volume 2*, Smart Innovation, Systems and Technologies 222, https://doi.org/10.1007/978-981-16-0119-4_32

32.1 The Elements of Design

32.1.1 Introduction of Categories of Elements of Design

The phenomenon of design involves a process that is dependent on the relationship between several complex factors and elements. It is not a clear linear process that begins with the project brief that culminates in a product/solution. At every stage, there is back and forth as the designer wrestles to layer the elements of influence in an appropriate manner. Design does not arise by a mere mechanical arrangement of tangible physical elements nor purely out of internal emotion and feeling. While there are many theories that touch on different individual and independent factors that influence architecture, "These theories ignore the fact that building form manifests the complex interaction of many factors ..." [1]. Design relies on aesthetic decisions (using physical elements), stemming out of external environmental facts (authoritative elements), refined by internal preferences and experiences (modifying elements). All design commences with an understanding of the authoritative elements that cause design to respond and engage with them to generate a creation through the use of the physical elements. To this creation is further applied a layer of modifying elements that work to create a sharper product/project by connecting strongly and personally with the designer or the client/user. This research paper aims to identify the primary design elements that authorize the existence of architecture and establish their importance as a source of architectural design. It was intended that through a survey questionnaire, data might be collected validating the importance of the authoritative elements. Through the analysis of this data, it would be possible to conclude whether these elements could facilitate design generation and enable place-making in the process (Fig. 32.1).

32.1.2 Place-Making in Design

The physical elements serve as tools that bring the essence of authoritative elements and modifying elements to the physical realm of being from the mode of imagination. Therefore, they operate at both levels. Modifying elements are those that while enabling the choice of materiality, form and space make that choice out of the existing anthropological context and milieu. The word "authority" has its base in the word author which means someone who has the power to bring something into existence and sustain it. In other words, with respect to design, it is these elements that bring richness to design or add the dimension of depth. The authoritative elements are history, typology and context. Through them, design is influenced and established through precedence set in facts from the past. With respect to architecture, the modifying and authoritative elements are key to enabling place-making. Place-making is vital to creation of spaces that resonate with the people and culture of a place in time (Fig. 32.2).

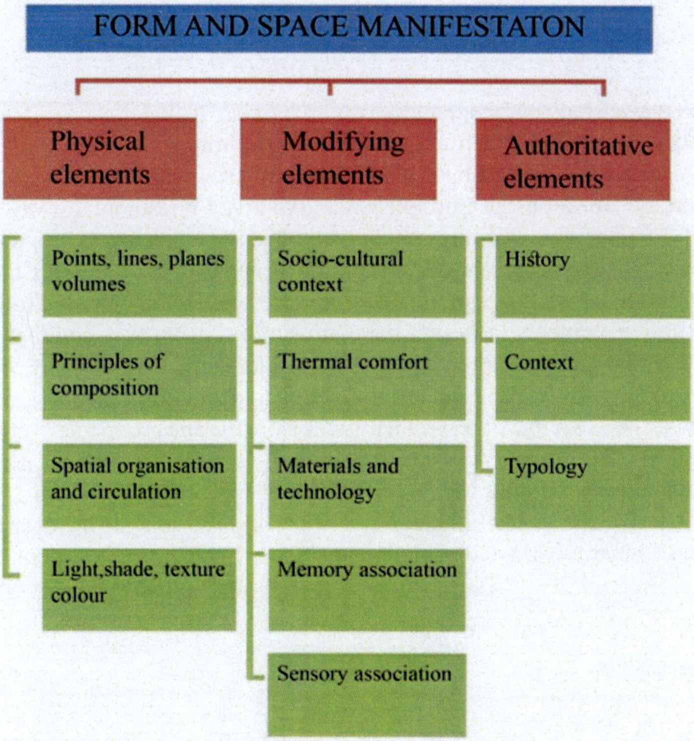

Fig. 32.1 Classification of elements that influence design

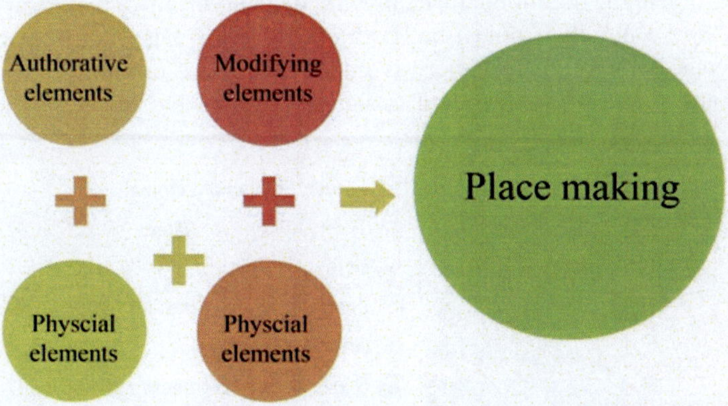

Fig. 32.2 Elements that contribute to place-making

32.2 Methodology

A comprehensive study of the literature was undertaken to understand different layers of elements that constitute the framework of design. Among these elements, the literature was reviewed to consolidate elements of authority which were instrumental in bringing architectural design into existence. A survey questionnaire was prepared to find out the importance of three of the authoritative elements. It was designed to analyze in detail, the degree of impact and influence of these elements on creating architecture and design generation. The objective was to establish the relevance of the authoritative elements in design generation. The survey was conducted among seasoned architects and designers, teachers of architecture and designers, fresh graduates from undergraduate program in architecture and others from a non-architectural background. The age range of respondents was from the age group of 23–68 inclusive of both genders. The survey was collected online, and the sample size was 90. The respondents hail from across India. The period of collection was during the month of April 2020. Pie charts and bar graph have been used to analyze the data collected.

The questionnaire consisted of three main parts: history, typology, context.

32.3 Analysis

32.3.1 History and Tradition

The first section of the questionnaire dealt with the relationship of history and design. In Fig. 32.3, the pie chart notes that both spatial principals, motifs and features are the means through which history is understood to be primarily referenced in contemporary works of architectural design as attested by 58.8% of the sample

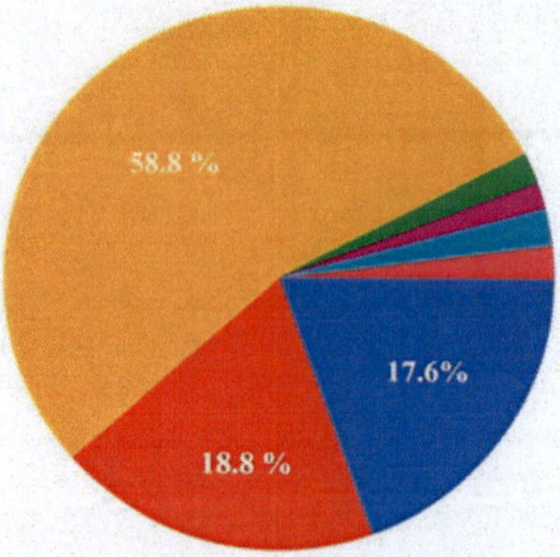

Fig. 32.3 Manner of referencing history in the present works of architectural design

respondents. Manner of referencing history through only spatial principles and only motifs is attested by a minority of 17.6 and 18.8% of the respondents, respectively, which is close to equal.

In Fig. 32.4, the pie chart depicts very clearly that history is a definite source of architectural design with 94% of the respondents attesting to this statement. This is because one always associates the present as the continuation of the past.

The question posed in Fig. 32.5 explores the possibility of evaluating developments against a design standard. From the pie chart, it is seen that 61.2% in the sample have expressed that there is a clear sense of aesthetics and right design. Therefore, there exists a frame of reference for a design standard, and there exists a design standard for developments against which they may be measured.

It is understood from the pie chart in Fig. 32.6 that there is only a small margin of disagreement, namely 4.7%, between the number of respondents in attributing the movement of design across the ages—on one hand as a cycle that begins, peaks and

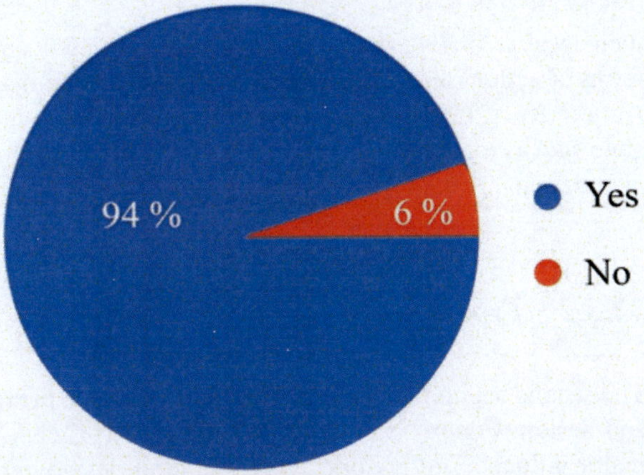

Fig. 32.4 History as a source for architectural design

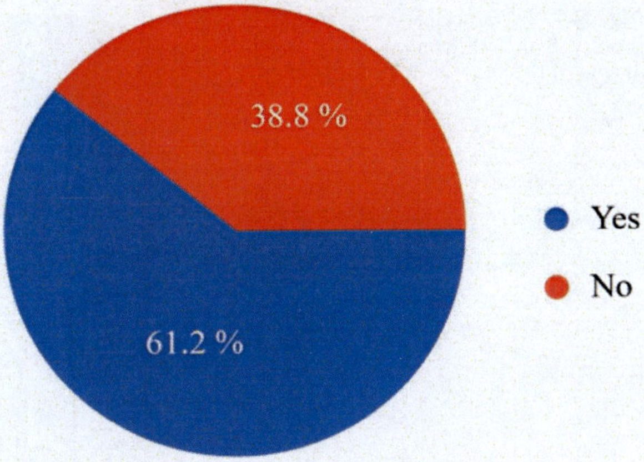

Fig. 32.5 Possibility of measuring developments against design standard

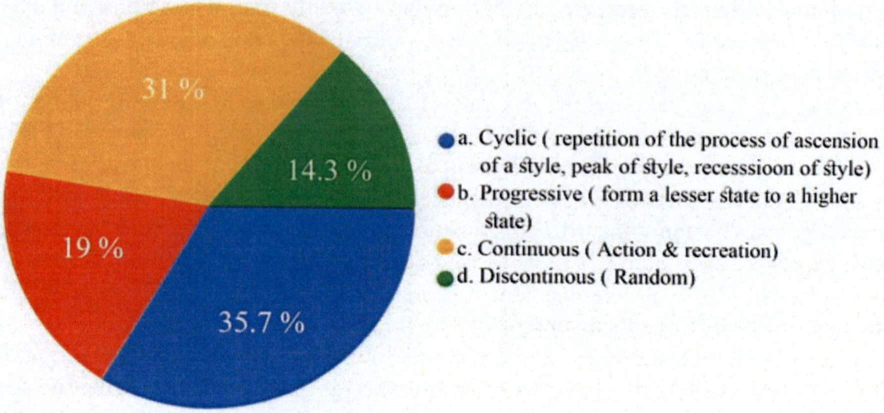

Fig. 32.6 Movement of design across the ages

then decays (35.7%) and on the other as a continuous movement that is responsive in terms of actions and reactions (31%). These two comprise the majority of responses. A minority of 19% hold that it is progressively moving from a lower state to a higher state, and a smaller minority of 14.3% hold that movement of design is just a random occurrence.

32.3.2 *Typology*

The second section of the questionnaire dealt with the relationship between typology and design. Figure 32.7 identifies prototype (primary pattern or model) as being a major influencer of a work of architectural design. 77.6% of the respondents have

Fig. 32.7 Influence of prototype on architectural design

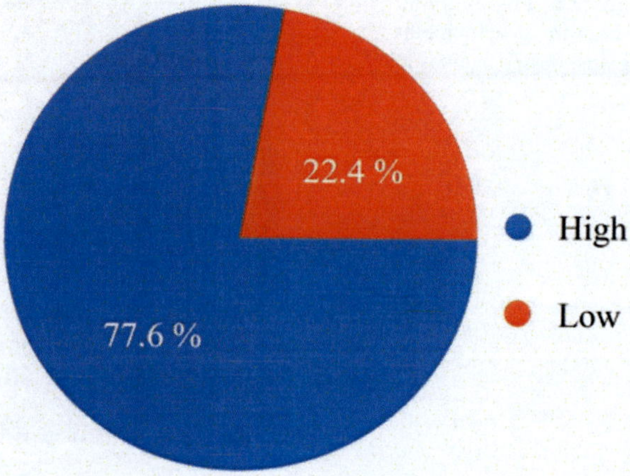

positively confirmed to its influence on design. The respondents identify prototypes as successful solutions to similar/identical challenges or briefs and hence a good place to start the process of design.

Apart from 12.9% of the respondents who are unsure about the relationship between typology and structural organization in the pie chart in Fig. 32.8, majority of the respondents of the sample, namely 50.6%, have stated that typology is only sometimes limited to the structural organization and 22.4% have stated that typology is never limited to the structural organization.

On questioned if prototypes would be useful to categorize and understand physical forms and/or spatial patterns while designing, 78.8% of the sample have agreed on its importance on the same as illustrated in Fig. 32.9. 17.6% have agreed on its usefulness sometimes. It can be observed that the number of respondents stating its unhelpfulness is negligible.

In the pie chart in Fig. 32.10, it is noted that the majority of the respondents of the sample, namely 92.9%, are in agreement that typology definitely helps a student/learner stay focused on his/her program or brief. It gives them

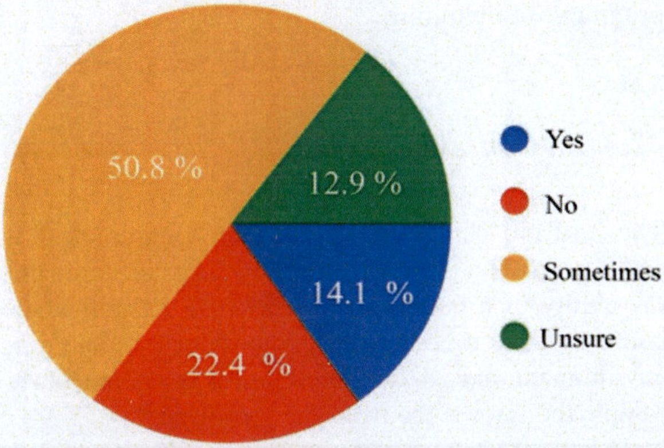

Fig. 32.8 Limitation of typology to structural organization

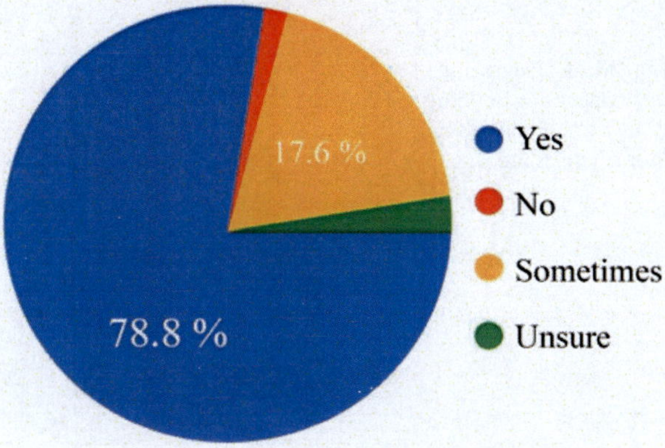

Fig. 32.9 Respondents on whether prototypes are helpful in categorizing and understanding physical forms/spatial patterns while designing

Fig. 32.10 Respondents on whether prototypes serve a student of design to maintain focus on his/her program

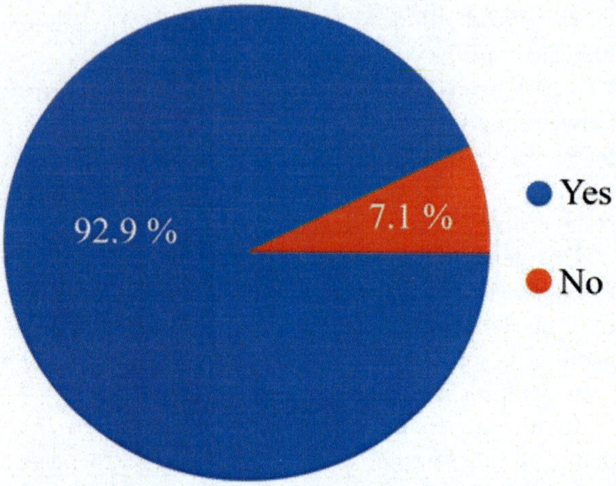

clarity by helping them understand the usual patterns for the same type of program/brief/building.

32.3.3 Context

The third and last sections of the questionnaire dealt with the relationship of context and design. The pie chart in Fig. 32.11 discusses the importance of conformity or the ability of a building to blend with its immediate context. 68.2% respondents confirmed that it is extremely important for a building to blend or conform with its environment and 28.2% stated that it is at least fairly important. Only 3.6% of the sample did not see the need for conformity.

In Fig. 32.12, the pie chart is a depiction of the idea of context and the importance of different types of context. It is seen that a physical micro-level context is as

Fig. 32.11 Degree of importance for building design to blend with its immediate context

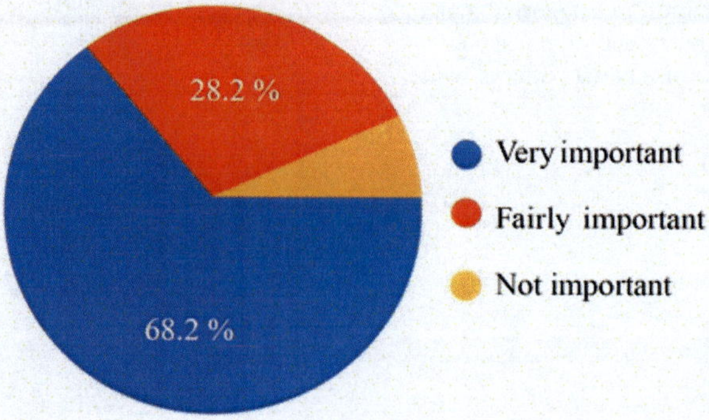

Fig. 32.12 Respondents on which factor is more importance while designing

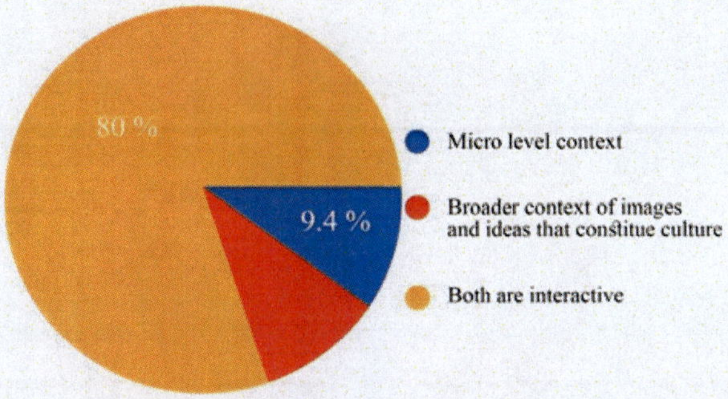

Fig. 32.13 Respondents on what determines their requirement to blend their design with context

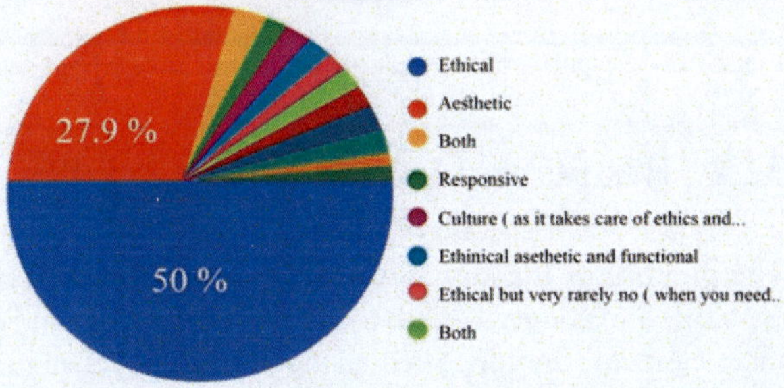

important as the broader cultural context, and they are both interactive with 80% of the respondents aligned in this favor.

While many designers automatically start with a study of the context and attempt to conform their designs to the context, the question posed in Fig. 32.13, the pie chart dives into the issue of what causes this requirement to conform. It is interesting to note that 50% of the respondents have stated that it is the ethical obligation of designers to conform and 27.9% of the respondents have stated that aesthetics can be achieved through conformation.

In Fig. 32.14 through the pie chart, it is observed that history, typology and cultural beliefs are all important while determining authority in design, of which history of a place is deemed the most important, namely 30.6%. It is apparent that these factors weigh in heavily while considering authenticity and validating context.

The respondents also stated that context helps to set out a precedent of the norms for the architect to make informed decisions in his approach toward planning, giving a set of design clues. At the same time, understanding and application of the knowledge of context helped with relevance which assists with an increased sense of project ownership.

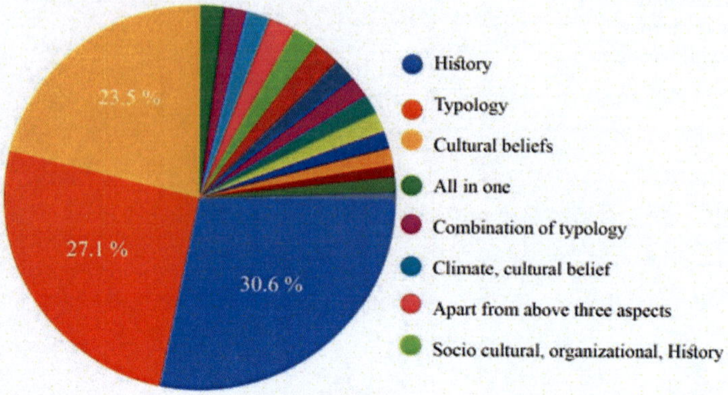

Fig. 32.14 Respondents on what gives context, authority in design

32.4 Results and Inferences

The purpose of the survey was to connect elements of authority—namely history, typology and context—with the creation of design and establish them as roots. These three elements are identified as the primary layer of design generators among many other layers of elements contributing to design.

"Design and Innovation builds on precedent, on ideas and concepts that have evolved over time. Architecture uses precedents from social and cultural history and applies these influences to contemporary buildings, forms and structures. Reacting against or responding to these ideas has been the basis of architectural evolution" [2]. History is instrumental in helping designers to make choices from the past by engaging with precedents/socio-cultural history as inspirational factors and choosing to improve/improvise on them or rejecting the past precedents for a justifiable cause by seeking to build afresh. Either way, it is observed that history is certainly a source of origin of design.

"Or are they merely a strategy, names invented to make tangible certain categories of coherence, … to maintain focus and authority and thereby help perpetuate architecture…?" [3]. This question is raised by author Paul Alan Johnson in his book Theory of Architecture while debating on the influence of typology on Architecture. However, this has been answered through Sect. 32.3.2 of this paper that discusses the relevance of typology in design, especially with respect to learners. It has been established that typology helps in our cognition and internalization of the layout of forms. It informs the user about the generic scheme of arrangement and is invaluable to someone who is attempting their first design at that "type." It helps us to make sense of buildings and is a vital point of design generation. This fact is particularly reflected in Fig. 32.10 where 92.9% of the sample respondents agreed that it helps a student of design maintain focus on the elemental framework of the program.

"Context ... has authority by virtue of its precedence over anything newly conceived that is desired to be placed within it" [3]. Architecture is always experienced as part of context. It is linked historically, physically, socio-culturally, geographically and visually to its immediate and broader context. "Every architectural work exists in the presence of a chorus of contexts that can impart meaning to and, in turn, derive meaning from their associations with the work" [4]. The requirement to design with conformation to context has been established in Sect. 32.3.3 as basically ethical. When an architect chooses to make his/her design respond to and engage with the context, it helps with extending the design narrative out of the boundaries of the site, connecting the project physically as well as intangibly/ephemerally in a cultural and social manner that is relevant to the context. Considering that no project is created to respond to isolation and all design arises from narratives that weave through people, experiences and places, it stands to reason that the context would lend authenticity to the narrative created on site. This authenticity is lent through the history, typology and cultural/social fabric of the context, to name a few. Therefore, it is established as pivotal point.

When the authoritative elements are arranged according to their degree of influence and importance on design, analyzed from Figs. 32.4, 32.7 and 32.11, the bar graph in Fig. 32.15 is derived. On a scale of 1–10, history values at a factor of 9.4, typology at a factor of 7.8 and context at a factor of 6.8. The average or mean of the importance factor of these elements is obtained as follows.

$$\begin{aligned}\text{Mean} &= \text{Sum of factors/no. of factors}\\ &= (9.4 + 7.8 + 6.8)/3\\ &= 8\end{aligned} \tag{32.1}$$

Considering the average influence factor for these elements is 8 on a scale of 1–10, where 1 implies very low impact and 10 implies very high impact, these elements maybe considered as those that empower the design with authenticity. Therefore, these elements author design.

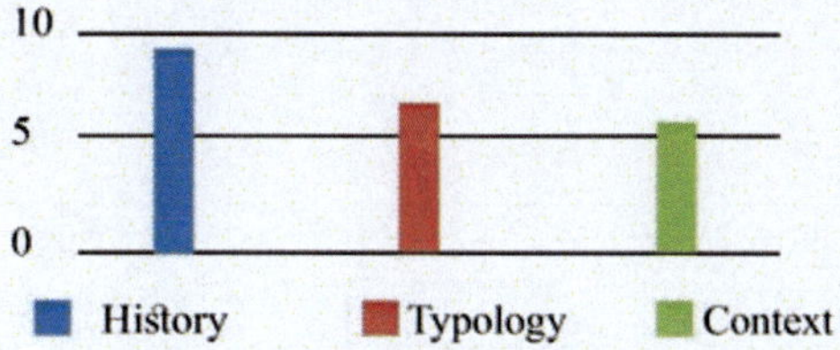

Fig. 32.15 Ranking of the degree of influence of the authoritative elements

32.5 Conclusions

The importance of history as a source for architectural design, the degree of influence of a typology on architecture and the importance of micro- and cultural context have been illustrated through this research.

Once the program for design has been established, for any project, these elements bring the work of architecture into existence to give it the basic form and spatial frame. The physical elements serve as tools to bring about this existence. Over this are superimposed the modifying elements to refine the design of space and environment in a precise manner while working with the physical elements.

With this framework in place, it is possible to derive a model to achieve design through the layering of elements that can be used in classrooms among students as well as among practitioners in design studios. It is imperative that the influences of these elements are taken into account, in the absence of which design remains superfluous, soulless and irrelevant.

References

1. Rapoport, A.: House Form and Culture. Prentice-Hall, Englewood Cliffs, NJ (1969)
2. Farrelly, L.: The Fundamentals of Architecture. AVA Publishing SA, Switzerland (2007)
3. Johnson, P.A.: The Theory of Architecture: Concepts, Themes and Practices. Wiley Inc., NY (1994)
4. Simitch, A., Warke, V.: The Language of Architecture: 26 Principles Every Architect Should Know. Rockport Publishers (2014)

Chapter 33
Social Connectedness in Online Versus Face-to-Face Design Education: A Comparative Study in India

Christy Vivek Gogu and Jyoti Kumar

Abstract The global pandemic has affected almost every aspect of life across the globe including medical services, essential services, production units and educational institutes. People have been forced to physically and socially isolate themselves. This study focuses on the sudden shift that design schools in India had to make from having traditional face-to-face (F2F) classes to starting virtual classrooms online. This paper reports the results from semi-structured qualitative interviews conducted with ten design tutors from ten different design schools across India. The authors attempt to understand how the design instructors compared the online classes with traditional F2F classes, and how they perceived the effects of instructor–student connectedness and interaction on teaching motivation and satisfaction in online design teaching. This study found that most instructors felt a lack of social connectedness with the students during the online classes. The paper discusses details of some of the reasons for this lack of felt connectedness and how it affects the teacher satisfaction and learnability of the design students. In addition, it reports the insights given by the instructors about some of the creative learning practices in design courses that they are currently finding challenging to replicate online.

33.1 Introduction

Inherently, human beings are social with a need to belong, feel connected with others, create and nurture social relationships [1]. It is assumed that positive personal and social attachments contribute substantially to the capacity of making sense of one's life [2, 3]. Social connectedness is one such attribute which has been argued to provide meaning in life for many people, impacting behaviour and life choices [4]. Students with better access to social support are better able to maintain a positive sense of

C. V. Gogu (✉) · J. Kumar
Indian Institute of Technology, Delhi, New Delhi, India
e-mail: christygogu@gmail.com

J. Kumar
e-mail: jyoti@design.iitd.ac.in

A. Chakrabarti et al. (eds.), *Design for Tomorrow–Volume 2*, Smart Innovation, Systems and Technologies 222, https://doi.org/10.1007/978-981-16-0119-4_33

social connectedness, reducing chances for academic and behavioural challenges [5, 6]. Social connectedness has been observed to be an important component of the academic settings. Social connectedness has been reported to be related to lower depression in students [7] and reduced risk-taking in youngsters [8]. The need for social connectedness is also influenced by many personal factors of students, such as sibling presence, economic strata of society, age of the person and time spent on the Internet [9].

The last two decades have seen a rise in online education, around the world. From increased computer literacy and Internet connectivity, growth in distance learning opportunities, to ease of learning while working part time, many reasons are attributed to this increase in popularity of online courses (or MOOCs). Countries are increasingly investing into multimedia-based courses [10]. In developing countries, online and distance learning gives better opportunities to people for whom either education is beyond reach or too expensive. For developed countries, multimedia education adds convenience in enhancing skills while at home or work. Because of this, teaching has become more innovative too, using technology to bring freshness and innovation to the course material, making courses accessible to multiple people at same time separated by geographical areas [11].

Similarly, the scenario in design education has changed with the growth in the IT sector. Some research has been done to inquire into the effectiveness and success of these courses but not much has been done to fully understand the impact on learning in online platforms [12].

When the global pandemic struck earlier this year, it affected almost every aspect of life across the globe including medical services, essential services, production units and educational institutes. Most universities were forced to suddenly shift to the online teaching space whether they were prepared for it or not. This study focuses on the sudden shift that design schools in India had to make from traditional face-to-face (F2F) classes to virtual classes.

There are many online learning and conferencing platforms developed by Institutes and IT companies, that are being used for virtual classes, for instance, Skillshare, Moodle, Blackboard, Google Classroom, MS Teams, Coursera, National Programme on Technology Enhanced Learning (NPTEL) a project funded by MHRD, Zoom, etc.

So far, these platforms are used for theory-based learning. How compatible these platforms are with the needs of design education is yet to be evaluated. There are specific features associated with design education which makes its pedagogy unique. The simultaneous combination of theory and practice in studio environments, focus on theories of art as well as science, classroom activities based on teamwork and collaboration, focus on problem-solving process with creative thinking and opportunistic usage of multimedia materials and contents are some of the unique aspects of design education [13].

Research shows that design and creative learning is often achieved by use of critical thinking and collaborative work, with a focus on hands-on work and creative problem solving [14–16]. Design education also brings together the know-how of a variety of other fields to uniquely solve problems. Studio culture is an integral part of design education. And this paper seeks to study the online design learning experience, especially in Indian design schools, since the COVID-19 pandemic having displaced normal F2F classes in most university campuses.

While many factors contribute to the effectiveness of an online design learning experience, in this paper, we focus on social connectedness and instructor–student interaction as a key in achieving an effective design learning experience. This study also attempts to understand how the design instructors compared the online classes with traditional F2F classes, and how they perceived the effects of instructor–student connectedness and interaction on teaching motivation and satisfaction in online design teaching.

33.2 Literature Review

33.2.1 Online Learning

With increase in technological and internet awareness, India is one of the largest growing consumers of online education. Statistics show that the number of users enrolled for various online learning courses is estimated to be 1.6 Million in 2016, which is expected to grow to about 9.6 Million by the end of 2021. Reports show that a blend of online and offline can be beneficial and productive [17]. Elson Szeto, while examining the effects of this blended synchronous learning in his paper, noted some of the differences and learning experiences of online and face-to-face learning [18] such as the instructor focused more on the online students, adopting a slow pace, clarity and repeated probing, thereby rendering the online students with clear explanations and ease of understanding of the topics. In contrast, the F2F students felt bored in some sessions because the instructor unintentionally paid too much attention to the remote online students.

33.2.2 Online Design Learning

While online learning has been around for a few decades, design education is still not readily available online [19]. Design uses extensive studio-based exercises which makes it challenging for design educators to transition to technology-driven changes into an online teaching and learning environment [20, 21]. With studio-based learning being at its core, design courses usually have small class sizes. They typically use project work and collaborative creative problem solving with many possible solutions [22]. Research shows that online collaboration in design can be successfully done only if student participation is high and instructor feedback is instantaneous [23]. The characteristics of studio-based teaching in art, architecture and design have been identified as supporting interaction, active learning, as well as social engagement [24, 25]. Peer learning and group discussion form an integral part of the learning process

[14, 26]. Keeping all these in mind, blended learning is seen as a possibility where certain courses are taken online, while others are studio-based [21].

33.2.3 Social Connectedness

Social connectedness is defined as 'a short-term experience of belonging and relatedness, based on quantitative and qualitative social appraisals, and relationship salience' [27]. Social connectedness is also defined as a personal sense of belonging to a group, family or community [28]. For the purposes of this research, the authors would like to define social connectedness as 'the experience of belonging and relatedness between people'. Social connectedness is recognized as a key motivator for social behaviour. It can also predict a successful life and result in a variety of social and health-related benefits [29]. In universities, a competitive environment, tech addiction and study workload can keep students preoccupied and often feeling lonely [30, 31]. Not only individuals but communities also benefit from the creating and nurturing of social connections. There is evidence that online students have a lower sense of university attachment [32] and do not feel as connected as regular students [33]. While in the west, youngsters are encouraged to be independent and individualistic at an early age, in eastern cultures, youngsters grow up in a more interdependent and socially interactive environment. So often, the move to university puts a strain on them and can result in feelings of loneliness [34]. Social interaction for students in traditional F2F college environments is instrumental in their learning and deep comprehension [35, 36]. Though students find it is easier to connect socially and make friends with students in traditional F2F classes, they think it is important to socially connect in virtual classroom settings as well [37]. Laffey et al. [38] claim education and various learning interactions, whether traditional F2F or virtual, to be social practices. A similar study done by Turki et al. [39] concludes that achievement motivation in university students significantly correlated with social connectedness, whereas another study indicated that students in both online and traditional programme formats had similar perceptions of their social connectedness and satisfaction with their learning experiences [40].

33.3 The Methodology

This study reports the preliminary section of a larger research and therefore includes the former of the stakeholder perspective of the two groups, design instructors and design students. The study surveyed 10 instructors teaching design courses at more than 10 architecture or design schools (including one visiting faculty member). Semi-structured interviews were used to pose the same questions to the ten design instructors, which provided an in-depth line of inquiry [41]. Some of the questions were 'How was your transition from traditional classes to online teaching?', 'Did you

Table 33.1 Coding template for the three dimensions

Dimension	Coding category
Instructional	1. Instructional management, 2. building understanding and 3. direct instruction
Communicative	1. Emotional expression, 2. open communication and 3. group cohesion
Learning	1. Triggering events, 2. exploration, 3. integration and 4. resolution

Adapted from Garrison et al. [43]

face any challenges? Please explain.', 'What do you think are the positives of online classes?', 'Did you notice any difference in the behaviour of the students online?' and 'Were you satisfied with the mutual communication in the online set up?'.

In-depth qualitative analysis was carried out on the inputs received from the instructors. The interview transcripts were broadly themed, coded and categorized using Strauss & Corbin's method [42] and the Garrison et al. [43] Community of Inquiry (CoI) framework. According to this well-received framework, quality education experiences are seen at the intersection of the teaching, social and cognitive presences. The three dimensions were instructional, communicative and learning, through which the learning and teaching effects were analysed. The following table shows the coding framework for the three dimensions (Table 33.1).

The responses were coded under three themes of 'instructional', 'communicative' and 'learning' phrases. Broad coding themes such as advantages and deterrents were established at the beginning of the analysis and sub-themes emerged by grouping responses. Some samples of the responses to questions about visual input/feedback were: 'students prefer to keep their video off … body language … many things that the mentor could see and give feedback accordingly', 'But if they are switching off their cameras, then it becomes very linear, very boring kind of a thing. Also, you don't know whether the students are really hearing you or are they multi-tasking', 'It's difficult to gauge so many students on one screen. Many don't even turn on their videos', 'Some students had a good knack, were able to master a particular medium really well compared to others, so the other students could watch… gain some insight … apart from just getting my feedback'. While some instructors listed the positives as 'I think virtual teaching is far more engaging from a creative perspective because there are tools and software that you can get the students involved in', 'students are a lot more professional (online) and I've seen an increase in attendance as well', 'they are definitely more disciplined and articulate (online)'. When asked about students' ease of access to instructors, some said 'I like being around the students, and being in campus also means that you know access to you is quite easy', 'Students can see you and reach out to you. When it is virtual, it also means that you have to send a message or an email to sir and see if he is available, students might hesitate to reach out'. Responses referring to colleagues included 'I like to also be around my peers, who come from a design background', 'small conversations over a cup of coffee about curriculum or anything on design are also enriching' and 'virtual classroom has also allowed is to have people from across the globe come and speak to my students'.

33.4 Findings

The findings are listed in Table 33.2, showing a comparison in the advantages of traditional F2F versus online teaching. They have been listed below in the order of their frequency of mention and emphasis by the instructors (from most frequent/emphatic to least).

Table 33.3 lists the ten most common deterrents that most of the instructors shared from their experience in teaching and interacting with design students online.

Table 33.2 A comparison of face-to-face/online student learning and instructor experiences

F2F teaching (percentage of mentions)	Online teaching (percentage of mentions)
Instructional dimension	
– Effective for teaching both theory and skill-based courses (80)	– Possibility of using online classes along with face to face (blended learning) (60)
– More personal engagement with each student (100)	– Can be taught from anywhere (guest lectures) (40)
– Can adjust content and speed based on students' verbal and non-verbal feedback (70)	– Saves commuting time and energy (30)
– Workshop facility and other tactile and sensory creative engagement available (60)	– Saves expenses on College amenities and overheads (20)
– More options and opportunities of engaging students based on their interests on campus (40)	– Exploring new ways to approach design teaching (20)
– More work done in office hours (30)	
Communicative dimension	
– Give immediate and accurate feedback for student work (80)	– Possibility of recording lectures and listening at convenience (50)
– Immediate feedback from body language and facial expressions of students (70)	– Higher attendance and professionalism (50)
– Discussions can continue even after class hours (50)	– More scope for improving articulation and online presentation skills (30)
– More vibrant communication, with jokes and appropriate examples (30)	
Learning (teaching) dimension	
– Greater engagement in class, more satisfaction and motivation (70)	– Can be attended from anywhere, anytime (70)
– Higher learnability—from more interaction with tutors/peers (60)	– Breakout groups option was useful in group work and discussions (50)
– Focused engagement between students and tutors without distraction (40)	– More online resources can be shared easily (40)
– More collaboration amongst faculty, sharing and discussing ideas, etc. (20)	– Learning online presentation skills and using interface and software (30)

Table 33.3 Deterrents in online design learning

Deterrents in online design learning
1. Unable to give immediate/accurate feedback; focus only on delivering lecture material
2. Convenient only for theoretical classes; prototyping and studio work not possible
3. Internet connectivity and power cut issues causes students' disinterest
4. Lack of connection with students due to lack of visibility and linear communication ('feels like talking to a screen')
5. Lack of engagement among students and between instructor and students (inversely proportional to the strength of class)[a]
6. Teamwork or collaborative projects are difficult to accomplish
7. Reduced discussion and critical thinking leading to feeling of lower learning efficiency
8. Personalized/individual attention not possible
9. Attention span of students is reduced, distracted by other things online or at home
10. Requires more effort from tutor and student to engage and ensure learning resulting in longer work hours for faculty

[a]More the number of students in class, lesser the interaction between students and instructor

33.5 Conclusions

When the pandemic struck, most schools and universities had to suddenly shift their on-campus teaching and interactions to the online realm. While some of the courses were seamlessly shifted online, some design courses suffered because of the lack of studio facilities and collaborative work. The aim of this study was to understand the design instructors' perspective on teaching satisfaction and effectiveness in virtual classrooms.

Ten instructors teaching in more than 10 design schools across India were interviewed about their experience in teaching design courses online. While most of the instructors were promptly able to adapt to the virtual classes, course content and duration of interaction had to be modified in considering compatibility of the online platform used and student attention spans for online learning.

Considering the findings in Tables 33.2 and 33.3, the prominent conclusions are:

1. The instructors experienced significant reduction in connectedness with the students online as compared to F2F interaction (ref. Table 33.2. They were unable to give immediate and accurate critique feedback on their work and in-class learning, unable to clarify doubts fully due to lack of time, unable to engage one on one with students. Often in traditional F2F classes, the instructors would slightly alter the pace or content of the class depending on the class vibe or response (perceived by their body language) to accelerate learning and

engagement. But this aspect is usually missing in the virtual classes as most of the students keep their videos OFF. One instructor, however felt no change in connectedness as their institute had mandated to keep the video option ON during classes.

2. The instructors had to put on hold courses or projects that needed workshop or studio use and skill-based learning. All of them agreed that such courses, though are uniquely integral to design education and thinking, cannot be done online. Group work that required collaboration and peer critique lacked student motivation and enthusiasm, though was possible online. Most instructors felt that the studio interactions allow students to enjoy and learn from each other and bring a sense of belonging and connectedness which reflected in their learning and class participation.
3. While virtual classes offered the convenience of teaching and learning from anywhere and save time and energy, allowing the instructors more time to prepare lessons and read, etc., the instructors missed the energy of the classroom and being socially connected on campus. Casual coffee time interactions and passionate discussions with colleagues are seen as an important social element that add to the instructor's experience and learning. Some instructors also felt that discussions and interactions with students on college campuses, whether on or off design subjects, also helped students engage better in the classroom, while most instructors felt that students hesitate to connect online outside of the virtual classes, leaving them frustrated or dejected to struggle with their doubts and questions on their own.

A consecutive study, to understand the students' perspective, is being done to gain a holistic understanding of the online teaching/learning experiences and for the triangulation of the findings here. A salient implication of the current study is the progressive need for institutions looking to engage in online education programmes to resolve to explore and implement practical steps to fully address connectedness and social interaction issues among the students and instructors. For design education in India, this study affirms the reports that a fully online design course is not yet feasible and that a hybrid or blended learning may be a more effective way forward.

References

1. Rettie, R.: Connectedness, awareness and social presence. In: Proceedings of 6th Annual International Workshop on Presence, pp. 1–7 (2003)
2. Lambert, N.M., Stillman, T.F., Hicks, J.A., Kamble, S., Baumeister, R.F., Fincham, F.D.: To belong is to matter: sense of belonging enhances meaning in life. Person. Soc. Psychol. Bull. **39**, 1418–1427 (2013)
3. Martos, T., Kopp, M.: Life goals and well-being: does financial status matter? Evidence from a representative Hungarian sample. Soc. Indic. Res. **105**, 561–568 (2012)
4. Stavrova, O., Luhmann, M.: Social connectedness as a source and consequence of meaning in life. J. Pos. Psychol. **11**(5), 470–479 (2016)

5. Henderson, D.X., Green, J.: Using mixed methods to explore resilience, social connectedness, and resuspension among youth in a community-based alternative to suspension program. Int. J. Child, Youth, Family Stud. **5**(3), 423–446 (2014)
6. Henderson, D.X., McClinton, J.: A qualitative exploration of suspended youths' social connectedness in a community-based intervention program. Child Youth Serv. (2016). https://doi.org/10.1080/0145935X.2015.1083402
7. Shochet, I.M., Homel, R., Cockshaw, W.D., Montgomery, D.T.: How do school connectedness and attachment to parents interrelate in predicting adolescent depressive symptoms? J. Clin. Child Adolesc. Psychol. **37**, 676–681 (2008)
8. Chapman, R.L., Buckley, L., Sheehan, M.C., Shochet, I.M., Romaniuk, M.: The impact of social connectedness on violent behavior, transport risk-taking behavior, and associated injuries in adolescence. J. Sch. Psychol. **49**, 399–410 (2011)
9. Mizan, T.B., Islam, M.A.: Factors influencing perceived community connectedness among university students. Dhaka Univ. J. Biol. Sci. **26**(2), 133–140 (2017)
10. Arya, U.: The rise of MOOCs (massive open online courses) and other similar online courses variants—analysis of textual incidences in cyberspace. J. Cont. Commun. Commun. **6**(3) (2017)
11. Chakravarty, R.: MOOCs in India: Yet to Shine. Int. J. Inf. Stud. Lib. 1(1) (2016)
12. Rieger, M., Turner, R.E., Barrick, R.K.: Student evaluation scores for courses delivered by interactive videoconferencing. NACTA J. **55**(1), 16–20 (2011)
13. Chen, W., You, M.: Internet mediated design course: A framework for using internet on design education. In: Bulletin of 6th Chinese Institute of Design Symposium on Design Institute Research (CD ROM), 6/16-17, Shu-Te University, Kaohsung, no. D06, pp. 243–248 (2001)
14. Park, J.Y.: Design education online: learning delivery and evaluation. Int. J. Art Des. Educ. **30**(2), 176–187 (2011)
15. Wood, A.: You Can't Learn Design Online. TNW—The Conversation. Retrieved from https://thenextweb.com/contributors/2018/03/24/cant-learn-design-online/ (2018)
16. Fleischmann, K.: A successive approach to multidisciplinary teamwork in undergraduate design education: from dysfunctional to functional teams. Arts Des. Stud. **37**, 25–35 (2015)
17. Hastie, M., Hung, I.-C., Chen, N.-S., Kinshuk.: A blended synchronous learning model for educational international collaboration. Innov. Educ. Teach. Int. **47**(1), 9–24 (2010)
18. Szeto, E.: A comparison of online/face-to-face students' and instructor's experiences: examining blended synchronous learning effects. Soc. Behav. Sci. **116**(2014) (2013)
19. Kumar, P., Kumar, A., Palvia, S., Verma, S.: Online business education research: system analysis and a conceptual model. Int. J. Mgmt. Edu. **17**, 26–35 (2019)
20. Bender, D.M.: Developing a collaborative multidisciplinary online design course. J. Educ. Online **2**(2), 1–12 (2005). https://doi.org/10.9743/jeo.2005.2.5
21. Fleischmann, K.: Online design education: searching for a middle ground. Arts Human. High. Educ. 1–22 (2018). https://doi.org/10.1177/1474022218758231
22. Blair, B.: At the end of a huge crit in the summer, it was "crap"—i'd worked really hard but all she said was "fine" and I was gutted. Art, Des. Commun. High. Educ. **5**(2), 83–95 (2006). https://doi.org/10.1386/adch.5.2.83_1
23. Bender, D., Vredevoogd, J.: Using online education technologies to support studio instruction. Educ. Technol. Soc. **9**, 114–122 (2006)
24. Crowther, P.: Understanding the signature pedagogy of the design studio and the opportunities for its tech enhancement. J. Learn. Des. **6**(3), 18–28 (2013)
25. STP: Curriculum Development in Studio Teaching: Volume One, STP Final Report. Studio Teaching Project (2009)
26. Blythman, M., Orr, S., Blair, B.: *Critiquing the Crit: University of the Arts London*. Retrieved from https://www.academia.edu/586074/Critiquing_the_Crit (2007)
27. Bel, D., Smolders, K., Ijsselsteijn, W., De Kort, Y.: Social Connectedness: Concept and Measurement, pp. 67–74 (2009)
28. Vancouver Foundation: Connections and Engagement: A Survey of Metro Vancouver (2012)
29. Riedl, C., Köbler, F., Goswami, S., Krcmar, H.: Tweeting to feel connected: a model for social connectedness in online social networks. Int. J. Human-Comput. Interaction **29**(10), 670–687 (2013)

30. Enez Darcin, A., Noyan, C., Nurmedov, S., Yilmaz, O., Dilbaz, N.: Smartphone addiction in relation with social anxiety and loneliness among University Students in Turkey. Eur. Psyc. **30**(S1), 1–1 (2015). https://doi.org/10.1016/S0924-9338(15)30398-9
31. Özdemir, U., Tuncay, T.: Correlates of loneliness among university students. Child Adolesc. Psyc. Ment. Health **2**, 29 (2008). https://doi.org/10.1186/1753-2000-2-29
32. Lane, F.C., Henson, R.K.: Differences between traditional, transfer, and online college students: measurement invariance of the university attachment scale. In: Educational Resources Information Center (ERIC) Document Number: ED530016. 35th Proceedings from the Southwest Educational Research Association. New Orleans, LA (2012)
33. Bolliger, D.U., Inan, F.A.: Development and validation of the online student connectedness survey (OSCS). Int. Rev. Res. Open Dist. Learn. **13**(3), 1–13 (2012)
34. Abdeldayem, M.M., Aldulaimi, S.H., Aldulaimi, M.L.A.: Virtual learning and students' connectedness in the time of coronavirus. Int. J. Adv. Sci. Technol. **29**(5), 12634–12645 (2020)
35. Anastasiades, P.S., et al.: Interactive videoconferencing for collaborative learning at a distance in the school of 21st century: a case study in elementary schools in Greece. Comput. Educ. **54**(2), 321–339 (2010)
36. Nortvig, A.M., Petersen, A.K., Balle, S.H.: A literature review of the factors influencing E learning and blended learning in relation to learning outcome, student satisfaction and engagement. Electron. J. E-Learn. **16**(1), 46–55 (2018)
37. Glisan, E., Trainin, G.: Online Community and Connectedness. Retrieved 5 Jan 2009, from University of Nebraska Digital Commons website: https://digitalcommons.unl.edu/cehsgpirw/7 https://eduxpert.in/online-education-india/ (2006)
38. Laffey, J., Lin, G.Y., Lin, Y.: Assessing social ability in online learning environments. J. Inter. Learn. Res. **17**(2), 163–177 (2006)
39. Turki, F.J., Jdaitawi, M., Sheta, H.: Fostering positive adjustment behaviour: social connectedness, achievement motivation and emotional-social learning among male and female university students. Active Learn. High. Educ. **19**(2), 145–158 (2017). https://doi.org/10.1771/14697841731202
40. Daves, D.P., Roberts, J.G.: Online teacher education programs: social connectedness and the learning experience. J. Inst. Pedag. 4."Elson Szeto / Procedia Soc. Behav. Sci. **116**(2014), 4250–4254 (2010)
41. Kvale, S.: Doing Interviews. Sage, London (2007)
42. Strauss, A., Corbin, J.M.: Basics of Qualitative Research: Grounded Theory Procedures and Techniques. Sage Publications, Inc. (1990)
43. Garrison, D.R., Anderson, T., Archer, W.: Critical inquiry in a text-based environment: computer conferencing in higher education model. Internet High. Educ. **2**(2–3), 87–105 (2000)

Part III
Design Teams, Collaboration, Communication

Chapter 34
The Infinity Process: A Design Framework for Interdisciplinary Problem-Solving

Sree Mahit Munakala, Chetan Manda, and Sharmistha Banerjee

Abstract Designers are posed with problems that typically require an approach involving knowledge and expertise sourced from multiple disciplines. They often rely on specific predefined design processes to tackle these complex, open-ended problems. These processes were conceptualized as a flexible methodology to approach design problems. However, these design processes are not explicitly designed to facilitate interdisciplinary problem-solving, which is fundamental in tackling most design problems. In this paper, we present an overarching framework for interdisciplinary problem-solving, as observed through an interdisciplinary project involving the hearing-impaired community while interacting with experts from different disciplines. The paper proposes the infinity process, a design framework that can be applied while working in interdisciplinary product design teams. It takes into account the facets of research, design, development, and testing while also addressing the gap of need validation and iteration. The infinity process breaks down the problem-solving process into four distinct phases. Each phase focuses on one aspect of product development. Designers can apply the four phases starting from any phase, thereby eliminating rigidity of the process. Designers may also use each phase of the framework independently, depending on their respective stages of product development. The cyclical nature of the infinity process captures the iterative nature of the design process.

S. M. Munakala (✉) · C. Manda · S. Banerjee
Indian Institute of Technology Guwahati, Guwahati, Assam 781039, India
e-mail: sreemahit@gmail.com

C. Manda
e-mail: chetm1998@gmail.com

S. Banerjee
e-mail: sharmistha@iitg.ac.in

A. Chakrabarti et al. (eds.), *Design for Tomorrow–Volume 2*, Smart Innovation, Systems and Technologies 222, https://doi.org/10.1007/978-981-16-0119-4_34

34.1 Introduction

34.1.1 Design Thinking and Models

Design thinking is an exciting new paradigm to tackle problems in the modern era, most notably in the fields of information technology and business [1]. With a sharp increase in the adoption of design thinking, there is a greater need to find suitable processes to apply this paradigm fruitfully. Several models and processes have been proposed over the years to redefine, formulate, and generalize the design process. There is also a lot of extensive management literature such as the planning and design process [2] and the third-generation process [3]. However, there is no method that can be applied universally, with most models having a relatively narrow focus [4].

Attempts have been made to categorize design models according to the processes or the strategies followed. One particular way to classify them is as analytical, procedural, or abstract models based on the strategy used to arrive at the solution. In addition to these, design models can be classified as problem- or solution-oriented models [4].

A majority of problem-oriented models such as Jones' model [5] and Cross' model [6] (Fig. 34.1) focus on structuring the problem statement instead of generating solutions. This requires the designer to frame a solution-neutral problem statement, which is crucial to arrive as a desired solution solely through logical deduction and not through subjective experiences. However, it is often highlighted that the designer would inherently structure a problem through prior experiences and knowledge, hence influencing the solution. This has been further explored by Hillier et al. [7]. This belief lays the foundation for most solution-oriented design models, some of which include Darke's model [8] and March's model [9] (Fig. 34.2). These models

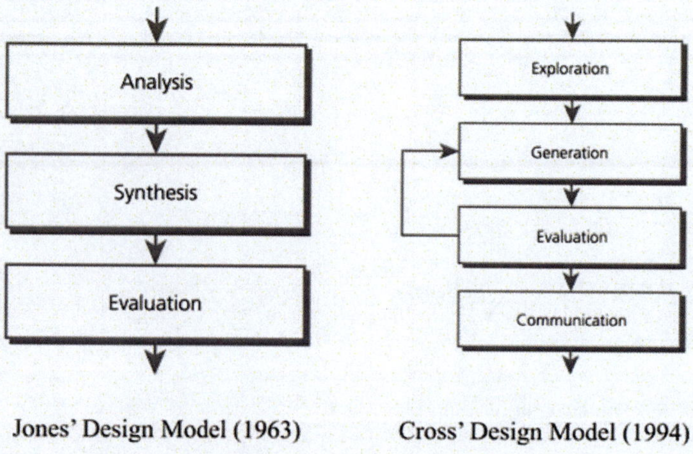

Fig. 34.1 Problem-oriented design models

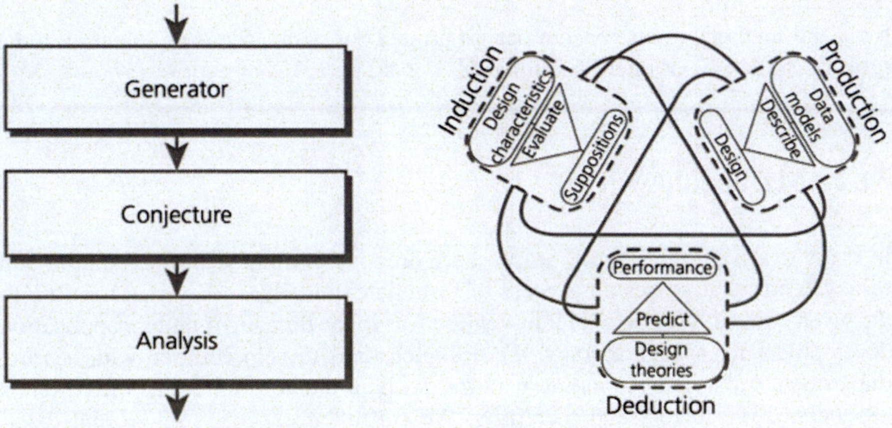

Darke's Design Model (1979) March's Design Model (1984)

Fig. 34.2 Solution-oriented design models

involve proposing an initial solution and modifying it as the problem statement is further explored. Solution-oriented models are considered to be more realistic and better describe the designer's thought process, unlike problem-oriented models.

34.1.2 Interdisciplinary Design

Within the field of design, collaborating with stakeholders across fields is often seen as a proven method to improve the overall outcome [10]. Drawing on skills across a variety of disciplines can lead to alternative approaches to problem-solving and unique design concepts. In a world with problems of rising complexity, the additional perspective can be crucial [11].

Considering the complex nature of interdisciplinary design and the ever-growing need to consider multiple disciplines, allowing room for integrating the nuances that come with interdisciplinary design is important. Design models that are solely problem-oriented or solution-oriented do not account for the multiple stakeholders that are typically involved in any interdisciplinary design task. A need for design models to be hybrid and cater to both linear and cyclic natures of the design process is apparent. Additionally, while design processes that follow both linear and cyclic nature of design do exist, they are not formulated with interdisciplinary design in mind.

There exists a need to formulate a design process that caters to the unconventional and complex nature of interdisciplinary design. In this paper, we propose a design framework called infinity process that enables a designer to use their preferred design tools and methods to approach a design project but with a focus on integrating the process challenges of interdisciplinarity.

34.2 Methodology

In order to design the infinity process, we drew upon a practical intervention and followed an iterative design process to formulate guidelines for generalization. In the given context, the practical intervention was in the domain of mobile application development for speech therapy. We designed tools to help children with hearing impairment train their intonation patterns, with an intent to minimize the patient's dependence on doctors and therapists. We consulted experts from several disciplines, including speech therapy, speech pathology, child care, psychology, computer science, and design throughout various phases of research, analysis, and design. As we completed each stage of the project, the underlying pattern and the process of consulting stakeholders were observed and documented. We incorporated the key insights into the design framework for interdisciplinary design projects, called the infinity process. To know about the project, visit https://youtu.be/ZfvRvBEGSWc.

34.3 The Infinity Process

The infinity process was the result of documenting and curating the process we followed when designing solutions for the hard of hearing community. We organized the process into four distinct and sequential phases. We describe each phase in the context of the project in the subsequent sub-sections.

34.3.1 Phase 1—The Requirement

The requirement phase lays the foundation for how an interdisciplinary project should be initiated by emphasizing an interdisciplinary value generation process. It is crucial to specify the exact problem area that will be tackled as part of the project.

Practical Intervention. Our intention to work with the hard of hearing community arose from our personal experiences of interacting with them. From past experiences [12], we noted various challenges people face within this community. These experiences led us to design solutions that created value for the hearing-impaired. As we were not experts from this field, we realized the need to interact with and consult other experts. We decided to take this project forward as an interdisciplinary project where we would involve several experts from multiple disciplines at different stages of the project. By doing so, we aimed to ensure our final solution would be better suited to the end-users.

What + **How** ⟶ **Result**

Fig. 34.3 Formalized process of problem-solving

Need	+	**Approach**	⟶	**Value**
Is there a problem to solve?		How would we approach it?		Does it generate value for my end user?

Fig. 34.4 Redefined equation for interdisciplinary design projects

Theory Building. Design thinking is fundamentally a formalized process of the underlying reasoning patterns that humans use to solve problems by comparing different known and unknown settings in an equation represented in Fig. 34.3 [13]. In this equation, *What* denotes the crux of the problem at hand. *How* denotes the processes used to solve the problem previously described. Together, they produce a *Result* that is desirable in the given context.

However, this equation best describes an abstract process for problem-solving. We reframed this equation to decrease ambiguity for the designer in the context of an interdisciplinary design project. In this redefined equation (Fig. 34.4), *Need* highlights the importance of the problem and ratifies whether the problem needs to be addressed, by involving stakeholders from multiple disciplines. *Approach* highlights the methods along with the processes when interacting with multiple stakeholders in an interdisciplinary project. Together, they produce the desirable value.

34.3.2 Phase 2—The Need

The second phase of the infinity process looks at how to arrive at a problem statement that will be tackled as an interdisciplinary design project. This phase covers a cyclic process of need identification and validation with the help of stakeholders from various domains.

Practical Intervention. With the intent to create value for the hearing-impaired, we began to highlight various problems that could be tackled with appropriate design interventions. This step involved using several research and analysis methods and interacting with various stakeholders, experts as well as end-users. The stakeholders consulted included children, parents, school instructors, doctors, speech therapists, and medical professionals. These interaction sessions highlighted several pain points that could be tackled. Taking into account the factors of time and available resources, we narrowed down on designing solutions for hearing-impaired children, to ease the process of speech therapy. In particular, we focused on designing tools to improve voice modulation. We re-consulted relevant experts to ensure this was an achievable target and validated the need for such a solution. We learned that voice modulation in itself is too broad and that we should tackle a section of voice modulation and intonation. Further, there was a need for suitable intonation trainers among children undergoing therapy. Having further refined the problem statement to tackle intonation training among hearing-impaired children, we got this problem statement ratified by experts before beginning the next phase.

Theory Building. From experience generated through the project at hand, we thus added the crucial stage of need validation along with need generation. Researching to frame a problem statement leads to a need statement biased towards the designers' preferences and inclinations. However, adding and prioritizing need validation in this phase eliminates this bias by seeking validation from experts within the domain.

Need generation is broken down into two primary objectives: understanding the target audience and identifying pain points (Fig. 34.5). It is suggested that research goals and analysis be geared towards these two objectives as they provide a firm foundation within the domain. Need validation is primarily achieved by involving subject–matter experts from the domain.

Another crucial addition to this process is to convert both need generation and need validation into a cyclic and iterative process (Fig. 34.6). The designer can ensure that the appropriate value is created for the target audience by frequently validating the need for the project. The need phase becomes a cyclic process of *Define* • *Validate* • *Learn* as the designer defines a problem statement that can be tackled, consults relevant stakeholders to validate the problem statement, and redefines it according

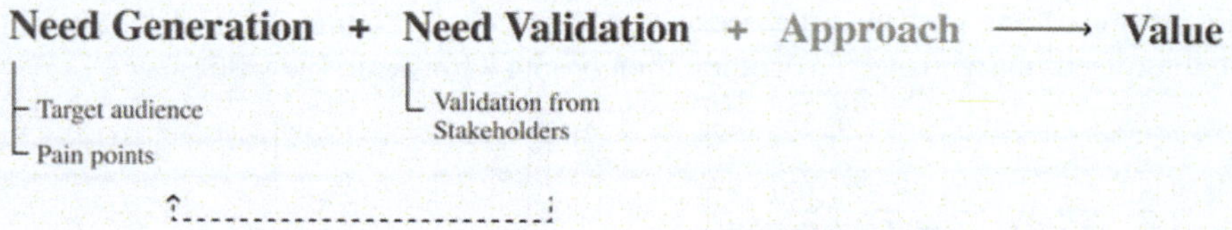

Fig. 34.5 Equation prioritizes need validation in the design process

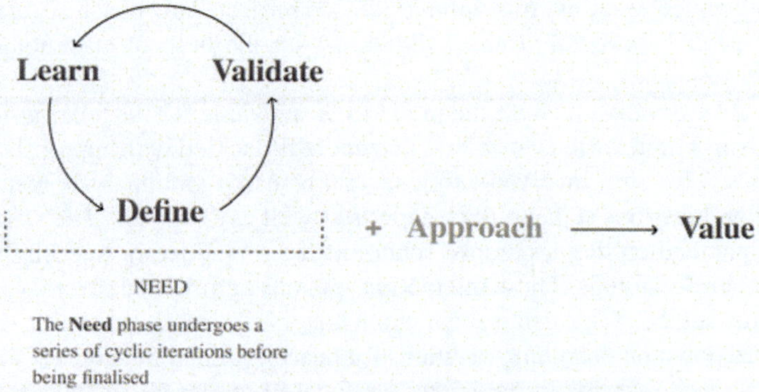

Fig. 34.6 Need generation and validation process becomes a cyclic process

to the feedback received. The cycle is continued until the designer is satisfied that real value will be created by tackling the decided problem.

34.3.3 Phase 3—The Approach

The third phase of the infinity process tackles the methods and processes used to design suitable solutions to the problem statement developed in phase 2. An essential aspect of this phase is the involvement of stakeholders from various disciplines repeatedly. This phase uses the cyclic process of *Build* • *Validate* • *Learn* to arrive at a desirable and feasible solution [13].

Practical Intervention. Designing solutions to tackle the problem involved interacting with experts from multiple disciplines sequentially. This was done to ensure the solution conceptualized is feasible and desirable. As designers, we narrowed down on the medium of designing a mobile-based application based on our understanding of the target audience and the need for improved accessibility. We involved front-end and back-end developers to refine our solution based on the technical feasibility of the idea. Subsequently, we involved speech therapists and speech pathologists in providing appropriate exercises that need to be administered to the children while also creating avenues for testing. As clinical testing was not feasible in the given time, the experts weighed in to refine the solution further. By following a cyclic process of design, development, and validation, the proposed solution was technically feasible and desirable for the end-user.

Theory Building. We broke down the *Approach* phase into two primary dependents: People and Process (Fig. 34.7). *People* include stakeholders from other disciplines, directly and indirectly, involved in the problem area. The domain of design involved and the testing parameters that will be considered directly influence the people involved in this phase. *Process* describes the steps that should be followed when interacting with various stakeholders. The process follows a *Build* • *Validate*

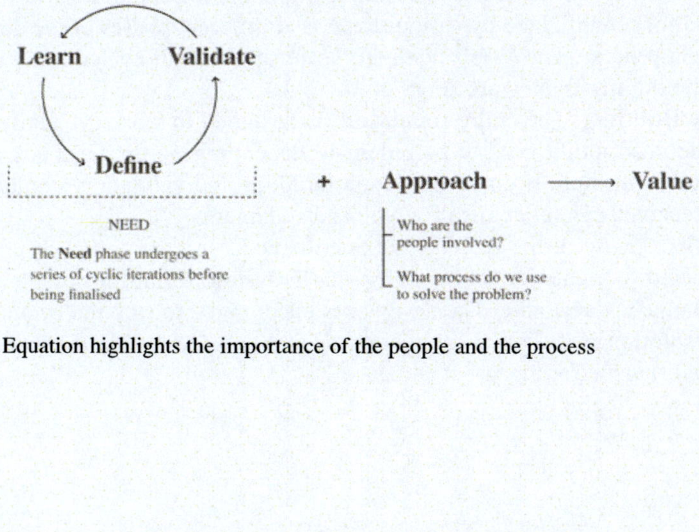

Fig. 34.7 Equation highlights the importance of the people and the process

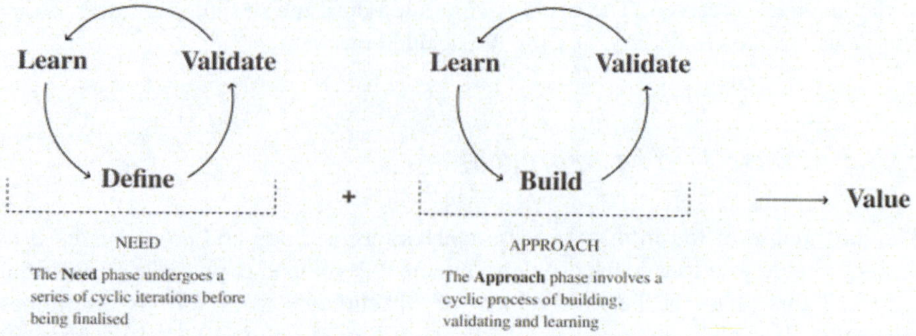

Fig. 34.8 Need and approach phases are both independently cyclic

• *Learn* sequence. We highly recommend this process be followed as it assigns all stakeholders equal importance.

The underlying cyclic process of *Build* • *Validate* • *Learn* is an iterative process, and we encourage the designer to follow this cyclic process (Fig. 34.8). Various stakeholders, across multiple disciplines, should be consulted to validate the solution. The feedback from these sessions is crucial in building a more suitable and robust solution.

34.3.4 Phase 4—The Interdisciplinary Cycle

In the final phase of the infinity process, the need and the approach phases are interconnected to create a cyclic process that is followed until a suitable solution, ready for launch, is crafted.

Practical Intervention. The project went through multiple cycles of need identification, validation, building, testing, and refining. In the process, the design decisions were influenced by the target audience and their pain points along with various technical limitations highlighted by various experts at different phases of the design and development process. The *Need* and the *Approach* phase thus cycled into each other to create the infinity framework (Fig. 34.9).

Theory Building. The entire framework is designed to work cyclically. This is because the need might need to be redefined depending on the feedback received from the other disciplines involved. Upon building and validating a solution, the feedback received can refine the original need statement.

In the final phase of the infinity process, the need and the approach phases are interconnected to create a cyclic process that is followed until a suitable solution ready for launch is available. The designer is encouraged to stop the cyclic process when the solution is designed to create a desirable and feasible (technically and economically) value for the target audience.

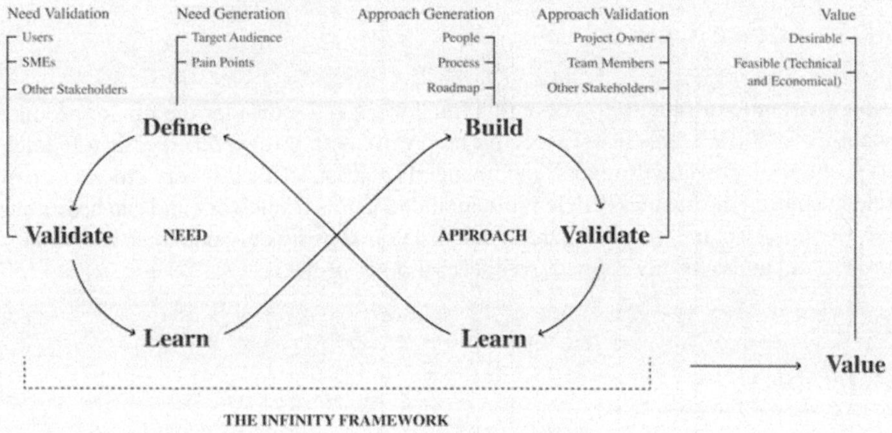

Fig. 34.9 Final equation for the infinity process

34.4 Discussion

The infinity process adds to the wider literature of abstract design frameworks by providing a tool for designers to tackle complex problems in an interdisciplinary setting. It provides a unique hybrid problem–solution-oriented approach that incorporates both the linearity of defining a solution-neutral problem seen in problem-oriented frameworks as well as the generative–iterative nature of solution-oriented frameworks. Furthermore, it allows the room for defining and redefining the need itself, which is the case when you have an interdisciplinary setting with multiple stakeholders redefining objectives and goals through various phases of a project.

Designers can choose to employ this framework starting from any phase, depending on the stage in which their project is currently in, eliminating the need to rigidly follow the process from phase 1. The framework also allows the user to apply each sub-framework individually as required. Finally, the infinity process was designed to be a cyclic and repetitive process with the ability to restart the process when required.

It is important to note that this framework was designed by following the product development process within the field of interaction design. All design decisions and prototyping involved digital interfaces exclusively. However, the framework is application for projects in other fields by suitably involving stakeholders from other disciplines as required. Further, it does not aim to elaborate on what design tools or methods need to be employed while conducting research or conceptualizing solutions. These decisions are left to the designer employing this framework.

34.5 Future Scope

An important consideration within this framework is to consider the role of product viability and how it fares in its respective market. It includes several other factors, such as profitability, revenue streams, and financial forecasts. This last very crucial step in understanding the business angle is not included in this framework and can be further developed. Also, the framework has been developed using only one interdisciplinary project. It can be further refined over several other projects.

References

1. IDEO Design Thinking | IDEO | Design Thinking. https://designthinking.ideo.com, last accessed 2020/04/18
2. Beitz, W., Pahl, G., Grote, K.: Engineering design: a systematic approach. MRS Bull. **71** (1996)
3. Cooper, R.G.: Third-generation new product processes. J. Prod. Innov. Manag. Int. Publ. Prod. Dev. Manag. Assoc. **11**(1), 3–14 (1994)
4. Wynn, D., Clarkson, J.: Models of designing. In: Design Process Improvement, pp. 34–59. Springer, London (2005)
5. Jones, J.C.: A method of systematic design. In: Conference on Design Methods, pp. 9–31. Pergamon Press Ltd. (1963)
6. Cross, N.: Engineering Design Methods: Strategies for Production Design. Willey, New York, Chichester (1994)
7. Hillier, B., Musgrove, J., O'Sullivan, P.: Knowledge and design. Environ. Des. Res. Pract. **2**, 1–3 (1971)
8. Darke, J.: The primary generator and the design process. Des. Stud. **1**(1), 36–44 (1979)
9. March, L.: The logic of design, i Cross, N (red). In: Developments in Design Methodology (1984)
10. West, R. E.: Communities of innovation: individual, group, and organizational characteristics leading to greater potential for innovation. TechTrends **58**(5), 53–61 (2014)
11. McDonald, J.K.: The creative spirit of design. TechTrends **55**(5), 53 (2011)
12. Munakala, S.M., Manda, C., Banerjee, S.: Tackling anger in children with hearing impairment. In: Accepted in the Designing for Children—With Focus on 'Play + Learn' Conference (2020)
13. Dorst, K.: The core of 'design thinking' and its application. Des. Stud. **32**(6), 521–532 (2011)

Chapter 35
Co-designing with Visually Impaired Children

Anupriya Gupta, Lokesh Fulfagar, and Pankaj Upadhyay

Abstract Co-design methodology takes user-centred approach a step ahead by including end-users directly in the design process. It ensures that their needs are well addressed by involving them as true experts of the experience. But when it comes to visually impaired (VI) children, it presents some unique challenges which mostly remain unaddressed with limited research in the domain. The paper addresses this gap by presenting a set of learnings which can assist the development of techniques for co-designing with VI children. We explore the co-design methodology for conducting ideation as part of a broader project centred around indoor navigation for VI children. We begin with a discussion of existing literature around co-design techniques for children and related work done for VI adults. Primary research was further done to understand the routine activities that VI children involve in and the challenges they face. Inspired by the insights gathered from this research, new co-design techniques were ideated. In all, five activities were chosen and conducted during the co-design session with six VI children. Findings and observations from this session are presented in the paper. Activities were found to be useful and thought provoking amongst children. It helped them in thinking and expressing new creative ideas which finally contributed in determining key directions for the solution. Finally, we present our learnings and discuss the best practices for conducting co-design sessions with VI children. These learnings can be further expanded and referred by fellow practitioners to build more innovative and accessible solutions for VI children.

A. Gupta (✉) · L. Fulfagar · P. Upadhyay
Indian Institute of Technology Guwahati, Guwahati, Assam 781039, India
e-mail: anupriya0028@gmail.com

L. Fulfagar
e-mail: lokeshfulfagar@gmail.com

P. Upadhyay
e-mail: pankaj.upadhyay@iitg.ac.in

A. Chakrabarti et al. (eds.), *Design for Tomorrow–Volume 2*, Smart Innovation, Systems and Technologies 222, https://doi.org/10.1007/978-981-16-0119-4_35

35.1 Introduction

The WHO estimates that there are 285 million visually impaired (VI) people in the world, of which 39 million are blind [1]. Also globally, only 1/10 disabled people have assistive products at their disposal [1]. Thus, there is a pressing need for more inclusive design for the future. Co-design is one such co-creation method of designing, where solutions are created in collaboration with the actual users. This hands-on collaboration has been proved to be an effective way of uncovering latent needs and developing effective design ideas for VI [2]. Moreover, involving children as active equal partners has shown to promote a meaningful exchange of ideas leading to better design outcomes [3]. There is a rich literature existing around co-design methodologies, but when it comes to designing along with VI, there are only limited tools available, especially for VI children. In this paper, we explore the co-design methodology with VI children as a part of a broader project concerning effective indoor navigation for VI. We focus on the extent to which existing co-design tools and techniques for children, that are not essentially designed with visual impairments in mind, could be adapted to be used for co-designing with VI children. We build on the existing work to find out effective ways of approaching the set-up, designing the activities and propose a set of learnings which can be used by practitioners seeking to include VI children in their design process. Over the course of the paper, we first discuss the existing literature in the domain of co-design techniques for children and expand it to include the same for VI adults. We conclude with the limited research available for co-designing with VI children and highlight the research gap that we aim to answer. Moving forward, we explain the methodology starting with our primary research followed by the five activities designed for the session. We then explain the details of the participants and set-up before describing the complete co-design session in detail. We summarise our research outcomes into two parts—project insights related to indoor navigation and the lessons learnt from the co-design session. Through this, we seek to empower users with visual impairments to actively participate in the design process and at the same time facilitate the designers and researchers to uncover insights from their experiences and develop more impactful solutions together with them.

35.2 Literature Research

Co-design has become an increasingly popular domain with its unique approach of involving end-users directly in the design process. Different methodologies have been developed by researchers to facilitate the involvement of varied user groups [4] across different phases of the design process including understanding, idea generation, prototyping and evaluation. There is a growing interest towards inclusion of children in the design process. Research has shown children, as co-designers, to possess great potential with their elements of natural playfulness and performativity

[5]. Practising co-design with children brings in a new set of dynamics and thus demands a different set of techniques. Practitioners have evolved varied methods to facilitate the inclusion of children in the design process [6]. The existing toolkit of child–computer interaction also lists specific methods evolved for the same. Nousiainen [7] classifies these methods into five categories—observation-based; (2) narrative; (3) documentation; (4) art-based and (5) game-like. Although there exists a rich research on the tools and techniques corresponding to each category of these methods like storyboarding, prototype modelling, journal reporting, etc., [8] but most of these are based on visual cues which limits its use to sighted children. Even the general framework for participatory design given by Sanders and Brandt [9] is largely based on visual capabilities [10]. However, recent years have seen an evolution of different methodologies for co-designing with VI, focused mostly on adults [11]. Different tools and material kits have been explored to create a tangible experience through different materials, textures and a range of interactions to tap their creativity, for example, e-textile weaving [12], exploring colours through sensory experience [13] and prototyping with methods like blank model prototyping [14]. Also, significant research shows that including visually impaired in the design process leads to uncovering their latent user needs, better design outcomes and more successful adoption with positive outcomes for the involved participants [15]. The scope of the research available specifically for co-designing with VI children is still relatively limited. Most of the research targets specific project areas like voice user interfaces in schools [16], motor rehabilitation [17], development of toys [18] and multi-sensory-assistive technology for education [19]. The research outcomes from these papers are greatly detailed in the context of their specific project domains but are limited to just that individual project. We could not find any set of generalised research outcomes that could be extended to other project areas and be used for conducting co-design sessions with VI children. We build on the existing research, focusing on addressing this gap by formulating a set of generalised learnings based on our co-design approach with VI children.

35.3 Methodology

The work presented in the paper forms part of the ideation phase of a broader project concerning indoor navigation with VI children. The study spans over three phases—research, design and implementation. The following sections discuss these in more detail.

35.3.1 Primary Research

After an initial literature research, a deeper understanding of behaviour and lifestyle of VI children was required for designing the co-design activities. With the same objective, a primary research was conducted with VI children of primary and medium sections of Guwahati Blind School to familiarise ourselves with their situation and the day-to-day activities they involve in, especially navigation. During the full-day visit to the school, observations were made around their activities, the problems they face with those and how they adapt to them with their special abilities. Their mental models and spatial understanding of the space were also understood. A key difference was noted in the same between children who were born blind and late blind. Overall, the interaction with VI people gave us a first-hand experience about who are the people we are designing for and provided an understanding required for the activity design.

35.3.2 *Activity Design*

Based on our literature review and understanding from primary research, we formulated three factors to keep in mind while designing for VI participants—first, ensuring accessibility, to keep them suitable for children with varying vision capabilities; second, using simplified robust tools and set-up considering the limitations of budget, mobility, and interaction dynamics; and the third around keeping the activities fun and still enabling enough freedom of expression and focus towards the end goal. Further, based on these, we conceptualised five activities for our co-design session.

Spatial Mapping (Individual). The spatial mapping activity involves the VI children in describing the surroundings they live in. In our context, this included their current school campus with various blocks such as mess, classrooms and hostel rooms. The aim of this activity is to understand their mental models and spatial understanding by gaining deeper insights into how they perceive their surroundings and in what ways do they remember and describe them.

Incident Sighting (Individual). The second activity revolves around narration, and it triggers the students to think and talk out aloud about their specific past experiences. For our project, we asked them to talk about their usual day in school premises, day out experiences, local visits and home trips. The aim of this activity is to use their anecdotes of past experiences to learn maximum possible in a short limited time.

Futuristic Scenario Ideation (Group). This is a group activity which involves kids in an ideation session in a futuristic scenario described by the facilitator. They are made to imagine their lives in a futuristic world and then probed to ideate on the solutions in the form of futuristic super gadgets that they would love to have. They are motivated to go wild by adding magical powers to those gadgets and further think of details in terms of its features and working. The aim of the activity is to push the limits and help them think out of the box. For our activity, the children were probed to ideate on super gadgets that would help them in navigation.

Magic box (Group). This activity continues from the previous one and aims to involve VI children to co-create solutions translating their ideas into tangible prototypes. A set of basic building tools are shared with the children using which they give shape and structure to their ideated concepts by creating a dummy prototype. In our activity, we used mechanics blocks [20] to build their ideated super gadgets with magical powers from the previous activity. To ease the activity for children, we prepared some elementary structures using those blocks which could be easily attached or detached in different orientations and combinations to create varied overall structures.

Ideas in Action (Individual). This is the concluding activity which takes the prototype developed in the previous activity and puts it into action. A task is framed by the facilitator which the children have to complete using their dummy prototype. Techniques like Wizard of Oz [21] can be put to use here to facilitate the working of the prototype. The core idea behind this activity is to see the prototype in action and observe the ways in which VI children interact with them.

35.3.3 Co-design Session

Participants and Environment. Students of the middle and secondary section of Guwahati Blind School were approached for voluntary participation. Three boys and three girls, of age 12–16 years, were chosen randomly from the set of students who volunteered. In conversations prior to the session, the contact described the participants as smart, active and mid-level tech-proficient. The contact also briefed us about their language limitations. All of the six participants were late blind with four partially sighted and two completely blind, as self-reported by them. Table 35.1 provides the participants details.

Henry [22] gives some recommendations while deciding an environment for a co-design session with VI. Participants need to adapt to different aspects of the environment like the structure of the room, the furniture present and their relative spatial

Table 35.1 Co-design participant details

ID	Age	Gender	Level of vision	Language	Hobbies
P1	15	Female	Very limited	Assamese	Singing
P2	16	Female	Limited	Assamese, Hindi	Singing, playing instruments
P3	13	Female	Limited	Assamese	Games, reading
P4	15	Male	Limited	Assamese, Hindi	Singing, technology
P5	13	Male	No vision	Assamese, Hindi	Computer games, cricket
P6	12	Male	No vision	Assamese	Cricket

positions. To keep the participants at ease and minimise any kind of difficulty in adaptation, we chose one of the classrooms in the school itself as the session location. This also helped in avoiding the inconvenience involved with transportation for the participants, as they reside in the school hostel inside the campus. The session did not involve any monetary incentive for participants, and chocolates were distributed at the end of the session. Due ethical approval was taken from Guwahati Blind School and each participant, in the presence of a familiar person with no visual impairments. Along with the participants, there were five volunteers in all to help in organising and recording the session. These volunteers were students of design department from Indian Institute of Technology, Guwahati. Volunteer V1 played the role of facilitator, V2 and V3 helped in organising the activities, V4 was responsible for taking notes, while V5 recorded the entire session. An additional sighted volunteer, V6, from the school authority helped with translations, wherever necessary, throughout the session. An initial briefing was also done to explain the volunteers about the project, their roles, the guidelines and code of conduct to follow throughout the co-design session.

Procedure. The co-design session was divided into six parts constituting an introduction session and five activities as discussed in Sect. 35.3.2. Once the participants were seated comfortably, volunteers introduced themselves in their own voice for their oral identification. It was followed by a project and session overview by V1. They were also explained about the recording apparatus and its importance, following which their consent was taken in the presence of a familiar sighted person. Though they were accustomed to the classroom already, a detailed outline of the room and relative positions of furniture and volunteers were explained. The entire session was recorded with a microphone integrated camera and two additional microphones.

The five co-design activities as explained in Sect. 35.3.2 were conducted sequentially. After the second activity, the groups were divided into two with three participants and one volunteer, each. A few tweaks were needed to be made for the magic box and ideas in action activities. We observed an initial difficulty faced by the participants while using the mechanics blocks. Thus, volunteers had to be actively involved in assisting the respective groups. While the participants described the structure and its physical features, volunteers undertook the task of attaching and detaching blocks to match their description. Participants were made to feel the evolving structure every now and then to ensure that their descriptions were effectively translated to the physical prototype. Both the groups built one prototype each (Fig. 35.1). After thorough explanations by both the groups, a voting session was conducted. Based on the voting, magic wand (Fig. 35.1b) was chosen as the favourite and selected for the final ideas in action activity. In this activity, the task was to navigate from the classroom to a defined point in the school garden, initially unknown to the participant, using the smart gadget prototype. Due to time constraints, the task had to be limited to two participants. Detailed instructions were given to P2 and P5, who volunteered for the task. Being a dummy prototype, Wizard of Oz technique was employed, wherein one volunteer enacted to substitute the super powers associated with the prototype while the other volunteer assisted the participant throughout the task. Participants

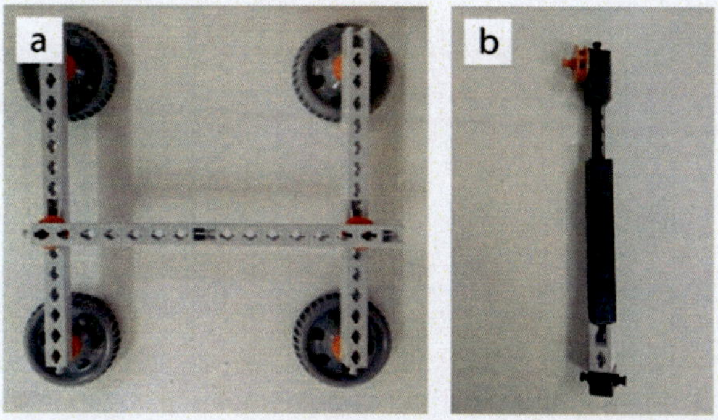

Fig. 35.1 Prototypes: **a** super ears by group 1: use sonar to predict route and obstacles. It can detect location of the person to whom the user wants to go to, by hearing their sound; **b** magic wand by group 2: guides the user by beeping when pointed in the right direction. User can also get the magic wand directly in their hands by calling it when it is not near

Fig. 35.2 Children participating in different activities of co-design session

considered the complete task as a mission and seemed to enjoy every bit of it. On completion of the task, each participant was given chocolates to mark their success. The activity took around 10 min each to complete, and the whole co-design session spanned around 3 h.

Following the activities, a small informal feedback session was conducted to receive participants' views focusing on the ideation and co-creation part. Detailed feedback was also taken from P2 and P5 for the last activity about what they thought of the magic box after putting it into use and what they would like to modify about it. After the session, all the volunteers were thanked for their valuable support, and detailed feedback was received from them (Fig. 35.2).

35.4 Research Outcomes

All the data recorded throughout the session was collated for a thorough analysis. The outcomes derived after complete analysis were broken into two major domains—project insights and research insights. The prior included the ideas and insights critical in conceptualising the final solution as part of the project, while the latter consisted of key learnings from the co-design session.

35.4.1 Project Learnings

Other than the ideas generated in the ideation process as part of the co-design session, there were some key insights observed from their behaviour and interaction methods that formed the basis for conceptualisation of an indoor navigation tool. An initially identified problem was the inability of the VI to walk in a straight line or any specific direction independently without any aid. It was observed during the activity that the VI with their heightened sense of hearing could interpret the position and direction of a sound source and could even follow the same correctly. In fact, it was also highlighted in their navigation anecdotes that they use this technique of snapping fingers or clapping to convey their position to each other while navigating together. This insight formed the basis of the solution which used the concept of 3D spatial sound input to provide directions to the visually impaired.

The other important insight was that the participants felt at ease when accompanied by a friend while navigating. They used to form chains by either holding hands or keeping one hand on the shoulder of the other person. In fact, they had their own navigation partners and mostly moved around the school in pairs. Even most of their ideas included their desire for independent travel, but still featured a presence of a partner. This highlighted the need for a companion to help them in navigation as well as build an emotional connection with them, which is something missing in all existing solutions. It directed the solution to incorporate a virtual persona which played a dual role of a guide and a partner.

These ideas and concept directions clearly originated from the insights discovered in the co-design session and would have been not possible otherwise. This highlights the value of including the end-users in the idea generation process, especially when those users have different and special abilities than the ones who are designing it.

35.4.2 Research Learnings

Based on the learnings from our study, we discuss here the best practices for designing activities and conducting co-design session with VI children.

Focus on active senses. While designing the activities, one should be very careful to not include any visual cues. This might seem like an obvious point at one glance but it is very crucial and any sighted designer or researcher might overlook this and miss it. The visually impaired often have enhanced abilities in other senses as compared to sighted people. Thus, activities should focus on interactions based on other senses like touch and sound. One could also explore taste and smell depending on the activity.

Understand the diversity of your participants. The key for designing the activities is to understand the abilities and limitations of the audience it is being designed for. The participants might have a varied range of visual impairment ranging from partially impaired vision to completely blind. Some might be born blind while others can be late blind. Differences like these bring differences in basic mental models and capabilities. Activities should adapt to these differences and align with the audience. Language is another important factor that should be considered beforehand. Be well aware of the participants' first language and make sure to have a translator if required.

Choose empowering tools, environment and dialogue. Throughout the session, conscious efforts should be given on avoiding any instance that can potentially highlight the differences in abilities between sighted and VI. Participants should feel empowered and in control in all aspects from tools to environment to language. The tools used should involve active senses and be familiar or easy to adapt by the participants. It is important that they feel at ease in the space they are in, and the best way to do it is by choosing an environment set-up familiar to them. All forms of communications should be kept simple and comprehensible for VI children. Constant communication about the time and status of the activity should be done to make them feel in control. Also, one should be careful in using appropriate terminologies and avoiding phrases related to sight like 'if you look at this', 'if you point it in this direction', etc.

Ensure equal participation. In case of a group activity, the dynamics in the group could significantly influence the results. The facilitator should ensure that everyone participates in the activities equally. In case of mixed audience with different capabilities, the distribution should be balanced well, so that everyone gets the opportunity to voice their opinions and thoughts. Volunteers and facilitators should ensure that their role in the activities is just passive and should avoid getting involved directly. One should be really cautious that they are neither directing the conversation nor interpreting the results only to confirm pre-existing beliefs. Also, keep the count of the participants and volunteers less, ideally somewhere between 5 and 8. Larger groups can cause confusion as it becomes difficult for the VI to identify speakers based on voice.

Enable them to speak and express. Through the activities, bring out their thoughts and perspectives in order to understand the actual needs. One needs to carefully draft the conversations and activities in such a way that it pushes them to communicate and express their thoughts constantly. The tools used in the activities should also invite active engagement and interaction during the activity.

Be observant and record meticulously. It is very crucial to be observant throughout the session and ensures a thorough audio–video recording of the entire session. It is also a good practice to ask volunteers to keep adding their own observations as notes to ensure all the observation points are recorded. Having said that, one should also ensure that the recording apparatus is explained to the participants, and their consent is taken before any type of data collection.

35.5 Discussion

Through this research paper, we explored the relatively unaddressed domain of engaging VI children in the co-design process. The exploration was conducted methodically with proper research and understanding and turned out to be fun and successful, where all the participants enjoyed the activities and designed creative experiences that led to development of an innovative solution and paved the way for the learnings as research contribution. Outcomes of the research include unbiased rationalised observations which would be helpful for practitioners engaging in a co-design session with VI children. An ethical dilemma that we faced over the course of the study was keeping the expectations realistic with the participants. During the session, they got overly enthusiastic about the technological prospects but the realisation that a research project might not always lead to a commercially available product, made them a bit disappointed. It can be an interesting course to dive deeper in understanding this ethical dilemma and the practices to address the same. Also during the prototyping session, we observed that the participants faced difficulty in interacting with the mechanics blocks and volunteers had to involve to help them specifically in building the prototype. This highlights the importance of using appropriate tools that the participants are already familiar with or are analogous to what they already use. The research was primarily focused around one centre and a limited set of children, all of whom belonged to the local community of Guwahati, India. It can be an interesting domain to explore how the cultural differences might influence the findings. Since it was an exploratory research with limited scope in terms of time and settings, the findings might not be completely exhaustive. However, it is still a significant starting point for enabling designers and researchers to conduct their co-design session with VI people, especially children. In turn, we hope this empowers users with visual impairments to voice their needs and motivations better, by becoming an active contributor in the development of ideas and products designed for them. We believe our contributions will help to address the gap in the domain and enable the community to build on this research to deliver more accessible and innovative design contributions in collaboration with the visually impaired.

References

1. World Health Organization: https://www.who.int/health-topics/assistive-technology#tab=tab_1. Accessed 16 Apr 2020 (2018)
2. Sanders, E., Stappers, P.: Co-creation and the new landscapes of design. CoDesign **4**, 5–18 (2008)

3. Hansen, A. S.: How to best communicate with and encourage children during a design process. In: Co-Design with Children (2019). https://pdfs.semanticscholar.org/ae0e/89081be616bfaea307b6de73c3660d9c75af.pdf Accessed 12 Apr 2020
4. Lindsay, S., Jackson, D.: Engaging older people using participatory design. In: Proceedings of the SIGCHI Conference on Human Factors in Computing Systems, pp 1199–1208 (2012). https://doi.org/10.1145/2207676.2208570
5. Vaajakallio, K., Mattelmaki, T., Lee, J.J.: It became Elvis: co-design lessons with children. Interactions **17**, 26–29 (2010). https://doi.org/10.1145/1806491.1806498
6. Guha, M.L., Druin, A., Chipman, G.: Working with young children as technology design partners. Commun. ACM **48**, 39–42 (2005)
7. Nousiainen, T.: Children's Involvement in the Design of Game-Based Learning Environments, Ph.D. Dissertation, University of Jyvaskyla, Finland (2008)
8. Giaccardi, E., Paredes, P., Díaz, P., Alvarado, D.: Embodied narratives: a performative co-design technique. In: Proceedings of the Designing Interactive Systems Conference, DIS'12 (2012). https://doi.org/10.1145/2317956.2317958
9. Sanders, E., Brandt, E.: A framework for organizing the tools and techniques of participatory design. In: Proceedings of the 11th Biennial Participatory Design Conference, pp. 195–198 (2010). https://doi.org/10.1145/1900441.1900476
10. Andrews, C.: Designing with people with visual impairments: an exploration of the value of extraordinary users in design. PhD Dissertation, Cardiff Metropolitan University (2016)
11. Branham, S., Kane, S.: Collaborative accessibility: how blind and sighted companions co-create accessible home spaces. In: Proceedings of the 33rd Annual ACM Conference on Human Factors in Computing Systems, pp. 2373–2382 (2015). https://doi.org/10.1145/2702123.2702511
12. Giles, E., Van der Linden, J., Petre, M.: Weaving lighthouses and stitching stories. In: Proceedings of the 2018 CHI Conference on Human Factors in Computing Systems (2018). https://doi.org/10.1145/3173574.3174044
13. Giugliano, G., Veneziano, R.: Open design approach, a project for inclusion of blind people. In: Cumulus Hong Kong Proceeding, pp. 237–240 (2016)
14. Arnowitz, J., Arent, M., Berger, N.: Effective Prototyping for Software Makers. Elsevier (2010)
15. Frauenberger, C., Makhaeva, J., Spiel, K.: Blending methods: developing participatory design sessions for autistic children. In: Proceedings of the 2017 Conference on Interaction Design and Children, pp. 39–49 (2017)
16. Metatla, O., Oldfield, A.: Proceedings of the 2019 CHI Conference on Human Factors in Computing Systems, pp. 1–15 (2019). https://doi.org/10.1145/3290605.3300608
17. Magnusson, C., Caltenco, H.: What do you like? Early design explorations of sound and haptic preferences. In: Proceedings of the 17th International Conference on Human—Computer Interaction with Mobile Devices and Services Adjunct, pp. 766–773 (2015)
18. McElligott, J., van Leeuwen, L.: Designing sound tools and toys for blind and visually impaired children. In: Proceedings of the 2004 Conference on Interaction Design and Children: Building a Community, pp. 65–72 (2004). https://doi.org/10.1145/1017833.1017842
19. Cullen, C., Metatla, O.: Multisensory storytelling. In: 17th ACM Conference on Interaction Design and Children (2018). https://doi.org/10.1145/3202185.3210775
20. Engino Discovering STEM Mechanics Kit: (2020). https://www.engino.com/w/index.php/products/stem. Accessed 12 Apr 2020
21. Dahlback, N., Jonsson, A., Ahrenberg, L.: Wizard of Oz studies—why and how. Knowl. Based Syst. 258–266 (1993). https://doi.org/10.1016/0950-7051(93)90017-N
22. Henry, S.L.: Just Ask: Integrating Accessibility Throughout Design, pp. 1–177. Breinigsville, PA, USA (2007)

Chapter 36
Design-Audited Mass Communication Model

Suresh Goduka and Amarendra Kumar Das

Abstract Assam is world's largest tea growing region, even as the tea industry labourers lag behind in human development parameters, suffering high magnitude of malnutrition, infectious diseases, superstition. This research paper tries to understand and analyse the communication process prevalent among them, whether various communication campaigns have effectively served necessary awareness messages to them, whether existing communication models are adequate for them and what design intervention can do for a better tomorrow of such population by exploring the scope of designing a new model. A usability study examines the characteristics of their interaction with media and processing of messages and meanings. The gap in communication message designs indicates the need for a new approach to deal with the unique challenges. An intervention study has been conducted. A new and exclusive model is proposed for them: a model of mass communication to be audited on the principles of design and contextual usability.

36.1 Introduction

[**Terminology**: *Communication Design*: Design efforts with communication objectives. *Visual Communication Design*: Design applications in communication using visual channels. *Mediated Communication*: Communication that mediates through interaction between screens, spaces and persons. *Mass Communication*: Communication directed to the mass of the people.]

About four million people live within and around *labour lines* or *coolie lines*, the designated settlements for labourers working in Assam tea industry in northeast India. They are descendants of 96 tribal communities migrated from Jharkhand, Chhattisgarh, Odisha, Telangana, and other regions, and collectively known as *tea*

S. Goduka (✉) · A. K. Das
Indian Institute of Technology, Guwahati, India
e-mail: sgoduka@iitg.ac.in

A. K. Das
e-mail: dasak@iitg.ac.in

A. Chakrabarti et al. (eds.), *Design for Tomorrow–Volume 2*, Smart Innovation, Systems and Technologies 222, https://doi.org/10.1007/978-981-16-0119-4_36

tribe at present. They pluck tea leaves for one-sixth of global tea production. There are visible gaps in standards of living of this tribe in comparison with other communities living in proximity.

The Assam Human Development Report has revealed that the health status of tea garden labourers is much below the state average [1]. Child Rights Situational Analysis Study by Save the Children [2] has stated that the rate of missing children especially girls and child trafficking is high in the tea gardens yet the reporting rate is less. 63% children in the tea gardens are reported to have dropped out from the school. Majority of the children reportedly start working at the age of 12–14 years. Girls are either involved in the tea gardens or engaged looking after their siblings while the parents are away for work. Child mortality is also observed, and reasons are not known to the labourers. 64% female and 40% male labourers are illiterate. Majority of the labourers are female, and 80% of them have more than 5 children [3]. Even after having a facility at home, 38.9% do not use toilets. 16.2% wash hands with ash, mud and plain water after defection. Only 4% do filtration and 37.5% boil water for drinking. Source of drinking water is accessible to domestic animal in 17.5% cases [4].

A study by Indian Council of Medical Research (ICMR) has revealed high magnitude of undernutrition and infectious diseases among tea garden population of Assam. The study revealed high burden of communicable, non-communicable and nutritional disorders among the tea tribe. Symptoms related to anaemia and associated morbidities are present amongst 75.8% adolescent girls [5]. Malaria, dengue, encephalitis and other epidemics kill them in hundreds every year. Superstition and witch-hunt have taken many lives. There have been several mob lynching acts by the tea industry labourers on the management or doctors serving in their tea estates. Around 69% of them spend more than 50% of their income on country liquor and gambling. Only 18% of them know about welfare schemes available for the labourers.

Posters, loud speaker announcements, printed leaflets, special programmes on radio and television are part of the communication campaigns created in the state language and disseminated by the governments and the tea industry managements. Such campaigns include those on sanitation, nutrition, maternal health, adult education, banking, drinking water filtration, family planning, waste disposal, educate girl child, prevention of alcoholism, skill development, disaster preparedness and such other themes.

This leads us to the following **research questions**:

(1) Whether existing models are adequate in community-context-based communication? (2) How design intervention can help mass communication models? (3) Whether a design-audited mass communication model can be proposed for improving conditions of communities such as the Assam tea industry labourers?

The design literature contains models that represent the relationship between how designers intend artefacts to be experienced and how they are subsequently experienced by the users. In de Souza's model of human–computer interaction, a distinction is made between two communicative roles of software products: they are messages sent from designers to users through the computational medium, but they are also "message senders and receivers at the immediate interface level". In addition to representing software systems as media, de Souza [6] represents users as interacting with those media by providing them with "input codes" and interpreting their "output codes". Thus, the users' interpretation of the artefact leads to action, and this action causes the artefact to exhibit some changes in it. This change is then perceived and interpreted to allow further action. Krippendorff and Butter's [7] more general design model situates the product within a context of use and represent consumers or users as manipulating not just the product, but the context too, and receiving feedback from both.

We can look at the communicative aspects of design from two different but related perspectives: first, from a perspective that views design as interpersonal communication and second, from a perspective that views design as mass communication. Research into mass communication has not been benefited by the knowledge and methods from neighbouring fields such as interaction design, human–computer interaction, service design and user experience (UX) design. Mass communication models give not much room for user involvement, contradicting much of the consensus in contemporary design research. Models remain ineffective unless the approaches are user-centric; unless they are understandable, meaningful and desirable by the users. There lies the gap. Information campaigns cannot achieve effects without understanding how audience members process messages [8].

The traditional communications models hold that mass communication is a one-to-many process, which presume the growth of a passive mass "audience" common in their tastes with respect to the information being transmitted rather than heterogeneous users seeking varied experiences. Later researchers brought attempts to isolate systematic receiver differences as the focal point of message design. We reviewed features of some of the communication models that are relevant to the discussion.

Shannon's [9] model of communication, with its representation of "communication as transmission", has strongly influenced communication theory generally and design theory specifically. It has encouraged scholars to represent artefacts as the transmitter of a message that is subsequently decoded by the users. "The designer transmits a message to the user by using the product itself as the device of transmission". Although Shannon's model is one of the most commonly cited representations of the communication process, its structure and language are often claimed to offer an inappropriate foundation for design thinking. In particular, the notion that artefacts act as transmitters has been strongly criticised for casting users of the artefact in an overly passive role [10]. Instead, users approach artefacts with their own motivations, experiences and expectations, and therefore artefacts will be interpreted in different ways by different people in different contexts, or as per the way that artefact relates to its surroundings.

Newcomb's model [11] provides a triangle with a simplistic indication of how people and systems in a society attempt to maintain equilibrium within the social system. This "fundamentally different" model was the first to introduce the role of communication in a society or societal relationship [12]. While subsequent studies have not supported the concept, the fact of the process continues in much of the communications literature albeit frequently now disguised as a "multi-step flow" [13]. Katz and Lazarsfeld in their two-step flow of communication were also

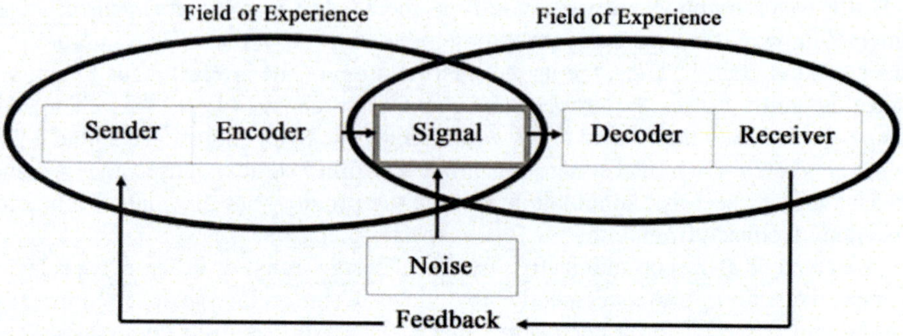

Fig. 36.1 Schramm's model. *Source* The process and effects of mass communication

concerned with the part played by people in the flow of mass communication in their study of the "diffusion of opinion". They first named the "personal influence" paradigm, codified it and brought it to the centre of the field [14]. This perspective recognized that informal social relationships played a significant part in modifying the manner in which given individuals acted upon a message that came to their attention via the mass media. They referred to these perspectives as interpretive frameworks on how people encounter the media. The psychosocial and other factors became incorporated in subsequent communication models which led to a proliferation of increasingly sophisticated versions of the linear, stimulus–response model or cause–effect interaction. With a concern for how mass audiences interpret these messages, Schramm [15] depicts many individual "receivers", each responding to one of many identical messages. Each of these receivers is connected to a group whose other members may or may not have received the same message, but who nevertheless interact with each other. As they interact, they respond to each other and to each other's interpretations, and this prompts reinterpretation of the message (Fig. 36.1).

The classical Westley and MacLean's model of mass communication [16], that applies to all artefacts to which people attach meanings, presents a "gatekeeper" phase, a one-way filtration for encoding from source, but no filtration for decoding at destination. The eventual users of the messages have no direct access to the message creators or designers. The processes by which the communicators' intentions are translated into artefacts are therefore only loosely connected to the processes by which users interpret those artefacts. It does not properly account for activities such as participatory or collaborative design.

"In Maletzke's model of the mass media, those who construct and those who interpret messages are each depicted as holding an image of the other party. This may influence the intentions that are held, the messages that are constructed, the media selected and the interpretations that are formed" [17]. The creative process of message design is depicted as cyclic; communicative intentions shape the message, but the message and the medium by which it will be expressed both inform intentions (Fig. 36.2).

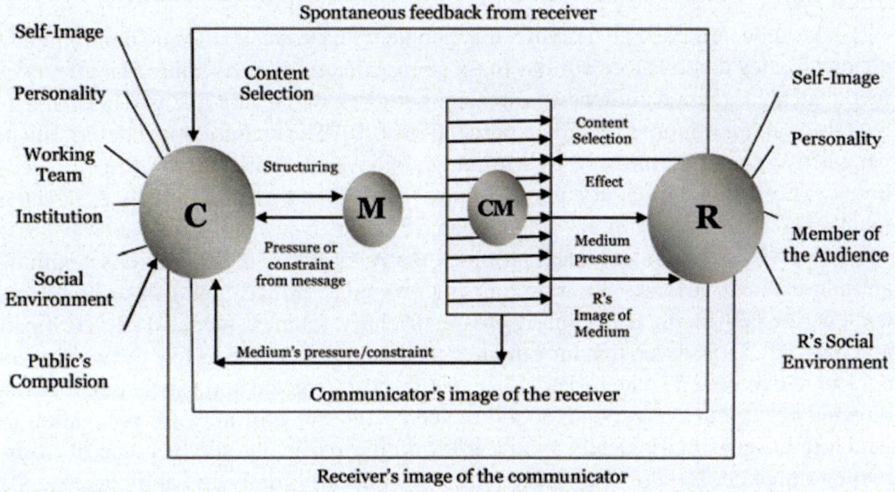

Fig. 36.2 Maletzke's model. *Source* Psychology of mass communication: theory and systematics

Lars' [18] alternative model is intended to highlight the essential communication entities and their interrelations, and potentially to cover all conceivable kinds of communication of meaning. It is designed to account for both verbal and nonverbal meanings, the different roles played by minds and bodies in communication, and the relation between pre-semiotic and semiotic media features. By avoiding the notion of code as a collective formula for meaning creation, this model claims to have the capacity to deal with all sorts of cognitive import. The visual diagram of this new model contains three entities of communication: (1) Something being transferred: cognitive import. (2) Two separate places between which the transfer occurs: producer's mind and perceiver's mind. (3) An intermediate stage that makes the transfer possible: media product. Additionally, it displays four essential interrelations among these entities: (1) An act of production "between" the producer's mind and media product. (2) An act of perception "between" the media product and the perceiver's mind. (3) Cognitive import "inside" the producer's mind and the perceiver's mind. (4) A transfer of cognitive import "through" the media product (Fig. 36.3).

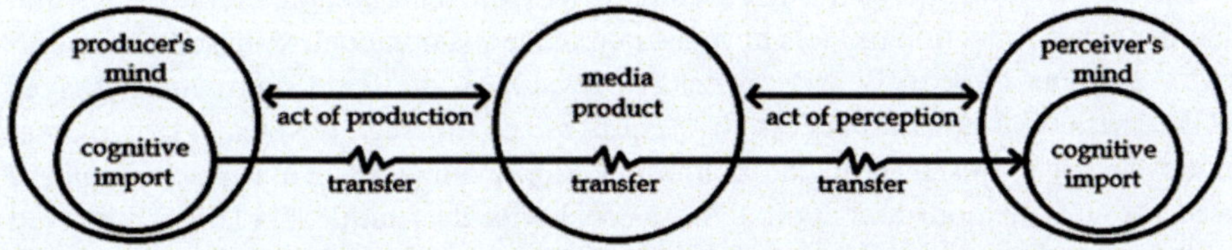

Fig. 36.3 Lars' model. *Source* A medium-centred model of communication. *Semiotica, 3* (2)

Blake and Haroldsen [19] predict that people with similar social and demographic characteristics demonstrate similar mass communications behaviour. The effects of mediated messages on audiences' intensions and behaviours are likely to be channelized through their more immediate perceptions [20]. The technological determinism perspective of Innis, followed by McLuhan, believes that the nature of media technology prevailing in a society at a particular point in time greatly influences how the members of that society think, act and behave [21].

With the rise of the Internet technologies, many "barriers" of the previous mediums are non-existent now. As the source is empowered to directly publish or broadcast content, the new media is becoming "a user-friendly yet mass media" [22]. Hoffman and Novak [23] observed that in computer-mediated environments like the web, users engage in two general categories of behaviour: goal-directed and experiential. Goal-directed behaviour corresponds to a directed search mode of network navigation in which the user is extrinsically motivated to find a particular site or piece of information on a site. Experiential behaviour is intrinsically motivated and corresponds to a non-directed, exploratory search mode. They suggest an alternative model of many-to-many mediated communication. In their new mediated model, information or content is not merely transmitted from sender to receiver, but instead, mediated environments are created by the participants and then experienced.

Internet, with its decentralized technical structure, facilitates interactivity and individuality among its users herald the arrival of what is described as the "second media age" [24].

With the advent of such overpowering new media, emerged concepts such as *communication design, media design, new media design, aesthetic design.* The concept communication design, closely related with *visual communication design* and *graphic design* includes the work of numerous scholars who use design as a mode of media research besides media innovation and development outside of academia, in particular in the media industries [25].

Sullivan and Carr [26] observed that with the emergence of powerful and flexible digital communication tools, individuals are using communication technologies in ways that expand the intersection of interpersonal communication and mass communication, calling for new frameworks. They introduced the mass personal communication model incorporating two dimensions—perceived message accessibility and message personalization—that link mass communication and interpersonal communication and redefine each independent of channel.

Having now considered the discussions so far from the available literature, we find that the communication models have remain inadequate in context-based communication that can effectively serve necessary awareness and empowerment messages to the users who have no access to the process of message designs. Their mental model is not represented in the communication campaigns. This gap indicates a need for new methods and approaches to deal with the unique challenges involved in designing for media. We therefore examine the scope of design intervention in mass communication models and explore a new model for improving conditions of communities such as the Assam tea industry labourers.

36.2 Methodology

36.2.1 Observation of Five Human Factors

For a primary research on how the tea industry labourers behave with the existing mass media tools as activity and as part of the society, we went into the tea garden areas in Charaideo District of Assam, India. We observed them in their own settings and took down notes and categorized the observations into five human factors: physical, cognitive, social, cultural and emotional.

36.2.2 Usability Study

We conducted a pilot survey among the tea estate managements. They are the owners, patrons or organizers of various communication channels in the tea industry areas. The survey examines the status of the tea industry labourers' mass communication consumption and processing, as well as the scope for a new model, from the management point-of-view. A questionnaire [27] was designed as a tool for the study. The selection of respondents is performed by means of purposive random sampling. Twenty-four managers and deputy managers from Assam took part in this survey.

36.2.3 Intervention Study

We experimented with a reception-based intervention among the tea labourers' community. For awareness on COVID 19 and lockdown, we prepared an audio, and subsequently a video [28]. The language was tea tribal. All visuals, colours, faces, ambience reflected the tea tribe. The narration was conversational, the way they talk among themselves, although it contained information prescribed by World Health Organization (WHO). The audio and the video were sent to tea industry labourer associations and groups for wider dissemination. They were sent an e-mail asking them to listen and view the audio and the video. They were asked to respond to a few questions about the same.

36.3 Results

36.3.1 Observation of Five Human Factors

Physical: The tea industry labourers and their families experience physical interaction with the media outlets as luxury items; most of the television remotes have either been missing or broken by kids or the cells were not replaced on time; most households have no newspaper or magazine. **Cognitive**: They use mass media outlets as entertaining objects, irrespective of whether they are watching, listening or reading them, as time-pass items, without provoking any new thought; they are largely indifferent to the outdoor advertisements like posters or banners on the street sides. **Social**: They are a close-knit society and behave each other with an affinity of same tribe, as most social interactions are informal and casual, and this reflects in their interaction with media as well; garden shows film screenings and theatre shows are social platform for them to meet and entertain. **Cultural**: During traditional festivals like Karam Parab and Jhumur, the radio plays their songs and television channels run special programmes. **Emotional**: They clap while watching a movie, talk in between, and exclaim together when an action or emotional scene is displayed on television screen.

36.3.2 Usability Study

The questionnaire returned interesting patterns for interpretation. The data collected are qualitative. Most of the labourers do not read text in the posters, and images and colours, not words attract them into a poster. The posters are in the state language, and not in the dialects of the tea tribe. Only a few of them listen to radio. Background music and songs attract them to tea labourers' special programme on radio, and not the message. Not all of them understand TV advertisements about government schemes and public interest appeals. Not all of them understand even the mic or loudspeaker announcements in the tea estates. There is no public awareness advertisement in the beginning or interval of the garden show cinema screenings. There is usually no plays/drama/theatre performance in the tea estate areas. The management think that existing mass communication medias are not sufficiently improving the social behaviours, knowledge, attitudes, practices of the labourers, and a redesigned mass communication may help.

36.3.3 Intervention Study

The audio prepared for intervention study on awareness of COVID 19 and lockdown was used for mic announcements in tea industry labourers' *coolie line*s. Tea Research

Association (Tocklai Tea Research Institute) sent the video prepared for intervention study on awareness of COVID 19 and lockdown to all tea estate managers of Assam, Bengal and Tripura for wider use. The responses to the mailers validated that this audio and video had helped to make tea tribe population aware on the subject of COVID 19. The message design principle was simple: the verbal and the visual language were not of the communicator, but of the audience, so that it connected the community with their own context. We received number of requests from various agencies for similar videos for their communities. The second video was designed [29] for Meghalaya state government targeting the Khasi community and well received. This design concept has thus demonstrated its usefulness in practice.

36.4 Discussion

By analysing the insight gained from observing five human factors of Assam tea industry labourers, we understood that for them there is either a luxury or an entertainment element associated with mass media tools and devices. The key insight is that tea industry labourers need a new design that will create an altogether different experience for their media use. The usability study shows that the labourers' media exposure is limited and not in tune with the technology. Whatever media they consume, the processing of messages and meanings is not making them informed. Their takeaway from the media is sensory, without the inherent communication. Existing mass communication medias are not sufficiently improving the social behaviours, knowledge, attitudes, practices of the labourers, and a redesigned mass communication may help. This is in fact validated by the simple intervention of context-based audio and the subsequent video designed exclusively for them.

36.5 Conclusion

Gaps in message designs have emerged in the case of Assam tea industry labourers. The solution lies in a new model of mass communication to be contextually audited on the basis of principles of design. Beyond involving design during production of media or message, this model necessitates the central role of design in decision-making for communication plans for the target audience. With respect to model development, this would involve taking into consideration the principles of visual design, interaction design and contextual usability. By varying the parameters in this design-based model of mass communication, we can produce a variety of responses with different end results in terms of awareness levels and behaviour changes. This will even require orientation of the end users of how they experience and interpret the representations of the intended communication. This will reconcile between the message sender's goal and the recipient's mental model. This may act as a step towards developing future models of design interventions in mass communication with scope for context-based adaptation.

Acknowledgements The author wishes to thank Tea Research Association of India for helping in carrying out the usability study, all tea planation labourers and tea estate management for helping the intervention study and district administration of West Jaintia Hills for taking initiative for Khasi version of the COVID 19 awareness video and also acknowledge debt to all the references listed below.

References

1. Sahoo, D., Konwar, K., Sahoo, B.K.: Health condition and health awareness among the tea garden labourers: a case study of a tea garden in Tinsukia district of Assam. IUP J. Agri. Eco. **7**, 50–72 (2011)
2. Save The Children Homepage: https://www.savethechildren.in/resource-centre/articles/child-rights-situational-analysis-in-tea-gardens-in-assam, last accessed 2020/04/10
3. Hazarika, K.: Tea tribes are lagging behind in the process of urbanization: a study on selected tea gardens of Jorhat district, Assam. Int. J. Trends Eco. Man. Tech. **1**(6), 3–4 (2012)
4. Bora, P.J., Das, B.R., Das, N.: Availability and utilization of sanitation facilities amongst the tea garden population of Jorhat district, Assam. Int. J. Community Med. Public Health **5**, 2506–2511 (2018)
5. Panyang, R., Teli, A.B., Saikia, S.P.: Prevalence of anemia among the women of childbearing age belonging to the tea garden community of Assam, India: a community-based study. J. Family Med. Prim. Care **7**(4), 734–738 (2018)
6. De Souza, C.S.: The semiotic engineering of user interface languages. Int. J. Man-Mach. Stud. **39**(5), 753–773 (1993)
7. Krippendorff, K., Butter, R.: Product semantics: exploring the symbolic qualities of form. Innovation **3**(2), 4–9 (1984)
8. Perloff, R.M.: Realism in alcohol messages. In: The Dynamics of Persuasion: Communication and Attitudes in the 21st Century, 2nd edn. Lawrence Erlbaum Associates Inc., New Jersey (2003)
9. Shannon, C.E.: A mathematical theory of communication. Bell System Tech. J. **27**, 379–423 (1948)
10. Crilly, N., Maier, A., Clarkson, P.J.: Representing artefacts as media: modelling the relationship between designer intent and consumer experience. Int. J. Des. **2**(3), 17–18 (2008)
11. Newcomb, T.M.: The prediction of interpersonal attraction. Am. Psychol. **11**(11), 575–586 (1956)
12. Fiske, J.: Introduction to Communication Studies, 2nd edn. Methuen & Co., New York (1982)
13. Budd, R.W., Ruben, B.D.: Beyond Media: New Approaches to Mass Communication, 2nd edn. Transaction Books, New Brunswick (1988)
14. Gitlin, T.: Media sociology: the dominant paradigm. Theory Soc. **6**(2), 205–253 (1978)
15. Schramm, W.L.: The Process and Effects of Mass Communication, 4th edn. University of Illinois Press, Urbana (1977)
16. Westley, B.H., MacLean, M.S.: A conceptual model of communications research. J. Q. **34**(1), 31–38 (1957)
17. Dervin, B.: Mass Communication: Changing Conceptions of the Audience. Public Communication Campaign. Sage Publications Inc., Beverly Hills (1981)
18. Lars, E.: A medium-centred model of communication. Semiotica **3**(2), 72–87 (2018)
19. Blake, R.H., Haroldsen, E.O.: A Taxonomy of Concepts in Communication. Hasting House, New York (1975)

20. Grube, J.W.: Alcohol portrayals and alcohol advertising on television: content and effects on children and adolescents. Alcoh. Health Res. World **17**(1), 54–60 (2004)
21. Nwanne, B.U.: Journalism and new media in Nigeria: issues, challenges and prospects. Int. J. Acad. Res. Reflect. **4**(3), 86–92 (2016)
22. Morris, M.: The internet as a mass medium. J. Comput. Med. Com. **1** (4) (1996)
23. Hoffman, D.L., Novak, T.P.: A new marketing paradigm for electronic commerce. Inf. Soc. **4**, 6–7 (1998)
24. Holmes, D.: Communication Theory: Media, Technology and Society, 1st edn. Sage Publications Ltd., Thousand Oaks (2005)
25. Lovlie, A.S.: Designing communication design. J. Med. Inno. **3**(2), 72–87
26. O'Sullivan, P.B., Carr, C.T.: Mass personal communication: a model bridging the mass-interpersonal divide. New Media Soc. **20**(3), 1161–1180 (2018)
27. Questionnaire on Google form. https://forms.gle/FKqYaGBdSyFBxkaN8
28. Tea Tribe video on YouTube: https://youtu.be/WEQJleLp_uU, last accessed 2020/04/10
29. Khasi video on YouTube: https://youtu.be/8GVX4kezSC4, last accessed 2020/04/10

Part IV
Design Theory and Research Methodology

Chapter 37
Impact of Culture on Design Versus Design on Culture

Lau Langeveld and Amarendra Kumar Das

Abstract Globalization has led to manufacturers selling their products all over the world and due to financial and manufacturing feasibility of designed products, these are mass manufactured and sold in different parts of the world. However, success of a product is impacted by the culture of a country/region. Therefore, it is important to consider cultural impact on design. Similarly, design can have significant impact on culture of a country/region. This paper studies impact of culture on design and vice versa and proposes a table for culture versus design, with qualitative parameters for culture and design and this is justified through case studies from two different culture. In these case studies, the impact of culture on design and design on culture in India and the Netherlands were carried out by considering the proposed table. The results help in understanding the impact of culture on design and vice versa.

37.1 Introduction

Culture versus design shows the difference in manifestation. Culture is about what happens in society and Design is about realizing the things that meet people's needs. Culture has impacts on design and design too impacts culture. The impact is not reversible from culture to design and from design to culture; both directions have different parameters to reach the goal. For the impact, see Fig. 37.1

The significance of culture and design must first be investigated, regardless of the impact. Then identify the influencing factors for achieving an effective result. The influencing factors are arranged in the order of importance. Of course, there are many other identifiable factors that need to be studied.

The research question is, what are the influences of culture on design on one hand and of design on culture on the other hand? Mutual influencing factors must be

L. Langeveld (✉)
Delft University of Technology, Delft, The Netherlands
e-mail: lau.langeveld@gmail.com

A. K. Das
Indian Institute of Technology, Guwahati, Assam, India
e-mail: dasak@iitg.ac.in

A. Chakrabarti et al. (eds.), *Design for Tomorrow–Volume 2*, Smart Innovation, Systems and Technologies 222, https://doi.org/10.1007/978-981-16-0119-4_37

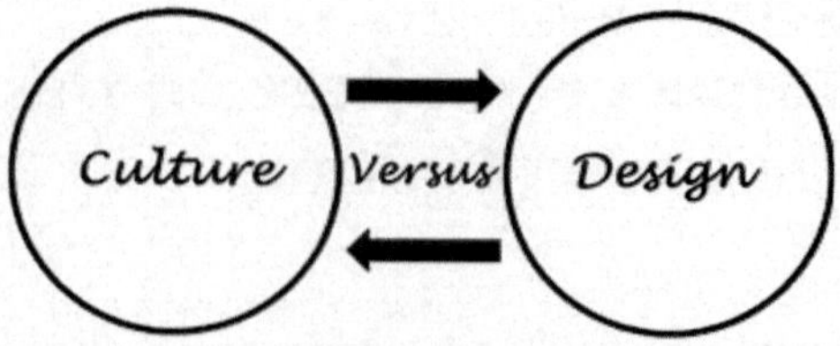

Fig. 37.1 Culture versus design

identified by means of a literature study. These influencing factors are determined as parameters, but due to the different language used in both domains also have a different accent. For example, happy as a fact in culture requires a design to evoke the concept of happiness, to be referred to as H_c (happiness in culture) and H_d (happiness in design).

After the exploratory study of the domains of culture and design, the influencing factors must be determined in a table with columns for culture and design. The rows are the subjects. The research question thus helps to understand what really means the impact of culture on design versus design on culture.

A literature review and the exploratory study for the domains of culture and design will provide the influencing factors with which the case studies are carried out. These case studies are studied and explained in two different environments:

1a. A design from India that changed the behaviour of the Indians,
1b. A cultural change in a part of India that changed the design.
2a. A design from the Netherlands that has changed the behaviour of the Dutch,
2b. The cultural change of Dutch society by the Delta Works design.

The discussion highlights the three most prominent parameters and examines their impact. The impact must be measurable from the effort consisting of knowledge, method, time and money. Subsequently, the conclusions can be drawn that the time over which it will be measured is the dominant factor.

37.2 Literature Review

In the literature review, mainstreams of the domains on the culture and design and also their impact features were explored. The impact feature must be researched on the properties for culture on design and design on culture. About the domains, culture and design have tremendous amounts of scientific literature available that we searched on known experts, institutions or universities, journal papers and blogs. For the impact features, different domains need to be searched including the Internet to arrive at a definition of the impact of culture on design and design on culture.

37.2.1 Culture

Culture starts with the definition of Sir Eward B. Tylor from 1871 [1], which reads as follows: "Culture … is that complex whole which includes knowledge, beliefs, arts, morals, law, customs, and any other capabilities and habits acquired by [a human] as a member of society". This definition of culture includes also the five basic characteristics as: learning, sharing, dynamic, existing and the whole (Material from MIT, MIT21A 01F12). In the book of the psychology of culture shock from Ward et al. [2] shows the psychological and social processes of learning, sharing and experiencing when intercultural contact take place. In general, it is the main culture aspect covered by the book and definition.

Organizational culture. The more specific organizational culture has been covered by Hofstede, Handy and Schein in their works. Three different visions are the most known at the moment; Hofstede with his five factors influencing culture, Schein with his three levels, Handy with his four types of culture. All the organizations have to design their own organization culture based on value, assumptions and artefacts [3]. Johnson's culture web embraced the organizational paradigm [4] from within the web. The six elements of the web are interconnected and have internal behaviours in the organization and outside the organization. The internal-related elements are symbols, stories and routine and rituals, and the external-related elements are power structures, organizational structures, control systems. The web offers many opportunities for human developments in communication, skills and knowledge. The relationships with the customers have internal and external behaviour in the organization.

Cultural mapping. Cultural mapping is a method to identify the identity of individuals, institutions, communities, associations, clubs, etc. This method finds out the strength against the weakness of the many cultural aspects of life: creativity, empathy, communication, social behaviour, morality, etc. Erin Meyer in her book "The Culture Map [5]" decoded how people think, communicate, lead and the things get done across many nations. The USA, China and India are so big, for example, that the cultures can be completely different from the north to south and west to east, metropoles, cities, urban areas and rural areas.

A quote by Pillai (2013) and Rashid (2015) from a University of the Philippines delivered lecture on Cultural Mapping [6] reads as follows: "Cultural mapping is a systematic approach to identifying, recording, classifying and analysing a community's cultural resources or cultural assets that traced the historical, economic, social, geographical significance of a site". This quote shows the culture's aspects of living, what is needed for the impact.

Culture profession. In our modern culture, a culture profession is a necessary evil to train high-level specialists for the future. But which professions are important in the future to indicate stewardship. Professional culture has a structure consisting of two elements: knowledge and value that must remain guaranteed for the future. The element of value includes the professional obligation which consists of personal and social matters and forms with the moral and civic duty, the profession of ethics.

However, the element of knowledge with the gigantic growth of information requires only a method to absorb the necessary knowledge for a profession and to share this knowledge [7].

37.2.2 Design

Design is caught with culture, material (tangible), non-material (non-tangible), process, nature, value and thinking. It takes great effort to fully explore these domains and to look for the impact of the domains on culture. The following examines the design culture, nature of design, material side of design, but also the non-material side of design, design process, value design and design thinking.

Design Process. Design is a creative process for solving problems of economic, social and cultural issues [8]. A design process goes through many paths that are laid down in a large number of methods, sequential, linear, circular, spiral, waterfall, etc. The common goal is to create or search for a solution. Creating a solution is often carried out by a designer and searching for a solution is often carried out by technocrats [9]. The designer is allowed to fully deepen his creative skills within the set problem limits with the aim of producing a result that enriches the community. The technocrats are looking for solutions that can be a new combination of partial solutions, the combinatorial power plays an important role in this [10].

Design Thinking. Design thinking can be carried out by anyone, provided the focus is on the design and the design process [11]. In order to discover the impact of design and culture, new paths must be taken in the way of thinking and demonstrating that it is a good way to map out the right influences. Being aware of the fact that researchers think in a solution-oriented way, while the solution lies in the space of problem-oriented thinking. In a literature review of the aforementioned design and cultural aspects, problem-oriented thinking investigates the possible impacts. These will guide the case studies.

Design Value. Design values are principles that create and shape our lives. The designers embody the design values through their qualities and characteristics to jointly develop the world that successfully distinguishes itself from others. According to Ivar Holm, five basic design values can be distinguished: aesthetic, social, ecological, traditional and gender. Individuals have their own design values that they use in their design work or practice. The personal design value is unique, such as empathy, learning experience, skills, creativity, behaviour, beauty, happiness, harmony and experience. Positive design values can be overshadowed by the negative design values, and these are the design restrictions imposed externally by clients or internally, the inner person himself.

Nature of design. The nature of design consists of a number of views, first, the way in which it is carried out, second, the nature of the engineering design and third, the coincidence of how the design is done. These views are considered in more detail, but certainly also design inspired by nature should be considered.

The execution of the design can be done in a natural way by adding more and more design information to the already existing. Facts are collected and arranged in logical order in order to develop something from abstract to something concrete, a gestalt [12].

Design Culture. Design culture, which must also be further investigated. None other than Manzini can be taken as a start with his vision of design culture and dialogic design [13]. For him, the design culture has changed considerably from "objects" to "thinking" and doing'. The design has become a means of approaching very complex design problems from a human perspective and often solving persistent social, ecological and even political problems. The design culture changes through our thinking, acting and doing such as the design methods, development methods, but also the production processes. Perhaps another step needs to be taken to integrate spiritual life and human aspects with design and production processes.

The design culture plays a vital role in the design process at design institutions, design studios, design departments, etc. The internal and external design aspects are essential here. For example, at design for the elderly, the external aspects are more important than the internal ones associated with culture and organization. Internal factors such as design, validation, production preparation, process communication often lead to a beautiful product. However, when the external factors are not involved in designing for the elderly, this often leads to design failure.

Design by Nature. Design by nature offers the elements of the sensuality of nature such as colour, structure, symmetry, harmony, beauty and form [14]. The whole gamut of colours can be experienced all day between light and dark. A day contains an amount of light and dark depending on the season, weather conditions and location. Man is dependent on light in his existence, and it also determines the mood of them. Little light gives them a chance of gloom, but bright light gives them an energetic life. Structure gives people dedication in their daily actions, in nature this dedication is based on the structure of the particles without any hindrance. A good structure of a problem often leads to a creative uncompromising solution.

Nature has a great wealth of observable value that teaches mankind to live in a wide variety of cultures and to come up with social and economic solutions.

Design of Material/Tangible Products. Design of a material/tangible object consists of observable elements and principles and contributes to the cultural identity of its environment [15]. A material/tangible object is a product that is used in its cultural context, for example a Frisian pink cake from Friesland, the Netherlands or Mysore sandal soap from Karnataka, India.

Complex products such as the car are distributed all over the world and cause cultural poverty in the world. Each country or region will have to make its own contribution to maintain its identity. But be independent of possible feasibility by others.

Design of Non-Material/Intangible Products. A non-material/intangible object has no physical expression, but the elements and principles are based on audio, visual, touch, taste and smell which appeals to our sensory capacity for our cultural identity and its existence.

Another example is the design of a fragrance from the natural environment to make spaces more comfortable that is usually enclosed in a fragrance carrier.

37.2.3 Impact

Impact cannot be captured in formulas because the results are unreliable due to the cultural interpretation, which cannot be generalized. Take the concept of design, which has general design aspects, but also the cultural characteristic aspects such as colour, appearance and interaction. But the designs are also often produced in a country or state where the business concepts are interpreted differently worldwide such as TQM total quality management, JIT just in time and BPR business process reengineering.

Impact factor. The impact factor for journals is a generally accepted concept in the publishing world, but how it is measured remains a hot topic. Namely, how many researchers are working on it, the domain of publication; the overlaps with other domains are sources that yield the unreliable or meaningless impact factors. So this model cannot be used for the purpose of impact assessment either.

The impact of thinking. The impact of thinking about the design has multiple facets, namely thinking is a mental activity that thoughts convert into physical objects. The mental activity is done by the individual in the appropriate manner. Thinking is a personal brain activity, which is unique, depending on knowledge, skills and motivation. The result of the entire design process is strongly influenced by the thinking and application of creativity. Thinking can have many starting points—rational, logical, intuitive, emotional, visual, design, practical, cosmopolitan, technical, etc. During the design process, these thinking forms are used, with the different skills and methods to achieve a successful result. A successful design can only be achieved through design thinking that determines the impact on design [16]. Design thinking is a cultural characteristic. This means thinking is one parameter of the impact from culture on design.

Social Impact. Social impact is created by the efforts of companies, institutions, individuals and activities on others that bring about change [17]. The change can also have an impact on cultural life. For example, a design of a curriculum for autistic children has a social impact and also a cultural impact on society. So social design has the goal to achieve a social impact and also a cultural impact for the individual, his environment and society.

37.3 Research Exploration

The exploration briefly examines culture, design and impact to come up with a table showing culture versus design with many facets that have been translated into parameters.

Culture is a collective characteristic of a group or category of people who differ from each other. The culture of a particular profession can be described with distinctive features. Culture as a concept encompasses a large number of knowledge areas in our digital age: education, learning, social, economic, industrial production, organization, politics, religion, art, lifestyle, habits, etc. Here only the knowledge that is more closely related to the design directly or indirectly is considered. Culture also has five basic characteristics [18] that all cultures have in common, such as dynamic, learning, sharing, based on expressions of the whole and a holistic approach.

Design has many facets, these are described in the literature study and some give cause for further explanation. Design culture is completely different from culture design, because one means under which conditions the design is created and the other means creating a new culture. Design thinking should not only focus on design, but there should also be space for other knowledge domains. Thought has so many facets that enrich design thinking, such as spirituality, intuition, concentration and mysticism.

Impact requires a certain approach that is worthy of culture and design and so a qualitative approach is suitable. Namely, a makeable world becomes unliveable, it is not a mathematical model. Social impact also needs a qualitative approach through each unique contribution.

Qualitative parameters for culture and design in Table 37.1 are used in the case studies. The parameters are always provided with a subscript c or d, but also capital letters C and D with subscript. These are used to make a good analysis of the cases.

37.4 Qualitative Impact Parameters

Qualitative means a described value assignment that will always be subjective: an actual assignment, a value indicated by a number. Happiness is a concept that is different for everyone, so a qualitative parameter that only produces a qualitative difference of more or less with another. Time and energy are described with bandwidths for the qualitative impact parameters. H_c from Table 37.1 has a holistic parameter description in the domain of culture with dimensions: time, energy, all-encompassing, happiness, flourishing and austerity. W_d from Table 37.1 represents the whole in the design domain with dimensions: content, volume, material, number, mass and quantity.

The culture parameter C_1 has many dimensions that are almost impossible to name, but they all relate to what takes place in community. The design parameter D_1 also has many dimensions, all of which relate to realizing people's needs.

Table 37.1 Culture versus design, with qualitative parameters for culture and design

Culture versus design			
Culture		Design	
Culture is about what happens in a society	C_1	Design realizes the things that fulfil the needs of people	D_1
The whole	W_c	Problem of a part	P_d
Social	S_c	Social solution	S_d
Religion	R_c	Spirituality	D_s
Family	F_c	Individual	D_i
Lifestyle	L_b	Needs	D_n
Elderly	E_c	Designs for and by elderly	D_{fe}
Art	A_c	Form giving	F_g
Behaviour	B_c	Functional fulfilment	D_f
Happiness	H_c	Happiness	H_d
Group or like-minded	G_c	Team	D_t
Profession	P_c	Value and knowledge	D_{vk}
Mapping	M_c	Identifying Design Task	D_{it}
Dynamic	D_c	Redesign	D_r
Learning	L_c	Experience	D_e
Sharing	S_c	Exchange	D_{ex}
Holistic	H_c	Whole	W_d
Impact	I_c	New design methodology	D_m
Changes	C_c	Impact	I_d

37.5 Case Study India

Culture impacting design: Defecation system in India historically till date

In many parts of India, there was open defecation and this was culturally acceptable and agricultural fields and adjoining forest land, etc. Defecation was considered as impuring act and after defecation, ablution was essential to visit any place of purity like temple or place of prayer. This practice was also difficult for elderly people. Later on as the population increased and in bigger settlements came up, design of service latrines came up outside the residences. These were serviced by lower caste people and the Father of the Indian nation, Mahatma Gandhi, popularly called as Gandhiji tried to eradicate this culturally degraded practice and called these people as *Harijan*, God's People.

In many parts of India, there was open defecation and this was culturally acceptable and agricultural fields and adjoining forest land. Defecation was considered as impuring act and after defecation, ablution was essential to visit any place of purity like temple or place of prayer. This practice was also difficult for elderly people. Later on as the population increased and in bigger settlements came up, design of service latrines came up outside the residences. These were serviced by lower caste people and the Father of the Indian nation, Mahatma Gandhi, popularly called as Gandhiji tried to eradicate this culturally degraded practice and called these people as *Harijan*, God's People. Later on Government banned this practice but country had difficulty in overcoming this practice. Over a period of time, as education propagated amongst people, and this practice became culturally unacceptable, service people unavailable, disposal of defecated matter difficult led to design of septic tank. Thus, the culture influenced design and this led to finally toilet being part of every household and attached to part of even bedroom. The cultural parameters and corresponding design parameters are followings:

$C_1, D_1, S_c, S_d, R_c, F_c, D_i, L_b, D_n, E_c, D_{fe}, B_c, D_f, C_c, I_d.$

Design impacting Culture: Portable LPG cylinder and LPG Gas stove

One design that has impacted the culture of India and the lives of women folk that is a portable LPG cylinder and a LPG gas stove. Indian population used firewood for cooking and for rural population, primarily women folk, it was a difficult, time-consuming but the routine task to go for collecting firewood from various sources. For this reason, girl children were also engaged and they had to forego their school. Burning of firewood indoor also had an adverse effect on the health of the people. For the rich people, they used to buy firewood and all these resulted in deforestation and resultant soil erosion, flush flood, etc. Design of safe portable LPG cylinder and stove led to non-use of firewood and clean cooking environment indoor and regeneration of forest and tree cover with a better environmental condition for the country. Impact of smoke from burning firewood too disappeared. In the absence of a requirement to go out for collecting firewood, women folk could take care of their home, cultivate vegetables and rear domestic animals. Girl child can attend their school. The impact of this was so strong on the culture of the country that, the central government of the country provided free LPG connections to more than 50 million households and 6 free refills per annum.

The design parameters and corresponding cultural parameters are followings:

$D_1, C_1, P_d, W_c, S_d, S_c, D_i, F_c, D_n, L_b, D_f, B_c, D_m, I_c, I_d, C_c.$

37.6 Case Study the Netherlands

Culture impacting design: Electric dry shaving

In the Netherlands, inventor, Alexandre "Sacha" Horowitz (1904–1982) developed the idea of the electric shaver with a rotary shaver system in the Philips research laboratory in Eindhoven, branded as the Philishave. Introduction of Philishave by Philips in 1939 encouraged dry shaving. This is a major cultural change in the man's personal care, which only took off after World War II. The rotary shaving principle is also used in various shavers for both men and women.

In 2003, Philips announced that it would move its shaver factory to China, while the annual production was at a level of approximately 11 million units. The 500 millionth Philishave rolled off the production line in 2007, an average annual production of 6 million units. The shaver design is now manufactured with mass production systems in China. The social impact of this shift of production facility was great in Drachten, a city of less than 50,000 inhabitants, Philips Drachten being the largest employer. However, the development facility of razors and means of production has remained. Naturally, a brain drain took place with regard to mass production knowledge, while the necessary renewal of processes and materials remains necessary.

The electric wet shave is also on the rise, and the renowned electric shaver manufacturers have come up with a high-end version here. Dutch culture is fast-paced and any time shaving in essential personal care are welcome, including electric wet shaving. The personal care of men and women has brought about many cultural changes over the centuries.

The relocation of production and later the sale of Philips Domestic Appliances and Personal Care unit has major cultural consequences for the community of Drachten, although in principle, it will change hands. Social relations are changing drastically, including employment that may decline, but it also offers new opportunities.

The social consequences are great: loss of a job, end of enhancement in disposable income, changes in lifestyle to a lower level has affected the entire population of Drachten and the surrounding areas. This is reflected in cultural quantitative parameters such as $\mathbf{C_1}$, $\mathbf{S_c}$, $\mathbf{L_c}$, $\mathbf{P_c}$, $\mathbf{C_c}$.

With the sale of Domestic Appliances and Personal Care unit, no more design and production means the following design parameter affected: $\mathbf{D_1}$, $\mathbf{D_r}$, $\mathbf{D_e}$, $\mathbf{I_d}$, $\mathbf{D_{vk}}$, $\mathbf{D_i}$.

Design impacting Culture: Delta Works

The struggle against water is a cultural characteristic of the Dutch population through the centuries. After first living on the mounds or weeding raised areas in the wetlands around 500 BC in the Iron Age, the dikes were built from the tenth century around the peat areas, which were later mined as peat areas, creating lakes and ponds. Dike areas could be drained with trenches and windmills that created the polders. The Delta Works is the third major water project in the Netherlands after the draining of the Haarlemmermeer and the Zuiderzee works. The Haarlemmermeer originated from Spieringermeer, Oud Haarlemmermeer and the Leidsemeer and threatened the cities with flooding due to a combination of spring tide and heavy storms. By removing the threat of flooding from in the cities, a ring dike has been built around the Haarlemmermeer in which the water could be drained into storage waters around the lake such as Westeindes plassen and Kaagerplassen. Three steam pumping stations were built for this purpose, De Cruquius, Leegwater and Lijnden for the draining instead of the well-known windmills. It took three years for the land to dry up. The Zuiderzee Works included an impressive number of sub-projects, first the impoldering of the Wieringermeer and then the closure of the Zuiderzee, now called the IJsselmeer. After the closure, reclamation of the Noord Oostpolder started, followed by the Flevopolders, Oost and Zuid.

Preparations for the Delta Works have already started for the flooding disaster in Zeeland. The entire project includes 5 storm surge barriers, 6 dams and two locks. After a flood of 1953, the coastline was accelerated from 700 km back to 80 km by closing off straits. In addition to safety, there is also a good infrastructure for recreation and nature development. The Oosterscheldekering is a unique solution to close the barrier during spring tide and heavy storm. This preserves the saltwater culture and ebb and flow. The shrinking of the coastline has brought safety, in which there is always a search for a balance between ecology, economy and culture. Translates to contemporary aspects, mussel cultivation, water sports, agriculture and horticulture, recreation and nature. Tourism is a major source of expenses in the summer season. The barriers were all closed only twice, the last of which was in the fall of 2019. The closure of the dams has also brought about ecological change, the temperature increased about one degree, salty freshwater, tidal energy extraction, and mussel cultivation have been created.

The realization has cost a lot of design energy, for this the Waterloopbouwkundig Laboratorium has been established, where the Delta of the Netherlands has been simulated on a scale model to study the currents under all kinds of circumstances. In this way, the civil engineers and designers searched for the necessary design data with the changed circumstances of closed sub-projects. The models have also been created for the last half-open Oosterschelde dam. Major design efforts have been made here to meet the needs of social and cultural demands. It contributed to the safety of the province of Zeeland and other parts of the Netherlands.

The great technological effort and sizeable government investment has brought much prosperity in the province of Zeeland and also in the rest of the Netherlands. The Delta Works has produced a large number of high-quality jobs, which means the qualitative parameters of the design $\mathbf{D_1, D_e, D_{ex}, D_m, D_f, D_{it}, D_{vk}}$. All knowledge and experiences are exported to all other countries that are also struggling against the oceans, seas, strait and river delta. The cultural changes are substantial to date, leading to a prosperous community. They have experienced all cultural changes dynamically and positively and now feel safe. This is expressed in the qualitative parameters changes of culture $\mathbf{C_1, L_c, A_c, B_c, D_c, P_c, S_c}$.

37.7 Discussion and Conclusion

First the next question must be raised, "how can the material (tangibles) be linked to the spiritual life (intangibles)". The coupling goes through the amount of energy from the sun that we receive and is converted into life energy on earth. This is used as cultural energy and material energy. Culture has a large energy content and an impact within this content is accompanied by changes between dimensions. So the impact of designs on culture can be seen in its dimensions. For example, gadgets are designed with energy and invested economically all in a very short time, which people purchase for a brief enjoyable experience. But a disaster can disrupt the energy balance that poses urgent questions to the design. The creativity absorbs the disaster energy and comes up with the solutions for the shorter term. However, the short term also requires creativity to meet basic needs.

The impact of culture on design versus design on culture is true under the condition that the energy balance remains fairly constant. In the Dutch case studies, Delta Works and Electric Shaving, that the impacts are not reversible when we look to the qualitative parameters of design and culture. Delta Works is an ongoing story for a high-quality lifestyle. The rise in sea level requires constant attention for the extreme conditions that may arise in the future. Using tidal energy and the ecological balance between nature and recreation requires a lot of attention and energy. Electric shaving has changed our personal care, but the concept is now electric wet shaving. When the shaving factory is sold, knowledge will disappear in a very short time. Personal care knowledge is very culture dependent.

In case of Indian case studies, culture impacting design: Defecation system in India, historically it took a long time but with education and changing lifestyle in modern housing and continued emphasis on hygiene, the existing culture is drastically changing and this changing culture is impacting design of sanitary products and systems even to luxury, leaving apart basic need. Against this in case of design impacting culture: Portable LPG cylinder and LPG Gas stove are being accepted by the larger population and people are also going for electric chimney to get rid of oily sticky fumes from LPG stoves and many new generation people are opting for electric induction plate where electricity supply is regular. These designs are also facilitating kitchen becoming part of living room or compact in size. So cultural changes of these designs are continuous and significant. In both these cases, these are dynamic.

Thus one can conclude that a designer need to consider both these impacts and cultural studies should be an essential part of design education for products to be successful in the market. In the name of globalization, one cannot produce one single product and be successful in all cultural landscape for long. Success may be until the novelty of the product exists but wane out sooner than later.

References

1. Anonymous: Sir Edward B. Tylor's definition of culture (1871). https://ocw.mit.edu/courses/anthropology/21a-01-how-culture-works-fall-2012/readings/MIT21A_01F12_Sir_Edwrd_cul.pdf (n.d.)
2. Ward, C., Bocher, S., Furham, A.: The Psychology of Culture Shock, 2nd edn. Taylor and Francis Inc., Philadelphia (2005)
3. Haworth Whitepaper: How to Create a Successful Organizational Culture: Build It—Literally. https://eu.haworth.com/docs/default-source/white-papers/how-to-create-a-successful-organizational-culture.pdf?sfvrsn=4 (visited on 25–4–2020), Manag. Sci. **29**, 363–377 (1983)
4. Sun, S.: Organizational culture and its theme. Int. J. Bus. Manag. **3**, 137–141 (2008). https://doi.org/10.5539/ijbm.v3n12p137
5. Meyer, E.: The Culture Map. Public Affairs, New York (2014)

6. Pillai, J.: Cultural Mapping: A Guide to Understanding Place, Community and Continuity. Strategic information and Research Development Centre, Malaysia (2013)
7. Karmazina, E.V.: Professional culture of the specialist of the future: moral aspects. In: The European Proceedings of Social and Behavioural Sciences (2018). https://doi.org/10.15405/epsbs.2018.12.02.87
8. Braha, D., Maimon, O.: The design process: properties, paradigms, and structure. IEEE Trans. Syst., Man, Cybern. Syst. Humans **27**, 146–166 (1997). https://doi.org/10.1109/3468.554679
9. Haik, Y., Shahin, T.: Engineering Design Process. Cengage Learning, Global Engineering, USA (2011)
10. Najar, N.A., Abdulluh, S.: Fundamentals of Engineering Design. J K E R T Foundation, Jammu & Kashmir (2017)
11. Owen, C.: Design thinking: notes on its nature and use. Des. Res. Q. **2**, 16–27 (2007)
12. Feucht, D.L.: Design in nature and the nature of design. Origins Des. **19**, 2 (n.d.). https://www.arn.org/docs/odesign/od192/designinnature192.htm
13. Manzini, E.: Design culture and dialogic design. Des. Issues **32**, 52–59 (2016)
14. MacNab, M.: Design by Nature: Using Universal Forms and Principles in Design. New Riders, Berkeley (2012)
15. Kari Kuutti, K.: Defining an object of design by the means of the cultural-historical activity theory. In: Proceeding of (CD-ROM), University of Arts Bremen. https://cc.oulu.fi/~kuutti/Julkaisukansio/KK_EAD06_final.PDF (2005)
16. Rashdan, W.: Impact of design thinking on conceptual design in interior design studio courses. People: Int. J. Soc. Sci. 3, 85–99 (2017). https://doi.org/10.20319/pijss.2017.32.8599
17. IF-World Design Guide: Social Impact and Design. https://ifworlddesignguide.com/design-articles/social-impact-and-design (2020)
18. Henslin, J.M.: Essentials of Sociology: A Down-to-Earth Approach, 13th edn. Pearson, Essex, UK (2019)

Chapter 38
Understanding the Dynamics of Emotions During the Design Process

Mritunjay Kumar, Satyaki Roy, Braj Bhushan, Ahmed Sameer, Swati Mittal, and Bharat Sarkar

Abstract Research on emotion and design literature has relied primarily on the product generated by the designers and the emotional experience felt by the users while using the product. A limited number of studies have addressed the dynamics of the designer's emotions during the design process. This exploratory study attempts to understand the emotional experience of designers during the design process with supporting empirical evidence. Twenty-five designers were asked to carry out a design task for a limited period of time. The data was analyzed using the FBS ontology framework, linkography, and PANAS ratings to establish the dynamics of the emotions during the design process based on the video and audio recordings of the task. This study demonstrates mostly positive affect throughout the design process with associated high entropy scores and high outcomes, where the affective states varied between different time intervals and at different phases of the design process.

38.1 Introduction

Present theories of emotions are a conglomerate of supporting pieces of evidence of several features and phenomena that constitutes an emotional episode [1]. These episodes activate or stimulate human behavior resulting in a complex act of decision making. Given the importance of emotions on high-level cognitive functions such as creativity, design, decision making, and reasoning [2, 3], scant attention has been paid to understand personal emotional experiences that a designer feels during the design process. Previous studies have demonstrated the importance of the product

M. Kumar (✉) · S. Roy · B. Bhushan · S. Mittal
Indian Institute of Technology Kanpur, Kanpur, India
e-mail: mjdesignfactory@gmail.com

A. Sameer
Indian Institute of Technology (ISM) Dhanbad, Dhanbad, India

B. Sarkar
One Plus, Hyderabad, India

A. Chakrabarti et al. (eds.), *Design for Tomorrow–Volume 2*, Smart Innovation, Systems and Technologies 222, https://doi.org/10.1007/978-981-16-0119-4_38

generated by the designers and the emotional experience felt by the users while using the product [4–6]. The design process is a complex phenomenon where designers often deal with wicked and ill-defined problems [7, 8], which generate a complex set of emotions to deal with the goal at hand. Different stages of design in the creative design process have also been explored and established [9]. Several tacit experiential decision making is applied at every stage of the design process, and designers keep jumping from one stage to another and several stages [10]. The emergence of an idea and insight during the creative design process is considered to be a highly emotional step that happens involuntarily [11]. Much has to be learned about designers and their design cognition from an emotional point of view. This gap can help us to understand and address the varieties of ways a designer faces an emotional experience and the challenges faced during the design process. Csikszentmihalyi et., al. believed that the designer's emotional experience would influence their decision making in the design process [12]. Some researchers have found the correlation of positive emotions with creativity during the design process [13]. Ho and Siu [14] proposed a conceptual model to understand emotions during the design process through two key concepts—*emotionalize design and emotional design.* The former describes the emotional experience of the designers during the design process, and the latter is about eliciting emotions on a specific set of users through a product. Remarkably few studies have tried to investigate and propose frameworks to understand the designer's emotions during the design process [15, 16].

38.2 Measuring the Design Process and Emotions

Protocol analysis (*concurrent and retrospective*) has been widely used to measure the cognitive process of the designers based on the verbal utterances by the designers [10, 17–20]. Diversity in coding schemes has been developed and used over these years [21, 22]. A widely accepted method, linkography, is used to measure the design process through protocol analysis. Linkography describes a design process by discerning the number of moves and the links produced between these moves [23]. Gero [24] proposed the FBS ontology of design to carry out the protocol analysis based on the goal of designing and transforming a set of functions (F) into a set of design descriptions (D). See Fig. 38.1.

The design starts with a set of requirements, and a designer then transforms the requirements into the Function (F), which is referred to as the teleology of design. A set of expected behavior (Be) tries to fulfill the function, and an abstract structure (S), which is referred to as the artifact's elements and their relationship, is generated. The derived behavior (Bs) from this structure is compared with the expected behavior to analyze the design. Once the structure is finalized, it is then documented (D). The syntactic design process explains the transformation of a design issue based on its previous preceding issue through Markov's chain [25]. Entropy scores represent creativity in the design process through the number of linked segments and the distribution of those links using Shannon's information theory model. A high number

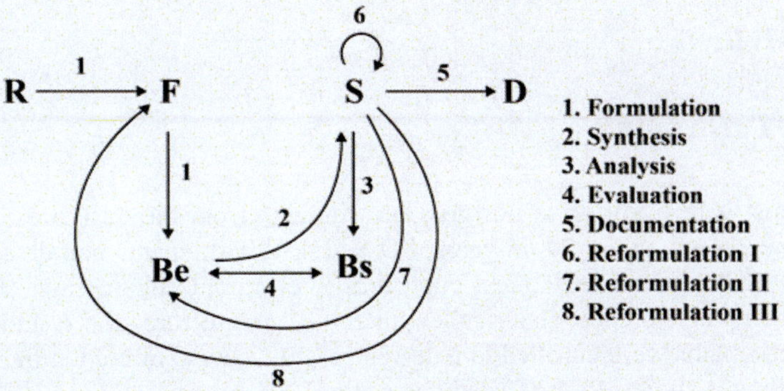

Fig. 38.1 FBS framework with the eight FBS processes [25]

of forelinks are associated with the divergent thinking process, and a high number of backlink explains the convergent thinking process. Horizon links are associated with incubation or cohesiveness during the design process [25].

Emotions are largely our feelings associated with physiological changes in the body. Since emotions are multidimensional, it becomes challenging to measure emotions on multiple factors. However, previous works have demonstrated methodologies to measure emotions on several dimensions [26]. Mauss and Robinson [27] did an extensive review on the measurement of emotions some of which are (a) self-report measures where participants report their moment to moment experiences, e.g., PANAS, mDES, SAM, etc. (b) physiological aspects of emotion measurement rely on physiological changes, e.g., autonomic nervous system responses from the subjects, eye blink rates, etc., and (c) behavioral elements rely on the facial expressions change, body expressions, and gestures [28].

38.3 Aims and Objective of the Study

This exploratory study aims to understand designers' emotional experience during the design process with supporting empirical evidence. This research took an inductive approach to investigate the variation and patterns of the designer's affect(s) during the eight syntactic FBS processes in the act of designing.

38.4 Method

38.4.1 Participants

Participants were twenty-five postgraduate students from the design discipline at IIT Kanpur (Mean age = 24.68 years, SD = 1.43). Participants had diverse backgrounds ranging from mechanical engineering, electrical engineering, civil engineering, computer science, electrical engineering, architecture, and fashion design. These participants were enrolled in a design practice course offered at the institute. They received a sum of Rs 200 as compensation.

38.4.2 Material

The Design Task To design a low-cost and efficient phototherapy unit for neonates who have neonatal jaundice that can be used in rural India. This problem was chosen as the students already had done their fieldwork in this area one week before sitting for this experiment as a part of the coursework.

Assessment of the design process Two Akaso V50 pro native cameras were fixed on the tripod to capture the ongoing ideation process. One camera captured the zoomed view, and the other master camera captured the overall scene. An android phone was used to capture the verbal utterances of the group members. Linkoder (www.linkoder.com) [29] software was used to analyze the design moves and the links between these moves based on utterances produced during the ideation process. Linkoder helps to produce several outputs, e.g., entropy scores, the ratio of forelinks, backlinks, horizonlinks, FBS issue distribution ratio, eight-design process syntactic distribution ratio, and link ratio.

Assessment of the affective states The psychology literature presents a distinction as well as an overlap with respect to emotions and affective states. The term 'affect' is a broader umbrella concept that encompasses moods and emotions [30]. We will be using the term 'affective state' and emotions interchangeably throughout this paper. Positive and negative affect schedule scale inspired by Peilloux et al. [31] was chosen for the study. Four positive affects (*interested, curious; inspired, stimulated; determined, decided; satisfied, blooming*) and four negative affects (*anxious, nervous; sad depressed; hesitant, doubtful; stressed, overwhelmed*) on a 5-point Likert scale ranging from 1 = not at all to 5 = definitively were used to capture the affective profiles of the designers. These eight affective profiles were alternately mixed and presented to the participant.

38.4.3 Procedure

Twenty-five participants were randomly divided into a group of 5 and were called to solve a design task mentioned in Sect. 38.4.2. These five groups participated alternately on the weekend (*three groups on Saturday and two groups on Sunday*) and were asked to think aloud while solving the tasks. This was a group design task where participants interacted and brainstormed together to come up with a solution. They were video and audiotaped throughout the entire session. Informed consent was taken from the participants prior to the task. The total time to finish the task was 30 min for each group. Preeceding with a short break, each participant was asked to rate their affective state retrospectively for the design process through video stimulated recall. Each participant was given a separate laptop to watch the video and rate their emotional profile (on PANAS) on a 30 page printed booklet. These participants were placed separately in 5 different rooms in the presence of five different volunteers. Each volunteer probed each participant every 60 s to rate on the parameters mentioned above. The video was paused, and participants then rated. Participants had the freedom to fast forward the video at their convenience. The 60 s interval was chosen for two reasons; first, this paper aims to find the emotional profile retrospectively for a long interval and not mood that happens in order of seconds and minutes, which can only be reported on that very moment it is happening, and the past research have used similar time interval to achieve good success [32]. The total administration time to complete the task and retrospective ratings together was approximately 75–90 min.

38.5 Analysis

38.5.1 Affective States During the Design Process

Combined mean scores of the positive and negative affects of each participant were calculated, providing us with thirty positive and thirty negative *time-affect* samples. To compute a single affect score, affect balance score was calculated by subtracting the mean negative ratings from the mean positive scores [33]. A total group affect balance mean was also computed (see Table 38.1).

38.5.2 Dynamics of the Design Process

Process refers to the transition or transformation of one design state to another. Protocols were segmented and coded twice using the Delphi method with two weeks of separation, as suggested by Gero and McNeill [20]. Five Linkographs of each session were produced through the Linkoder software (see an excerpt here, Fig. 38.2).

Table 38.1 Groupwise affect, linkography, entropy, and the design outcome scores

Group	Mean_affect balance score	Total segments	Total links	Link ratio	Forelink entropy	Backlink entropy	Horizonlink entropy	Total entropy	Mean outcome score (novelty and creativity)
1	9.42	270	654	2.42	0.4332	0.4985	0.142	1.073	9.66
2	9.27	236	405	1.72	0.3803	0.3828	0.087	0.850	6.66
3	8.05	239	483	2.02	0.3832	0.4246	0.101	0.909	7.33
4	8.78	202	461	2.28	0.4213	0.4289	0.092	0.942	8.33
5	7.82	272	511	1.88	0.3617	0.4133	0.102	0.877	6.33

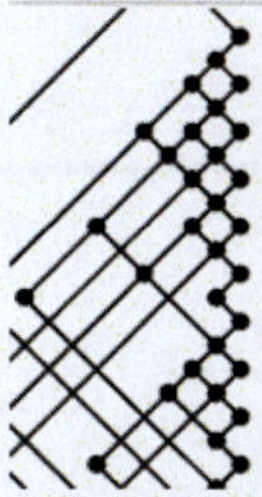

Linkograph with links	Moves	Participant	FBS code	Verbal Utterances
	186	P1	Be	the airflow can be circulated by ....
	187	P1	S	...small electrical vents inside the fabric...
	188	P1	Be	that can be controlled from outside
	189	P4	S	yeah...may be we can have air ducts made...
	190	P4	Be	...to control the airflow...
	191	P2	S	...I have seen something similar ....the project done by MIT media lab....Let me have a look
	192	P2		[searches the internet]
	193	P2	S	yeah....see this.....these ducts...here
	194	P3	Be	yeah...that looks great...we can have similar mechanisms like this.....
	195	P1	Bs	yeah....I mean......this seem like a breathable fabric........

Fig. 38.2 Design activity excerpt details of Group 2, where column one is the visual Linkograph with the links, column two is the number of moves, column three is the participant number, column four is the FBS code assigned to verbal utterances in column five. In this heavily linked segment, participants were brainstorming to figure out the design prototype. The design process sequence was very rapid, where participants were jumping between the behavioral and structural aspects of the design prototype. Move 192 is blank as it fell under the 'other' category

Three independent experts in the domain of product and mechanical engineering design rated the final output on the criteria of novelty and creativity using a 5-point Likert scale (from 1 = 'less novel; uncreative' to 5 = 'very novel; very creative'). Overall summary of the quantitative results is produced for each group in Tables 38.1, 38.2, and 38.3. Also, for the purpose of this study, early phases of the design process are referred to as *the conceptual design phase* and the later phase as the *embodiment design phase*, as proposed by Howard et al. [9]. Based on the time-affect sample, 30 time-based linkograph statistics (*Entropy scores and FBS processes*) for each group were computed. Afterward, 3-axis graphs were plotted for the FBS processes, and affect, over a 30 min timeline, combined for all the groups. This cumulative mean computation was done to visualize the overall affect change during the design process and not just on individual groups.

Table 38.2 Groupwise distribution of the FBS issues

Group	Mean_affect_balance score	*R*%	*F*%	Be%	Bs%	*S*%	*D*%
1	9.42	14.2	8.5	17.7	12.3	34.2	13.1
2	9.27	10.3	3.6	24.7	18.8	34.1	8.5
3	8.05	3.4	5.6	22	15.1	40.1	13.8
4	8.78	10.7	4.1	17.7	13.7	31.5	22.3
5	7.82	5.2	1.5	25.4	20.9	32.8	14.2
	Mean	8.76	4.66	21.5	16.16	34.54	14.38

Table 38.3 Groupwise distribution of the eight FBS syntactic processes

Group	Mean_affect balance score	Formulation %	Synthesis %	Analysis %	Evaluation %	Docmentation %	Reform ulation1%	Reform ulation 2%	Reformulation 3%
1	9.42	3.3	20.7	17.3	8.3	11.6	25.6	9.9	3.3
2	9.27	2.9	25.7	23.8	7.6	7.6	13.3	18.1	1
3	8.05	1.5	20.3	16.5	11.3	12	17.3	17.3	3.8
4	8.78	2.5	16.5	12.7	6.3	19	27.8	13.9	1.3
5	7.82	0	21.1	21.8	15	11.3	18	12.8	0
	Mean	2.04	20.86	18.42	9.7	12.3	20.4	14.4	1.88

38.6 Results

The overall design process for every group was associated with positive affects and high arousal. In the first half of the design process, i.e., *conceptual design phase*—(1 to 12 min), participants reported having positive affects. They were interested/curious; inspired/stimulated during this phase, as reported in PANAS. Toward the second half of the process, i.e., *the embodiment design phase,* participants reported being anxious/nervous, hesitant/doubtful (12–25 min), and more determinant/decided; satisfied/blooming at the end (25–30 min). These results are based on the mean affect balance score for all the groups. However, we are losing important data by ignoring the negative ratings. Studies have reported the occurrence of both the affects parallelly [34, 35]. So, we present a separate graph to illustrate the occurrence of positive and negative affect parallelly based on the ratings (see Fig. 38.4). It can be inferred from the graph that the beginning part of the design process (1–7 min and 8–12 min) was associated with positive affects and high arousal. The process at the end (12–25 min) was associated with negative affects and high arousal. Though the affect balance score is overall positive (above the X-axis), we still can observe several peaks and lows signifying the affect size variability.

As visible in Table 38.2, the majority of the cognitive efforts were dedicated to discerning the structural issues (34.54%), followed by expected behavior issues (21.5%). The groups majorly worked on understanding the elements and their relationships of the artifact.

High entropy scores (1.073), high link ratio (2.42), and high mean outcome scores (9.66) were associated with a high affect mean balance score of 9.42 for group 1 (see Table 38.1). High forelink (43.47) and backlink entropy (49.85) show that this group generated many ideas and then later built upon them. The second group to score higher on the mean outcome score (8.33) was group 4 with an affective balance score of 8.78 and an entropy score of 0.942. Interestingly, the lowest-scoring group no 5 with a mean outcome score of 6.66 and a total lowest entropy score of 0.850 scored high on the affect balance score (9.27). In general, high entropy scores were associated with positive affects.

We present the following observations related to eight FBS processes and the associated affective state based on Fig. 38.3 and Table 38.3. Student's cognitive efforts were majorly expended upon the syntactic processes of synthesis (20.86%), where students were engaged in deriving the structure to fulfill the expected behaviors. During reformulation I, student's were indulged in modifying the proposed structure space by introducing new components and elements in their design (20.4%) followed by the analysis process (18.42%) where students derived the actual behavior (Bs) from the proposed structure (S) (Fig. 38.4).

(1) Formulation ($R \rightarrow F \rightarrow$ Be) and the associated affective state(s)—A sharp peak of formulation happens at (2–5 min), and multiple peaks can be observed between (15–26 min). The participants were majorly reading and discussing the design requirements and assigning a function (purpose) to the artifact that was to be designed during this process. The affect balance score shows peak during

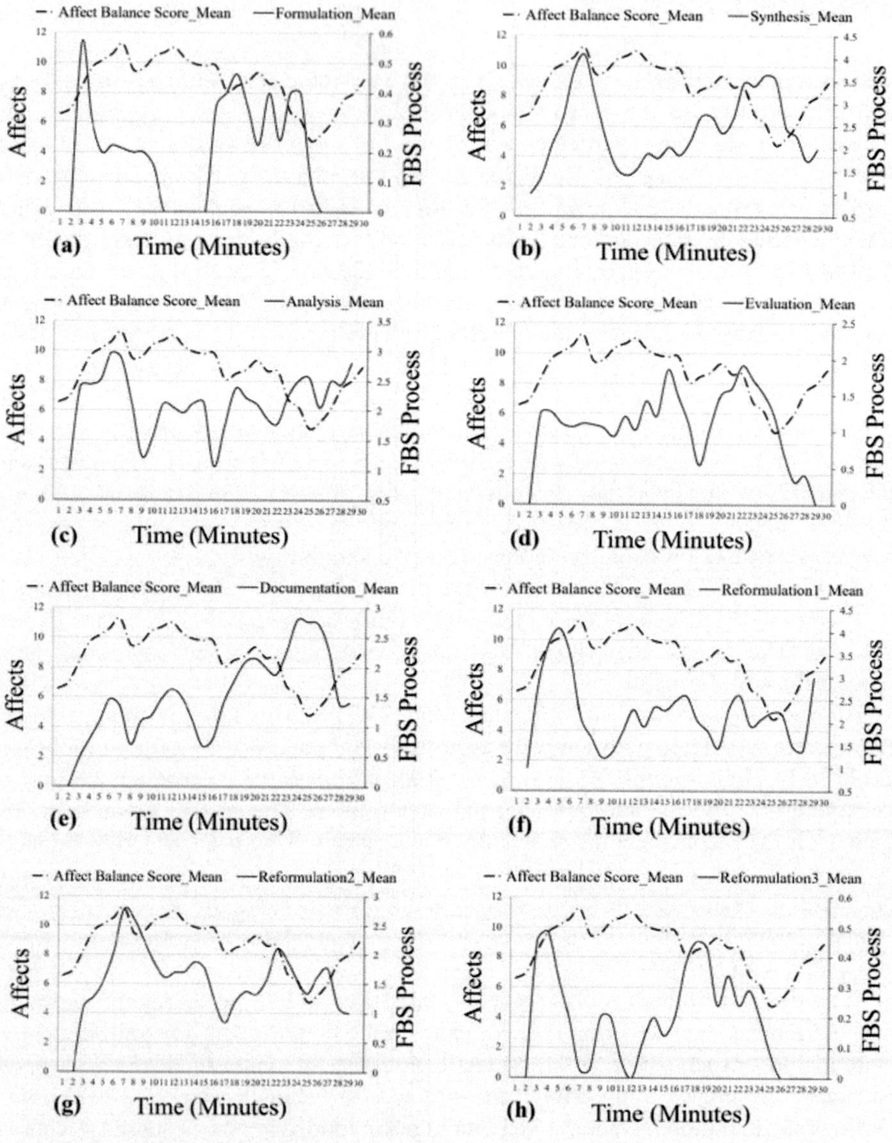

Fig. 38.3 Mean of the eight FBS syntactic processes and associated mean affect balance score combined for all the five groups

the beginning, where participants reported to be more excited and curious (2–5 min). However, formulation peaks between (15–26 min) were associated with low positive affects (16–25 min). Rather than discussing the artifact and relationship (*S*) at this point in time, designers were still trying to either define the

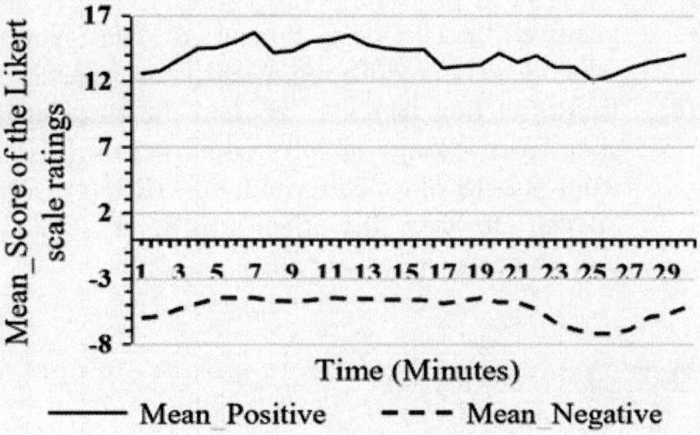

Fig. 38.4 Mean positive and negative affect balance score combined for all the five groups

artifacts' purpose or discuss the expected behavior (Be), where they reported being nervous and anxious.

(2) Synthesis (Be → *S*) and the associated affective state(s)—The synthesis process showed a peak at (2–7 min) and was associated with the positive affect (2–7 min) and at (11–25 min). However, the synthesis process declined from (7–11 min) and again after the 25th minute. The affect mean score was high during this period (11–22 min). This indicates the designers' expending the majority of their cognitive efforts in proposing a structure (*S*) by transforming a set of expected behavior (Be) and reported being inspired and stimulated.

(3) Analysis (*S* → Bs) and the associated affective state(s)—Multiple peaks can be seen at (2–6 min), (9–11 min), (16–18 min), and from (22–29 min). This indicates that the designer's expended most of their cognitive efforts in deriving behavior from the structure at different intervals. The affect balance score declined at (23–25 min) and grew back again (25–30 min).

(4) Evaluation (Bs ↔ Be) and the associated affective state(s)—Evaluation process (2–15 min) was majorly associated with positive affect. Also, the graph drops sharply (22–29 min) with which the mean affect balance score also drops (22–25) and increases at (25–29 min). Designers reported being stressed and overwhelmed before the end, and as they sorted and adjusted their concept they reported being satisfied and blooming.

(5) Documentation (*S* → *D*) and the associated affective state(s)—the graph shows a growth in the documentation process from (2–26 min). This makes sense as the documentation requires partial description of designs through doodles, sketches, and specifications that usually happened once the structure was finalized by the groups. However, in the end, groups reported being stressed and overwhelmed (24–29) for a few minutes.

(6) Reformulation I ($S \rightarrow S'$) and the associated affective state(s)—During Reformulation I, a very short peak in the graph (2–5 min) can be seen during which the mean affect scores also shows a growth for this timeline.

(7) Reformulation II ($S \rightarrow Be'$) and the associated affective state(s)—the graph shows a progression from timeline (2–8 min) which was mainly associated with

positive affects. During this process, designers reframed the behavior space, which leads to modifying the structure space.

(8) Reformulation III (S → F′) and the associated affective state(s)—Function state space change or reformulation is seen in the beginning (2–5 min) and from (12–18 min). Reformulation III has also been associated with positive affects. However, the affect size decreases at the end for reformulation III (18–26 min).

38.7 Discussion

This exploratory study investigated the dynamics of emotions during the design process using the FBS ontology and linkography. We observed that high affect balance scores were associated with high entropy scores of the individuals. Also, positive affects dominated the conceptual phase of the design process, and negative affects were more dominant in the embodiment stages of the design process. However, the syntactic processes differ in their emotional profiles during the early and the later phases of the design process. This is an interesting finding, as it sheds light on the designer's cognitive efforts distributed to the same FBS processes at different time frames resulting in different affective profiles. During the initial phase, designers were interested, curious, and inspired to solve a new task in the beginning, and as time progressed, they got anxious and nervous regarding the deadline. The later stages also demanded designers to converge to a single concept, which was associated with negative affect profiles (stress, nervousness, hesitant, and doubtful). The intensities of these negative affects were moderate as compared to the high activated positive affects. In conclusion, this study was an attempt to understand the designer's emotional experience in with supporting empirical pieces of evidence. However, the affective ratings taken were retrospective, and one can argue regarding the memory loss to recall the specific emotional experiences during the subjective ratings. Despite this fact, this study is the first of its kind to capture emotions using linkography and FBS ontology of design with empirical evidence. Another limitation of the study is the limited design experience of the designers. Studies have shown a difference between the expert and novice designer's design process approach. Studying emotions during the design process is of great importance for understanding the creator's cognitive and behavioral aspects. Several design support systems could be developed to help designers self-regulate their emotions to avoid fixation and improve their design process. For example, Blom [36] proposed a system *(video game)* that adapts to the individuals' emotional state based on facial expression change. Several such systems could be designed and developed.

Extending this research, we have conducted a behavioral experiment with 85 samples using the ML algorithm to capture facial action units in real time while solving multiple creativity tasks. We are still in the process of data analysis. We also wish to do emotional profiling based on the creative stage using time-series analysis.

References

1. Ekman, P., Davidson, R.J. (eds.).: Series in affective science. In: The Nature of Emotion: Fundamental Questions. Oxford University Press (1994)
2. Blanchette, I., Richards, A.: The influence of affect on higher level cognition: A review of research on interpretation, judgement, decision making and reasoning. Cogn. Emot. **24**(4), 561–595 (2010)
3. De Dreu, C.K., Baas, M., Nijstad, B.A.: Hedonic tone and activation level in the mood-creativity link: toward a dual pathway to creativity model. J. Pers. Soc. Psychol. **94**(5), 739–756 (2008)
4. Norman, D.A.: Emotional Design: Why We Love (or hate) Everyday Things. Basic Civitas Books (2004).
5. Singh, A.: Managing Emotion in Design Innovation. CRC Press (2013)
6. McDonagh, D., Hekkert, P., Van Erp, J., Gyi, D.: Design and Emotion. CRC Press (2004)
7. Simon, H.A.: The structure of Ill structured problems. Artif. Intell. **4**(1), 181–201 (1973)
8. Buchanan, R.: Wicked problems in design thinking. Des. Issues **8**(2), 5–21 (1992)
9. Howard, T.J., Culley, S.J., Dekoninck, E.: Describing the creative design process by the integration of engineering design and cognitive psychology literature. Des. Stud. **29**(2), 160–180 (2008)
10. Dorst, K., Cross, N.: Creativity in the design process: co-evolution of problem–solution. Des. Stud. **22**(5), 425–437 (2001)
11. Eastman, C.: On the analysis of intuitive design processes. In: Moore, G.T. (ed.) Emerging Methods in Environmental Design and Planning: Proceedings of the Design Methods Group First International Conference, pp 21–37. Cambridge, Mass (1970)
12. Csikszentmihalyi, M., & Rochberg-Halton, E.: The Meaning of Things: Domestic Symbols and the Self. Cambridge University Press, Cambridge (1981)
13. Desmet, P.M.A.: Inspire and desire. In: Desmet, P.M.A., van Erp, J., Karlsson, M.A. (eds.) Design and Emotion Moves, pp. 108–127. Cambridge Scholars Publishing, Newcastle upon Tyne (2008)
14. Ho, A.G., Siu, K.W.M.G.: Emotion design, emotional design, emotionalize design: a review on their relationships from a new perspective. Des. J. **15**(1), 9–32 (2012)
15. Hutchinson, A., Tracey, M.W.: Designers' own emotions and the practice of designing: a literature review and preliminary research agenda. J. Des. Res. **15**(3–4), 197–213 (2017)
16. Biagioli, M., Grimaldi, S., Ali, H.: Designer's emotions inthe design process. In: Design Research Society 2018: Catalyst, pp. 25–28. Limerick, Ireland (2018)
17. Cross, N., Cross, A.C.: Observations of teamwork and social processes in design. Des. Stud. **16**(2), 143–170 (1995)
18. Dorst, K., Dijkhuis, J.: Comparing paradigms for describing design activity. Des. Stud. **16**(2), 261–274 (1995)
19. Olson, G.M., Olson, J.S.: Distance matters. Human-Comput. Interaction **15**(2–3), 139–178 (2000)
20. Kan, J.W., Gero, J.S.: Acquiring information from linkography in protocol studies of designing. Des. Stud. **29**(4), 315–337 (2008)
21. Gero, J.S., Mc Neill, T.: An approach to the analysis of design protocols. Des. Stud. **19**(1), 21–61 (1998)
22. Hughes, J., Parkes, S.: Trends in the use of verbal protocol analysis in software engineering research. Behav. Inf. Technol. **22**(2), 127–140 (2003)
23. Goldschmidt, G.: Linkography: Unfolding the Design Process. MIT Press (2014)
24. Gero, J.S.: Design prototypes: a knowledge representation schema for design. AI Mag. **11**(4), 26–26 (1990)
25. Kan, J.W., Gero, J.S.: Quantitative Methods for Studying Design Protocols. Springer, Dordrecht, The Netherlands (2017)
26. Meiselman, H.L. (ed.).: Emotion Measurement. Woodhead Publishing (2016)
27. Mauss, I.B., Robinson, M.D.: Measures of emotion: A review. Cogn. Emot. **23**(2), 209–237 (2009)

28. Lang, P.J., Bradley, M.M., Cuthbert, B.N.: Motivated attention: affect, activation, and action. Attent. Orienting Sens. Motivation. Process. **97**, 135 (1997)
29. Pourmohamadi, M., Gero, J.S.: LINKOgrapher: An analysis tool to study design protocols based on FBS coding scheme. In: DS 68–2: Proceedings of the 18th International Conference on Engineering Design (ICED 11), Impacting Society through Engineering Design, Vol. 2: Design Theory and Research Methodology, Lyngby/Copenhagen, Denmark (2011)
30. Ekkekakis, P.: The Measurement of Affect, Mood, and Emotion: A Guide for Health-Behavioral Research. Cambridge University Press (2013)
31. Peilloux, A., Botella, M.: Ecological and dynamical study of the creative process and affects of scientific students working in groups. Creat. Res. J. **28**(2), 165–170 (2016)
32. McMakin, D.L., Santiago, C.D., Shirk, S.R.: The time course of positive and negative emotion in dysphoria. J. Pos. Psychol. **4**(2), 182–192 (2009)
33. Koydemir, S., Şimşek, Ö.F., Schütz, A., Tipandjan, A.: Differences in how trait emotional intelligence predicts life satisfaction: the role of affect balance versus social support in India and Germany. J. Happiness Stud. **14**(1), 51–66 (2013)
34. Cacioppo, J.T., Berntson, G.G.: Relationship between attitudes and evaluative space: a critical review, with emphasis on the separability of positive and negative substrates. Psychol. Bull. **115**(3), 401 (1994)
35. Botella, M., Zenasni, F., Lubart, T.: A dynamic and ecological approach to the artistic creative process of arts students: an empirical contribution. Emp. Stud. Arts **29**(1), 17–38 (2011)
36. Blom, P.M., Bakkes, S., Tan, C.T., Whiteson, S., Roijers, D., Valenti, R., Gevers, T.: Towards personalized gaming via facial expression recognition. In: Tenth Artificial Intelligence and Interactive Digital Entertainment Conference (2014)

Chapter 39
Design of Multifunctional Artefacts as Perceived by Potential Users: Findings from a Preliminary Investigation

Vishal Singh and V. Srinivasan

Abstract Multifunctional designs such as the Swiss army knife, sofa-cum-bed and smartphones are commonplace, but there is limited research to assist designers in determining what factors to consider while designing such products. Therefore, findings from a survey-based empirical study that seeks to understand how potential users perceive the desirability of multifunctional products in which the functions of two or more existing products are combined into one product, are presented. The findings from this study can be used by designers to build better multifunctional designs. The empirical study builds on related theoretical work on sharing in designs, especially structure sharing and resource effectiveness, where factors identified from previous work such as relative importance (RI) of the different functions performed by the product, the quality of functions (QoF) and the emergent negative functions (NF) in the multifunctional product, are found to be relevant. The results and insights from the responses to the survey are presented, and the implications for opportunities for further research are discussed.

39.1 Introduction

Multifunctional designs such as Swiss army knives, sofa-cum-beds and smartphones are commonplace. Intuitively it appears that such products are more desirable because by achieving more functions with relatively fewer structures, one can achieve resource and cost efficiency. In contrast, combining more functions within the same design may negatively influence the core functions and their performance, increase effort required for maintenance, (dis)assembly, etc. Nonetheless, given that such

V. Singh (✉)
Centre for Product Design and Manufacturing, Indian Institute of Science (IISc), Bangalore 560012, India
e-mail: singhv@iisc.ac.in

V. Srinivasan
Department of Design, Indian Institute of Technology Delhi (IIT Delhi), New Delhi, Delhi 110016, India
e-mail: srinivenk@iitd.ac.in

A. Chakrabarti et al. (eds.), *Design for Tomorrow–Volume 2*, Smart Innovation, Systems and Technologies 222, https://doi.org/10.1007/978-981-16-0119-4_39

multifunctional designs are commonplace, it is safe to assume that designers face questions such as: Should two or more functions be combined in the same design? Which functions are better suited together in the same design? How do potential users perceive the desirability of such multifunctional designs? This paper presents findings from a survey-based empirical study on related questions.

The reported empirical study builds on previous theoretical work on structure sharing and resource effectiveness [1, 2]. Structure sharing is one of the four types of sharing in design [3, 4], in terms of the mapping between the functions (*F*) and structures (*S*) in a product [5]. The other three types of sharing include multi-modal integration, function sharing and structural redundancy (see Fig. 39.1). A function can broadly be defined as the objective or purpose of a design as intended by designers. A structure can be defined as a discrete identifiable element and feature in a design to which one or more (partial or full) functions are attributed. Accordingly, a window-pane is an example of structure sharing where the glass panel (S) serves three functions—allow light, enable visibility and provide protection and barrier from outside particles—all at the same time. The pencil shaft (*S*) also serves multiple functions—cover lead, provide grip to hold the pencil to write and provide grip to pencil to erase—but not all at the same time. That is, only one of writing or erasing can be used at any instant in time. In contrast, the four legs (*S*) of the chair collectively serve the same functions—support the weight of the person who is seated on the chair and provide balance—at the same time. That is, neither of the two functions can be independently performed by one of the legs by itself. And finally, the stepney wheel in a vehicle provides a backup for a function that is already being performed by other wheels.

As Chakrabarti and colleagues [1, 4, 6] suggested, it appears in structure sharing and multi-modal integration, sharing modes that collectively represent multifunctional products, tend to be more resource-effective, and hence, are more desirable. However, not all multifunctional designs are likely to be equally desirable. So, what are the factors that designers may need to consider in determining the desirability of a potential multifunctional design?

Ghazanfari and Singh [2] claimed that some of the factors that should be considered in determining the desirability of a multifunctional product (though they erringly only write about structure sharing even though their examples include solutions with multi-modal integration) are: the number of functions and structures in the product;

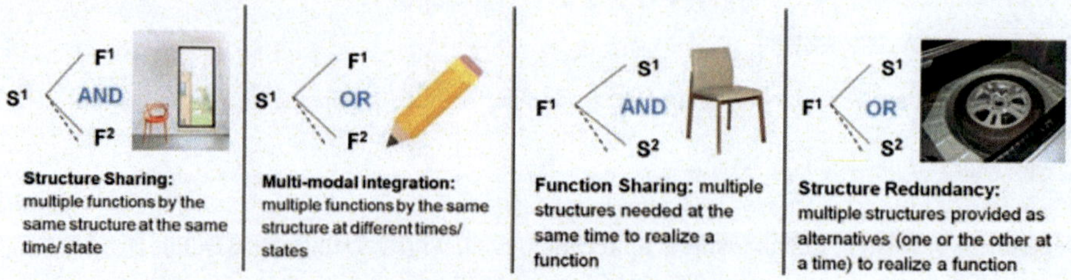

Fig. 39.1 Different types of sharing in design with an example of each

the relative importance (RI) of the different functions performed by the product; the quality of functions (QoF) and; the emergent negative functions (NF) in the multi-functional product that does not exist when the separate functions are not combined in the same product.

Ghazanfari and Singh [2] also proposed an equation to evaluate the admissibility/desirability of such multifunctional solutions using the above variables and provide a preliminary validation of the usefulness of the equation through limited examples. They also claim that the values for variables such as RI and QoF can also be obtained through feedback from potential users via methods such as quality function deployment [7].

Therefore, this paper specifically focuses on understanding the perceptions of potential users about the desirability of multifunctional designs. In particular, this empirical study seeks to test whether it is possible to get an acceptable estimate of values such as RI, QoF and NF through user feedback. Since questionnaire-based surveys are one of the most commonly used and familiar approaches for both researchers and respondents, this study is based on a questionnaire survey.

39.2 Survey Design and Data Collection

A survey is designed to collect data relevant to the perceived desirability of the sample multifunctional products and approximate estimation of RI, QoF and NF. Following are the main considerations in the design of the survey: (1) both direct and indirect questions are considered to assess RI, QoF and NF, and (2) a cross-validation of survey responses is planned by considering three different complementary surveys, each with a separate multifunctional product (see Fig. 39.2). Products *A*, *B* and *C* are chosen such that they include different combinations of—a pen, a USB and a Swiss army knife. The main questions included in each of the three complementary surveys are listed in Table 39.1.

The questions are designed such that it was easier for the respondents to answer the questions and have the same set of factors in cases where it would have been

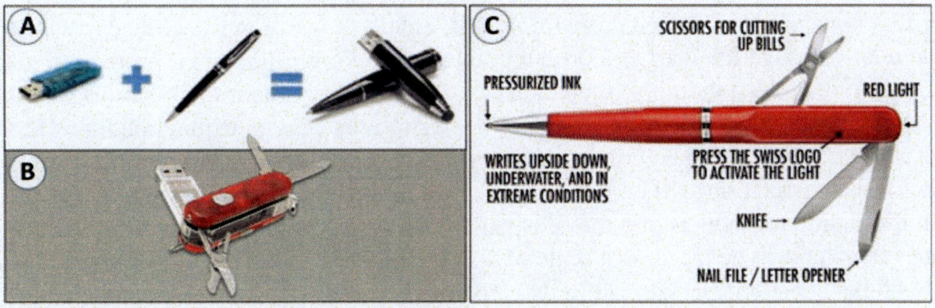

Fig. 39.2 Three multifunctional products used for the three surveys

Table 39.1 Template for questions to be used in the three complementary surveys

Question types (X = A, B, or C)	Description
How would you rate the importance of each sub-product in multifunctional product X?	Direct question related to RI
How important are each of the listed requirements in multifunctional product X?	Direct question related to RI
If you are given a choice to remove one of the functions (least important) from Product X, which would it be?	Direct question related to RI
How much would you be willing to pay for X, given what constituents cost separately?	Question related to the desirability of the multifunctional product
How would you rate the likely effect on the QoF due to the listed changes in the listed attributes?	Directly related to QoF and NF
How frequently do you use each sub-product in a multifunctional product?	Indirect question related to RI, via the frequency of usage
How detrimental is it if you do not own a sub-product but you need one at a given instance?	Indirect question related to RI, via criticality
How readily is a sub-product available if you are not carrying one?	Indirect question related to RI, viathe availability
How long do you use each of the sub-products before replacing it with another one?	A question added to understand relative lifespan compatibility

possible to have wide variations in responses. For example, rather than asking the respondents to conjecture what could be the likely NFs that could emerge in a multi-functional product, a list of potential NFs is given, such that respondents could assess how the attributes associated with NFs could likely influence the QoFs. In addition, various factors that could influence the perception of RI are included. For example, how frequently one uses a product can be an indirect indicator of RI in a multifunctional product. Similarly, if a product is extremely critical even though it is not used frequently, it may be deemed important, e.g., a fire extinguisher is expected to be rarely needed but it is a must-have in every building. Also, if an alternative is readily available when a specific product or function is needed, one may not consider it important enough to be included in a multifunctional product where other products or functions may be deemed more important. Finally, a question is included to assess the relative usage lifespan of a product included in a multifunctional product to get an intuitive understanding of their lifespan compatibility. The implicit conjecture is that the respondents are likely to judge the desirability of a multifunctional product also based on the lifespan compatibility of the sub-products. That is, it is expected that some correlation will be found in the other indicators of the desirability of the multifunctional products and the lifespan compatibility of the sub-products. All the questions have to be rated on a scale of 1–5 or 1–7.

All the questionnaires are administered online via Google forms. All respondents (undergraduate and postgraduate students) are sent a link to all three questionnaires, expecting that they will respond to all three. Since the functions of the three products in the surveys are related to common activities that students pursue, it is assumed that these students are potential users of such products.

39.3 Results and Findings

The results for the survey on Products A, B and C are based on 52, 43 and 32 responses, respectively. The results for the three surveys are separately reported, which is followed by a brief discussion on the observed patterns in the responses.

39.3.1 Survey Results for Product A

Figure 39.3 shows the results from three separate questions. As shown in Fig. 39.3, the RI of the USB in product A is greater than the pen. Among the QoFs available in product A, the capacity of the pen is deemed not as important as the other requirements. Most respondents (mode) are willing to pay only $6 for the combined product, which is the same as the price of the USB. However, the median value is $7, while the mean is $6.71. Median indicates that more than half of the respondents are willing to pay $7 or more.

When a multifunctional product is conceived, combining functions from two or more existing products, it may have other intended and/or unintended effects on the QoFs performed by the combined product. To assess how the respondents perceive the effects of potentially emergent NFs on the QoFs, a list of attributes associated with NFs is given. As shown in Fig. 39.4, the users had no severe concerns about the compromise on the QoFs, as nearly all mean values are close to the neutral value of 4. However, the mode values are particularly more extreme.

Figure 39.5 shows the responses to questions related to the usage of the different sub-products. In Fig. 39.5a, the scales for the frequency-related question is 1–10, while for the other two questions is 1–5, even though the charts are plotted on the same graph. The perceived frequency of usage for both the pen and the USB is high, their criticality is also on the moderate to the high side (scale of 5), while

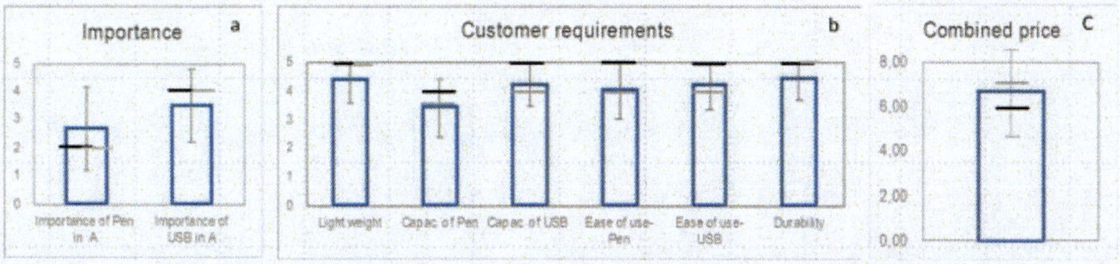

Fig. 39.3 **a** RI of each in A (5 highest), **b** RI of each requirement in A (5 highest), **c** If pen costs 2$, USB cost 6$, how much will you pay for A [Box–Mean; Grey–Median; Black–Mode]

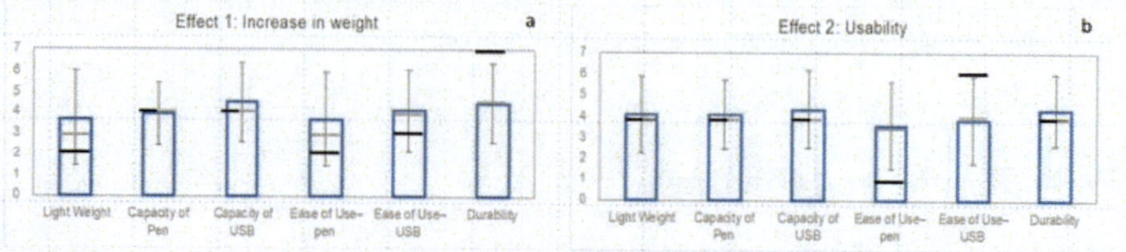

Fig. 39.4 Rate the likely effect on QoF due to **a** increased weight of the combined product and **b** usability concerns in the combined product (High Negative Impact-1 to High Positive Impact-7, 4 = neutral) [Box–Mean; Grey–Median; Black–Mode]

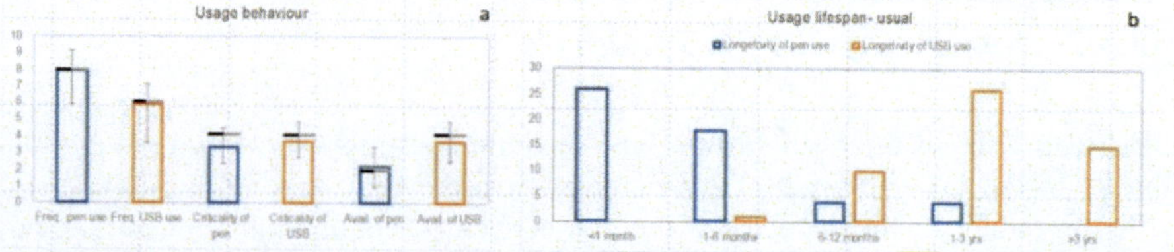

Fig. 39.5 **a** How frequently do you use each? How detrimental is it if you do not own but need one (1: No worries, 10: very critical)? How readily is it available if you need one? (1: easily available, 5: very difficult), **b** how long do you use each product before replacing it with another one [Box–Mean; Grey–Median; Black–Mode]

their availability shows much difference. While a pen is easily available, a USB is relatively less readily available when needed. Similarly, as shown in Fig. 39.5b, there is remarkable difference in the usage lifespan of a pen and a USB.

39.3.2 Survey Results for Product B

As shown in Fig. 39.6, the RI of the USB in product B is high while that of a file is low. File is the most dispensable sub-product in product B (see Fig. 39.6b).

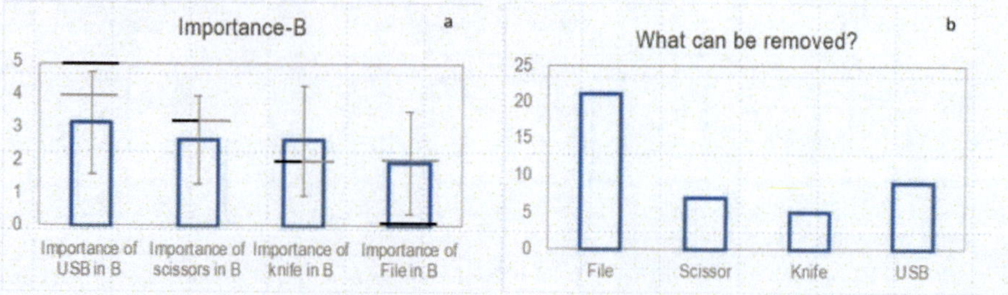

Fig. 39.6 **a** RI of each in B, **b** if you are given a choice to remove one of the functions (least important) from B, which of these would it be? [Box–Mean; Grey–Median; Black–Mode]

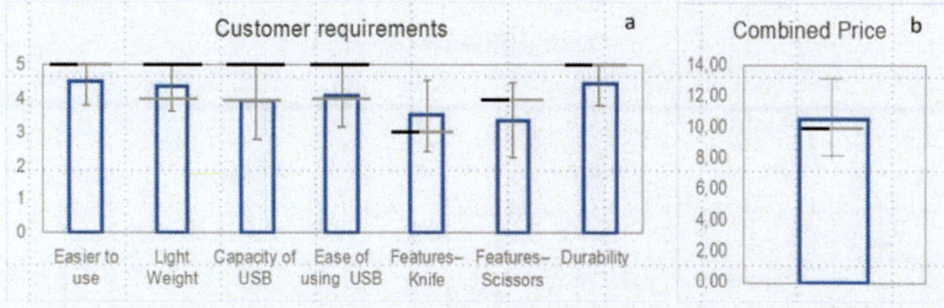

Fig. 39.7 **a** RI of each requirement in B (5 highest), **b** if Swiss knife costs 8$, USB cost 6$, how much will you pay for B [Box–Mean; Grey–Median; Black–Mode]

Figure 39.7a shows that all the functional requirements are rated high, with a slightly lower and moderate value for features associated with the knife and a scissor. Figure 39.7b shows that the mean of the price that the respondents are willing to pay from the combined product is more than that of either sub-products taken separately but less than the sum of their individual prices.

As shown in Fig. 39.8, the mean values across all the three charts are close to neutral, indicating that the respondents did not perceive the QoFs to be significantly impacted by the listed NFs.

The frequency of usage of the USB is much higher than that of a Swiss army knife (see Fig. 39.9a). The perceived criticality of a USB is also higher than that of a Swiss army knife, while a USB is more readily available than a Swiss army knife. As shown in Fig. 39.9b, though there are some notable differences in the perceived usage lifespan of a USB and a Swiss army knife, the two products can broadly be considered to have reasonable lifespan compatibility.

39.3.3 Survey Results for Product C

Figure 39.10a shows that the RI of the different sub-products, with a slightly lower value for the file. This is further confirmed in Fig. 39.10b where most respondents choose the file as the most dispensable sub-product.

Figure 39.11a shows that the RI of the different requirements, based on the mean values is comparable, with a slightly lower value for the LED light. In Fig. 39.11b, based on the mode ($8.00) and mean ($8.27) values, respondents are willing to pay only marginally more for the combined product than what the Swiss knife costs ($8.00). However, the median value suggests that more than half of the respondents are willing to pay at least $9, which is more than the price of the Swiss knife but less than the sum of the prices of the Swiss knife and the pen.

Figure 39.12a shows that the respondents do perceive that the pen's QoF will be negatively impacted by the increase in weight when put together in a combined

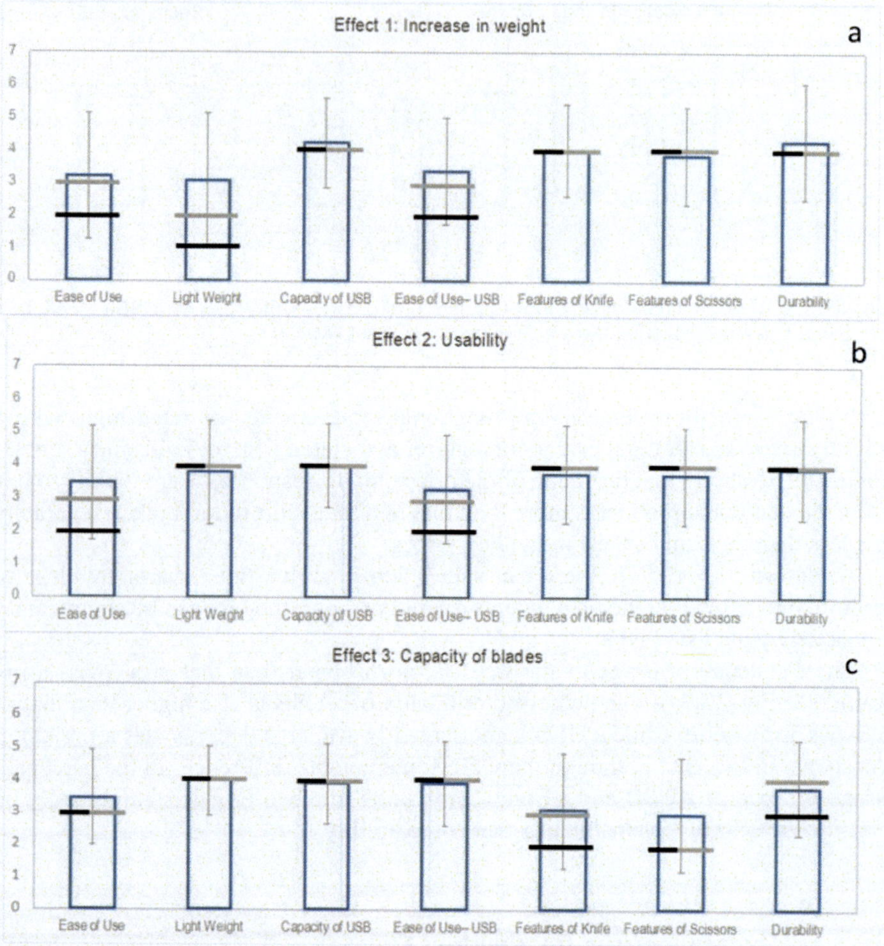

Fig. 39.8 Rate the likely effect on QoF due to **a** increased weight of the combined product, and **b** usability concerns in the combined product **c** capacity/strength of the knife blade (High Neg. Impact-0 to High Pos. Impact-7, 4 = neutral) [Box–Mean; Grey–Median; Black–Mode]

product with a Swiss army knife. Besides the reduced QoF of the pen, there is some concern for the reduced QoF of the knife blade, as shown in Fig. 39.12c.

Figure 39.13a shows that the frequency of use of a pen is relatively much higher in product C. Similarly, the criticality of the pen and its availability, both are noticeably different. Figure 39.13b shows that the pen and the Swiss army knife also do not seem to have a usage lifespan compatibility, whereas the usage lifespan of the Swiss army knife and the LED light are relatively closer.

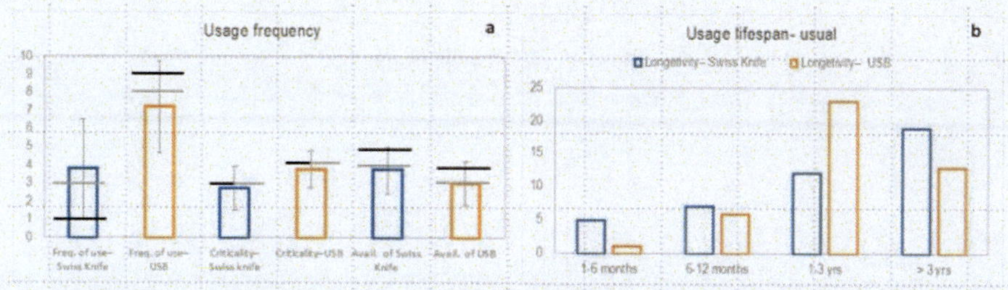

Fig. 39.9 **a** How frequently do you use each product (scale 1–10)? How detrimental is it if you do not own but need one (No worries: 1 to very critical: 5)? How readily is it available if you are not carrying one (easily available: 1 to very difficult: 5)? **b** How long do you use each product before replacing it [Box–Mean; Grey–Median; Black–Mode]

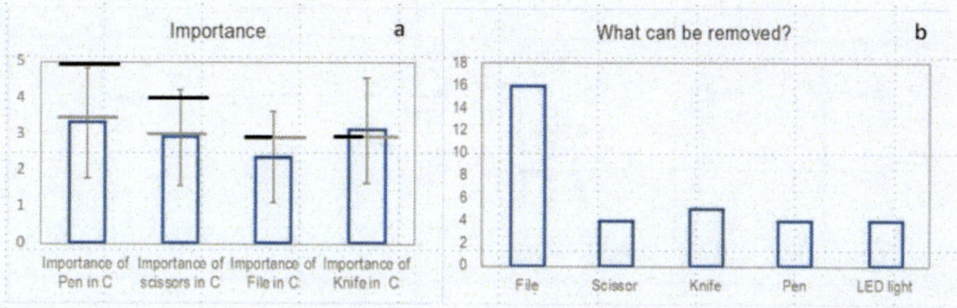

Fig. 39.10 **a** RI of each in C, **b** If you are given a choice to remove one of the functions (least important) from C, which of these would it be? [Box–Mean; Grey–Median; Black–Mode]

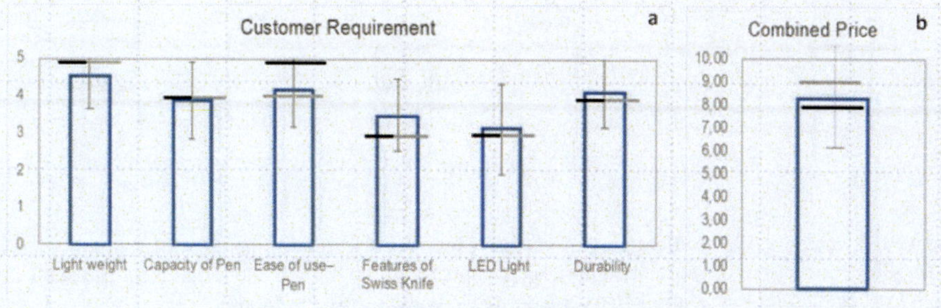

Fig. 39.11 **a** RI of each of the requirements in C (5 highest), **b** If pen costs 2$, Swiss knife cost 8$, how much you will be willing to pay for C [Box–Mean; Grey–Median; Black–Mode]

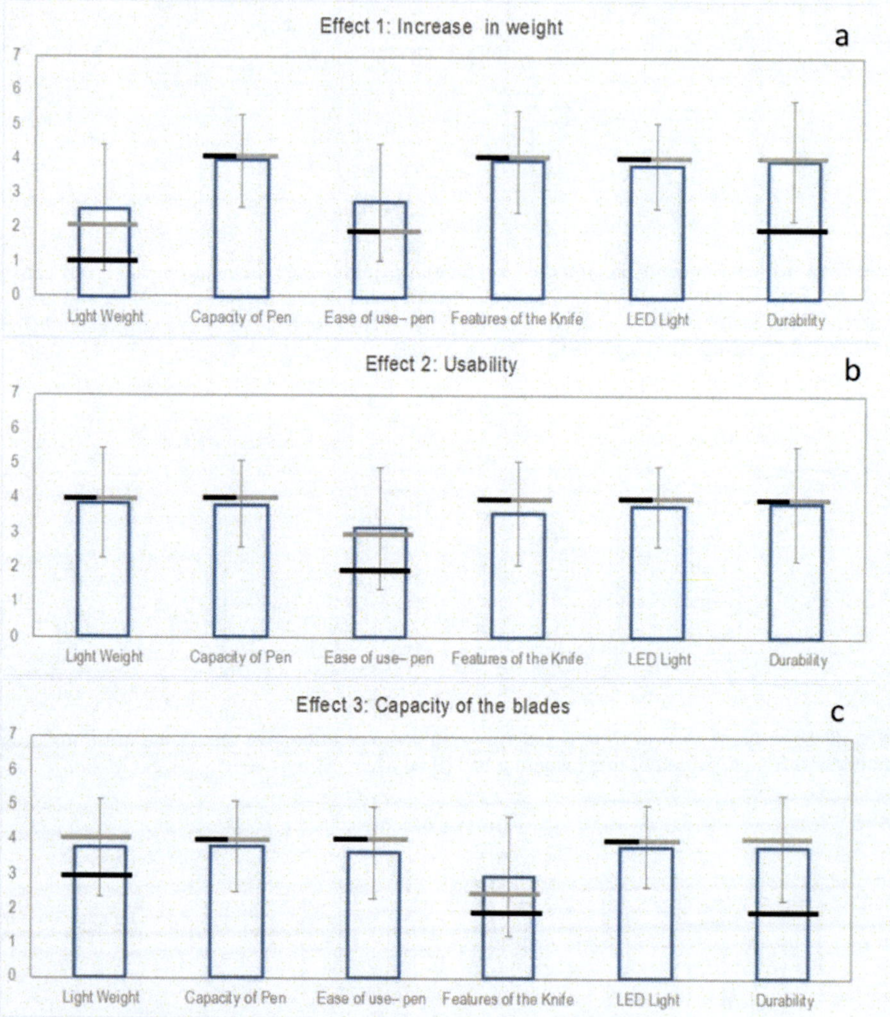

Fig. 39.12 Rate the likely effect on QoF due to **a** Increased weight of the combined product, and **b** usability concerns in the combined product, **c** capacity of the knife blade (High Neg. Impact-1 to High Pos. Impact-7, 4 = neutral) [Box–Mean; Grey–Median; Black–Mode]

39.4 Discussion and Conclusion

The reported empirical study aimed to assess whether potential user perception of the desirability of a multifunctional product can be obtained through a questionnaire survey or not. This includes an attempt to get an intuitive assessment of values for variables such as RI of the functions in multifunctional products, QoFs in such multifunctional products, and the impact of NFs. Both direct and indirect questions

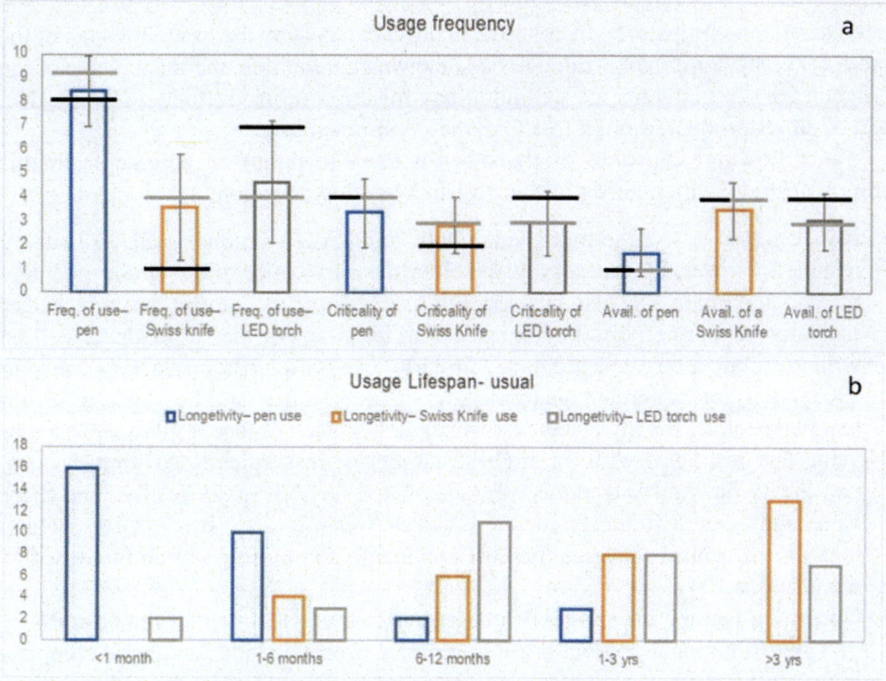

Fig. 39.13 **a** How frequently do you use each (scale 1–10)? How detrimental is it if you do not own but need one (No worries: 1 to very critical: 5)? How readily is it available if you are not carrying one (easily available: 1 to very difficult: 5)? **b** How long do you use each product before replacing it [Box–Mean; Grey–Median; Black–Mode]

were conceived to obtain the desired values. The survey results provide encouraging patterns and indicators that such an empirical approach can complement the theoretical work on developing measures that can support decision making on the design of multidisciplinary products.

In all the three complementary survey results, clear patterns of RI values emerge both through direct and indirect measures. In product A, it is clear that the USB has higher RI than a pen, and an acceptable value can also be derived from the ratings. In product B, the USB, scissor and knife have comparable RI within a range, but the file is clearly rated to have much lower importance. In product C, once again the file is rated much lower than all the other sub-products. Similarly, in all the three surveys, the indirect indicators of RI—usage frequency, the criticality of functions and usage lifespan—seem to provide clear patterns. In particular, usage lifespan compatibility appears to be a particularly useful indicator with apparent congruence with perceived value (via price) of the multifunctional product. For instance, in product A and product C, the mean values of what the respondents seem to be willing to pay for the combined product is nearly no more than the price of one of the constituent products, and in both these cases, the usage lifespan of the main products being combined are rated to be non-compatible. In contrast, in product B, where the usage lifespan of the products being combined is rated to be somewhat compatible, the mean value of the price that the respondents are willing to pay for the combined product is more than that of either product, though less than their combined price.

The following are some of the other noticeable points to emerge from this empirical study, which needs further research and investigation:

- It is common to use the mean value with standard deviations to analyse survey results. However, for this survey, the other measures of central tendency, namely median and mode, can also be used. For instance, the median can provide a clear indicator of a price point beyond which at least half of the respondents will be willing to buy a targeted product. Similarly, most often if a product succeeds to appeal to a certain percentage of the target market, it may still be a success. From that perspective, the mode can also be a good choice because it gives a particular value that resonates with the highest number of respondents. All three values, i.e., mean, median, and mode were calculated for the survey results, and there were significant differences for these values in many cases. It is not clear which measure of central tendency is most applicable and more research is needed to identify this.
- One of the factors that should be considered in the assessment of the desirability of a multifunctional product is the 'functional proximity and functional compatibility' of the products/functions being combined. For example, the survey results show that the file is the least desired sub-product/function but it is still common in a Swiss army knife. The most likely explanation is that a Swiss army knife is a collection of basic tools, and a file belongs to that family of functions. Thus, even though it is not deemed particularly useful, it still fits in because of functional proximity and functional compatibility with the rest of the tools.
- The results seem to indicate that the price that the potential users will be willing to pay for a multifunctional/combined product is likely to be less than the sum of the prices of the constituent products taken separately

In summary, this paper provides preliminary findings from a survey-based study to advance our understanding of factors to consider in the design of multifunctional products. Further work is needed to test and validate how to translate these indicators as inputs into the theoretical equations [2], as well as in modifying and updating the equations to account for factors that have not been considered so far.

To conclude, the empirical responses can complement the theoretical findings and help in the creation of usable and useful multifunctional products. Such responses from potential users can be feedback to designers to take into consideration users' needs while designing multifunctional products.

References

1. Chakrabarti, A., Singh, V.: A method for structure sharing to enhance resource effectiveness. J. Eng. Des. **18**(1), 73–91 (2007)
2. Ghazanfari, E., Singh, V.: Structure sharing for resource effective solutions: improving measures to account for importance and quality of functions. In: DS 87–7 Proceedings of the 21st International Conference on Engineering Design (ICED 17) Vol 7: Design Theory and Research Methodology, Vancouver, Canada (2017)
3. Ulrich, K.: Computational and Pre-Parametric Design, Ph.D. Thesis, Artificial Intelligence Laboratory, MIT Cambridge (1988)
4. Chakrabarti, A., Regno, R.: Sharing in design: categories, importance and issues. In: Proceedings of Int'l Conference on Engineering Design (ICED01), pp. 563–570, Glasgow (2001)
5. Gero, J.S.: Design prototypes: a knowledge representation schema for design. AI Mag. **11**(4), 26–36 (1990)
6. Chakrabarti, A.: A new approach to structure sharing. ASME J. Inf. Sci. Eng. (JCISE) **1**(1), 1–78 (2004)
7. Akao, Y. (ed.): Quality Function Deployment: Integrating Customer Requirements into Product Design. Productivity Press, Cambridge, MA (1990)

Chapter 40
Patterns and Insights From a Design Group's Mobile Chat–Preliminary Findings and Commentary on Participation in Online Communities

Vishal Singh and V. Srinivasan

Abstract The discussions in online groups and communities that share a common identity and interest can provide useful insights into various attributes and behaviors of the group as well as their beliefs and thinking. Such discussions may also be useful to make preliminary conjectures and hypotheses about the wider community they identify with. Accordingly, this paper seeks insights from an extended group chat among the alumni of a design school of a reputed institute over a logo rebranding exercise commissioned by their alma mater. An almost collective sense of dissatisfaction among the design alumni with the approved rebranded logo of their alma mater ensured that the discussion extended over a month, providing a rich volume of data and discussion to analyze and reflect on. While the situation was somewhat unique because of the sentiments attached to their alma mater, the range of discussions covered several comments on the outcomes (designs), process, and people (stakeholders). Therefore, this paper presents a preliminary analysis of the discussion. This analysis sought to gain insights into their thinking and beliefs, both as individuals as well as a group. Implications of the findings for design education as well as participation in online communities are discussed.

40.1 Introduction

Ever since mobile devices and instant messaging social platforms have become commonplace, most people are part of one or more chat groups, ranging from small

V. Singh (✉)
Centre for Product Design and Manufacturing, Indian Institute of Science (IISc), Bangalore 560 012, India
e-mail: singhv@iisc.ac.in

V. Srinivasan
Department of Design, Indian Institute of Technology Delhi (IIT Delhi), Delhi, New Delhi 110016, India
e-mail: srinivenk@iitd.ac.in

A. Chakrabarti et al. (eds.), *Design for Tomorrow–Volume 2*, Smart Innovation, Systems and Technologies 222, https://doi.org/10.1007/978-981-16-0119-4_40

family-based groups to professional chat groups and communities. Such social platforms also facilitate ongoing engagement and continuity of groups and communities such as college batchmates and alumni who transition into a connected online community that retains a common history of shared physical experience and identity. Consequently, it is reasonable to claim that online communities, including closed communities that share a common ground and interest, are increasingly becoming part of our social life [1], which in turn may contribute to our social cognition and learning [2]. For instance, the discussions in such groups may reveal who has been doing what, what are the trending professional or social issues that are directly or indirectly relevant to the specific group or community, and what are the group dynamics and the social network structure within the community [3, 4]. Another particularly notable point with these group chats is that unlike the face-to-face conversations, these conversations automatically get documented, making it easier to retrospectively follow the sequence and flow of the conversation. Therefore, this research investigates a specific sequence of discussion in one such closed mobile chat group—an extended group chat among the alumni of a design school of a reputed institute over a logo rebranding commissioned and adopted by their alma mater.

The paper presents preliminary patterns identified in the design group discussion around the logo rebranding that was done by their institute. The discussion on this topic extended over more than a month and a half, providing a rich dataset that can be analyzed both qualitatively and quantitatively from various perspectives, ranging from the content of the discussion to the patterns of interaction among the group members. Assessment of such conversations or discussion is a well-established research approach that is used regularly in design research [5] as well as several other fields such as psychology, organizational studies, media studies, education, nursing, healthcare, and so on [6–8]. A qualitative—interpretive and subjective—analysis can be applied at both higher and lower levels of granularity such that the volume of data is not particularly critical. In contrast, for a structured and objective content analysis that relies on statistics, the volume of available data can be critical to have reasonable confidence in the identified patterns. Therefore, in the dataset available for this research, both the approaches can be adopted, as discussed and demonstrated in this paper. Accordingly, this paper provides preliminary insights into broad qualitative and quantitative patterns. This paper serves as an introductory article from research where the further in-depth analysis will follow in later articles.

40.2 Context

The design school that the group is associated with is part of a well-known research-driven science and technology institute that primarily offers a postgraduate program in product design, with a particular focus on a balanced approach to technology and design. Consequently, the students in the program typically come from architecture or engineering backgrounds and graduate as trained product designers who work across diverse design fields such as the design of consumer goods and products, furniture design, automotive design and styling, user experience, and user interface design, healthcare product design and medical device design, industrial product design, design education, and so on. This diversity of the professional working background and experience of the group members is important to highlight for multiple reasons. First, despite their training as product designers, the group members work across a wider range of design disciplines, and some a bit closer to the areas such as visual communication. Second, since the formal training of all the group members is in product design, besides the introduction to fundamental elements of design and some training on visual aspects of design, they are not trained as specialists in visual communication. It is likely that this pattern is not unique to this group, and it appears from general observation in the field that designers trained in one specific sub-area of design do end up working across diverse design fields, often acquiring further insights and specialization into these complementary fields through practice and experience [9]. Consequently, at a finer level of detail and from the acquired design expertise and experience perspective, the group members in this case represent a heterogeneous group even though it appears to be a homogeneous group by training. In addition, the group includes more than 250 active members representing more than 20 different academic batches (yearly), a few design research alumni with doctoral degrees, and a few faculty members, which further adds to the group's heterogeneity in age and experience.

Furthermore, even though this online chat group has been active for more than five years with regular discussion on various topics, this event, that is, the rebranding and redesign of the institute logo, generated a much more engaging and relatively longer-lasting discussion around a single topic. Multiple reasons can be posited for this overwhelming response. First, the institute logo is part of the institute identity and brand, which is the fundamental common ground and shared identity that binds this group together. Second, as trained designers, many of the group members feel qualified as well as responsible to respond to a logo redesign action, which has also raised muted concerns in the general media. Third, as part of the institute's alumni, some of the group members also believe that design school and the design school alumni should have been consulted as stakeholders in the decision process. And, perhaps one of the most significant factors, which seems to also implicitly reflect in some of the comments in the group, is the scope of the design activity. That is, logo design is seemingly viewed as a challenge that does not require any specialization and training. Consequently, unlike some of the other domain-specific design problems that have previously been discussed on the same forum, the logo design problem appears to be an inclusive challenge, within everyone's reach and competence.

Therefore, this unique case context provides an opportunity to not only get case-specific insights but also an opportunity to understand patterns and observations that are potentially generalizable to (1) design, designers, and design communities, and (2) closed online groups and communities.

40.3 Related Literature and Approaches

Critical analysis of online chats and discussion is not a new approach. There have been several studies that have looked at discussion patterns across a diverse range of online groups, communities, and forums ranging from closed chat groups to open microblogging sites to professional and social discussion forums on the Web [3, 4, 10]. Once again, while the text data on these online platforms lend themselves readily suitable for content analysis and research on the patterns of interaction, similar methodological approaches and research objectives have existed for analysis of transcripts of verbal data and conversations using techniques such as verbal protocol analysis [5, 11] and conversation analysis [12, 13]. That is, there are numerous research approaches such as content analysis [14], thematic analysis [7], protocol analysis, and conversation analysis that broadly fall in a similar category of largely qualitative research methods that can also have a complementary quantitative approach in some cases, such as content analysis [7]. Accordingly, most of these techniques have been applied across diverse disciplines, but the terms have also occasionally been used interchangeably, especially content analysis and thematic analysis. While content analysis and thematic analysis are equally usable for analysis of synchronous and asynchronous communication such as analysis of documents and monographs, the other two techniques, especially verbal protocol analysis is used typically for synchronous verbal data, which may also reveal the thinking process and cognitive aspects via a think-aloud process. This is not to say that content analysis or thematic analysis may not reflect the thinking, but the content written after due deliberation and thinking may not be the same as content obtained through a think-aloud process. Nonetheless, without going into the details of the differences between these seemingly similar techniques in this paper, it suffices to state that content analysis is more suitable for the dataset available in this research.

40.4 Research Methodology

This research combines both qualitative and quantitative research methods. While the qualitative approach can adopt a purely interpretative and subjective analysis of the conversation, a more structured content analysis approach also allows some quantitative analysis even though it is also primarily a qualitative method [7]. A content analysis approach often requires data segmentation and a coding scheme that can be applied to the collected data.

Figure 40.1 shows the research framework and research steps. One of the advantages of working with chat messages is that the data is already in form of a segmented text, with each message as a separate segment. As the first step, the messages are copied and transferred to a spreadsheet. This data is anonymized before further processing. Once the data is anonymized (removing personal details and other

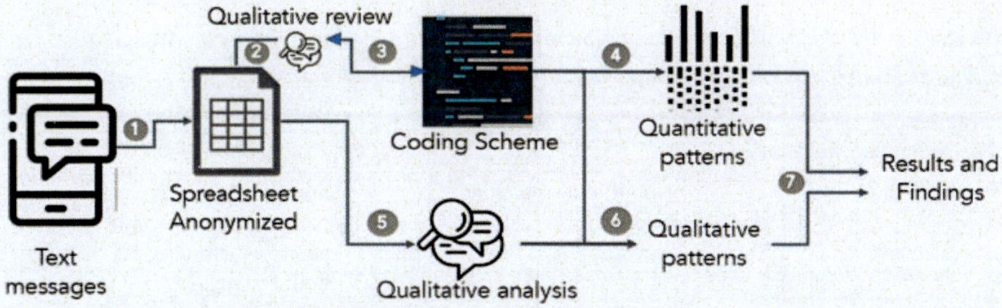

Fig. 40.1 Research framework and the research steps

specific information that can breach the privacy of individuals), an initial qualitative review of the data is conducted as a preliminary assessment of the content (step 2 in Fig. 40.1). This preliminary assessment helps to devise the coding scheme that may allow more structured and in-depth analysis of the data and the patterns that seem apparent from the initial qualitative review. However, this step (step 3 in Fig. 40.1) is an iterative process, which is also informed by related classifications and terms that may be relevant to the targeted coding scheme. In addition, it may involve going back and forth between the qualitative review and the coding scheme to modify and update the choice of codes and terms in the coding scheme as it is applied. Once the coding scheme is applied to the data, it allows a qualitative analysis of the quantitative patterns in the data (step 4 in Fig. 40.1). Besides the quantitative patterns revealed through the codification process, an in-depth qualitative analysis is possible by adopting a subjective and interpretive approach to understand the discussion at various levels of granularity (steps 5 and 6 in Fig. 40.1). For example, at the broadest level, an entire discussion session can be taken as a unit of analysis, and in contrast, even a single message or segment could be dissected further for interpretive analysis. The final step (step 7 in Fig. 40.1) involves putting together the observations from the quantitative patterns as well as the qualitative interpretations to make sense of the findings.

This paper presents preliminary findings from these research steps. More research is needed for further validation of the coded segments and findings. For instance, it is desirable that the coding of the data is independently done by multiple researchers and analysts for cross-validation. The current set of findings presented in this article is based on data coded by a single researcher, and hence, it is prudent to consider these findings preliminary.

The details of the coding scheme are shown in Table 40.1. Based on the initial qualitative review, three categories of codes are found relevant:

- The subject of the comment, which includes products, processes, and people.
- Tone and tenor of the comment, which includes appreciation, encouragement, objectivity, sympathy/empathy, nostalgia, criticism, questioning/skepticism, defensiveness, disappointment, judgment, and sarcasm.

Table 40.1 Coding scheme applied in the research—description with illustrative examples

Coding scheme	Description	Representative comments (random samples)
The subject of the comment		
Product	About a design or its properties,	For me, this shape and composition resemble a leaf, a petal, a surfboard, etc. if I go literal way…
Process	About an activity and how it is done	Hence let us conceptualize as far as we believe that the options have enough scope to improve
People	About stakeholders and actors, both individual as well as organizational	This logo is a stunning example of how highly talented people think they know what design is and how clueless they are about design and aesthetics
Tone and tenor of the comment		
Appreciative	A positive comment on something or someone	A picture speaks a thousand words
Encouraging	A comment to encourage a further continuation of an activity or idea	Really appreciate the effort. But I think it should be much simpler and clean
Objective	An attempt for balanced comment, often citing a source or theory	And as in product design, it is very important that we think more about the fundamentals and the questions and not how to simply come up with the answers
Sympathetic/ empathetic	Comment with an effort to try and understand a situation or an actor	And thus even logo design is in itself a very, very thought intensive and intricate job …
Nostalgic	Referring to memories	Lot of memories here!
Critical	Toward pointing shortcomings and limitations	If one does want to leverage a metaphor, then abstraction and composition ought to support it. Both are lacking in this case
Questioning/Skeptical	A comment raising doubts and suspicion	I'm also strongly hinting at their competence …
Defensive	Statements to defend an idea/action/actor, occasionally preemptive	Not necessarily … both do different things and have different skills … both need each other

(continued)

Table 40.1 (continued)

Coding scheme	Description	Representative comments (random samples)
Disappointed	Expressing dissatisfaction implicitly or explicitly	If it's such a high budget project, as the costs which have been discussed I am super disappointed by the …
Judgemental	Passing or pronouncing a verdict	Finally, the buck should stop at the top heads of the two entities
Sarcastic	Expressing a strong negative sentiment in a lighter note	The triple rings of the new logo are costlier than wedding rings
Attributes/elements of design being referred to		
Visual	About visual attributes	For me this shape and composition resembles a leaf, a petal, a surfboard etc. if I go literal way …
Cultural	About cultural and historical context	The crest already has brand recognition and reflects the heritage of the Institute
Psychological	About emotional and psychological aspects	Something about the shape has visual pleasantness
Philosophical	About beliefs or principles that allude to an explanation of why and how	A logo is difficult to put a price on …
Rhetorical	A comment intended to impress, persuade or influence, often beyond apparent and explicit attributes	The new identity is inspired by the perpetual motion and dynamism of the atom, the wisdom and transcending capabilities of the lotus and the flame of knowledge. It embraces change while also maintaining a level of continuity
Ideological	About beliefs or principles that are taken as a guiding rule or position	Here is an attempt to find a middle ground between a traditional institutional logo and a modern one

- Attributes and elements of design are referred to, which include visual, cultural, psychological, philosophical, rhetorical, and ideological.

Table 40.1 provides a brief description of each of the codes, including a sample of a coded segment relevant to each of the codes. It must be noted that all the segmented data are coded across all three categories wherever applicable. In addition, many segments, including some listed as examples in Table 40.1, are found to be relevant to multiple codes within the same category. For example, the statement "*Really appreciate the effort. But I think it should be much simpler and clean*" is not only encouraging, but it is also critical. Nearly 2800 messages are available in the dataset, of which there are several diversions on topics that are not related directly to the topic. Thus, out of the 2800 segments, only around two-third of these segments are coded. Nonetheless, while these diversions and interruptions (non-relevant segments) could have been removed altogether from the dataset, these are retained and marked such that the number of discussion sessions could be identified, and assessed on why and how the discussions on the specific topic of interest resumed despite repeated diversions, and how the patterns across these sessions changed, if they did.

40.5 Preliminary Results and Findings

Figure 40.2 shows the pattern of discussion as revealed through the code-based content analysis. The top part of Fig. 40.2 shows that there were at least sixteen different discussion sessions that were interrupted mostly by other discussion topics during the one and half months of discussion. While some of the discussion sessions were short, especially toward the end (14–16 in Fig. 40.2), there were some intensive sessions that were packed within a much narrower timespan. For instance, sessions 2–5 were spread between 3 days, and similarly, sessions 6–14 were spread between 7 days. A breakup of each of the sixteen sessions in terms of the number of times each code appears in the discussion in the specific session is shown in Fig. 40.2.

For example, the discussions in the brief first-session primarily focused on the logo and the people involved—the consultants who designed the new logo and the decision-makers at the institute who approved the redesigned logo. Most of the comments were critical, objective, and sarcastic, referring to the visual, ideological, cultural, and philosophical aspects of the design. Similarly, the discussions in the second session were more substantial, including comments about the product (logo), the process, as well as the people (involved stakeholders). The discussions continued to be critical, questioning, and skeptical, revealing disappointment, sarcasm, and a judgmental tone. These discussions covered all the elements of design—visual, cultural, ideological, psychological, rhetorical, and philosophical—in that order. There was also appreciation and encouragement for group members who proposed the need to raise the issue further and explore design alternatives that could be proposed to the institute to demonstrate and convince how a better redesign of the logo was still possible.

Nonetheless, without going any further into the specifics of each of the sixteen sessions in this paper, it can be seen from the overall patterns in Fig. 40.2 that the discussions gradually moved from criticism, disappointment, and sarcasm that was directed toward the redesigned logo, the process, and the stakeholders to discuss on the alternative designs proposed by the group members. Nonetheless, even for the design proposals from the group members, there were plenty of critical, skeptical, and judgmental comments, largely related to visual, cultural, ideological, and rhetorical aspects of the proposed concepts. There were occasional comments of appreciation and encouragement as well, but as the discussions progressed, those who volunteered

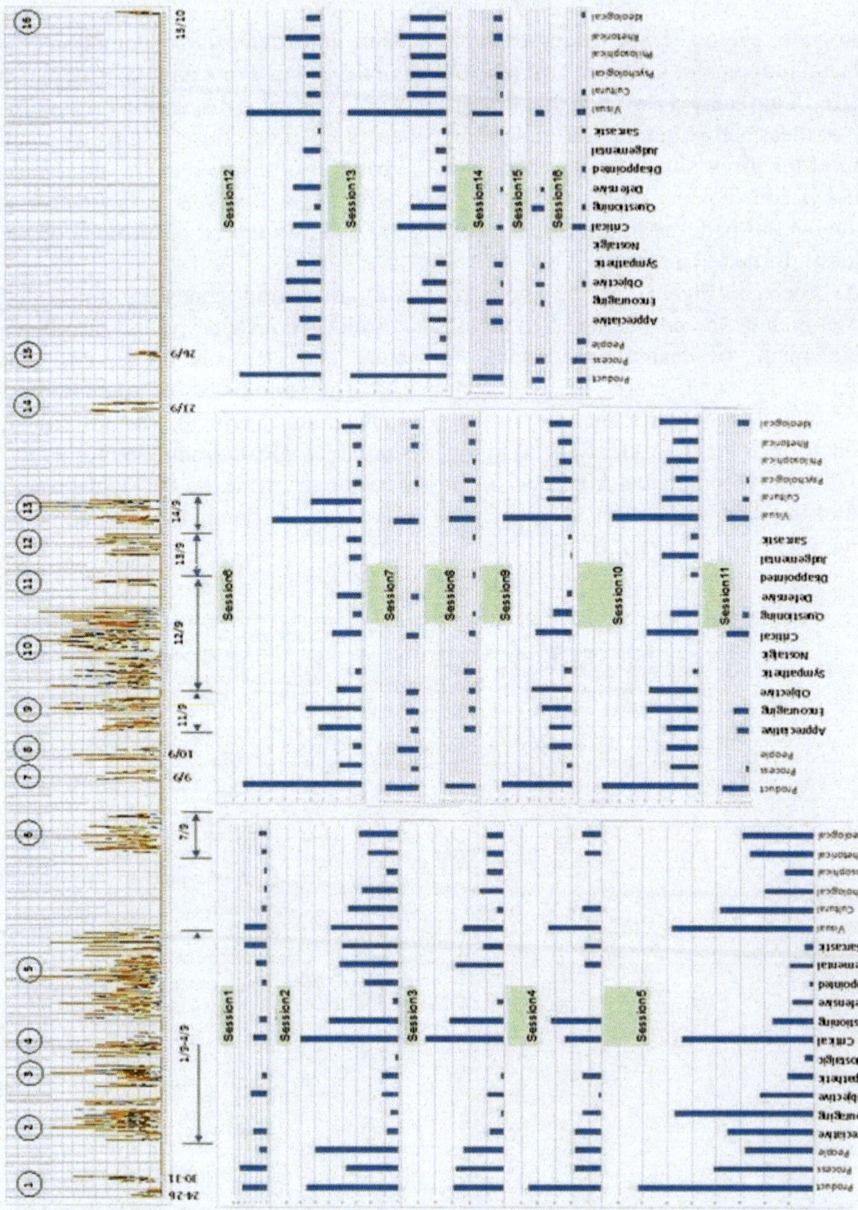

Fig. 40.2 Quantitative patterns of chat discussion revealed through the coded and segmented data

to propose design alternatives increasingly became more defensive, preempting critical, and judgmental feedback from the others in the group. For example, statements such as "I am sorry, I assumed that this group would welcome discussion on a multitude of ideas before getting into designing concepts …" reveal the changing tone and tenor of the discussion. Similar changes in the patterns of discussion are observable across various aspects, which will be reported in future publications. As an overview of the overall patterns in the discussion, Fig. 40.3 shows the co-occurrence of the codes in the coded dataset.

As seen in Fig. 40.3, the most frequently occurring codes in the coded dataset include the product (design solutions or related concepts), visual (attributes and elements of design), ideological (attributes and elements of design), and critical (tone and tenor of the comment). The top 10% of the highest co-occurring code-pairs include product-visual, product-critical, visual-critical, product-ideological, product-encouraging, visual-ideological, product-cultural, and visual-cultural comments. All the co-occurring code-pairs in the top 10% and the next 10% of the dataset are marked in enclosed cells and underlined cells, respectively, in Fig. 40.3.

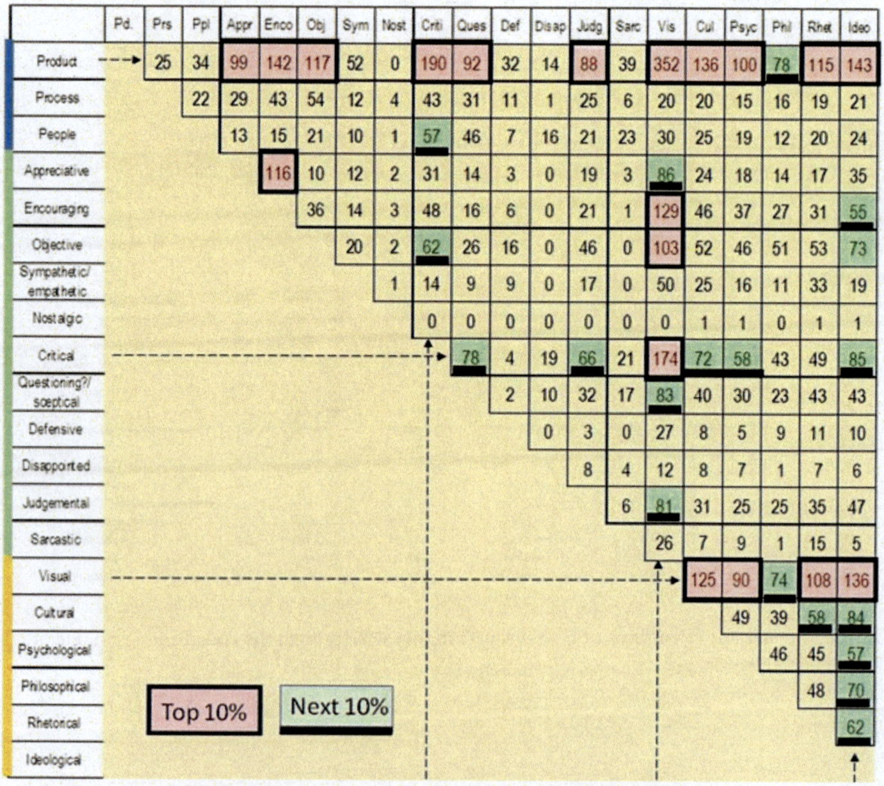

	Pd.	Prs	Ppl	Appr	Enco	Obj	Sym	Nost	Crti	Ques	Def	Disap	Judg	Sarc	Vis	Cul	Psyc	Phi	Rhet	Ideo
Product		25	34	99	142	117	52	0	190	92	32	14	88	39	352	136	100	78	115	143
Process			22	29	43	54	12	4	43	31	11	1	25	6	20	20	15	16	19	21
People				13	15	21	10	1	57	46	7	16	21	23	30	25	19	12	20	24
Appreciative					116	10	12	2	31	14	3	0	19	3	86	24	18	14	17	35
Encouraging						36	14	3	48	16	6	0	21	1	129	46	37	27	31	55
Objective							20	2	62	26	16	0	46	0	103	52	46	51	53	73
Sympathetic/empathetic								1	14	9	9	0	17	0	50	25	16	11	33	19
Nostalgic									0	0	0	0	0	0	0	1	1	0	1	1
Critical										78	4	19	66	21	174	72	58	43	49	85
Questioning?/sceptical											2	10	32	17	83	40	30	23	43	43
Defensive												0	3	0	27	8	5	9	11	10
Disappointed													8	4	12	8	7	1	7	6
Judgemental														6	81	31	25	25	35	47
Sarcastic															26	7	9	1	15	5
Visual																125	90	74	108	136
Cultural																	49	39	58	84
Psychological																		46	45	57
Philosophical																			48	70
Rhetorical																				62
Ideological																				

Fig. 40.3 Co-occurrence matrix for the different codes for the overall coded dataset

40.6 Discussion and Conclusions

The preliminary results presented in this paper demonstrate that the social interactions in an online chat group can reveal various aspects of the groups and group members' beliefs and thoughts about their areas of common interest, and how they engage and interact with each other.

For instance, the findings that the coding scheme developed specifically for this dataset turned out suitable for coding and revealing the patterns of interaction in the group suggests that the group members implicitly or explicitly recognize that designs and designing comprise visual, cultural, psychological, philosophical, rhetorical and ideological elements and attributes. Similarly, the criticism of the redesigned logo was often coupled with questioning and skepticism of the design consultant's capabilities as well as that of the entire process of design management and design review. This suggests that the group members not only recognize the intertwining of the products, processes, and people but at the same time, it appears that it is often difficult to review and judge a design artifact independently, without making attributions and judgment toward the designers and their emotions.

Similarly, the pattern of discussion also revealed various aspects of group dynamics. For instance, while some of the group members readily questioned the capability of the design consultant (an outsider) based on their designs, there was never any attempt to question a group member's capability even though the concepts proposed by the group members were equally criticized, if not more. A clear pattern emerged in how group members typically responded to a design concept proposed by one of their own. The first few responses were almost always appreciative and encouraging until one of the members mixed encouragement with criticism. Once a hint of criticism broke the ice, the tone and tenor of responses turned negative toward criticism, skepticism, and judgmental comments. There is clear evidence that group members were initially guarded and careful not to offend anyone within the group, but their position changed once someone else created an opportunity. Such a pattern seems to indicate that the group can be characterized as a semi-formal closed group where group members are neither absolutely frank nor too formal to not be openly critical. For example, referring to the pattern of discussion itself, one of the group members stated the following in relief to a critical comment from another member—"Thank you XX! I was wondering how to say this thing in this august forum".

The pattern of interactions also provides clear evidence of social cognition and learning. For instance, as the discussions progressed, the group members who volunteered to propose alternative concepts adopted a defensive approach and qualified their new proposed concepts with an explanation such as "Ideation on the logo … pls ignore the text contrast/ positioning/height for the time being…the feeling to be conveyed was boldness, presence of the institute, 'crowning' of science …". Such a qualified introduction of the new concept shows how this member recognized the pattern of criticisms and preempted the kind of critical comments he may receive if not explained upfront. Thus, this member not only learnt what kind of criticism can be expected but also learnt and developed his own strategy on how to engage with the group and how to present his ideas for a more favorable response.

In summary, though this paper presents only preliminary findings from an ongoing research, the paper demonstrates that online group chats and discussions can be insightful when appropriate qualitative and quantitative research methods are applied in conjunction.

References

1. Shove, E., Pantzar, M., Watson, M.: The Dynamics of Social Practice: Everyday Life and How it Changes. Sage Publications, London (2012)
2. Shum, S.B., Ferguson, R.: Social learning analytics. Edu. Tech. Soc. **15**(3), 3–26 (2012)
3. de Wever, B., Schellens, T., Vallcke, M., van Keer, H.: Content analysis schemes to analyze transcripts of online asynchronous discussion groups: a review. Comput. Educ. **46**(1), 6–28 (2006)
4. Dong, H., Cheung Hui, S., He, Y.: Structural analysis of chat messages for topic detection. Online Inf. Rev. **30**(5), 496–516 (2006)
5. Gero, J.S., McNeill, T.: An approach to the analysis of design protocols. Des. Stud. **19**(1), 21–61 (1998)
6. Blake, C., Scanlon, E.: Analysing online discussions in educational and work-based settings. In: Proceedings's of the 9th International Conference on Networked Learning 2014, pp. 25–32 (2014)
7. Vaismoradi, M., Turunen, H., Bondas, T.: Content analysis and thematic analysis: implications for conducting a qualitative descriptive study. Nurs. Health Sci. **15**(3), 398–405 (2013)
8. Gu, N., Singh, V., London, K., Lijiljana, Taylor, C.: BIM: Expectations and a Reality Check. In: Unknown, (ed.) 12th International Conference on Information Technology in Construction, 2008, Beijing, China: QUT Digital Repository (2008)
9. Ball, L., Pollard, E., Stanley, N.: Creative Graduates, Creative Futures. Report for Council for Higher Education in Art and Design, London (2010)
10. Chew, C., Eysenbach, G.: Pandemics in the age of twitter: content analysis of tweets during the 2009 H1N1 Outbreak. PLoS ONE **5**(11), e14118 (2010)
11. Ericsson, K.A., Simon, H.A.: Protocol Analysis: Verbal Reports as Data. The MIT Press (1984)
12. Heritage, J.: Conversation analysis: methodological aspects. In: Quasthoff, U.M. (ed.) Aspects of Oral Communication, pp. 391–418. Walter de Gruyter, Berlin/New York (1995)
13. Giles, D., Stommel, W., Paulus, T., Lester, J., Reed, D.: Microanalysis of online data: the methodological development of "'digital CA."' Dis. Context Media **7**, 45–51 (2015)
14. White, M.D., Marsh, E.E.: Content analysis: a flexible methodology. Lib. Trends **55**(1), 22–45 (2006)

Chapter 41
Our Machines

Louise Colbourne

Abstract Our relationship with machines has evolved to the point where it dislocates our connection to the natural world. *Our machines* are a series of curated exhibitions and events which include a selection of video art and objects that explore emerging technologies and reflects on our lives as cyborgs. Artists and designers often work at the forefront of new technological eras, exploring, utilising and adapting new apparatus, materials and systems. By looking at the evolution of some of the work of these practitioners, it is possible to trace a trajectory of the development of new means to express transformation, empathy and affinity to one another and with technology. This text considers how the selected works within these exhibitions enhances our post-human existence by offering imaginative and positive experiences of our lives as cyborgs.

41.1 Introduction

41.1.1 Initial Concept

> Our Machines are disturbingly lively, whilst we ourselves frighteningly inert [1].

Artists and designers whose pioneering work with and about *machines* over the last century have been far from *inert*. By testing and pushing the boundaries of new technologies, their work engages with our experiences of technoscience. This paper looks at the first in a series of curated exhibitions which brings together a selection of work made throughout the last five decades. The *Our Machines* project has been on-going for the last 5 years bringing into focus a wide range of cross-discipline and inter-generational art and design work. The curatorial concept was to give a collective voice to the creative works that have at times broken apart devices and systems in order to both question and re-humanise the latest technological developments. Some

L. Colbourne (✉)
University of Brighton, East Sussex, UK
e-mail: lncs@springer.com

A. Chakrabarti et al. (eds.), *Design for Tomorrow–Volume 2*, Smart Innovation, Systems and Technologies 222, https://doi.org/10.1007/978-981-16-0119-4_41

of the frequently probed questions within the exhibition are about the necessity for limitless progress, or to point out that references to the natural world are fast disappearing. The endeavours of the artists and designers are carried out in a spirit of openness and freedom. Freedoms that have often been afforded by the arrival of new technologically advanced tools, which have offered the opportunity for these practitioners to seize the means of production [2].

Most of the works included in the exhibition are time-based media; considered illusionary and non-existent, yet instrumental in asking the viewer to question a world, where multiple views show reality in a state of flux [3]. The focus of this presentation will be on specific examples of work from the exhibition that offered experiential, immersive interactions for the viewer, challenging the notion of the illusionary and non-existent and refusing to be experienced as *inert*. Our relationship with machines has indeed evolved, and in the age of the Anthropocene, does the brining together of this selection of work communicate a restorative affinity with nature, each other and machines?

41.1.2 Exhibition Programme Design

For the purposes of this paper, I will be concentrating on the first exhibition held at observer arts, at the observer building in Hastings, UK in 2016. The design of the exhibition and the supporting printed material are important aspects of this project. The curation aims to both reflect the concepts of the overall exhibition and to offer a platform for the individual works within it to resonate in their own right (Fig. 41.1).

Starting with the reading of Donna Haraway's *Cyborg Manifesto* (1985), the title for the exhibition came out of the above quote. Conversations with peers initiated the momentum to challenge the notion of Haraway's statement that our relationships with machines are 'disturbingly inert'. Haraway's statement is in many ways true and

Fig. 41.1 *Our Machines*, exhibition poster, L. Colbourne 2016

prophetic; now that our dependency on devices in the current age is essential, and our data is gathered and stored without a full understanding of why or how. However, the statement by Haraway quickly bought to mind a list of artists and designers who have challenged this assumption of *inertia.* The curatorial gambit became one that would attempt to plot a trajectory of the pro-active interventions with technologies and machines that artists and other creatives have established through their working practices and within recent history.

As well as reflecting on this particular statement from *A Cyborg Manifesto* [1], it soon become relevant to reflect on the relationships that the various art works also had on the overall call of the manifesto; to reject rigid boundaries between the human and non-human machines, and to an assert an affinity between them. So exuberantly does Haraway proclaim this concept that in last line of the manifesto, she states that she 'would rather be a cyborg than a Goddess.' [1].

The *Our Machines* exhibition includes a selection of work that often require dark, quiet spaces, the challenges of this resulted in some surprising out-comes. The occasional bleed between the sights and sounds of the works as they were played out in the exhibition space, were experienced as unique and serendipitously engaging. This evocation of 'conversations' between the different voices of the artists, eras and mediums, echoed across the vast industrial space, asserting the notion of affinity between the works and their collective purpose (Fig. 41.2).

The location used for the exhibition in Hastings was at a former print works for a regional newspaper. It was a very large 600 m^2 space, which once housed vast printing beds. Some of the work by the artists uses the latest in new digital media, and it was evocative that this exhibition was placed in this once industrious, mechanical factory floor. The publication for the exhibition was printed as a small six page spread on tabloid-sized newsprint, which was printed locally to align with and reflect on the previous occupancy of the building.

Fig. 41.2 *Our Machines*, exhibition installation view, L. Colbourne 2016

41.2 Themes

41.2.1 Overview

The exhibition (and associated events) was put together around three areas for discussion or as quasi themes which either romanticised, humanised or criti-cised the subject of technology.

For purposes of this exhibition, it helped to ground some of the similarities or indeed differences, between the works chosen within these themed groups. This helped to establish a structure for dialogue and to emphasise certain frames of reference. However, these themes were not articulated in the publication or even in the curatorial journey as such, they were predominantly set out as points for symposium discussion. Most art does not seek to centre itself on the communication of a single concept or theme, traditionally the work of a designer seeks to come up with definitive answers to a brief or problem. Yet more frequently designers are initiating projects as self-authorial endeavours, often aimed at positive change. Designers and design collectives are more frequently being associated with the endeavours of what we might traditionally experience as a work of art; exampled in the work of Forensic Architecture or Theaster Gates.

41.2.2 Romanticise

Historical interactions with spaces, places and things can offer romantic associations to new technologies. As new machines emerge in the present and have done so in the past, they inform how we might imagine a future [4]. Early cinema offered a magical yet challenging experience for the viewer as the depicted scenes enthralled audiences. This was also experienced earlier with the Magic Lanterns of the seventeenth century; audiences were as perplexed as they were mystified by the sequences of slides depicting scenes of the surreal and mythical, played out in darkened rooms with the aid of candle light. The arts are quick to explore and utilise the mechanisms of new technological epochs, and in doing, so many stories can be told about the possibilities of other worlds and new horizons. *Our Machines* considers various points of reference that resonate with these imagined futures that at times attempt to re-instate the sublime through new aspects of nature and landscape or beyond (Fig. 41.3).

Band 9 is an installation that considers aspects of nature through a techno-scientific lens. This image shows the installation of nine light boxes, which each show a digital image of cloud data. This data has been captured by a remote sensing satellite, which orbits the Earth [5]. Semiconductor (the artist duo based in Brighton) has used optical sensors to collect the reflected light in various wavelengths of the electromagnetic spectrum. By focusing on very thin slices of these, scientists can pinpoint individual

Fig. 41.3 *Band 9* 2015, installation view, semiconductor

phenomena such as the bands we see in this installation. The machine used to collect this data is designed to reveal high-altitude clouds called cirrus.

Another piece of work in the exhibition also reflects on the natural world is *Voice Windows*, 1990 by Steina and Woody Vasulka and Joan La Barbara. This piece included vocal elements and voice patterns by the singer Joan La Barbara that were then intercut with images of desert landscapes. Steina and Woody Vasulka then added speed changes, directional variations, and additional electronic voice processing to the final piece. The sounds of La Barbara's voice become a visually rendered rhythm of movement resonating within an electronic land/sound scape (Fig. 41.4).

Saskia Olde Wolbers' films are painstakingly crafted to appear as if they were digitally rendered but are in fact made of analogue processes. Wolbers, carefully films model sets underwater, they have been dripped in paint and lit to disorientate the sense of scale. The underwater action of dripping paint appears to be happening in slow motion but it is playing out in real-time. These lo-fi sculptural and chemical

Fig. 41.4 Paredolia 2011, installation view, Saskia Olde Wolbers

processes are beautifully rendered. *Paredolia* (2011) is accompanied by a voice-over of a reading of Eugen Herrigel's book *Zen and the Art of Archery*. The work, however, subverts any notion of the truth of filmed reality and creates instead a dream-like surreal experience for the viewer.

41.2.3 Humanise

This group if work seeks to communicate the complexity of the human experience from the domestic to the global. The works often imbue a sense of empathy and connectivity through the use of teleporting contraptions, humour and scientifically rendered voice patterns.

The cinematic genius of Buster Keaton entered into the audience's experiences of the spectacular illusionism of cinema at the beginning of the last century. His antics were thrilling and engaging but also offered a humorous distraction to what was challenging new technological apparatus; the cinema view of large out-of-scale images. Keaton reflecting humanity through his physicality and use of the cinematic frame as devise to communicate with audiences and draw them into the narrative with slapstick humour.

Senses of scale and a call to see the bigger picture inspired Stewart Brand to ask the question; 'Why have not we see an image of the whole Earth?' In his campaign to do just that he ran in 1965. When NASA did release a photograph of the whole Earth in 1967, Brand used it for the cover of his magazine the *Whole Earth Catalogue* [1968–72]. Brand's catalogue was full of tools and tips to live greener and sustainable lives, making it a forbearer of Internet search engines, the opening lines of the catalogue stated:

> We are as gods and might as well get good at it.

This was all part of the groundswell of positive thinking of the time in reaction to the Vietnam War and civil unrest [8]. Other designers also had dreams to change the world for the better such as Buckminster Fuller, and Ken Garland's the *First Things First* manifesto of 1964 called for more meaningful design. These designers and innovators saw a way forward from advertising campaigns and consumer capitalism of the time; towards a more empathic, humane future.

Later, there was growing opportunity for women in the field of design through the new technological developments of apple computers in the 1980′s. With digital technology emerging, the designer Muriel Cooper was inspired to set up the visible language workshop at MIT. She wanted to see what happens when designers are given direct access to the means of production:

> I was convinced that the line between reproduction tools and design would blur when information became electronic and that the lines between designer and artist, author<> and designer, professional and amateur, would also dissolve [9].

Within the work on display as part of the *Our Machines* exhibition was a curious machine by the design team: David Chatting, Paulina Yurman, Jo-Anne Bichard and

David Kirk. It was called *Family Rituals 2.0 Ritual Machine 'Anticipation.'* This bespoke technology was made for a busy couple that lived apart from each other during the week. The machines placed in each of their homes meant that they could share moments of reflection, allowing them to communicate resonances about their work/life balance with each other. The operational functions of the machine are based around the everyday rituals of the home, which are missed in their day-to-day separation from one another (Fig. 41.5).

'Screen Machine' is a device created by Paul Sermon and Charlotte Gould to offer public audience participants interactive play between different spaces and places. The audience have the opportunity to co-create chance encounters between friends or other random participants and to then initiate self-direct spontaneous performances between two separate installation locations. Inspired by urban and cultural surroundings and the re-contextualized of locations and spaces 'Screen Machine' which aims to allow audience members the agency over the outcomes of this intervention, akin to a tele-present flux happening (Fig. 41.6).

Vesna Petresin's piece '*explores the topology, boundaries, immersion through embodiment and the notion of gaze. It situates the space of the body in relation to its environment, and the sound of human voice in relation to technology. Particle fields informed by the body geometry interact with the space that surrounds the body; their transformation is driven by acoustic parameters of the piece of music performed by the artist. The pattern shaped by the body and the voice becomes a crystallised sonic pattern*' [3].

Fig. 41.5 *Screen Machine* 2016, installation view, Charlotte Gould and Paul Sermon

Fig. 41.6 *Audiomorph* [Can I Wear My Song] 2016, video still, Vesna Petresin

41.2.4 Criticise

Key moments in history such as the industrial revolution created spiritual shifts as much as they heralded a practical force for change. In the current era, machine learning and generative processes are becoming the normative practices of interactive design. Artists such as Hito Steyerl use AI to make informed art work such as *Power Plants* 2019, but to what end? Has the evolution of technology as a grand utopian gesture, in fact, become a passive extension of the human experience; perhaps reinforcing our creative powers by amputating our natural ones [3]. In this section, we look at some of the more critical voices of the selected artists. They may communicate a call to action for technology to become less invasive in our lives or to utilise it for expressing the politically urgent (Fig. 41.7).

Dara Birnbaum makes use of the video editing software as soon as it becomes possible to carry out VHS post-production on a small budget. By editing excerpts from the 1970's TV series *Wonder Woman*, Birnbaum's expresses a feminist viewpoint as well as her skill as an editor and montage artist. This pro-active assemblage

Fig. 41.7 *Technology/Transformation: Wonder Woman*, 1978/9, video still, Dara Birnbaum

come into full play with *technology/transformation: Wonder Woman 1978/9*. The explosive bursts of fire open the sequence, the repetitive action throughout the video works to deconstruct the ideology embedded in the television icon. Wonder Woman is trapped in her metamorphic state as she endlessly turns, full circle through a myriad of chopped edits. The radical manipulation of the female icon re-appropriates the female forms and asserts a new strength and power both in the scenes she has edited and in her own command of the medium.

For *Disambiguation* 2009 James Richards and Steve Reinke exchanged discs of music, stray footage and fragments of existing works for each other to remix, add alternate soundtracks to and re-order [15, 16]. The work used and edited together for this work had been collected and downloaded from years of trawling the Internet by James Richards. The earlier analogue sound and video recordings had been collected by Steve Reinke. The result of editing these two archives together is becomes something between a compilation, a new work and a curated programme. Like an exquisite corpse forged from their attempt find a common voice that speaks of the dissolution of self. The dissolution of meaning, dissolution of image, dissolution of the original, ends up being a raw, disjointed, free, abstracted, visceral, retinal, discordant, close, distant, disembodied, liquid, implacable, democratic, dub montage.

In Luke Pendrell's *Rictus* 2016, a similar fracturing occurs to Richards and Reinke. This work '*inhabits a broken time populated with dislocated fragments of the digital Sargasso. Stuttering jpegs, feedback loops, glitches and fractures, thwart our navigation through a world choked within its own visual detritus* [14] (Fig. 41.8).

In Matthew Noel-Tod's piece, a blazing inferno turns a 3D image of the world orbited by four cartoon children. Encircling the globe is the Latin palindrome; *In girum imus nocte et consumimur igni* (We go in circles into the night, we are consumed by fire) [15]. The phrase is taken from the title of Guy Debord's final film and was used because it is attributed to the behaviour of moths around fire. In Noel Todd's, *A Season in Hell 3D* the children are in constant movement, never landing, never leaving, far from innocent the other-worldliness of the game environment is far from playful. *Purgatory is overwritten by the ecstasy of the spectacle* [17].

Fig. 41.8 *A Season in Hell 3D* (2014), 3D video loop still, Matthew Noel-Tod

41.3 Conclusion

In conclusion, *Our Machines* has bought into focus the affinities and complexities between disciplines, processes and communicative means of our experience of the cyborg. Artworks that reflect on the sublime within nature such as *Band 9* by semiconductor may seek to communicate the human experiences of nature as reflective and enlightening, offering extraordinary perspectives of the world around us, whereas traditionally, design has a more problem-focused approach to the here and now. However, by considering the trajectory of brand, fuller and cooper, we have witnessed their call to humanise technoscience and work towards long-term change. More frequently now and with a growing sense of urgency the role of the designer has shifted to one whereby their projects are evolved by research-active practitioners who intend to add to the restoration of our world, such as the extraordinary work of *Forensic architecture*; the research group based at Goldsmiths University in London who have unearthed architectural evidence in cases of war crimes and human rights abuses. Design for change, does not focusing purely on making *better* things for *better* futures but sticks with and addresses current problems.

Whilst looking back to the past, we are alerted to what needs changing now, for a brighter future. Aspects of some of the critically engaged works included in the exhibition speak directly of the urgency of our over-reliance on social media and the accumulation of big data. Richards and Reinke show us that the dissolution of the 'self' is fragmentary as we all become a series of images of infinite abundance, now ever more prevalent via digital uploads. Birnbaum's transformative video of Wonder Woman speaks directly to Haraway's call in her manifesto that we move beyond gender and transform into cyborgs. This point is echoed both in the reading of the video and in the making of it; that these women have become one with technology.

Many of the works in the exhibition do begin to pin point certain aspects of Haraway's manifesto, from transformation to affinity and pro-activity in the face of the possible inertia. These practitioners not only accept technological advances, but they embrace, question and utilise them to communicate aspects of our techno-scientific lived experiences. Gathering these works together in this series of exhibitions and events has shown a light on the possible affinities we have with each other, nature and technology. Our human condition in relation to technology is communicated in an attempt to negotiate and articulate a vision of a brighter future, by being fully engaged with and immersed in the urgent present.

References

1. Haraway, D.: A cyborg manifesto. Soc. Rev. 65–107 (1985)
2. Marx, K.: Capitol, A Critique of Political Economy, vol. 1. Penguin Books, New York (1976)
3. Petresin, V.: Our Machines. Available at: ourmachines.blogspot.com. Accessed 7 June 2020 (2016)
4. O'Kane, P., Sun, L.: Technologies of Romance Part 1. Eeodo, London (2017)

5. Semiconductor.: Band 9 2015. Available at: https://semiconductorfilms.com/art/band-9/#:~:text=Band%209%20is%20an%20installation,sensing%20satellite%2C%20orbiting%20the%20Earth.&text=The%20angle%20the%20light%20boxes,been%20captured%20and%20archived%20at. Accessed 6 June 2020
6. Olde Wolbers, S.: Pareidolia (2011). Available at: https://www.saskiaoldewolbers.com/works/pareidolia-stills-synopsis/. Accessed 6 June 2020
7. Vasulka, S. La, W., Barbara, J.: Voice Windows (1986). Available at: https://www.foundation-langois.org/html/e/page.php?NumPage=466. Accessed 6 June 2020
8. Kirk, A.: Appropriating technology: the whole earth catalog and counterculture environmental politics. Environmental History **6**(3), 374 (2001)
9. Cooper, M.: Computers and design. In: Armstrong, H. (eds.) Digital Design Theory; Readings from the Field. Princetown Architectural Press, New York (2016), pp. 64–71
10. Sermon, P., Gould, C.: Container: research.brighton.ac.uk. Available at: https://research.brighton.ac.uk/en/publications/bdf-screen-machine. Accessed 6 June 2020
11. Chatting, D.: Family Rituals—Family Two (2015). Available at: https://davidchatting.com/familyrituals/two/. Accessed 6 June 2020
12. Petresin, V.: Our Machines.: Homepage, ourmachines.blogspot.com. Last Accessed 6 June 2020
13. Birnbaum, D.: Technology/Transformation: Wonder Woman (1976/7). Available at: https://eai.org/tiotles/technology-transformation-wonder-woman. Accessed 6 June 2020
14. Pendrell, L. Our Machines
15. Available at: ourmachines.blogspot.com. Accessed 7 June 2020 (2016)
16. Richards, J., Reinke, S.: Disambiguation (2009). Available at: https://lux.org.uk/uk/work/disambiguation. Accessed 6 June 2020
17. Noel-Tod, M.: A season in Hell 3D. Available at: https://www.bannerrepeater.org/matthew-noel-tod-a-season-in-hell-3d. Accessed 6 June 2020

Chapter 42
A Qualitative Study of Global Design Practices to Build Directions and Opportunities for Indian Social Design

Tanishqa Bobde and Raman Saxena

Abstract The design industry including design knowledge and practice is constantly evolving itself to fulfill the evolving needs of society in order to deliver a larger social impact. Social design is a field that stresses the positioning, responsibility and collective societal impact of designers and designed products. The underdeveloped and developing world can benefit largely from social design interventions. India is a developing country with social problems like poverty and poor health care—giving a large scope for social design work. However, major development of social design is limited to other countries. A qualitative study was conducted with 14 global design practitioners to understand the social design theory, knowledge and practice existing globally and identify directions and opportunities for the Indian context, by using a cultural lens. The study suggests creating open dialogue, simplifying vocabulary, bias-busting, fostering cross-cultural social design and encouraging mindfulness as the main opportunities. These opportunities can be applied both in social design practice and design education.

42.1 Introduction

Social design has gained rapid traction over the years and can be traced to writers like Victor Papanek [1] and Victor Margolin [2]. As design thinking has progressed, there has been a growth of socially centered designers tackling critical social issues like mental health and poverty [3]. These issues are associated with the social sector which is central to the overall socio-economic development of a country [4].

India is a country of over 1.2 billion people [5]. A large chunk of society is deprived of basic needs including education and health care. 67 million Indians comprise the poorest half of the population [6]. India is now at the low rank of 112 in terms of

T. Bobde (✉) · R. Saxena
USID Foundation, NCR Region, Noida 201301, India
e-mail: tanishqa@usidfoundation.org

R. Saxena
e-mail: raman@usidfoundation.org

A. Chakrabarti et al. (eds.), *Design for Tomorrow–Volume 2*, Smart Innovation, Systems and Technologies 222, https://doi.org/10.1007/978-981-16-0119-4_42

gender equality due to poor economic opportunities for women among many more reasons [7]. 25.6% of Indian population is illiterate [8] and India also faces social problems like fake news [9] and poor mental health [10]. Despite these social issues, social design in India is limited and a slowly developing practice. There is a lack of innovative reflection on the diverse perspectives emerging from this field, especially compared to other countries [11].

Current social impact work in India is dominated by social enterprises and NGOs who lack knowledge and skills of human-centered design [12]. There are 1.5 million NGOs in India, 21.3% work in the area of community and/or social service. About one in five NGOs work in education, 17.9% in sports and culture and 6.6% in the health sector [13]. They reflect social development work but not necessarily social design work.

In other countries, social design projects have emerged for similar problems, e.g., The Good Kitchen, a government kitchen service for the aging population in Denmark. The social problem is that 60% of Denmark's seniors in assisted living facilities or residential care units have poor nutrition, and 20% are malnourished. The result is health problems, a low quality of life and a greater economic burden on the government. Design firm Hatch & Bloom did rigorous ethnographic design research and created a solution which involved improving the working experience of kitchen staff, increasing interaction between the kitchen staff and the people they serve, redesigning the menu and changing the name from 'Hospitable Food Service' to The Good Kitchen. This saw a 500% increase in food orders in the first week itself [14].

Design methodologies, like participatory design, that are used in the process of social design have been primarily developed in the West [15] and have gained little attention in terms of being built for the culturally different developing world.

Culture plays an important role in design and has an influence on social behavior and hence social design [16]. Cross-cultural design gives designers a greater possibility to develop their own intentions and design languages, as well as more creative and relevant outputs—which are produced by observing seemingly 'hidden' dimensions of a particular societal culture [17]. There is a need to take a closer look into this topic and the globally emerging standards and design tools like citizen participation from a cultural lens [11].

Outside India, there is not only the development of social design-related theories and tools but also various social design education initiatives [18–20] against which there is only one full length Master in Social Design course in India (School of Design-Ambedkar University, Delhi) [21]. There is a need to blend design study with other domains like the social sciences [22]. As Meyer and Norman [23] have stated, tackling social issues through design requires a holistic design education.

It is clear that social design can play a critical role in solving social issues and in spite of having a large scope for such social design interventions in India, most of the interventions are being carried out by Non-Governmental Organizations/NGOs which lack skills and expertise in social design. There is a need for exploring new approaches, theories, conversations and concepts of social design from the lens of

Indian culture [11] keeping in mind the current nascent state of Indian social design and the simultaneous development of social design in other parts of the world.

Determined from the above, this study explores the following research question: Based on the social design knowledge and practices existing among designers globally, what are some theories and opportunities for development of social design in India, from a cultural lens?

42.2 Research Methodology

The research was conducted using a qualitative research methodology with semi-structured interviews applying a cultural lens. This methodology facilitated the understanding of global social design practices and knowledge from designers while contextualizing it to India with the cultural lens. This cultural lens was guided by an understanding of three widely recognized models, the Iceberg Model of Culture proposed by Hall [24]; the six dimensions model of national culture proposed by Hofstede [25] and the cultural dimensions proposed by Hall [26, 27]. These models supplement each other well [28].

42.2.1 Cultural Models

The Iceberg Model of Culture by Edward T. Hall (1976). Hall [24] proposes that in any culture, there are some parts that are visible (above the iceberg) but a larger portion is hidden (see Fig. 42.1).

The external or conscious part of culture is visible and includes beliefs and behaviors. The internal or subconscious part consists of values and thought patterns that underlie this behavior. He suggests that the only way to understand this hidden, internal part of any culture is by actively participating in that culture.

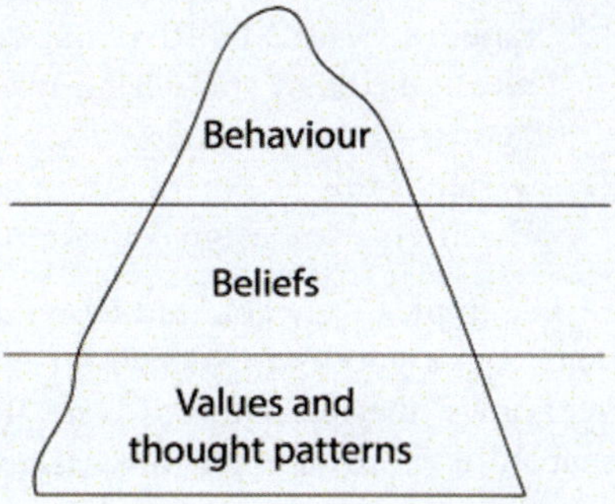

Fig. 42.1 Visual depiction of Hall's Iceberg Model of Culture

Six dimensions model of national culture by Hofstede. Cultural dimensions [25] are overarching independent characteristics that describe a culture's behaviors and preferences like the relation to authority, for example. These are used to understand cultures at a national level [29]. Hofstede's model proposes six cultural dimensions: power distance (to what extent people tolerate unequal distribution of power), masculinity (the extent to which personal achievement and material reward are valued), long-term orientation (ability to adapt to rapid innovative changes, i.e. creativity, curiosity and liberalism), uncertainty avoidance (tolerance to ambiguity), indulgence (extent to which people are optimistic and experience joy in simple things) and individualism (how society members identify themselves, i.e., 'I' versus 'We'). Hofstede defined this widely recognized model based on an extensive worldwide survey of employee values from 1967 to 1973 done by IBM.

Cultural dimensions by Edward T. Hall. Edward T. Hall proposes the cultural dimensions of space perception (comfortable physical distance) [26], context of communication (implicit and high context communication or explicit and low-context communication) [24] and time perception (preference to do one task at a time meaning monochronic cultures or more tasks at a time meaning polychronic cultures) [27]. This model supplements the model of Hofstede by giving directions to help understand communication and task management by different cultures.

42.2.2 Semi-Structured Interviews

Semi-structured interviews were conducted using video calls individually with 14 designers who were geographically distributed (ranging from Australia and Europe to Asia and Africa) and all had work experience in India. Semi-structured interviews were an effective tool in the context of this study because they help gain in-depth data of samples' perspectives, experiences and stories [30] and enable better expression of interviewee viewpoints compared to other tools [31].

The samples were selected using the following criteria:

1. Experience working in social design in varied design domains (experience ranging from 2 to 10 years; domains ranging from service design, communication design and design research to product design, exhibition design and human–computer interaction).
2. Varied levels of experience in design (student to expert).
3. Experience in design education (experience ranging from 3 to 15 years).

14 samples were selected based on the above criteria. Guest et al. [32] reported that 12 is a good sample size for a relatively homogenous sample set. Before the interviews, the interviewees were familiarized with the cultural models since this enabled a more in-depth discussion about social design from the lens of Indian culture. Semi-structured interviews enabled open expression of the interviewees' narratives around the topics: what is social design, what processes and practices are

followed in social design, does working in different cultures affect these processes and what opportunities exist for social design development in India's cultural context.

The qualitative data obtained from these interviews was analyzed through a thematic analysis [33] wherein common themes were identified from topics and ideas that came up repeatedly. These themes led to a set of results which were used to answer the research question.

42.3 Results

Table 42.1 shows the themes that emerged and the number of interviewees who brought them up. The remaining interviewees who did not bring up a certain theme does not indicate disagreement, but simply that they did not bring up this theme during the interview. This quantification will help determine the priority and relevance of the themes in regards to the research question.

While discussing global social design, all the interviewees mentioned that social design is about the collective and is a complex process which should be simplified through use of simple and inclusive language. It is speculative and critical, but grounded in reality. It starts with education and is not necessarily tangible.

On how social design can be practiced in India, all of them focused on the constant flux and heterogeneity in India. They stated that due to this heterogeneity, cultural restraint, hierarchies [34], and high context, implicit communication [35], there's a need for modification of flat design processes like participatory design and even standard design research interviews which seem relatively simpler in other more

Table 42.1 Summary of themes that emerged and number of interviewees who brought them up

Theme	Number of interviewees who brought it up ($n = 14$)
Questioning and dialogue about the role, scope and point of entry of social design should be a part of Indian social design	13
Reexamine jargons and use simpler language	11
Informal education and dialogue with non-design communities about social design	10
Sensitize designers towards their own personal issues	9
Cross-cultural design work and partnerships with Indian firms who already have rapport with the target communities	8
In India, social design should not be used just for vulnerable/'poor' communities	6
Encourage open student dialogue and questioning in Indian design schools	5

homogenous and low-context countries. Co-creation sessions and interaction with the non-design community in India currently requires a long period for breaking the ice. To tackle this, lateral thinking and quick adaptation is needed.

> There is a lack of beautiful solidarity in Indian slums. (interviewee, design anthropologist)

Before diving into the problem areas with existing design methods, time should be spent questioning what role design is playing here and why it is required in this context. Informal education and dialogue about social design among the non-design community in India was pointed out as an opportunity. Collaboration and two-way knowledge exchange are important for Indian social designers due to India's cultural diversity—which makes it harder to understand the larger social community that one is designing for, and the designer's positioning within the same. In this sense, the non-design community members should also be seen as a form of social designers.

> No matter how well we think we know people, they know themselves better. (interviewee, service designer)
>
> Professionals in environmental planning and urban development used design as a tool to protect the ecological systems of Guayaquil while managing an integrated provision of both housing and ecological infrastructure during the Designing Inclusion Summer School: Guayaquil, Ecuador [36]. It involved engagement with economically, environmentally and socially vulnerable people. We presented the designed strategies in the following manner to the target audience (see Fig. 42.2).
>
> This manner of presentation failed due to miscommunication and translation errors. After doing some more iterations, we presented the concepts again in a poster format where the audience could engage 1 to 1 and discovered that this was very well received. (interviewee, urban designer)

Incubation time and handing the baton to the target audience are very important when working in new areas with new people who are unfamiliar with the design process. The word target audience should be used for designers as well, since the real social design activity and learnings are two-way.

> Avoid design indigestion. One cannot swallow a hamburger whole; we must chew it slowly. (interviewee, urban designer)

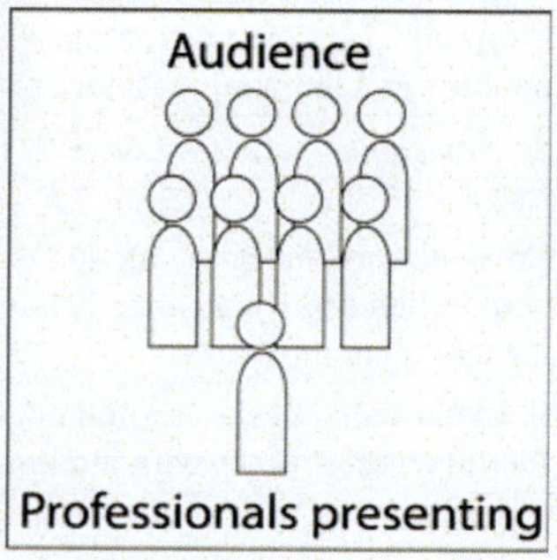

Fig. 42.2 Visual depiction of the professionals presenting their designed outputs to a group of target audience from the local community

Many interviewees pointed out cross-cultural social design as an opportunity for developing social design in India. This can be done through partnership with Indian firms who already have a rapport and understanding of the target communities.

> Students from different institutions in the West visited India for 2 weeks and immersed themselves in complex societal challenges in an interdisciplinary team during EPFL + ECAL Lab's India Switzerland Social Innovation Camp (INSSINC) [37]. This was done in partnership with SELCO, a for profit social enterprise based in Bangalore. It was intimidating at first, but SELCO trained us with workshops in Switzerland. Due to SELCO's familiarity with the context, we had no trouble in understanding their needs. Since we came from a different cultural context, we thought about unique solutions. We sometimes came with solutions that would not be possible in the Indian context and other solutions looked silly to us but were actually the best for the camp. Our occidental perspective along with SELCO's expertise led to a fruitful output. (interviewee, designer and researcher)

Traditional design jargons and vocabulary need to be simplified and contextualized to wipe out the communication barrier—this would make them inclusive.

Some interviewees questioned the stereotype that social design in India should only help tackle topics like poverty and waste. Social design can also include designing a restaurant's system to increase employee happiness, reduce plastic waste and chances of food contamination. Frugal innovation is already being practiced by many Indians who are not formally trained designers. The positioning of the same must be identified during any social design activity.

> Are poor people really genuinely unhappy? If yes, then why? And what can we, as social designers, really do for them, if anything? Jugaad or frugal innovation is very common in India. How are people practicing it? And is there a need for any more intervention? We must ask these questions. (interviewee, service designer).

Lack of dialogue and presence of stereotypes was traced by the interviewees to a lack of open dialogue, debate and questioning in Indian design schools. Some interviewees mentioned that this is due to the hierarchical Indian culture.

> When I taught at an Indian design school, I set up conversation spaces with the students for moments of reflection and questioning. This was very new to them but it enabled a critical approach necessary for social design. (interviewee, urban designer and design educator)
>
> I noticed that students who have suffered from depression are intrinsically motivated to design for the social issue of mental health. (interviewee, communication designer and design educator)

Making students aware of themselves can help to naturally and informally create a more socially oriented pedagogy in Indian design schools.

Despite the varying professional level of the interviewees, there was a repetition and clear pattern of insights received and this shows a common line of thought among the design audience regarding context-sensitive social design in India. The results already clearly indicate directions for developing theory and action and have been discussed in the next section.

42.4 Discussion

In an Indian social design context, the themes of the results can be put into two broad categories.

Under **social design methodology**: conversation and questioning as a formal social design methodology; re-examining vocabulary and replacing jargons with simpler language; using social design for all communities; cross-cultural work and partnership with local bodies.

Under **design education**: encouraging open student dialogue and questioning in Indian design schools; sensitizing people toward their personal (social) issues and scaling design education informally to non-design communities.

The study findings are discussed below in light of the research question.

42.4.1 Based on the Social Design Knowledge and Practices Existing Among Designers Globally, What Are Some Theories and Opportunities for Development of Social Design in India, from a Cultural Lens?

The opportunities have been arranged in order of priority, based on the number of interviewees who brought up the connecting themes (refer to Results section Table 42.1). Since this is an exploratory design research, each opportunity ends with a how might we statement to spark possible future research and design activity toward implementation.

Embed dialogue in the social design process. The social design process in India must have dialogue woven into it. India has a high context culture, meaning there is a lack of open dialogue and direct conversation compared to some Western cultures [35]. However, this open dialogue needs to be built up bit by bit in Indian social design practice. Prasad [11] has mentioned the need for conversation and dialogue among diverse actors for the advancement of social design and innovation in India. The findings of this study add on to Prasad, by stating that this dialogue needs to be imbibed in the definition of social design itself. This dialogue should be among design teams, between designers and the non-design community, among the non-design community and between the designer and themself. Mindful dialogue of the designer with themself is an integral part of social design because understanding ourselves helps us understand others better [38]. The non-design community can also be seen as a form of social designers. This is complemented by a large presence of frugal design in India which, as stated by Khan [39], has a practical contribution towards pressing social needs.

How might we facilitate conversation and dialogue throughout the social design process?

Simplify vocabulary. Design jargons like co-creation, democratic design and empathy are used widely by designers. These words and their meanings are important, but over the years, have we retained their true meaning in the context of varying design activities? India has high context communication [35] and a plethora of languages [40]. When engaging in social design, we not only deal with this but also with diverse communities who have diverse literacy backgrounds [8]. Hence, going back to the design vocabulary and understand the real meanings of jargons in a particular context should be a formal activity of the Indian social design process.

How might we simplify the design vocabulary we use in our daily practice to make it easy to understand in varying contexts during social design work?

Scope of social design in India and bias-busting. Bias-busting should be a formal activity in the Indian social design process. Behavioral biases affect us all, especially while engaging with different cultural groups [41]. Saying that social design for India is involved with only helping poor people living in slums is a very biased and incorrect statement. These statements should be made only after doing research through conversation and understanding the people (as mentioned in one of the previous points). Social design can tackle the needs of other communities in India too. People all over India suffer from fake news [9] and mental health issues [10]. The social design example of The Good Kitchen in Denmark [14] tackles issues that exist in India too. Social design can be used to tackle systemic issues related to food service, aging population and alleviation of overall quality of life of Indian people. Social design can also be used to promote desired pro-social behavior implicitly in people. Behavior impacts social design and social design impacts behavior [42]. Biases have a direct relation to our quality of work as designers [43], more so here as social designers. These biases must be addressed as a part of design education curriculum as well as design practice.

How might we tackle our biases so that they do not interfere judgmentally into the social design process?

Cross-cultural social design as an opportunity. This means that designers from different cultures intervene in Indian social design. Lee and Bain [17] have stated that cross-cultural design practice leads to larger design discoveries. The findings of this study supplement this. In an Indian social design context, large design discoveries can be made through cross-cultural design, but this must be done in partnership with local bodies. Partnership with local bodies that are familiar with the subcultures can help warm up and begin conversation. People should be aware of cultural rules, stigmas, communication, etc. Social design practices should be moved from big cities to smaller towns [11] in partnership with local bodies to increase sensitivity and familiarity towards the subcultures in India.

How might we nurture cross-cultural collaboration and global exchange for social design projects in India?

Dialogue in design schools. Formally built safe, trustable conversation spaces where students can comfortably share and question concepts, projects and problems should be present in design schools. These conversation spaces should not be open only to members of the design school but also to the non-design community and design professionals. Awareness of design students about themselves should be encouraged as well through encouraging mindfulness about one's emotions and the world around them. Making students aware of certain social problems they have personally experienced, like mental health issues or periods of debt and poverty could sensitize them to work on social problems for others. Following these cues can catalyze the social design pedagogy in Indian design schools without going through the rigid hierarchy and large time gaps which are caused due to high power distance in Indian culture [25].

How might we facilitate open conversation and mindfulness in Indian design schools?

The strength of this study lies in the wide geography and background of the samples. The study suggests that there is a common thread of thinking among this audience about the topic of social design in regards to Indian culture. This diversity of samples facilitates any designer curious about social design to utilize the directions provided and apply them within their own domains and contexts. The findings are not relevant only to design practice but can also aid Indian design education. They can catalyze the process of building more socially motivated Indian designers, which, in the big picture, can help in the growth of Indian society, economy and culture.

42.5 Limitations and Future Scope

Ideally, in-person interviews should have been conducted. However, since most of the interviewees were in varying geographies and there was time and resource limitation, interviews were conducted through video calls. Due to this, the study may have missed body language cues which could have given better insights. However, the use of video calling software in such situations is justifiable [44].

A topic brought up by one of the samples when discussing social design 'needs' was Maslow's hierarchy of needs [45] which proposes five sets of basic human needs arranged in a hierarchy—starting with physiological needs up to self-actualization needs. Future scope could involve going deeper into whether this hierarchy can be connected to Indian social design theory, and if yes, how?

Future research can involve testing the scope and validity of the proposed opportunities in different Indian subcultures and identifying what works where and when, and creating solutions that can be directly implemented in Indian social design practice and Indian design education based on the guidelines provided by this study.

42.6 Conclusion

The study enabled understanding of global social design knowledge and practices and helped build theories and opportunities for social design within the context of

India. A qualitative study was done with designers from varied backgrounds who have all worked on social design on social design in different geographies including India. Using a cultural lens contextualized the findings to India. Creating open dialogue, simplifying vocabulary, bias-busting, fostering cross-cultural social design and encouraging mindfulness emerged as the main opportunities. These opportunities can be applied both in social design practice and design education. Existing literature has mentioned the need for conversation but not gone in-depth about whom this conversation should be between and why. This study gives a clearer view of the same. The literature also stressed the need for developing theory and knowledge specific to Indian social design, considering the Indian socio-cultural phenomena. This study has taken a step forward to build this theory and created knowledge for present and future design studies and social design work.

References

1. Papanek, V.: Design for the Real World: Human Ecology and Social Change. Van Nostrand Reinhold Co, New York (1984)
2. Margolin, V.: The Politics of the Artificial: Essays on Design and Design Studies. University of Chicago Press, Chicago (2002)
3. Stickdorn, M., Schneider, J.: This Is Service Design Thinking: Basics, p. 84. Tools, Cases (2010)
4. Mishra, R., Singh, P., Sarkar, S.: The Indian Social Sector—Trends and Issues. (2015), p. 137
5. The World Bank in India.: World Bank. https://www.worldbank.org/en/country/india. Accessed 11 Apr 2020
6. India: Extreme Inequality in Numbers | Oxfam International. In: Oxfam International. https://www.oxfam.org/en/india-extreme-inequality-numbers. Accessed 13 Apr 2020
7. World Economic Forum: Global Gender Gap Report **9**, 11–14 (2020)
8. Education Statistics: All Indicators | DataBank. In: Databank, World Bank (2018). https://databank.worldbank.org/source/education-statistics-%5e-all-indicators. Accessed 16 Apr 2020
9. Banerjee, A., Haque, M.: Is Fake News Real in India? (2018). doi: https://doi.org/10.31620/JCCC.12.18/09
10. The burden of mental disorders across the states of India: The Global Burden of Disease Study 1990–2017. Lancet Psychiatr. **7**, 148–161 (2019). https://doi.org/https://doi.org/10.1016/S2215-0366(19)30475-4
11. Prasad, S.: Thinking through social innovation and social entrepreneurship in India. In: Social Innovation in Social Entrepreneurship: Strengthening the Ecosystem for Scaling Social Innovation, pp. 1–27. Villgro & IDRC, Bhubaneswar (2020)
12. Sharra, R., Nyssens, M.: Social Innovation: An Interdisciplinary and Critical Review of the Concept (2011)
13. PRIA: Invisible, Yet Widespread: The Non-Profit Sector in India. New Delhi (2002)
14. Liedtka, J., King, A., Bennett, K.: How design thinking improved the lives of a group of Danish seniors and inspired the people that serve them.
15. Smith, R., Iversen, O.: Participatory design for sustainable social change. Des. Stud. **59**, 9–36 (2018). https://doi.org/10.1016/j.destud.2018.05.005
16. Cronk, L.: Culture's influence on behavior: Steps toward a theory. Evolut. Behav. Sci. **11**, 36–52 (2016). https://doi.org/10.1037/ebs0000069
17. Lee D, Bain J (2016) Developing a Cross-Cultural Design Methodology: A study of UK-South Korean Collaboration in Higher-Education Design. THE GLOBAL eLEARNING JOURNAL 5:

18. M.A. in Social Design (MASD). In: MICA. https://www.mica.edu/graduate-programs/social-design-ma/. Accessed 2 May 2020
19. Design Studies MA: Design Theory for Social Change | The New School. In: Newschool.edu. https://www.newschool.edu/parsons/ma-design-studies-research/. Accessed 3 May 2020
20. BSc Industrial Design Engineering. In: TU Delft. https://www.tudelft.nl/en/ide/education/bsc-industrial-design-engineering/. Accessed 6 May 2020
21. Masters in Social Design. In: Sod.aud.ac.in. https://sod.aud.ac.in/programme/postgraduate-studies/21-mdes-(social-design). Accessed 11 Jun 2020
22. British Council: The Future of Design Education in India. (2016), p. 29
23. Meyer, M., Norman, D.: Changing design education for the 21st century. She Ji: J. Des. Econ. Innov. **6**, 13–49 (2020). https://doi.org/10.1016/j.sheji.2019.12.002
24. Hall, E.: Beyond culture. Anchor Press, Garden City (1976)
25. Hofstede, G.: Culture's consequences. Comparing Values, Behaviors, Institutions, and Organizations Across Nations. Sage, Thousand Oaks, California (2001)
26. Hall, E.: The Hidden Dimension. Doubleday, Garden City, NY (1969)
27. Hall, E.: The Dance of Life: The Other Dimension of Time. Anchor Press, New York (1989)
28. Lachner, F., von Saucken, C., 'Floyd' Mueller, F., Lindemann, U.: Cross-cultural user experience design helping product designers to consider cultural differences. Lect. Notes Comput. Sci. 58–70 (2015). doi: https://doi.org/10.1007/978-3-319-20907-4_6
29. Chakraborty, J., Norcio, A.: Cross cultural computer gaming. Lect. Notes Comput. Sci. (2009). https://doi.org/10.1007/978-3-642-02767-3_2
30. Balushi, D.: The use of online semi-structured interviews in interpretive research. Int. J. Sci. Res. (IJSR) **7**, 726–732 (2018)
31. Flick, U.: An introduction to qualitative research, 4th edn. SAGE Publications, London (2009)
32. Guest, G., Bunce, A., Johnson, L.: How many interviews are enough? Field Methods **18**, 59–82 (2006). https://doi.org/10.1177/1525822x05279903
33. Braun, V., Clarke, V.: Using thematic analysis in psychology. Qualit. Res. Psychol. **3**, 77–101 (2006). https://doi.org/10.1191/1478088706qp063oa
34. Hofstede, G.: Country Comparison—Hofstede Insights. In: Hofstede Insights (2020). https://www.hofstede-insights.com/country
35. Nishimura, S., Nevgi, A., Tella, S.: Communication Style and Cultural Features in High/Low Context Communication Cultures: A Case Study of Finland, Japan and India (2008)
36. Back to Front: Envisioning an alternative for the Estero Salado. In: Designing Inclusion. https://designinginclusion.wordpress.com/. Accessed 13 Jun 2020
37. Understanding reality and shaping solutions. In: India Switzerland Social Innovation Camp. https://inssin.camp/program/. Accessed 14 Jun 2020
38. Böckler, A., Herrmann, L., Trautwein, F., Holmes, T., Singer, T.: Know thy selves: learning to understand oneself increases the ability to understand others. J. Cogn. Enhancement **1**, 197–209 (2017). https://doi.org/10.1007/s41465-017-0023-6
39. Khan, R.: How frugal innovation promotes social sustainability. Sustainability **8**, 1034 (2016). https://doi.org/10.3390/su8101034
40. Kulkarni-Joshi, S.: Linguistic history and language diversity in India: views and counterviews. J. Biosci. (2019). https://doi.org/10.1007/s12038-019-9879-1
41. Understanding Bias When Working with Multi-Cultural Populations. In: Ncda.org. (2020). https://www.ncda.org/aws/NCDA/pt/sd/news_article/278313/_PARENT/CC_layout_details/false. Accessed 12 Oct 2020
42. Tromp, N.: Social Design: How Products and Services Can Help Us Act in Ways that Benefit Society (2013)
43. Hallihan, G., Shu, L.: Considering confirmation bias in design and design research. J. Integr. Des. Process Sci. **17**, 19–35 (2013). https://doi.org/10.3233/jid-2013-0019

44. Krouwel, M., Jolly, K., Greenfield, S.: Comparing Skype (video calling) and in-person qualitative interview modes in a study of people with irritable bowel syndrome—an exploratory comparative analysis. BMC Med. Res. Methodol. (2019). https://doi.org/10.1186/s12874-019-0867-9
45. Maslow, A.: A theory of human motivation. Psychol. Rev. **50**, 370–396 (1943). https://doi.org/10.1037/h0054346

Chapter 43
Anecdotes to Animation: Role of Oral History in Visual Adaptation Study

Swayamsiddha Panigrahi, Hans-Martin Rall, and Mazhar Kamran

Abstract This paper investigates the qualitative research method of oral history collection through in-person interviews in the context of visual adaptation studies. For this discourse, two academicians cum animation scholars across two countries, India and Malaysia, are interviewed. Personal testimonies are validated to understand the process and decisions for designing traditional visual art informed animation styles. The study evaluates the research method through relevant, practice-based examples and context analysis. This scholarly approach is nascent in the field of visual art studies and animation. Hence, the findings of this study hold significant value for designers, creative practitioners and academicians.

43.1 Introduction

This paper examines the visual adaptation of traditional art into animation styles through oral history. This study investigates how oral history can help gather rich data and enhance the interpretative value of visual adaptations. Through in-depth interviews, the paper explores the benefits (and limitations) of using oral history to interpret and map the methods adopted by creative practitioners to select elements from traditional art for animation. The interview examples are cross-cultural. In the conventional sense, cross-cultural interviewing refers to the collection of interview data across cultural and national borders [1]. The interviews for this study are conducted with participants across two different countries, India and Malaysia. It is a lengthy and planned process of data collection.

S. Panigrahi (✉) · M. Kamran
IIT Bombay, IDC, Mumbai, India
e-mail: swayamsiddhapanigrahi90@gmail.com

S. Panigrahi · H.-M. Rall
NTU Singapore, ADM, Singapore, Singapore

A. Chakrabarti et al. (eds.), *Design for Tomorrow–Volume 2*, Smart Innovation, Systems and Technologies 222, https://doi.org/10.1007/978-981-16-0119-4_43

"In tandem with oral history, which also seeks to uncover hidden, marginalized aspects of the past, the interview appears to privilege first-hand narratives and experience". Interviews provide the circumstance and opportunity for retrospective reflection [2] and a means to close the gap between the interpretations and facts. Within the visual arts, oral history has been used extensively with several archive building collections. However, here the study goes beyond only archive building. In this study, oral history collection serves as a significant component as literature for animation styles informed by traditional art. Visuals evoke varied emotions, and the experience of it is different for each individual. Hence, it was essential for this project to know and distinguish between the individual perspectives of lived experiences while engaging with visuals and visual mediums.

43.2 Oral History: A Research Tool in Visual Adaptation Studies

43.2.1 Background

Use of personal testimony in the investigation of society is not new, but the mechanical recording of the in-depth interview is not so old [2]. Oral history as a formal term came into use in the nineteenth century. But the widespread use of tape-recorded interview happened in the twentieth century only after World War II, and the digital revolution has made the process much more accessible [2–4]. Academics have defined oral history since the 1940s as a research method that asks eyewitnesses to talk about their lives, and their own experiences in usually very extensive life history interviews or event-focused interviews [5]. Oral history is most frequently used to refer to the recorded in-depth interview, which can be further defined as the recording of personal testimony delivered in an oral form which is a combination of taped memoir, type-written transcript and a research method that involves in-depth interviewing [3].

James Bennett, an oral historian, has referred oral history with a string of terms such as *life history, self-report, personal narrative, life story, oral biography, memoir and testament.* [2, 3]. Here, the context of oral history pertains to interviewee's back story of building the visual language for the animation. The terms such as *in-depth interview, recorded memoir, taped memories and life review* suggest that there is an involvement of a second party who frames the topics and inspires the narrator to begin the act of remembering, jogs memory, and records and presents the narrator's words [3]. Hence, the role of both the stakeholders, interviewee and the interviewer, is significant as well as consequential in this evidence-building process. Oral history in visual studies can be a multivalent, diverse, co-constructed practice that challenges conventional autonomous production and identities [2]. The great oral historian, Alessandro Portelli, has also written eloquently on the responsibility of oral historians and their work in creating narratives of meaning constructed by its two interlocutors, the interviewer and the interviewee [2].

43.2.2 Narratives from the Visual Practitioners and Experts

Animation professionals and scholars are interviewed for this research. The interviews are based on their work experiences in taking inspiration from traditional visual art to develop contemporary animation styles, and these serve as theoretical and practical evidence of a visual transformation process of traditional visual forms. The animation is the final product, but insight into the creative process can only be given by its creator(s). The decisions of the animator are crucial as they shape the visual language of an animation film. From this study, it is deduced that only analysing an animation film cannot unpack the method of the transformation. The analysis of the researcher, along with the testimony of the animator, completes the bigger picture and information on what informs the visual language. This paper argues that oral history can be a method to create the first account evidence of visual transformation.

Use of audio and video recordings, interview transcripts and note-taking has become essential tools for modern oral history collection, archiving and further works. This study explores the significance of this data gathering and data building method for visual transformation studies and discourses. Owing to the relevance of this method in contributing to the scholarly visual studies, it comes with several challenges too. In this, there are challenges like legitimacy, accuracy, language barrier, comfort and willingness of the participants to engage with researchers. This study considers all the advantages and disadvantages of adopting this method in scholarly practices as opposed to other prevalent methods of data collection.

Designers and scholars in the field of visual art media use the oral history method of data collection. One example is the doctoral research by Timothy Jones. He used the method of interview and collected oral history to study the Indian animation community in the context of reflexivity and identity in Indian animation production culture [6]. Dr. Nina Sabnani, an animation scholar, has widely used oral history method in studying indigenous cultures and their art forms [7]. However, a context review in this topic showed that the use of oral history as a research method is popular and advanced across humanities and social science subjects. The use of this method for scholarly visual design study is sparse. In qualitative research subjects, researchers go for the primary form of data collection, which involves conducting one to one or focus group interviews—finding participants. But it has been less articulated from the perspective of conducting an interview, collecting a personal testimony to study visual adaptations and transformations. The scarcity of such information and written texts is one of the primary reasons for the researcher to opt for the oral history method for data collection.

This paper poses a question—what is a personal testimony adding, and how? Visuals are the predominant language of both traditional art and animations. It uses and at the same time evaluates the method of oral testimonies by creative practitioners. The in-depth interview enables the researcher to give the subject leeway to answer as he or she chooses, to attribute meanings to the experiences under discussion and to interject topics. In this way, new hypotheses may be generated [2]. The following discourse is based on two cross-cultural interviews conducted in person by the interviewer. Traditionally content gathering is the primary objective of oral history [2]. Oral history collection from creative practitioners is based on validated data, professional experience and scholarly work.

The following section discusses the in-depth interviews with a personal point of views of interview participants.

43.3 Oral History: Anecdotes to Adaptations

The oral history method was conducted through in-person interviews across India and Malaysia. The paper focuses on one interview example from each country to examine the role of oral history. The research happened in three stages, namely secondary study, data collection, and interpretation and analysis.

43.3.1 Secondary Study: Literature Review and Interview Preparation

This stage is the most crucial step as it determines what data needs to be collected and how. The need for oral history was visible in the literature review stage of the research. There were limited data on the processes of adaptation of traditional art into animation styles when the researcher (later interviewer) started the study in India. Even in the digital library, scholarly works and theoretical explanations were sparse on this subject. The study initially relied on analysing available animation films, but later it was essential to validate the interpretations and observations made from the film analysis. Many animation films were also not openly accessible for research. Hence, oral history was found suitable as the next step to discover first-hand accounts of the creative practitioners who have worked on this topic. It was inferred from the film analysis that multiple factors determine the degree of reference that an animator takes from an existing art form to develop an animation style. Each animator's approach could be individualistic and personal on many accounts. Hence, it is important not to generalize and understand individual takes to form an overall opinion on the processes. Therefore, oral history collection through one to one interview is an excellent way to know the creative practitioner's vision. Also, the data collected from interviews can be further verified through existing literature and other forms of measurable and verifiable means.

An especially helpful way to begin preparation is to read and listen to other oral histories [4]. It is necessary to do a background survey, study existing literature about the interviewee and their works as that is the only way to determine what questions to ask [4]. Research also equips the interviewer and supplies information that the interviewee might forget to mention while they backtrack through their pieces of work. Adjunct to the tape recorder to collect the stories, an oral history interviewer is a co-author to create the accounts [8].

A structure was prepared to conduct the interviews. A set of questionnaire was designed according to the need of the research project. The questionnaire was divided into four parts, e.g. visuals, story, motion and identity. Total 26 common questions were prepared in a structured manner but with the option to expand towards more specific and individualized questions. As the interview is dealing with living persons, there is an opportunity to find out what is important to them in their history and their present lives [2]. A standard questionnaire was used for both interviews with small contextual, geographical and case-specific tweaks.

The first example is from the interview with Indian animation scholar and filmmaker, Dr. Nina Sabnani. Dr. Sabnani has more than two decades of experience on traditional art informed animations and has received national and international recognition for her contributions in this field. So she was one of the most suitable candidates for the interview. Her animation films, such as *Hum Chitra Banate Hai* (*We make drawings,* 2016), *Tanko Bole Chhe* (*Stitches speak,* 2009), *BaatWohi Hai* (*It's the same story,* 2011), *Bemata* (2012) and *Shubh-Vivah* (1984), are all inspired from indigenous Indian folk art forms.

As rightly said, the oral historian has to play detective [9]. As word-of-mouth referrals unearth many potential interviewees. So was the finding of the second interviewee, the Malaysian animator and educator, Siti Hajar Aznam. For her master's project, she made an animation film called *Ulek Mayang: Spirits of the Sea* which went on to win many international awards and screened at multiple prestigious film festivals around the world. She took inspiration from two prominent traditional Malay art, *Wayang Kulit* (Shadow Puppet) and Batik. She is amongst the few animators of Malaysia who has been working in this subject area. Hence, knowing her story was essential. After figuring out the suitable interview partners, the researcher (interviewer) placed contact with them through emails stating the purpose and nature of the interview.

43.3.2 Data Collection: Interview and Personal Testimonies

The interview for both participants happened in person in their home countries. Dr. Sabnani was interviewed at her residence in Mumbai, India, and Mrs. Hajar was interviewed in a museum library in Kuala Lumpur, Malaysia (Refer Figs. 43.1 and 43.2). The location of the interview is vital as the participant needs to feel comfortable in their surroundings, and the interview should also be uninterrupted from any external influences. The interviews were planned in a semi-formal manner for a two- to three-hour session. The sessions were audio recorded and not video recorded as a camera might make the interviewee conscious, and also it becomes an additional physical object to handle for the interviewer. The interviewer initiated the conversation by warming up to the objective of the interview and slowly introduced the planned questions. The order of asking questions was not chronological at all

Fig. 43.1 Interview with academician and animator Hajar Aznam

Fig. 43.2 Interview with academician and animator Hajar Aznam

points as it allowed room for new stories and discoveries. Some open-ended questions were asked to enable interviewees to come forward with their accounts, reflect on matters and include all the material that they think is relevant. In framing open-ended question, oral historian, Charles Morrissey, postulates that the two-sentence format always works best. The first sentence should state the problem; the second poses the question [4]. One such question's example from Mrs. Hajar's interview: "Unlike the original Wayang kulit your animated characters don't have perforations in their body design. Was that a conscious decision?".

Empathy helps greatly in conducting interviews but with scholarly scepticism to avoid personal biases or admiration for the interviewee [4]. The interviewer also made handwritten notes to maintain the flow of the conversation and not to miss out on essential questions. Another aspect that the interviewer kept in mind was not falling into the researcher's bias. It is very likely to happen as the interviewer comes prepared with a perception about the interviewee and their work from prior research from secondary sources. Thick data were collected from both interviews and further analysed. The scholar and the subject collaborate: "They come together and provide a good advantage for understanding the meaning of the experience". The above notion that Frisch calls a "shared authority": the grounds of authority are very different and have a different meaning, but there is a kind of sharing in the process of the interpretive authority, which is one of the exciting things about conducting oral history [4, 10].”

43.3.3 Interpretation and Analysis: Extracting Information from the First-Hand Narratives

After the interview, the audio recordings were transcribed into text format. The next stages involved another method, thematic analysis for data analysis and interpretation. The oral history method gave insights into the working styles of the creative practitioner as well as into the decisions that informed the adaptation of the traditional visual imagery into animation. Observations from the two interviews are as follows:

Dr. Nina Sabnani: Dr. Sabnani was interviewed to gain anecdotes of her experiences and choices that resulted in a distinct style of scholarly work as well as animation films. The distinctiveness in Dr. Sabnani's work is her method of data collection through ethnographic methods and using animation film making as a research tool. She uses a collaborative approach where the story emerges from the conversations with artist communities. Her process starts with choosing an artist community, and to narrate the community's story, she uses that community's art form as the visual language of the film. Collaborative film making and capturing the stories of artist communities through their visual style are her approach. She conceptualizes that the artist prepares the drawings in the traditional method, and later it is taken for processing. Her ethnographic animation *Tanko Bole Chhe* (The stitches speak, 2009) (Refer Fig. 43.3) is an animated documentary. The film celebrates the art of kutch artisans through conversations and memories of their evolution of a craft tradition [11]. The visual analysis of the film *Tanko Bole Chhe* gave an understanding of the style of representation, scale, alignment and almost linear movement of individual elements in the film. All the characters are shown inside profile and houses in top view cross sections (Refer Fig. 43.4). The rudimentary ways of cloth stitching have

Fig. 43.3 Scenes from the animation film *Tanko Bole Chhe*. A complete scene (left)

Fig. 43.4 Profiles of characters from *Tanko Bole Chhe* (Right)

been preserved. However, the meaning of elements, icons and decisions of visual compositions came to the fore through the personal interview.

Dr. Sabnani gave examples of the visual decisions to convert static to dynamic images in her film *Tanko Bole Chhe*. She mentioned that the fascinating part in the overall frames is the way the images are composed with the knowledge of spaces between objects. She started with static characters composed and stitched on to a cloth. She scanned and digitized those elements from the cloth. So when she animated the component on the stitched cloth, things moved, and they created blank spaces;

Fig. 43.5 Still from *Ulek Mayang*

Fig. 43.6 Pohon Beringin/shadow puppet prop

they had to be filled. The aesthetic has to be precisely the way they plan it. So she not only looks at the forms but also at the spaces between the forms and neighbouring elements, and those dynamics have to remain. Another example she gave of the prominent borders used across the animation. Those borders mean actual boundaries surrounding houses and communities. Those same boundaries have been used as grids and movement paths of the characters. The researcher had a similar experience of facts unravelling through interviews conducted with the Malaysian animator Hajar Aznam's film Ulek Mayang.

Hajar Aznam: Along with methods used for Dr. Sabnani's interview, additional pre-interview preparations were made as this was a cross-cultural interview. The interviewer was new to the Malay culture and people. Hence, the interviewer did a secondary study to understand the nuances of Malaysian culture and art. Background study of the interviewee, Mrs. Aznam, and her films were carried out in detail. Before the interview, the interviewer connected with third parties like other Malaysian and non-Malaysian animation scholars who have known her professionally and could talk about her work. Mrs. Aznam has also used fabric (Batik) patterns in her film *Ulek Mayang* (Refer Fig. 43.4) like Dr. Sabnani. However, the designs are not as basic and direct as the film *Tanko Bole Chhe*. The creator has decided to use nicely cut Batik cloth patterns as the skin on the shadow puppet characters. For example, the sequence of the narrative and visual placement in the animation is inspired by the actual Wayang Kulit performance. Unlike the original Wayang Kulit silhouette style, Mrs. Aznam has experimented with colour light, while keeping the core principles of shadow Wayang Kulit intact. When asked is she has used a silhouette style for opening and closing sequences, her answers helped in interpreting the unknowns. She mentioned that it's just to show that the Wayang Kulit element, the silhouette from the beginning and then she ends with the silhouette as well. In Wayang Kulit performance, they have a ritual, Bukapango; it is how you open the stage. They usually have a ritual for opening the scene and so on, and they will start with the *Pohon Beringin*[1] (Refer Fig. 43.4) in the centre which is why she has a similar shadow puppet prop at the first opening scene. Even though the character is a Javanese character, she wanted to associate with Malaysian Wayang Kulit as well". (Figs. 43.5 and 43.6).

43.4 Findings from the Cross-Cultural First-Hand Narratives

The oral history method enhanced the interpretative value of the adapted visuals. It also validated the findings from the literature review and film analysis. This study opened doors to a broader study of comparing works across nations. For instance, Hajar Aznam's film *Ulek Mayang* and Nina Sabnani's film *Tanko Bole Chhe: Stitches*

[1] *"Pohon Beringin"* (Banyan tree) is a traditional leaf shaped prop used for the beginning, changing and ending of scenes in the Malay Wayang Kulit performance.

speak, both animation films have adopted a similar style of animation. They have taken traditional cloth patterns and used them as visual patterns in their respective animation films. They both have used the original designs and colours of the traditional fabric. But both their approach to executing the visual language is very different. For Dr. Sabnani, it starts on a rudimentary level. She collaborates with artisans to come up with their characters in hand-stitched cloth applique. She then digitally manipulates the cloth character to come into action. However, Hajar has worked on her animation independently though they also relied on a field visit to understand the process of Malaysian shadow puppetry for her film *Ulek Mayang*. Mrs. Hajar draws the characters digitally and superimposes the batik cloth patterns on the surface on the character. Both the filmmakers stay true to the traditional art form and try to bring in the forms and functions of the traditional visuals into their animations.

The decision to adopt an oral history interview method seemed right as personal and circumstantial choices can influence the visual decisions for animation. Even though the interviewer asked for personal anecdotes and testimonies, the interview process was pre-planned and semi-structured. The interview method is used not just for data gathering but also a validation of secondary data from literature review and analysis of animation films. It was done to find answers for particular questions of visual adaptations of traditional art for animation. It is indicating that to understand transformation and reason of the visual decisions, it is essential to adopt the method of interview and gain a picture of the creative choices and symbolic meanings. The interesting fact about this study is that oral history also covers the interviews conducted by interviewees for designing their visuals. It became a means of not just accessing information but also a way to find signification, interpretation and meaning [12]. Oral history interview method enhances the interpretative value of visual adaptations designed for traditional art informed animation styles.

Oral history helped in finding other potentials interviewees as there was no such database to find the specific people working in this type of work. This is how the oral narratives helped in the construction of the traditional art transformation process. The cross-cultural narrative came together to show patterns and parallels—the similarities and differences in the oral history collection process in two countries. The oral history collection process in the two countries was mostly followed by a standard structure. The cross-cultural angle gave the research a comprehensive perspective. The interview participants were from different cultures and communities, which gave the study diversity and fewer chances of overlap and repetition. It also brought out similarities in practices across cultures and research models. It was observed that creative professionals give importance to primary studies collecting first-hand data from the field. They rely on the narratives and methods of the traditional art practitioners and emulate them in animation. The role of the interviewer proves to be essential too as they steer the thoughts in a direction that brings out meaning in this whole process. The content of the narrative is also not just built based on the interviews but also from other participant's views on their contemporaries and colleagues' work. The contribution of personal testimony is adding to the research my creating literature. "A good oral history can present and preserve convincing evidence and put it in quotable, first-person prose that enlivens historical narratives. But oral history should not stand alone as a single source. Researchers need to seek out available material to substantiate both written and oral evidence. If written and oral information contradicts each other, then the researcher must dig even deeper to determine which is more accurate". [4].

43.5 Conclusion

From the above discourse, it is deduced that the oral history method is quite apt to build literature as well as validate existing secondary information on adaptation studies. It is also found that each interview and its subject are independent individuals and so are their backgrounds and circumstances. Hence, there cannot be a standard method of conducting an interview or posing questions. However, a semi-structured approach and pre-planned context review go a long way in having a successful interview. Especially, cross-cultural interviews require thorough understanding and sensitivity to the nuances of a new culture and its specifics with context to the art form and people being studied. The subjectivity incurred in the interviews is validated through relevant contextual analysis and visual analysis of selected animation films.

The possibility of discovering something not even thought of before is an advantage of this method. However, in-depth interviews are time-consuming, and so the qualitative researcher cannot examine the number of cases that other methods (quantitative research) can. Generalizations about a wider population have to be held tentatively. The visual design decisions are influenced by their life experiences, events which make it valid as oral history. They are being interviewed as subject experts, but they also have life memoirs included in it. Many personal experiences lead to professional choices is seen in these oral histories collected from animators. It needs to be sieved and scanned through secondary and primary validation measures and tools. Oral narratives are subjective and may be influenced by external variables. But once verified using supporting data, they become reliable facts and literature for studies of adaptation of traditional visual styles for animation.

References

1. Hajek, A: Oral History Methodology (2014)
2. Bennett, J.: Human values in oral history. Oral Hist. Rev. **11**, 1–15 (1983)
3. Morrissey, C.T.: Why call it "oral history"? Searching for early usage of a generic term. Oral Hist. Rev. **8**(1), 20–48 (1980)
4. Richie, D.A.: "Doing oral history" A Practical Guide: Using Interviews to Uncover the Past and Preserve it for the Future (2003)
5. Freund, A.: https://www.youtube.com/watch?v=Xk3gb9xCTFo
6. Jones, T.G.: Animating Community: Reflexivity and Identity in Indian Animation Production culture, PhD dissertation, University of East Anglia

7. Sabnani, N.: Sharing Stories: Learning with Stories, International Conference on Usability and Internationalization
8. Portelli A.: Speaking of Oral History. https://www.youtube.com/watch?v=vETоq3T_LZQ
9. Partington, M., Sandino, L.: Oral History in the Visual Arts. Bloomsbury, London (2013)
10. Frisch, M.: A Shared Authority: Essays on the Craft and Meaning of Oral and Public History
11. Wang, Y.: Ethnographic animation: participatory design with the Longhoen Maio. Int. J. Film Media Arts **3** (2018)
12. Abrams, L.: Oral History Theory. Routledge, Milton Park (2016)
13. Yow, R.V.: Recording Oral History: A Practical Guide for Social Scientists. Sage Publications, Thousand Oaks (1994)
14. Sharpless, R.: The History of Oral History. W Thinking about Oral History. Theories and Applications, red. Thomas L. Charlton, Lois E. Myers, Rebecca Sharpless, 7–32 (2008)

Chapter 44
Design Thinking as a Strategy to Inculcate Problem-Based Learning (PBL) in Undergraduate Education Across South Asian Universities

Shakuntala Acharya, Apoorv Naresh Bhatt, Amaresh Chakrabarti, Venkata S. K. Delhi, Jan Carel Diehl, Nelson Mota, Andrius Jurelionis, and Riina Subra

Abstract The Bologna declaration states that, "*successful learning and studying in higher education should involve students in deep learning*". However, a survey of faculty across institutes in Nepal and Bhutan highlights that the undergraduate students in engineering and management lack skills needed to be industry-ready. They face difficulty in getting employed after graduation and if placed, then struggle during their employment due to insufficient practical experience, lack of good communication skills and unawareness of broader socio-economic contexts. The Erasmus +

S. Acharya (✉) · A. N. Bhatt · A. Chakrabarti
Indian Institute of Science, Bangalore, India
e-mail: shakuntalaa@iisc.ac.in

A. N. Bhatt
e-mail: apoorvbhatt@iisc.ac.in

A. Chakrabarti
e-mail: ac123@iisc.ac.in

V. S. K. Delhi
Indian Institute of Technology Bombay, Mumbai, India
e-mail: venkatad@iitb.ac.in

J. C. Diehl · N. Mota
Delft Institute of Technology, Delft, The Netherlands
e-mail: j.c.diehl@tudelft.nl

N. Mota
e-mail: n.j.a.mota@tudelft.nl

A. Jurelionis
Kaunas Institute of Technology, Kaunas, Lithuania
e-mail: andrius.jurelionis@ktu.lt

R. Subra
Aalto University, Aalto, Finland
e-mail: riina.subra@aalto.fi

A. Chakrabarti et al. (eds.), *Design for Tomorrow–Volume 2*, Smart Innovation, Systems and Technologies 222, https://doi.org/10.1007/978-981-16-0119-4_44

funded project, "Strengthening Problem-based learning in South Asian Universities" (PBL South Asia) is an endeavour to address these pressing concerns in education quality, employability and overall sustainable development of the region and to imbibe deep learning capabilities. Therefore, as an empirical study to clarify and in turn, inculcate PBL in South Asian undergraduate education, the young faculty of the inexperienced institutes from Nepal and Bhutan, alongside the students from the experienced institutes from India and Europe, were mentored by faculty and researchers from the latter to undertake multidisciplinary case studies. The strategy of "design thinking" was employed to methodologically guide the cases and keep it consistently problem-based, i.e. the learning is driven by the problem with no one correct solution. Results showed that the participants reflected improvement in problem-solving skills and increased motivation, apart from enhanced collaboration and improved communication ability. Based on these findings, further development of curricula to imbibe PBL in its existing courses and guidelines to train the trainers for implementation of the same, are currently in progress.

44.1 Introduction

The World Economic Forum 2016 [1] recognises critical thinking, complex problem solving, self-learning, collaboration and people management, and communication as top skills for 2020. Problem-based learning (PBL) methods support these specific skills [2] by positioning the onus of learning on the student through "investigation and resolution of messy, real-world problems" [3].

Globally and in South Asia, Engineering is the preferred undergraduate programme of study, followed by Management. Ideally, these programmes should enable the graduates to "design effective solutions to meet social needs" [4]. However, the Bologna process identifies "employability of graduates" and "shortage of skills in key areas" as major challenges in higher education and outlines objectives to promote employability [5]. Its declaration states that "*successful learning and studying in higher education should involve students in deep learning*" [5]. Unfortunately, literature reports that students are not industry-ready as they lack communication skills and teamwork experience, as well as awareness about social, environmental, economic and legal issues, and programmes, being content-driven instead of need-driven, do not provide sufficient design experiences to these students [6].

A survey of five South Asian Universities, from Nepal and Bhutan, corroborated the same and revealed that the undergraduate curricula across the region are predominantly instructional and are not adequately hands-on, due to several constraints, such as:

- University directed lesson plans with heavy syllabi to cover and restricted time for practical activities,
- Dearth of motivation in students to self-learn and innovate during the stipulated practical hours within a course,

- Poor critical thinking ability due to a general lack of awareness on sustainable development goals and their local implications in the students,
- Less number of co-instructors to guide in practical, real-world issues that can be addressed in courses,
- Fewer collaborations in these programmes/courses, and
- Poor communication skills.

Therefore, to make students industry-ready and motivate them towards development of the five essential top skills for the future [1], inculcating PBL approach into existing undergraduate courses is proposed. However, the lack of experience in practising PBL gave rise to the need of clarifying; (i) what is PBL?, and (ii) how it can be contextually adapted for South Asia, prior to being incorporated as a pedagogical method in undergraduate studies at South Asian Universities.

The aim of this paper is to present the unique strategy of employing design thinking to (i) clarify and, in turn, (ii) inculcate problem-based Learning through experiential case studies where young faculty and students collaboratively address complex and real-world problems.

44.2 Literature

44.2.1 Problem-Based Learning: A Means to Develop Top Skills

PBL is a learner-centred approach [7] where students learn through "*facilitated problem solving that centres on a complex problem that does not have a single correct answer*" [8].

While earlier studies stated that students who experienced PBL showed; (i) improvement in problem-solving skills [9, 10] and (ii) increased engagement and motivation to learn, as they preferred PBL to the traditional methods of teaching [3, 11]. Recent empirical findings for engineering discipline reveals that conceptual understanding is higher through PBL than traditional lectures [12] and peer discussion enhances it [13]. This, in turn, nurtures critical thinking skills as sharing of opinions, analysing situations through different perspectives and thinking of more possibilities to solve a problem [14] are results of PBL approach. It is also reported that students perform better in both interpersonal skills as well as practical domain skills [15].

A creative thinking approach has been used for implementing PBL in the classroom and resulted in enhancement of both creative skills and technical abilities [16]. This approach focusses only on "problem solving", whereas PBL takes an overarching view beginning with the problem, and design thinking enables problem finding [17]. Several similarities between the two, such as large number of stages, requirement of motivation, organisation skills and capability for group work and collaboration [18] exists, while the latter is known for its "potential to scaffold the development of skills" [19].

44.2.2 Design Thinking

Design thinking is a cognitive process [20] of identifying ill-defined or "wicked" problems [17] and developing solutions through iterative steps or activities, supported by several methods and tools.

There are several design thinking models, of which the most popular are: Stanford's school design thinking process [21] and IDEO human-centred design model [22] for generic design, and Pahl and Bietz [23], Hubka and Eder [24], Cross [25], Dieter and Schmidt [26], Eppinger and Ulrich [27] for systematic design process stemming from engineering. These models are either descriptive or prescriptive and use varied terminologies to guide the design process.

However, the common approach of all these models may be summarised into four steps as follows [28]:

1. ***Identify Requirements*** from abstract needs and observations upon analysis of ill-defined problems (*terms: Understand, Observe, Define, Empathise*);
2. ***Ideate solutions***, which involves generating several ideas and combining these into solution alternatives (*terms: Ideate, Visualise, Synthesise, Co-create*);
3. ***Consolidate solutions*** *into feasible solutions* through development and analysis (*terms: Build, Prototype, Simulate, Model*); and
4. ***Select concept***, i.e. the most promising solution upon evaluation of alternatives through testing (*terms: Test, Validate*).

44.2.3 Inculcating PBL Through Design Thinking Methods

Upon assessment of the needs of the partner South Asian Universities, the following shortcomings and drawbacks were identified.

(a) Lack of clarity in the understanding of PBL approach.
(b) Lack of knowledge of PBL processes and methods.
(c) Lack of experience in practise of PBL courses.
(d) Lack of trained faculty to facilitate PBL courses (and its problems).

Thus, two needs were broadly identified from the above, that:

i. There is a need to clarify "what is PBL?" and provide understanding, know-how of processes and methods and hands-on experience of the same; and
ii. There is a need to contextually appropriate the PBL experience with respect to both, the domain of study and the socio-cultural that determine the priorities and problems of a community or region, so that the untrained faculty can adopt it in the future and facilitate the proposed PBL courses in their home universities.

44.3 Descriptive Study

To address these lacking aspects and familiarise the faculty participants from the partnering South Asian Universities with PBL, a two-week programme at IIT Bombay (IITB), Mumbai, was conceptualised with the aim to "*provide a tangible introduction to PBL casework, engaging in cases with external partners/stakeholders and moving through a full PBL inquiry and design process in a simplified, condensed form*". It was envisioned as a platform for teacher training, reflection and debate on the hands-on implementation of PBL cases through field studies.

44.3.1 Workshop Participants and Programme

The workshop had about 70 registered participants, divided into 7 teams with 7–8 members each. Every team was assigned a mentor and had access to other resources. The participants were as follows;

- Faculty from five Universities across Nepal and Bhutan;
- Advanced students from three European and two Indian Institutes of Eminence;
- PBL project mentors from European and Indian universities;
- Case owners and subject experts; and
- PBL South Asia project managers and coordinators.

The two-weeks comprised of:

- Lectures on PBL—literature survey on definitions, terms and roles in PBL; history of PBL; PBL process and present application in technical domain with focus on Design for Sustainability and well-being.
- Interactive sessions on a few variants of design thinking, prototyping, etc.
- A dedicated session on design thinking, with problem identification, ideation and decision-making (consolidation and selection) methods, as a framework to guide the casework and enable the application of the same during the assignments; and
- Case studies that involved fieldwork for data collection and validation from local communities and stakeholders, as well as development of their "problem statement", analysis, conceptualisation and final presentation.

44.3.2 Workshop Cases: Problem Briefs and Teams

The cases were as follows:

a. *Liveability in Slums* (2 teams)

The teams were required to take up challenges surrounding the theme of "50% of the urban population of Mumbai residing in slums, chawls and squatter settlements" and work on possible interventions to improve the conditions surrounding the living in the slums in Mumbai. The study would aim to observe and analyse the liveability of people residing in these units. The study would involve analysis of the neighbourhoods, interaction spaces, built-up area and so on. The teams were expected to define the parameter for living conditions and how their intervention will improve that particular parameter.

b. *Affordable Housing* (2 teams)

The theme of this case was to understand the affordability of housing in Indian conditions. The teams would define opportunities related to how to design housing that can create a trade-off between efficiency and resilience, being affordable as well as adequate for people to live in. Teams visited the organisation's sites to understand the execution of projects in Indian conditions. To arrive at possible interventions, the participants focused on site-specific conditions to use energy and thermal comfort simulations, vehicular and pedestrian traffic studies, materials and construction techniques, amongst other aspects.

c. *Construction Demolition Waste* (CDW) (1 team)

The challenge in this theme was for the team to identify the creative uses for the use of construction demolition waste (CDW) in the country. The teams visited sites of construction demolition as well as some recycling plants and laboratories to acquaint themselves with the dynamics of CDW, and worked on how to creatively utilise the CDW in the society. The students were given access to the materials laboratory of IIT Bombay to cast and test some materials/prototypes made out of CDW.

d. *Accessible Healthcare* (1 team)

Parents of hearing-impaired children need to consistently deliver goal-oriented speech therapy at home; otherwise, the child's progress in learning to understand language and speak will be hampered. The team was expected to address the accessibility of such a healthcare need and propose how to motivate parents and children to conduct therapy at home.

e. Net-zero energy development—rehabilitating sustainably: *Redevelopment of Artist Village in Belapur* (1 team)

Based on the details, the team needed to develop a prototype model for incremental housing in the area, focusing on energy efficiency and sustainability without compromising on living standards and affordability. The prime focus was more on the possibility of renovation and densification of the existing neighbourhood rather than planning the area from ground zero. This case was explored to test the possibilities to develop brownfield projects sustainably in contrast to greenfield projects that cause environmental problems.

f. Net-zero energy development—rehabilitating sustainably: *Modular Housing scheme for Bhiwandi locality* (1 team)

Bhiwandi, an area located in the Eastern suburbs of Mumbai, is predominantly a textile hub which has the largest number of power looms in the country. A significant chunk of the population is employed in this sector, but the residential area is lesser developed to the Industrial Area. In order to end this disparity, a number of housing schemes have been coming up in the area recently. The aim was to create a prototype for modular housing which would fall under the affordable category. The focus would be on multi-utility areas including modularisation, affordability, energy usage, etc., based on the user group.

44.4 Case Study: An Example

As an exemplar, the PBL process of the above case study *(f)* on "*Modular Housing for Bhiwandi locality*", using design thinking as a framework, is presented.

44.4.1 Problem Identification

- "Problem" extracted from given brief: *How to provide affordable and modular housing in Bhiwandi?*
- The team observed and demarked the system boundary as "Bhiwandi locality" and enlisted a number of probable problems and concerns to be considered for both, the locality and modular housing in general, enlisted in Table 44.1 (Col. I).
- Based on the above, a preliminary list of requirements was generated, as shown in Table 44.1 (Col. II).
- Field visit was conducted to clarify the requirements, as well as, secondary research on climate and temperature data, wind flow pattern and population.
- Stakeholder interviews were conducted with questions categorised under the three pillars of sustainability, namely
- Social aspects (*What works—what do they like? What doesn't—what should be improved? Does it feel safe? Comfortable? What would they want if they moved to a new home?*),
- Environmental aspects (*What kind of annual environmental changes take place? How do they adapt to them? How does water management work? What about waste management?)*
- Economic aspects *(Is it affordable? Would they prefer to buy or rent?)*

Insights were drawn from the responses of the interviews, and the problem statement was redefined to "How to provide affordable, safe and spacious housing for Bhiwandi's low-income labour/workers through modular solutions?"

Prioritisation of requirements using checklist [22] are shown in Table 44.2.

Table 44.1 List of probable problems and preliminary requirements

I. Probable problems	II. Preliminary list of requirements
• For **locality**; – **Accommodating** low-income group – Risk of **flooding** in monsoon – **Adaptation** capacity – Risk of **high temperature and humidity** – Scarcity of pure **drinking water** – Improper **waste management** system – Ground/land **strata** – Availability of **space** for construction – Risk of **earthquakes** • For **modular housing;** – Environmental sustainability – Socio-cultural adaptation – Economic feasibility	1. **Need low-cost** rental/own housing 2. Modular system should withstand flooding 3. Modular system **should withstand high temperature and humidity** 4. Need provision for pure **drinking water** 5. It should inhibit **waste management** system 6. It should facilitate construction **in different ground strata** 7. It should withstand **mild to severe earthquakes** 8. Modules should be **scalable** 9. Modules should be **lightweight to facilitate transportation** 10. Should **facilitate expansion** in the future 11. **Materials and construction** process should be **sustainable** 12. Modular system should **encourage socio-cultural interaction** 13. Should **provide flexibility** for personalisation

Table 44.2 Prioritised list of requirements

Requirements	Demand (D)/wish (W)	Priority
Need low-cost rental/own housing	D	1
Modules should be scalable and flexible	D	2
Should facilitate expansion in the future	W	3
Materials and construction process should be sustainable	W	4
Need provision for pure drinking water	D	5
It should inhibit waste management system	D	6

44.4.2 Ideation

Several ideas were explored for the prioritised requirements (Fig. 44.1).

44.4.3 Solution Consolidation

Of the ideas and piecemeal solutions generated, some were merged to develop alternatives.

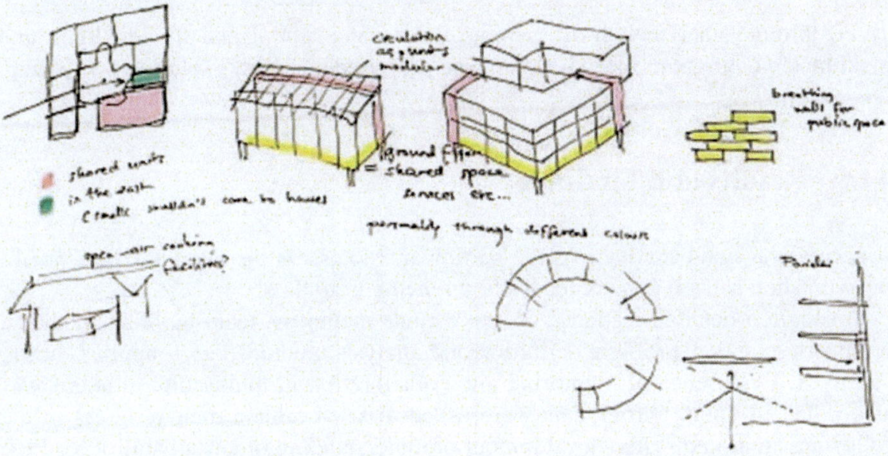

Fig. 44.1 Ideation: solution exploration A

44.4.4 Concept Selection

The concept:"*Shared spaces and facilities for different scales* (Fig. 44.2), was selected as the most promising solution, based on its modular flexibility for different configuration possibilities and breathable walls that take advantage of prevailing wind flows. The concept also entailed a smart energy and waste management systems, with facilities for recycling, earning opportunities from energy production through small-scale biogas production and solar PV system integrated to the grid, and energy

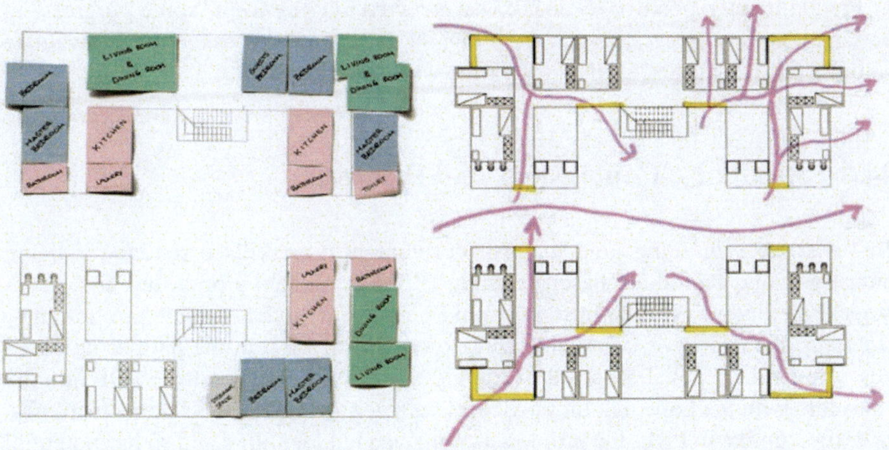

Fig. 44.2 Selected concept 1

saving through smart switch-off systems, Solar water heating and efficient lights and maximising daylight usage. The team also proposed a "policy" solution for renting.

44.5 Results and Findings

A survey was conducted across the participants to assess the overall experience of the workshop as well as seek feedback on methods applied.

It sought a detailed response on the various methods—team-building methods, fieldwork methods, problem definition and analysis methods, etc.—applied during casework. Experience of mentoring and collaboration in multi-cultural teams was also reflected upon. Participants were asked to self-evaluate their progress in the following areas and skills: development of understanding of sustainability for PBL and apply design thinking, aimed at improving "*critical thinking, self-learning and complex problem-solving*"; skills of team working, leadership, project management and increased motivations and interest, aimed at building "*collaboration and people management*"; and improvement in English as well as technical language under "*Communication skills*".

The results and findings were as follows:

- Most participants found the lecture sessions to be "fulfilling their expectations", as in Fig. 44.3a, whereas the relevance and value of the methods sessions were not adequately understood and drew contrasting responses.
- A majority, i.e. 56.4% of the participants responded positively about their experience of PBL through the PBL test cases, as shown in Fig. 44.3b. However, 41% stated that the experience only "partially fulfilled their expectations.
- Predominantly, positive feedback was received ranging from "some progress" to "significant progress", as shown in Fig. 44.3c, upon being asked to self-evaluate on specific areas and skills through the workshop experience.

44.6 Summary, Conclusions and Discussion

To foster deep learning, and in turn, develop the top skills essential for being industry-ready, a problem-based learning (PBL) approach is proposed as a pedagogical methodology to be introduced in South Asian undergraduate programmes. However, the dearth of know-how and experience of PBL in the partner universities prompted the novel proposal of using design thinking to inculcate PBL into the educators with the hope that they will disseminate the same at their home institutes. In design, unlike in PBL, the problem is not given but identified and so has potential of reflecting the contextual nuances of the domain of study and socio-cultural fabric of South Asia. Thus, a two-week programme was hosted at IITB to clarify what is PBL and contextually appropriate it for South Asia, through lectures, hands-on

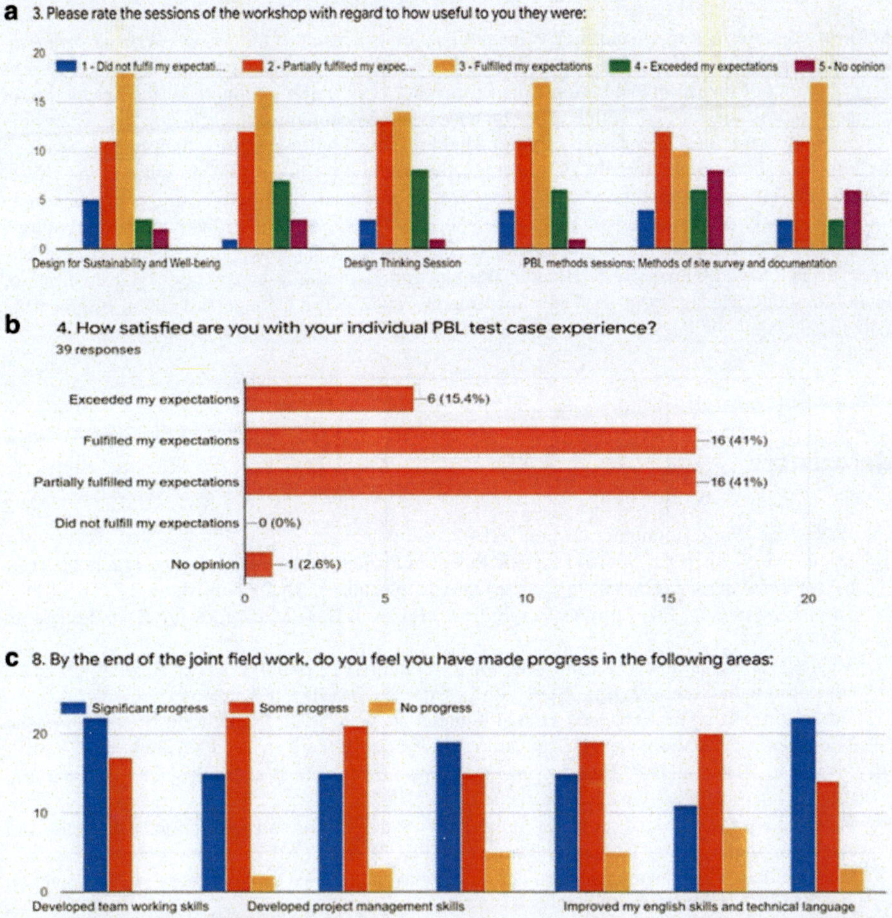

Fig. 44.3 **a–c** Survey results

sessions, and case studies addressing challenges unique to the region, so that the participants without prior knowledge or training receive a first-hand experience of PBL.

It was noted that the participants successfully completed the PBL case by following the design thinking methods were overall satisfied with the case experience and reported that they had made some to significant progress with respect to skills.

In conclusion, the two-week programme showed promise in inculcating PBL through design thinking in undergraduate level and became a pilot for the further development of PBL courses. Currently, the re-designed course with PBL approach is being implemented across the five partner institutes in Nepal and Bhutan, and a MOOC is being collaboratively produced by all partners.

Acknowledgements and Disclaimer This publication is a result of the "***Strengthening Problem-based Learning in South Asian Universities***" (PBL South Asia) project, co-funded by the Erasmus + programme of the European Union (The European Commission's support for the production of this publication does not constitute an endorsement of the contents, which reflect the views only of the authors, and the Commission cannot be held responsible for any use which may be made of the information contained therein.). This study would not have been possible without the participation of over seventy members from across five countries and ten partner universities, namely Aalto University, Finland; Indian Institute of Science (IISc) and Indian Institute of Technology Bombay (IITB), India; Kaunas University of Technology (KTU), Lithuania; TU Delft, The Netherlands; JNEC, Royal University of Bhutan, Bhutan; and Kathmandu University, Asian Institute of Technology and Management (AITM), Sagarmatha Engineering College and Nepal Engineering College, Nepal.

References

1. WEF.: The World Economic Forum (2016)
2. Duch, B.J., Groh, S.E., Allen, D.E.: The Power of Problem-Based Learning: A Practical "How to" for Teaching Undergraduate Courses in Any Discipline. Stylus Publishing, LLC (2001)
3. Torp, L., Sage, S.: Problems as Possibilities: Problem-Based Learning for K-16 Education (2002)
4. McNeill, B.W., Evans, D.L., Bowers, D.H., Bellamy, L., Beakley, G.C.: Beginning design education with freshmen. Eng. Educ. **80**(5), 548–553 (1990)
5. Declaration, B.: The European Higher Education Area, Joint Declaration of the European Ministers of Education, convened in Bologna, 19 June 1999. Retrieved November 8 (1999)
6. Mills, J.E., Treagust, D.F.: Engineering education—Is problem-based or project-based learning the answer. Australas. J. Eng. Educ. **3**(2), 2–16 (2003)
7. Savery, J.R.: Overview of problem-based learning: definitions and distinctions. Exploring and extending the legacy of Howard S. Barrows. IJPBL **9**, 5–15 (2015)
8. Hmelo-Silver, C.E.: Problem-based learning: what and how do students learn? Educ. Psychol. Rev. **16**(3), 235–266 (2004)
9. Albanese, M.A., Mitchell, S.: Problem-based learning: a review of literature on its outcomes and implementation issues. Academic Medicine-Philadelphia **68**, 52–52 (1993)
10. Vernon, D.T., Blake, R.L.: Does problem-based learning work? A meta-analysis of evaluative research. Acad. Med. (1993)
11. Denton, B.G., Adams, C.C., Blatt, P.J., Lorish, C.D.: Does the introduction of problem-based learning change graduate performance outcomes in a professional curriculum. J. Excell. College Teach. **11**(2–3), 147–162 (2000)
12. Yadav, A., Subedi, D., Lundeberg, M.A., Bunting, C.F.: Problem-based learning: influence on students' learning in an electrical engineering course. J. Eng. Educ. **100**(2), 253–280 (2011)
13. Spronken-Smith, R., Harland, T.: Learning to teach with problem-based learning. Act. Learn. High Educ. **10**(2), 138–153 (2009)
14. Rideout, E., Carpio, B.: Learning model of nursing education. In: Transforming Nursing Education Through Problem-Based Learning (2001), p. 21
15. Schmidt, H.G., Van der Molen, H.T., Te Winkel, W.W.R., Wijnen, W.H.F.W.: Constructivist, problem-based learning does work: a meta-analysis of curricular comparisons involving a single medical school. Educ. Psychol. **44**(4), 227–249 (2009)
16. Awang, H., Ramly, I.: Creative thinking skill approach through problem-based learning: pedagogy and practice in the engineering classroom. Int. J. Hum. Soc. Sci. **3**(1), 18–23 (2008)

17. Melles, G., Anderson, N., Barrett, T., Thompson-Whiteside, S.: Problem finding through design thinking in education. In: Blessinger, P., Carfora, J.M. (eds.) Inquiry-Based Learning for Multidisciplinary Programs: A Conceptual and Practical Resource for Educators, pp. 191–209. Emerald Group Publishing Limited (2015)
18. Williams, A., Williams, P.J., Ostwald, M., Kingsland, A.: Problem based learning: an approach to teaching technology. Res. Dev. Probl.-Based Learn. **2**, 355–367 (1994)
19. Petray, T., Taylor, P., Otto, T., Anderson, N., Adam, R., Melles, G., Kueh, C., Wright, N. (2014). Design thinking frameworks as transformative cross-disciplinary pedagogy: Final OLT Seed Grant Report
20. Cross, N.: Design cognition: results from protocol and other empirical studies of design activity. In: Design Knowing and Learning: Cognition in Design Education, pp. 79–103. Elsevier Science (2001)
21. Plattner, H., Meinel, C., Weinberg, U.: Design-thinking. Mi-Fachverlag, Landsberg am Lech (2009)
22. Brown, T., Wyatt, J.: Design thinking for social innovation. Dev. Outreach **12**(1), 29–43 (2010)
23. Beitz, W., Pahl, G., Grote, K.: Engineering Design: A Systematic Approach (1996)
24. Hubka, V., & Eder, W. E. (2012). Theory of Technical Systems: A Total Concept Theory for Engineering Design. Springer Science & Business Media
25. Cross, N.: Designerly ways of knowing: Design discipline versus design science. Des. Issues **17**(3), 49–55 (2001)
26. Dieter, G.E., Schmidt, L.C.: Engineering Design. McGraw-Hill Higher Education, Boston (2009)
27. Eppinger, S., Ulrich, K.: Product Design and Development. McGraw-Hill Higher Education (2015)
28. Bhaumik, R., Bhatt, A., Kumari, M.C., Menon, S.R., Chakrabarti, A.: A Gamified Model of Design Thinking for Fostering Learning in Children. In: Research into Design for a Connected World, pp. 1023–1036. Springer, Singapore (2019)

Part V
Design Management, Knowledge Management and Product Life Cycle Management

Chapter 45
Capturing Knowledge Transfer Using Zachman Framework in Bio-inspired Design Process

Sunil Sharma and Prabir Sarkar

Abstract Natural entities are sources of inspiration for designers as they exhibit nature's unique and novel strategies and help them design products. For understanding and abstracting principles from nature, the knowledge of a biological entity is a must. Knowledge capture and representation in bio-inspired design is a complex process. Firstly, the biological entities are documented in different places such as books, articles, blogs, etc. A designer might need to spend considerable time while shuffling through these documents without coming to any conclusion. Secondly, biological knowledge is intricate and from a different domain. This complexity at the entity level and retrieving captured knowledge can lead to confusion and frustration. Thirdly, the roles of various stakeholders are not clearly defined. Hence, the knowledge collection, compilation, usage, and representation process are vague. We provide an organizational approach to represent bio-inspired design knowledge collection, compilation, and usage and representation process using the Zachman framework. Zachman framework provides the descriptive representation of enterprise architecture, uses six primitives, and can handle complex systems. In this research, we understand and apply the Zachman framework to capture and represent bio-inspired design knowledge with an organizational viewpoint. This research aims to organize and represent captured knowledge transfer, building, and representing processes and making it readily available for designers who can use it to make design decisions.

Bio-inspired design takes inspiration from natural entities to solve human design problems. Many revolutionary products and applications such as Lotuson, Velcro, Sharkskin inspired swimsuit, etc. have been developed in the past. However, realizable and repeatable novel designs and applications need multiple tools, methods, people, resources, databases, and process knowledge. The facilitation, collection,

S. Sharma (✉) · P. Sarkar
IIT Ropar, Rupnagar, Punjab, India
e-mail: 85sunilsharma@gmail.com

P. Sarkar
e-mail: prabir@iitrpr.ac.in

A. Chakrabarti et al. (eds.), *Design for Tomorrow–Volume 2*, Smart Innovation, Systems and Technologies 222, https://doi.org/10.1007/978-981-16-0119-4_45

and processing of knowledge also become a critical part of bio-inspired design organizations. Since organizations act as social systems, they have to deal with problem-solving, coordination, and uncertainty in assigned tasks. With a countable number of bio-inspired design organizations, research on organizational processes can discover significant organizational structures. In this research, we intend to find (1) an appropriate work configurations and links between systems and actors (2) and recognize stakeholders and their specialized tasks in bio-inspired design organization.

The answer to these questions can help streamline the process and optimize the bio-inspired design organization's structural set. Defining the organization form possesses numerous difficulties. One issue in organizational design is knowledge management. Organizational learning is improved if the knowledge management processes are effective and efficient. Knowledge management is a business process that identifies knowledge from previous experiences and selectively applies it to current decision-making processes [1]. Modern-day technology enterprises are highly dependent on knowledge management. The emergence of bio-inspired design products has led to the development of bio-inspired design organizations, where the knowledge management process can be implemented [2]. The success of design projects is facilitated by design communication in a structured organization. Design communication and performance are affected by the team organization [3]. Knowledge management is also strengthened by defining knowledge flows in an organization. Firms have no formal, explicit representation of knowledge, and their knowledge is processed in the unstructured form [4]. However, many models exist that are used to describe the organizational structure and knowledge management in an enterprise. Co-ordination theory [5], enterprise modeling using IDEF (Integration Definition) modeling method [6], knowledge mapping to identify knowledge flows in the organization [7], ontology-based framework such as Core Organizational Knowledge Entities (COKE) for specifying organizational knowledge using representation language [8], modeling languages such as Web Ontology Language (OWL), ArchiMate [9], and enterprise modeling technique like Generalised Enterprise Reference Architecture and Methodology (GERAM)[10] has been used in the past. Many of them are effective, detailed and efficient but there are some limitations. They may not be suitable for current bio-inspired organizational knowledge structuring. GERAM requires details of several building blocks for its development. It may not be appropriate to represent the bio-inspired design organization, where little is known about the knowledge architecture, management, organization structure, and stakeholders. COKE, OWL, and ArchiMate require an understanding of modeling language first. The modeling is often too elaborative, complicated and requires knowledge of the organization's functioning. We assume that bio-inspired design organizations' structure for knowledge management can be represented using the Zachman framework. Zachman framework is developed to be independent of any methodology or process within an organization but can be implemented using a rational selection of variables [11]. Similarly, the knowledge transition in bio-inspired design is a complicated task involving multiple stakeholders [12]. Therefore, a biological knowledge process comparable to information system architecture has been developed using ZF.

45.1 Introduction

Zachman framework was developed by John Zachman in 1987 [13, 14]. Zachman framework (referred to as ZF in this paper) is used to model an organization, information, and knowledge architecture. The representation of ZF is shown in Fig. 45.1. Primarily, it is developed by involving multiple stakeholders and their perspectives [15]. Thirty-six categories are provided in ZF to specify complex objects such as manufactured products, structures, and enterprises. A complicated enterprise can also be represented by ZF like a periodic table exhibiting elements in an orderly fashion of increasing atomic numbers. As complicated entities, such as information, enterprise, and architecture, are visible to the observer in one go, the ZF brings order to a complex world. Some of the crucial definitions related to the ZF are as follows. *Abstraction:* One dimension of classification of descriptive representation. *Perspectives:* Second dimension of classification of descriptive representation. *Artifacts:* Each cell is populated with artifacts, which describe entirely the cell—for example, design documents, specifications, models, etc.

It is essential to understand how bio-inspired design occurs in an organization. The customer usually walks with a design problem in hand. The organization reformulates the problem in terms of specifications, functions, constraints, performance, and processes. The natural world is explored, and the transition of selected solutions is performed by various stakeholders using multiple tools and techniques. Finally, solutions are generated.

45.1.1 Background

ZF has been extensively used commercially and by researchers to represent US-CIPP, US government's infrastructure protection program [16], as a diagnoses model [17],

Abstractions (Column) →

Perspectives (Row) ↕

Zachman Framework	DATA (1) (what) *Entities*	FUNCTION (2) (how) *Activities*	ORGANIZATION (3) (who) *People*	NETWORK (4) (where) *Locations*	SCHEDULE (5) (when) *Time*	STRATEGY (6) (why) *Motivation*
SCOPE (A) (Contextual) *Planner*	List of things important to business	List of processes the business performs	List of organization important to business	List of location in which the business operate	List of events significant to business	List of business goals/strategies
BUSINESS MODEL (B) (Conceptual) *Owner*	Semantic model	Business process model	Work flow model	Business logistics system	Master schedule	Business plan
SYSTEM MODEL (C) (Logical) *Designer*	Logical data model	Application architecture	Human interface architecture	Distribution system architecture	Processing structure	Business rule model
TECHNOLOGY MODEL (D) (Physical) *Builder*	Physical data model	System design	Presentation architecture	Technology architecture	Control Structure	Rule design
DETAILED REPRESENTATION (E) (Out of Context) *Sub contractor*	Data definition	Program	Security architecture	Network architecture	Timing definition	Rule specification
FUNCTIONING (F) *Enterprise*	Actual business data	Actual application code	Actual business organization	Actual Physical network	Actual business schedule	Actual business strategy

Fig. 45.1 Zachman framework for an enterprise

and decision making in business [18]. Academically, it has been applied in computing classrooms for improved student learning [19] and representation of digital libraries [20]. It has also been used to analyze sustainability standards [21], for reverse engineering and integrated with other frameworks as well [22–24] and has been used to define knowledge management architecture in an enterprise [4].

Knowledge management in the design organization is in the nascent stage. The knowledge asset is to be preserved by the design organizations. This design knowledge can be in the form of design artifacts, design intuition, design rationale, and design theories. The bio-inspired design can be categorized under design artifacts. It is established that biological entities' attributes are mimicked and transformed into commercial applications [25]. However, the input knowledge and output idea in the bio-inspired design process also utilize design rationale, design theories, and design intuition. In the last twenty-five years, enormous literature on the biomimetic process and its applications has been researched [26]. The actual process involves different cross-disciplinary teams, where each actor performs multiple roles. This involvement has created a need for knowledge management in the bio-inspired design process and has not been explored to date. Traditionally, many formal, informal methods, and tool categories of design knowledge representation exist. One research has classified the knowledge representation of the tools as pictorial, symbolic, linguistic, virtual, and algorithmic [1]. The knowledge capture and representation can be pictorial, symbolic, and linguistic for the bio-inspired design process. Computational tools can support the bio-inspired design problem-solving process by integrating biological knowledge and multimodal representation that can help rationalize ideas and generate concepts [27].

The knowledge collection and compilation process in the bio-inspired design process rarely have been focused on. An online database of biological strategies, AskNature.org, has described its data collection process. In the data compilation process, trained biologists compiled 1600 pages of biological information. They have explored scientific journals and books to assess and include the strategy in the online database. Enthusiastic individuals and paid staff appended the database. Taxonomy has been used to categorize functions [28].

45.2 Methodology

The compiled knowledge transition for biological entities (single or more) by different stakeholders has been considered equivalent to an organizational workflow process. For developing this ZF for bio-inspired design organizations, we follow these steps. *Step I*: Studying the Zachman framework for understanding enterprise architecture and its representational elements. All dimensions of the ZF are explored for complete understanding. The objective is to use the understanding of developing the ZF. *Step II*: The stakeholders in the enterprise are identified, and their roles are defined. A syllogism is drawn between the enterprise architecture and the knowledge transition process. *Step III*: Perspectives and descriptions are identified and realized, and the ZF is developed. The knowledge transition stages in the bio-inspired design process are decomposed for each function of the cell. The roles and related processes are recognized for populating the ZF. *Step IV*: The developed ZF is revised.

Step I: Understanding Zachman framework [29]

Zachman framework descriptive representation of the organization is shown in Fig. 45.1. Here, initially, perspectives and later abstractions are detailed. The cells of ZF are populated based on rules, as described further.

(1) *Planner perspective (Row 1):* This establishes inner and outer limits and lists relevant constituents for remaining perspectives. (2) *Owner's perspective (Row 2):* This provides a conceptual view of the end product, for what the owner will do with the end product. (3) *Designer's perspective (Row3):* Intermediary between what is desirable and what is feasible. They reflect laws of nature, system, or logical constraints for designing the product. (4) *Builder's perspective (Row 4):* Reflect the physical constraints of applying the technology in developing the product. (5) *Subcontractor's perspective (Row 5)*: Detailed descriptions that dissociate the parts or pieces of complex products for manufacturing. (6) *User's perspective (Row 6):* View of the functioning system in its operational environment.

The columns represent the following interrogatives for the enterprise.

(1) *What*—This interrogative is a representation of data. (2) *How*—This interrogative is a representation of the function. (3) *Where*—This interrogative is a representation of the network. (4) *Who*—This interrogative is a representation of people. (5) *When*—This interrogative is a representation of time. (6) *Why (motivation)*—This interrogative is a representation of motivation.

Step II: Identification of stakeholders.

The biological world has a plethora of information for observation. From this living world, knowledge can be collected, compiled, and then applied. Knowledge dissemination is essential to make it available to the necessary stakeholders [30]. Naturalists describe their critical observations on nature and pass them to other stakeholders such as biologists and researchers. These stakeholders closely to observe, dissect, and compare the literature. Biologists also act as stalwarts to abstract attribute-based knowledge, experiment, modify, and design various bio-inspired applications based on strategies such as the Lotus effect (inspired Lotuson) and Gecko's feet (inspired Gecko tape). Publishers further publish the information as generalized theories with supporting images or adapted images provided by naturalists and biologists. This information is available for public use for designers and engineers.

Researchers may team up with designers and engineers to abstract analogy leading to revolutionary applications and products. Similarly, designers and engineers also explore causal mechanisms in nature published in the scientific research documents. A 1D knowledge process flow for knowledge transition and analogy abstraction from nature, delivered to designers and engineers, is given in Fig. 45.2. Designers also explore the physical and digital databases, web, eBooks, textual documents, and digital media to identify the natural entities. Then, they reformulate the question by biologizing to seek the answers to trigger inspiration and solution generation.

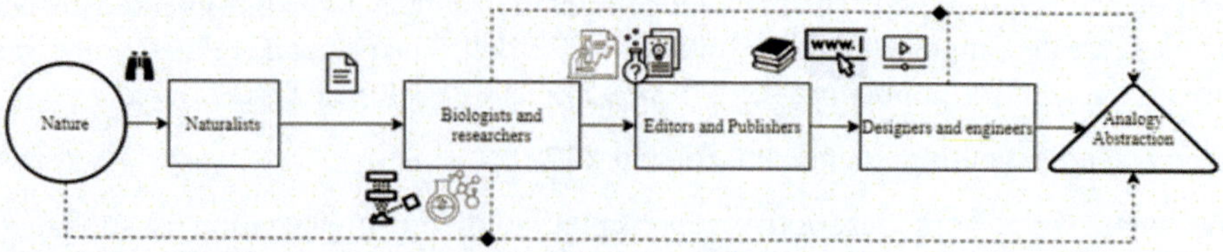

Fig. 45.2. 1D process to capture knowledge in bio-inspired design

The sources are often scattered; modes of representation are vague and biological terms different from engineering. Engineering designers with queries also approach peers for information. This is a usual practice as assessing information present in documents is difficult [31]. These stakeholders perform many activities for generating bio-inspired designs. As already described, this knowledge transition through the stakeholders is shown in Fig. 45.2. However, the 1D process does not explain how different stakeholder processes the information. Therefore, the Zachman framework, a 2D-based ontology, is proposed to represent this knowledge process, which shows a flow of information with logic for processing at each different level. However, a complete description of all the artifacts represented for individual cells is outside this research's scope. As described earlier, the ZF's standard stakeholders are planner, owner, designer, builder, and sub-contractor. These are replaced by stakeholders for bio-inspired design organizations, such as knowledge planner, knowledge compiler, knowledge builder, tool selector, knowledge fitter, and designer. The description entities in abstraction are kept the same. This proposed framework of bio-inspired design knowledge organizational structure is represented in Fig. 45.3.

Figure 45.3 provides the various titles of the column's descriptive representation as numbered from 1 to 6. The rows from A to F detail descriptive representations. Cell A1 is the intersection of Ath row and first column, which describes 'List of things important to the business.' Cell F5 is the Fth row intersection with the fifth

Zachman Framework	DATA (1) (what) *Entities*	FUNCTION (2) (how) *Activities*	ORGANIZATION (3) (who) *People*	NETWORK (4) (where) *Locations*	SCHEDULE (5) (when) *Time*	STRATEGY (6) (why) *Motivation*
SCOPE (A) (Contextual) *Knowledge Planner*	A1	A2	A3	A4	A5	A6
BUSINESS MODEL (B) (Conceptual) *Knowledge compiler*	B1	B2	B3	B4	B5	B6
SYSTEM MODEL (C) (Logical) *Knowledge builder*	C1	C2	C3	C4	C5	C6
TECHNOLOGY MODEL (D) (Physical) *Tool Selector*	D1	D2	D3	D4	D5	D6
DETAILED REPRESENTATION (E) (Out of Context) *Knowledge Fitter*	E1	E2	E3	E4	E5	E6
FUNCTIONING (F) *Designer*	F1	F2	F3	F4	F5	F6

Fig. 45.3 Classification structure for knowledge transition of bio-inspired design process artifacts

column, which defines the 'Actual business schedule.' The row-column numbering system does not belong to the ZF. It has been provided for ease to pinpoint the exact cell locations using row and column combination.

Stakeholders: Usually, the bio-inspired design research details the role of designers, biologists, and engineers [32]. The functions of knowledge compiler, knowledge builder, and others supporting knowledge inclusion are rarely described. The roles of different stakeholders are defined as follows. (1) *Knowledge planner:* The stakeholder goes though design question and describes boundaries of knowledge and details the planning. (2) *Knowledge compiler:* This stakeholder collects and manages knowledge about different biological strategies from multiple sources and reasons them with logic. (3) *Knowledge builder:* The compiled knowledge is developed to form meaningful information supported by external representations with logic. (4) *Tool selector:* The modal representations for nature's strategies are comprehended and selected depending upon the designer requirement and explained to knowledge transposer. (5) *Knowledge transposer or fitter:* This stakeholder transposes compiled knowledge from knowledge builder in tool provided by tool selector. (6) *Designer:* This stakeholder obtains knowledge in final representation and provides feedback, inputs in knowledge to other stakeholders. Different professionals can play multiple roles in an organization. Therefore, we do not personalize the roles in this research work.

Step III: Decomposing organizational hierarchy for individual roles of each cell.

In this step, each stakeholder's role is decomposed at each interrogative level to populate the cells. Ideally, each cell is supported by relevant modes, diagrams, rules, models, or processes. In this research, we only describe an indicative function and not an exhaustive list. Here, each row represents data (what), function (how), organization (who), network (where), schedule (when), and strategy (why). Table 45.1 describes the proposed ZF for bio-inspired design organization.

Step IV: Revision of the proposed Zachman framework.

As already described, Fig. 45.3 gives the review and revised version of the final ZF. These cells can be further described in great detail. The above list is only indicative and not exhaustive. Diagrams, texts, and rules, etc., further detail each cell.

Applications: The developed ZF has the following applications. (1) Improve design communications within bio-inspired organization leading to efficient knowledge management. (2) The framework can be integrated with data management systems to support knowledge management. (3) Help in planning and managing the design projects. (4) Integrated with other frameworks such as TOGAF for detailed organization representation and management. (5) Knowledge pathways can be easily tracked using digital representation of this framework.

Limitations: We represent organizational structure using ZF and understand that the actual practice can differ from what is described in this research. The framework is structural and does not specify any method to develop the framework itself. We

Table 45.1 Proposed Zachman framework with possible cell descriptions

Stakeholder	Cell position	Cell detail
First row: knowledge planner	A1 (Goal list)	Decides knowledge capture and representation form; responsible for complete management of knowledge. Determines output content and communication
	A2 (Process list)	Who to include and knowledge management support
	A3 (Organizational unit)	Allocation of the task to the responsible person
	A4 (Organizational identification)	The storage location can be detailed
	A5 (Event list)	Noting knowledge collection time
	A6 (Knowledge project goals)	Purpose is to deliver the designer with knowledge in a selective form based on biologized question
Second row: knowledge compiler	B1 (Entity relationship model)	Whether biological strategy is suitable for designer's question?
	B2 (Process model)	How to abstract strategy to capture and represent?
	B3 (Workflow model)	Work details and suitability rationale; knowledge compiling, keeping it annotated, filtered, supported by source notes
	B4 (Business flow model)	Initial and final location of trimmed knowledge
	B5 (Master schedule)	Keeping time notes of compiling and editing
	B6 (Business plan)	Understand concept behind biological entity and its relationship; list sources, knowledge of biological entity
Third row: knowledge builder	C1 (Logical data model):	Know source input and output, confirmation? Describes logically organized information for answering the designer's question
	C2 (Application architecture)	Information presenting technique can be digital, physical. Knowledge presentation in multiple formats with capabilities
	C3 (Human interface architecture)	Stakeholders adding, deleting, and editing details
	C4 (Distribution system architecture)	Where is relevant knowledge located for access?

(continued)

Table 45.1 (continued)

Stakeholder	Cell position	Cell detail
	C5 (Processing structure)	Knowledge duration final total time
	C6 (Business rule model)	Complete knowledge packages represented in many modes
Fourth row: tool selector	D1 (Physical data model)	Includes various representation techniques and their modes, models, tools, etc
	D2 (System Design)	Capabilities of different models and their needs; Elemental understanding of various frameworks and requirements
	D3 (Presentation architecture)	Physical form of frameworks or representational form with details that are needed to fill
	D4 (Technology architecture)	Database and details of frameworks stored at locations
	D5 (Control structure)	Time filling details, duration
	D6 (Rule design)	To know the technical tools list
Fifth row: knowledge fitter	E1 (Data definition)	Appropriately fit the knowledge gathered in tool selected
	E2 (Program representation)	Use knowledge and tool selected and fit the knowledge in tool
	E3 (Security architecture)	Knowledge mappers understanding tool and understanding knowledge
	E4 (Network architecture)	Storing knowledge at appropriate places
	E5 (Timing definition)	Time notes for this task
	E6 (Rule specification)	Fit knowledge to capture and represent
Sixth row: designer	F1 (Actual data)	Usage of delivered knowledge captured forms to designer. Requires input and feedback
	F2 (Actual function)	Linking using formats and answering question
	F3 (Actual organization)	Read formats and understand them
	F4 (Actual network)	Particular locations, links
	F5 (Actual timing)	Designer reflection times can be recorded
	F6 (Actual motivation)	Trigger inspiration and analogy selection

do not formally test the ZF for representing organizational structure or conduct an actual examination of bio-inspired design organization. The architecture of information capturing and representation encompasses tools and methods to deliver the representation to the consumer of this information, typically the designers. A detailed ZF version can be developed using software but is complex and rigorous. Often, this framework is difficult to use.

45.3 Conclusion and Future Work

This research has examined the ZF for the organizational representation of the bio-inspired design knowledge transfer process. We also identify different stakeholders for knowledge transfer in the bio-inspired design process. We also populate the roles of these stakeholders to define their attributes. The characteristics of this knowledge transition are based on the classic ZF template. Representing knowledge as design knowledge in an organizational entity makes it easier to manage and handle bio-inspired design knowledge. This proposed framework results in a comprehensive perspective of bio-inspired design knowledge capture and representation. This can be studied to determine the efficacy of bio-inspired design knowledge mapping for knowledge capture. Thus, using the ZF can help the design organizations transfer captured and represented information to the designer. It can also be useful in knowledge management, developed to capture and reuse the organization's knowledge. It must be noted that awareness of other stakeholder tools and techniques can cast off many doubts about knowledge processing by other stakeholders. This can ease up communication across teams. If there is any need for changing tools and techniques, the same can be modified, recorded, and represented in the ZF. Currently, the ZF representation is based on interpretation, as proposed by the author. The future versions of this work can be further refined as per the agreed views of the stakeholders and the analysts. Furthermore, the organizational processes can be studied and presented to represent the bio-inspired design organization better.

References

1. Chandrasegaran, S.K., Ramani, K., Sriram, R.D., et al.: The evolution, challenges, and future of knowledge representation in product design systems. Comput.-Aided Des. **45**, 204–228 (2013). https://doi.org/10.1016/j.cad.2012.08.006
2. Salgueiredo, C.F.: Modeling Biological Inspiration for Innovative Design (2013)
3. Chiu, M.-L.: An organizational view of design communication in design collaboration. Des. Stud. **23**, 187–210 (2002). https://doi.org/10.1016/S0142-694X(01)00019-9
4. Farazmand, E., Moeini, A.: A framework for knowledge management architecture. In: Proceedings of the 13th International Conference on Enterprise Information Systems. SciTePress - Science and and Technology Publications, Beijing, China, pp. 425–430 (2011)
5. Crowston, K.: A coordination theory approach to organizational process design. Organ. Sci. **8**, 157–175 (1997). https://doi.org/10.1287/orsc.8.2.157

6. Liles, D.H., Presley, A.R. (1996). Enterprise modeling within an enterprise engineering framework. In: Proceedings of the 28th conference on Winter simulation—WSC '96. ACM Press, Coronado, California, United States, pp. 993–999
7. Jonker, C.M., Sharpanskykh, A., Treur, J., Pi, Y.: A framework for formal modeling and analysis of organizations. Appl. Intell. **27**, 49–66 (2007). https://doi.org/10.1007/s10489-006-0030-2
8. Gualtieri, A., Ruffolo, M.: An ontology-based framework for representing organizational knowledge. In: Proceedings of I-KNOW '05. Graz, Austria, pp. 71–78 (2005)
9. Pereira, D.C., Almeida, J.P.A.: Representing Organizational Structures in an Enterprise Architecture Language (2014)
10. Vernadat, F.: Enterprise Modeling and Integration: Principles and Applications. Chapman & Hall, London, New York (1996)
11. Vail, E.F.: Causal architecture: bringing the Zachman framework to life. Inf. Syst. Manag. **19**, 8–19 (2002). https://doi.org/10.1201/1078/43201.19.3.20020601/37165.2
12. Morganwalp, J., Sage, A.: A system of systems focused enterprise architecture framework and an associated architecture development process. Inf.-Knowl.-Syst. Manag. **3**, 87–105 (2002)
13. Noran, O.: An analysis of the Zachman framework for enterprise architecture from the GERAM perspective. Annu. Rev. Control. **27**, 163–183 (2003). https://doi.org/10.1016/j.arcontrol.2003.09.002
14. Zachman, J.: The Framework for Enterprise Architecture: Background, Description and Utility. In: Zachman Int. https://www.zachman.com/resources/ea-articles-reference/327-the-framework-for-enterprise-architecture-background-description-and-utility-by-john-a-zachman(1996)
15. Pienimäki, T.: A Business Application Architecture Framework in Manufacturing Industry. Tampere University of Technology (2005)
16. Tatar, U., Karabacak, B., Katina, P.F., Igonor, A.: A Complex Structure Representation of the US Critical Infrastructure Protection Program Based on the Zachman Framework, p. 14
17. Dantu, B., Smith, E.: Medical process modeling with a hybrid system dynamics Zachman framework. Procedia Comput. Sci. **6**, 76–81 (2011). https://doi.org/10.1016/j.procs.2011.08.016
18. Danny, J., Shanlunt, W.G., Alianto, H.: The application of Zachman framework in improving better decision making. In: 2018 Indonesian Association for Pattern Recognition International Conference (INAPR), pp. 245–249. IEEE, Jakarta, Indonesia (2018)
19. Mani, M., Uludag, S., Zolinski, C.: On evaluating the use of Zachman framework in computer science and information systems classes. J. Comput. Sci. Coll. **31**, 9 (2015)
20. Abdullah, A., Zainab, A.N.: The digital library as an enterprise: the Zachman approach. Electron. Libr. **26**, 446–467 (2008). https://doi.org/10.1108/02640470810893729
21. Rachuri, S., Sarkar, P., Narayanan, A., et al.: Towards a methodology for analyzing sustainability standards using the Zachman framework. In: Hesselbach, J., Herrmann, C. (eds.) Glocalized Solutions for Sustainability in Manufacturing, pp. 543–548. Springer, Berlin Heidelberg (2011)
22. Chen, Z., Pooley, R.: Rediscovering Zachman framework using ontology from a requirement engineering perspective. In: 2009 33rd Annual IEEE International Computer Software and Applications Conference, pp. 3–8. IEEE, Seattle, Washington, USA (2009)
23. Kurniawan, H., Salim, A., Suhartanto, H., Hasibuan, Z.A.: E-cultural heritage and natural history framework: an integrated approach to digital preservation. In: Proceedings of CSIT. IACSIT Press, Singapore, pp. 177–182 (2011)
24. Urbanic, R.J., ElMaraghy, W.: A design recovery framework for mechanical components. J. Eng. Des. **20**, 195–215 (2009). https://doi.org/10.1080/09544820701802261
25. Müller, R.M., Thoring, K.: A typology of design knowledge: a theoretical framework. In: AMCIS 2010. Lima, Peru, p. 12 (2010)
26. Sharma, S., Sarkar, P.: Biomimicry: exploring research, challenges, gaps, and tools. In: Chakrabarti, A. (ed.) Research into Design for a Connected World, pp. 87–97. Springer Singapore, Singapore (2019)
27. Vattam, S.S., Goel, A.K.: Foraging for inspiration: understanding and supporting the online information seeking practices of biologically inspired designers. In: Volume 9: 23rd International Conference on Design Theory and Methodology; 16th Design for Manufacturing and the Life Cycle Conference, pp. 177–186. ASMEDC, Washington, DC, USA (2011)

28. Deldin, J.-M., Schuknecht, M.: The AskNature database: enabling solutions in biomimetic design. In: Goel, A.K., McAdams, D.A., Stone, R.B. (eds.) Biologically Inspired Design, pp. 17–27. Springer, London, London (2014)
29. Ertaul, L., Sudarsanam, R.: Security Planning Using Zachman Framework for Enterprises, p. 11
30. Beck, S., Mahdad, M., Beukel, K., Poetz, M.: The value of scientific knowledge dissemination for scientists—a value capture perspective. Publications **7**, 54 (2019). https://doi.org/10.3390/publications7030054
31. Culley, S., International Conference on Engineering Design, Institution of Mechanical Engineers: Design Applications in Industry and Education. Professional Engineering Publ, Bury St. Edmunds (2001)
32. Bhasin, D., McAdams, D.: The characterization of biological organization, abstraction, and novelty in biomimetic design. Designs **2**, 54 (2018). https://doi.org/10.3390/designs2040054

Chapter 46
MCDM-Based Decision Support System for Product Design and Development

Prabhat Kumar and Ayan Tiwari

Abstract The process of product design and development (PDD) consists of various sequential stages. Each stage requires a complex evaluation and the right decision to attain a successful product. Decision making in these product design stages often is involved with multiple criteria and it is important to use multiple criteria decision making (MCDM) to assist design practitioners for more appropriate decisions. Nowadays, various MCDM methods are available and applied in various areas. The objective of this paper is to identify the types of decision-making problems that may creep during different design stages and possible MCDM methods that might be applicable to solve them. This paper presents comparative analysis and gives information about some of the most popular MCDM methods with the design decision applications as per the available literature. This knowledge can help enterprises make better decisions in a particular design stage to ensure the success of their PDD.

46.1 Introduction

Product design and development (PDD) is a problem-solving and knowledge-accumulation process. This process passes through several key design stages before it gets a final design. Each design stage requires valid inputs and complex evaluation with clear decision making to obtain the desired output(s). Hence, PDD is a complex, interdisciplinary, sequential, uncertain, and risky process of extensive planning and activity. The right decisions in each design stage are vital to the success of the 'design.' A good product design ensures, product working as per customer requirements (CR), within optimal cost. To take the right decision by identifying, a feasible combination of customer requirements and satisfy the conflicting requirements is a tough task for both the design practitioners and the manufacturer [1]. Nowadays,

P. Kumar (✉) · A. Tiwari
Department of Fashion and Lifestyle Accessories, National Institute of Fashion Technology, Bhopal, India
e-mail: prabhat.kumar1@nift.ac.in

A. Tiwari
e-mail: ayan.tiwari@nift.ac.in

A. Chakrabarti et al. (eds.), *Design for Tomorrow–Volume 2*, Smart Innovation, Systems and Technologies 222, https://doi.org/10.1007/978-981-16-0119-4_46

product design has grown rapidly in the last few years. Increased competition for better product (or system) functionality, quality, and cost along with shorter delivery time presents remarkable challenges for any product manufacturing enterprise [2]. Over the past decades, the complexity of product design has increased rapidly.

The complex and dynamic nature of design leads to uncertainty and risk, which makes taking the right decision critical [3]. Most of the design problems often include the necessity to identify the best optimal design solution from a large number of potentially good alternative solutions. The selection of one good alternative among a large number of potential alternatives is a very critical managerial task. It is also challenging due to, design decision involves multiple criteria, both quantitative and qualitative in nature with dependent and independent variables.

Decision making in product design often is involved with multiple criteria. Therefore, it is advantageous to use multiple criteria decision making (MCDM) methods to find an appropriate assessment. During the past few years, there have been tremendous attempts on using MCDM techniques for decision making in product design. The implementation of MCDM could pave the way for a new horizon in the decision support system for product design. MCDM techniques improve the quality of decisions by creating the development more efficient, rational, and explicit [4]. As per the Wang, an optimal design scheme not only improves the performance of the product but also lead to the greatest satisfaction of customers [5]. In recent decades, the MCDM techniques and approaches have received a great deal of attention from design practitioners.

Nowadays, a large number of MCDM techniques have been proposed, which are diverse in their theoretical process, the type of input required, and the type of obtained results. Thus, it is important to identify the types of different MCDM techniques applicable in the product design and document the exponentially grown interest in the MCDM techniques and provide a state-of-the-art review of the literature regarding the MCDM applications for decision making in product design. This work gives an idea to identify suitable MCDM techniques for accurately and efficiently decision making in product design.

The remaining sections of this paper are organized as follows. Section 2 provides a summary of the literature review on decision making in product design. Section 46.3 describes the research methodology and the procedure of this study. Section 46.4 presents the discussion. Finally, Sect. 46.5 presents the conclusion and shows the prospects for future research framework, opportunities, and challenges.

46.2 Summary of the Literature

PDD process is a sequence of steps or stages to conceive design and commercialize a product. The decision making in PDD is considered as a typical MCDM problem. Due to the great interest of academics and practitioners, many efforts have been devoted to develop various types of MCDM methods. As per the literature, methods which are mostly used in design evaluation and decision making are: analytical hierarchy process (AHP) [6], analytic network process (ANP) [7], elimination and choice translating reality (ELECTRE) [8], goal programming (GP) [9], multi-attribute utility theory (MAUT) [10], preference ranking organization method for enrichment evaluation (PROMETHEE) [11], technique for order preference by similarity to ideal solutions (TOPSIS) [12], and weighted sum method (WSM) [13]. However, it is very challenging to decide which MCDM method is the best for the particular design problem. To date, several articles carried out a systematic review of the literature [14–17] on MCDM methods. Siskos and Spyridakos (1999) presented a survey of the history and the recent status of the multiple criteria decision support systems [14]. Zavadskas & Turskis presented a panorama of decision-making methods in light of the recent developments of multiple criteria decision-making methods [15]. Recently, Mardani et al. [16] reviewed a total of 393 articles published from 2000 to 2014 and investigated the developments of various methods of MCDM and their applications. Renzi et al. [16] presented a review of decision-making methods focused on the selection of decision support methods for automotive industry design problems. Nowadays, one MCDM method with a combination of other methods of MCDM methods has been proposed in the literature extensively and solved decision problems. Similarly, Mayyas et al. [17] proposed a combined quality function deployment and analytical hierarchy process; Peng and Xiao [18] proposed combined ANP and PROMETHEE; Zhou et al. [19] proposed AHP and TOPSIS for the suitable decisions. A method, which is often combined with the MCDM methods, is the fuzzy sets theory [16]. Fuzzy logic employed approximate modes of reasoning for decision making in imprecise and uncertain environment over the entire design cycle [20]. A further AHP method is most popular among the MCDM methods and frequently combined with the other MCDM [21]. AHP is a simple method that focuses on prioritizing the criteria by capturing the degree of importance of criteria to assist enterprises in the product development phase [22]. AHP is simple to use, flexible, effective, versatile, and transparent methods. This has made AHP extremely important and useful tools in solving design decision-making problems.

As per the literature, numerous MCDM methods are available, and no single method is considered the most suitable for all types of decision-making situations [22–24]. Moreover, different MCDM methods can yield different results when applied to the same design decision problem. Further, the selection of an appropriate decision-making method leads to an MCDM problem itself. Thus, it is felt that there is a serious need to classify the MCDM method as per the product design stages.

46.3 Research Methodology

The methodology of the work includes:

Step 1: Identification of strength, weaknesses, and application area of MCDM methods.

Step 2: Classification of various design stages and decision problems.

Step 3: Mapping of seven design stages and applicable MCDM methods.

46.3.1 Identification of Strength, Weaknesses, and Application of MCDM Methods

For the aforementioned goal, the article reviews the literature published in popular journals on decision making in product design. An extensive search was carried out to find MCDM in titles, abstracts, keywords, and research methodologies of the article. Currently, research on MCDM continued and found many applications in different fields. Each of these methods has its own features and application area. After a comprehensive analysis of journal articles, it was found that the eight most frequently used techniques in PDD are: WSM, MAUT, AHP, ANP, TOPSIS, ELECTRE, GP, and PROMETHE. Based on the extensive literature reviewed, the observed strength and weaknesses, as well as application areas of the popular MCDM methods, are compiled in Table 46.1. As several types of MCDM methods are available and applied in various areas, this work considers only the major application areas.

46.3.2 Classification of Various Design Stages and Decision Problems

As shown in Table 46.1, MCDM methods have their own strength and weakness as well as application fields; none of the methods dominate the other methods. PDD process involves a systematic series of design stages that design practitioners follow to develop an appropriate design solution. During this process, there are numerous decisions made in every design stage to attain a successful design. Inappropriate decision making during any design stage may lead to product failure [3]. Therefore, in this section, the major stages of product design are classified and associated decision-making problems are identified. Many PDD process has been published over the years. Therefore, it is important to establish the comprehensive PDD process to conceive, design, and commercialize a product. As per the inputs gained from Ulrich and Eppinger [35], Pahl and Beitz [36], Anderson and Pine [37], Montagna [38], the seven major stages of PDD and associated decision-making problem are established as shown in Table 46.2 (column 1 to column 3). As per the available literature, the relationship between the features of the reviewed methods presented in Table 46.1, and the decision-making problems that characterize the design process in Table 46.2. There are many decision-making activities involved in each of the seven design stage. However, this work considered only the major decision-making job involved in the associated design stage.

Table 46.1 MCDM methods, their application, strength, and weakness

Methods	Area of application	Strength	Weakness	References
WSM	• Structural optimization • Product assessment • Fund allocation	• Risks and uncertainty are considered • Mechanism of the method is straight forward	• Needs a lot of input • Preferences need to be precise	[13, 25]
MAUT	• Supplier selection • City planning • Testing and robustness assessment	• Represent the uncertainty directly to decision model	• Needs a lot of input	[10, 26]
AHP	• Concept selection • Buyers selection • Customer needs selection • Resource management	• Flexible, intuitive appeal to the decision makers • Ability to check inconsistencies • Capture both subjective and objective evaluation measures	• Allows less number of alternatives • Expressed in different measurements	[1, 22, 27]
ANP	• Concept selection and evaluation • Supplier selection • Material selection • Site selection	• Allows complex interrelationships among decision levels and attributes • Allows tangible as well as intangible criteria in decision making	• Data acquisition is a time intensive process • Subjectivity of the comparisons is not considered • Requires a lot of calculations	[7, 16, 29]
TOPSIS	• Technology forecasting • Concept selection • Competitive benchmarking • Plant layout design	• Simple, rational, comprehensibility • Good computational efficiency and ability to measure the relative performance in a simple mathematical form	• Basically works on Euclidian distance and so does not consider any difference between negative and positive values	[1, 12, 28]
ELECTRE	• Business policy and strategy • Risk and financial management • Facilities planning	• Takes uncertainty and vagueness into account • Deals with heterogeneous scales	• Less versatile	[8, 30, 31]

(continued)

Table 46.1 (continued)

Methods	Area of application	Strength	Weakness	References
GP	• Product planning • Production planning • Cost estimation	• Capable of handling large-scale problems • Effective in combination with other MCDM methods	• Demands substantial information from decision makers on their objectives	[9, 32, 33]
PROMETHE	• Plant location selection • Concept selection • Resource evaluation	• Deals with qualitative and quantitative information • Incorporate uncertain and fuzzy information	• Output depends on the decision maker to assign weight	[11, 18, 34]

Table 46.2 PDD stages and associated MCDM methods

Product development stage	Involve decision	Methods
1. Planning	Business policy and strategy, fuzzy and conflict customers' requirements, risk and financial management	ELECTRE [31, 39], GP [32, 33]
2. Concept generation and selection	Technology forecasting, concept and functional analysis	AHP [22, 27] ANP [18, 29], TOPSIS [1, 19, 28], PROMETHE [11, 34]
3. Embodiment and detail design	product architecture, configuration, parametric, and material	TOPSIS [1, 28], WSM [13, 25], AHP[22, 27]
4. Testing and refinement	Performance, reliability, and durability	PROMETHE [11, 34], WSM [13, 25], MAUT [10, 26]
5. Production ramp-up	Resource management and utilization	AHP [22, 27], TOPSIS [1, 19, 28], GP [32, 33]
6. Approval and launching the product	Product demands, logistics, and transportation	AHP [23, 28], TOPSIS [1, 19, 28], PROMETHE [11, 34]
7. Planning for its retirement	Business management	ELECTRE [31, 39], GP [32, 33]

46.3.3 Mapping of MCDM Methods and Seven Stages of Design

Nowadays, a several MCDM methods available and each method are appropriate for some specific types of decision-making situation. The last phase of this work involves the mapping of the product design stage along with the associated MCDM methods (as obtained in Table 46.1). Based on the areas of application, the popular MCDM methods that are applicable to the particular product development stage are summarized in Table 46.2.

46.4 Discussion

An explicit decision-making process is one crucial aspects of efficient project execution in PDD. As shown in Sect. 46.3, WSM, MAUT, AHP, ANP, TOPSIS, ELECTRE, GP, and PROMETHE methods are well known, often cited, and commonly used MCDM methods for the decision making in PDD.

The first stage of the design process is the planning. This step undertakes to need assessment and viability of the project idea along with customer requirements decision. Further, designing a new product requires extensive planning. The early works had indicated that ELECTRE [31, 39] and GP [32, 33] are useful tools for handling decision making in the planning stage. The subsequent step of the design process begins with concept generation and concept selections. This stage is associated with knowledge and information processing to generate concepts, and thereafter, technical and economic feasibility investigation to select the most optimal concept. The excellence of the decision in this stage greatly impacts on the quality, cost, and desirability of the final product. In this stage, if the wrong concept is chosen, the design may be said to suffer from conceptual weakness. The available literature indicated that AHP [22, 27], ANP [18, 29], TOPSIS [1, 19, 28], and PROMETHE [10, 34] are some of the popular decision-making tools that can be employed in determining the most appropriate design concept.

Embodiment design and detail design are the subsequent step in the design process. Decision-making activity involves in this stage are: arrangement of the physical functions, selection of materials and size of parts, selection of final dimensions/parameters and tolerances, etc. As per the literature, methods that are mostly used for decision making during the embodiment design and detail design stage are TOPSIS [1, 28], WSM [13, 25], and AHP [22, 27]. During the testing and refinement stage, a number of prototypes are built and tested to examine the desired functionality. This stage involves compatibility with mating components, product performance, reliability, durability, etc. Literature proposes that PROMETHE [11, 34], WSM [13, 25], and MAUT [10, 26] are the most suitable MCDM methods to improve decision making during the testing and refinement stage. In the consequent stage, production ramp-up is the beginning of commercial production. The decision-making activity involves in the production ramp-up stage are: quality, equipment, technique, process, personnel, training, materials supply, volume, yield, cost, etc. As per the available literature, the techniques AHP [22, 27], TOPSIS [1, 28, 29], and GP [32, 33] offer enough scope for the decision making in the production ramp-up design stage. The next stage in the design process is the approval and launching. Decisions regarding, launch date, pricing and packaging, positioning, communications plan, marketing content, media relations, infrastructure, sales strategy, and training, are some of the crucial factors in this stage. Literature proposes AHP [23, 28], TOPSIS [1, 20, 29], and PROMETHE [11, 34] have been successfully applied for decisions related to product approval and launching. The last step in the design process is the product retirement. This stage is the end of the product life cycle. The decision for product retirement mostly depends on the product performance, technology progress, sales growth, market, competition, etc. Literature highlights that ELECTRE [31, 39] and GP [32, 33] techniques have been successfully applied for the decision making in the product retirement stage.

46.5 Conclusion

This paper carried out a unique literature review to identify the decision-making problem that may creep during different design stages and the MCDM methods that might be used to solve the decision problem. This work found that MCDM methods have rapidly developed and have been applied to support strategic decisions in various stages of PDD. It is also highlighted that each MCDM methods have their own strength and weakness as well as application fields. Accordingly, this work categorizes the associated MCDM techniques for optimum decision making in different product design stages. It is felt that the design practitioners may benefit from this work as they can identify what type of decision-making activity involves during a particular design stage, along with the MCDM methods that can be used for optimum decision making. In the literature review, it is observed that AHP was the most regular technique followed by TOPSIS and then PROMETHEE in PDD.

The study is limited to find most individual MCDM methods and their applications as per the design stages. However, there is a lack of empirical evaluations of the different MCDM methods. Nowadays, the development of hybrid and modular methods is becoming significant increasingly. Therefore, in order to help decision makers for more suitable decision making, it is necessary to publish reviews on hybrid MCDM methods and approaches in future. Further, the paper identifies a need for further research into decision-making methods in global product development, in order to develop effective decision support tools for manufacturing companies involved in global product development.

References

1. Kumar, P., Tandon, P.: A paradigm for customer-driven product design approach using extended axiomatic design. J. Intell. Manuf. **30**(2), 589–603 (2019)
2. Kumar, P., Tandon, P.: Bionic knowledge and information reuse methodology for uncertainty minimization in product design. Knowl. Inf. Syst. **57**(2), 287–309 (2018)

3. Kumar, P., Tandon, P.: Classification and mitigation of uncertainty as per the product design stages: framework and case study. J. Brazilian Soc. Mechan. Sci. Eng. **39**(11), 4785–4806 (2017). https://doi.org/10.1007/s40430-017-0822-9
4. Mardani, A., Jusoh, A., Nor, K., Khalifah, Z., Zakwan, N., Valipour, A.: Multiple criteria decision-making techniques and their applications–a review of the literature from 2000 to 2014. Econ. Res. Ekonomska Istraživanja **28**(1), 516–571 (2015)
5. Wang, C.H.: Incorporating customer satisfaction into the decision-making process of product configuration: a fuzzy Kano perspective. Int. J. Prod. Res. **51**(22), 6651–6662 (2013)
6. Saaty, T.L.: A scaling method for priorities in hierarchical structures. J. Math. Psychol. **15**(3), 234–281 (1977)
7. Saaty, T. L., Decision Making with Dependence and Feedback: The Analytic Network Process. RWS Publications Pittsburgh (1996)
8. Roy, B.: Classement etchoixenprésence de points de vue multiples (la méthode ELECTRE). La Revue d'Informatiqueet de RechercheOpérationelle (RIRO) **8**, 57–75 (1968)
9. Charnes, A., Cooper, W.W., Ferguson, R.: Optimal estimation of executive compensation by linear programming. Manage. Sci. **1**, 138–151 (1955)
10. Churchman, C.W., Ackoff, R.L., Arnoff, E.L.: Introduction to Operations Research. Wiley, New York (1957)
11. Brans, J.P., Vincke, P.: A preference ranking organization method: the PROMETHEE method for MCDM. Manage. Sci. **31**(6), 647–656 (1985)
12. Hwang, C.-L., Lai, Y.-J., Liu, T.-Y.: A new approach for multiple objective decision making. Comput. Oper. Res. **20**(8), 889–899 (1993)
13. Afshari, A., Mojahed, M., Yusuff, R.M.: Simple additive weighting approach to personnel selection problem. Int. J. Innov. Manage. Technol. **1**(5), 511–515 (December 2010)
14. Siskos, Y., Spyridakos, A.: Intelligent multicriteria decision support: overview and perspectives. Eur. J. Oper. Res. **113**(2), 236–246 (1999). https://doi.org/10.1016/S0377-2217(98)00213-6
15. Zavadskas, E.K., Turskis, Z.: Multiple criteria decision making (MCDM) methods in economics: an overview. Technol. Econ. Dev. Econ. **17**(2), 397–427 (2011)
16. Renzi, C., Leali, F., Di Angelo, L.: A review on decision-making methods in engineering design for the automotive industry. J. Eng. Des. **28**(2), 118–143 (2017)
17. Mayyas, A., Shen, Q., Mayyas, A., Abdelhamid, M., Shan, D., Qattawi, A., Omar, M.: Using quality function deployment and analytical hierarchy process for material selection of body-in-white. Mater. Des. **32**(5), 2771–2782 (2011)
18. Peng, A.-H., Xiao, X.-M.: Material selection using PROMETHEE combined with analytic network process under hybrid environment. Mater. Des. **47**, 643–652 (2013)
19. Zhou, L., Zheng, P., Xu, X., Zhou, W.: A novel AHP-TOPSIS integrated method for case-based retrieval in mechanical product design. Int. J. Prod. Dev. **22**(3), 212–229 (2017)
20. Kumar, P., Tandon, P.: Uncertainty and decision making in product design: a fuzzy approach. In DS79: Proceedings of The Third International Conference on Design Creativity, Indian Institute of Science, Bangalore (2015)
21. Darko, A., Chan, A.P.C., Ameyaw, E.E., Owusu, E.K., Pärn, E., Edwards, D.J.: Review of application of analytic hierarchy process (AHP) in construction. Int. J. Constr. Manage. **19**(5), 436–452 (2019)
22. Kumar, P., Tandon, P.: Improvised concept development process in design through product ingredients. In International Conference on Research into Design, pp. 453–463. Springer, Singapore (2017, January)
23. Guitouni, A., Martel, J.M.: Tentative guidelines to help choosing an appropriate MCDA method. Eur. J. Oper. Res. **109**, 501–521 (1998)
24. Roy, B., Słowinski, R.: Questions guiding the choice of a multicriteria decision aiding method. EURO J. Dec. Proces. **1**, 69–97 (2013)
25. Rehman, S., Khan, S.A.: Multi-criteria wind turbine selection using weighted sum approach. Int. J. Adv. Comput. Sci. Appl. **8**(6), 128–132 (2017)
26. Xiao, Z., Jianhua, Y., Xin, S.: Multi-criteria decision model for imperfect maintenance using multi-attribute utility theory. Int. J. Perform. Eng. **14**(12), 3014–3024 (2019)

27. Razi, P.Z., Ramli, N.I., Ali, M.I., Ramadhansyah, P.J.: Selection of method in construction industry by using analytical hierarchy process (AHP). In IOP Conference Series: Materials Science and Engineering, Vol. 712, No. 1, p. 012015. IOP Publishing (2020, January)
28. Kumar, R., Singh, K., Jain, S.K.: A combined AHP and TOPSIS approach for prioritizing the attributes for successful implementation of agile manufacturing. Int. J. Product. Perfor. Manage. (2020)
29. Kheybari, S., Rezaie, F.M., Farazmand, H.: Analytic network process: an overview of applications. Appl. Math. Comput. **367**, 124780 (2020)
30. Rogers, M.: Using electre III to aid the choice of housing construction process within structural engineering. Constr. Manage. Econ. **18**(3), 333–342 (2000)
31. Govindan, K., Jepsen, M.B.: ELECTRE: a comprehensive literature review on methodologies and applications. Eur. J. Oper. Res. **250**(1), 1–29 (2016)
32. dos Santos Rubem, A.P., de Mello, J.C.C.S., Meza, L.A.: A goal programming approach to solve the multiple criteria DEA model. Eur. J. Oper. Res. **260**(1), 134–139 (2017)
33. Munoz, D.A., Nembhard, H.B., Camargo, K.: A goal programming approach to address the proposal selection problem: a case study of a clinical and translational science institute. Int. Trans. Oper. Res. **25**(1), 405–423 (2018)
34. Vinodh, S., Girubha, R.J.: PROMETHEE based sustainable concept selection. Appl. Math. Model. **36**(11), 5301–5308 (2012)
35. Ulrich, K.T., Eppinger, S.D.: PDD. McCraw-Hill. Inc., New York (1995)
36. Pahl, G., Beitz, W.: Engineering Design: A Systematic Approach, 2nd, rev Springer, New York (1996)
37. Anderson, D.M., Pine, J.: Agile Product Development for Mass Customization: How to Develop and Deliver Products for Mass Customization. Irwin Professional Publishing, Chicago, USA (1996)
38. Montagna, F.: Decision-aiding tools in innovative product development contexts. Res. Eng. Design **22**(2), 63–86 (2011)
39. Certa, A., et al.: ELECTRE TRI-based approach to the failure modes classification on the basis of risk parameters: an alternative to the risk priority number. Comput. Industr. Eng. **108**, 100–110 (2017)

Part VI
Lexicon, Taxonomy, Ontology, Machine Learning and Data-Driven Design

Chapter 47
Mindmaps-Based Tool for Planning and Structuring Design Research Papers

Mashahib Nawaz Hassan and Sharmistha Banerjee

Abstract Research papers are a tremendous source of cutting edge knowledge. Writing a good research paper requires proper planning and structuring of paper, which often comes from experience. Planning and structuring the research paper is tedious for novice researchers due to the lack of a process that could assist novice researchers in planning and structuring their paper. We conducted workshops on writing research papers based on a generic framework to understand the mindset and problems faced by novice design researchers. Workshops provided us with helpful insights about how and what novice researchers think while writing research papers. We came up with a concept of using mindmaps that allows novice researchers to plan and structure the content of the research paper before writing the paper. It was observed that it reduced stress and anxiety around organizing the contents of a paper while writing it and researchers could focus more on the language of the paper while writing it. This directly increases the quality of a paper in terms of its language and narration.

47.1 Introduction

According to L. Bruce Archer: "Design research is systematic inquiry whose goal is knowledge of, or in, the embodiment of configuration, composition, structure, purpose, value, and meaning in man-made things and systems." [1, 2]. According to Cross, the purpose of performing design research is to "extract" and "construct" knowledge from the natural or the artificial world and, after that, to make it available to the community at large. This knowledge should be reproducible and reusable [2]. The research should be purposeful, inquisitive, methodical, and communicable.

Research papers are the medium through which new knowledge is disseminated amongst peers and the masses in general. A general guideline for writing a research

M. N. Hassan (✉) · S. Banerjee
Indian Institute of Technology Guwahati, Guwahati 781039, Assam, India

S. Banerjee
e-mail: sharmistha@iitg.ac.in

A. Chakrabarti et al. (eds.), *Design for Tomorrow–Volume 2*, Smart Innovation, Systems and Technologies 222, https://doi.org/10.1007/978-981-16-0119-4_47

paper suggests sequencing the paper into the following sections: Title, Abstract, Introduction, Methodology, Observations, Results, Conclusions, Discussions, Future Work, and References [4]. Structuring the paper appropriately helps in conveying the knowledge effectively as well as in a reproducible manner. A survey among design undergraduate, post-graduate, and Ph.D. students at the Department of Design, IIT Guwahati, revealed that novice researchers, irrespective of their level of study, usually find it difficult to chalk out an effective structure while writing a paper. Cross also mentions in his editorial, “A frequent reason for rejecting a paper is that the research aims and objectives, and the research methods and techniques are unclear and reported with insufficient clarity and detail.” [3].

Also, multiple research papers can be chalked out of a design research project. However, we observed that novice design researchers lack an understanding of this art of research reporting. In this paper, we present our experiments using a mindmap-based design tool for novice researchers to assist them in planning and structuring the research paper. We developed and tested the tool with design undergraduate, post-graduate, and Ph.D. students at the Department of Design, IIT Guwahati.

47.2 Methodology

We interviewed 30 design undergraduate (third and fourth year), post-graduate, and Ph.D. students at the Department of Design, IIT Guwahati. The interviews were unstructured and helped us to understand the problems that the participants face while writing a research paper. Next, we conducted workshops for these researchers in writing a research paper. Each workshop had four to six participants. We followed the general sequencing guideline (Title, Abstract, Introduction, Methodology, Observations, Results, Conclusions, Discussions, Future Work, and References) for writing the research paper. This workshop aimed to observe the participants while they wrote the different sections of their research paper and identify their pain points in structuring. We followed up the observation with a one-to-one discussion and further defined the difficulties they faced. Next, we designed a mindmap-based tool to help the participants in structuring their paper. We then conducted one-to-one sessions with the participants to test the efficacy of the mindmap-based tool.

47.3 Observations—Unstructured Interview and Paper Writing Workshop

We identified several problems that novice researchers face. Some of the significant issues they face while writing the title are:

- Trouble articulating their research into an appropriate title.
- Using unnecessary words in the title or making it very concise.

- Unclear if the title conveys the message of the paper.
- Miscommunicate the scope of the paper while framing the title.

Some of the major problems they face while writing the abstract are:

- Over or under explanation of the paper.
- Miscommunicate or miss out on explaining the scope, objectives, research problem, methodologies, and the results of the research.
- Less thought to the suggested word limit.

Some of the significant problems they face while writing the body of the paper are:

- Missing out on mentioning the key concepts and research work used in the presented research work.
- Missing out on establishing the context of the research problem, the research question(s), and the hypothesis.
- Writing the right content in the wrong section. For example, some of the participants wrote the contents of methodology in the introduction.
- Inadequately defined methodology.
- Unclear on the differences between observations, results, conclusions, inferences, and discussions. As a result, unable to establish well-argued connections between them.

We, thus, observed that the novice researchers faced confusion in organizing (content as well as order) their paper sections. They also faced challenges in deleting unnecessary information. As a result, they stressed out with the idea of writing their paper leading to delays and procrastination. Also, the final paper churned out to be less effective in communicating the research to the world.

47.4 Mindmaps-Based Tool

Like a building needs to be planned before beginning construction, a research paper needs planning too. To prepare the architecture of a research paper, we used inspiration from mindmaps and developed a tool. The aim of the mindmap-based research paper planning tool is to help a researcher in planning and to organize a research paper. We assume that the tool will help in improving the quality of research reporting and also reduce the stress and anxiety of novice researchers in research reporting.

Mindmap helps in connecting information and planning. All the necessary information can be laid out in one place, and the essential connections between them can be made, allowing the researchers to put their thoughts together regarding the content of the paper. We present the tool below using screenshots.

As shown in Fig. 47.1 on the next page, the tool consists of two major screens: mindmap and roadmap. The mindmap screen consists of generic cards, namely introduction, methodology, results/observation, and conclusion/discussion. An author can

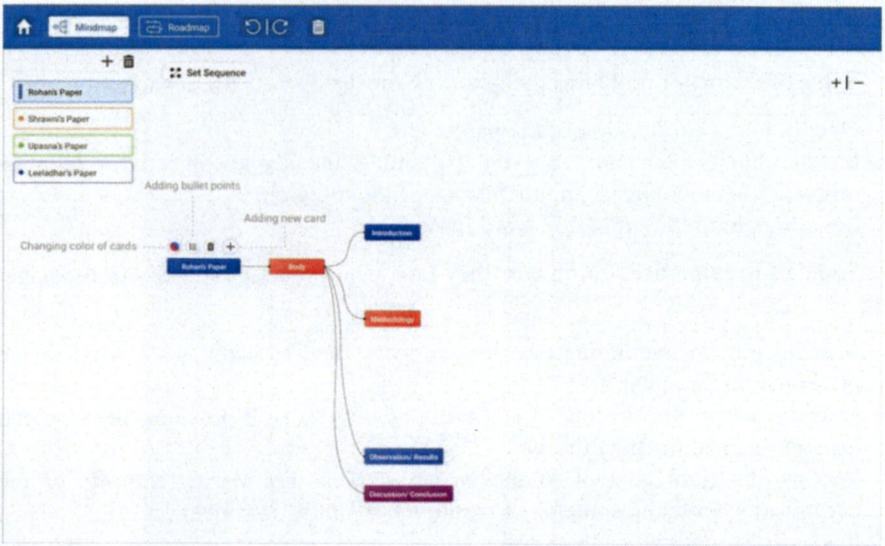

Fig. 47.1 Mindmap default UI

add or remove cards based on her requirements. Next (Fig. 47.2), the author proceeds to create sub-branches. This helps her to organize her thoughts and work. Here, she can add bulleted contents to each sub-branch. She can also move content from one branch to another. Completing the mindmap will give the author an outline of the

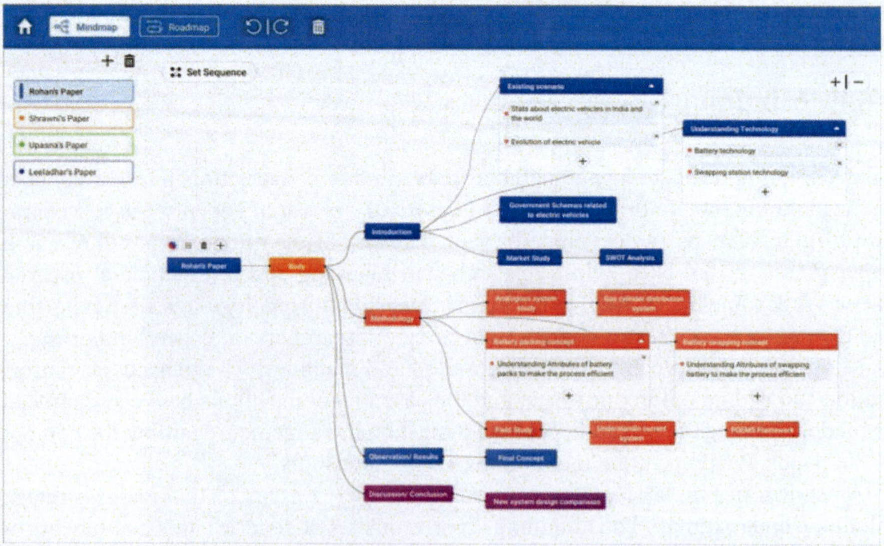

Fig. 47.2 Mindmap UI after completing a mindmap

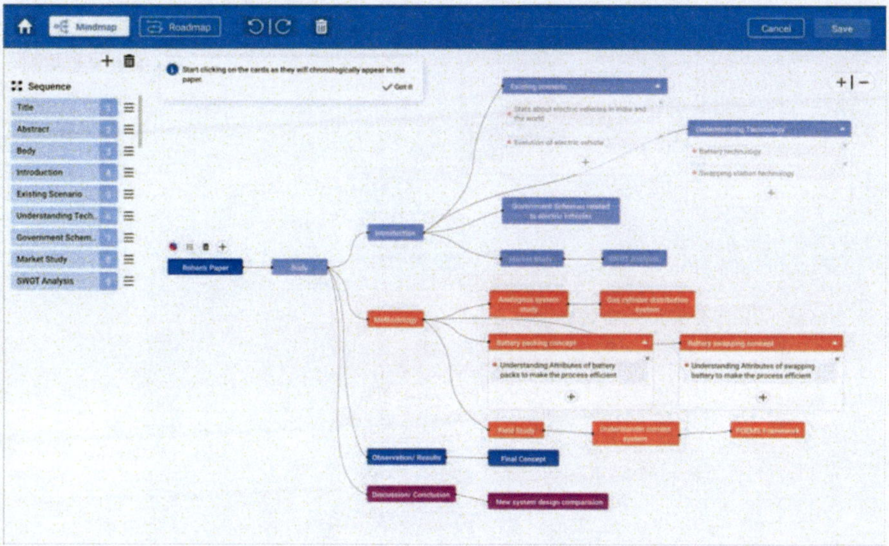

Fig. 47.3 UI for sequencing the content of the paper

paper and its content. The author can now reflect upon this map and see that there is no overlapping content, and correct content has been put in the right sections of the paper.

The next step is to sequence the content as they will appear in the paper. It is, thus, not necessary to add contents in the mindmap in the order as they will appear in the paper. After completing the mindmap, the author has to click on the "set sequence" button (Fig. 47.3). The left panel of the UI will show the current sequence as per the selection. The sequence can be changed from this panel. The author will save the sequence after setting it in the correct order.

Mindmaps will help in selecting and planning the content of the paper. However, that itself will not be enough. Contents of the paper appear in a linear fashion whereas mindmap is not linear. This might lead to disconnects between the paper and the outline provided by the mindmap. So the point of creating a chronological sequence (see Fig. 47.3) is to convert mindmap into a linear roadmap (see Fig. 47.4). Thus, roadmap will provide a linear outline of the paper. The writer can switch between mindmap and roadmap from the top panel.

47.5 Testing the Mindmap-Based Tool with Novice Researchers

The mindmap and roadmap were tested with four novice researchers. The tool helped the researchers in better organizing the content of their paper. Figures 47.1, 47.2, 47.3,

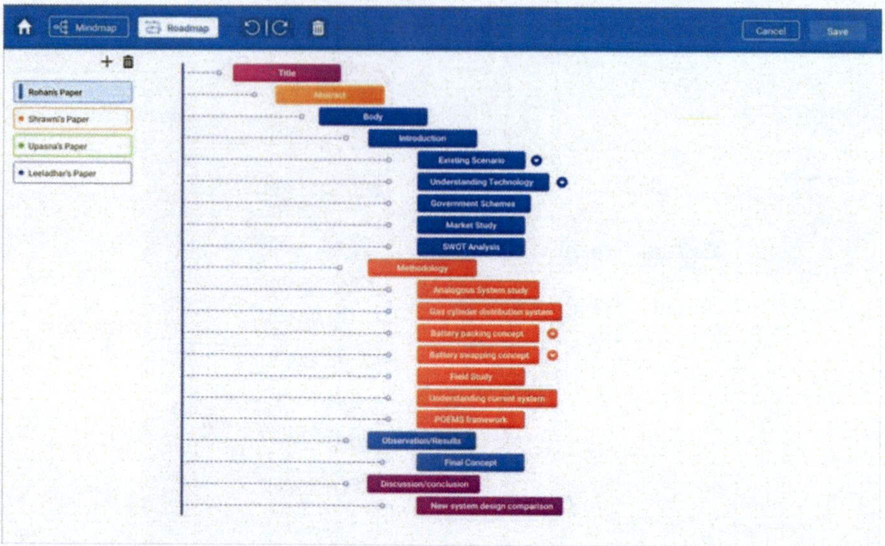

Fig. 47.4 Roadmap UI

and 47.4 shows one of the sample researchers' work using the tool. This paper has been written using the tool too. Some of the papers that are completed using this tool are cited below:

- Munakala, S.M., Manda, C., Banerjee, S. The Infinity Process: A Design Framework for Interdisciplinary Problem-Solving (Submitted to ICoRD'21).
- Sehji, U. & Banerjee, S. Campaign design to nudge men in order to reduce the crime rate against women (Submitted to ICoRD'21).
- Hassan, M.N. & Banerjee, S. Mind Maps-based tool for planning and structuring design research papers (Submitted to ICoRD'21).

47.6 Future Scope

A complete HCI-based tool for novice design researchers to assist them while writing research papers is under progress. The tool consists of the mindmap and the roadmap followed with a step-by-step guidance to a novice researchers in framing the title, writing the abstract and the body of the paper. The tool is a text editor and helps in better structuring. Once the basic structuring is complete, the author can use any of the existing grammar, plagiarism, and formatting applications to complete the authoring process.

References

1. Archer, B.: A view of the nature of design research. De. Sci. Method **1**, 30–47 (1981)
2. Cross, N.: Design research: A disciplined conversation. Des. Iss. **15**(2), 5–10 (1999)
3. Cross, N.: Editing design studies—and how to improve the likelihood of your paper being published. Des. Stud. **63**, A1–A9 (2019). https://doi.org/10.1016/j.destud.2019.06.001
4. Lester, J.D., Lester, J.D.J.: Writing Research Papers—A Complete Guide. Pearson (2015)

Chapter 48
Inference Variable—A Human Cognition and Design Perspective on Knowledge Graph Visualization

Rani Joseph, Abhilasha, Prithvi Raj Ramakrishnaraja, Ramesh Manickam, K. Vineetha, and Srinjoy Ghosh

Abstract In this paper, we expound our theoretical hypothesis covering effective visualization factors in the area of visual analytics for a knowledge graph (KG) considering human cognition aspects of inferencing. The paper reflects on the first and foremost challenge with visualizations—which is effectively harnessing the potential of a knowledge graph visually. While there is a progress on the technical front in allowing an inference to be queried, there is limited scope to do the same visually. The second challenge identified is when the user has to use his mental space and cognitive reserve to conflate and schematize knowledge to infer. There are certain constraints with this as to how the human attention is efficiently oriented toward the relevant information and the visualized information is not cognitively loaded for the user to infer from it. As far as the visualization aspect of the knowledge graph is concerned, the aim is to reduce the cognitive resources used for building a new schema of the task inside long-term memory. Hence, the proposed inferencing variable in the paper enables the release of space and conserves the cognitive reserve for further higher cognitive process of decision making. It also includes the elements of motivation for the user to visualize knowledge in knowledge graphs and an overview of the existing visualizations of KG.

R. Joseph (✉) · Abhilasha · P. R. Ramakrishnaraja · R. Manickam · K. Vineetha · S. Ghosh
Siemens Technology and Services Private Limited, Bangalore, India
e-mail: rani.joseph@siemens.com

Abhilasha
e-mail: abhilasha@siemens.com

P. R. Ramakrishnaraja
e-mail: prithvi.ramakrishnaraja@siemens.com

R. Manickam
e-mail: ramesh.manickam@siemens.com

K. Vineetha
e-mail: vineetha.k@siemens.com

S. Ghosh
e-mail: ghosh.srinjoy@siemens.com

A. Chakrabarti et al. (eds.), *Design for Tomorrow–Volume 2*, Smart Innovation, Systems and Technologies 222, https://doi.org/10.1007/978-981-16-0119-4_48

48.1 Introduction

KG represents a collection of interlinked descriptions of entities—real-world objects, events, situations, or abstract concepts achieved by interconnecting large network of data. It is perhaps important to suggest that such generic structuration of general world knowledge lays the cornerstones for both domain-independent and domain-specific information systems, especially in this digital age. In words of Paulheim [1], a knowledge graph mainly describes real-world entities and their interrelations, organized in a graph and it defines possible classes and relations of entities in a schema. In the journey of data turning into knowledge, a process of rationalization takes place where—(a) data is refined into information, (b) schematized into correlated information, and (c) transformed into high-quality, conflated, and schematized knowledge ready for inferencing. Figure 48.1a, b showcases the latter two points. In this entire process, there are two integral aspects that this paper will focus and expand on: visualization of the schematized knowledge and the requirement of an '*Inference Variable*' in the KG to aid the cognitive processing of the user.

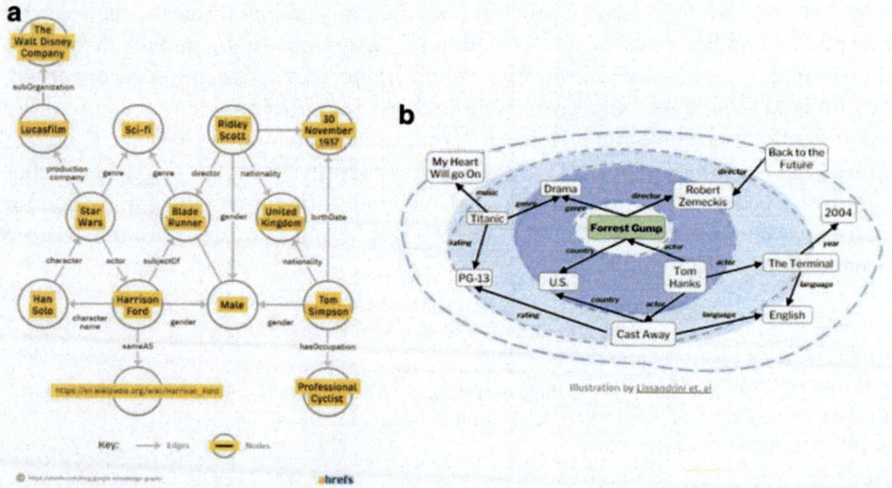

Fig. 48.1 **a** Sample Google Search KG (*Source* https://ahrefs.com/blog/google-knowledge-graph/) **b** Sample Netflix Movie Recommender KG (*Source* https://towardsdatascience.com/movie-recommendations-powered-by-knowledge-graphs-and-neo4j-33603a212ad0)

48.1.1 *Why is Sensible Visualization Important in a KG Ecosystem?*

"*A good sketch is better than a long speech...*"—a quote often attributed to Napoleon Bonaparte. We live in the information age where every second large sets of data get generated from which insights are drawn constantly either with an automated data pipeline or by data scientists and data analysts trying to make sense with that data. In a more visual analysis-based approach where you want to discover the insights with a visual representation of data, visual data exploration and thus visual analytics is particularly useful when little is known about the source data and the analysis goals are vague. In this context, visual data exploration can be viewed as an evolving hypothesis-generation process [2], during which hypotheses can be validated or rejected on a visual basis, and new ones can be introduced. According to Schneiderman [3], visual data exploration is a three-step procedure encompassing data overview, facet zoom and filter, and on-demand detail request. To give dimension to the size of the challenge associated with visual data exploration, a new forecast from International Data Corporation (IDC) estimates that there will be 41.6 billion connected IoT devices, or 'things,' generating 79.4 zettabytes (ZB) of data in 2025. So, it is natural to migrate toward superior visualizations to cope with this information ocean.

48.1.2 *Factors of Visualization*

Users need information to make effective decisions. In the era of humongous information, the more the data, the more challenging it is to connect the dots and come to logical conclusions. Considering the aspects of human comprehension, the below section outlines a few key visual elements from a data visualization standpoint which can be extrapolated to knowledge graph visualization as well.

Ben Schneiderman's [3] paper offers a task by data-type taxonomy with seven data types (one-, two-, three-dimensional data, temporal and multi-dimensional data, and tree and network data) and seven tasks (overview, zoom, filter, details-on-demand, relate, history and extracts). The paper analyses the necessary factors to be addressed for a data visualization for a better comprehension and decision making. The elements put together help in forming an effective understanding of the visualization.

Fayyad et al. [4] provide an overview of importance of visual perceptual elements in design. These elements (listed below) would not only just apply to information visualization but also to visualize knowledge graph.

1. Point with position, color, and size
2. Line with location, length, width, color, and style
3. Polygon with location, shape, size, orientation, color, texture, and translucency
4. Image with pattern, range of values, colormap, size, and scale

The above-mentioned first four are the basic building blocks of visualization, while the following elements build over these to add different dimensions of comprehension to them

5. Point plots with position coordinates
6. Wire frames with connection components (edges)
7. Icons and glyphs, such as arrows, spheres, boxes, stars, stick figures, and faces
8. Parallel coordinates taking a multivariate point and mapping it to a polyline across N vertical axes
9. Multiple simultaneous views.

Considering the above factors of visualization, graph-based visualizations benefit by less time spent to assimilate information, higher chance to discover hidden insights, better understanding of a problem by visualizing patterns and context, effective communication, and applicable for technical and non-technical users.

48.1.3 Related Work

There are several knowledge graph visualization tools in the market that brings out information and hidden insights and helps one to understand the problem better to solve it. A few focus on exploration and analysis of linked data and emphasize on browsing data rather than showing the semantics of it. In this section, we provide an overview of five known knowledge graph visualization tools (Table 48.1).

Table 48.1 Overview of knowledge graph visualization tools

Tool	Visualization characteristics	Example visualization
PoolParty [5]	Variety of KG visualization features to visualize nodes and relationships in data to support visual analytics. Can be used by technical and non-technical users with advanced natural language-based queries and search/filtering	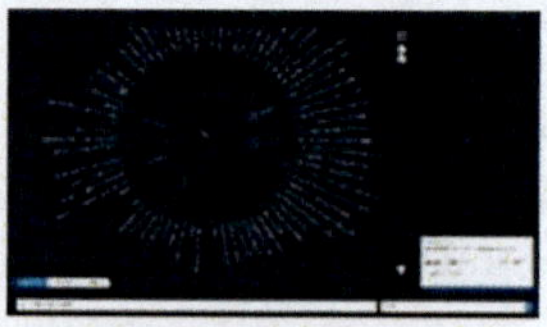
Protégé [6]	Open-source tool used for ontology and knowledge graph creation having four major visualization possibilities based on nodes and relationships. (a) indented list, (b) node-link and tree, (c) zoomable and d. focus + context. Visual analytics capability is not very advanced, however, provides search and filtering options	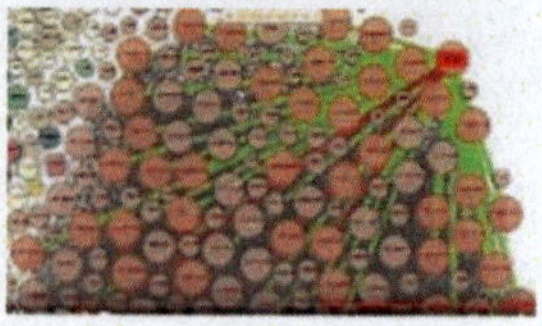

(continued)

Table 48.1 (continued)

Tool	Visualization characteristics	Example visualization
Neo4j [7]	Open source and paid tool that supports different visualizations like timeline, dataset, graph2d, graph3d, and network and is mostly based on network visualizations on nodes and relationships. It supports visual analytics with natural language queries and advanced search and filtering	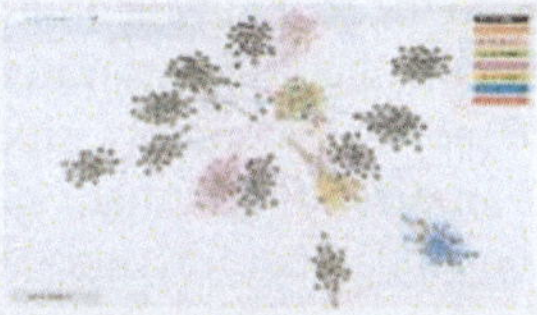
GraphXR [8]	A Web-based KG visualization tool that supports different interactive visualization formats in 2D, 3D, geospatial, timeseries, scatter plots, network. It enables visual analytics with advanced nodes and relationship visualizations and is apt for technical and non-technical users alike with natural language queries and search and filtering options	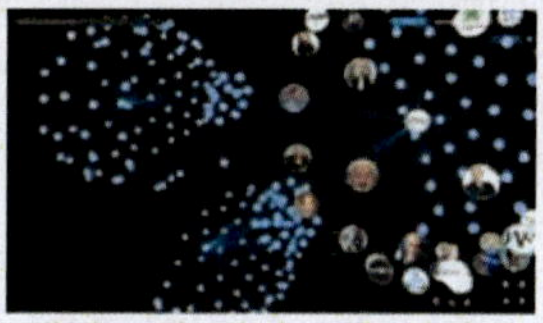
Gephi [9]	Open-source tool that can visualize KG nodes and relationships in the forms of networks. It aids in visual analytics by search, and filtering and visual cues of colors, etc.	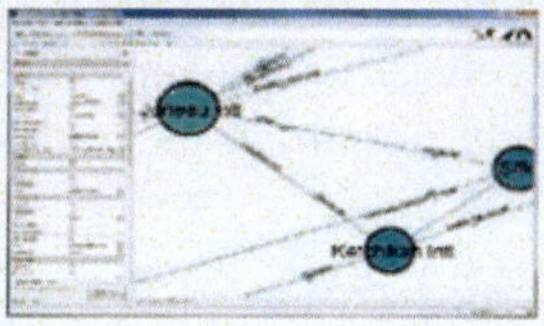

48.1.4 *Human Cognitive Aspects of Inferencing—The Need for an* Inference Variable

With large amount of data that is visualized in a knowledge graph, the challenge is with respect to an efficient usage of user's cognitive reserve to conflate and schematize knowledge to be inferred. Two types of mental processing occur when a person decides—Type 1 (outside of user's conscious awareness) and Type 2 (thoughtful decision while also dependent on some rules of the thumb). Type 1 processing may utilize just small amount of working memory to function. However, Type 2 process relies heavily on a deliberate and conscious application of working memory to drive the process. Although both these processes are not mutually exclusive and rather interdependent. In the context of knowledge graph visualization, we are referring to a subset of visuospatial reasoning, i.e., visualization cognition. It refers to the cognitive process of deriving meaning from external representations of visual information that maintain consistent spatial relations [10]. This happens broadly in two distinct ways either by perceptually focused frameworks or by considering the influence of prior knowledge.

Pinker's model [11] details an integrative structure with a distinction between top-down and bottom-up encoding mechanisms in understanding such graphs. It suggests that a visual description is constructed using bottom-up mechanisms. After the mental encoding of the visual stimulus, a long-term memory search for relevant knowledge is initiated and represented in form of a graph schema. The graph schema is then matched, and the most similar visual array is retrieved. Once, the match is found the schema becomes instantiated and the visualization conventions associated with that helps the user in interpreting the respective visualization. And this denotes the message assembly process.

Studies suggest that a mismatch between the visualization and a decision-making component requires more working memory resources in order to do corrective mental transformations as compared to scenarios where the visualization is aligned with the mental schema and task [12]. If the fit is poor, larger working memory resources would be required to cover corrective mental transformations. Visualizations that do not match the mental schema require cognitive transformations to make the visualization and mental representation aligned.

Considering the above visual and human cognitive challenges and its interaction between the inferencing of a KG and decision making, the research explored the approaches that can be used to aid visual inferencing so that one can effectively harness the potential of knowledge graph through visual analytics. It is further hypothesized that introducing a new '*Inference Variable*' attribute in knowledge graph visualizations would eventually enhance the visual inferencing of a knowledge graph.

Inference variable (IV) is a cognition led design exploration toward unearthing intelligent way of acquiring knowledge from raw data in a structured manner. Given the massiveness of a knowledge graphs and the user's limited information needs, inference variable distills the available information and presents in a concise and intuitive manner to the users for visual inferencing. It is dynamic in nature and can change according to the context and specific user needs.

48.2 Research Perspective

In a knowledge graph, the basic elements are represented by Subject, Object, and Predicate (Fig. 48.2a) where, the Subject (ABC) is represented in a circle, Object (1A) is represented in a circle, and the Predicate (Works with) is the connecting element shown with an arrow.

Based on the outcome from our research, we introduce '*Inference Variable*' which can enable the visualization of a fourth element in the visual analysis of a knowledge

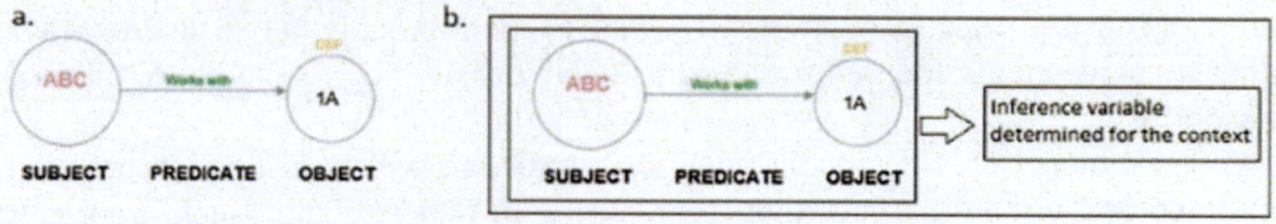

Fig. 48.2 **a** Basic elements of a knowledge graph 2. **b** Inference variable as per the visual analysis task

graph. An inference variable is the goal of your visual analysis task while exploring the data in the knowledge graph. Example:

a. If one wants to know all the employees who have joined and left an organization in a particular time frame, in this case, the goal of inferencing is based on time, hence time can be considered as the inference variable.
b. If one wants to know the movement of a package within a factory location, in this case, the goal of inferencing is based on movement, hence location can be considered as the inference variable.

This translates to a new element in knowledge graph visualizations (Fig. 48.2b), where the Subject (ABC) is represented in a circle, the Object (1A) is represented in a circle, and the Predicate (Works with) is the connecting element shown with an arrow and the inference variable as per the visual analysis task within the context of Subject, Object, and Predicate.

48.2.1 Case Study

48.2.1.1 Inferencing from a Typical Node-Relationship Knowledge Graph Visualization

Let us take an example of a query performed by visual analysis: If someone wants to know how many people work in FIN had joined as Intern after 2015 from college 'IDC' and are still working within the organization.

Figure 48.3 represents a typical graph visualization from a snippet of data of an organization FIN which employs full-time employees, interns, and contractors. In the visualization, E1, E2, E3, E4, and E5 represent full-time employees, I1, I2, I3, I4, and I5 represent interns, and C1 and C2 represent contractors.

One of the approaches to visually analyze the data to arrive at the result is to:

1. Filter and visualize all interns in the organization after 2015.
2. Narrow down the resultant data using a sub-filter to extract all interns who are still with the organization.
3. Further filter this narrowed list based on the relationship to find the data with a link to college 'IDC'
4. Finally, you arrive at the count of interns for the query as '3 interns'.

The maturity of the visualization tool defines the number of steps required to arrive at the result for visual inferencing.

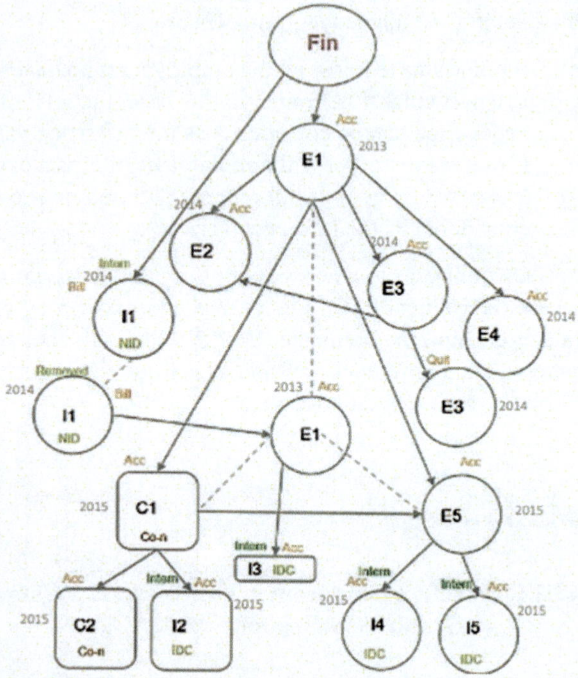

Fig. 48.3 Traditional knowledge graph visualization

48.2.1.2 IV Based Knowledge Graph Visualization: Time as the *Inference Variable*

Figure 48.4 illustrates the core differentiation in the visualization proposed which lies in how a node and its relationship with other nodes are represented.

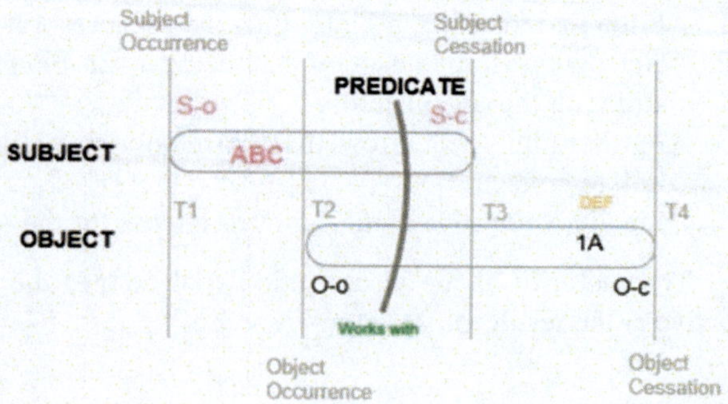

Fig. 48.4 Basis of IV visualization

Node: A node in this context is visualized as an elongated capsule that spans across the inference variable (in this context the inference variable being 'time').

Relationship: The relationships between respective nodes are represented by the overarching 'arc' which indicate the connecting relationships between these nodes.

The basis of IV visualization consists of:

1. The Subject (ABC) which is visualized as an elongated capsule across the IV instances (T1, T2, and T3) starts at Subject Occurrence (S-o) and ends at Subject Cessation (S-c) in the focus area of visualization.
2. The Object (1A) which is visualized as an elongated capsule across the IV instances (T2, T3, and T4) starts at Object Occurrence (O-o) and ends at Object Cessation (O-c) in the focus area of visualization.
3. The Predicate (Works with) is the relationship between ABC and 1A established in the IV instances between T2 and T3 in the focus area of visualization.
4. The different instances of IV (T1, T2, T3, and T4) are visualized across the corresponding Subject–Object–Predicate (S-O-P) contexts.

A sample visualization of the knowledge graph based on IV for the same visual analysis query is highlighted in Fig. 48.5 consisting of:

1. Full -time employees (E1, 2, 3, 4, and 5) and interns (I1, 2, 3, 4, and 15) are visualized as elongated capsule across the inference variable instances (2013, 2014, 2015, 2016, and 2017) with the nodes in the periphery indicating the attributes like the college or sub-department information with filled colors indicating 'current presence' and unfilled colors indicating 'past presence'
2. Now by glancing through this visualization, you can narrow your visual search to the capsules that extend after the 2015 instance.
3. You can then simply count from left to right all intern capsules with the IDC-filled colored attribute in the periphery.

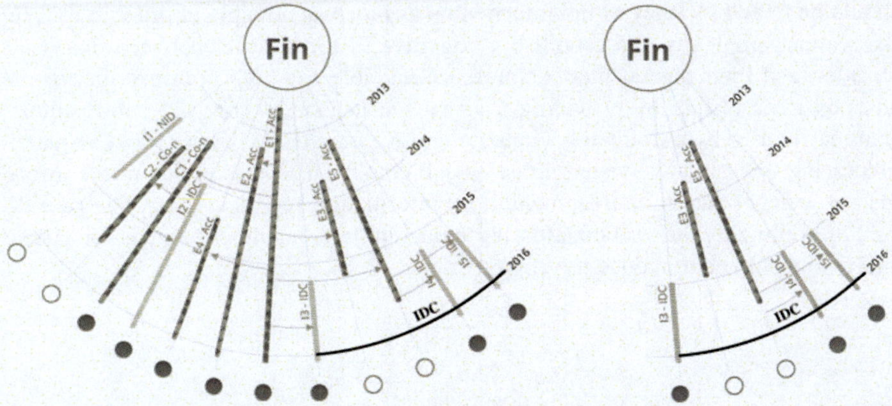

Fig. 48.5 Visualization of the of knowledge graph based on inference variable.

48.2.2 Visual and Cognitive perspective

The traditional approach of showcasing knowledge graphs with nodes and relationship represent the node clusters and their immediate relations. When a new parameter of inference is brought, there is a cognitive requirement from the user to imagine/extrapolate how this inference variable can link in this context with the traditional nodes and relations. The concept of inference variable in this context by using an example visualization showcases how such an inference variable can be introduced enable the user in quickly inferring in the context.

Figure 48.3, traditional way of representing nodes and relationships is comparatively difficult for the user to infer as such it misses the important element of visual affordances. If the user wanted to find out the time-based relationship ('time' being the inference variable here) for how E1-A is related to E3-A, the user does not have any visual affordance to bring out this relationship in Fig. 48.3. It only has nodes and relationships presented visually in the raw linked format.

While in Fig. 48.5, since the nodes are represented as elongated capsules—keeping in context the inference variable—the user can easily infer visually the time-based relationships between E1-A and E3-A, wherever they are in their corresponding arc range. Inference variable tailors the visualization from the raw linked data format to a form that is much organized and matches the human mental models during a decision-making process.

This visual affordance here is enabled by the factors referred to by Fayyad et al. [4], in terms of 'Line with location and length'. The location of the 'elongated capsule' brings in the 'Length' component. Its proximity with other elongated capsules brings in the aspect of 'Location'. With the cognitive alignment in terms of Gestalt principle of proximity, these two parameters can help the user easily enable him/her to correlate the aspects of time with the corresponding nodes. The number of steps in arriving at the result is influenced by the visualization maturity of the tool one is using.

The research analysis has led us to conclude that the aim of this visualization should be to add saliency to the information as much as possible in order to reduce the search complexity and establish a cognitive fit for the user between the visualization and their mental model. Inference variable can help optimize the visual analysis steps further by providing a visual mechanism to compare sub-resultant steps easily thereby introducing a better cognitive fit in terms of user's visuo-spatial processing (pre-attentive visual processes), hierarchical representation of the information, categorization and segmentation, information relevance, etc. The overall goal of optimizing the visualization (by reduction in the number of steps for visual analysis) would be to reduce the cognitive load.

48.3 Discussion and Conclusion

We argue that with respect to human cognitive aspects of inferencing, both the types of information processing (Type 1 and 2) have its own advantage and disadvantage. It is thus important to understand which specific task would call for what type of processing on the user's part to make the visualizations optimal. There are studies that show that capitalizing on Type 1 processing helps the users to make accurate decisions when less is known as it is more intuitive. However, if the visualizations are designed inappropriately and require surplus mental transformation, then the visualizations prove to be detrimental.

Three distinct sources generate the level of cognitive load on part of the user: intrinsic load, extraneous load, and germane load [13, 14]. Intrinsic load is related to the challenge of the task itself and the level of difficulty, while the extraneous load is related to the complexity of the task represented and the germane load is associated to the cognitive resources used for building a new schema of the task inside long-term memory. All these combined together offer an overall experience.

When it comes to the visualization aspect of the knowledge graph, the aim to reduce the Germane load so that the cognitive reserve is relatively free for other higher cognitive activities of decision making. The inferencing variable enables the release of space and conserves the cognitive reserve for further usage. We suggest the graph visualizations begin with basic perception where the raw data can be denoted by visual attributes and hence would easily distinguishable by the user. There can be two levels of information in the form of, a. Nodes, connections, relationships, attributes and b. Timeline and versioning. One can also have zoning and prioritizing which would in turn enhance the saliency of the information by making it visible when needed. This would direct the user's attention by categorizing (edge filters), grouping based on priority or timeline and clustering based on criticality (pattern formed with priority groups).

Here, the designers can play with the bottom up features of the visualization to direct user's attention. It consists of involuntary shifts in attention to salient features of a visualization and does not utilize working memory [2].

For subsequent levels of information inferencing, one could introduce a categorical drill down and delineating relationships. This is to implicate the notion of user's mental model such as (a) Do the mental representations belong or differ from a set of propositions? (b) What rules of inferencing the mental processes embody underlying reasoning? and (c) Whether the representation and the questions depend on a de-compositional dictionary or set of meaning postulates?

The visual representations can thus synthesize the new knowledge and know-how by enabling a connection between the new and existing knowledge base. Reflective cognition would lead to problem solving, planning, reasoning, and decision making where the visualizations allow further data cleansing. This inclusion of Type 2 processing in our proposed perspective suggests that there is a variation in the cognitive fit between the visual description, graph schema, and conceptual model.

It is a knowledge -driven process and it interacts with the user's visual encoding technique.

We put forth the view that in an era of data explosion and its automation through technical advancements, it is crucial to consider the user experience elements of visualizing knowledge graph (human cognitive aspects). This would not only make the knowledge graph visualizations more intuitive but also aid in the advancement of visual analytics. We have added to the state of the art in visual analysis of KG through tailored visualizations that are 'compact summaries' of KGs containing only the facts most relevant to individuals' interests. These are well aligned with the user's mental model of decision making by drawing inferences. In addition to the existing work, we intend to continue future exploration with a more concrete use case to arrive at a different variation of the possible visualizations with an inference variable.

References

1. Paulheim, H., Ristoski, P., MItichkin, E., Bizer, C.: Data mining with background knowledge from the web. In: Proceedings of the 5th Rapid Miner World, Boston, USA, 1–14 (2014)
2. Connor, C.E., Egeth, H.E., Yantis, S.: Visual attention: bottom-up versus top-down. Curr. Biol. **14**(19), R850–R852 (2004)
3. Shneiderman, B.: The eyes have it: a task by data type taxonomy for information visualizations. In Proceedings 1996 IEEE Symposium on Visual Languages (1996)
4. Fayyad, U.M., Wierse, A., Grinstein, G.G.: Techniques for graphical data presentation. In: Fayyad, U.M., Wierse, A., Grinstein, G.G. (eds.) Information Visualization in Data Mining and Knowledge Discovery, Morgan Kauffman Publishers, San Fransisco (2002)
5. PoolParty—Semantic Web Company GmbH. https://www.poolparty.biz/poolparty-graphviews
6. Sivakumar, R., Arivoli, P.V.: Ontology visualization protégé tools—a review. Int. J. Adv. Inform. Technol. (IJAIT) **1**(4) (2011)
7. Neo4j, Inc. https://neo4j.com/developer/graph-visualization/
8. GraphXR—KINEVIZ, INC. https://www.kineviz.com/
9. Mathieu, B., Sebastien, H., Mathieu, J.:Gephi: an open source software for exploring and manipulating networks. In: Proceedings of the Third International ICWSM Conference, AAAI Publications (2009)
10. Tversky, B.: Visuospatial reasoning. In: Holyoak, K., Morrison, R.G. (eds.) The Cambridge Handbook of Thinking and Reasoning, pp. 209–240. Cambridge University Press, Cambridge (2005)
11. Pinker, S.: A theory of graph comprehension. In: Artificial Intelligence and the Future of Testing, pp. 73–126 (1990)
12. Joslyn, S., LeClerc, J.: Decisions with uncertainty: the glass half full. Curr. Direct. Psychol. Sci. **22**(4), 308–315 (2013)
13. Chen, F., Zhou, J., Wang, Y., Yu, K., Arshad, S.Z., Khawaji, A., Conway, D.: Robust multimodal cognitive load measurement. In: Tan, D., Vanderdonckt, J. (eds.) Human-Computer Interaction Series. Springer International Publishing, Switzerland (2016)
14. Cîmpanu, C., Dumitriu, T., Ungureanu, F.: Instructional design "based on the assessment of cognitive load and working memory load. In: The 14th eLearning and Software for Education Conference—ELSE 2018, Bucharest, Romania (2018), April 19–20.
15. Anonymous: Graph Visualization: Why It Matters (2017). https://linkurio.us/blog/why-graph-visualization-matters/

Part VII
Innovation Management and Entrepreneurship

Chapter 49
Store Atmospherics, Shopping Motives, and Buyer Behavior—An Indian Consumer Perspective

Ramchandra Alias Ameet Chate and S. R. Bharamanaikar

Abstract In the modern era of marketing, creating pleasurable experience by enhancing store atmospherics, which we coin as ambience, has emerged as a deciding factor for profitability of firms. Hence, retailers have been investing a lot to create positive experience and identifying the factors that stimulate the conative, affective and cognitive factors. Current increased usage in information and communication technology, free knowledge sources has made the consumer more aware of substitutes and better options at better prices. Hence, studying the impact of store atmospherics, which include lighting, visual merchandising, color that stimulates positive experience has been studied to increase the profitability of retailers and improve customer experience from user's perspective. The study undertaken will help to understand the dimension in designing the store atmospherics, which are most effective in attracting the customers and affecting their conative, cognitive and affective replies. This study is based on the user's experience in buying the product and evaluating the factors that stimulate the buying intention through the usability audit. The findings of this investigation will help the retailers to develop better in-store atmospherics, helpful in increasing the sales and consumer loyalty that contribute to the profitability.

49.1 Introduction

Store atmospherics was first coined by Philp Kotler in the year 1973 [6]. It was defined as "The effort to design buying environments to produce specific emotional effects in the buyer that enhance his purchase probability". However, several studies were conducted to evaluate the impact of atmospheric variables on purchasing behavior considering external variables (exterior signs, exterior display windows, color of

R. A. A. Chate (✉) · S. R. Bharamanaikar
Research Resource Centre, Visvesvaraya Technological University, Belagavi, Karnataka, India
e-mail: ameet.chate@gmail.com

S. R. Bharamanaikar
e-mail: im_srb@yahoo.com

A. Chakrabarti et al. (eds.), *Design for Tomorrow–Volume 2*, Smart Innovation, Systems and Technologies 222, https://doi.org/10.1007/978-981-16-0119-4_49

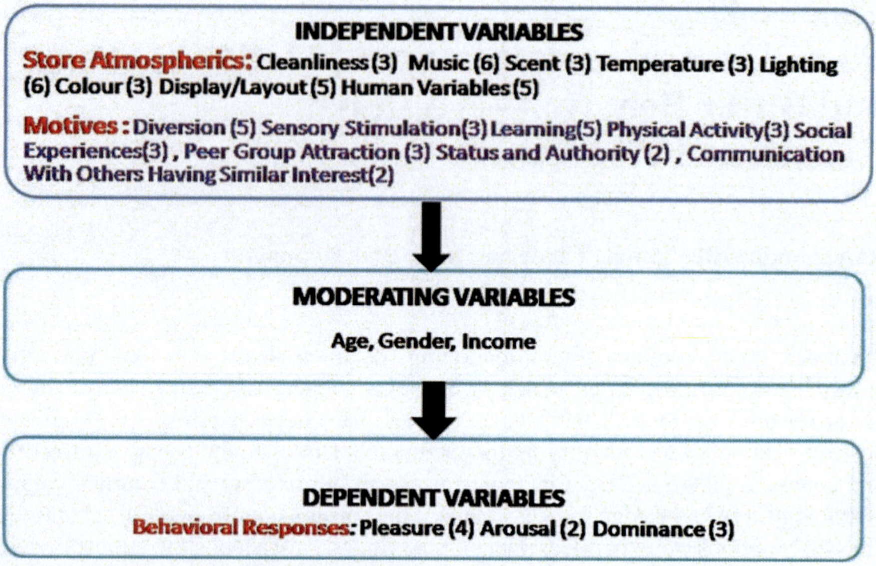

Fig. 49.1 Proposed model for evaluating impact of store atmospherics on buyer behavior

building, lawns and gardens, surrounding stores, parking availability, and architectural style), general interior variables (flooring and carpeting, color schemes, lighting, music, scents, temperature, and cleanliness), layout and design variables (space design and allocation, placement of merchandise, waiting ques) ,point-of-purchase variables(point-of-purchase displays, wall decorations, usage instructions) and human entities (employee characteristics, customer characteristics, and privacy) as independent variables and consumer behavior as dependent variables.

The research papers aims to cater only to physical store settings, to study the hedonic outcomes in terms of behavior. Taking into consideration the cultural and behavioral aspects of Indian consumers, physical store settings play a major role in shaping ones shopping behavior. However, the literature studied during the research work for online platforms indicated all together, a different set of variables that influence the shopping behavior, that are no way related to physical shopping variables. Hence, the research paper is restricted only to physical store environment.

A revised, comprehensive, and conceptual frame work for the study can be designed as shown in Fig. 49.1, and the numbers in the parentheses represent the number of questions (in the questionnaire) pertaining to the variable under study.

49.2 Literature Review

A great deal of connect between store atmospherics and behavior exists. There are several proposed methods to establish, connect, and evaluate the relationships between the store atmospherics and behavior, also various dimensions of store atmospherics that are discussed below. Right from conducting a controlled experiment to $2 \times 2 \times 2$ factorial design to devising methods to collect data by verbal inputs have been discussed. Kotler in the year 1973 defined atmospherics as "Efforts initiated by retail stores to create an environment that stimulates a buyer to generate a desired emotional response to enhance the possibilities of purchase and thus contribute to profitability" [6]. Donovan and Rossiter [3] briefed about the experiment conducted with in-store environment as independent variable and attitudes and intentions - emotions during the shopping experience in terms of extra time spent in-store and impulsive purchasing as dependent variables and age as a moderating variable. The outcomes of the study revealed that pleasantness experienced by visitors of the store due to store environment is the factor responsible for visitors spending more time in the store and also making impulsive and unplanned purchases [3]. Gardner [4] briefed about the research conducted with mood states as independent variable and consumer behavior as dependent variable to prove that consumer behaves in accordance to the mood while availing a service or during service encounter. It is easy to please consumers in a good mood as they display helpful behavior compared to consumers in neutral moods. Also consumers in good mood are likely to perform the expected task at the point of purchase with positive outcomes compared to consumer in negative mood where the outcomes or results are not as expected [4]. Baker et al. [1] conducted a research on 147 graduate students to find out that architecture, style, and layout, in-store decor cues interrelate with social cues to persuade respondent's contentment and social cues persuade provocation in-store environment. For the study, lighting and music were categorized as in-store decor cues and number/friendliness of employees was categorized as social cues considered as independent variables and retail patronage decisions (Pleasure, arousal, and willingness to buy were examined) as dependent variables [1]. Bitner [2] said that dimensions in the physical environment have an impact on employees and customers behavior in terms of cognitive, emotional, and physiological response. Here ambient conditions space/function, signs, symbols and artifacts were considered as independent variables and behavior-approach and avoidance behavior and social interactions were considered as dependent variables [2]. Turley and Milliman [12] put together atmospheric variables-layout and design, point-of-purchase and decoration variables, human variables, external variables, general interior as independent variables and shopping behavior: approach and avoidance behavior as dependent variables to prove that retail environments has influence on consumer's approach-avoidance behavior [12]. Hussain and Ali [5] studied the effect of cleanliness, scent, lighting, music, color display/layout on buying behavior. The outcomes stated that atmospheric variables such as cleanliness, scent, lighting, and display/layout have a positive influence on consumer purchase intention whereas music and color have insignificant impact on consumers purchase intention. The temperature has almost no impact on the purchase intention of the consumers. Table 49.1 represents interpretation of pleasure arousal and dominance [5].

Table 49.1 Interpretation of pleasure arousal and dominance

S. No	Author	Year	Title/Journal	Outcomes
Interpretations of pleasure				
1	Mehrabian, A., & Russell, J. A	1974	"An approach to environmental psychology"	Pleasure was defined as a affective (emotional) responses
2	Russell, J. A., & Mehrabian, A	1977	Evidence for a three-factor theory of emotions	Feelings of pleasantness and un pleasantness are dimensions of evaluating pleasure
				Pleasure is defined as feelings ranging on a spectrum of pain toward happiness
3	Mehrabian	1996	Balanced Emotional Empathy Scale (BEES)	Positive versus negative affective states (e.g., enthusiasm, respite, feeling of acceptance, serenity versus cruelty, disgrace, apathy, and tedium)
Interpretations of Arousal				
4	Thayer G.D and Russell	1967 1979	Circumflex model of marital and family systems	Arousal described in terms such as: watchful, aroused, enthusiastic, heartfelt, watchful, roused
5	Mehrabian, A., & Russell, J. A	1974	An approach to environmental psychology	Feeling state ranging along a single dimension ranging from sleep to anxious excitement such as expectant, serene energized, and somnolent
		1977	Evidence for a three-factor theory of emotions	Arousal ranges from slumber through transitional states of lethargy and then preparedness to hyperactive stimulation at the opposite extreme
Interpretations of dominance				
6	Mehrabian, A., & Russell, J. A	1974	An approach to environmental psychology	Dominance is associated to behavior such as controlling, influencing others and self-governing

(continued)

Table 49.1 (continued)

Interpretations of pleasure				
7	Russell, J. A., & Mehrabian, A	1977	Evidence for a three-factor theory of emotions	Feelings of being deficient in controlling or influencing on events and surroundings to the contrary extreme of feeling influential and in control

49.3 Summary

Several research papers were considered for the study; however, in all the papers that have been considered, the independent variables considered are either single variable indices or multiple variable indices restricted to 2, 3, or maximum 4 variables that are related. Also, most of the studies have adopted an experimental approach limiting the sample size due to availability of resources or time at disposal. The studies have been carried out in a controlled environment as per the availability of resources, which gives a fabricated outcome, restricting a holistic outcome for consideration and hence the findings cannot be generalized.

The study undertaken has taken into consideration eight dimensions of store environment that are relevant to make a considerable impact on buying behavior and eight dimensions of pre defined motives, put together which give a holistic outcome for consideration. Hence the research contributes to the body of knowledge by giving a broader perspective of store environment variables and antecedent states that drive shopping behavior. This study will surely help retail store owners to design an environment for inducing favorable shopping behavior and better shopping experiences.

49.4 Research Methodology/Questionnaire Framework

Based on the literature study carried out, the following gaps have been identified.

Research Gap 1: Majority of the studies carried out have been controlled experiments with a limited sample size, studied in a controlled environment. Hence, the outcomes of the study cannot be generalized and applied to a larger mass or real-time situation.

Research Gap 2: Off the few research studies carried out in Indian context, the data collection from samples has been limited to a particular city or town, also the behavior of people varies from jurisdiction or varied geographical locations. Hence, generalizing the findings of a particular geographical location to broader Indian context is not feasible.

Research Gap 3: To a greater extent, mood or other antecedent states of individuals are the key determinants of driving behavior. The literature studies conducted so far have not indicated any studies to prove the impact of store atmospherics on mood. Hence, this gives rise to an opportunity to study the impact of store atmospheric elements on mood and in return, the impact on behavioral outcomes of consumers.

The framework of the questionnaire is as mentioned below in Table 49.2. A sample size of 1000 visitors and shoppers of organized retailer outlets was decide, but eventually data was collected from 1028 samples from the five zones of the country, namely north, south, east, west, and central Indian. A structured questionnaire was administered both online and offline. The stores considered for the study were supermarket (Big Bazaar, Dmart—46%), departmental stores (Westside, Pantaloons, and Shoppers Stop—24%), convenience store (More, Reliance fresh—17%), malls (Central Mall—12%) and specialty stores-1%. The percentage numbers represent the sample unit. The sampling method adopted for the study was stratified random sampling, as the whole idea was to generalize the overall behavior of India consumers. A questionnaire consisting of 77 was designed. The responses were recorded on a 5-point Likert scale with the following ratings (1—Always, 2—very often, 3—sometimes, 4—rarely, and 5—never). A correlation, regression, chi square test, and ANNOVA test were carried out to establish a relationship between the variables of the study and also to find the strength of association and degree of variance. The insights gained from the study would help to develop a framework for designing store atmospherics to suit the target customers psych to induce an expected behavior. Also developing a store environment to suit the target customers need would help to increase the store patronage and loyalty.

49.5 Findings and Discussion

See Tables 49.3, 49.4 and 49.5.

Table 49.2 Questionnaire framework

Layout	Variable	Adopted from	Cronbach's Alpha	Citations
Part A Independent variable-34 Questions	Atmospheric variables (Cleanliness, music, lighting, scent, color, temperature, display/layout)	Hussain and Ali [5]	0.937	20
Part B Independent Variable-26-Questions	Buyers motives (Diversion, sensory stimulation, learning, physical activity, social experiences, peer group attraction, status and authority, communication with others	Mooradian and Olver [10]	0.76	84
Part C Dependent variable 09-Questions	Behavioral intentions —Pleasure (P), arousal (A), dominance(D)	(P) Oliver [11] (A) Zeithaml [13] (D) Hussain and Ali [5]	0.76 0.77 0.77	84 6352 20

Table 49.3 Statistical representation of chi square test

Relationship between store atmospheric and consumer behavior				
Store atmospheric dimensions	Consumer behavioral dimension	Chi-Square value	Cramer's V value	Discussion
Cleanliness	Pleasure	0.0	0.23-strong	Clean stores drive people to spend more time in the store
	Arousal	0.0	0.17-strong	Clean stores motivate people to make unplanned purchases
	Dominance	0.0	0.24-strong	Store cleanliness increase store patronage
Music	Pleasure	0.0	0.21-strong	Music creates a relaxed atmosphere while shopping and also encourages customers to spend more time in the store
	Arousal	0.0	0.21-strong	Music also tends to motivate people to buy more
	Dominance	0.0	0.21-strong	Music played in the store increases the customers well-being and comfort that induces store recommendations and store revisits
Scent	Pleasure	0.0	0.19-strong	Scent has an considerable impact on customers spending time in the store
	Arousal	0.0	0.20-strong	Scent has a very strong relationship with customers making unplanned purchases
	Dominance	0.0	0.18-strong	Scent in the store is responsible for consumers revisiting the outlet

(continued)

Table 49.3 (continued)

Relationship between store atmospheric and consumer behavior

Store atmospheric dimensions	Consumer behavioral dimension	Chi-Square value	Cramer's V value	Discussion
Lighting	Pleasure	0.0	0.32- very strong	In-store lighting makes customers comfortable and is one of the reasons for them to spend more time in the outlet
	Arousal	0.0	0.25-strong	Store lighting makes things visually attractive and also generates an interest in the product. Lighting is also one of the parameters that help consumer evaluate the quality of the product
	Dominance	0.0	0.32-very strong	Store lighting may also increase store patronage and recommendations
Temperature	Pleasure	0.0	0.26-very strong	Store temperature has very strong relationship with customers comfort
	Arousal	0.0	0.19-strong	Store temperature affects unplanned purchases
	Dominance	0.0	0.25-strong	Store temperature increases store patronage and recommendations
Color	Pleasure	0.0	0.31-very strong	Color of the store attracts customers towards the store
	Arousal	0.0	0.29-very strong	Store color creates a positive perception
	Dominance	0.0	0.29-very strong	Store color induces store patronage and recommendations
Display and layout	Pleasure	0.0	0.31-very strong	Display induces interest to pose an enquiry

(continued)

Table 49.3 (continued)

Relationship between store atmospheric and consumer behavior

Store atmospheric dimensions	Consumer behavioral dimension	Chi-Square value	Cramer's V value	Discussion
	Arousal	0.0	0.26-very strong	Attractive and impressive displays motivate customers to buy more
	Dominance	0.0	0.32-very strong	Displays induce revisit and store loyalty
Human variables	Pleasure	0.0	0.27-very strong	Employees at the store form a crucial element to enhance customer shopping experience
	Arousal	0.0	0.25-strong	Employees and crowding at the store are dominant factors in inducing unplanned purchases
	Dominance	0.0	0.27-very strong	Employees motivate customers to revisit the outlets and display store patronage

Table 49.4 Statistical representation of ANOVA analysis

Analysis of behavior variance with respect to store atmospherics			
Atmospheric variables	Consumer behavior	Significance value	Discussion
Cleanliness	Pleasure	0.229	Although overall store atmospherics have an impact on pleasure dimension of consumer behavior; music, lighting, color, display, and layout dimension of store atmospherics have emerged as dominant dimensions that drive customers to spend more time in the store
Music		0.003	
Scent		0.524	
Temperature		0.189	
Lighting		0.000	
Color		0.000	
Display and layout		0.000	
Human variables		0.091	
Cleanliness	Arousal	0.218	Results prove that overall atmospherics are instrumental in affecting the arousal dimension of consumer behavior but music, scent, color, display, and layout and human variables have emerged as dimension that are responsible for customers buying more and making unplanned purchases
Music		0.003	
Scent		0.002	
Temperature		0.603	
Lighting		0.129	
Color		0.000	
Display and layout		0.006	
Human variables		0.000	
Cleanliness	Dominance	0.003	Store cleanliness, lighting, color, human variables, display, and layout are the dimension that induce customers revisits and store recommendations
Music		0.015	
Scent		0.731	
Temperature		0.969	
Lighting		0.001	
Color		0.000	
Display and layout		0.000	
Human variables		0.000	

Table 49.5 Statistical representation of AMOS modelling (regression analysis)

Direct and indirect impact of store atmospherics and motives on consumer behavior						
Atmospheric variables	Direct effect			Motives as mediating variables		
				Indirect effect		
	Pleasure	Arousal	Dominance	Pleasure	Arousal	Dominance
Cleanliness	**0.225**	**0.151**	**0.264**	**0.181**	**0.154**	**0.162**
Discussion: Desire to make unplanned purchases will increase with cleanliness, irrespective of the pre-defined motive with which the consumer visits the store. But cleanliness as an atmospheric variable seems to have a reduced impact in terms of desire to spend more time and display store patronage because the predefined motives dominate over cleanliness						
Music	**0.134**	**0.121**	**0.135**	**0.251**	**0.257**	**0.227**
Discussion: Music in the store tends to have a dominating impact on the motives of the customers and in turn on the behavior because from the statistics above, desire to spend more time in the store make unplanned purchase and recommend the store to others and display store loyalty seems to have increased with the presence of music irrespective of the pre-defined motive with which the consumer visits the store						
Scent	**0.097**	**0.136**	**0.076**	**0.169**	**0.168**	**0.153**
Discussion: Store fragrance or scent is instrumental in changing the pre-defined motives of customer as the statistics above show that desire to spend more time in the store, make unplanned purchase and recommend the store to others and display store loyalty seems to have increased with the presence of store fragrance						
Temperature	**0.267**	**0.180**	**0.250**	**0.200**	**0.191**	**0.193**
Discussion: Desire to make unplanned purchases will increase with temperature, irrespective of the pre-defined motive with which the consumer visits the store. But temperature as an atmospheric variable seems to have a reduced impact in terms of desire to spend more time and display store patronage because the predefined motives dominate over temperature						
Lighting	**0.346**	**0.245**	**0.343**	**0.219**	**0.209**	**0.202**
Discussion: Store lighting is not instrumental in changing the predefined customer motives. Hence, the customer will act and behave in the way that he has decided						
Colour	**0.289**	**0.268**	**0.287**	**0.245**	**0.249**	**0.223**
Discussion: Color is not instrumental in changing the predefined customer motives. Hence, the customer will act and behave in the way that he has decided						

(continued)

Table 49.5 (continued)

Direct and indirect impact of store atmospherics and motives on consumer behavior						
Display/layout	**0.292**	**0.242**	**0.338**	**0.241**	**0.211**	**0.211**
Discussion: Display/layout is not instrumental in changing the predefined customer motives. Hence, the customer will act and behave in the way that he has decided						
Human variables	**0.175**	**0.187**	**0.259**	**0.278**	**0.276**	**0.229**
Discussion: Human variables are instrumental in changing the pre-defined motives of customer, as the statistics above show that desire to spend more time in the store make unplanned purchase and recommend the store to others and display store loyalty seems to have increased with the presence of human variables						

References

1. Baker, J., Levy, M., Grewal, D.: An experimental approach to making retail store environmental decisions. J. Retail. **68**(4), 445 (1992)
2. Bitner, M.J.: Servicescapes: the impact of physical surroundings on customers and employees. J. Market. **56**(2), 57–71 (1992)
3. Donovan, R.J., Rossiter, J.R., Marcoolyn, G., Nesdale, A.: Store atmosphere and purchasing behavior. J. Retail. **70**(3), 283–294 (1994)
4. Gardner, M.P.: Mood states and consumer behavior: a critical review. J. Consum. Res. **12**(3), 281–300 (1985)
5. Hussain, R., Ali, M.: Effect of store atmosphere on consumer purchase intention. Int. J. Market. Stud. **7**(2) (2015)
6. Kotler, P.: Atmospherics as a marketing tool. J. Retail. **49**(4), 48–64 (1973)
7. Mehrabian, A.: Manual for the Balanced Emocional Empathy Scale (BEES), Alta Mesa. Mehrabian, Monterrey, CA (1996)
8. Mehrabian, A., Russell, J.A.: An Approach to Environmental Psychology. MIT Press, Cambridge (1974)
9. Michon, R., Chebat, J.C., Turley, L.W.: Mall atmospherics: the interaction effects of the mall environment on shopping behavior. J. Bus. Res. **58**(5), 576–583 (2005)
10. Mooradian, T.A., Olver, J.M.: "I can't get no satisfaction:" the impact of personality and emotion on postpurchase processes. Psychol. Mark. **14**(4), 379–393 (1997)
11. Oliver, R.L., Rust, R.T., Varki, S.: Customer delight: foundations, findings, and managerial insight. J. Retail. **73**(3), 311–336 (1997)
12. Turley, L.W., Milliman, R.E.: Atmospheric effects on shopping behavior: a review of the experimental evidence. J. Bus. Res. **49**(2), 193–211 (2000)
13. Zeithaml, V.A., Berry, L.L., Parasuraman, A.: The behavioral consequences of service quality. J. Mark. **60**(2), 31–46 (1996)

Chapter 50
Creation of Value in Ecosystems—A Perspective from Design Innovation

Nikhil Ravindranath Zope

Abstract With design becoming more and more explicitly recognized and used as part of the innovation in businesses, design innovation as a discipline is also maturing. Design innovation is helping recognizing value creation explicitly in all stages of innovation. Innovation itself is moving from offering innovation to experience innovation and from a single business firm driven to ecosystems driven. In this paper, we review literature from design innovation discipline and business ecosystems discipline. From the literature review, we establish that design innovation needs to go beyond offering innovation and address value creation within multiple firms from the ecosystem perspective. We also discuss business models of ecosystem participants and their influence on each other as first step in this research direction.

50.1 Introduction

50.1.1 Introduction—Design and Innovation

Design has become language of innovation [5]. Various design methods and design thinking as a discipline have become integral to innovation processes practiced by innovation teams across startups and corporates alike. While management and innovation literature has adopted design discipline in terms of design thinking methods and using it as language of innovation, "design innovation" has emerged as discipline within design community [15].

Design methods and design processes have been aiming at identifying the value, and an offering must provide to its users and design the offering for that value. Erstwhile product design and in last decade service design has successfully applied design methods for the respective offerings to create value for intended customers. Several authors have successfully demonstrated through various cases how design

N. R. Zope (✉)
Tata Consultancy Services Limited, Mumbai, India
e-mail: nikhil.zope@tcs.com

A. Chakrabarti et al. (eds.), *Design for Tomorrow–Volume 2*, Smart Innovation, Systems and Technologies 222, https://doi.org/10.1007/978-981-16-0119-4_50

thinking methods can be used along with other management concepts to design business models for firms [16]. In last two decades, business models are conceptualized as one of the essential steps to go from idea of an offering to successful innovation. Further in recent efforts, other authors have shown that design can also help in ensuring value creation along the innovation process with help of business model canvas [12].

50.1.2 Research Direction—Ecosystem-Driven Innovation

Nature of innovation itself is transforming over last two decades. Innovation has moved from centrality of an individual firm to an organizational ecosystem. It is the organizational ecosystem together that provides compelling value to the customer which cannot be provided by any individual firm by itself. The organizational ecosystem for an offering or set of offerings may still be coordinated primarily by an anchor organization. Organizational ecosystem literature identifies actors that can be part of an ecosystem and different types of ecosystems based on how organizations in the ecosystem work together [2, 8, 13]. Various roles that an individual firm can play in ecosystem has also been discussed in the literature [8]. In recent literature, it has also been recognized that the role a firm plays in ecosystem can change over time and dynamics of organizational ecosystem is considered as one of the important research agenda for the future [17].

In this paper, we focus on ecosystem-driven innovation. The literature identifies three types of organizational ecosystems—business ecosystem, knowledge ecosystem, and innovation ecosystem [19]. However, such distinction is not followed widely in the literature and often the term "business ecosystem" or "organizational ecosystem" is used to represent an ecosystem in any context. Different value related methods like "value modeling," "value blueprint," etc., are used in the literature to identify and define organizational ecosystems [2]. Well-known design schools have also identified business ecosystem design as key capability for growth and innovation [10].

In this paper, we establish perspective based on literature review to identify the area in which that design innovation discipline can complement.

50.1.3 Objective and Proposed Contribution

The objective of this paper is to explore how design innovation can complement in ensuring value creation in an innovation ecosystem. We review organizational and innovation ecosystem literature and key papers from design innovation literature to present our views. We also leverage ecosystems design canvas and steps toward ecosystems design from literature.

50.2 Design Innovation and Value Creation

The term "design innovation" is interpreted in multiple ways by various authors. From innovation perspective, it is useful to first understand how design can contribute to innovation. Hernández et al. [5] recognize following roles played by design from innovation perspective: design to differentiate; design for introduction and adoption of innovations in the market; design to transform ideas into concepts; design (as) research, design as a (creative, generative) thinking process; design as techniques to articulate ideas and to integrate concepts, people, and functions; designers' contributions to innovation.

Na et al. [15] articulate role of design from innovation perspective in the form of design innovation spectrum. The spectrum includes different design theories and relationship of these theories to innovation at different levels of business (activity, strategic and organization level).

Hobday et al. [7] in their "coupling model" of innovation with design as a bridging function identify different types of innovation activities, relationships between them, and the innovation activities in which design practice is utilized the most. The authors argue that design practice is utilized primarily in idea generation, concept development and prototypes, while its potential in the other innovation activities remains underutilized.

The Danish Design Ladder represents different levels of influence or integration of design within a business [9]. The four steps of the framework are: (1) no design, (2) design as styling, (3) design as process, and (4) design as strategy. The authors propose that in the fourth level—design as strategy—design is integral part of the innovation process in the business and designers take part in the strategy of the business.

Heskett [6] has discussed design and value creation comprehensively. Heskett in his book has argued that design community should explicitly study value creation.

Lokku et al. [11] further link creativity, design, and innovation with focus on value creation. Lokku [12] further utilizes value creation stages with respect to business models (and hence innovation) and argue that design must understand value creation beyond just value proposition and explicitly address the concern of ensuring value creation for the stages—value creation, value delivery, and value capture.

Value creation is the central concept for innovation. Any offering (whether service or physical product) must create value for its customer. With modern offerings including both service and products, trying to fulfill needs of potential customer together, the modern discussion on innovation has shifted in terms of value than service/product and their features.

Boztepe [4] studies the notion of value through three approaches toward user value: value as exchange and use; value as a sign; and value as experience. The author further identifies different categories of user value and finally discusses how these different approaches and categories can be applied in design practice. Value as exchange and use is recognized from economic paradigm where exchange value is considered as monetary or other material value that can be sacrificed for acquiring the offering while use value is the value realized by the user through the benefits of using the offering. Value as a sign refers to the symbolic meaning attached to the offering. It may be driven primarily by social and cultural aspects. Value as experience refers to the benefits including feelings derived from the offering consumption experience. Smith and Colgate [18], similarly in their customer value creation framework also refer to four types of values—functional/instrumental value, experiential/hedonic value, symbolic/expressive value, and cost/sacrifice value. They further offer examples and refer to related literature for each type of value.

From the literature we note that, design innovation is focused on offering and value that can be delivered to the customer of the offering. As more and more innovations are being conceptualized and executed at ecosystems level, it is important for design innovation practice to address value creation for all the ecosystem participants.

50.3 Ecosystems and Innovation

50.3.1 Understanding Ecosystems

Moore [13, 14] first presented concept of ecosystem in business context. Moore [14] defines a business ecosystem as "an economic community supported by a foundation of interacting organizations and individuals—the organisms of the business world." The author describes business ecosystems in three layers: core business layer, extended enterprise layer, and business ecosystem layer [13]. The author uses analogy of biological ecosystems and argues that an organization's capabilities co-evolve around innovations (Fig. 50.1).

Iansiti and Levien [8] developed the concept further using the analogy of biological systems used by Moore [13] highlighting both similarities and differences between biological ecosystems and business ecosystems. The authors emphasize that collective behavior of organizations in the ecosystem determines how well organizations in the ecosystem do; health of an individual organization in an ecosystem depends on the health of the ecosystem. Iansiti and Levien [8] proposed four different roles that an organization can play in the business ecosystem they are part of: Keystones, Niche players, Dominators, and Hub Landlords.

In this paper, we are focusing on ecosystems from innovation point of view.

Adner [2] considers innovation ecosystem as "a world in which the success of value proposition depends on creating alignment of partners who must work together in order to transform a winning idea to a market success." Valkokari [19] classified ecosystem into three types of bring clarity to different perspectives with which various researchers studied ecosystems. The three types are: business ecosystems, innovation ecosystem, and knowledge ecosystems. Valkokari [19] argues that innovation ecosystem is understood with the objective of co-creation of innovation.

Lewrick et al. [10] present business ecosystem design as capability for growth and innovation. The authors propose an ecosystem design canvas that can be used

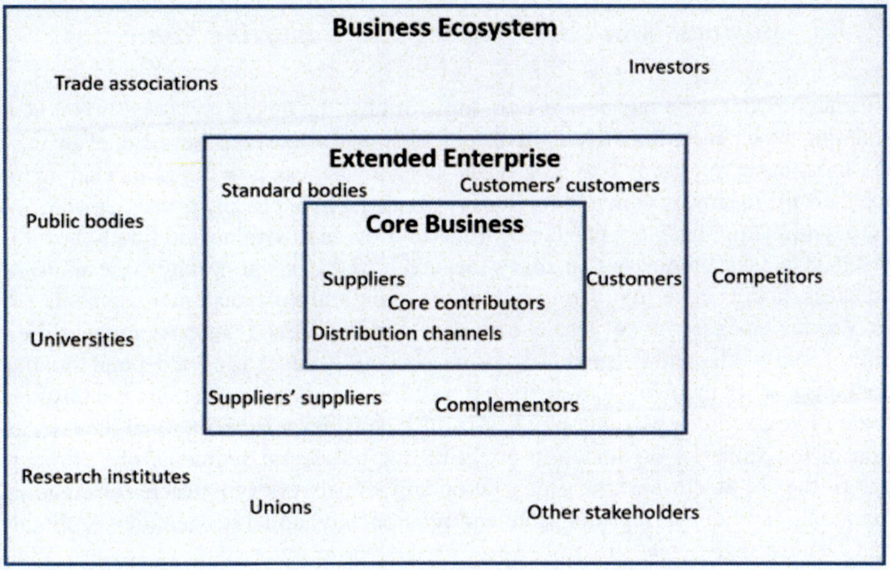

Fig. 50.1 Business ecosystem layers (Adapted from: [14])

to design or improve an ecosystem. The elements of the ecosystem design canvas highlight importance of considering business models of each actor, how business models of actors influence each other and the resultant overall business model for the ecosystem.

50.3.2 *Value Creation in Ecosystems*

Value creation is not a linear process [3]. Organizations participating in ecosystem cooperate and interact with each other in various ways to create value for each other and for the customer. Ecosystem allows organizations to create value that no individual organizations could have created alone [3].

Value creation is the central concept that is used to innovate at ecosystem level. Researchers have used business models to express and analyze ecosystem-driven innovation as they capture the value creation and relationships between ecosystem participants [3, 10, 20].

50.3.3 Business Models—Creating and Capturing Value

Business model for a business is core logic of how the business works. It is understanding the business holistically in its most abstract form. A business model answers the most basic questions like what value does a business provide to its customers, what are the means by which this value is provided (channels, key process, resources, etc.), with whom business needs to partner to provide this value and finally how the business makes money out of this endeavor [16]. A business may have multiple business models covering its multiple offering and multiple customer segments.

Various business model researchers have conceptualized business model differently. Osterwalder and Pigneur [16] define a business model as a conceptual tool that contains a set of elements and their relationships and allows expressing the business logic of a specific organization. They further consider business model as description of the value an organization offers to one or several segments of customers and of the architecture of the organization and its network of partners for creating, marketing, and delivering this value and relationship capital, to generate profitable and sustainable revenue streams.

50.3.4 Ecosystem Design Canvas

Lewrick et al. [10] have proposed ecosystem design canvas to design and analyze ecosystem like business model canvas for business models. However, unlike business model canvas the ecosystem design canvas is not a single view but consists of multiple views for some of the elements of the ecosystem design canvas. Table 50.1 shows list of elements of the ecosystem design canvas. From the user/customer side, the canvas contains two elements: *user/customer needs* and *core value proposition.* As any ecosystem consists of multiple organizations, such organizations are recognized as actors in the ecosystem design canvas. The canvas has two elements for the actors,

Table 50.1 Elements of ecosystem design canvas (*Source* [10])

Category	Element
	Core value proposition
User/customer	User/customer needs
Actors	Describe the actors
	Analysis of the advantages and disadvantages of each actor
	Multidimensional view of business models
	Definition of the value streams
	Design/redesign, build/test, explore
	Prototype, test and improve the business ecosystem

one element includes *description of all the actors* while the other analyzes *advantages and disadvantages of all the actors* from the ecosystem perspective. Actors also feature in the element—*Multidimensional view of the business models.* In this element, business models of all the actors are analyzed along with their contribution to the core value proposition for the ecosystem. One of the important elements of the canvas is the *definition of value streams* in which the value streams across multiple actors are recognized and articulated. The other two elements represent the activities in ecosystem design. One element is *prototype, that tests and improves the business ecosystem,* while the other element represents the cyclical stages of ecosystem design: *design/redesign, explore, and build/test.*

The ecosystem design canvas and the approach proposed along with it connects design thinking with innovation and ecosystem perspective. However, it does not go deeper into how value is accrued to all the actors in the ecosystem. The element—multidimensional view of business models—would be key to analyzing this value perspective along with value streams.

50.4 Leveraging Design Innovation

We note from the literature for ecosystems discipline that current research focuses on bringing out the importance of ecosystems view and need to embrace it. There are many methods that help identify ecosystem participants and value exchange within the ecosystem participants. However, a systematic approach toward how complex interrelations within these ecosystem participants influence each other collectively creating value for each other and for customer has not yet matured. We believe that multidimensional view of business models mentioned in ecosystem design canvas is the view that will help organizations most in finding and working on this synergy.

Design Innovation starts with user needs and has matured as a discipline in designing to deliver value for user/customer through offering or experience of the offering. However, similarly explicitly working on value creation across multiple business entities is an area that is not adequately addressed by design innovation. We believe that this gap can be addressed by enhancing design practice to formulate approach toward designing and analyzing multidimensional business models (a set of business models that influence each other and align in a way to deliver through an overall business model). We can refer to overall business model as "ecosystem business model." Some researchers like Bahari et al. [3] have tried to define and formulate approach to design such ecosystem business model. Design innovation can further strengthen such approach along with focus on identifying influence of business models of organizations (ecosystem participants) on each other. Figure 50.2 shows pictorial representation. This research direction can benefit both the design community as well as business community.

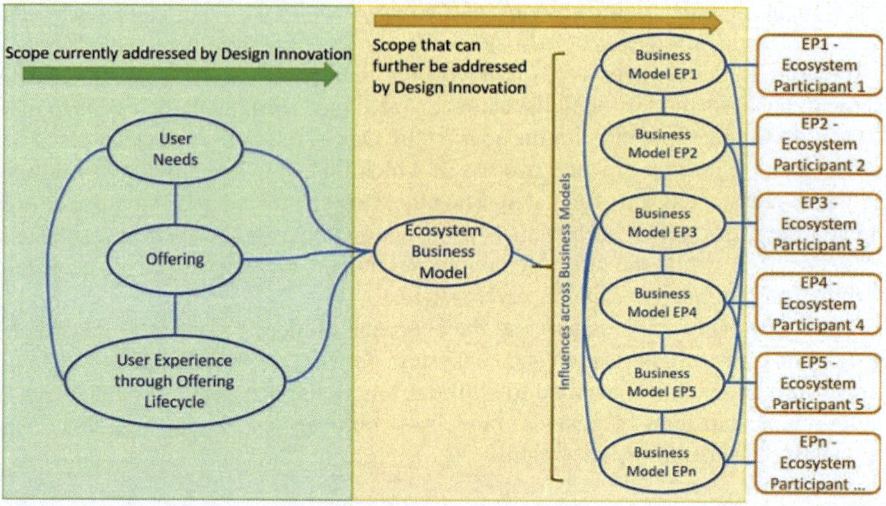

Fig. 50.2 Leveraging design innovation for ecosystem context

50.5 Conclusion

Innovation is increasingly becoming ecosystem-driven with ability of individual organizations to compete and innovate alone in a sustaining manner is decreasing. Design innovation as a discipline has matured in addressing innovation processes with respect to offering innovation and delivering value through experience throughout offering lifecycle. Ecosystem-driven innovation processes require going beyond offering paradigm and address value creation among ecosystem participants along with value creation for customer/user.

Business models serve the purpose of understanding and designing value creation in all stages for an organization. Multidimensional view of business models—business models of all ecosystem participants and their influence on each other—is an important area where we can systematically approach ecosystem-driven innovation. Value networks can further help in explicitly identifying value exchange across ecosystem participants. Design innovation can address this value creation paradigm using business models. We observe this area of research is not adequately addressed in design practice. We believe conducting research in this direction will benefit both design community and business community. We are undertaking research to study influence of business models of multiple organizations on each other as part of our further work.

References

1. Adner, R., Kapoor, R.: Value creation in innovation ecosystems: how the structure of technological interdependence affects firm performance in new technology generations. Strateg. Manag. J. **31**(3), 306–333 (2010)
2. Adner, R.: The Wide Lens: A New Strategy for Innovation. Penguin UK (2012)
3. Bahari, N., Maniak, R., Fernandez, V.: Ecosystem Business Model Design. In: XXIVe Conférence Internationale de Management Stratégique, pp. 1–18 (2015)
4. Boztepe, S.: User value: competing theories and models. International Journal of Design **1**(2), 55–63 (2007)
5. Hernández, R.J., Cooper, R., Tether, B., Murphy, E.: Design, the language of innovation: a review of the design studies literature. She Ji: J. Des. Econ. Innov. **4**(3), 249–274 (2018)
6. Heskett, J.: Creating economic value by design. Int. J. Des. **3**(1), 71–84 (2008)
7. Hobday, M., Boddington, A., Grantham, A.: An innovation perspective on design: Part 1. Des. Issues **27**(4), 5–15 (2011)
8. Iansiti, M., Levien, R.: The keystone advantage: what the new dynamics of business ecosystems mean for strategy, innovation, and sustainability. Harvard Business School Press, Boston (2004)
9. Kretzschmar, A.: The Economic Effects of Design. National Agency for Enterprise and Housing, Copenhagen, Denmark (2003)
10. Lewrick, M., Link, P., Leifer, L.: The Design Thinking Playbook: Mindful Digital Transformation of Teams, Products, Services, Businesses and Ecosystems. Wiley, Hoboken (2018)
11. Lokku, D.S., Onkar, P.S., Mathew, D.J.: Leveraging design innovation for ensuring creation of value: an approach to identify the corresponding design concerns towards enabling the design practice. In: Research into Design for a Connected World, pp. 133–142. Springer, Singapore (2019)
12. Lokku, D.S.: Design Innovation and Creation of Value: A Study based on Business Models for an Approach to enable Design Practice, Working Paper (2020)
13. Moore, J.F.: Predators and prey: a new ecology of competition. Harvard Bus. Rev. **71**(3), 75–86 (1993)
14. Moore, J.F.: The death of competition: leadership and strategy in the age of business ecosystems. Harper Business, New York (1996)
15. Na, J.H., Choi, Y., Harrison, D.: The design innovation spectrum: an overview of design influences on innovation for manufacturing companies. Int. J. Des. **11**(2), 13–24 (2017)
16. Osterwalder, A., Pigneur, Y.: Business Model Generation: A Handbook for Visionaries, Game Changers, and Challengers. Wiley, Hoboken (2010)
17. Rong, K., Lin, Y., Li, B., Burström, T., Butel, L., Yu, J.: Business Ecosystem Research Agenda: More Dynamic, More Embedded, and More Internationalized (2018)
18. Smith, J.B., Colgate, M.: Customer value creation: a practical framework. J. Mark. Theor. Pract. **15**(1), 7–23 (2007)
19. Valkokari, K.: Business, innovation, and knowledge ecosystems: how they differ and how to survive and thrive within them. Technol. Innov. Manage. Rev. **5**(8) (2015)
20. Weill, P., Woerner, S.L.: Thriving in an increasingly digital ecosystem. MIT Sloan Manage. Rev. **56**(4), 27 (2015)

Chapter 51
Understanding and Evaluating the Needs of a Respiratory Assessment Device for Community Health

Nibedit Dey and Priyabrata Rautray

Abstract In public health settings, most of the patients with respiratory-related complaints come with chronic conditions, where the disease has already developed to a chronic stage that requires immediate care. As per the Global Asthma Report (2018), more than one billion patients suffer from chronic respiratory diseases (CRD). Pneumonia is considered the single most significant cause of mortality in children worldwide with over 2 million deaths. Factors such as malnourishment, low zinc intake, overcrowding, exposure to parental smoking, indoor air pollution play an essential role in increasing children's susceptibility to respiratory infections. Due to the lack of pulmonologists in primary care centres and lack of time, patients remain undiagnosed, which leads to a rise in the number of chronic cases. The problem is worse for babies suffering from pneumonia, where any delayed diagnosis can be fatal. In a crowded and noisy atmosphere, it is challenging to hear the sound of congested lungs through a regular stethoscope. To improve our understanding of the screening process and gather clinicians' perspectives, several face-to-face interviews and online surveys were conducted. We also documented the existing diagnosis methods and tests prescribed by clinicians to detect pneumonia. This research paper documents and highlights the need for a respiratory assessment device. The research paper is divided into three segments; firstly, it highlights the existing diagnosis methods for respiratory disease and issues related to it, and secondly, recording and analysis of clinicians' perspectives and challenges. Finally, conceptualizing a novel solution to assist clinicians in the diagnosis of respiratory diseases.

N. Dey
Ibrum Technologies, Bangalore, India
e-mail: nibedit@ibrum.tech

P. Rautray (✉)
Indian Institute of Technology, Hyderabad, India
e-mail: md17resch11001@iith.ac.in

A. Chakrabarti et al. (eds.), *Design for Tomorrow–Volume 2*, Smart Innovation, Systems and Technologies 222, https://doi.org/10.1007/978-981-16-0119-4_51

51.1 Introduction

In the twenty-first century, respiratory disease is the foremost cause of death and illness worldwide, affecting all age groups and especially infants are more vulnerable. The respiratory of the disease varies from severe infections to prolonged non-communicable diseases. Acute respiratory infections (ARI), tuberculosis (TB), asthma, chronic obstructive pulmonary disease (COPD), and lung cancer are five prevalent respiratory diseases identified by the Forum of International Respiratory Societies (FIRS) [1]. Every year, nearly 1.3 million children die due to pneumonia, and with early detection, most of the death could be averted. One in five children below the age of five dies due to pneumonia each year [2]. Currently, asthma is one of the widespread non-communicable diseases in case of children worldwide [3]. In 2016, it was assessed that more than 339 million people were affected by asthma globally [4]. According to WHO estimates, there were 417,918 deaths due to asthma at the global level and 24.8 million Disability-Adjusted Life Years (DALYs) attributable to asthma in 2016 [5]. Globally, around 5% of death is from COPD compounding to more than 3 million in number. As per WHO estimates, COPD will be the third leading cause of death in the year 2030 [6].

India, with its massive population, chronic respiratory diseases are adding to an already overly burdened healthcare system. As a country, we lack an overall understanding of the prevalence of chronic respiratory diseases for individual states [7]. As per the global asthma report 2018, about 2% of adults and 6% of children have asthma in India [4]. Asthma affects over 30 million and with a high mortality rate of 18.1 death per 100,000 people. ARI accounted for 69% of the total cases of infectious diseases and caused 23% of deaths in 2017 [8]. In 2018, the National Health Profile (NHP) of India recorded 41,996,260 ARI cases and 3740 deaths.

This research paper on understanding and evaluating the needs of a respiratory assessment device for community health is divided into four sections. Section 51.1 introduced the background for the paper and established the need for the research project. A detailed literature review to understand the existing respiratory assessment for diagnosis of different lungs conditions was done as elaborated in Sect. 51.2. Section 51.3 documents the methodologies used for clinical immersions, interviews and online survey. It also includes illustrated data sets, data interpretations and finding. Finally, Sect. 51.4 gave a summary of the problem and methods to overcome it. This paper is an effort to initiate further research and development of novel medical devices.

51.2 Literature Review

Literature related to physical observation, respiratory assessment and computerized lung sounds analysis (CLSA) to detect respiratory disorders were studied. The traditional way of physical observation and examination has a lower accuracy in identifying pneumonia cases. Research says that even with a specialist doctor, physical diagnosis sensitivity is between 47 and 69% and specificity is between 58 and 75% [9]. A respiratory assessment is a physical examination that includes evaluation of the respiratory rate, chest movements and breathing pattern. The assessment consists of inspection, palpation, percussion and auscultation along with the patient health history as explained in Fig. 51.1 [10]. Once a doctor assesses the patient's medical history and general appearance, physical assessment is done. Palpation provides vital information about respiratory health conditions. The clinician measures body temperature looks for abnormal sounds, checks alignment of trachea, excursion and percussion. Auscultation is the final critical component of a respiratory assessment, i.e. listening to chest sounds [11].

A stethoscope is used to listen to respiratory sounds, also known as lung sounds or breath sounds. Normal lung sounds can be heard throughout the chest area, which includes the area below of the rib cage and above the collarbones. The clinician may listen to the normal, decreased or absent, and abnormal chest sounds using a stethoscope [12]. The four most common types of abnormal breath sounds are rales,

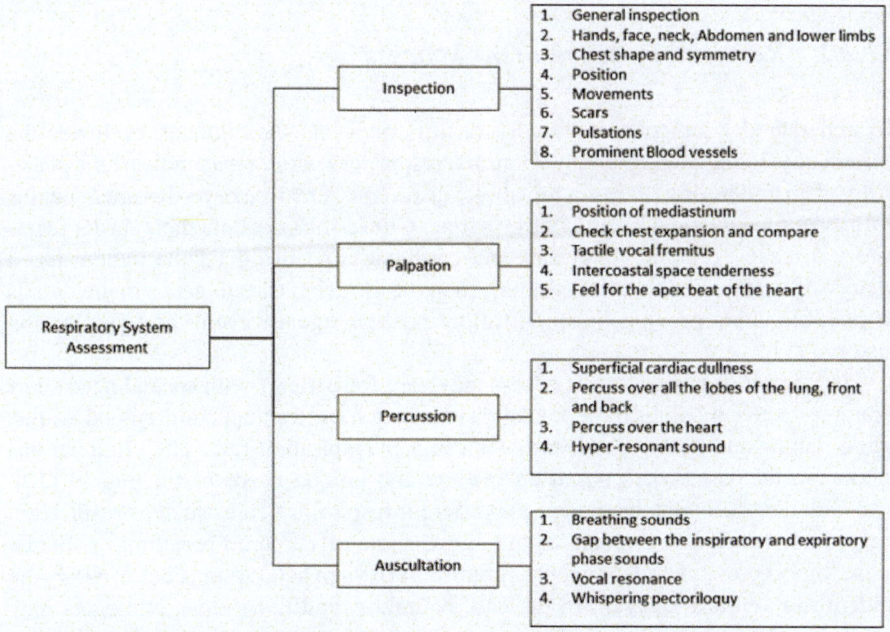

Fig. 51.1 Respiratory assessment procedure

rhonchi, stridor and wheezing [13]. Abnormal chest sounds may indicate underlying health conditions like asthma, bronchitis, pneumonia, emphysema, interstitial lung disease, bronchiectasis, etc.

Auscultation is a very skilful procedure and gained by practice. Therefore, most of the clinicians require confirmation using blood tests and medical imaging modalities like X-rays or CT scans [14]. The user's listening skills limit the capabilities of traditional stethoscopes for chest auscultation. Auscultation using electronic stethoscopes and machine learning-based classification of lung sounds may bridge these shortcomings. In one of the studies, the sensitivity of CLSA for the detection of crackles or wheezes was at 80%, and specificity was 85% [15]. 2D visualization of lungs sounds has the edge over traditional stethoscopes. 3D visualization of lungs sounds may help in localizing the infection. Unlike traditional stethoscopes that provide access to only one site information, sensor arrays can provide a 2D or 3D visualization of the lungs. However, existing sensor array-based solutions are not preferred by clinicians as they are not user friendly, expensive and bulky [16]. As per the study conducted by Marques et al., computerized respiratory sound analysis has shown potential in monitoring respiratory health in COPD patients. The researchers found that the quality of lungs sounds from standardized anatomic locations was excellent and reliable. The usefulness of computerized respiratory sound assessments in evaluating disease severity and response to treatment may be improved by using these standardized anatomical locations [17].

51.2.1 Respiratory Assessment Process

To understand a patient's respiratory health condition, the clinician evaluates the current physiological status. Any abnormal respiratory pattern may indicate the possibility of an underlying illness or metabolic disorders. A respiratory assessment begins with a detailed medical history of the patient. Clinicians ask about their earlier respiratory diseases, chronic conditions and cardiovascular health. If the patient has a history of cardiopulmonary conditions, then the clinician tries to get as many details as possible about previous hospitalization, post-emergency events and vaccination history [11].

The respiratory assessment process may vary for patients with special needs. For example, preterm babies have weaker respiratory muscles than children and adults, while babies and young children have a higher respiration rate. The clinician has to change the assessment based on the age and unique needs of the patient [10]. The clinician observes the patient for different respiratory clues such as respiration rate, abnormality in the shape of the chest, signs of laboured breathing, patient's pulse, blood pressure and oxygen saturation. If oxygen saturation is below 90%, the patient may require oxygen. In the case of infants and new-borns, clinicians may check for flaring nostrils, which indicates breathing problems. Any retractions or bulging of the muscles between the ribs may indicate an emergency condition with lack of enough air [18]. Respiratory assessment can be challenging for patients with pneumonia as the symptoms are similar to patients with colds or the flu. People often ignore pneumonia until the illness lasts longer than the other conditions. The clinician diagnoses pneumonia based on the patient's medical history, a respiratory assessment examination, chest X-ray and laboratory test results. Pneumonia has four stages, namely consolidation, red hepatization, grey hepatization and resolution. Most of the pneumonia cases are reported at the later stages.

Clinicians usually follow below parameters for pneumonia diagnosis:

- CXR (Chest x-ray) to confirm the presence of pulmonary infiltrate or two or more of the following:
- Fever > 100.4 °F (Oral)
- SpO2 < 92% in normal room condition
- RR > 24 breaths per minute (bpm)
- Evidence of focal pulmonary consolidation on respiratory assessment, including auscultation results.

Similarly, clinician usually follows the below parameters for COPD/Asthma diagnosis:

- CXR showing COPD with hyper-inflated lungs and no infiltrates and two or more of the following:
- Symptoms of shortness of breath, wheezing or increased sputum production
- SpO2 < 92% in normal room condition
- Acute reduction in peak flow or FEV1 in spirometry
- Respiratory rate > 24 bpm.

51.3 Clinical Survey

This section includes our efforts in clinical immersions and needs validation following Stanford Biodesign methodology. After our initial personal interviews with doctors from the selected hospitals, we conducted an online survey to increase the data set and to have a broader perspective.

51.3.1 Personal Interviews

Initial clinical studies were conducted in 15 hospitals, and the current methods of diagnosis were studied. We identified a few focus clusters based on our observations during phase 1 clinical immersion. To gain further insights, we discussed our views with doctors during phase 2 focused immersion. It was found that it is difficult for an undergraduate doctor or alternative medicine practitioners to identify abnormalities in lungs sounds in the noisy environment of primary care. There were many patients with complaints of sneezing and cold, which are diagnosed as allergic rhinitis.

Following are the observations documented during our interviews with doctors. They mentioned that chronic allergy cases convert to asthma. Asthma is prevalent, and patients require regular treatment. Doctors regularly persuade asthmatic patients to stop smoking to avoid asthma attacks. Most of the days, people come early in the morning for nebulization. People do not keep inhalers and depend on PHC. The condition can be critical at midnight. Villages are far away and no proper transport available at night. Patients do not visit hospitals until the situation goes out of control and believe that allopathic treatment will make them dependent on drugs. The doctor said the early treatment could cure and reduce the progress of asthma. In case of patients coming with breathing difficulty, the doctor does not have tools to differentiate between asthma, COPD, pneumonia and other disorders. In one of the PHCs, there were 2275 outpatients during the month, and 1579 laboratory tests were done. Tuberculosis was common among the villagers. Twelve patients were getting treated for TB with DOTS. From the clinical immersion, we found that pneumonia is prevalent among children and often treated late causing the casualties. From the discussions with the doctors, we concluded that there is a need for a tool for quick diagnosis of patients in primary care.

51.3.2 *Online Survey*

After the discussion with primary care doctors, we discussed with doctors from different specializations to increase the unique data sets and to have a better understanding of the issue. An online survey was conducted where 29 doctors responded from various specialties.

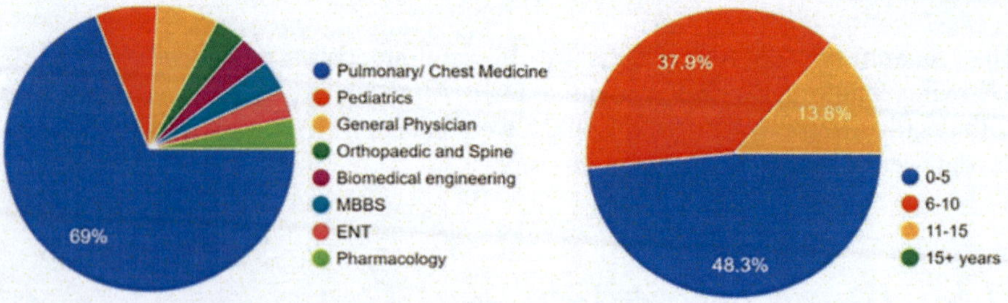

Fig. 51.2 Specifications of participant and years of experience

51.3.3 Data Visualization

Figure 51.2 highlights the specializations of the doctors who participated in the online survey and their experience in years. Around 69% of the doctors were specialized in pulmonology and with a fair amount of experience, which gave us in-depth insights.

The next figure explains the frequency of most common lungs' diseases and the number of asthma case handled by participants every month, as shown in Fig. 51.3.

Figure. 51.4 shows the number of pneumonia cases handled every month by doctors and the methods used to detect them.

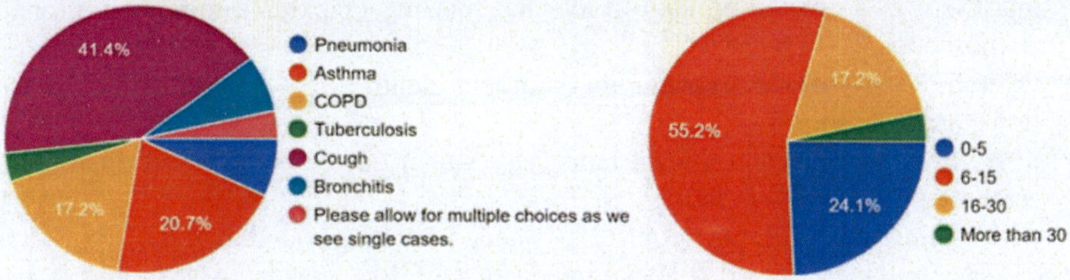

Fig. 51.3 Most common lungs diseases and the number of asthma cases per month

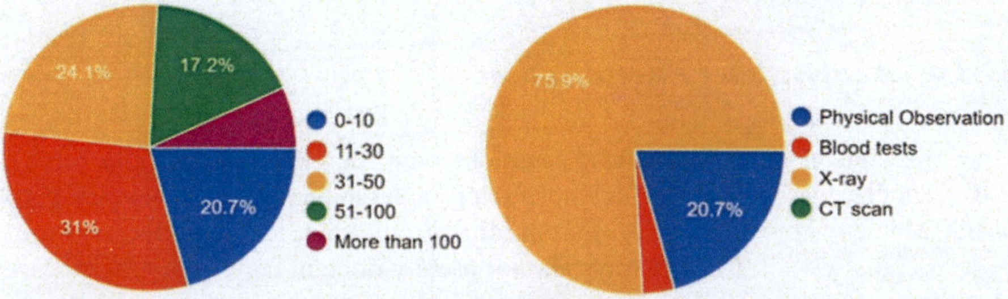

Fig. 51.4 Number of pneumonia case per month and methods used in the diagnosis

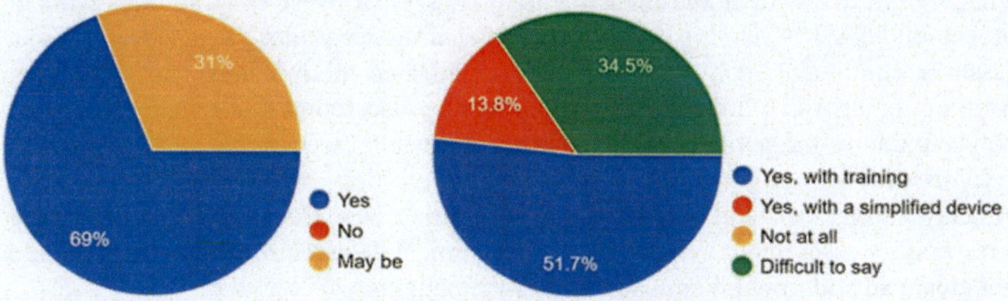

Fig. 51.5 Need for assistive respiratory assessment system and ease of use by semi-skilled healthcare user

In Fig. 51.5, validate the need for an assistive respiratory assessment system for primary care and ease of use by semi-skilled healthcare user.

Thus, after analysing the data from our online survey, we found that most of the cases, the participants came across were cough-related problems followed by asthma and COPD. The important findings are mentioned below:

- 75.9% of participants responded that they rely on X-rays for pneumonia diagnosis.
- 75.9% clinician also mentioned that they come across more than five pneumonia cases per month.
- 20% of doctors with specialization in pulmonology and chest medicine mentioned that there is a chance of identifying and missing crackles through traditional stethoscopes.
- Overall, 24.1% of doctors were not confident identifying a crackle with conventional auscultation.
- 69% clinicians said that a respiratory assessment device could assist doctors in screening pneumonia in PHCs.
- 44.8% clinicians said they require assistance with lung sound analysis.
- 51.7% clinicians agreed that paramedical staff like nurses, ANMs and Asha workers could identify childhood pneumonia cases with a skilling program or with the help of a simple respiratory assessment system.

51.3.4 Analysis and Interpretation

From the clinical immersion and discussions with the clinicians and the survey, it is clear that there is a need for a respiratory assessment system for community health. Majority of respiratory cases come at a chronic stage and require assessment by a specialist doctor. Chest X-rays are not usually done in the primary healthcare centres due to unavailability of the X-ray machine or the technician and due to poor economic condition of patients. In several cases, the rural healthcare centres are managed by ayush doctors or by paramedical staffs who are not trained to do a respiratory assessment and prescribe treatment. Over usage of antibiotics without proper diagnosis is leading to anti-microbial drug resistance. It is observed that essential equipment such as pulse oximeter, infrared thermometer and spirometer were not present in primary care centres. It was also found that telemedicine was not available to the patients in government-run healthcare centres and life support systems such as ventilators are only available in tertiary care centres. It is leading to delay in diagnosis and treatment in most of the cases. Hence, there is a need for a respiratory assessment system to enhance clinical decision making, which should not create an additional diagnostic cost to the patients.

51.4 Conclusion

Due to lack of access to the healthcare facilities, vaccines and medicines, millions of people die every year. These deaths are preventable, and community spread can be stopped. Availability of a respiratory assessment system can help in identifying underlying respiratory conditions early, and patients can be treated before it becomes an emergency. It can also assist in decision making about the next treatment steps and avoid any guesswork. Children can be protected from pneumonia with a simple respiratory health assessment device and can be treated with low-cost medication and care. Thus, from our literature reviews, clinical and online surveys, we found that there is an unmet need for the development of a respiratory assessment system that can be used with existing telemedicine platforms. These devices will become the first line of defence in preventing respiratory disease related to mortality. Due to remote locations and a massive rural population of our country, these innovative devices based on latest technologies will go a long way in providing early detection and last-mile healthcare services to the people. We hope our research findings will help to inspire and push for the development of such novel medical devices that are not only cost effective but easy to use.

References

1. Ferkol, T., Schraufnagel, D.: The global burden of respiratory disease. Ann. Am. Thoracic Soc. **11**(3), 404–406 (2014)
2. Wardlaw, T.M., Johansson, E.W., Hodge, M.J.: Pneumonia: The Forgotten Killer of Children. Unicef (2006)
3. Zar, H.J., Ferkol, T.W.: The global burden of respiratory disease—impact on child health. Pediatr. Pulmonol. **49**(5), 430–434 (2014)
4. Network, G.A.: The global asthma report 2018 (2018)
5. WHO.: Asthma (2020). https://www.who.int/news-room/fact-sheets/detail/asthma
6. WHO.: Chronic respiratory diseases (2020). https://www.who.int/respiratory/copd/burden/en/
7. Salvi, S., Kumar, G.A., Dhaliwal, R., Paulson, K., Agrawal, A., Koul, P.A., et al.: The burden of chronic respiratory diseases and their heterogeneity across the states of India: the global burden of disease study 1990–2016. Lancet Global Health **6**(12), e1363–e1374 (2018)
8. Central_Bureau_of_Health_Intelligence.: National Health Policy 2019. In: Services, D. G. o. H. (ed.). Government of India, New Delhi (2019)
9. Wipf, J.E., Lipsky, B.A., Hirschmann, J.V., Boyko, E.J., Takasugi, J., Peugeot, R.L., et al.: Diagnosing pneumonia by physical examination: relevant or relic? Arch. Intern. Med. **159**(10), 1082–1087 (1999)
10. Mehta, M.: Performing a respiratory assessment. Nurs. 2020 Crit. Care **5**(3), 45–47 (2010)
11. Say, S.D.: Three Critical Components of a Respiratory Assessment Checklist (2020)
12. Auscultation, E.: Respiratory Sounds Reference Page (2011). https://www.easyauscultation.com/respiratory-sounds2020
13. MedlinePlus.: Breath Sounds (2020). https://medlineplus.gov/ency/article/007535.htm2020
14. Abbas, A., Fahim, A.: An automated computerized auscultation and diagnostic system for pulmonary diseases. J. Med. Syst. **34**(6), 1149–1155 (2010)
15. Gurung, A., Scrafford, C.G., Tielsch, J.M., Levine, O.S., Checkley, W.: Computerized lung sound analysis as diagnostic aid for the detection of abnormal lung sounds: a systematic review and meta-analysis. Respir. Med. **105**(9), 1396–1403 (2011)

16. Rao, A., Huynh, E., Royston, T.J., Kornblith, A., Roy, S.: Acoustic methods for pulmonary diagnosis. IEEE Rev. Biomed. Eng. **12**, 221–239 (2018)
17. Jácome, C., Marques, A.: Computerized respiratory sounds are a reliable marker in subjects With COPD. Respirat. Care **60**(9), 1264–1275 (2015). https://doi.org/10.4187/respcare.03922
18. Say, S.D.: What is Involved in a Respiratory Assessment? (2020)

Chapter 52
Context of Design Innovation: Development of Practice Support in Accordance with Design Research Methodology for Enabling Creation of Value

Doji Samson Lokku and Deepak John Mathew

Abstract The context of design innovation refers to harnessing human creativity for 'creation of value'. With innovation pointing to transforming the ideas for success in the market place, it is a reasonable expectation for creativity-based design practice to enable this particular transformation. The review of design innovation literature suggests that the practice ought to be explicit with regard to 'creation of value'. The research study looks at this scope towards developing a practice support for enabling 'creation of value'. The literature on business model refers to the rationale behind creation of value, which may be adapted into design practice. The development of practice support adopts the various perspectives from the literature on business model and introduces the resulting approach as part of standard design phases, in order to address 'creation of value' explicitly. The stages in design research methodology are adhered with, while developing the practice support. An initial study with reference to the context of design startups indicates validation of the developed practice support.

52.1 Introduction

The context of design innovation refers to combining the practice of creativity-based design with innovation. The multidisciplinary definition of innovation [1] refers to transforming the ideas in order to differentiate successfully in the market place. On the other hand, creativity-based design refers to shaping the ideas to make things that are meaningful in our lives. Thereby, there exists a common theme between design and innovation in the form of shaping and transforming of the ideas. The publication by Boztepe [3] succinctly describes notion of value and design.

D. S. Lokku (✉)
Tata Consultancy Services, Hyderabad, India
e-mail: doji.lokku@tcs.com

D. J. Mathew
Indian Institute of Technology, Hyderabad, India
e-mail: djm@iith.ac.in

A. Chakrabarti et al. (eds.), *Design for Tomorrow–Volume 2*, Smart Innovation, Systems and Technologies 222, https://doi.org/10.1007/978-981-16-0119-4_52

The phrase 'creation of value' in this paper refers to wealth or value creation as it happens to be the case with respect to the context of startup firms. The notion of value is both qualitative (which refers to perception of value by a stakeholder) and quantitative (which refers to net benefits accruing to a stakeholder).

The scope of this paper aims at identifying the design concerns that enable 'creation of value' explicitly. Development of practice support to address these concerns is part of the research scope. The research questions pertain to state of the art of the design practice and 'creation of value'. The paper describes the possible approaches for it with an outline of content as per design research methodology [2]. Each of the methodology stages is substantiated with exclusive figures prepared by this paper authors themselves.

52.2 Research Methodology Based upon DRM Stages

The study in this paper adheres with the stages as per design research methodology (DRM) that include research clarification (RC), descriptive study I (DS I), prescriptive study (PS), and descriptive study II (DS II). Research clarification describes the argument for making the practice explicit with regard to 'creation of value'. Since the scope of research includes identifying the concerns that pertain to 'creation of value', a review of literature on business model is undertaken as part of DS I. The understanding gathered is sufficient to undertake development of practice support as part of a comprehensive prescriptive study. The resulting actual support is subjected to an initial DS II for evaluation purpose.

The research clarification consists of literature review in the area of design innovation to highlight the gap in the state of the art of design practice while addressing 'creation of value'. The descriptive study I captured the literature review on business model in an attempt to describe the perspectives on 'creation of value'. The resulting understanding lends a handle towards developing the practice support. The subsequent prescriptive study aims at developing the practice support for enabling 'creation of value'. Lastly, the descriptive study II evaluates the developed practice support with reference to the context of design startups, as these firms have an agenda for 'creation of value' aimed at intended stakeholders.

52.3 DRM Stage: Research Clarification—A Review Study

The study for this stage includes review of literature pertaining to both design and design innovation as well. The point of view while studying these literatures is on 'creation of value' in terms of whether the practice is implicit or explicit in addressing 'creation of value'. Generally, the scope of design includes business offering and the associated value proposition, whereas the scope of design innovation extends further to Value Realization. Accordingly, the former refers to implicit nature of practice,

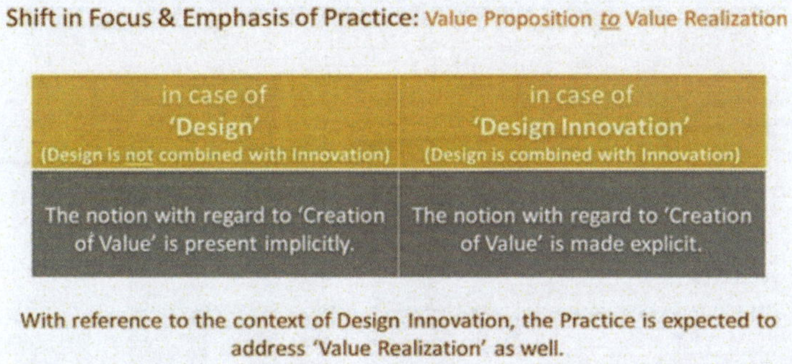

Fig. 52.1 Design versus design innovation—shift in focus and emphasis

whereas the latter is about being explicit with regard to 'creation of value'. The recent publication [6] points to the need for design practice to be explicit with regard to 'creation of value' but it does not put forth an approach for it. Figure 52.1 depicts the shift in focus and emphasis of the practice with respect to 'creation of value'. Most of the literature articles were fetched thru a study during the years 2017–2019.

52.3.1 Current Understanding and Expectation

Most of the authors and researchers who have explored in the area of design innovation have worked on either value of design as a practice, or the definition of value from design point of view, or they have looked at the connection between design and creation of value, etc. As such, these authors and researchers have not explored 'creation of value' exclusively and how design (as a practice) can enable 'creation of value' explicitly.

52.3.2 Success Criteria, Research Questions, Hypothesis

Since the scope of study is enabling 'creation of value', the criteria for success points to effective translation of 'value proposition' into 'value realization'. Also, since the timeline for value translation stretches until the phase 'Go to Market' by startup firms, which is generally of the order of a couple of years from the time business offering is developed, there exists a need for measurable success criteria.

The research questions include what it means for the design practice to be explicit with regard to 'creation of value'. And how the design practice can be explicit with regard to 'creation of value'. These two questions are addressed as part of review-based studies in RC and DS I stages. Additionally, the questions pertaining to support

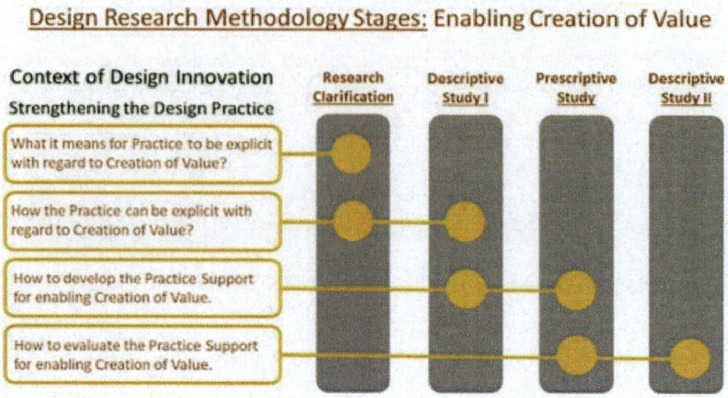

Fig. 52.2 Research questions in each of the design research methodology stages

development and evaluation are also included as shown in Fig. 52.2. These questions are addressed as part of PS and DS II stages.

Towards hypothesis, it is stated that the design practice can be explicit with regard to 'creation of value' by addressing the stages or building blocks in 'creation of value'. The subsequent study in adherence with DRM has aided greatly in undertaking this particular design research as part of doctoral studies in design.

52.3.3 Areas of Relevance and Contribution

The topic of 'value' is a subject matter in economics. Heskett [7] who fervently talked about design and creation of value is an economist himself. His notes on design and creation of value has been published by his colleagues Dilnot and Suzan [8]. Sharon [15] suggests that it is initially up to the designers to understand economics as lay people. In addition, perhaps some designers in collaboration with economists to take Heskett's work further, to model more fully the 'creation of value' from a united economic and design perspective. Sharon advocates for design to take the initiative to remedy economic blind spots, as well as for economics and design to work out collaboratively to yield an integrated, useful understanding with regard to 'creation of value'.

On the other hand, a business model refers to the rationale behind creation and capture of value. Several authors have addressed this subject matter of whom business model canvas [14] is most popular. Subsequent authors [5] have organized this canvas elements in the form of stages or building blocks for 'creation of value'. Accordingly, these stages offer a handle for design practice by which it can enable 'creation of value'.

Chesbrough [4] describes a business model in terms of six functions. He states that a business firm with a superior business model and inferior business offering would do well compared to a business firm with an inferior business model and superior business offering. One can infer from his writings that the function of a business offering is to carry certain potential value, whereas the function of a business model is to aid in realizing this value.

Johnson and Christensen [12] describe a business model in terms of four key components. The authors describe value proposition in combination with the intended customer segment. Their emphasis on business viability reflects in the suggested effort to detail the overall profit formula. The perspective broadly covers all the important elements that point towards 'creation of value'.

52.3.4 *Type of Research and Research Plan*

DRM refers to several types of research with various combinations of study as part of its stages, namely RC, DS I, PS and DS II. Among these, RC is always a review-based study. For the research topic that is chosen here which is about enabling 'creation of value', the theoretical basis is found in business model literature. Accordingly, a review-based study on business model is undertaken as part of DS I. The resulting understanding is sufficient to undertake support development as part of PS consisting of a comprehensive study. The DS II includes an initial study towards evaluation of the developed support.

Overall this amounts to a research type wherein both RC and DS I are review-based studies, and PS is a comprehensive study, accompanied by DS II as an initial study. As part of DRM, this amounts to research type 3.

Accordingly, the research plan accounts for review-based studies to arrive at the theoretical basis for support development. This is followed by a comprehensive study that includes relating the understanding on design innovation to the insights from business model literature. Keeping in view of the topic of study, the evaluation of the developed support is carried out with reference to startups, whose generic mandate includes 'creation of value' for the targeted stakeholders. The scope of evaluation is limited to design startups because of their ability to undertake creativity-based designs of their respective business offerings in adherence with standard design phases.

52.4 DRM Stage: Descriptive Study I—A Review Study

The review-based study undertaken as part of DS I includes both the design innovation literature and business model literature.

52.4.1 *Review of Literature on Design Innovation*

An elaborate description of innovation perspective on design, including the policies, is published by Hobday [9–11]. An overview of design influences on innovation is covered in the publication on design innovation spectrum [13]. A key reference is a review publication by Ricardo [6]. As per these authors, design as a practice needs to be explicit with regard to 'creation of value'. Another key publication is a book review by Sharon [15] on 'design and creation of value'. Based upon these review-based studies, it is inferred that the current practice of design has focus and emphasis predominantly on 'value proposition' of a given business offering. Accordingly, the designed offer is likely to have chances for 'creation of value' that are taken for granted. Which means, the actual translation (between value proposition and value realization) towards 'creation of value' may or may not happen.

52.4.2 *Review of Literature on Business Model*

Business model represents the rationale behind 'creation of value'. The key reference is business model canvas [14]. It consists of 9 elements, detailing of which will result in description of the business model. Further to it, Gunzel and Holm [5] have organized these 9 elements into four blocks or stages for 'creation of value', namely value proposition, value creation, value delivery, and value capture. Of which, the last 3 stages combined together refer to the scope on 'value realization'.

Based upon this review-based study, it is inferred that the extended context of design innovation expects the practice to address 'value realization' as well. Thereby, the designed offer to have better chances for 'creation of value'. These scenarios are captured as part of reference model (shown in Fig. 52.3) and impact model (shown in Fig. 52.4), respectively. The proposed practice support is shown in Fig. 52.4 in a hexagon.

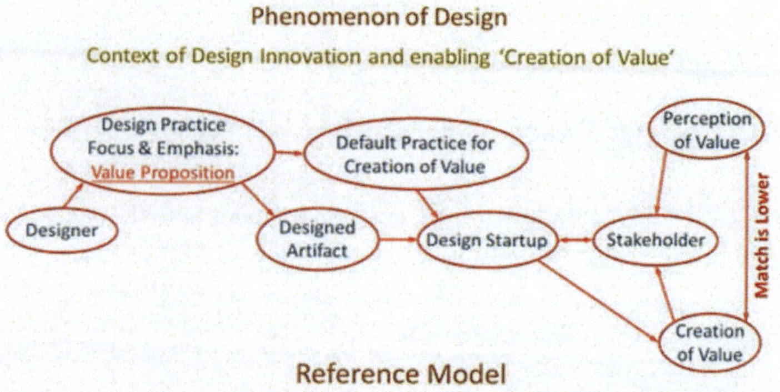

Fig. 52.3 Reference model—context of design startups aimed at creation of value

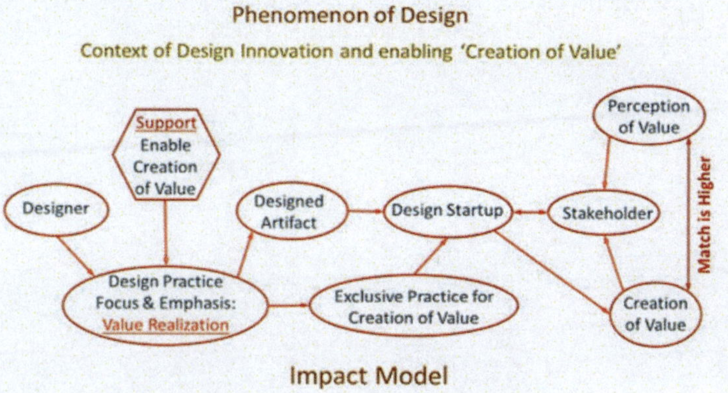

Fig. 52.4 Impact model—context of design startups aimed at creation of value

Both these models refer to the context of design startups whose generic mandate is 'creation of value' to an intended stakeholder. The match between 'perception of value' and 'creation of value' is expected to be better if the focus and emphasis of design practice extends to 'value realization' as well.

A better match between perception of value (by the stakeholder) and creation of value (for the stakeholder) is the goal behind enhancing the design practice.

52.4.3 Overall Summary of DS I

The extended shift from design-to-design innovation can be enabled by enhancing the practice for its shift in focus and emphasis towards 'value realization', in addition to 'value proposition'. The corresponding design concerns are identified in business model literature. These include the stages in 'creation of value', namely value creation, value delivery, and value capture.

52.5 DRM Stage: Prescriptive Study—A Comprehensive Study

The insights from DS I are adapted and made use of to develop the support for strengthening the practice towards enabling 'creation of value'.

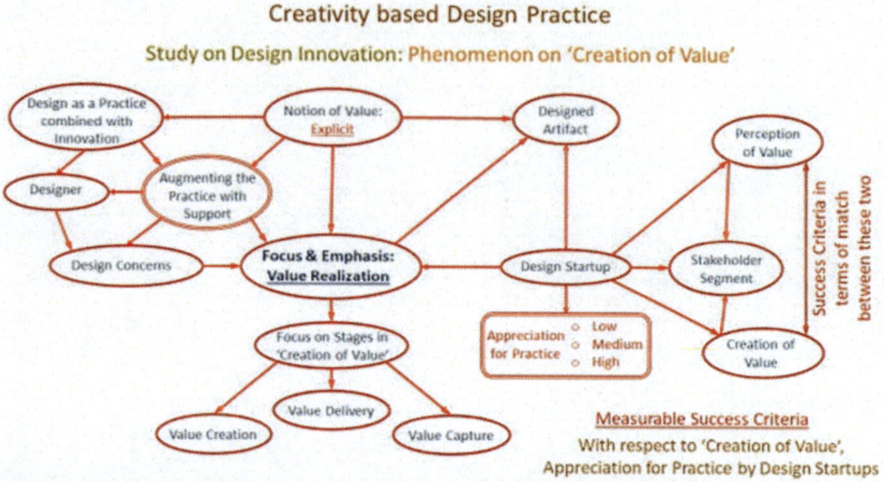

Fig. 52.5 Support model—context of design startups aimed at creation of value

52.5.1 *Task Clarification and Conceptualization*

The stages in 'creation of value' are adapted from business model literature towards the design concerns that need to be addressed by the practice for enablement purpose. Accordingly, the practice focus and emphasis extends to include the stages, namely value creation, value delivery, and value capture as shown in Fig. 52.5.

52.5.2 *Support Realization and Introduction*

Towards introducing the support as part of practice, the stages in 'creation of value' are put through the standard design phases as shown in Fig. 52.6. Accordingly, each

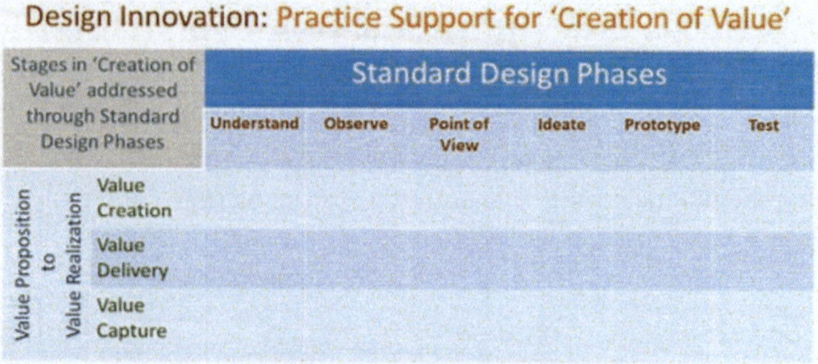

Fig. 52.6 Practice support—introduced through standard design phases

of the stages, namely value creation, value delivery, and value capture, is addressed by the practice to arrive at corresponding designs that enable 'creation of value' explicitly.

52.6 DRM Stage: Descriptive Study II—An Initial Study

The measurable success criteria stated in Fig. 52.5 has been made use of as part of DS II. This includes interaction with design startups to measure the effectiveness of the developed Support in terms of their appreciation for enhanced practice.

52.6.1 Evaluation Focus and Plan

Since the topic is about enabling the practice for 'creation of value', the research plan includes initial validation of the Support with respect to the context of startups whose agenda include 'creation of value'. Also, since the core of design practice involves human creativity, design startups were chosen which are capable of undertaking designs in adherence with standard design phases. Accordingly, the appreciation for design practice by the Startups is measured towards initial validation of the support.

Startups usually get involved with presentation pitches to prospective investment firms in their efforts to raise funds. The preparation for the startup pitches can benefit from leveraging the developed Support. Accordingly, the appreciation for practice is measured through an interaction with design startups as they resort to investor pitches. These initial interactions have pointed towards medium to high appreciation for the enhanced design practice consisting of the developed support.

52.6.2 Evaluation Studies

A series of interactions are planned with design startups whose outcome is yet to be published in the form of case studies. Overall, the initial evaluation indicates good prospects for the developed Support for use by startups.

52.7 Summary and Conclusion

This study has covered the extended shift from design-to-design innovation, which has resulted in an expectation that the practice ought to be explicit with regard to 'creation of value'. This particular expectation translates into a shift in focus and emphasis of the practice on not only 'value proposition' but also on 'value realization' as well. A handle from business model literature is identified and adapted towards design concerns that are to be addressed by the practice in order to enable 'creation of value' explicitly. These include the subsequent stages in 'creation of value' in addition to value proposition, namely value creation, value delivery, and value capture, which have become the basis for support development. As part of introducing the support, these stages are put through standard design phases in order to arrive at the corresponding designs that would enable 'creation of value' explicitly. Further, the support is subjected to initial validation by applying it with reference to the context of design startups and measuring their appreciation for the enhanced practice. The initial results are indicative of better appreciation. Elaborate case studies in this regard are planned for future publication.

References

1. Baregheh, A., Rowley, J., Sambrook, S.: Towards a multidisciplinary definition of innovation. Manage. Decis. **47**(8), 1323–1339 (2009)
2. Blessing, L.T., Chakrabarti, A.: DRM: A Design Research Methodology. Springer, London (2009)
3. Boztepe, S.: The notion of value and design. In: Journal of the Asian Design International Conference, Vol. 1 (2003)
4. Chesbrough, H.: Open Business Models: How to Thrive in the New Innovation Landscape. Harvard Business Press (2006)
5. Günzel, F., Holm, A.B.: One size does not fit all—understanding the front-end and back-end of business model innovation. Int. J. Innov. Manage. **17**(01), 1340002 (2013)
6. Hernández, R.J., Cooper, R., Tether, B., Murphy, E.: Design, the language of innovation: A review of the design studies literature. She Ji: J. Des. Econ. Innov. **4**(3), 249–274 (2018)
7. Heskett, J.: Creating economic value by design. Int. J. Des. **3**(1), 71–84 (2008)
8. Heskett, J.: Design and the Creation of Value. Bloomsbury Publishing (2017)
9. Hobday, M., Boddington, A., Grantham, A.: An innovation perspective on design: Part 1. Des. Issues **27**(4), 5–15 (2011)
10. Hobday, M., Boddington, A., Grantham, A.: An innovation perspective on design: Part 2. Des. Issues **28**(1), 18–29 (2012)
11. Hobday, M., Boddington, A., Grantham, A.: Policies for design and policies for innovation: contrasting perspectives and remaining challenges. Technovation **32**(5), 272–281 (2012)
12. Johnson, M.W., Christensen, C.M., Kagermann, H.: Reinventing your business model. Harvard Bus. Rev. **86**(12), 57–68 (2008)
13. Na, J.H., Choi, Y., Harrison, D.: The design innovation spectrum: an overview of design influences on innovation for manufacturing companies. Int. J. Des. **11**(2), 13–24 (2017)
14. Osterwalder, A., Pigneur, Y.: Business Model Generation: A Handbook for Visionaries, Game Changers, and Challengers. Wiley, New York (2010)
15. Poggenpohl, S.H.: Blindspots in economics and design: a review of John Heskett's design and the creation of value. She Ji J. Des. Econ. Innov. **3**(4), 251–261 (2017)

Chapter 53
Designing Brand-Led Cultures in Small and Medium-Sized Enterprises (SME's) Struggling Through Midlife Crisis

Isha Patel

Abstract Small and medium-sized enterprises are engines of employment and innovation, contributing to 95% of businesses in most countries. The world of branding is dominated by large businesses, but there exists a lack of SME perspective on the subject (Berthon et al. in J Small Bus Manage 25–45, 2008, [1]; Krake in J Prod Brand Manage 14:228–238, 2005, [8]; Wong and Merrilees in A brand orientation typology for SMEs: a case research approach, 2005, 15). This study investigates into challenges of SMEs that limit their understanding and use of brand-led strategies and states how SMEs can use their own brand to enhance longevity and sustenance for their companies. It will therefore benefit SMEs across the globe, irrespective of what sector they belong to. As observed, SMEs have successfully challenged large corporations by developing market-focused brands in the last decade. The UK Design Council has also noted that SMEs adopting a design-led approach are more successful than those who do not. It therefore becomes essential to identify the challenges and overcome them through further research. This research follows a qualitative approach and expresses a metaphorical relation between humans and SME brands. Brands are personified and their life cycle is illustrated. SME struggles are compared to midlife crisis in the human life cycle. It can be said that SMEs are in the same ambiguous state of mind as humans facing the advent of midlife crisis. The recommendations made in this study will equip SME leaders to understand their organization from a grassroots level and benefit through development of a brand-led culture within their organization, making them competitive through strategic use of their own brand. This study is attending to identify a much larger gap in the industry, raising a conversation that has potential for further research.

I. Patel (✉)
ISDI Parsons Mumbai, Mumbai, India
e-mail: isha9194@gmail.com

Brunel University, London, UK

A. Chakrabarti et al. (eds.), *Design for Tomorrow–Volume 2*, Smart Innovation, Systems and Technologies 222, https://doi.org/10.1007/978-981-16-0119-4_53

53.1 Introduction

The world of branding is dominated by large businesses and brands play an important role in consumers' daily lives. Small business is big business and has been for the past four decades as the category has been experiencing a conducive market environment that is highly competitive [9]. Successful SMEs are capitalizing on new opportunities of technological advancement, adapting to changing trends and meeting consumer demands in a rapidly changing world. In the heydays of branding, from 1950 to 1970, mass manufacturing economy rewarded big brands with profits and big companies did well. This remains a perceived notion even today as the best company rankings we see today comprise of only big brands. In the recent years with advancement in privatization, globalization, and technology, mass manufacturing has been replaced by customization, leading to growth in SME businesses. With a shift in focus from big brands to SMEs, it becomes important to investigate their positioning in the world of big brands. Branding has been studied by economists, marketers, designers, psychologists, philosophers, and cultural critics as research and practice. Yet, very few disciplines have understood the concept of branding and leveraged it to its full potential. Branding is much more than just the identity of a product. It gives people a reason to believe in something larger than themselves: an idea, an experience, and a perception of a personality to identify with. If a brand is what a company stands for, branding is the activity of conveying the company's product or service. A brand is the cause and branding is the effect [7]. Although brands are traditionally measured in numbers, to the consumer, it is the brand's value that makes him attract or repel it. The rise in power of markets and materialism has increased the demand for meaning in people's life, as consumers and as employees. This has led to a shift in meaning of branding from a product to a concept. Today, brands enrich cultures, give people new sense of belonging and identity, and add value to businesses and lives of people, both socially and culturally [7]. Although this kind of brand adoption is common in large organizations as corporate culture, SMEs have been resistant to adopting it. While large organizations are brand-led and are adopting an approach that is termed brand orientation, most SMEs see brand as one of the many factors in business. In recent years, only few SMEs have successfully challenged large corporations by developing market-focused brands. Burson-Marsteller, published its annual 'Crisis Survey' that we are living in a disruptive era with communication facing stormy challenges. Airbnb, Uber, and Deliveroo are the biggest success examples of disruptors as they have acquired leadership and consumer appreciation in the marketplace. With design-led branding at their cores, these are successful examples of small business turned big, realigning the market in their respective industries to gain leadership. SMEs are engines of employment and innovation. However, as few as 50% of firms founded in 2011 actually survived beyond the five-year mark. It is therefore vital to understand the real challenges and opportunities that can shape the future of SME businesses. There exists a lot of literature on SMEs and branding, respectively, but the integration of both is relatively new subject [15].

53.2 Methodology and Research Process

Adapting the double diamond framework, a balance in strategic and creative thinking is achieved [3]. Since SME branding is an emerging area of research, a qualitative approach allows for the development of insights into the phenomenon to be studied. Every SME is different and will have its own viewpoint and understanding of branding. Bearing this in mind, a mix of qualitative primary and secondary methods was selected to gain an unprejudiced view on the subject (Fig. 53.1).

Pilot Survey

A well-structured and designed questionnaire was shared with 50 SMEs globally to gain an insight in the subject and understand the current situation of the concept of branding in SMEs across the globe. The sample group was handpicked, comprising of SME owners, founders, and employees, equally balanced. However, the responses were negative in nature, with just few word answers that represented confusion in understanding terms within business branding marking clear evidence of lack of interest, understanding, and awareness of branding in small and medium businesses, motivating further research.

Literature Review

A total of 18 books on branding, 26 journal papers, three reports, four blogs, five lectures, and multiple Web searches (expert blogs and company websites) were reviewed to give a rationale to this study with focus on the subjects of branding, small and medium-sized enterprises, and theories on brand culture.

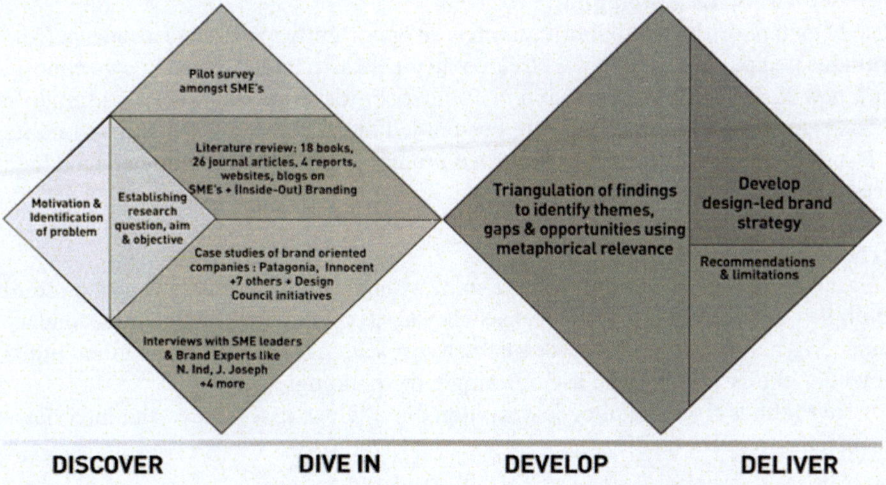

Fig. 53.1 Research plan. Adapted from Design Council's Double diamond, 2015

Case Studies

In this research, 12 case studies were carried out and analyzed. Whilst most of the case studies were carried out through secondary data, interviews for four of the SME companies were further conducted and insights from SME leaders were gathered through semi-structured interviews. The case studies were chosen on the basis of their use of branding in business as an example of success or failure for topics discussed in the study. They included companies such as Innocent, Heck Sausages, Patagonia, WeWork, Ka Sha, Airbnb, and Warby Parker.

Semi-structured Interviews

Four SME leaders and seven experts were interviewed from across the globe. It was a carefully handpicked list of experts, professors, consultants, and authors in the field of internal branding and brand culture. They were Nicholas Ind (professor and author of over six books on branding), Jerome Joseph (also known as the Guru of Branding in South Asia), Allan Steinmitz (founder, Inward consulting, USA, with over 20 years of experience on the subject), Ian Ferris (Design Associate, Design Council, UK and Innovation Director for Co-Innovate at Brunel University, UK), Alejandro Masferer (Faculty at various universities and founder, Trigger Cards), Rajan Shah (Brand Consultant and CEO, Shah & Partners, NYC), Sahil Sachdev (Strategy Director with over 10 years of experience globally and Brand Head, Quiqup), Karishma Shahani Khan (Fashion Designer and founder, KaSha, India), Dhruv Paknikar (Founder, Dominix, India and Dubai) and Peter Van Groos (Techpilot, Germany). This mix of experts' insights on various subjects is discussed in the study.

Personification as a Metaphor

Use of metaphors in qualitative research is an opportunity to examine a concept from a unique perspective. It provides structure to the data, to understand a known concept in a new light [2]. In this research, a comparison of struggling SMEs and midlife crisis is made. Struggling SMEs are personified in the context of the human midlife crisis, with an aim to evoke non-creative professionals to identify, understand, and act on the problem at hand.

Triangulation

This method establishes validity in the study by analyzing the research question from multiple perspectives to arrive at consistency across gathered primary and secondary data. Originating in the 1950s as a qualitative research technique, this method limits potential biases arising from use of a single methodology [5].

The findings from the pilot survey, literature review, case studies and interviews are discussed in Sect. 53.3.

53.3 Findings

53.3.1 Design and Branding Are Low Priorities in Most SMEs

SMEs are an outcome of a great idea, but over time there is a lack of clarity in translating the idea to its implementation. In technical terms: the idea is innovation, the implementation is design, the selling process is marketing and advertising, and the experience it creates for its prospects is branding. The confusion in understanding the differences in these terms leads to fewer successes as companies cannot leverage the great assets they hold. Design is often perceived in an aesthetic sense at product level and not as creative thinking technique, also known as design thinking (IDEO n.d.). There has always been a survival mentality in SMEs. However, with globalization and emerging technologies, there has also been a move toward adopting design and branding in SME businesses. This movement has been slower than the industry demands. Even though large organizations and a few SMEs have been able to leverage this change and have become disruptive brands, most SMEs have not been able to understanding branding and its power. Branding is considered an investment to advertise the company. Branding is perceived as a logo and collaterals to be circulated in order to increase sales and awareness, which can be noted as a marketing activity since it follows an outside-in approach building an outer image, without identifying and leveraging the brand's core purpose. It can be concluded from the research findings and interviews conducted in this study that SMEs perceive design and branding as a short-term cost and not as a long-term investment of time, finances, and resources. It is considered a luxury if a brand can spare finances for brand building [11]. Organizational management and culture are poorly strategized and designed in SMEs and are often based on the leader's vision and gut feeling. The focus in completely on making profits by sale and marketing and is outside-in, which makes for poor or weak communication strategies and is limited to an exterior image— often different from what the brand really stands for. Most of the focus, talent, and time is spent on short-term activities. Branding, in most SMEs is not prioritized and less or no funds are available for investing in design and branding practices like brand strengthening, positioning, building an integrated internal culture, recruitment, mentoring, and conducting regular audits to keep the firm's equity maintained (Fig. 53.2).

53.3.2 Lack of Focus on the Fundamentals of Business

Secondary sources from reviewed literature have clearly stated that SMEs have a poor articulation of their business, leading to poor delivery. This is a circular process and needs further attention. In his book *Start with Why* (2009), Simon Sinek explains how the human brain works. Companies are in business to deliver to their customers and to improve business performance, they need to understand the mental model of

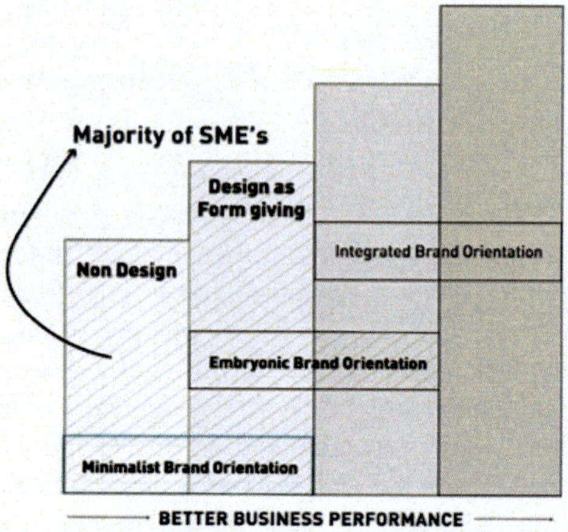

Fig. 53.2 Adapted from Danish Design Ladder (2001) and Wong and Merillees Brand Orientation Ladder (2005)

consumers. Hence, the articulation of why a brand's purpose and core values are important and the need for further development in the subject is required as they are closer to their consumers than most corporations, owing to their smaller size. Sinek's concept discusses a simple yet powerful principle of success. It all boils down to the leader's mentality and ability to inhibit change. If the leadership is functional and fundamentally strong, internal branding is strong and employees are happy to work at the company, increasing productivity and efficiency. This also enhances collaboration and cultivates a creative culture that aims to solve problems innovatively rather than superficially. The decision making of the company that follows in inside-out approach will be vested in the employees and the leaders collectively, with holistic rather than individual benefits and leader's vision alone. An inside-out approach is open, flexible, and keeps relevance to changing trends and competition as it is not just the leader or leaders but also the employees' strengths that are responsible for the brand's success and failure. The employees in the company will therefore work because they believe in the brand and not solely for financial benefit (Fig. 53.3).

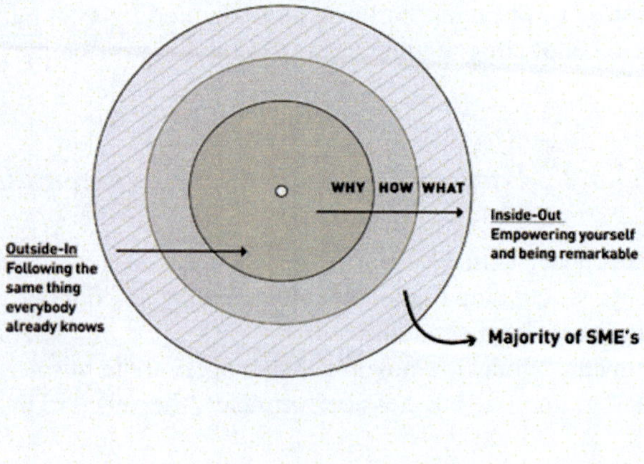

Fig. 53.3 Adapted from Simon Sinek's 'Start with Why' (2009). The Golden Circle poses the question 'why?' to identify the purpose, 'how?' To implement the purpose, and 'what?' To sell. Most SMEs remain on the outer ring of this circle, unable to articulate the brand's purpose

53.3.3 Poor Leadership

Every source of research for this study has widely identified leadership as a root cause for an SME to accept or resist change, impacting its ability to grow and succeed. Due to the small size of the company, an SME will aspire from the vision, perception, and leadership of the managing person unlike larger organizations where a corporate culture is developed as a discipline for one and all. As the business grows, it becomes the leader's job to personify the company as an extension of his personality to spread the 'why' of the brand to his employees for the company to grow [10–12]. He is responsible for conducive settings to ensure that employees are at the core of the company's decision making as it is their implementations that produce an outcome and their productivity depends on their engagement and motivation. Appreciation and participation play a huge role in efficiency and can only be initiated by leadership of the company. Adaptation is key. It is up to the leader to take his employees with him and not beneath him to build a closely knit, well-functioning, coherent team which is culturally brand-led to be innovative, disruptive, and successful in a competitive world.

53.3.4 Majority of SMEs Do Not Refer to Theory and Research

Nicholas Ind mentioned that he has not come across any SMEs that read research or published articles or follow any resources or frameworks. In the case studies and SME interviews conducted in this study, it was found that SMEs rely on the leader's directions, which are an extension of his gut feeling. Leaders and employees do not immerse in research activities and existing theoretical and practical tools that can give them a push in the right direction. They find academic tools 'tedious'. Owning to the limited time, resources, and talent in the organization, an ignorance is exhibited toward constantly evolving consumer demands, advancing trends, technologies, and relevant case studies that could benefit and ease their business' operational and functional activities, therefore making short-sighted decisions.

53.3.5 Brands Are Personified Humans

From the findings, it could be said that SME brands, if metaphorically compared to humans, behave in similar ways. They can be personified as humans as they have a character and personality of their own, unique from any other brand, which is the reason for their existence. Like humans, brands tend to have a life cycle, a voice of their own to communicate, and an intuitive and intrinsic ability to grow which is led by character and personality. They have an emotional, physical, cognitive,

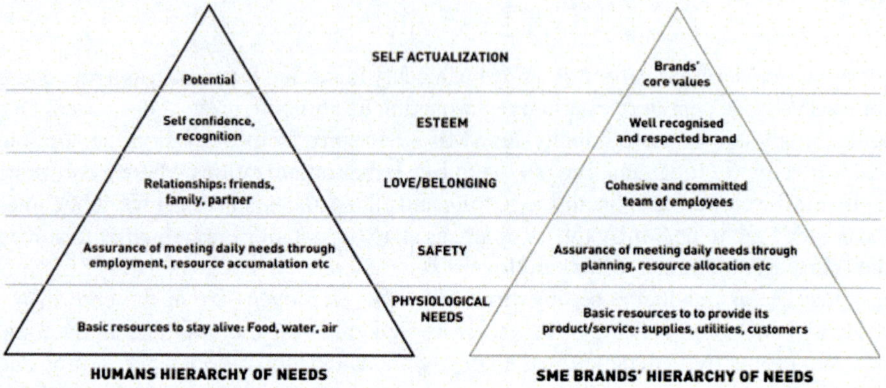

Fig. 53.4 Adapted from Maslow's Hierarchy of needs: individuals versus small businesses. *Source* New venture mentor (2018)

and intellectual aspect that is determined by the values they believe in. Recently, it has been noted that few successful brands have started to display more human characteristics and sound less like businesses [4,13]. Brands are made by humans for humans and for human survival. The common factor that connects brand to humans is the human connection. Like a person has a story to tell that is based on his experiences, brands have their own story to tell based on their values and purpose. Humans make connections with each other by connecting emotionally over stories, experiences, and meanings that their relationship silently creates for themselves to be able to evolve in that space. Similarly, brands can only connect with humans through experiences and storytelling that provide meaning and accumulate a high degree of loyalty by becoming like a religion or a culture. The stronger a person's personality, the stronger is his impression amongst people. In the case of brands, the stronger the vision of the business, the more distinguished its culture becomes. For this research, the personification of brands is limited to SME brands. Due to their smaller size and non-corporate existence, they have more human qualities than large organizations which are evolved entities on their own, having their own corporate personality. An SME is an extension of a leader's personality and vision and can therefore be more closely related to human behavior (Figs. 53.4 and 53.5).

53.3.6 *Midlife Crisis*

Dr. Elliott Jaques (1965) explains midlife crisis as a psychological stress occurring in the middle age of a human's life, triggered by a physical, occupational, or domestic event. It tends to be a triggered realization that his life has passed the halfway point and the experience of change. It is an introspection of oneself, taking into account life until now and the future ahead, often in a negative light. Psychologist Carl Rogers explains that midlife crisis occurs when the gap between the congruence

SME CRISIS		MID LIFE CRISIS
Blur vision and mission of brand	⟷	Lost sight of goal
Lack understanding of 'why' they do business	⟷	Lost purpose
Lack articulation of brand ethos	⟷	Confused and unclear on direction
Confused with 'how' to achieve goal	⟷	Debating on what action to take
Lack of empathyand consumer feedback	⟷	Working on autopilot mode
Unclear marketing and advertisement	⟷	Making big changes that arent relevant
Dragged culture within the company	⟷	Everything feels like a chore
Survival mentality	⟷	No longer playing To win, But to not lose

Fig. 53.5 SMEs are in the same ambiguous state of mind as a human facing the advent of mid-life crisis

in a person's actual behavior and his ideal self grows larger. In simpler words, it is a state of disorientation in a person's life. As much as it is a turning point in a persons' life, it is perceived as negative because of the use of the word 'crisis', but it can be turned into an opportunity to take responsibility for learning about oneself and by making conscious decisions rather than impulsive ones [14]. Having established a similarity in humans and brands, it can be concluded that brands and human follow life phases which can be quite similar. As a typical phase, midlife crisis is not only confined to humans, but also brands too. Findings suggest that it becomes difficult for a company to sustain itself and evolve after being established into a structural organization. A reason for this is an unarticulated organizational and operational approach that conveys unclear messages to prospects. As an SME matures, it too suffers from midlife symptoms like anxiety, disorientation toward their future, denial, depression, and anger in its own way. SMEs specifically suffer from an ambiguous state of mind as they reach the tipping point from which they either progress or begin to wither. It is in this phase of introspection into the company's brand that it needs to become aware of the crossroads the company is at and how gaps can be filled holistically.

53.4 Triangulation

Gathering findings found from primary and secondary sources, discussing and analyzing them, it can be concluded that:

1. SMEs need a well-defined and refined meaning of branding that is clearer than the many definitions available. In simpler terms, they need to understand the purpose of their business and how neatly and innovatively they can communicate this to their audience through experiences.

2. Available research, theory, and books do not focus on SME branding, but on branding for large organizations, also termed as 'corporate culture'. Also, SMEs do not refer to research as it is time-consuming and tedious. It requires expertise to understand the technical use of language that is most often outside their field of knowledge. They need low-investment, user-friendly, and easy to adopt tools that can be a combination of theory and practice, owing to their dynamic nature.
3. SMEs are incompetent of and unwelcoming to radical change. The solution should therefore be gradual in nature, bringing a slow transformation, which can be more easily acceptable by all in the organization.
4. Leadership is the first point of adaption and evolving for SMEs, as the company is often an extension of a leader's vision. It is important for leadership to be persuaded that he needs to be introspective, identifying and dealing with problems and gaps in the existing process in order to have the employees follow. Change trickles top-down in SMEs.
5. SME leadership and employees are not always creative-minded professionals and need to be triggered to think distinctively, not analytically but creatively. The solution should therefore be provocative in nature for more efficacy and efficiency amongst the company.
6. The solution needs to be simple but immersive in its content, targeting every aspect of human and brand decision making.
7. SMEs are similar to humans in behavior and can find practical solutions that lead to progressive theories if personification as a theory is used to seek answers.

It can be concluded that if SMEs understand the relationship between the functioning of brands and humans better, they will be able to make more impactful decisions. It can also help to bridge the gap between understanding the vast amount of research on branding on the one hand and the less available information on SME branding and culture on the other hand. From this point onwards, this paper will therefore focus on explaining the human brain before making any recommendations that link together humans and brands. The circular framework in below in Fig. 53.6 represents the human brain and how it thinks and reasons, and its adaptations to humans and personification to brands, which, again impacts the human brain to be decisive on the experience it has had with the brand. Hence, a human-centered approach needs to be the basis for developing brands and evoking a brand culture around it. As established earlier, culture is formed as a reflection of the brand and it governs the brand space of that company. Everything that a company buys or sells is for consumers to buy. Therefore, the understanding of human behaviors, needs, demands, reactions, and actions is important for companies to be brand-oriented in their strategies. Brands play an important role as a differentiator and should translate the uniqueness with a human-centric approach. By doing so, they also add value and meaning to the offering that a consumer can relate to, influencing their buying decision. The product/service is then used or not by the consumer, depending on the experience he has with the brand. It is a chain reaction, by humans, for humans with a human personification (brand). This circular ecosystem is a continuous process and it is important to target existing gaps in this ecosystem, here, the midlife crisis.

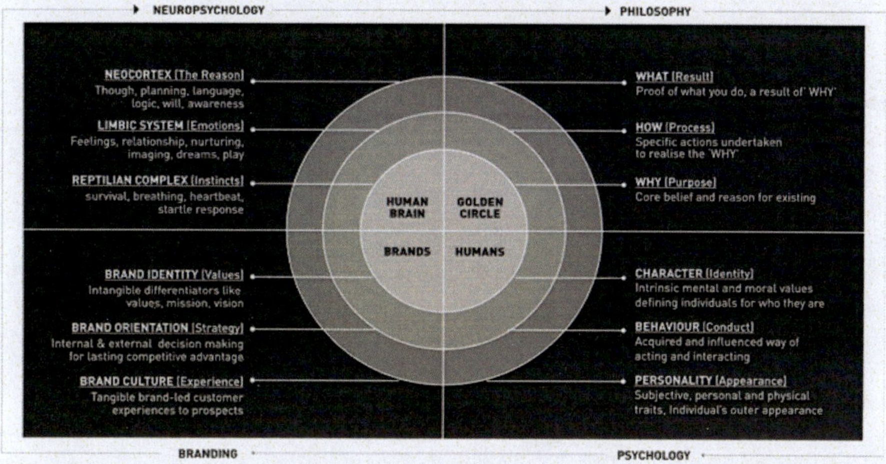

Fig. 53.6 Illustrating the relation between branding and humans by analyzing the relationship of how human brain works and can affect and be affected by branding approach

53.5 Recommendations and Conclusion

53.5.1 Recommendations for Toolkit Development

Based on the previous discussion and the synthesized framework showing the translation of values from the human brain to brands and back to the brain, recommendations for a strategically designed toolkit have been established. This toolkit should aim to foster brand-led cultures in SMEs struggling through midlife crises. Branding is a holistic process and focusing on only one aspect risks overlooking others. Therefore, the solution is a strategy toolkit with distinct phases, each connected and translating seamlessly to a recommended strategy. Each aspect of the strategy is a metaphor to midlife crisis. The terminologies, content, and visual outlook are aimed at identifying, understanding, and acting on midlife crisis. The toolkit should consist of thought-provoking, discussion-evoking questions for team members within the SME. Exercises should act as triggers for rediscovering the company's core values subconsciously. It aims to take the pressure off hierarchy and to bring members within the organization closer, to answer questions that can help all identify the organization's personality, and be at the same level of understanding. It will build, evolve, and define the brand for the SME. The toolkit is a work in progress but the flow of the toolkit would be as shown in Fig. 53.7. The metaphor of a midlife crisis which is commonly understood by most people is strongly used. The tone of voice of this toolkit will reflect that as shown in Fig. 53.7.

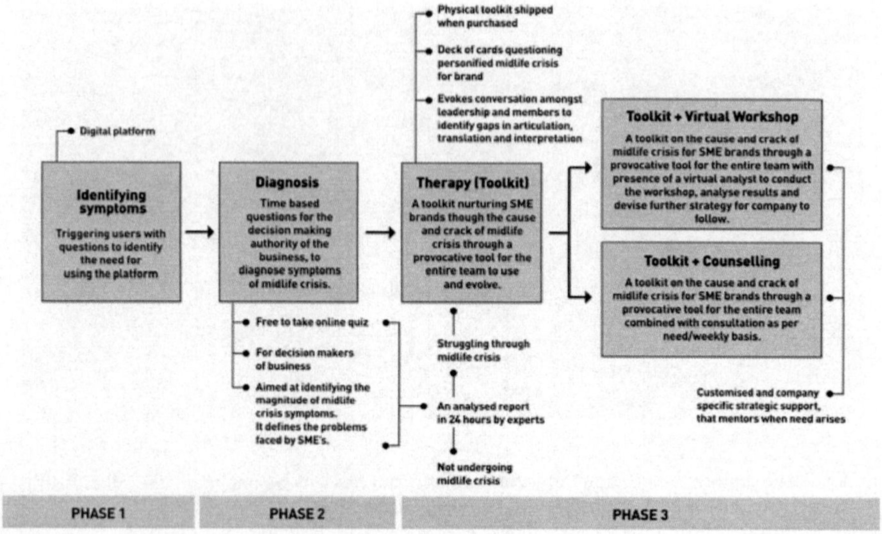

Fig. 53.7 Work in progress flow of the toolkit that is currently being expanded into tools and resources. Furthermore, a system will be developed

53.5.2 Recommendations for Future Research

In future research building on findings from this study, the following recommendations can be taken into consideration:

1. The study can be further developed, especially the recommendation section as it has potential to be a fully functional practical solution to integrate the use of design and branding in SMEs.
2. A psychologist's opinion can further be implemented to develop strategic questions for the recommendation.
3. The recommendation strategy needs to be validated further by experts and an experimental try-out can be conducted.
4. The concept of midlife crisis in SME brands is relatively new and has potential to become a trend and a solution for businesses to realize the importance of being brand-led and culture-driven.

53.5.3 Conclusion

This study is a result of lack of study in the subject of SME branding, despite the growing importance. Through primary and secondary sources of data collection, it can be concluded that there exists a lack of research in the sector and that the concepts of branding are ill defined for SMEs. They do not base their business' decision making on existing research, but it is an extension of the leader's vision and personality. The study has used metaphorical relevance to develop a recommendation strategy personifying SME brands to identify their struggles with midlife crises. The recommendation is a conversational strategic toolkit provoking SMEs to identify the situation and take action, first as a leader and then as a team to evoke discussions and creatively develop a brand for themselves that can become a culture for employees and consumers to co-exist.

Referencess

1. Berthon, P., Ewing, M.T., Napoli, J.: Brand management in small to medium sized enterprises. J. Small Bus. Manage., 25–45 (2008)
2. Carpenter, J.: Metaphors in qualitative research: shedding light or casting shadows? Res. Nurs. Health, 274–82 (2008)
3. Design Council.: The Design Process: What is the Double Diamond? Accessed 2018 (2000). https://www.designcouncil.org.uk/news-opinion/design-process-what-double-diamond
4. Divakaran, S.: 3 Great examples of brands that acted like humans. Digital Uncovered. Accessed Aug. 2018 (2016). https://digitaluncovered.com/3-great-examples-brands-acted-like-humans/.
5. Heale, R., Dorothy, F.: Understanding Triangulation in Research. ON (2013)
6. IDEO.: ideo (n.d.). Accessed Sept. 2018. https://designthinking.ideo.com/
7. Jones, R.: Branding: A Very Short Introduction. Oxford University Press, UK (2017)
8. Krake, F.: Successful brand management in SMEs: a new theory and practical hints. J. Prod. Brand Manage. **14**, 228–238 (2005)
9. Razeghi, Y., Roosta, A., Alemtabriz, A., Gharache, M.: The role of corporate brand in the entrepreneurial SMEs total brand. Int. Bus. Manage. **8**(2), 120–130 (2014)
10. Sachdev, S.: Interview by Isha Patel. Head of Brand+Innovation, Quiqup (2018)
11. Shah, R.: Interview by Isha Patel. Brand Consultant (2018)
12. Steinmitz, A.: Interview by Isha Patel. Founder, Inward Consulting (2018)
13. Techshu.: 6 Ways World's Best Brands Are Acting Like Humans. Accessed Aug. 2018 (2017). https://www.techshu.com/blog/6-ways-worlds-best-brands-are-actinghumans
14. Witt, K.: Mid-Life Crises: Disaster or Turning Point? December 5. Accessed Aug. 2018 (2015). https://drkeithwitt.com/mid-life-crises-disaster-turning-point/
15. Wong, H.Y., Merrilees, B.: A brand orientation typology for SMEs: a case research approach (2005). https://www.researchgate.net/profile/Ho_Yin_Wong/publication/29455908_A_brand_orientation_typology_for_SMEs_A_case_research_approach/links/5590e4b408ae1e1f9bae36e9.pdf

Chapter 54
Understanding Bodo Identity in Their Handloom

Maneswar Brahma and Utpal Barua

Abstract According to S. Endle, the Bodos are one of the largest ethnic communities among the Indo-Mongoloid origin of Tibeto Burmese language family and the largest single speaking ethnic community in North-east India. The community has significant traditions of visual culture, particularly of wood carving, handloom and textile, craft works, horticulture, and folklores. The community also had wide knowledge in weaving garments from grasses, fibers and cotton, what they grew from silk, both wild and domesticated. It is significant that the silk of sericulture was introduced in north-eastern states of India, many centuries before the arrival of Vedic Aryans by the ancestors of Bodos (Endle 1997). However, the new generation of the Bodo community rarely follows the ancient cultural practices due to lack of preservation, documentation, and research of about the art and artefacts of this community. "Considerable amount of changes has taken place in the dress outfits not only the male section but the female section as well. Similarly, the art of dying yarn and cloths which was a community secret of the tribe is gradually fading out". ("Tribe of Assam Part- I" by Bordoloi et al. [Bordoloi B.N., Sharmah G.C., Saikia M.C.: Tribe of Assam. Part-I. B. N. Bordoloi, Guwahati (1987)]) So, this paper evolves from the research gap through a critical review of the limited literature available in the field of Bodo handloom design, understanding of Bodo community, and philosophy of the culture.

54.1 Introduction

As an ethnic community, the Bodo is one of the largest of the Indo-Mongoloid origin of Tibeto Burmese language family. The Bodo language is single largest speaking language among of the other communities of northeastern states of India.

M. Brahma (✉) · U. Barua
Indian Institute of Technology Guwahati, Guwahati, Assam 781039, India
e-mail: maneswar67@gmail.com

U. Barua
e-mail: utpalb@iitg.ac.in

A. Chakrabarti et al. (eds.), *Design for Tomorrow–Volume 2*, Smart Innovation, Systems and Technologies 222, https://doi.org/10.1007/978-981-16-0119-4_54

Bodos are considered to be the sons of the soil in Assam and culturally rich [1]. The Kacharis are different in many ways from their Hindu and neighbors in things like material and moral living of the life. He said that in general appearance, they bear some resemblance to the Nepali, and as a rule shorter and stouter than the people of North-West India, and well-shaped to bear up against physical fatigue and hardship [1]. The physiognomy and figure show a distinct approximation, which known as the Mongolian type. Bodos are well fitted for all forms of outdoor labor that require for agriculture and other cultural works capabilities and they are more reasonable to be regarded as the 'navvis' of Assam. Bodos are known as Kachari too. Kachari are the aboriginal or earliest known inhabitants of the Brahmaputra valley. In Brahmaputra valley, Kacharies are known as Bodo or Boro 'fisa' (sons of the Bodo) and in North Cachar Hill, they are called themselves Dimasa [2]. Again, Bodos are known as Kirata, Kacharies, and Miecchas. Sir Edward Gait has written that "the earliest mentioned king of Kamrupa was named Mahirang Danab who was succeeded in turn, in the direct line by Hatak Asura and Ratna Asura" [2]. The earlier kings of Assam were belonged to Non-Aryan tribes and known as Danavas and Asuras [3]. Ancient Indian literature like Mahabharata and Ramayana also narrates the story of Indo-Mongolian and the Sino-Tibetan peoples as 'Kirats'. In Kalika Purana, there is beautiful representations of Kiratas and describes them as short stature, golden color, shaven head, rough skin, and addicted to meat and local drink. Vedas like Yajurveda and Artharveda also describe about the Kiratas.

Thus, the community continues following with rich cultural heritages and cultural dialects. The cultural dialect and cultural heritage are subsequently a subject of recollection having new approach of research. Here, we are trying to do such approach of research by collecting designs of motifs and patterns of handloom. Design and pattern of weaving material establish the cultural and community identity. Besides the cultural identity of the community, it speaks the understanding of the socio-economic state. The discussion in an intended manner of this paper is to understand the Bodo identity in their handloom.

54.2 Literature Review

54.2.1 Bodo Religion

Bodo is an aboriginal community; its origin of religiosity and religion is 'Bathou' which is based on five philosophical principles and, respectively, the five principles are 'Ong', 'Hring', 'Khling', 'Fwt', and 'Che' [4]. In Bodo language, the 'Ba' means five and 'Thou' means deep spiritual things, and it indulges the consciousness of individual's 'self' and introspection. Subsequently, the five principles consider the consciousness of self-realization and self-introspection. Scholars like S. Endle, Dr. Liladhar Brahma, Dr. Kameswar Brahma, and others also have same comprehensive opinion considering the study on Bodo religion. There again they have deciphered and researched by saying that the Bodo religion is animistic. The animistic religion, the Bodo had the first imaginary gods, who worshiped in earlier age. These five gods are named as Ailong (the god of earth), Agrang (the god of water), Khwila (the god of air), Sanjabwrlee (the god of light), and Rajhkumbree (the god of sky). There is a historical trace and folk narratives which are of local narrative to narrating in colloquial lingo that represents and saying of the primitive Bodo called Mwnchinchin. The Mwnchinchin is considered of five states of consciousness. These five states of consciousness are Saithi (peace), Laothi (control of mind), Nalathi (cooperation of the mind and body), Maothi (activeness), and Sonathi (dearer) [4].

Bathou dharma speaks about the creation of the universe, the philosophy, and its belief. In Bathou dharma, there is a belief that the absolute vacuum was the prior state to the creation of the universe and there was an existence of vacuum, the only Almighty, the Anan Gosai (the eternal God). Who became of His formless existence and so he desired Himself by taking a form of flesh and blood. The following spiritual words are surged in His consciousness: [4].

- Laoba laoswm
- Khaoba Khaoswm
- Ada Gwswm
- Dwiao Barswm

The verse connotes the meaning of 'oh mind the great!' Take to you the five organs of perception and control the organs of action to jump into the worldly ocean. In the reference, 'Lao' means organs of perception and 'Ba' means five. 'Laoba' means five organs of perception. Similarly, 'Khaoba' means 'five organs of action' and 'Khaoswm' means 'to tie with' i.e. 'Khaoba Khaoswm' means 'to tie or control the five organs of action'. 'Ada' means elder brother or the great and 'Gwswm' means Mind, hence the mind is elderly addressed to tie or control the five organs of action. 'Dwiao' means 'into water' while 'Barswm' means 'to jump into' [4].

The important shifting of the social and religion of the Bodo- Kacharis' can be seen since the nineteenth century. Since the nineteenth century, the shifting of Bodo religion has transgressed into two ways; first one is Christianity and second is the Brahma Dharma. Though the conversion was going on in later periods, but the Brahma Dharma also was on reachable path for preaching to till modern days. And the traditional animistic belief that continued to guide the religious philosophy of Bodo-Kacharis began to change by the preaching of Guru Kalicharan Brahma who was inducted to the Brahma belief [5].

The conversion to the Brahma cult paved the way to bring the larger sections of followers to the wider fold of the Vedic Sanatan Dharma and it was possible to carry out a social revolution among the Boro-Kacharis residing in and around the district of Goalpara with epicenter at Kazigaon in Dhubri Subdivision (it was then comprised of the present Kokrajhar district as well). It was a movement for total change of outlook in all fields- social, religious, political, economic and educational.

The living hood or basic survival of Bodo community, since its inception of human culture has been based on agriculture and been followed that way since. Thus, they were settled agriculturist, empowering two ways like broadcasted and transplanted of variable of paddy which accounted the skill of irrigation by the dong system. The community also was comprehensive in weaving garments from grasses, fibers, and cotton, what they grew from silk, both wild and domesticated. It was significant that the skill of sericulture was introduced in North Eastern India, many centuries before the arrival of Vedic Aryans by the ancestors of Bodos [1]. The Bodo community is predominant in the commercial production of silk and Endi cocoons and finished products from Endi yarn. Spinning of Endi yarn and weaving is household industry and every woman of the community is skillful enough to express in the finer art of handloom craftsmanship. According to him, yellow color is favorite choice of Bodo women which is significantly visible in their dresses, specifically in their 'Dokhona'. And other dresses they wear during ritualistic dance/performance like 'Bagarumba' and 'Kherai' which shown very artistic taste and decor of the tribes as a whole. Observing the present context, he mentioned that in the course of fashion designing and globalization, the considerable changes have been happening in the dress outfits not only of the male section but in the female section as well [5].

54.3 Research Gap

The major gap of the subject (handloom design, visual element and symbolical representations and analysis, etc.) is its cultural interpretation and propositional analysis of dialectical paradigm and cultural value. It is intended from this research gap that the researcher would like to focus only on study and understanding of visual elements related to Bodo handloom.

54.4 Aim

A study to understand the cultural identity and the meaning of its visual representations, through the visual motifs present in the handloom designs of the Bodo Community by having the example of Bunduram design (it is a significant form of design being weaved on various cloths of the Bodo Community).

54.5 Methodology

The basic methodology will be ethnographic research and descriptive research. We have followed primary documentation process to collect the data for abovementioned subject. To do so, we have gone through many villages, few particular families, weaving centers and NGOs to make visual documentation of the old cloths and newer cloths of the Chirang and Kokrajahar districts. In this case, we have had photographic documentation, video documentation, and individual interviews (Semi-structured) with older people and weavers of the community. During our data collection, we have found few significant old collections of cloths (like Aronai, Dokhona, Pasra, Pali, Sadar and Gamocha) which are indeed rare collection and preservable cultural product. We met weavers of Manas weaving center of Nichima village, Chirang district, The Ant NGO of Rowmari, Chirang district, Weaving center Bijni Khabli Bagan Chirang district and Still Rout Weaving Center, Bagan Chali, Kokrajhar district. And had talked and interviewed of shopkeepers, weavers of Kokrajhar and Chirang district. The interviews conducted with these people were intended to understand the seven tradition of Bodo weaving culture and acquire knowledge of the heritage. And how does the same culture survive in this age? Many of them had enlightened us by saying its present scenario and considerable condition for survival.

54.6 Analysis and Discussion

"The art of spinning, weaving, and handloom embroidery of the Bodo women are never inferior to another race of people all over India. They know the art of swelling embroidery in their clothes and costume," said Dr. Kameswar Brahma. The Bodo women weave various kinds of clothes for household use and business purposes. Few of them are—Zi Gidir (big cloth), Sima (use as bad cover and general use), Endi Si (endi cloth), Dokhona, Gamsa, Alon, Aronai (wrapper on neck), Zwmgra (women wrapper on body), Sadri, Phali (hand-car-ship), Aranga Si (table cloth), Sopha Si, Si Gudung (Wool cloth), Gandu Si (pillow cloth), Daothu Godo (design of doves neck), Phareo Megon (design of pigeons eye), Pahar Agor (design of hill), etc. [6]. They weave ancient traditional motif designs in dresses of the Bodo community that consist of—Mwider Agan (design of elephant foot prints) and Mwcha Agan (design of Tiger foot prints) which are currently not seen in present motif design of their practice [7].

The new generation of the Bodo community rarely follows the ancient cultural practices due to lack of preservation, documentation and research of those Art and Artefacts of the community. In their choice of dress, the Bodo women show a certain fondness for colored garments and acquainted with the art of dyeing (blue and various shades of red) being supplied from leaves or roots of tree. He said that the male persons of young and old generation put on Gamcha, woven at home. During winter, they use to wrap the body with a wrapper of cotton or Endi spun and woven at home. This wrapper is known as Jumgra or Madamni Gamcha. They also use a banian, a sort of coat of cotton or Endi. According to him, such dresses of the Bodo male persons are now found rare. In the same book, author said that the women of the community formerly and even now tie a cloth round the chest just below the arm that hangs to the toe, which is called Dokhona. If it is plain, then it is called Sala Matha and if it is ornamented, it is called Dokhona Thawsi. Dokhona Thawsi is essential dress of the bride to wear during marriage ceremony. He observed that this type of ornamented dress is now not easily found. The popular designs of the ornamented Dokhonas are many in number. The most popular Dokhona are Daothu Godo (designs of dove's neck), Phareo Megon (designs of pigeon's eye), Pahar Agor (designs of hills scenery), and Moider Agan (designs of elephant's foot print). The orange, yellow, and sky colors are their favorite colors.

In the view of [1] "one would be astonished to know about the art, craft and weaving skill of traditional Boro-Kacharis. For example, a kind of aria pachara, made of endi thread, which gets from eri worm still has a great demand among the people of Assam. But with the advent of modern textile technology their primitive weaving process is being swept away. Nowadays, many Boros are attracted toward the power loom. So far as agricultural economy is concerned, it is difficult to ascertain the era of traditional hoe cultivation (jhum cultivation) of the Boros." The Bodo dancers use Dokhona, Aronai, Rege-Regang, Janji Khanai Fali (wide belt made of cloth) and blouse. And latest are attractively hand-woven and beautifully embroidered. Embroidery is made by handloom.

The cultural representations are not only of the products of handloom, besides that there are lots of products or representations. But in truth, such products are getting important place in market. The market value or economic value is becoming eye catching part in today's living. So, the same cultural-economic value of the past is yet refuting and transforming on a new format of homogeneity.

Weaving in the Bodo community bears a significant role. In the ancient time, Bodo woman without weaving expertise was not give due respect in the society. From ancient times, Bodo women have been using dresses and materials which are hand-woven by them. They create splendid design on their traditional dress materials which are woven with the help of handloom and some other tools (Figs. 54.1, 54.2, 54.3 and 54.4).

This type of design represents high quality of creativity. While producing dress material and creating magnificent design on their dress materials, Bodo women use

Fig. 54.1 Samples of Sadar collected from Kokrajahar Musuem (BTAD). Name of the object: Sadar (with Bunduram design)

Fig. 54.2 Samples of Sadar Collected from Kokrajahar Musuem (BTAD). Name of the object: Sadar (with Bunduram design)

Fig. 54.3 Name of the object: Aronai, designed of the hill on Bunduram

Fig. 54.4 Name of the object: Aronai, designed of the hill on Bunduram

certain type of tools which are made of bamboo and wood. And those design motifs represent naturalist forms and eclectic aesthetic significance. The design Bunduram weaved in a Sadar represents the symbol of tree leaves and flower leaves with various colors. For doing such, Bodo women recollect their traditional narratives which they found in their lively culture. Even they weave Bunduram design on various cloth and narrate other naturalistic representations like hill, bank of the river, etc. The artistic notion is that they represent through the weaving, their familiar surroundings in which they live. When we look at these types of designs, then we find their cultural and geographical identity which are been portrayed in their designs. Color which they used are all hue, what always speaks and narrates the ethnic beauty and eclectic. Besides of all, there is another significance that is about of the season, in which they live for weaving and render the following season of spring, summer, and autumn.

Aronai (Fig. 54.5.) is a small scarf used by both man and woman. Aronai symbolized the Bodo tradition and is used to felicitate guests with honor as a gift. During their traditional performance and dance, it has been used. In ancient time, Bodo warriors used Aronai as a belt in the battle field. At the time of war, Bodo women weaved Aronai within a single night and presented it to the warriors as they set out for the battlefield. And it represents the belief of victory. The use of hue color represents the ethnic eclecticism and purity (Fig. 54.6).

Again, this subject will lead to understand the present cultural alienation, identity politics, suppression of gender politics, and other cultural over empowerment. The technological progress and upheaval are another threat to ethnic cultural unity and

Fig. 54.5 Different designs on the cloth piece called Aronai

Fig. 54.6 Jwmgra (Rege-Regang), Bodo women use Jwmgra (scarf) to cover upper portion of the body (length around 2.5 m, width around 1 m

eclecticism. Hybrid culture and globalized market is ceasing day by day the local products and productivities.

54.7 Conclusion

This work will signify multidimensional subjects of ethnic community and homogeneity of cultural plurality. Ethnic community study is in modern and contemporary academic discourse should open up the idealism and concept of open culturalism which is in truth inherited into the ethnic community, culture, and living hood. And need to pave a way to sustain this heritage stepping up an alternative cultural synthesis.

References

1. Endle, S.: The Kacharis (Bodo). Low Price Publications, Delhi (1997 reprint)
2. Gait Sir, E.: A History of Assam. Bina Library, Guwahati (2001 reprint)
3. Barua, B.K.: A Cultural History of Assam, 4th edn., p. 05. Bina Library, Guwahati (2003)
4. Brahma, L.: Religion and Dances of the Bodos, p. 11. Onsumoi Library, Kokrajhar (1991)
5. Bordoloi B.N., Sharmah G.C., Saikia M.C.: Tribe of Assam. Part-I. B. N. Bordoloi, Guwahati (1987)

6. Basumutary, S.: Sukumar Basumatary's Swrji Bihung, 1st edn., p. 256. Bodo Publication Board, Bagsa, Mushalpur (2012)
7. Brahma, K.: A Study in Cultural Heritage of the Boros. Bahniman Printers, Guwahati (1998)

Chapter 55
Entrepreneurs Motivation for Selecting Homestay Businesses: Special Reference to Ella, Sri Lanka

C. P. Danthanarayana, T. G. A. H. C. Amarawansha, and P. G. M. S. K. Gamage

Abstract Sri Lankan tourism industry is one of the key important sectors to develop within the country. Ella is one of the destinations which has recently become popularized with its attractive and unique geographical features, chilling climate, and authentic local lifestyle. Homestay accommodation is playing a major role in fulfilling an excess demand for accommodation in Ella and it is contributing to the economy of small and medium entrepreneurs (SMEs). The main objective of this study is analyzing push and pull motivation factors of entrepreneurs to select homestay business in Ella. Primary data has collected from fifty homestay entrepreneurs and researcher used semi-structured questionnaire to collect data. Confirmatory factor analysis has been used to analyze the data. The study revealed that, to generate additional income is the primary push factor which is pushing entrepreneurship by negative external factors. Enjoyable things to do in the homestay accommodation business can be identified as main pull factor which is attracting entrepreneurs to homestay businesses. Apart from that, this study identified some issues and challenges which entrepreneurs have to face when they are doing homestay businesses such as lack of security, lack of facilities, competition with unregistered homestay operators, lack of service orientation, difficult to maintain the standards, not enough training to manage homestays, language barriers, lack of cross-cultural understanding, lack of brand image and marketing activities, lack of accessibility for transportation and communication, and lack of monitoring from the ministry and Sri Lanka Tourism Development Authority (SLTDA). Basically, this study reveals the reasons of entrepreneurs to start homestay businesses in Ella, Sri Lanka. Relevant authorized parties can be involved and analyze the challenges and issues of homestay entrepreneurs and required to give solutions for them as well. Finally, this study suggests innovative practices for homestay entrepreneurs to face identified challenges and problems and become more unique business in the industry.

C. P. Danthanarayana (✉)
University College of Matara, Matara, Sri Lanka
e-mail: chathupiumika@gmail.com; chathupiumika@ucm.ac.lk

T. G. A. H. C. Amarawansha · P. G. M. S. K. Gamage (✉)
Uva Wellassa University of Sri Lanka, Badulla, Sri Lanka
e-mail: Sewwandikagamage20@gmail.com

A. Chakrabarti et al. (eds.), *Design for Tomorrow–Volume 2*, Smart Innovation, Systems and Technologies 222, https://doi.org/10.1007/978-981-16-0119-4_55

55.1 Introduction

55.1.1 Background of the Study and Problem Statement

Motivation is playing a major role in new business creation [1]. Lack of empirical research on entrepreneurial motivation can be seen among scholars [2]. An entrepreneur, a person who is self-employed, taking personal challenges, innovative and creative, starting, organizing and managing the business, and taking business responsibilities. Always they are taking the risks related to the business such as financial and market risks. Selecting entrepreneurship as a career is a challenge itself because the entrepreneur has to face uncertain everyday life and work situations which can be caused to failure and frustration toward the new business creation [3]. A number of researches have conducted researches on investigating the motivation which drives people to take risks and uncertainties to become entrepreneurs. For that, this study is investigating the component of motivation which is contributing toward the individual decision to become an entrepreneur.

The tourism industry in Sri Lanka is the most important sector for the country as the third-largest foreign exchange earner in 2018 [4]. The number of tourist arrivals for the country is increasing yearly and in 2018, foreign exchange earnings increased 15.9% compared with the year 2017 [4]. Sri Lankan government has identified that tourism as one of the main sectors which contributing economic growth, generate employment opportunities, earn foreign exchange earnings, attract foreign direct investments, and enhance the livelihood of the community. As one of the key tourism destinations in South Asia, the Sri Lankan government has collaborated with the private sector and local communities in destination marketing and promotional activities.

Sri Lanka is enriching with a great culture and local communities are playing a major role in actively promoting traditional culture among tourists. The government has encouraged local authorities to make efforts to introduce community-based tourism, rural tourism, and homestay tourism by developing and diversifying tourism products and services. Homestay tourism is a kind of cultural tourism based on the rural culture, which has developed by diversifying cultural tourism and to contribute to the economic well-being of the local community. Homestay business directly brings benefits to the local community by providing employment opportunities and improving the quality of work-life of the local people [5]. Past researcher [6] stated that homestay businesses are providing accommodation opportunities for budget travelers while making an income source for tourism operators. It is an important method of tourism since tourists can directly contact with the local community, living style, and traditional culture while keeping financial benefits within Sri Lanka without any leakages to the other countries.

Homestay can be categorized under supplementary establishments and in Sri Lanka, total capacity is including of 1855 supplementary establishments by the year 2018. It has increased the accommodation capacity by 9.56% while having 13,457 rooms in 2018 [4]. Homestay accommodation will be provided opportunities for tourists to experience the rural lifestyle and authentic culture in the host community while closely relating to nature, traditions, and local customs. Since the homestay sector is contributing toward economic development, micro-, small-, and medium-scale entrepreneurs are trying to participate with the homestay businesses. Therefore, there is a question of what motives these small-scale entrepreneurs to attract homestay accommodation businesses in Sri Lanka. Still, there is a lack of studies that examined the motives of entrepreneurs in these kinds of businesses [7]. In the Sri Lankan context, different studies have focused on homestay accommodation services, restricted to homestay potentiality, homestay accommodation supply, difficulties, and challenges related to the homestay and restricted to some areas such as Meemore [8, 9, 10, 11]. Therefore, this study is to overcome the restrictions and limitations of identifying real motives to select homestay business as entrepreneurs by considering push and pull motives. Moreover, it is discussing challenges and real-world issues for homestay businesses which are restricting the attractiveness of entrepreneurs toward homestay businesses and how innovative practices can be used as entrepreneurs to face these challenges. According to the researchers' perspective, this study greatly contributes to the hospitality and tourism sector since it is providing a huge opportunity for the local entrepreneurs to make innovative marketing and strategic decisions based on information provided by the study.

55.1.2 Research Objectives and Questions

In the study, it is focusing on small and medium entrepreneurs in homestay businesses and seeking whether motivators most frequently suggested by empirical researches employing the push–pull model of entrepreneurs in different business sectors can be applied to the homestay businesses in Sri Lanka. The main objective of the paper is to identify the motivation factors of entrepreneurs to select homestay business by confirming factors through push and pull factor analysis. Apart from that, the research study is focusing on the problems and challenges that entrepreneurs are facing in operating homestay businesses as well as how entrepreneurs should face these challenges with innovative practices. Therefore, to achieve the objectives, this study addressed the research questions of, what are the push and pull motivational factors of homestay accommodation entrepreneurs? What are the challenges and problems homestay accommodation operators have to face when they are doing business specifically in this industry? And how entrepreneurs should face these challenges with innovative practices.

55.2 Literature Review

55.2.1 *Hospitality and Tourism Industry in Sri Lanka*

The tourism and hospitality sector considered as the major and it among the world's largest industries, employing 250 million people around the world and generating more than 15% of global GDP. Tourism is a highly varied sector that includes tour operators, accommodation providers, transport, caterers, leisure activities, and entertainment. According to a United Nations report [12], compared to other sectors of the global economy, the hospitality industry is one of the fastest-growing industries, which has marked more than one-third of the total global services trade. Moreover, the hospitality industry is much broader than most other industries [13]. The majority of business positions are collected of only a handful of different businesses, although this industry applies to nearly any company that is focused on customer satisfaction and meeting leisure/recreation needs rather than basic ones.

Furthermore, this industry is identically broader than other conventional industries. The number of tourist arrival and tourism activities is affecting by making a huge impact on national Gross Domestic Production. When concerning Sri Lankan tourism contribution to the economic development of the country, it offers new job opportunities and creates new source of income by developing infrastructure. According to Sri Lanka Tourism Development Authority (SLTDA), they have expected around 4 million tourists arrive in 2020. When concern about tourist arrival patterns tourists are arriving on specific seasons of the year. This refers to the seasonality effect on tourism. Moreover, this seasonality effect in tourism can done a huge impact on the host community which maintains their daily expenses via tourism-generated revenues. This seasonality effect can be seen in the southern area regularly and this study is going to discuss the issues related to the seasonality effect of southern area tourism in Sri Lanka. Due to this seasonality effect within Sri Lanka, a tourism-related business cannot be expected a continuous revenue. Due to this reason, most of the business is moving away from the business and some are running the business even at the period of losses. Moreover, when considering the nature of the tourism business, most of the time, it based on the motivational factors which affected or faced by the entrepreneur or business owner.

55.2.2 *Entrepreneurial Intention*

According to the past researchers [14], intentions play a key role in explaining human behaviors. Rosli [15] has added to the above statement by mentioning intentions play a key role to develop entrepreneurship activities and the capability of being an actual entrepreneur. Past researcher [16] mentioned that entrepreneurial intention as a state of mind that underlines a person's attention and experience to create a new venture. Past researchers [17, 18] have identified in the same way that entrepreneurial intention is the commitment to performing entrepreneurial behavior. There are many studies in the literature related to the antecedents of entrepreneurial intentions. Even though there are more models, it can be identified a widely recognized or accepted model. That model has been explained by the past scholar [17] and it is called as theory of planned behavior (TPB). Many scholars have validated the above model. According to this model, there are three factors to predict entrepreneurial intention. They are attitude toward the behavior, social norms, and finally perceived behavioral control. If a person has a favorable or unfavorable evaluation of behavior, it is called an attitude toward the behavior. Basically, it may depend on the person's assessment of the expected outcomes of the behavior. The social pressures alleged by a person to perform or not to perform the behavior is called a subjective norm. Social pressure may come from friends, the family, peers, networks, or mentors. On the other hand, a person will suffer a subjective norm that influences them to stop carrying out the behavior. The final component is the intention, perceived behavioral control. Perceived behavior can be identified as the perception of easiness or difficulty which is encounters in a particular behavior. As per the theory of planned behavior accomplishments or the results depend on the combination of intention or motivation and ability. This theory of planned behavior may depend on non-motivational factors like accessibility of opportunities and different resources such as money, time, a cooperation of others, etc. A person's actual or real control over the behavior will be represented by the above-stated factors.

There is another model that will help to the formation or establishment of the entrepreneurial intention. It is called Shapero and Sokol's Entrepreneurial Event (SEE) and founded by Shapero and Sokol [19]. This theory of entrepreneurial event proposes entrepreneurial intention based on perceived viability and alleged interest of the prospect of starting a business along with the propensity to act. In the available literature, there is another model for the intention. It describes intentions are based on a mixture of both personal and contextual factors [16]. In addition to that, the development of Bird's model was prepared to include the concept of self-efficacy taken from the social learning theory by Boyd and Vozikis [20]. Finally, past scholars [17] have developed a model based on the theory of planned behavior. It has been done with some alterations to adapt to an entrepreneurial environment. Consequently, three critical perceptions defined as personally desirable, supported by social norms, and feasible have predicted intentions toward pursuing an opportunity. By having a look at the above illustrations, entrepreneurial intention can be identified as follows. It is a mental representation of actions for exploiting an opportunity by applying entrepreneurial knowledge and skills. This entrepreneurial knowledge and skills can be identified as entrepreneurial learning. Researchers have confirmed that entrepreneurial intention effectively predicts entrepreneurial behavior. That is, factors influence entrepreneurial behavior through influencing intention, which is derived from attitudes. When concern about these being and motivational to run a business basically deals with the push and pull factors. These push and pull factors can be a concern as both in a negative and positive aspects.

55.2.3 *Pull Factors*

Theory of planned behavior well explains about the attitude inculcate in the mind set of the business owners through being engage in self employment. According to past researchers [19], the initiation of a business is dependent based on the different and important changes that happen in the life of an individual. This movement can be negative or can be positive. Moreover, they explain that the entrepreneurs' business or their life perception will decide the willingness to continue the entrepreneurial business career. Moreover, according to the past researchers [21], there are six categories of motivation as pull factors in positive mode can be considered as an innovation, independence, recognition, roles, financial success, and self-realization. Moreover, according to the past researcher's [22] argument, that women are pulled into self-employment by the promise of independence, flexibility, autonomy, and opportunity to escape employment barriers. Furthermore, according to the researchers [23], is providing a number of pull motivations, including the need for approval, independence, personal development, improved welfare and wealth, and following role models within his research study.

55.2.4 *Push Factors*

Past researchers [19] explain that the claim that jobless individuals or individuals with low interest in wage employment may become self-employed. Moreover, according to the researchers [24], push motivation is typically associated with unemployment. Moreover, there are other factors that might push persons into the way of new venture creation, like family pressure to transfer the business to the new generation of job dissatisfaction, people who are removed from their jobs, individuals who decide to leave wage employment because their boss does not want to commercialize their ideas or inventions, and individuals who are lack of education or skills or having poor backgrounds.

55.3 Research Methodology

55.3.1 *Research Design*

Ella is a well-known tourist destination that has been successfully implemented the homestay model in Sri Lanka. There are 6% of tourists arrival has been recorded in Ella from the total tourists arrived in the country annually according to the Sri Lanka Tourism Development Authority (SLTDA). There are 17% of tourists visit Ella to experience rural life and 24% to enjoy the unpolluted air and 30% to experience natural environmental beauty. Hence, the researcher selects Ella to in investigating the push and pull factors of entrepreneurs to select homestay business in the Ella area under these tourist arrivals.

The sampling unit of the study was a homestay entrepreneur. Fifty homestay entrepreneurs were selected for the study based on the convenience sampling method. The primary data was collected through a semi-structured questionnaire by using five-point Likert scale questions.

55.3.2 Data Analysis Method

Confirmatory factor analysis was applied to investigate the main objective of the study, to analyze push and pull factors of entrepreneurs to select homestay business in the Ella area. Descriptive statistics such as mean and standard deviation has used to select factors and challenges faced by entrepreneurs. The demographic factors were analyzed based on the descriptive analysis which represents the count and the percentages. The analysis was supported by SPSS statistical package.

Accordingly, this study has been investigated the factors, which are related to the push and pull factors of homestay entrepreneurs and challenges and problems faced by these entrepreneurs when operating homestay businesses. Then, it has been deliberated the result of the study and it has been organized the discussion and conclusion related to the result.

55.4 Data Analysis and Interpretation

55.4.1 Demographic Profile of Respondents (N = 50)

Table 55.1 represents the demographic profile of the respondents. According to the table, the majority of homestay accommodation providers in Ella are female (66%). As per the literature [24], it is consistent with the entrepreneur gender breakdown in the homestay sector recorded as most of them as female.

The majority of the respondents (72%) have completed up to their primary and secondary education and 8 percent of respondents have completed their Bachelor's degree. Some of them (18%) do not have a formal education background.

Most of the respondents (74%) are more than 40 years old and the rest of the 26% is at 39 years or below. The majority of the respondents are indicating that they have passion and energy to operate the business effectively.

Most of them (54%) do not having previous experience in working at small businesses. However, about 34% had less than 5 years' experience in managing small and medium businesses.

Table 55.1 Demographic profile of respondents

Variable	Group	Count	%
Gender	Male	17	34
	Female	33	66
Education	No formal education	9	18
	Primary level	17	34
	Secondary level	19	38
	Tertiary level	4	8
	Other	1	2
Age	30 years or younger	4	8
	30–39 years	9	18
	40–49 years	26	52
	50 years or old	11	22
Previous working experience in small businesses	No working experience	27	54
	5 years or less	17	34
	5–7 years	4	8
	8–10 years	2	4
	More than 10 years	–	–
Years in business operation	3 years or less	29	58
	4–6 years	16	32
	7–9 years	3	6
	10 years or more	2	4
Number of employees	Less than 5	48	96
	5–10	2	4
	More than 10	–	–

(N = 50)

The majority (58%) of homestay entrepreneurs are operating homestays business for less than 3 years and 42% operating for more than 4 or more years. When considering the number of employees in the homestay business, majority (96%) is having less than five employees.

55.4.2 *Motivational Factors for Homestay Accommodation Entrepreneurs in Ella, Sri Lanka*

55.4.2.1 Quantitative Analysis

Table 55.2 indicates the motivational factors for being homestay entrepreneurs in Ella. A higher value indicating the stronger motivation factors to being an entrepreneur

Table 55.2 Mean and standard deviation of motivation factors of entrepreneurs to select homestay business

Item	Mean	SD
Desire to generate additional income	4.79	1.65
Need for supplementary family income	4.72	1.63
Unemployment	4.65	1.67
Personal satisfaction	4.58	1.59
Enjoyable things to do in the homestay accommodation business	4.42	1.63
Flexibility of lifestyle	4.23	1.72
Family encouragement to start homestay business	4.13	1.69
To work for myself	3.86	1.64
Current market demand in the tourism industry	3.54	1.71
Dissatisfaction with previous job	3.49	1.55
To achieve something and get recognition in this business	3.45	1.62
Community members' motivations to participate	3.38	1.60
Desire for independence	3.31	1.66
New market potential and opportunity	3.14	1.59
Ability to make own decisions	3.02	1.71
Preparing for retirement	2.93	1.53
Availability of resources in strategic area for homestay Business	2.76	1.60
The homestay business had tremendous opportunity	2.31	1.57
To apply experience, knowledge, and expertise in the field	2.31	1.57
Continuation of family businesses	1.74	1.55
Job insecurity	1.74	1.55
Gain a personal sense of accomplishment and grow business	1.41	1.52
Previous job experience in similar business	1.22	1.59
Inspired by friends' success in the business	0.98	1.61

in a homestay business. The researcher has selected factors which are having higher mean values for further analysis. Factors which are having more than 3.00 mean value have been selected to conduct factor analysis.

The researcher has conducted confirmatory factor analysis by using SPSS software and outer loading values (OLVs) should be are more than 0.5 and significant at 0.05 significant statistical level to consider as influential factors for the variable [25]. Table 55.3 shows the outer loading values of selected factors. Factors with less outer loading values did not consider the influential factors for motivation to select homestay accommodation among small- and medium-scale entrepreneurs. Accordingly, to work for me, new market potential and opportunity and ability to make own decisions have removed from consideration.

As most of the studies, economic motivation to generate an extra income is the primary factor for selecting homestay business among entrepreneurs. Mostly, when

Table 55.3 Push and pull factors

Item	Outer loading values	
Desire to generate additional income	0.922	
Need for supplementary family income	0.828	
Unemployment	0.888	
Personal satisfaction	0.815	
Enjoyable things to do in the homestay accommodation business		0.859
Flexibility of lifestyle		0.726
Family encouragement to start homestay business		0.610
To work for myself	0.453	
Current market demand in the tourism industry		0.713
Dissatisfaction with previous job		0.606
To achieve something and get recognition in this business	0.806	
Community members' motivations to participate		0.753
Desire for independence	0.736	
New market potential and opportunity		0.431
Ability to make own decisions		0.465

the husband is doing a job somewhere and the wife is tending to start homestay to generate extra income for their home [26, 27]. In Ella, most of the homestay entrepreneurs have selected this sector with the desire to generate additional income as the primary push factor. Homestay tourism is providing additional economic opportunities for rural communities as many of them required additional income sources [28].

Unemployment and need for supplementary family income, personal satisfaction, to achieve something and get recognition in this business, and desire for independence can be seen as other push factors.

Enjoyable things to do in the home-stay accommodation business, community members' motivations to participate, the flexibility of lifestyle, current market demand in the tourism industry, family encouragement to start a homestay business, and dissatisfaction with the previous job, can be seen as pull factors which attract entrepreneurs to homestay businesses in Ella. This study reveals that involving homestay business can be influenced by community members' motivation to participate and it indicates that social awareness about the benefits of homestay businesses, social and affiliation needs, and personality of entrepreneurs are playing a major role in considering entrepreneurs' participation in homestay businesses. Community participation in the homestay business is contributing to enhancing the intrinsic value of society [29].

55.4.3 Challenges and Problems Faced by Homestay Entrepreneurs

Challenges and problems are not common for all businesses and entrepreneurs. It can be changed based on the industry, business, business size, and geographical location also. Especially, micro, small, and medium businesses are experiencing different challenges when starting the business and different stages of the business life cycle.

As per Table 55.4, the high mean value is representing the most critical challenges, and the low mean value is representing less critical challenges. Lack of facilities, not enough training to manage home-stay, lack of monitoring from the ministry and state government, and competition with incompliance unregistered home-stay operators can be seen as the most critical challenges faced by homestay entrepreneurs. Because of the lack of training and experience in the industry, homestay entrepreneurs facing difficulties in managing the standard of cleanliness, safety, and hygiene. Homestay businesses do not require many financial capabilities to start the business as other industries [30]. Therefore, lack of financial abilities is not an important challenge for these entrepreneurs. Apart from that, after discussing with homestay business entrepreneurs, they have stated that, managerial incapability, insufficient resources, and failure to manage service quality.

Table 55.4 Challenges and problems faced by entrepreneurs in homestay business

Item	Mean	SD
Lack of facilities	4.97	1.63
Not enough training to manage home-stay	4.52	1.60
Lack of monitoring from the ministry and state government	4.39	1.53
Competition with incompliance unregistered home-stay operators	4.03	1.58
Lack of cross-cultural understanding	3.96	1.58
Difficult to act-and-think tourist attitude	3.85	1.55
Difficult to maintain standard/optimum levels of cleanliness	3.84	1.64
Lack of safety	3.72	1.55
Lack of accessibility for transportation and communication	3.60	1.57
Lack of service orientation	3.56	1.71
Lack of brand image and marketing programme	3.32	1.70
Not enough guest bedrooms	3.19	1.60
Difficult to maintain the requirement and compliance	3.01	1.59

55.4.4 How Innovative Practices Can Be Used as Entrepreneurs to Face the Challenges

Innovativeness is a key trait of entrepreneurs and with the innovative practices, it can reduce lot of challenges and problems faced by entrepreneurs when doing business. In Ella, most of the homestay businesses are operating by following traditional methods and they are not much focusing on technological advancements. They are not promoting their business through online platforms such as Web sites and social media. Most of the times, walking guests are arriving to the homestays and they are not taking online reservations for the homestay accommodation. Comparatively, there is less demand and awareness of tourists toward these types of homestay businesses. Therefore, it can use modern technological advancements such as mobile apps, Web sites, social media, and reservation systems for their business. Apart from that, payment methods such as credit/ debit card payments, PayPal payment methods can be accepted as most of the tourists are not using local currency to pay bills. As Sri Lanka enriched with traditional culture, cuisine and arts homestay businesses can promote local businesses for ornaments and products related to culture, traditional cuisines, cooking classes for traditional and unique food items to make more profits from them. Apart from the mere bedroom and meals, it will give another advantage for the business. They can create homestays as traditional Sri Lankan houses by using natural material as clay walls, coconut leaves roofs, and wooden furniture as well as natural water pools, without making them as luxury buildings, comfortable beds, and attached bathrooms with amenities. It will give totally new experience for tourists as well as it can earn high income as other countries (Japan homestays at Beppu using traditional Onsen pools for tourists' bath purpose without offering bathrooms). At homestay businesses, they can offer Sri Lankan traditional cuisines as main meals for the tourists with some modification of taste. They can promote Sri Lankan tea and tea-related products since Ella is having different tea estates around the area. Homestay entrepreneurs can become tour guides for homestay tourists to educate tourists about history of Ella, traditional values, motivate them to engage at different adventure activities at Ella and encourage them to purchase local products to support local businesses. Apart from the homestay business, they can operate adventure park, spa and wellness centers to promote traditional Ayurveda treatments and products and flower shops for tourists. Innovatively, it should give some practical education for homestay entrepreneurs about tourism industry and tourism businesses and local government should give their guidance for the entrepreneurs to maintain proper standards within the business.

55.5 Conclusion

This study was conducted to analyze the push and pull factors of entrepreneurs to select homestay business in the Ella area based on 50 homestay entrepreneurs.

As per the confirmatory factor analysis, to generate additional income unemployment, need for supplementary family income, personal satisfaction, to achieve something and get recognition in this business and desire for independence have been revealed as the push factors which are pushing entrepreneurship by negative external factors. Further, the study found evidence to support pull factors which are attracting entrepreneurs to homestay businesses as enjoyable things to do in the home-stay accommodation business, community members' motivations to participate, the flexibility of lifestyle, current market demand in the tourism industry, family encouragement to start home-stay business and dissatisfaction with the previous job. Moreover, when concerning on the tourism business, redesigning of tourism business is essential feature to enhance the entrepreneurial practices among them. Therefore, based on the analysis, it has identified that enhancing the collaboration and networking may facilitate the entrepreneurial culture. Moreover through stakeholder collaboration, cluster creation of the SMEs can be enhanced. Through that it will be easy to develop new innovative practices and new opportunities for tourism SMEs. Moreover, through that it can easily develop a new system or process to capture a greater value from tourism SMEs in Ella.

In addition, the study revealed the issues and challenges which are faced by homestay entrepreneurs. They are lack of security, lack of facilities, competition with unregistered homestay operators, lack of service orientation, difficult to maintain the standards, not enough training to manage homestays, language barriers, lack of cross-cultural understanding, lack of brand image and marketing activities, lack of accessibility for transportation and communication and lack of monitoring from the ministry and Sri Lanka Tourism Development Authority. Apart from that, study discussed some innovate practices to overcome the challenges and problems face by homestay accommodation entrepreneurs.

References

1. Herron, L., Sapienza, H.J.: The entrepreneur and the initiation of new venture launch activities. Entrepreneurship Theory Pract. **17**(1), 49–55 (1992)
2. Kuratko, D.F., Hornsby, J.S., Naffziger, D.W.: An examination of owner's goals in sustaining entrepreneurship. J. Small Bus. Manage. **35**(1), 24 (1997)
3. Campbell, J.J., Dunnette, M.D., Lawler, E.E., Weick, K.E.: Managerial behavior, performance, and effectiveness (1970)
4. Annual Statistical Report—Sri Lanka Tourism Development Authority (2018)
5. Bhuiyan, M.A.H., Siwar, C., Ismail, S.M., Islam, R.: The role of home stay for ecotourism development in east coast economic region. Am. J. Appl. Sci. **8**(6), 540 (2011)
6. Liu, A.: Tourism in rural areas: Kedah. Malaysia. Tourism Manage. **27**(5), 878–889 (2006)
7. Li, L.: A review of entrepreneurship research published in the hospitality and tourism management journals. Tourism Manage. **29**(5), 1013–1022 (2008)
8. Jayasooriya, S.S.W., Perera, K.J.T., Niroshini, S.: Growth of home stay entrepreneurs with reference to Meemure village, Sri Lanka (2019)
9. Jayawardena, C.C., Teare, R., Bandara, C.: Engaging the rural communities of Sri Lanka in sustainable tourism. Worldwide Hospitality and Tourism Themes (2013)

10. Ranasinghe, R.: Evaluation of homestay accommodation supply in Sri Lanka. Int. J. Multidiscip. Res. Dev. **2**(2), 442–447 (2015)
11. Wijesundara, N., Gnanapala, A.C.: Difficulties and challenges related to the development of homestay tourism in Sri Lanka. Tourism, Leisure Global Change **3**, 90–104 (2016)
12. Un.org..: Millennium Development Goals Report 2008. Latest Major Publications—United Nations Department of Economic And Social Affairs (online) (2020). Available at: https://www.un.org/en/development/desa/publications/millennium-development-goals-report-2008.html. Accessed 29 June 2020
13. Karmarkar, A.: A Study on Career Shift of Women from Hotels to Other Fields. Atithya: *A* J. Hospit. **1**(1), 1–5 (2015)
14. Tubbs, M.E., Ekeberg, S.E.: The role of intentions in work motivation: implications for goal-setting theory and research. Acad. Manag. Rev. **16**(1), 180–199 (1991)
15. Mohd, N., et al.: A study on entrepreneurial intention among engineering technology students. Mediterranean J. Soc. Sci. **6** (2015)
16. Bird, B.J.: The operation of intentions in time: the emergence of the new venture. Entrepreneurship Theory Pract. **17**(1), 11–20 (1992)
17. Grossman, G.M., Krueger, A.B.: Economic growth and the environment. Q. J. Econ. **110**(2), 353–377 (1995)
18. Reynolds, P., Miller, B.: New firm gestation: conception, birth, and implications for research. J. Bus. Ventur. **7**(5), 405–417 (1992)
19. Shapero, A., Sokol, L.: The social dimensions of entrepreneurship. In: Kent, C., Sexton, D., Vesper, K.H. (eds.) The Encyclopedia of Entrepreneurship, pp. 72–90. Prentice-Hall, Englewood Cliffs (1982)
20. Boyd, N.G., Vozikis, G.S.: The influence of self-efficacy on the development of entrepreneurial intentions and actions. Entrepreneurship Theory Pract. **18**(4), 63–77 (1994)
21. Carter, N.M., Gartner, W.B., Shaver, K.G., Gatewood, E.J.: The career reasons of nascent entrepreneurs. J. Bus. Ventur. **18**, 13–39 (2003)
22. Kirkwood, J.: Motivational factors in a push-pull theory of entrepreneurship. Gender in Manage. Int. J. (2009)
23. Scheinberg, S., MacMillan, I.C.: An 11 country study of motivations to start a business. In: Kirchoff, B.A., Long, W.A., McMullan, W.E., Vesper, K.H., Wetzel, W.E. (eds.) Frontiers of Entrepreneurship Research, pp. 669–687. Babson College, Wellesley (1988)
24. Giacomin, O., Janssen, F., Pruett, M., Shinnar, R.S., Llopis, F., Toney, B.: Entrepreneurial intentions, motivations and barriers: differences among American, Asian and European students. Int. Entrepreneurship Manage. **J. 7**(2), 219–238 (2011)
25. Barclay, D., Higgins, C., Thompson, R.: The partial least squares (PLS) approach to casual modeling: personal computer adoption ans use as an Illustration (1995)
26. OECD: OECD Employment Outlook. Organization for Economic Cooperation and Development, Paris (1998)
27. Dahles, H.: 10 Tourism, small enterprises and community development. Tourism Sustain. Commun. Dev. **7**, 154 (2002)
28. Embacher, H.: Marketing for Agri-tourism in Austria: strategy and realisation in a highly developed tourist destination. J. Sustain. Tour. **2**(1–2), 61–76 (1994)
29. Razzaq, A.R.A., Hadi, M.Y., Mustafa, M.Z., Hamzah, A., Khalifah, Z., Mohamad, N.H.: Local community participation in homestay program development in Malaysia. J. Mod. Account. Audit. **7**(12), 1418 (2011)
30. Bitzenis, A., Nito, E.: Obstacles to entrepreneurship in a transition business environment: the case of Albania. J. Small Busin. Enterp. Dev. (2005)

Chapter 56
Measuring Design in Businesses and Government—A Framework to Measure Design Impact

Ashreya M. Venkatesh, Kristin L. Wood, and Arlindo Silva

Abstract Private and public sector organizations are realizing significant and lasting economic and social value through the adoption of design (Sheppard et al. in The business value of design. McKinsey & Company (2018), [1]); (2015 dmi: Design value index results and commentary—design management institute (2015), [2]); (Brown in Design thinking. Harvard Business Review, [3, 4]). As such, projects and initiatives that are design focused are increasingly being undertaken by businesses and governments. However, there is only a nascent understanding of how the impact of design should be measured. A comprehensive framework has been developed and tested with businesses and the government of Singapore to measure the outcomes of their designs. The framework consists of four impact areas—enabling economic transformation, raising quality of life, advancing brand and culture, and making ground-breaking achievements in design. The design impact framework considers different types of impact and shows that while certain types of "impact" can be quantified, we also now have other ways for measuring impact and widening its definition. The design impact framework provides a basis and common language for discussing designs across disciplines and scale. The framework and tool jointly create a comprehensive and robust structure that may assist designers and organizations communicate their designs more effectively.

A. M. Venkatesh (✉) · K. L. Wood · A. Silva
SUTD-MIT International Design Center (IDC), Design Innovation (DI) Program, Singapore University of Technology and Design, Tampines, Singapore
e-mail: ashreya@mymail.sutd.edu.sg

K. L. Wood
Comcast Media and Technology Center (CMTC), College of Engineering, Design, and Computing, University of Colorado Denver, Denver, USA

A. Silva
Program Director for the MEng in Innovation By Design, Tampines, Singapore

A. Chakrabarti et al. (eds.), *Design for Tomorrow–Volume 2*, Smart Innovation, Systems and Technologies 222, https://doi.org/10.1007/978-981-16-0119-4_56

56.1 Introduction

56.1.1 Purpose and Scope of This Paper

According to Simon [5], design is "devis[ing] courses of action aimed at changing existing situations into preferred ones." In this context, "Good Design" is, on the one hand, subject to the eye of the beholder (the art or craft), and, on the other hand, an indisputable outcome (the science). In either case, "Good Design" comes about through interdisciplinary design innovation and comprises the integration of people, process, methods, principles, and mindsets [6–8]. Products, services, systems, or combinations are the result of Good Design, creating a transformative design impact on community, society, and people's lives. Design impact can take several forms. Some of these may be quantitative, such as an increase in economic value after a rebranding campaign, improved sustainability, or a greater employee satisfaction after re-designing a workspace. Other forms of impact may be qualitative, such as increased positive sentiments, judgments in ethical sensitivities, and sense of well-being. In the end, Good Design emerges from the designers' well-articulated ethos and creates a "Wow" through its impact. Measuring design impact can be highly subjective. Currently, there is no comprehensive and universally accepted approach to measure the impact of design. The purpose through this paper is to present an assessment framework that can aid in measuring design impact comprehensively and accurately across disciplines and scale. The goal is also to use the framework to assess designs and present the results using a common language.

The design innovation process is governed by a set of methods. These methods guide activities carried out by the people, leading to desired outcomes. The selection of methods and transition between specific methods is governed by a process or approach [9, 10]. Finally, a culture of design is grown organizationally through ubiquitous mindsets and "principles." These principles are cherished and adopted into organizations and their design communities via execution of the methods and approach.

56.2 Measuring Design Impact

56.2.1 Value of Design

Design is a powerful force that has the potential to add and boost competitiveness to organizations, companies, communities, and teams. Design has the potential to add value across many different business areas. The business value of design [1] has been studied extensively. McKinsey, for example, performed a study regarding the value of design. A Mckinsey Design Index (MDI) was rated with each company that was analyzed and its financial performance was observed. Higher MDI scores correlated with higher revenue growth and financial outperformance of top-quartile companies across all three industries studied, albeit product and service-based sectors. The Design Danish Centre in their 2018 Design Delivers report, found that 67% of the companies who use design find their competitiveness improve, 60% of them to a high or very high degree sell more products and/or services and 92% report that design has a positive impact on their bottom line [11], An example of such a company is matter. By using basic design principles, they were able to create a digital platform on the market that allows younger customer segments to follow specific environmental effects of their pension, and more importantly making pension an engaging and relevant topic. Similarly, The Design Management Institute's 2015 Design Value Index (DVI), based on a portfolio of 16 publicly traded stocks from companies considered to be "design-centric" contingent on a set of criteria that reflects best practices in design management, shows a 211% return over the S&P 500, continually for three years in a row [2], 35]. In 2018, IBM was able to cut costs by $20.6 million accelerating projects and increased portfolio profitability by $18.6 million using design thinking [12].

The economic value of design is clear. Design, however, also contributes positively to non-economic or non-financial gains. In one of the largest design maturity studies conducted by Invision, with 2200 companies across 24 industries of various scales across the world, it was found that design enabled 81% of the companies to have improved product usability, 71% delivered higher customer satisfaction, and all of the companies demonstrated an improved employee productivity by an average of 33% [2, 13].

While a number of studies clearly demonstrate the multi-faceted value of design, it is clear that measuring the overall impact of design is not as straightforward as measuring purely financial returns [2, 12]. This limitation is problematic because leaders who have a desire to practice design innovation in an organizational context are often assessed solely on their ability to perform via traditional metrics. Therefore, they have less incentive to work in creative and innovative ways that crosses the boundaries of these traditional metrics [14]. Ultimately, the key goal is to be able to answer questions such as "How do we know that a design is successful?" and "How can we demonstrate to managers that design innovation has a significant impact on the project's or company's outcomes?"

56.2.2 *Why is design important?*

Design innovation and design thinking are important foundations for understanding the value of design [3, 6, 7, 9, 15], and thus companies are turning to design innovation (or innovation by design) to connect deeper with their users-customers and find and/or boost their competitive advantage. When top innovative companies "use" design in their businesses to drive value [16] and when organization adopts design as a norm to stay ahead, design becomes a strategic tool. As more jobs in design become available, as organizations encourage design archetypes across personnel, and as more designers move into leadership roles, leaders in design become responsible for both design and business decisions [17]. This means that designers, and those taking on design archetypes, need to effectively communicate the impact and value of design and how it bridges the gap with the pre-existing functions such as engineering, operations, and business [16].

"You can't manage what you can't measure"—this is the traditional management mindset that executive leaders and managers use to track and measure performances of their teams, departments, and companies [18]. However, when interviewed, more than 50% of business and design leaders "admitted that they have no objective way to assess or set targets for the output of their design teams" [1]. Those who do not measure often do not know how to or lack the resources to do so [19]. Sometimes, measuring the impact of design takes a long time, especially in the case of service design [18]. Because of this characteristic, it becomes extremely challenging to show how and to what extent design work contributes to the overall meaningful value for users-customers, and how effective transformative change may be brought to organizations and companies.

56.3 Design in Singapore

56.3.1 Overview

According to Prime Minister Lee Hsien Loong, as part of his address to the Ministerial Forum at the Singapore University of Technology and Design (SUTD), April 5, 2018 [20]: "At the national level, design is similarly a core element of our nation-building. Singapore is a nation by design. We didn't call it design thinking then, but with each of these major policies, our founding fathers had to understand the issues, define the problem, come up with creative ideas and solutions, prototype the idea, test out the innovations, and constantly review the thinking and solutions… It needs a deep understanding of human beings, their emotions and psychology—how individuals behave, how society works." The ideas expressed by PM Lee focus on Design Thinking. Singapore's pathways in design began with the establishment of the first national design school in 1968. Subsequent advancements in the design ecosystem include emphasis on industrial design business by the Economic Development Board (EDB) in the 1980s, founding of the Design Business Chamber Singapore (DBCS) in 1985, the creation of the Design Singapore Council in 2003, the development of Nanyang Technological University's (NTU) School of Art, Design and Media In 2005, the establishment of the President's Design Award (P*DA) in 2006 [21], the creation of SUTD (sutd.edu.sg) as Singapore's fourth national university focused on Technology and Design, and the launch of the Singapore Good Design Mark Award in 2013. Design Innovation initiatives and Centers are cropping up across the Singapore public and private industry landscape.

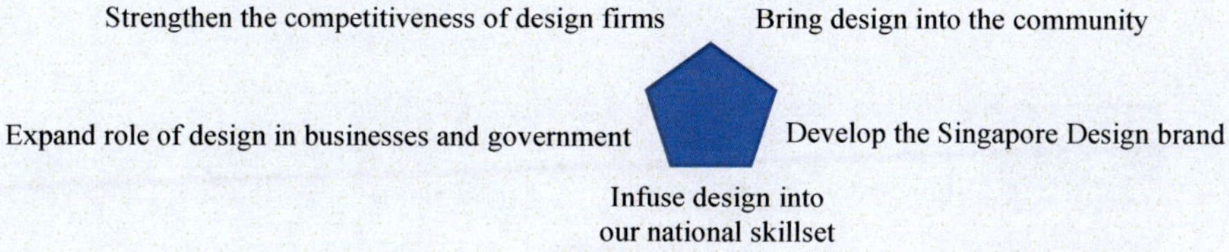

Fig. 56.1 Design 2025 Masterplan

In 2015, a Design Masterplan committee was created where 16 experts from the industry, academia, business, and government were put together to devise the "Design 2025" Masterplan for Singapore. Design 2025 envisions for Singapore to be an innovation-driven economy and a loveable city [22]. The committee drew up five (5) strategic thrusts with a total of 15 recommendations to be implemented as shown in Fig. 56.1. The plan ranges from infusing design as a national skillset to expanding the role of design in businesses and government, strengthening the competitiveness of design firms.

With almost every aspect of the nation adopting design, and from all of the developments in the pathways that Singapore has already created as shown above, it becomes even more pertinent and critical for there to be a framework that allows for the measurement of design impact. This would then translate to the effectiveness of the decisions made to achieve the above goal of the 2025 framework. Singapore being a developed nation served as the initial test bed to test the framework. However, the applications of the framework are for beyond Singapore.

56.4 Design Impact Assessment Framework

56.4.1 Introduction to the Framework

Literature study was carried out on how various international design award programs use their chosen criteria that they do use to assess designs. Design literature across disciplines that focused on measuring design, such as design research and impact of design research on practice (IDRP) literature, were also carefully studied. Some of the exemplar sources that were reviewed include Red Dot Design Award [23], metrics for measuring ideation effectiveness in design studies [24] and The Design 100, Time Magazine [25]. Through interviews with expert designers, a careful study of design projects across companies, an understanding of design award programs across the world, and the concept of a design impact framework was developed. The overall methodology is shown in Fig. 56.2.

Design impact categories, outcomes and metrics (demonstrators) formed the framework. This framework, as a unification of the concept of design impact across disciplines and scales, places emphasis on outcome-based design impact alongside

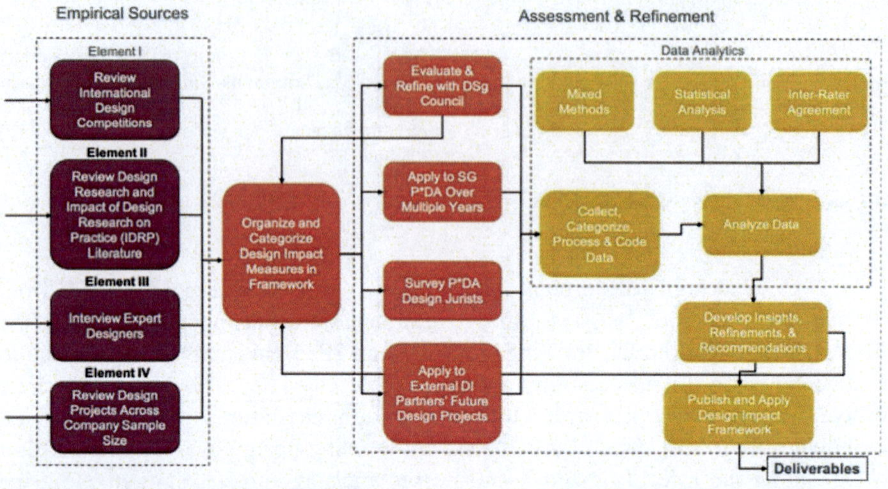

Fig. 56.2 Research study and methodology

recognition of excellence in design craftsmanship and aesthetics. We worked with Singapore and the DesignSingapore (Dsg) Council to explore, define, and test design impact evaluation strategies as part of Singapore's most prestigious national design award program, the P*DA [26]. Through the systematic literature review and interviews with design experts, four impact areas were identified to constitute the framework, initially tested, and utilized as part of the P*DA: Enabling economic transformation, raising quality of life, advancing Singapore's brand and culture, and making ground-breaking achievements in design.

Outcomes and demonstrators were developed to help define and assess each impact area. The outcomes represent sub-categories within each impact area and the demonstrators are the tangible results of designs that may be measured quantitatively or qualitatively. Examples include enhancing employee experience, enhancing living experience, empowering and facilitating positive user learning, increasing revenue, simplifying usage and increasing efficiency and effectiveness, increasing identity and bonds with the community, and innovating processes and functionality. The end result is a table, Fig. 56.3, with the resulting design impact framework.

56.4.2 Design Impact Assessment Methodology

To test the developed design impact framework, a submission portal was developed, through the DesignSingapore Council's Web site, for designers and organizations, across disciplines and across project scales, to submit their designs as part of the Singapore P*DA. Over a three-year period, and across two cycles of submissions for

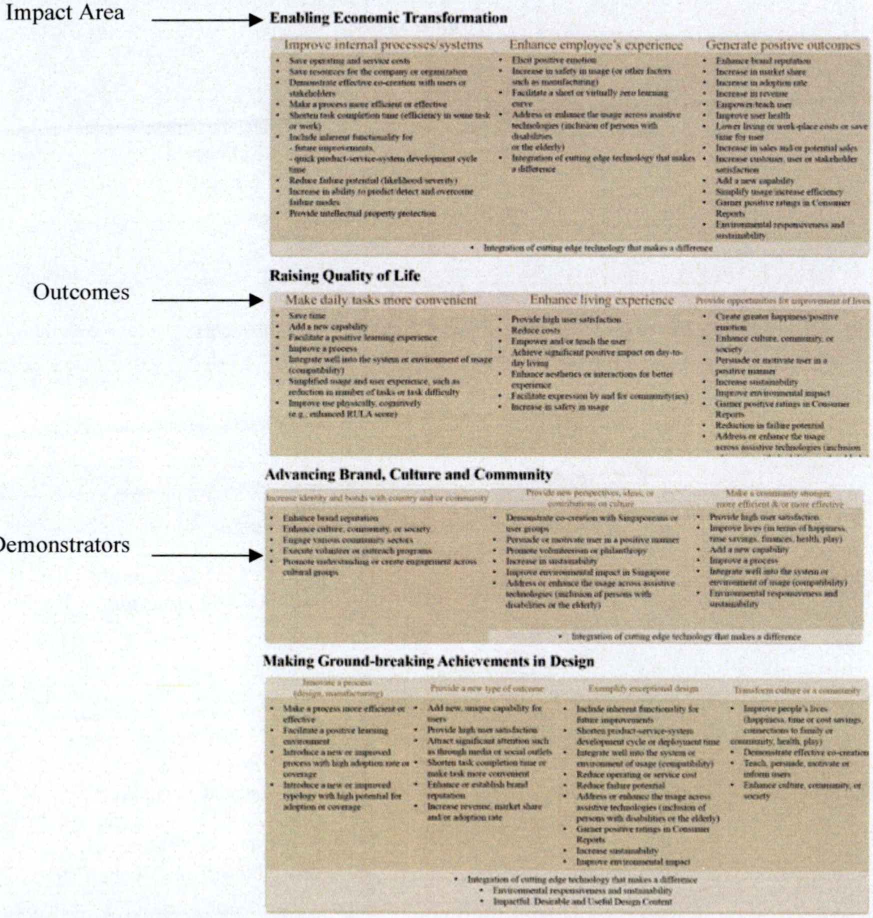

Fig. 56.3 Design impact framework

the P*DA, i.e., P*DA 2018 (design projects only) and P*DA 2020 (design and architecture projects), hundreds of designs were submitted and juries of design experts shortlisted the submissions.

Applicants used the impact assessment framework while submitting their designs through the portal. We then assessed the shortlisted designs using the framework and ultimately to document the extent of impact of the design submissions. The team assessing each submission extracted the area(s) of impact along with their corresponding set of outcomes and demonstrators listed. In parallel, the evidence provided by the applicant was also analyzed. The declared evidence and data were then connected to the claims and checked for their validity. This process was carried out via several methods—online searches, site visits, interviews with designers/design teams, surveys of users and customers, calls to referrals, economic studies, etc.

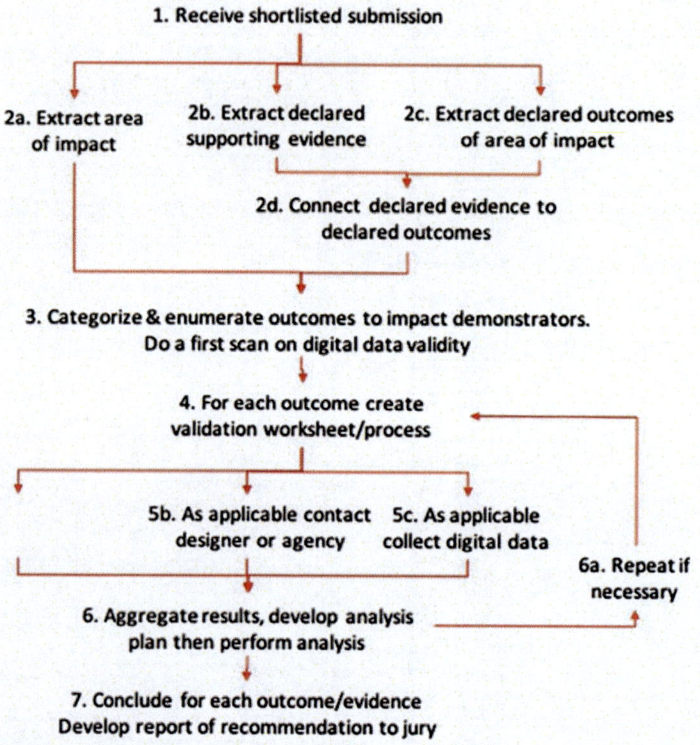

Fig. 56.4 Design impact assessment methodology and workflow engaged

Each project was also independently allocated a "rating" with an overall "Design Impact Rubric" table with five (5) ranks on the impact level—no impact, low impact, moderate impact, evident impact, and extensive impact. Figure 56.4 shows the design impact assessment methodology and workflow used.

A final report documenting all evidence and impact assessment was generated for all shortlisted submission. The final reports went through three to four iterations across assessment teams and members with increasing seniority and design experience levels to ensure that there was consistency throughout all projects. Once sufficient relevant data were collected, the design impact results were collated, summarized, and reported back to the design juries.

56.4.3 *Effectiveness of the Design Impact Assessment framework*

So far the framework has been used through 2 rounds of competitions in 2018 and 2020, with more than a 150 design applications submitted each year. 58 projects

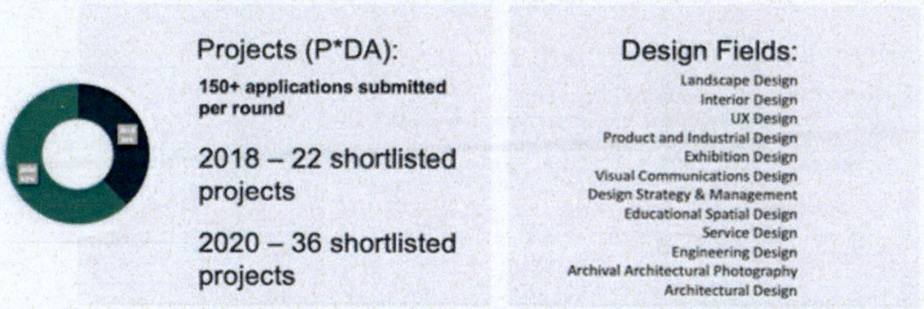

Fig. 56.5 Profile of shortlisted projects from P*DA 2018 and 2020

have been shortlisted using this framework. The design fields of these projects range anywhere from landscape design to UX design to architectural design as shown in Fig. 56.5.

The team assessing each submission extracted the area(s) of impact along with their corresponding set of outcomes and demonstrators. The evidence provided by the applicant was also analyzed. The declared evidence and data were then connected to the claims and checked for their validity. This was done via several methods—online searches, site visits, interviews with designers/design teams, surveys of users and customers, calls to referrals, economic studies, etc. Each project was also independently allocated a "rating" with an overall "Design Impact Rubric" table with five (5) ranks—no impact, low impact, moderate impact, evident impact, and extensive impact—as part of the conclusion of the report, depending on the how strongly the evidence provided back the claims up. A final report documenting all evidence and insights from the impact assessment was reported back to the design jury. Figure 56.6 shows an example of a single claim from a landscape design project from P*DA 2020. The exact is anonymized but shows sufficient detail to understand the connection the design impact framework.

Analyzing the 58 projects from 2018 and 2020 submissions revealed the following insights:

- 4 out of 4 (100%) impact areas, 13 out of 13 (100%) outcomes, and 53 out of 57 demonstrators (93%) were used at least once.
- There are a total of 57 unique demonstrators that have been used a total of 442 times.
- The top 10 primary demonstrators are shown in Fig. 56.7 with "increase customer, user or stakeholder satisfaction" being most commonly used.

Another insight revealed that there is no correlation between the number of demonstrators and impact areas and the assessment of the impact as shown in Fig. 56.8. There were projects that had 1 impact area, 1 outcome and a few demonstrators, with an equivalent rating to projects that had multiple of all. This is because designers tend to think their design has a greater coverage of impact than the truth.

Feedback was collected from the jury members who were part of the P*DA 2018 competition. Around 80% of them found the assessment reports to be extremely

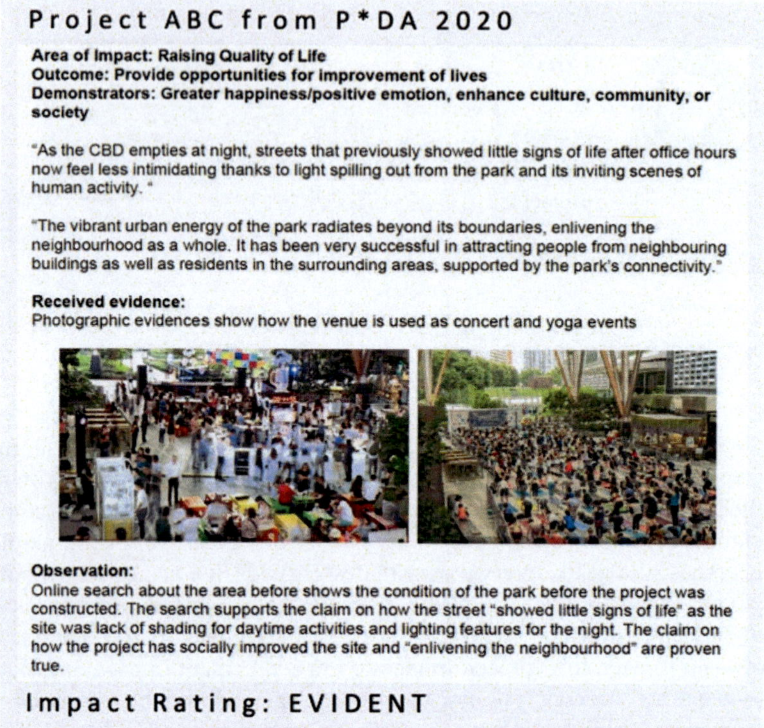

Fig. 56.6 Single claim from a P*DA 2020 landscape design project being assessed

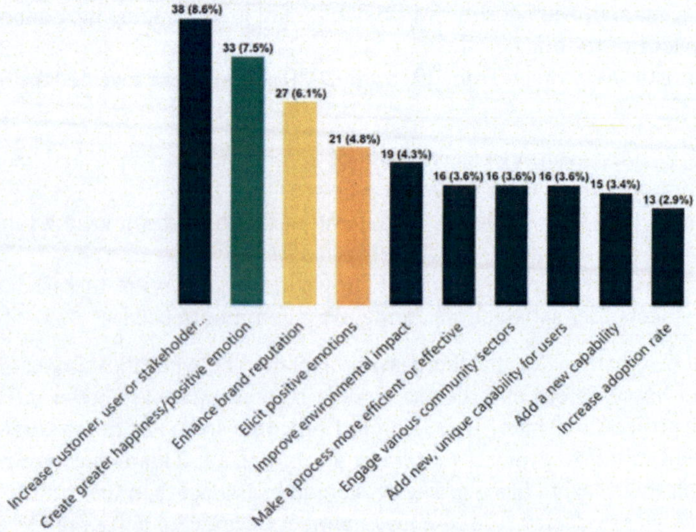

Fig. 56.7 Top 10 demonstrators used most commonly across all projects from P*DA 2018 and 2020

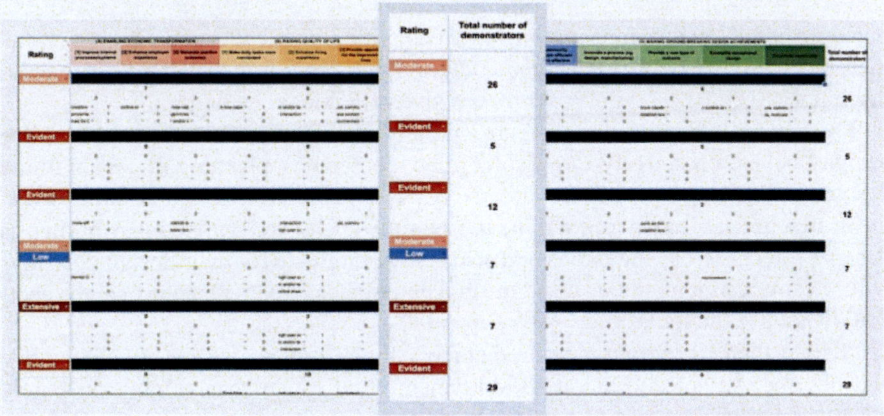

Fig. 56.8 No correlation between the overall impact rating of a project and the number of demonstrators claimed

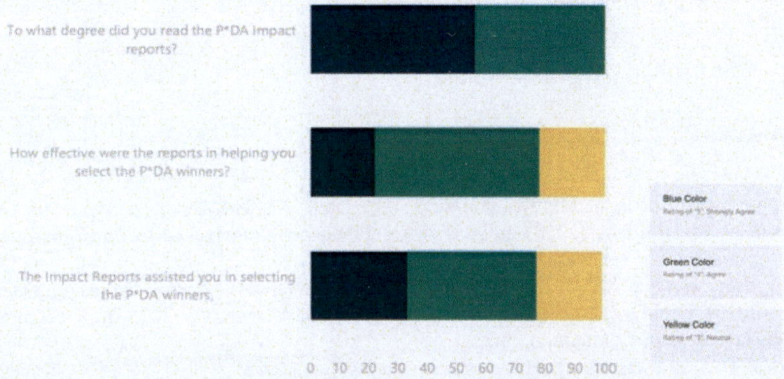

Fig. 56.9 Jury members' feedback on the reports using the impact framework

helpful and in fact assisted them in choosing the winners of the PDA competition as shown in the graphs below in Fig. 56.9.

56.5 Conclusion and Future Work

Overall, these insights and conclusions include: the design impact framework is effective and comprehensive in representing, understanding, and measuring design impact. The design impact framework provides a basis and common language for discussing designs across disciplines and scale. This framework places emphasis on outcome-based design impact alongside recognition of excellence in design craftsmanship. Designers who bring together diverse skills and work collaboratively across disciplines to address complex issues are given significant recognition.

There could also be value in representing design impact for not only assessing designs, but for proactively forecasting project outcomes, planning impact in future designs and driving teams more effectively. As such, looking ahead at future work, the design impact framework will be used jointly and tested with teams working on design projects across industries and scales. The rationale for their design decisions will be captured along the way and the end output measured with teams and projects that do not use the tools.

"Rather than something measured at the end of a project, it needs to be iterative and become part of continuous development. And rather than relying on purely quantitative data, it needs to become more experience and practice-led." [27]. We believe this is exactly what we are able to do.

Acknowledgements This research was sponsored by the DesignSingapore (Dsg) Council, as part of the President's Design Award (P*DA). SUTD-MIT International Design Center (IDC) served as the official Knowledge Partner for the P*DA program.

References

1. Sheppard, B., Sarrazin, H., Kouyoumjian, G., Dore, F.: The Business Value of Design. McKinsey & Company (2018). https://www.mckinsey.com/business-functions/mckinsey-design/our-insights/the-business-value-of-design
2. 2015 dmi: Design Value Index Results and Commentary—Design Management Institute. (2015). Dmi.Org. https://www.dmi.org/page/2015DVIandOTW/2015-dmiDesign-Value-Index-Results-and-Commentary.htm
3. Brown, T.: Design Thinking. Harvard Business Review (2008)
4. Brown, T.: Change by Design: How Design Thinking Transforms Organizations and Inspires Innovation. (Revised, Updated edition (March 5, 2019)). Harper Business (2019).
5. Simon, H.A.: The Sciences of the Artificial. MIT Press, Cambridge (1996)
6. Tushar, W., Lan, L., Withanage, C., Sng, H.E.K., Yuen, C., Wood, K.L., Saha, T.K.: Exploiting design thinking to improve energy efficiency of buildings. Energy **197**, 117141 (2020)
7. Tushar, W., Basnet, S., Yuen, C., Xu, Y., & Wood, K. (2017). Design innovation for a smarter Singapore: a case study on smart energy innovation. The Singapore Engineer Magazine, 18–25.
8. Red Dot Design Award.: www.Red-Dot.Org (n.d.). Retrieved 20 Oct. 2020, from https://www.red-dot.org/
9. Camburn, B.A., Auernhammer, J.M, Hui K., Mignone, P.J., Arlitt, R.M., Perez, K.B., Huang, Z., Basnet, S., Blessing, L.T., Wood, K.L.: Design innovation: a study of integrated practice. In: ASME 2017 International Design Engineering Technical Conferences and Computers and Information in Engineering Conference (IDETC), pp. V007T06A031-V007T06A031. American Society of Mechanical Engineers (2017)
10. Amarsy, N.: Why and How Organizations Around the World Apply the Business Model Canvas. (online) Strategzer (2015). Available at: blog.strategyzer.com/posts/2015/2/9/why-and-how-organizations-around-theworld-apply-the-business-model-canvas. Accessed May 2020
11. Design Delivers 2018: How Design Accelerates your Business (2018)

12. Brown, B.: The Total Economic Impact TM of IBM's Design Thinking Practice How IBM Drives Client Value and Measurable Outcomes With Its Design Thinking Framework (2018). https://www.ibm.com/design/thinking/static/Enterprise-Design-Thinking-Report-8ab1e9e1622899654844a5fe1d760ed5.pdf
13. Design Maturity Model by InVision: The New Design Frontier. Design Maturity Model by InVision: The New Design Frontier (2019). https://www.invisionapp.com/design-better/design-maturity-model/
14. Koch, C., Meinel, C., Leifer, L.: Introduction: The HPI-Stanford Design Thinking Research Program. Understanding Innovation, 5–12 (2016). https://doi.org/https://doi.org/10.1007/978-3-319-19641-1_2
15. Lauff, C., Menold, J., Wood, K.L.: Prototyping Canvas: Design Tool for Planning Purposeful Prototypes. In Proceedings of the Design Society: International Conference on Engineering Design, vol. 1(1), pp. 1563–1572. Cambridge University Press (2019)
16. Rumsey, R.: Connect design & business. Business Thinking for Designers—DesignBetter (2020). https://www.designbetter.co/business-thinking-for-designers
17. SSG Design - 24102019
18. Cisero, C., Ji, V., Marcoli, S., Diana, C.: SDN|You Can't Manage What You Can't Measure (2017). Www.Service-Design-Network.Org. https://www.service-design-network.org/touchpoint/vol9-2-measuring-impact-and-value/you-cant-manage-what-you-cant-measure
19. Plattner, H., Meinel, C.: In: Meinel, C., Leifer, L., (eds.) Design Thinking Research: Making Design Thinking Foundational. www.springer.com. Springer International Publishing (2016). https://www.springer.com/gp/book/9783319196404
20. Singapore, P.M.O.: PMO|PM Lee Hsien Loong at the SUTD Ministerial Forum. Prime Minister's Office Singapore (2018). https://www.pmo.gov.sg/Newsroom/pm-lee-hsien-loong-sutd-ministerial-forum
21. President's Design Award Singapore.: Pda2020.Sg (2020). https://www.pda2020.sg/judgeImpact Assessment Framework
22. Design 2025.: Base (n.d.). https://www.designsingapore.org/what-we-do/resources/design-2025.html
23. Shah, J.J., Smith, S.M., Vargas-Hernandez, N.: Metrics for measuring ideation effectiveness. Des. Stud. **24**(2), 111–134 (2003)
24. The Design 100, Time Magazine
25. Borges, L., Soares, P.: McKinsey Design Index. Solutions.Mckinsey.Com (n.d.). Retrieved 21 Oct. 2020, from https://solutions.mckinsey.com/design-index/
26. https://www.designsingapore.org/presidents-design-award)
27. Drew, C.: SDN | You Can't Manage What You Can't Measure. www.Service-Design-Network.Org (2017). https://www.service-design-network.org/touchpoint/vol9-2-measuring-impact-and-value/you-cant-manage-what-you-cant-measure

Chapter 57
Design, Frugal Innovations, and Low-Resource Settings: An Analysis of Five Contextual Aspects

Santosh Jagtap

Abstract Designing frugal innovations is crucial to alleviate problems faced by people living in low-resource settings. Many design studies have been undertaken in such low-resource settings. These studies are discussed using a variety of names like 'frugal innovations', 'appropriate technology', 'design for the Base of the Pyramid (BOP)', 'product service systems for BOP', 'community development engineering', 'design for development', etc. There is an important need to know in what context these studies were undertaken. In order to gain an in-depth understanding of the contextual aspects of these studies, we review a wide range of literature, focussing on design studies in this field. The review findings show a multifaceted picture, revealing a large variety in examination and presentation of contextual aspects such as income, design sectors, countries, rural–urban, and gender. Based on the review findings, we offer recommendation for practice, education, and research of designing frugal innovations in low-resource settings.

57.1 Introduction

People living in low-resource settings in developing countries face numerous problems. Their income is irregular and unpredictable [38]. They generally live in shanty towns, urban slums, or rural villages, and have little or no formal education. They typically lack access to basic facilities including, among others, clean water, infrastructure, sanitation, public health, and energy. They are generally not good at saving and investing, and prefer immediate consumption. This short-term approach can be due to their weak access to various resources and assets (e.g. [16]).

Design, with its core idea of changing an existing or undesired state into a desired state, has a prominent role in satisfying unmet or under-served needs of people living in low-resource settings at the base of the world income pyramid, typically known as the base of the pyramid (BOP). Alleviating poverty or satisfying needs of BOP people demands the design of frugal innovations, which can manifest in several forms

S. Jagtap (✉)
Blekinge Institute of Technology, Karlshamn, Sweden
e-mail: santosh.jagtap@bth.se

A. Chakrabarti et al. (eds.), *Design for Tomorrow–Volume 2*, Smart Innovation, Systems and Technologies 222, https://doi.org/10.1007/978-981-16-0119-4_57

such as products, services, or product service systems (e.g. [15]). Some examples of such frugal innovations are medical devices, educational devices, income-generating products or services, vaccination plans, systems supporting agriculture and water distribution, or any other products and services that support social and human development of BOP individuals and communities. Design of these frugal innovations is undertaken, for example, by NGOs, governments, for-profit companies, and BOP people as well (e.g. [15, 28]).

Whilst design research has been predominantly undertaken in Western countries or relatively affluent regions Jagtap [13], many design studies have been undertaken in the context of BOP societies in developing countries. These design studies in the BOP context are discussed using names like 'frugal innovations', 'appropriate technology', 'design for the Base of the Pyramid', 'product service systems for the BOP', 'community development engineering', 'design for development', etc. (e.g. [7, 16]). Furthermore, several universities have begun to offer projects or courses in this field, while undertaking design investigation to understand and support design for supporting social and human development of the BOP (e.g. [17]).

Whilst several design studies have been undertaken in this field, there is absence of their analysis, making it difficult to gain an understanding of context in which they were carried out. Since design is a context-specific activity, it is crucial to gain an overview of contextual aspects (e.g. countries, rural–urban, design sectors, etc.) of these studies. Therefore, we aim at analysing contextual aspects of these studies. This is addressed by reviewing the relevant literature, focusing on contextual aspects of the studies. Analysing and scrutinising contextual aspects of the design studies are crucial in undertaking further design research in this field and in developing and testing methods and tools to support practice and education of designing frugal innovations. We identify crucial areas that are still not explored, offering recommendations for design research this field. Our purpose is to explain some of its current issues and to suggest areas for scholarly exploration of this field.

Following this introduction, the rest of the paper is organised as follows. Section 57.2 presents an overview of design research in BOP societies. Section 57.3 presents the research methodology employed. Section 57.4 discusses five contextual aspects as revealed in this review. Finally, Sect. 57.5 discusses the review findings, offering suggestions for future research in this field.

57.2 Design, Frugal Innovations, and Low-Resource Settings

'Design for the Real World' and 'Appropriate Technology' concepts, articulated in the 1970s, by Victor Papanek and E. F. Schumacher, respectively, can be considered as early attempts of employing design to support development of BOP societies [27, 31]. Papanek's movement suggested designers to undertake projects for satisfying needs of people in developing countries. The appropriate technology movement, motivated by the failures in transferring products and technologies from industrialised to developing countries, suggested designing suitable technologies, contextualised for societies in developing countries. The appropriate technology movement could not deliver expected impact in addressing problems of marginalised societies in developing countries.

In the case of the 'Design for the Real World' and 'Appropriate Technology' movements, the role of NGOs in undertaking design projects can be recognised. On the other hand, the roles of multinational corporations (MNCs) are explicit in the BOP concept articulated by Prahalad and his colleagues in 1998–1999 [29]. In parallel to the BOP concept, a complimentary concept, known as 'subsistence marketplaces' approach, is articulated by Viswanathan and his colleagues evolved (e.g. [35]). In this approach, they have developed a bottom-up approach for designing solutions to address needs of BOP individuals and societies.

In recent years, research on the design of affordable products and services using several terms such as 'frugal innovation', 'jugaad innovation' and 'grass root innovation' is developing (e.g. [10, 30]). These innovations generally consider creating highly affordable and low-cost innovations in a low-resource setting. Furthermore, similar concepts using names such as community development engineering, humanitarian engineering, engineering for development, and design for extreme affordability are growing (e.g. [25]). Engineering and design departments in many universities support and carry out design projects, while offering courses in this field [17]. This has resulted into many design studies undertaken in the BOP context.

57.3 Research Methodology

Design investigations in the BOP context have been carried out in several contexts in many developing countries. Since design is a social process (e.g. [3, 22]), with close connection between design practice and the context of the practice, it is important to examine contextual aspects of the studies. This analysis of contextual aspects is useful in understanding similarities and differences between contexts researched in various studies. We searched databases such as Scopus, Web of Science, and Google Scholar to search for design studies in this field. This paper discusses a broad range of studies, with close analysis of 30 papers, focusing on their contextual aspects (e.g. income, design sectors, countries, rural–urban, and gender). Of the 30 papers, 25 are journal and five are conference papers. The journal papers are published in outlets such as 'Co-Design', 'Design Studies', 'Engineering Studies', 'Research in Engineering Design', 'Technology in Society', 'International Journal of Design', 'Design Management Review', 'Journal of Product Innovation Management', 'Design Issues', 'Journal of Mechanical Design', etc. The 30 papers present studies that have been undertaken from design perspective, for example, studies aimed at understanding and supporting how products, systems and services are designed in the BOP context. This selection of articles allowed a focused approach, permitting their close analysis. Due to the constraint on the number of pages allowed in this paper, we have not included the list of 30 papers. This list is available in our previous study [11].

57.4 Contextual Aspects in Designing Frugal Innovations

The reviewed studies consider contextual aspects such as rural or urban region, income of BOP people, design sectors, specific developing countries, and gender aspects. These contextual aspects ought to be considered in undertaking design research in BOP societies, and in developing and evaluating methods and tools to create intended impact on design practice in this field (e.g. [2, 16]). These contextual aspects are discussed in the following subsections.

57.4.1 Income

Income is a quantifiable measure of poverty. Income enables people to buy products and services and permits them to fulfil basic needs, reducing their vulnerability and powerlessness. Therefore, income-based measurement of poverty has received attention [20]. Some of the reviewed studies have defined or reported income. For instance, income level of $4 a day is reported in Jagtap et al. [16]. Murcott [23] and Thomas [33] define the level of $1 a day in their paper. Whilst there is diversity and inconsistency in reporting and defining income, these studies provide some information on the examined contexts. Studies not mentioning or defining income threshold leave much room for interpreting the context. This also makes it difficult to compare contexts examined in such studies.

In addition to income, qualitative aspects of poverty also provide important insights into the examined contexts. Many studies qualitatively report other dimensions of poverty. For example, poverty has been described as social, cultural, and political exclusion, in terms of various constraints and deprivations, in terms of problems of gender inequality, or in terms of vulnerability due to effects of climate-changes (e.g. [15, 24]). Such qualitative aspects expand the concept of poverty measurement from income-based to multidimensional construct. Overall, it is crucial to describe qualitative characteristics of the context and define or report income.

57.4.2 Rural and Urban

Using the multidimensional poverty Index (MPI), Alkire et al. [1] compared rural and urban poverty, revealing differences in various characteristics of poverty in these areas. The three dimensions—education, health, and standard of living—are taken into account in the estimation of the MPI. These three dimensions are assigned equal weights. Whilst greater prevalence of poverty is seen in rural than urban areas, poor people from rural regions are migrating to urban areas. Deprivations in electricity, water, and flooring contribute more to MPI in the case of rural poverty. On the other hand, deprivations in school attendance, child-health, and nutrition contribute more in the case of urban poverty [1]. Furthermore, social networks of poor people in urban areas are relatively weak, and they face problems of crime and social unrest.

Design studies are undertaken in both rural and urban areas of developing countries. For example, Whitney and Kelkar [36] have undertaken an ethnographic design study in urban slums in India and Guimarges et al. [9] studied how small firms in urban areas in Brazil design products. In addition to studies in urban areas, many studies have been undertaken in rural areas. For example, Nieusma and Riley (2010) examined how engineering designers consider cultural and technological aspects in designing products for marginalised communities. As a part of their study, they undertook a case study in a rural village in Sri Lanka. Their findings suggest that there is a tendency to consider mainly technological factors, with less emphasis on the issues related cultural and social aspects. Design studies have also dealt with the design of pumps and irrigation systems to support farmers in their agricultural activities (e.g. [14]). It is important to direct design efforts for alleviating poverty in both rural and urban areas, taking into account specific requirements and problems in these areas (e.g. [11]). For example, since rural village in developing countries can be geographically dispersed, issues of marketing and distribution need to be handled carefully in designing frugal innovations for rural areas. Whilst studies have been undertaken in both rural and urban areas of developing countries, the literature has just presented general guidelines to design frugal innovations for people in these areas.

57.4.3 Design Sectors

Although design studies have been carried out in several sectors like healthcare, water, energy, etc., several cases are drawn from some specific sectors, in particular, from sectors such as artisanal-goods, information and communication technologies (ICTs), healthcare, and water. For instance, in healthcare sector, some studies have aimed at supporting designers to generate concepts for healthcare devices or to develop holistic understanding of target community (e. g. [2]). In a similar fashion, studies in artisanal-goods and ICTs sectors, for example, have developed methodologies to support design activities of craftspeople or to support the development of ICTs for marginalised societies (e.g. [5, 8]).

Studies such the above have been carried out in one of the sectors. On the other hand, some studies have explored several design sectors. For example, by analysing data from the Growing Inclusive Markets (GIM) initiative of the United Nations Development Programme (UNDP), Jagtap et al. [16] examined strategies used by businesses in designing and developing frugal innovations for the BOP. Likewise, by analysing studies from several sectors, guidelines have been developed to aid the design of integrated frugal innovations for BOP societies [12].

57.4.4 Countries

There is variation in poverty rates across developing couriers. In some developing countries, poverty rates have increased [26]. The Development Assistance Committee (DAC) of the Organisation for Economic Co-operation and Development (OECD) has categorised developing countries into four types—'least developed', 'low income', 'lower middle income', and 'upper middle income'. The classification of developing countries into 'low income', 'lower middle income', and 'upper middle income' categories is based on the Gross National Income per capita. The 'least developed' countries are 'the poorest and weakest segment of the international community', representing approximately 12% of the global population and supporting not more than 2% to the world GDP [34].

Several studies are carried out in, or use or refer to secondary data, from India—a 'lower middle income' country (for example, see Subrahmanian et al. [32]). From the 'least developed' category, studies are carried out mainly in Nepal, Bangladesh, and Cambodia (e.g. [19, 23]). Similarly, from the 'upper middle income' category, studies focus primarily on South Africa and Brazil. From the 'low income' category, studies are undertaken in Zimbabwe and Kenya (e.g. [7, 33]). Findings of a few studies are applicable to developing countries in general (e. g. [2, 37]). However, there is a need of category-related or country-related design guidelines.

57.4.5 Gender

As compared to men living in poverty, women living in poverty face numerous problems, weakening their access to, for example, healthcare and education (e.g. [4]). Biased allocation of resources within households also weakens their capabilities such as health and literacy level. Poor women are more susceptible to biases in labour markets Kabeer [18] In addition, they suffer more from the lack of public services (e.g. problems due to lack of access to toilets). As gender equality is crucial for social and human development of societies, one of the sustainable development goals (SDGs) focuses on gender equality.

The reviewed studies reveal a broad variation in presenting details about gender aspects. While most of the studies have not thoroughly dealt with gender aspects, a few studies have focused on such gender aspects. These studies, for example, have developed guidelines, prescribing designers to give attention to the needs of poor women, have developed design methodologies for craftswomen, or have presented details of how women in a community can contribute towards the implementation of frugal innovations (e.g. [8, 15]. Overall, a broad range exists in dealing with gender aspects in the studies, stressing the need for considering these crucial aspects in design research in this field.

57.5 Discussion and Recommendations

Many design investigations have been carried out to examine and support the design of frugal innovations in the BOP context. The reviewed literature presents a great diversity in terms of contexts examined. There is also a diversity in how details of contextual aspects are (not) reported.

The review of contextual aspects of the reviewed papers allows suggesting areas for further research in this field (see Fig. 57.1. Some suggestions focus on reporting details of contextual aspects. Overall, there is a need to report detailed information on the characteristics of the context in which studies are undertaken. Researchers can add such detailed information in scientific articles, for example, by adding supplements as scientific journals typically allow supplements. Detailed information on contextual aspects can help in understanding commonalities and differences between various contexts, providing a basis for making potential generalisations or explaining variations in findings obtained from similar or dissimilar contexts. Konda et al.'s [21] concept of 'shared memory' in design also supports the need of providing rich and detailed information on contexts examined. In design, shared memory helps in developing better understanding of design situations and related contexts. Providing

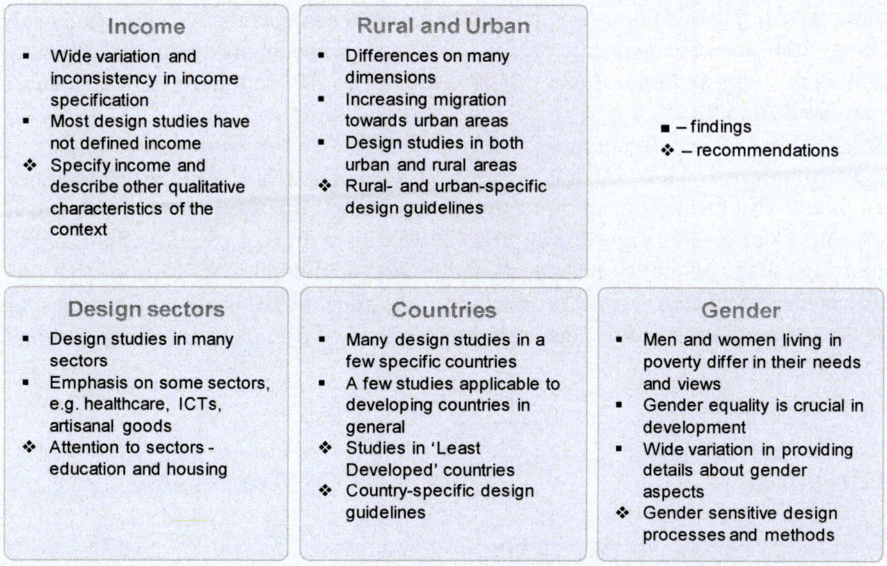

Fig. 57.1 Contextual aspects of the BOP and design of frugal innovations

detailed contextual information about marginalised communities and influence of frugal innovations on those communities is essential to determine if those innovations can be applicable and used in other communities. In order to avoid reuse of failed innovations, it is crucial to have access to rich and detailed information on related contexts of marginalised settings in which those innovations were unsuccessful. Reporting the details of contextual aspects is also crucial to establish external validity of a study—that is—to assess the degree to which the conclusions of a study can be applicable to other marginalised settings. Various contextual aspects, as identified and examined in the present paper, can usefully provide an initial basis for reporting a study carried out in marginalised contexts. It is expected that this review of contextual aspects will motivate scholars to identify further contextual aspects (if any) of marginalised communities.

In addition to providing thorough information on contextual aspects, future research in this field can benefit from using appropriate research methods to glean information on such contextual aspects. For example, since men and women living in marginalised societies largely differ in their needs and views on problems, it is important to adopt gender-sensitive research methods in collecting information on contextual aspects.

Future research in this field can also benefit from examining areas that are not explored. For instance, because many design studies are undertaken in a few specific countries, further research ought to be undertaken in other countries, specifically in countries from the 'least developed' category of the DAC's categorisation of developing countries [6]. Furthermore, the diversity across developing counties, rural and urban areas, and sectors can provide promising and fascinating opportunities for future research in this field. As such, future research can aim at developing and evaluating rural- and urban-specific design guidelines or at supporting design in sectors such as housing and education. It is also crucial to develop and evaluate gender-sensitive design methods and processes to support design of frugal innovations for both men and women living in BOP societies.

To summarise, whilst design research in the context of marginalised societies can potentially create impact on design practice and education in this field, and it can gain from a more rigorous approach. This involves, for example, thoroughly reporting details of contextual aspects and related methodologies. We hope that this work can motivate and support researchers to examine the recommended avenues, as they are crucial for practice, education and research of designing frugal innovations in this field.

References

1. Alkire, S., Chatterje, M., Conconi, A., Seth, S., Vaz, A.: Poverty in Rural and Urban Areas: Direct Comparisons Using the global MPI 2014 (2014)
2. Aranda Jan, C.B., Jagtap, S., Moultrie, J.: Towards a framework for holistic contextual design for low-resource settings. Int. J. Des. **10**(3), 43–63 (2016)
3. Bucciarelli, L.: Designing Engineers. MIT, Cambridge (1994)

4. Cagatay, N.: Gender and Poverty. UNDP, Social Development and Poverty Elimination Division (1998)
5. Cai, J., Yang, D., Li, D.: Designing the right IT services for the bottom of the pyramid. Commun. Assoc. Inf. Syst. **19**(1), 22 (2007)
6. DAC, https://www.oecd.org/dac/stats/documentupload/DAC%20List%20of%20ODA%20Recipients%202014%20final.pdf. Last accessed Jan. 2016.
7. Donaldson, K.: The future of design for development: three questions. Inf. Technol. Int. Dev. **5**(4), 97 (2009)
8. Girón, J.D.L.P., Domínguez Hernández, M.L., Jiménez Castañeda, J.C.: Participatory methodologies and the product development process: the experience of Mixtec craftswomen in Mexico. Dev. Pract. **14**(3), 396–406 (2004)
9. Guimarges, L.E., Penny, J.E., Moody, S.: Product design and social needs: the case of North-East Brazil. Int. J. Technol. Manage. **12**(7–8), 849–864 (1996)
10. Gupta, A.K.: Grassroots innovation: minds on the margin are not marginal minds. Random House India (2016)
11. Jagtap, S.: Design and poverty: a review of contexts, roles of poor people, and methods. Res. Eng. Des. **30**(1), 41–62 (2019)
12. Jagtap, S.: Key guidelines for designing integrated solutions to support development of marginalised societies. J. Cleaner Prod. **219**, 148–165 (2019)
13. Jagtap, S.: Design creativity: refined method for novelty assessment. Int. J. Des. Creat. Innov. **7**(1–2), 99–115 (2019)
14. Jagtap, S., Kandachar, P.: Representing Interventions from the base of the Pyramid. J. Sustain. Dev. **3**(4), 58–73 (2010)
15. Jagtap, S., Larsson, A.: Design of Product Service Systems at the Base of the Pyramid. In: Chakrabarti, A., Prakash, R.V. (eds.) ICoRD'13, pp. 581–592. Lecture Notes in Mechanical Engineering, Springer India (2013)
16. Jagtap, S., Larsson, A., et al.: Design and development of products and services at the Base of the pyramid: a review of issues and solutions. Int. J. Sustain. Soc. **5**(3), 207–231 (2013)
17. Jagtap, S., Larsson, A., Hiort, V., Olander, E., Warell, A., Khadilkar, P.: How design process for the base of the pyramid differs from that for the top of the pyramid. Des. Stud. **35**(5), 527–558 (2014)
18. Kabeer, N.: Agency, well-being & inequality: reflections on the gender dimensions of poverty. IDS Bull. **27**(1), 11–21 (1996)
19. Kang, L.: Social design as a creative device in developing Countries: the case of a handcraft pottery community in Cambodia. Int. J. Des. **10**(3), 65–74 (2016)
20. Karnani, A.: Fighting Poverty Together—Rethinking Strategies for Business, Governments, and Civil Society to Reduce Poverty. Palgrave Macmillan, New York (2011)
21. Konda, S., Monarch, I., Sargent, P., Subrahmanian, E.: Shared memory in design: a unifying theme for research and practice. Res. Eng. Des. **4**(1), 23–42 (1992)
22. Minneman, S., Leifer, L.: Group engineering design practice: the social construction of a technical reality. In: Proceedings of the International Conference on Engineering Design (ICED). The Hague (1993)
23. Murcott, S.: Co-evolutionary design for development: influences shaping engineering design and implementation in Nepal and the global village. J. Int. Dev. **19**(1), 123–144 (2007)
24. Narayan, D., Chambers, R., Shah, M.K., Petesch, P.: Voices of the Poor: Crying Out for Change. The World Bank—Poverty Division (PRMPO, New York (2000)
25. Nielsen, C., Samia, P.M.: Understanding key factors in social enterprise development of the BOP: a systems approach applied to case studies in the Philippines. J. Consum. Market. **25**, 446–454 (2008)
26. Olinto, P., Beegle, K., Sobrado, C., Uematsu, H.: The state of the poor: where are the poor, where is extreme poverty harder to end, and what is the current profile of the world's poor?. Econ. Prem. **125**(2) (2013)
27. Papanek, Victor, Buckminster Fuller, R.:. Design for the Real World. Thames and Hudson, London (1972)

28. Prahalad, C.K.: The Fortune at the Bottom of the Pyramid: Eradicating Poverty Through Profits. Wharton School Publishing, Upper Saddle River (2004)
29. Prahalad, C.K., Hart, S.: Strategies for the Bottom of the Pyramid: Creating Sustainable Development (Working paper). University of Michigan, Ann Arbor (1999). Retrieved from https://www.bus.tu.ac.th/usr/wai/xm622/conclude%620monsanto/strategies.pdf
30. Radjou, N., Prabhu, J., Ahuja, S.: Jugaad Innovation: Think Frugal, Be Flexible, Generate Breakthrough Growth. Wiley, New York (2012)
31. Schumacher, E.F.: Small is Beautiful: Economics as if People Mattered. Harper and Row, New York (1973)
32. Subrahmanian, E., Eckert, C., McMahon, C., Reich, Y.: Economic development as design: insight and guidance through the PSI framework. In: International Conference on Engineering Design, ICED17, Vancouver, Canada (2017)
33. Thomas, A.: Design, poverty, and sustainable development. Des. Issues **22**(4), 54–65 (2006)
34. UN (2017), https://unohrlls.org/about-ldcs/. Last accessed July 2017
35. Viswanathan, M., Rosa, J.A.: Product and market development for subsistence marketplaces: Consumption and entrepreneurship beyond literacy and resource barriers. In: Cheng, J., Hitt, M. (eds.) Advances in International Management Series, pp. 1–17. Elsevier, Oxford, UK (2007)
36. Whitney, P., Kelkar, A.: Designing for the base of the pyramid. Des. Manage. Rev. **15**(4), 41–47 (2004)
37. Wicklein, R.C.: Designing for appropriate technology in developing countries. Technol. Soc. **20**(3), 371–375 (1998)
38. World Bank: World Development Indicators. World Bank Publications, Washington (2010)

Part VIII
Fablabs, Makerspaces and Design Spaces

Chapter 58
Journey of Product in Maker Spaces—A Case Study

Arun Soman and Neha Chourasia

Abstract India being a country of variety, there is a variety of innovative products, available and made. Here is a journey of a small product and its development with the help of different maker spaces. The interest to use bamboo as a product designer has always pushed the imagination after experiencing the material. Bamboo is a very common material and has been a major source of living for many tribes. In past decades, mainly south Asian countries have brought bamboo as an industrial material. Just research and development in bamboo is not enough as the world is changing rapidly. An emerging trend of design collaboration is the key to our problems. As our environment and resources are changing, it puts us into a great threat. As the technology could not grow beforehand, the growth of such technology depends on research and development of materials and product and user experiences. The small-scale industry could not afford such research and development, designers, and engineers do not have laboratories to innovate. Thus, a makers culture brought in some revolution were in, making is more focused then designing or innovation. Making something automatically inculcates innovation based on the requirement and comfort of the product. With this maker's culture, many makers were inspired but not everyone can avail space as per their need. In India, such space might be seen especially in the art institute, but these spaces are restricted to students only. Such a concept was taken forward and some maker spaces and design spaces were evolved. The paper will have a comprehensive and case study of a product. This product which was incepted by a maker's space and went through different maker spaces until its first prototype. The further study shall include the development requirement of a maker's space and survey on some other maker spaces. Post research will be examined on the need of such maker space and its boon to the society.

A. Soman (✉)
ImaginXP, Pune, India
e-mail: arun.piads@gmail.com

N. Chourasia
SNAP Associates, New Delhi, India
e-mail: ar.nehajc@gmail.com

A. Chakrabarti et al. (eds.), *Design for Tomorrow–Volume 2*, Smart Innovation, Systems and Technologies 222, https://doi.org/10.1007/978-981-16-0119-4_58

58.1 Introduction

A variety of creativity and innovation is carried out in India as a lot of natural materials are available here in the woods. One of the abundantly found materials here is bamboo. Bamboo is one of the fastest-growing plants in the world and is found mainly in Southeast Asia and South America where the climate is best suitable for its cultivation. It is a renewable and extremely versatile resource with multi-purpose usage. Bamboo is the most common and easily available material. In past decades, bamboo is been largely acknowledged as an industrial material mainly in South Asian countries. Nevertheless, just research and development in bamboo is not enough as the world and its environmental conditions are changing rapidly (See Fig. 58.1).

It is very important to make students aware of such new age materials and wonders it could make. To understand a material, one has to work with it hands-on to get a good idea about the material. Bamboo has been part of Indian tradition for long, but consumers and industry do not use it because consumers were not aware and well versed with the material. If it were introduced to students, future generations would keenly be interested in making or buying bamboo products. Educational institutes always are big investors in knowledge transfer. But it is very difficult for them to provide students with such hands-on activities, and the gap between bamboo artists and customer needs, is a major problem in bamboo development. An emerging trend of design collaboration is the key to our problems. On the other hand, the small-scale industries could not afford such research and development, plus designers and engineers do not have laboratories to innovate. Thus, a maker's culture brought in some revolution, making it more focused then designing or innovation. Makers culture evolved the concept of "making in a space" now commonly known as maker space. It has also become to be known as places where people can pursue their creativity by making; this has manifested the talent and interest of like-minded people which has thus implemented maker spaces. The study will contemplate the making of a product from naturally available material (bamboo) in different types of maker spaces since it is directly proportional to the growth and need of maker spaces. Detail literature and practical examples are referred to create a database. Discussion in the

Fig. 58.1 Bamboo as an industrial material (*Source* Author)

Fig. 58.2 Word cloud showing different terms in a maker space (https://www.google.com/search?q=makerspaces+wordcloud&rlz=1C1MIMX)

paper may lead to a collaborative approach and a step towards future—"Atmanirbhar Bharat".

58.2 Maker Spaces as an Idea

Maker spaces are performed by a community of makers, and in such places, peer learning and collaborations are highly valued. The old proficiencies such as woodworking and architecture to as modern as 3D printing, robotics and Internet are featured in spaces as making attracts people from different backgrounds such as entrepreneurs, students, senior citizens, and children. Thus, from being a hobby, making now serves many causes: a vector for teaching and learning skills, a driver of citizens' empowerment, a start-up kick-starter, and a platform to meet like-minded or different people. Maker spaces evolved as the right platform to physically anchor the do it yourself (DIY) momentum providing materials, tools, and machinery for the execution of diverse projects, especially in the USA. The thriving making movement was soon exported from the USA to four other continents. Switzerland and India, two technology-oriented nations, have welcomed it enthusiastically in the last 8 years (See Fig. 58.2).

58.3 Evolution of maker spaces

It was very difficult to track the very first maker space as many standalone spaces already exist. The creation of the *making* movement started in the USA at Massachusetts Institute of Technology (MIT) and the campus was already a nest of maker spaces which was attracting a lot of engineers and technology enthusiasts. The official release of the first Make magazine in 2005 followed by the organization of the first Maker Faire in 2006, coined the term "*Maker*". Fittingly, as early as 2009, USA President Obama acknowledged the benefits of *Making* during his "Educate to Innovate" campaign, recognizing the importance of the *Maker* movement in the modern society and the opportunities it represented for the world. Following the success in the USA, the concept of *Making* was also implemented in other countries. In India, for instance, the incorporation of *Making* was largely driven by Indian expats returning from the USA, in the early 2010s. As with other countries, in India, *Making* has been incorporated in resonance with its socio-economical background and cultural heritage. Many stakeholders such as companies or institutions have been investing in *Making*. Recently, the country has also seen Government initiatives arise such as **Atal Tinkering Lab2**, designed to introduce children to *Making* in schools.

Maker space has come in the existence in around 2005 with "Maker Movement" launched by Make Magazine which was the promotion of creating work and methods of making.

58.4 Journey of the product

58.4.1 The Making Stage-1 (Idea Inception)

The idea all started on a detail study of the material bamboo. Bamboo being a versatile material is explored in every part of the world. A Geodesic dome constructed which was covered with canvas nick the attraction towards a bamboo workshop (Bamboo Maker space) which is affiliated by the Government of India (See Figs. 58.3 and 58.4). That place was allotted and developed by Bamboo Board of India. It was a bamboo shop, which displays handcrafted bamboo products by different organizations and bamboo artists from different parts of India. It was also equipped with all the machineries used for bamboo processing, which included different types of saw, planners, lathe machine, etc. The latest discussion with the officials there was about a laser engraving machine, which was soon to be installed at the premises. Generally, laser machines work on straight plain surfaces, but here the discussion was about a laser cutting machine which could work on curved surface of bamboo. Until now, laser cutting machine was used on bamboo ply and bamboo timber, which are flat surface so the engineering student decided to design a machine with rotary

Fig. 58.3 Local craftsmen working on bamboo at a government affiliated bamboo workshop (*Source* Author)

Fig. 58.4 Local craftsmen working on bamboo at a government affiliated bamboo workshop (*Source* Author)

function which would work on curved bamboo profile. This discussion thus proved that persons with the same interest meet to share their thought and boost the local craftsman to keep the art alive and become a bridge to connect to real market.

58.4.2 The Making Stage-2 (Fab-Lab)

Once decided about the product to be made, there were a series of discussions which boosted between the designers and engineering students about the shape, form, compatibility between the materials, comfortable usage of the product, etc. The project started with generating requirements for the machine. The machine has to be designed for a designer or bamboo product manufacturer. Peer learning is a common practice in maker spaces or workshops. It thus happened that the group of engineering students met an architect cum product designer and they shared their thoughts regarding the same product. Exchange of thoughts directly helped in improving the product design (laser- engraving machine) and also manifested about the need of the market. This interaction proved very friendly and a collaboration between those like minds started which thus resulted in the working of one more innovative product for the market. Maker spaces are thus promising venues for supporting agency and endeavours that are personally meaningful to the maker.

It was at this stage another maker space was introduced which was Fab-Lab where all the machines and guidance was available for the prototyping of the project. Fab-Lab was equipped with all woodworking tools like table saw, circular saw, zig zag saw, sander, planners, files, etc., it was also equipped with rapid prototyping tools like CNC, laser cutting, 3D printer, and MCB printer. Fab-Lab thus proved an ideal place to develop the project. The prototype was designed and after a practical research it was finalized that the bamboo will rotate as per the design and the laser could move in single plan, i.e. parallel to the ground. The conceptualization started through sketches, but that was not enough as imagining the product and its working was not that easy for the producers (See Fig. 58.5).

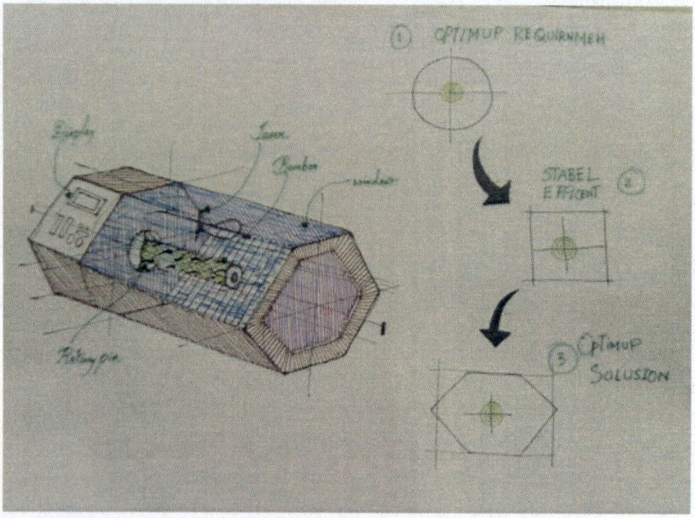

Fig. 58.5 Conceptual sketches of laser cutting machine (*Source* Author)

Prototyping

Mentors at Fab-Lab, Nagpur guided students to use "Fusion 360", an integrated software built to bridge designer and producer. The project was first created virtually in the software clearing the details in the design of outer body and the machine (See Fig. 58.6).

From major working like access to working space machine, bamboo fixing, etc. To smaller details like display and interaction panels, the 3D model became the reference

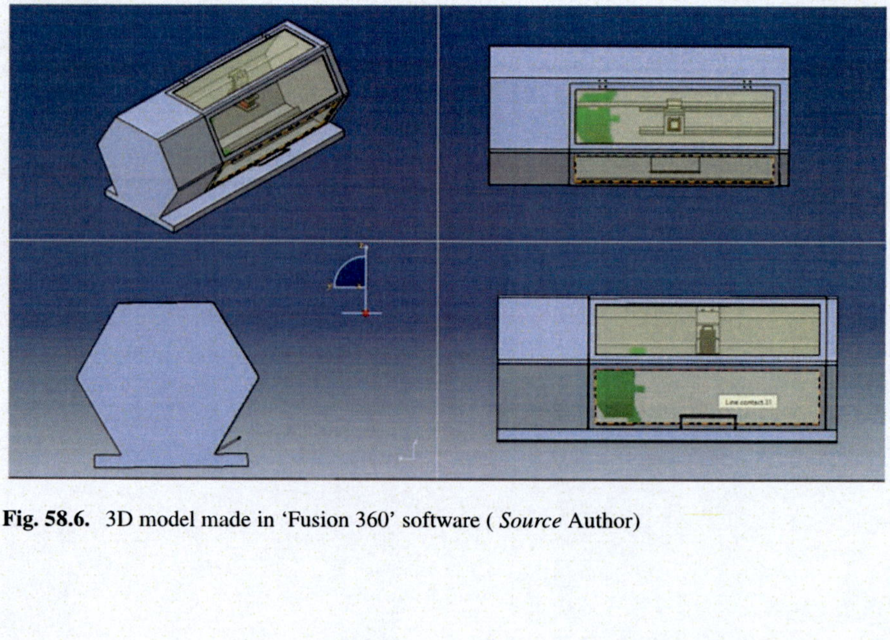

Fig. 58.6. 3D model made in 'Fusion 360' software (*Source* Author)

for prototyping. With use of power tools, the body of machine were made with ply and acrylic. Outer body and compartment for machine was made with plywood, whereas doors to access the working area was designed with plywood and acrylic for easy governess of laser. Power tools were used to cut the plywood and laser to cut acrylic.

In the machine, bamboo is fixed and rotated as per design. A clutcher attached to gears was designed in fusion 360 which was then 3D printed for better finish and efficiency. Fusion 360 is a software by Autodesk which is easily compatible with maximum rapid prototyping tools like CNC, laser cutting, 3D printer, etc. Fab-Lab is a worldwide organization which has tie up directly to Autodesk allowing access to such advance software. Soon other parts such as display and interaction panels were also designed and made with help of PCB printers and laser cutting of MDF for supports and structuring.

58.4.3 The Making Stage-3 (Viprush Laboratory)

After the prototype is ready, the working of the machine needs to be done. For this, a proper coding required and so such type of maker space is needed where all the electric circuit work can be completed with proper guidance. The Viprush laboratory was a perfect laboratory where all the young minds play with their innovations and designs to make it real. Here, the working circuit of the machine was developed which gave life to the machine and brought it to working stage.

58.5 Maker Spaces—A Discussion

Maker space has their origin in maker culture and hackerspace. Maximum times advanced technologies have developed in a small laboratory or may be some garage space, which when used by group of innovators becomes a hackerspace. Most of these spaces have started from closed community of similar interest, much similar to clubs. Hackerspace basically was an IT nerd's club. Here, these engineers experimented with new knowledge and skills in their inventions and this generated a new wave of makers. Every product made is not always by a craftsman or an artisan with all the perfection and artistry looks, and it can also be made by a person who is just interested in making by simple tools and technology. Thus, a significant term *Maker* stood out. Some maker spaces were created to share the resources to the needful makers and artists. Sharing of resources led to sharing of knowledge and with friction of knowledge innovation sparked out.

58.5.1 Contribution to the Project

The laser engraving machine made by the engineering students was thus modified in different maker spaces and was displayed at Makers Mela 2020. Hence, maker spaces are important for such projects on different levels to modify the product. As discussed in the stage 1, the inception of the product was merely visiting a local government bamboo workshop (maker space) where the local artisans work. Many times different experts and designer come there to learn from the local craftsman and innovate as per market need. This process helps local craftsman to adapt to time (See Fig. 58.7).

Once the idea generated of a laser engraving machine with rotary function, the process of making started in the second maker space (discussed in stage 2), which was a technical one, equipped with all the machineries for making the same. It was an institutional maker space which is associated with many companies and institutes like IIT. They help institutes develop their maker space for school or college students. The product was then taken to the third maker space (discussed in stage 3), which was a private maker space. These maker spaces were very specific to the maker's interests. So they tend to have experts hanging around. They seem to be very useful for budding engineers and artists.

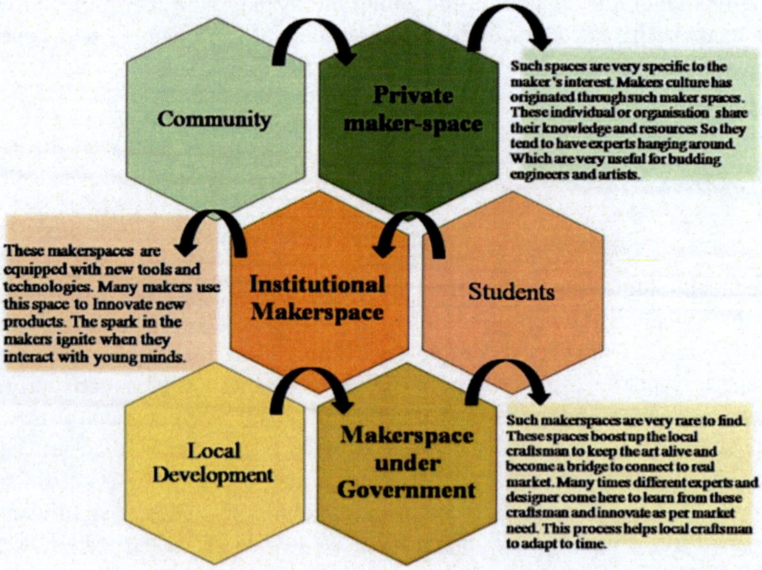

Fig. 58.7 Graphical representation of types of maker spaces (*Source* Author)

58.5.2 Role in Society and Education

Although the importance of maker spaces has been discussed many times in the past for technical field, but there are many non-technical field as well, which includes the school education where the students are not so sure about their future endeavours. With this changing period of time, it is inevitable that the college education is not enough to stand in the current market. With expanding career opportunities, bunch of unique qualities are also needed to stand out individually. Moreover, educational institutes could not just be the only source of recent updated knowledge as they are information centres, rather than these maker spaces as they are centres for innovative research and better. That means students need to explore the new trends on their own, which many times need some resource or technology to truly get a deep understanding.

On the other hand, it is very difficult for educational institutions to provide practical exposure to students along with theoretical knowledge. Many institutes do take students on different field visit, hands-on workshop, etc. To make students aware about the application of theoretical knowledge to practical, keen students find such maker spaces to apply the knowledge they grasp in academics, or to make projects for their assignments. Maker space provides them with infrastructure according to the type required. As we discussed in the case study engineering students used different makers spaces at different stages of their project. Different maker spaces create specific practical experience in different streams. For new trend of collaboration, students should peak into as many streams as possible (Figs. 58.8 and 58.9).

A maker space in Nagpur has become a torch of this maker culture. They introduced Nagpur to makers culture with MAKER ADDA. A passionate family of artists and designers nurturing infant designers and artists. They have woodworking, sculpting, painting, etc. Art supplies, immense amount of energy, experience and enthusiast makers.

Fig. 58.8 Image from Makers ADDA, Nagpur (*Source* Author)

Fig. 58.9 Output of bamboo workshop (*Source* Author)

58.5.3 Need of Maker Space

Many problems can be found but solutions cannot be dreamt or searched without a problem. They have to be worked out by experimenting, research, feedback and prototyping. On the same lines, all the problems cannot be solved at a single maker space, as problems related to specific material or technology would need spaces with specialized tools and experts. This would be very difficult and financially inappropriate for a single maker space to cover all the possible maker tools. Institutions like Fab-Lab provides with updated technology for prototyping with CNC or 3D printing, etc. But what if the innovator needs to use fabric, or clay work in their product. There would be a very less ratio of people interested in such activities. But for Fab-Labs to maintain the equipment for such small no. of people would not be logical. At this point, organization with such equipment could step up to create their maker space available for people interested in fashion or clay work. This could provide extra revenue to organizations and unique opportunity for community to learn different skills.

"India has always been known for '*Jugaad*'—innovation on a shoestring budget. But we are also typically scared of failures. A maker space allows for 'the weird' and encourages experimentation", says Maker's Asylum founder Vaibhav Chhabra.

If we look at the maker spaces most of them are technology and rapid prototyping-based spaces. Designed for all students and technology-driven makers. It is seen in very few maker spaces which are focused on particular skills or material.

58.5.4 Understanding the Gap

Maker's culture is catching up a quick pace in India with private players like Fab-Lab and Viprush Laboratory. These Laboratories are building up new opportunities for curious students and designers. Social media influencers are motivating new generations to try out different ways to innovate. But there is an invisible gap between maker and maker spaces which is holding back makers potential to innovate. Makers

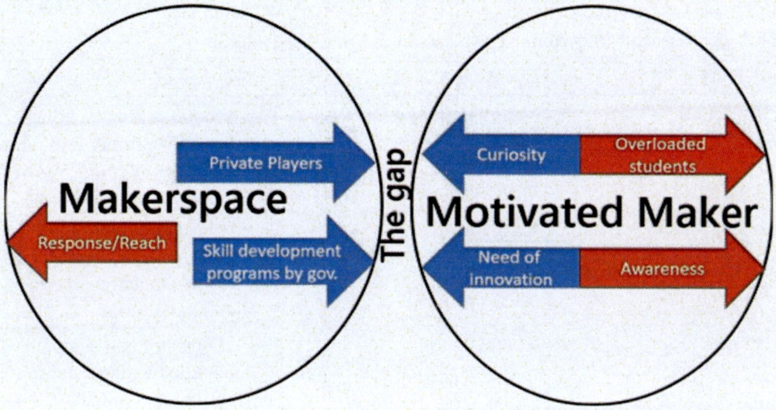

Fig. 58.10 Image explaining the gap between maker space and motivated makers (*Source* Author)

especially students are curious to try out but they are overloaded with academic work on other side makers and designers are innovating but they are not aware about such spaces of work. On the other hand, private players like "Technoventor Innovations" are getting maker spaces like Fab-Lab to makers in small cities like Nagpur. Government skill development programmers are adapting to maker's culture through places like "Bamboo Shop" at Nagpur by MBDB. Yet they are not able to reach out to makers (See Fig. 58.10).

Let us try to understand the reasons for the gap with the help of B. J. Foggs behavioural model. "B = MAP(T)".

"Three elements must converge at the same moment for a behaviour to occur: motivation, ability, and a prompt (trigger)".

In our case,

B = use of maker spaces

M = curiosity of user

A = availability of maker spaces

P(T) = prompt (trigger)

But that prompt (trigger) is missing which limits use of maker spaces.

58.6 Present Scenario

58.6.1 Maker Space in India

Currently maker's culture is spreading in metro cities with a great speed. Due to booming, maker spaces and the gift to more skilled-based thinking of younger generation have boosted the makers culture into smaller cities too (See Table 58.1).

Table 58.1 List of well-known maker spaces in metro cities in India

Name of space	Location	Type
Maker's Asylum	Mumbai and Delhi	Woodworking, robotics, bicycle building and repair, rapid prototyping, open-source computer-aided design
Workbench Project	Bengaluru	Fabrication laboratory, power tools, CNC's, 3D printers, sewing machine, welding kit and hand tools
Maker's Loft	Kolkata	3D printer and scanner, Google Cardboard, laser cutting machine, etc., for creative children
Chennai Makerspace	Chennai	Fabrication laboratory, power tools, CNC's, 3D printers, etc.
IKP-EDEN (Engineering, Design and Entrepreneurship Network)	Bengaluru	Private room, meeting room, classrooms, cafe, co-working space, Internet, locker facility, mentorship services, funding access
CEPT FabLab	CEPT University, Ahmedabad	Circuit production, 3D printing, CNC milling, laser cutting/engraving, precision milling and vinyl plotter
Project DEFY (Design Education for Yourself)	Bangalore and Mangalore	Self-learning for schools and other institutes

58.6.2 Working of Maker Spaces

The idea was to set up spaces with a variety of tools and promote an open culture that would encourage people to experiment and share ideas so they could innovate whatever they want to.

Maker space serves not only students but also professionals to design new products using the tools and knowledge sharing at the maker spaces. This is well explained by an article which discussed about a young BITS-Pilani graduate was working in research and development of home appliances. He wanted to make something for his elderly parents which could be easily moveable, bendable and stretchable with less efforts. He went to "mahadevpura", an area in Bengaluru city (India) where many workshops are there. As he approached the workshop owner for building his idea into reality, he was not really welcomed by the same as the machineries were not easily accessible for the outsiders as well as the owners were not really keen in the interest of making a prototype. On the other hand, it is difficult for the designer/innovators to explain his thoughts and ideas to the craftsmen for the making the same. This gave the individual an opportunity to make the prototype by himself at a start-up known as IKP-EDEN, a maker space in Bengaluru's Koramangala area. Within few months, the prototype was ready by using the tools and technology available to him as an employee in IKP-EDEN. He finally left his job to establish a start-up for his new product. This example manifested that maker spaces can become a birth/foundation place for future "Antmanirbhar Bharat". Maker space provides the ground to work out the solutions.

58.7 Conclusion

Human has collected uncountable information and it is ever growing. This information is transferred to the new generation through academics. It becomes impossible to teach so much theory in any stream of academics that it loses the practical approach of the knowledge students gain. Institutes can become the trigger for students by creating a practical focused curriculum then a theory-based curriculum. Accordingly, students will find the need for such spaces where they can work on their assigned projects to learn more. Here, the maker space can play a major role in creating such practical platforms. Maker spaces are becoming the playground for art and innovation which helps grow a self-sustaining society. Maker spaces offer a rich and diverse pool of people and skills, community-driven approach, and easy access to machinery and tools. Maker space could prove to be the missing key between education and market. Artists with functional set up need to open their space to aspiring and budding artist. Hence, a collaborative approach is very much needed in the society. Here, the process could be the "give and take" one, as in collaboration could be from both sides for a better functioning. Educational institute will help society to build a platform for the future generations and to update the market with new job opportunities. Educational organizations and private organizations have to take steps in coordination to grow this making culture.

References

1. Avneet, H., Morgan, M.: People, Means, and Activities: A Conceptual Framework for Realizing the Educational Potential of Makerspaces. Hindawi Education Research International, vol 2018, Article ID 6923617, pp. 1–2 (2018). Retrieved from https://doi.org/https://doi.org/10.1155/2018/6923617. February 2020
2. Swissnex India.: Making in India- A Hardware Approach to Drive Progress and Innovation. Article published in Swissnex India (2017) innovation@swissnexindia.org
3. https://www.fablabs.io/labs/fablabnagpur, May 2020
4. https://viprushtech.in/techlabs/, June 2020
5. https://www.makersadda.in/about, June 2020
6. https://theconstructor.org/building/bamboo-as-a-building-material-uses-advantages/14838/#:~:text=Bamboo%20as%20a%20building%20material%20has%20high%20compressive%20strength%20and,%2C%20bridges%20and%20structures%2C%20houses, June 2020

7. https://m.economictimes.com/small-biz/startups/how-makerspaces-are-helping-startups-and-a-new-breed-of-entrepreneurs-to-innovate-new-products/articleshow/58741461.cms, June 2020
8. https://www.hindustantimes.com/more-lifestyle/have-you-used-a-3d-printer-yet-india-s-got-a-host-of-new-maker-communities-and-you-can-join-in/story-lPl01I4myfZ6YIXfcS82tM.html, June 2020
9. https://www.google.com/search?q=makerspaces+wordcloud&rlz=1C1MIMX_, June 2020
10. https://behaviormodel.org/, September 2020

Part IX
Enabling Technologies and Tools (Computer-Aided Conceptual Design, Augmented/Virtual Reality, MEMS, Haptics, Smart Technologies, IoT, Cobotics, etc.)

Chapter 59
Digital Health Interventions to Enhance Patient Care for Indian Nurses

Vydianathan Ramaswami and Raj Arjunan

Abstract The health care and nursing profession in India has suffered neglect both historically and in present times. We propose the use of information and communication technology-assisted learning (ICTAL) for creating a large capacity of qualified, skilled, and competent health care workforce in a short period. The "Indian Nursing Knowledge System (INKS)" an online platform uses a blended learning approach with, including, but not limited to customized, individualized digital health content, maximum utilization of smartphones, Web portal, SMART products, simulations, with inter-sectoral coordination with sectors of society addressing health and well-being. We envisage improving nurses' clinical skills, competence, critical thinking, decision making, and the ability to work in teams.

59.1 Objectives

- Identify the problems confronting the Indian nursing education, training, particularly at smaller institutions, without their own hospitals.
- Understand the prevailing conditions of the nursing profession in the Indian healthcare system, particularly from the patient and family point of view.
- Propose digital health intervention (DHI) for efficient care management.
- Integrate information and communication technology in nursing education, training, and practice for enhancing personalized patient care.

V. Ramaswami (✉)
V-SIGN, Vellore Institute of Technology, Vellore, Tamil Nadu, India
e-mail: vydianathan.r@vit.ac.in

R. Arjunan
Indian Institute of Technology, Madras, Chennai, Tamil Nadu, India
e-mail: am15d029@smail.iitm.ac.in

A. Chakrabarti et al. (eds.), *Design for Tomorrow–Volume 2*, Smart Innovation, Systems and Technologies 222, https://doi.org/10.1007/978-981-16-0119-4_59

59.2 Research Methodology

Three subject matter experts consulted, from the health sector in the research for this paper are: Dr. K. V. Arulalan; Nurse Mohana: Practicing in Singapore, Nurse Renuka: Practicing in the USA, possessing vast experience in using technology for patient care. The authors have years of experience taking care of their parents with chronic comorbidities at their end of life. The process of this study are:

Step 1: Study relevant content through online published sources, and understanding the subject through secondary data [1].

Step 2: Analysis of existing nursing practice through semi-structured interviews with experienced people from medium-sized hospitals (bed strength not less than 150), telephonic interviews with nursing practitioners in India, Singapore, and in the USA about nursing in global contexts' [2]. Learned about realistic issues related to the implementation of health technologies in pilot project from written notes of health activists on of Jan Swasthya Abhiyan (JSA) in Tamil Nadu.

Step 3: The author's core skills and competencies are in designing solutions using digital technologies; we realize its immense potential for the seamless integration in point of care systems in Indian nursing education and its practice.

Persons involved in **Steps 2 and 3** had personally given care to at least one bedridden patient at home for more than 12 weeks. This process enabled in conceiving solutions for Indian nursing education, training, and practices, in underexplored area.

Background to Indian nursing: The way to transform the Indian healthcare system is by educating, enabling, and empowering nurses; the nurses are the last mile deliverers of patient care. There are many published literature about the types of problems and possible solutions for the nursing profession in India [3–11].

59.3 Indian Nursing Education, Production of Nurses

Some worrying trends appear in the landscape of nursing education; 91% of nurses come from the private sector [3], a large number of these institutions do not have their own hospitals with patients. Nursing institutions lack periodical audits and quality control systems. Spelling mistakes in case sheets of hospitals are an easily available example. DHI is unheard of in the majority of nursing institutions.

59.4 Emerging Technologies Transforming Nursing Practices

The transformation of nursing practices using emerging technology is happening at warp speed, i.e., conceptually, "warp speed is faster than light." We have developed Table 59.1, an adaptation from an article by Huston C, "The Impact of Emerging

Table 59.1 Status of seven emerging technologies transforming nursing in India

A. Technologies already available in large Indian hospitals
1. Non-invasive/minimally invasive tools for diagnosis, treatment, and patient care
Benefits: Detection of diseases, providing care, minimally invasive, and non-invasive tools being developed to reduce patient risk at less price **Challenges**: The rapid penetration of non-invasive and minimally invasive tools, the nurses need to use in their practice effectively is a serious challenge
2. Digitization of data enable informed clinical decision making and support
Benefits: Digitization of health care has made services affordable, error reduction, and improved evidence-based practices **Challenges**: Integrate technology in health care requires a change of mindsets. The learning/training requires investments for implementation
B. Technologies nascent in India need time to develop and implement
3. Electronic Healthcare Records (EHR)
Benefits: Digitization of health records enable service providers to access patient information from anywhere at any time **Challenges**: EHR is expensive systems, ownership of patient data, proper usage of data, privacy protection, and security
4. Biometrics
Benefits: Biometrics increases the security of information and eliminates the costs of managing user access credentials **Challenges**: Impacting costs due to frequent change in the workforce, updating biometric identifiers in challenging situations
C. Nursing technology present only at theoretical level in India
5. Genetics and Genomics
Benefits: Most disease risks, conditions arising due to hereditary elements with the influence of the environment, lifestyle, affecting the entire care **Challenges**: Nurses have little or no knowledge on genetics and genomics. Educate and counsel patients and train nurses on latest fields of medicine
6. 3D printing
Benefits: 3D printing enables the creation of human organs as a replacement **Challenges**: At present, many biomaterials limitations in 3D printing
7. Robotics
Benefits: Robotics perform routine tasks; do accurate diagnosis, less invasive procedures, small interventions are precise **Challenges**: Robots lack emotions, difficult to replace human caregivers

Technology on Nursing Care: Warp Speed Ahead" [4]. According to this article, seven emerging technologies will change the practice of nursing; emphasizing three competencies that nurses have to develop to learn, practice, apply, and four challenges nurse leaders will face in integrating new technologies.

Hospitals in future will not use paper to store patient data. Use EMR to manage patients using types of data (images or reports). Share these discreet dynamic data among care providers via standardized electronic transactions for analysis to make informed decisions enabling continuity of care [5] among stakeholders to improve the quality and safe care of the patient.

59.5 Overcoming Challenges in Indian Nursing

Indian health care faces five main challenges namely awareness, accessibility, absence, affordability, and accountability in the healthcare system [6]. Let us look at existing challenges and ways to overcome using existing and emerging technologies as digital health interventions.

59.5.1 Awareness

Most of the Indian public health campaigns are unsuccessful due to ineffective communication. There is a need to design innovative and sustained public health campaigns to change attitudes and behaviors of the citizens. Create awareness regarding the potential of DHI and its benefits among health workers, policymakers, administrators, academicians, and health field workers.

59.5.2 Accessibility

Peter F. Drucker, in his "Technology, management and society", explains the *total marketing approach as increasing the profit by decreasing the cost for the customer*. Peter F. Drucker asserts cost reduction cannot happen without eliminating the activity, which produces the cost. Hospital and health center visits are the source of rising health costs. If the cost of health needs to reduce, then the hospital visits will have to be eliminated or minimized [7]. To do this, we need the nurses to perform complicated tasks easily with high accuracy and speed. This is possible only through DHI.

59.5.3 Absence of Healthcare Workforce

Indian health care is facing a severe shortage of nurses. Need to reframe the "scope of work of Indian nurse practitioners," improve the retention, define career paths, and assign responsibility.

59.5.4 *Affordability by Making Nurses Deliver Primary Care*

The use of technology can reduce the workload stress on the nurses helping them to perform their tasks in an easier, accurate, and speedy manner. From the time, a patient enters the hospital or community care center, and the nurse can provide complete attention to the patient for eliciting information about their condition, perform series of stipulated tests, record data for the physician's reference to treating, and prioritize appointment depending on severity. This requires training and practice for nurses to acquire new skills, knowledge, and competence for efficient technology integration.

59.5.5 *Accountability*

A nursing professional is accountable for their actions, consequences of their actions, and behaviors [8]. Ensuring proper documentation of patient health and treatment data is the first step in making bringing accountability factor in healthcare services. After DHI implementation, the confidential patient health information in-hospital case sheets only available to the patient and potential approved health service providers (Table 59.2).

Table 59.2 Factors influencing accountability and its enhancement after DHI

Lack of accountability	DHI enhances accountability
Job description not available	Clearly defined job description provided
Lack of guidelines, standards, and control	Easy access to policies and standard operation procedures documents
Authority not specified	Competent authority to conduct nursing audit for ensuring compliance
Shortage of staff leading to overwork, fast burn out, and high attrition	Assign roles and responsibilities as per norms and guidelines
Lack relevant on job training and experience to improve efficiency	Provide advance training based on individual's need/aspiration
Do not have knowledge of best practices, no opportunity to upgrade skills, knowledge, and competence	Provide access to acquire skills, knowledge, and evaluate for accreditation; recognize performance through financial incentives and promotion
Obsolete policies, protocols, and procedures	Review, redefine policies, and procedures aligned to present needs
Poor working environment, resulting in high stress levels	Encourage, incentivize performance in good work environment
Inadequate staff, supervision by ward managers	Ensure adequate personnel and resources available for patient needs
Lack of decision making	Training on leadership, critical thinking, and decision-making skills

59.6 Need for Continuous Revision of Nursing Curriculum

In present times, the majority of nurses train for hospital care and not for community care. Even in-hospital care, technology, let alone digital plays minimum role now.

59.6.1 *Bridging the Academia and Industry Gap*

ICTAL will enable bridging this critical gap in nursing education and practice, by training novice nurse students through various case studies as best practices in the use of technology relevant to the Indian healthcare context.

59.6.2 *Nurses Need to Adapt and Adopt New Technologies*

"Technology is not about gadgets, but how a person organizes his or her work"—Peter F. Drucker. Technology has innovatively disrupted a physician's job in India and their service is productized [9]. Technological solutions need customization based on local needs across Indian subcontinent. Secondly, medicine is culture-based, which means the food, lifestyle, approach to health; compliance with health advice is variable. New technologies can bring about changes in healthcare accountability.

The components of public health system classified into points of care as primary, secondary, and tertiary care centers depending on their complexity and the service. Presently, many "point of care technology"-based gadgets starting from measuring temperatures; vitals to blood sugar are used. As the health industry matures, nurses should know the principles of operation for communities proactively participate by adopting emerging technologies and putting them into practice. To employ DHI meaningfully in health services, we must know the activities that take place in a health center. It is ideal to use a medium-sized hospital for exploring the needs of the patients, lacunae in the provision of services, and the niche for DHI.

i. Medium-sized hospital provides secondary and primary care services.
ii. They have a strong link with the community.
iii. Continuous sharing of learnings from medium-sized hospital with tertiary care hospitals and adapted to primary healthcare centers.

National Rural Health Mission in 2005 started to create an architectural correction in the health system [10]. Of the five pillars created, one was community participation, and thus Community Action for Health was born in 2007 [11]. Tamil Nadu with advanced health indices was one of nine chosen states for comparison. Jans Swasthya Abhiyan was the nodal NGO at the national level and TNSF, Tamil Nadu. Science forum at the state level. During 2007–2011, the project done at five districts, Dr. K. V. Arulalan, was the Vellore district secretary covering 117 village panchayats in three blocks. The outcome of this project resulted in creating over 1000 trained health activists in Village Health Water Sanitation Committee. The felt needs of field workers derived from their experiences incorporated in this paper (Table 59.3).

The DHI solutions designed and developed by Dr. K. V.Arulalan Health Center with Helyxon, a Healthcare Information Technology Company at IIT Madras Research Park as listed below:

Personal Health Records (mobile application): A pediatric consultation has 10 components.

(1) Need for child's hospital visit, (2) details of the complaints, (3) past health history of the child, (4) physical growth, (5) brain development, (6) immunization, (7) drugs, (8) advice apart from drugs, (9) when to come back, and (10) storage of laboratory data. Once a patient visits, the summary is available under major headings at a glance, such as the date of visit, major problems action is taken and formerly known as problem-oriented medical record (POMR) [12].

Sensor-based Internet of Things instruments—Checking vitals; temperature, oxygen level, and pulse.

Cloud-based Individual CHATS—A parent and child need not come every time to the hospital. Particularly, under two situations.

- Already visited once. The patient needs to follow up care
- In case, the problem is a minor one and a mere text message is enough. The key difference between SMS and WhatsApp messages, this CHAT is in conjunction with the health record of the child.
- Telemedicine effectively and commonly used during the COVID-19 times.

59.6.3 Leveraging ICTAL and DHI for Nursing Education and Remote Care

Nurses in many instances are required to provide care outside of the hospital setting, which is complicated for coordinating care between clinicians and community care providers; managing chronically ill patients to prevent worsening of the ailment by using technological tools to deliver improved quality of effective care. This would entail the use of mobile-based electronic health records, applications powered CHAT messages, emails, and answering traditional voice through telephony [13].

59.6.3.1 Online Healthcare Community Platform

As part of the National Skill Development Program, design "Indian Nursing Knowledge System (INKS)" an online learning platform, where nursing trainers, students, and practitioners can learn, train, and practice from anywhere and anytime. This

Table 59.3 Tasks of nurse and opportunities for DHI intervention

1 Casualty
Tasks: Take vital signs: Blood, temperature, pulse rate, respiration rate, and blood pressure **DHI intervention**: Transfer of patient data digitally to patient health record system for doctor's actions. Use of sensor-based instruments for the same
2 Out patient department (Male and Female)
Tasks: Patient health records: enter patient complaints and inform them about follow up care **DHI intervention**: Use of EMRs: at point of care, replace had written notes with digital technology. There is a realization; nurses need to use technology
3 Patient waiting halls
Tasks: Classify patients' disease status into mild moderate severe and initiate treatment **Intervention**: Electronic health record specific for the common ninety conditions is available; this data used for managing patients and it is triaging
4 Pharmacy
Tasks: In many places, ANM's work as pharmacists because the salary for a trained pharmacist is very high **DHI intervention**: Design a medicine dispenser with a tray with prescribed medicines for the patient. Create multimedia medication procedures to educate patients for ensuring compliance, avoiding prescription errors
5 Nursing superintendent office
Tasks: This is a crucial place often forgotten. Depending on the caseload, trained people should go where there is a need. For example, a nurse trained in intubation needs to be posted in the ICU ward; a nurse trained in putting drips for a child should be posted in the pediatric ward. Normally there is rotation **DHI intervention**: Nurses complain that they cannot work in "tension-filled" ward daily; a valid point. Presently, the decision is a non-clinical one. We need to consider multiple parameters before a nurse assigned to a particular ward. Using DHI, the daily work allocation done by considering their skills and experience of the nurses' before assigning duties
6 Intensive care
Tasks: How to hand over the patient problems to the next duty sister consisting of the patient list, injections to be given, tests to be done, results to be collected and communicated with doctors for further care and informing family on developments and requirements **DHI intervention**: A study has shown a neonatal intensive care unit produces 1300 pieces of data per bed. This puts many cognitive loads that are difficult to handle without proper computing tools. Need for designing specific mobile apps for this purpose
7 Medical and surgical wards
Tasks: In OPD, in the ward also we have to classify people who need attention first, who have the potential to become sick and discharged patients. Additionally, provide indent of drugs and injections for the day. Visit of numerous consultants. They may not write down, but say orally and walk off **DHI intervention**: Nurses will have to record, assign, share, and ensure the accuracy of all these various types of information to that and report it to the concerned department for record maintenance. Design simple digital devices with tools that will facilitate performing these regular repetitive tasks
8 At Anganwadi center

(continued)

Table 59.3 (continued)

Tasks: The team has tools to assess the growth and development of the baby. It is very time-consuming to fill by referring to many available paper-based charts **DHI intervention**: DHI will provide simple design tools to record, store, and retrieve information about the growth and development of the baby quickly
9 For Asha worker or village health nurse
Tasks: ASHA workers have to maintain 70 plus patient records be it immunization or giving drugs to Tuberculous patients **DHI intervention**: DHI will deliver a comprehensive workflow that will help them to prioritize their work accessing through their mobile phones

platform will help individual nursing trainers and institutions to "design, implement, and evaluate the effectiveness of learning and training of their student's proficiency."

59.6.3.2 Revise Indian Nursing Curriculum

Constitute a core team of healthcare experts to formulate standards, guidelines for nursing educators, review content of the curriculum, assess accreditation requirements, and revise norms to allow practice for credits consideration using INKS. Practicing nurses and aspiring trainers will benefit by using INKS for their career development.

59.6.4 Benefits of Using INKS in Nursing Education and Practice

INKS will enable nurses working in a hospital environment and community nurses serving the underserved and marginalized communities living in inaccessible places in urban areas or rural areas in the following ways:

1. The majority of tasks performed toward patient education, automation of this content, and its distribution would greatly help to reach quickly.
2. Individualized procedures and protocols to follow, given to patients as audio-visual presentations, with customization of patient's language choice.
3. Multimodal communication happens with the nurses apart from one to one communications: Messages from phone, emails, Web portals, and handles of social media can be refined and using standard templates, chatbots created, thus, greatly reducing the workload of the nurses. This saves the nurse's quality time to provide humane care, which is only possible by nurses.

59.6.5 Need for Nurse Researchers and ICTAL Niche

This glaring gap between nursing academia and the needs of the healthcare industry reinforces the compelling need to conduct continuous research in nursing education with adequate funding to support evidence-based teaching and facilitating the sharing of transdisciplinary medical professional knowledge. Nurse researchers need to focus on essential improvements that are the key aspects in the delivery of nursing care to improve the quality and safety of the patient [14]. INKS application in nursing requires exploring three domains:

(a) Customizing leaning to individual at institutions without additional costs.
(b) Classify nursing duties to reduce their non-clinical work time, e.g., linking thermometer and weighing machine to the electronic health record. Orally given physician orders are handwritten notes as records in short time with the intention for 100% accuracy.
(c) Leveraging communication channels: Use TV, radio, and Internet using mobile phones for nursing education and practice in the community.

59.6.6 Broadening of Nursing Competence

In future, the Indian healthcare system will become complex with the increasing demand of patient care expectations from the health service providers, and registered nurses will have to possess all-round competence such as use of digital technology, genetics, quality improvement, and geriatrics. We will need a lot more nurses than physicians to deliver care; directly, over the phone, and through multiple gadgets specifically during crisis like COVID-19. Nursing education and practice will have to be in line with changes in clinical requirements. The nursing coursework must align to real-life situations to provide care by inheriting legacies with great responsibilities; particularly for serving the marginalized and underprivileged. The pioneer in Nursing Florence Nightingale in the 1870s wrote, "It would take 150 years before we would achieve the kind of nursing I envision." That has arrived both metaphorically and chronologically.

59.7 Conclusion

The Indian healthcare system is facing a severe shortage of 2.4 million nurses [15]. To build large qualified, competent and skilled nurses is only possible by digital technological intervention like "Indian Nursing Knowledge System" (INKS), in a short time. Future nurses must be competent in using digital technologies while providing patient care. INKS platform will provide simple to complex healthcare simulations accessible through mobile devices. In India, the government need to frame relevant healthcare policies and guidelines with methods for ensuring proper implementation with periodical audits to enumerate data on the effectiveness of social healthcare programs. As an example: Let us take diarrhea, ensure regular hand washing, drinking boiled/hot water, along with the intake of prescribed medicines by individuals supported by local healthcare service providers. The external factors such as regular cleaning, sanitization of the living habitat/environment, and subsequent periodical chlorination are issues that are to decide upon by the policy makers in coordination with respective organizations.

Use DHI to disseminate policy information among stakeholders for execution and monitoring, sharing audit results for improvement. The transformation in providing enhanced healthcare service using technology must reach primary health centers and village hospitals for making timely, accessible, affordable, and quality health care reaches the unreached and marginalized urban and rural poor. The pioneer in Nursing, Florence Nightingale in the 1870s wrote, "*It would take 150 years before we would achieve the kind of nursing I envision.*" That time has arrived both metaphorically and chronologically.

Acknowledgements Dr. Arulalan, MD, DCH, DNB, a community-based Pediatrician and Founder of AA Child Care Center at Vellore with 35 years of experience, shared his experience under Dr. Miss. P. Chandra, MD DCH FIAP, 87 years former Director, Institute of Social pediatrics, Stanley Medical College, Chennai, from 1981-till date, provided timely support, patient guidance, encouragement, and insightful critiques for this paper. His constant advice directly and Dr. Chandra's support, indirectly, helped us.

References

1. Reed Geertsen, H.: Rethinking Thinking About Higher-Level Thinking. American Sociological Association, vol. 31(1), pp. 1–19 (19 pages) (2003). https://doi.org/https://doi.org/10.2307/3211421
2. Fetters, M.: Getting started in primary care research: choosing among six practical research approaches. Family Med. Commun. Health **7**, e000042 (2019). https://doi.org/10.1136/fmch-2018-000042
3. Worldometers, 17 06 2020. (Online). https://www.worldometers.info/world-population/india-population/#:~:text=India%20population%20is%20equivalent%20to,(and%20dependencies)%20by%20population. 26 06 2020
4. Press Trust of India, "Home page; Times of India," 23 03 2018. (Online). https://timesofindia.indiatimes.com. 26 06 2020
5. PTI, 8 05 2019. (Online). https://economictimes.indiatimes.com/news/economy/policy/still-a-long-way-to-go-to-raise-public-health-expenditure-to-2-5-per-cent-of-gdp/articleshow/70131459.cms?from=mdr#:~:text=The%20report%20said%20following%20projections,20%20(BE)%20to%20the%20health
6. Nivedita Rao, "Health," 14 April 2018. (Online). https://thewire.in/health/who-is-paying-for-indias-healthcare (26 06 2020)
7. Akshaya K, Lalu A, Sathyaprakasan D, Antony S.: SERO—The nurse robot. IJIRT **6**(1), p. 5, Unique Paper ID: 148326, **6**(1), 357–361 (15 06 2020)

8. Walton-Roberts, M., Irudaya Rajan, S.: Migration Policy Institute, 23.01.2020, **Global Demand for Medical Professionals Drives Indians Abroad** Despite Acute Domestic Health-Care Worker Shortage. https://www.migrationpolicy.org/article/global-demand-medical-professionals-drives-indians-abroad
9. Mishra, S.: Respect for nursing professional: silence must be heard. Indian Heart J. **67**(5), 413–415. https://doi.org/https://doi.org/10.1016/j.ihj.2015.07.003. Epub 2015 Aug 5. PMID: 26432724; PMCID: PMC4593846
10. Gill, R.: Scarcity of nurses in India: a myth or reality? J. Health Manage. **18**(4), 509–522 (2016). https://doi.org/10.1177/0972063416665932
11. Issac, A., Syam, N.: Migration of Health Care Professionals from India: A Case Study of Nurses. Working Papers id:2403, eSocialSciences (2010). https://ideas.repec.org/p/ess/wpaper/id2403.html
12. Nair, K.S.: Health workforce in India: opportunities and challenges. Int. J. Commun. Med. Public Health **6**(10) (2019). http://dx.doi.org/https://doi.org/10.18203/2394-6040.ijcmph20194534
13. Huston, C.: The impact of emerging technology on nursing care: warp speed ahead. OJIN: Online J. Issues Nurs. **18**(2), Manuscript 1 (2013). https://doi.org/https://doi.org/10.3912/OJIN.Vol18No02Man01
14. Stephen, S., Vijay, V.R.: Metamorphosis of nursing profession: an Indian perspective. J Glob Health **9**(2), 020314 (2019). https://doi.org/10.7189/jogh.09.020314
15. Kasthuri, A.: Challenges to healthcare in India—the five A's. Indian J Commun. Med **43**, 141–143 (2018)
16. Drucker, P.F.: Technology, Management & Society; Essays. Harper & Row New York (1970)
17. Kumar, H., Gokhale, Jain, K., Mathur, D.R.: Legal awareness and responsibilities of nursing staff in administration of patient care in a trust hospital. J. Clin. Diagn. Res.: JCDR **7**(12), 2814–2817 (2013). https://doi.org/https://doi.org/10.7860/JCDR/2013/7737.3886s
18. Chellaiyan, V.G., Nirupama, A.Y., Taneja, N.: Telemedicine in India: where do we stand? J. Family Med. Prim. Care **8**(6), 1872–1876
19. National Rural Health Mission, Meeting people's health needs in rural areas. Framework for Implementation 2005–2012 Ministry of Health and Family Welfare, Government of India, New Delhi-110001
20. Pilot Phase, Community Action Health, Population Foundation of Indian (Online). https://nrhmcommunityaction.org/pilot-phase/. Accessed 15. 06. 2020
21. Weed, L.L.: Medical records that guide and teach. New Engl. J. Med. **278**(11), 593–600 (1968). https://doi.org/10.1056/NEJM196803142781105
22. Zamanzadeh, V., Jasemi, M., Valizadeh, L., Keogh, B., Taleghani, F.: Effective factors in providing holistic care: a qualitative study. Indian J. Palliat Care **21**, 214–224 (2015)
23. Gulliford, M., Figueroa-Munoz, J., Morgan, M.: What does 'access to health care' mean? J. Health Serv. Res. Policy **7**(3), 186–188, (Issue published: July 1 2002)
24. Oda, H., Tsujita, Y., Nil, I.R.: An analysis of factors influencing the international migration of Indian nurses. J. Int. Migr. Integr. (2018). https://doi.org/10.1007/s12134-s018-0548-2

Chapter 60
Enhancing Creative Learning Methods by Immersive Virtual Reality: A Pilot Study in Classroom Environment

Sumana Som, Deepak John Mathew, Shivanes Phillipson, and Indranil Saha

Abstract Technology is increasingly essential in engaging millennials. Over the past few decades, education pedagogy has struggled to develop new methods and tools to create better and higher quality education. Virtual reality (VR) technology is one of the solutions for new education pedagogy that can solve the restriction of space, time, and the act of learning. The VR inclusion into learning pedagogy also facilitates self-learning and problem-solving as well as assist students to explore new ideas. This study aims to develop and trial lessons of art and design using VR technology instead of traditional art and design classes. The participants are 6th standard school students. Within traditional teaching of art and design, depth of learning is insufficient. To challenge the traditional environment, VR technology-based art and design lesson are applied in the core curricula of art education in this study. Two sets of tests are conducted with 6th graders, one is pre- and post-test with VR-based lesson, and the other is pre- and post-test with non-VR-based lessons. Overall, this study hypothesizes that VR will make for better learning and upskilling of students' knowledge of art and design. This study found that the teaching outcome with VR and conventional teaching outcomes has a number of significant differences. Besides, it shows the pathway of VR-based teaching methodology in classroom environment.

S. Som (✉) · D. J. Mathew · I. Saha
Indian Institute of Technology Hyderabad, Hyderabad, India
e-mail: md16resch11005@iith.ac.in

D. J. Mathew
e-mail: djm@des.iith.ac.in

I. Saha
e-mail: md18resch01001@iith.ac.in

S. Phillipson
Swinburne University of Technology, Melbourne, Australia
e-mail: sphillipson@swin.edu.au

A. Chakrabarti et al. (eds.), *Design for Tomorrow–Volume 2*, Smart Innovation, Systems and Technologies 222, https://doi.org/10.1007/978-981-16-0119-4_60

60.1 Introduction

Teaching and learning are the core and a rather complicated part of any education system and forms the activity between teachers and students. The teacher is one who plays a vital role in teaching subjects, developing skills and concepts. To a certain extent, the teacher's role is not only teaching the subject but also to develop the effectiveness of teaching on the topic [1]. Effective teaching includes a set of criteria, such as proper instructional design and delivery by using particular subject content and recourse, classroom management, process of action in the classroom such as cultivate thinking skills and stimulate interest in the topic [2]. Effective teaching and creative teaching are mutually exclusive and vice-versa [3]. On the other hand, creative evolutionary teaching requires three process; generating a new idea, a variation on original subject, and the procedure must be original and unique [4]. According to Jerome Burner (1962), the act of creativity produces 'effective surprise.' Hence, the effective and creative teaching apparently intertwined in the teaching–learning scenario. From the last few decades, educators from all over the world have been concerned and are continually evolving the new pedagogy methods. Nevertheless, the applicability of enhancing creativity in the pedagogy is one of the challenging tasks [5]. In the Indian educational context, educational policymakers are reinforced creativity from a few years before [6], but it is diversely accomplished in a modern school setting.

Due to the Indian government initiative on creative education, but there are many ambiguities and misconception between creative education and teaching of creativity [7]. Part of the study claims that the school's busy curriculum schedule kills creative skills of students [8]. Few article and report stating that 50% of educators believe that current education system does not foster creativity and 34% of them shows insufficient recourses and tools [9]. In the present situation, with the advent of new technology, teaching methodology is seeking a new technology aid [10]. However, there is no technology and simple framework prescribed for the application of effective pedagogy. Therefore, it is clear that the concept of creative education is not widespread in the school education scenario in India.

One source of solution to enhance creative learning is virtual reality-based (VR) pedagogy. Several kinds of research have been showing its possibility in teaching and training sectors [11]. The characteristic features of VR technology have opened up many possibilities such as enabling visualization of abstract content, providing information with enhanced clarity, and allowing students to take part physically in the learning environment actively [12]. Within these possibilities of pedagogy assisted by VR, specific characteristics are an invocation of creative phenomena in the field of pedagogy of school education. For instance, immersion, that nourishes the idea of problem-solving and instills persistence toward the topic [13]. Rich visual information connected with similar topics, co-presence, that changed/develop the concept of space [14]. In addition, using 'Avatar' for self-representation, multimodality that can help to encapsulates the functionality the virtual worlds to connect with different communication systems [15].

In this small quasi-experimental pilot study, researchers took initiative to create a process of VR-aided teaching for art and design as a cross-curricular activity. According to Bill Lucas creativity, it can be demonstrated in any subject and any aspect of life [4]. The method of experimental teaching of other subjects such as physics, biology, and geography, VR tools have already been developed and successfully implemented in school. But the study of art and design assisted by virtual reality for the school level is rare and is not implemented in an obvious way in India. Therefore, this study examined the possibility and effects of VR-assisted art and design programs for the Indian school. Consequently, this research introspects the results in light of a qualitative discussion to identify changes in the creative learning platform.

60.2 Literature Review

In order to review the existing knowledge based on the effectiveness of virtual reality in education, the trajectory of the virtual reality and education sector meets in a common factor; the creative phenomena. This review identified some crucial points to make connectivity between virtual reality, creative teaching, and art and design study.

60.2.1 Awareness of Technology-Based Pedagogy

Currently, digital technology is more personalized and has come in various forms, such as augmented reality, virtual reality, and the holograph [16]. The government of India's initiatives on the digitalization of education systems is promising. Although there are many inequalities of digital technology, adoption issues in government school across countries. Gurumurthy (2010) presented a case study of information and communication technology (ICT) programs in school education in terms of private–public partnership models versus the integrated approach [17]. In 2013, National Council of Educational Research and Training (NCERT) published a curriculum for ICT training [18]. On the contrary, some studies show the imbalance of technology-integrated learning such as recourses, connectivity, and storage-related challenges, maintenance, up-grading software, content developer, and the trained teacher are significant challenges across Indian school and universities [19]. After that, a report by TATA Trust and IT for change published in 2018 in which it was vividly described ICT implementation procedure in school [20]. In a very recent national policy of education 2019, it has actively focused on technology-based education [21].

Throughout the reports and study, it is very evident that Indian schools' proceeds into technology-based education forum. Although emerging technologies are not yet adopted widely, it is under experiment that how much technology could be applicable for classroom teaching and how it could be applicable for self-learning.

60.2.2 Virtual Reality and Teaching–Learning Process of Creativity

Jerome Dinet et al. says that immersive feelings are an intersection of perceptual-cognitive motor process and external environments, although immersive-ness of every single people could be different from each other [22]. Likewise, full-body interactive learning environments are very impactful to develop cognitive insight where the body is the mediator of making meaning [23]. According to author M. Glenberg conceptual knowledge grounded on sensory motors representations [24]. A VR environment took an active part to triggered imagination with miscellaneous visualization. The imagination as a prevalent technological metaphor connected students with situated activity [25]. Felipe Becker Nunes et al. explored a comparative study on the OpenSim virtual technology versus Moodle learning platform on science class on 6th standard students, and the result shows the VR-based learning performance is better than Moodle platform [26]. A quasi-experimental study on 10th standard students by Tiwi Nur Asthuti et al. shows that VR integrated learning improves critical thinking and scientific attitude [27].

60.2.3 Creative Corner of Virtual Reality Technology Within Art and Design Study

Many VR researchers have explained how VR impacts learning experiences and outcomes concerning knowledge acquisition [28]. However, is this virtual reality-based technology equally useful for learning the art or innovative design learning? The relevance of art and design in educational practice is 'hands-on' or practice-based learning [29]. The hands-on learning provides an opportunity for authentic sensory experiences and encourages empathy, self-reflection, and a tangible connection to the real world. Likewise, virtual reality provides students with a three-dimensional interactive environment, which is very effective for perceptual understanding [30]. However, this technology is rarely used for teaching art and design, such as using 3D rendering in virtual reality to develop sketch leaning [31], 3D visualization of virtual scene simulation for illustration learning, digital art teaching system with interactive virtual technology. Koun-Tem Sun et al. (2010) show improvements of sketch capability with VR-aided training systems of elementary students [31]. Yan Song (2018) says that to demonstrate the design concept, VR technology help to simulate the sense effect and also increase fidelity, vividness, and perceptual knowledge of design or artwork and consolidate the experience [32].

60.2.4 Summary of Review

The literature review shows that there is continuous effort to develop a tech-savvy pedagogy in school education. Especially, the study indicates the creative aspects of VR-based pedagogy in overall education and potential aspects for art and design study. There is an enormous attempt to apply VR in artistic activity, whereas VR application on art and design teaching is comparatively less. Besides, a technology-based training procedure includes embodied interaction that amplifies a hands-on learning procedure, which is a crucial aspect of art and design study. Hence, the study presumption is how differently effective VR-based art and design learning than traditional studio-based learning (H1).

However, there are many advantages of digital technology-based pedagogy for art and design. Still, there is no detailed instruction to develop the curriculum. Specifically, we need to identify whether or not the part of the syllabus endorses digital technology integration. The school not yet applied broadly to VR-based art and design pedagogy. Therefore, this empirical study (pilot) attempts to discover the effectiveness of VR application on art and design learning concerning skill (H_2), concept, and perception development (H_3, H_4). The four assumptions (H1, H2, H3, H4) focuses on a trial of VR-based teaching methodology for art and design study in classroom environment.

60.3 Hypothesis

- **H1**: There is statistically significant difference in mean without and with VR learning performance of 6th grade students.
- **H2**: There is statistically significant difference in mean value of learning artistic skill of 6th grade students with VR and non-VR
- **H3**: There is statistically significant difference in mean value of achieving creative capability of 6th grade students with VR and non-VR.
- **H4**: There is statistically significant difference in mean value of accumulation of knowledge of creating art and design and development of cognition of 6th grade students with VR and non-VR.

60.4 Methodology

A quasi-experience follows the study with a quantitative approach followed by a qualitative discussion on the performance of students and VR-based teaching method in classroom environment. The methods contain pre-test and post-test observations based on the tasks. The entire test procedure obtained three stages; first, the pre-test where only the question of the task provided without any explanation. Second, the intervention (without VR and VR) explains the concept and teaches them the traditional method and, later, the VR-assisted method. Third, a post-test where students have performed with the same question.

60.4.1 Sample

The setting of this study is a residential, public educational institution—the residential school named Telangana Social Welfare Residential Educational Institutions Society (TSWREIS) based in the city of Hyderabad. A convenience sampling is carried out. Students have limited experience in technology-based teaching–learning such as mobile, tablet, and desktop. They have no experience in a virtual reality environment. Certain factors are taken into account when selecting the target participants and school for the test. First of all, for research, students are selected from 6th standard students, age group of around 11–14 years. A small-scale pilot study conducted with six students to figure out the VR impact on particular art and design lesson in classroom scenario.

60.4.2 Data Collection Strategies

A task-based pre- and post-test are performed with the intervention of non-VR and VR tools in which students are instructed to create a design. Three variables are defined, and 33 items have been constructed as a tool to collect the responses using 10 points Likert scale in which the score 1–2 poor, 3–4 reformative, 5–6 moderate, 7–8 good, and 9–10 excellent.

The main observation deals with significant aspects; (1) each student's artistic skill capacity (ASC), the skills are considered as substantial characteristics of studio art practice [33]. (2) Overall creative capability (CC), it follows standard art assessment criteria prescribed by Donna Key Beattie [34], and (3) topic-wise knowledge and cognitive development (KOC). The items are developed by following Henrik Hagtvedt's (2008) scientific framework of perception and evaluation of visual art [35] and Anderson and Krathwohl's (2001) higher-order level of thinking skills framework [20]. The tool was reviewed by art and design educationist and VR expert. The participants were suitably coded to ensure confidentiality. The evaluation of the test done by three schoolteachers individually.

60.4.3 Pre-test—Intervention—Post-test Design

The test will include two procedures; First, without a technical device and, secondly, with technological tools (VR-based teaching). The entire group of students assigned

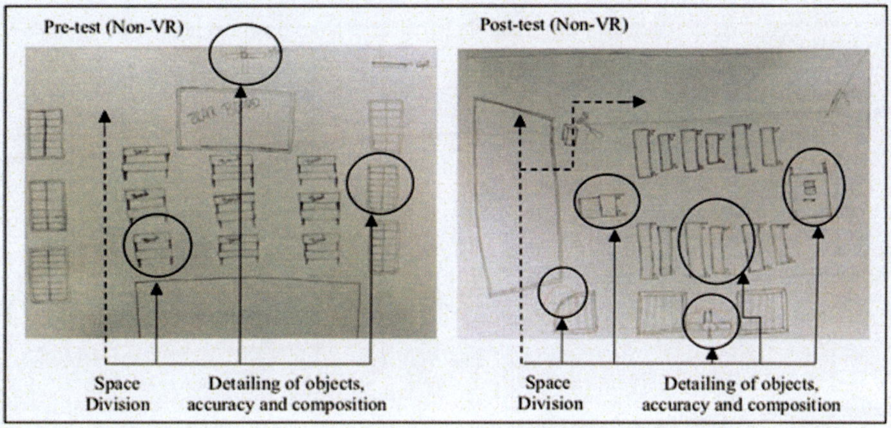

Fig. 60.1 Pre-test and post-test with non-VR teaching (Task: interior view of classroom)

for the first pre-test 30 min, and 15 min intervention with traditional teaching methods, then for the post-test for 30 min. The same group of students assigned for first pre-test 30 min and intervention with virtual reality tool-based teaching, then post-test for 30 min. The VR-based intervention took 30 min; each student gat 5–6 min to experience in the VR environment. Two days gap has maintained between traditional teaching-based tests and digital tool-based tests.

60.4.4 Limitation of This Study

This study is limited to art and design pedagogy. The VR content is developed based on NCERT art and design syllabus instruction and 3D models have taken from NCERT other subjects' books (History, Geography and General Science). Therefore, this content is customized who are following NCERT books. This particular VR environment can run only in Unity 2019.1.0 Beta 1 and viewer can experienced with HTC VIVE pro head mounted set and hand controller (Figs. 60.1 and 60.2).

60.5 Results and Data Analysis

The study applied three stages to analyze the quantitative data of this study. Firstly, the mean values of assigned marks based on the measurement items were calculated. Mean values were categorized in two subgroups for each student: pre-VR demonstration and post-VR demonstration. Secondly, the mean values of assigned marks were classified based on the dimensions of the research framework, which are ASC, CC, and KOC, keeping the pre-VR and post-VR categorization in place. Finally,

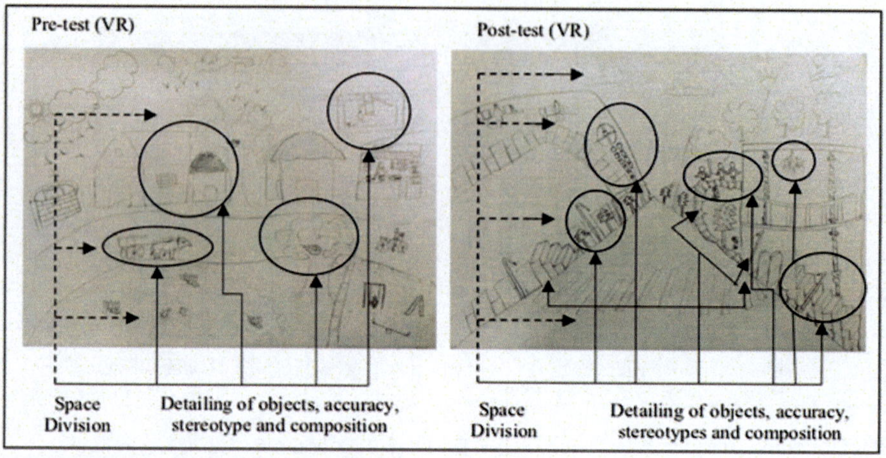

Fig. 60.2 Pre-test and post-test with VR-based teaching (Task: outside view of your school)

the overall and dimension-based mean values were analyzed through paired two samples t-test, which is a statistical method to conclude if there is any mean difference between two sets of observations [36] Pre- and post-VR observations, in the context of the current study. Tables 60.1, 60.2, 60.3, and 60.4 illustrate the findings of this analysis.

There was a significant difference in the overall scores for post-non-VR (M = 4.01, SD = 0.34) and post-VR (M = 5.69, SD = 0.48) observations; t (5) = −8.3, p = 0.00. For the variable artistic skill capacity of individual artistic items (ASC),

Table 60.1 t-test: paired two sample for means (overall)

	Mean	SD	Variance	Observation	df	T stat	$P(T \leq t)$ two-tail
Post-non-VR	4.01	0.34	0.12	6.00	5.00	−8.38	0.00
Post-VR	5.69	0.48	0.23	6.00			

Table 60.2 t-test: paired two sample for means (ASC)

	Mean	SD	Variance	Observation	df	T stat	$P(T \leq t)$ Two-tail
Post-non-VR	3.60	0.20	0.04	6.00	5.00	−7.34	0.00
Post-VR	5.20	0.46	0.21	6.00			

Table 60.3 t-test: paired two sample for means (CC)

	Mean	SD	Variance	Observation	df	T stat	$P(T \leq t)$ two-tail
Post-non-VR	4.67	0.35	0.12	6.00	5.00	−6.01	0.00
Post-VR	6.78	0.86	0.74	6.00			

Table 60.4 t-test: paired two sample for means (KOC)

	Mean	SD	Variance	Observation	df	*T* Stat	$P(T \leq t)$ Two-tail
Post-non-VR	4.33	0.89	0.79	6.00	5.00	−5.15	0.00
Post-VR	5.89	0.91	0.82	6.00			

there was a significant difference in the scores for post-non-VR (M = 3.60, SD = 0.20) and post-VR (M = 5.20, SD = 0.46) observations; t (5) = −7.34, p = 0.00. For the variable creative capability of entire work (CC), there was a significant difference in the scores for post-non-VR (M = 4.67, SD = 0.35) and post-VR (M = 6.78, SD = 0.86) observations; t (5) = −6.01, p = 0.00. For the variable knowledge orientation and cognition with topic (KOC), there was a significant difference in the scores for post-non-VR (M = 4.33, SD = 0.89) and post-VR (M = 5.89, SD = 0.91) observations; t (5) = −5.15, p = 0.00.

From the above results of each variable, it is evident that the mean post-VR ratings were significantly greater than zero. The two-tail p-values (0.00 in all cases) were less than the significance level 0.05, providing evidence that teaching through VR technology was effective in students' art and design learning performance. The summarization of the findings says as the two-tail p-values for all four cases came out to be less than the significant level 0.05, we accept all four hypotheses (H1, H2, H3, H4).

60.6 Discussion and Conclusion

The purpose of this pilot study to develop and trial the effectiveness of non-VR and VR-based art and design teaching in set up of classroom environment. The study utilized VR-based pedagogy to improve the learning of drawing and design skills, and creative capacity to develop a complete artwork. The study triangulated findings from the observation of participants' activity, and a quantitative analysis produced several highlights. This small-scale indicative data utilized to test VR-based teaching methodology.

The results showed that VR-based learning performance indeed offers a significant benefit. The study examined the changes in drawings before and after VR-based intervention. The evaluation started with assessing the skill of drawing and how students observe and reimagine the objects and space. The findings of artistic skill capacity (ASC) and creative capability (CC) assessment came out with some fact that indicates embodied learning with VR tools gives them a new insight to break the stereotype imagination (appendix). The ASC assessment follows holistic rubrics for standard-based assessment in the visual art 6th standard prescribed by Dorn [33] and NCERT fine arts evaluation criteria [37]. It is mentioned that the work of art must represent complex conceptual ideas, inventiveness, imagination, risk-taking, and compositional skills. The test result of post-VR work carries several changes, such as introducing a scientific perspective, detailing each object, adding new objects, making variations in lines, and making accurate shapes of objects.

Overall creative capability (CC) assessments follow suggestions for creativity assessment prescribed by Beattie [38]. As per Beattie's suggestion, the work should show intuitive skills and show the formulation of a new idea, a heuristic search on method, and the application of abstract relation to generating a new understanding. In the analysis of post-non-VR and post-VR tests, the study responds to significant changes that lead to the effectiveness of immersive VR environments and embodied learning. The works are showing a continuous process of adjustment the overall composition, manifestation of a different viewpoint, an intense attempt to create a new style, breaking the stereotype patterns, introducing folk elements in a new style. Evaluation on knowledge orientation and cognition with the topic (KOC) followed by Henrik Hagtvedt's (2008) scientific framework of perception and evaluation of visual art [35] and Anderson and Krathwohl's (2001) higher-order level of thinking skills framework [20]. Here, the study assessed conceptual, factual, and procedural knowledge applied with cognitive dimensions, such as remembering crucial points, understanding the facts, problems, and agendas, analyzing the topic heuristically, creating the final work of art, or formulating a design. In this study post, the VR test shows many changes in their work (appendix). Their retention is higher than the post-non-VR test. Every single point, they are pointed out by annotation and symbols. As students are first experiencing VR, they are taking little extra time to get accustomed to the virtual environment. Once they habituated, the level of understanding is more profound. It reflects on their verbal and gesture reciprocation. The work also shows they analyze the complete work and emphasizing particular objects by detailing, bold line, and size.

In the endeavor of creativity teaching in the twenty-first century, VR-based art and design teaching is not a substitute for the traditional teaching method. It is a tool for making effective, versatile teaching and enriches students understanding and invigorating their abstract ideas. It is a parallel method of teaching. Although the integration of VR technology in art and design education is still challenging for teachers, it requires a profound knowledge of content and software knowledge.

60.7 Scope for Further Study

The pilot study obtained a positive result. Consequently, this research will be applied to a more significant number of students in order to establish a much more reliable art and design effective pedagogy strategy based on virtual reality.

Appendix

Variable 1: Artistic skill capacity (ASC). Items: fluency of line pattern, stroke strength, variations of line quality, accuracy of couture, level of confidence, naturalism of each object, detailing of each object, ability to render complex image, originality of innovative form making, adorance to one-point perspective, consideration of gravitational parameter, space management within the frame, use of design/art principles

Variable 2: Creative Capability (CC). Items: shape accuracy of elements, structural arrangement of various elements within this frame, linear variations of different objects, accuracy of each objects and proper placing within the frame, awareness of art and design principle, ease of making

Variable 3: Knowledge & Cognitive process dimension (KOC). Items: Factual(remember): recalling all the important information, provide annotation. Conceptual and Procedural (understand, apply, analyze): Annotation with objects, clear idea, came out with specific visuals, Adoption of images from teaching visuals, develop image with Juxtapose, imaginary image, demonstration their idea critically

References

1. Tateo, L.: Education as "dilemmatic field." Integr. Psychol. Behav. Sci. **52**, 388–400 (2018)
2. Layne, L.: Defining effective teaching. J. Excellence Coll. Teach. **23**, 43–68 (2012)
3. de Sousa, F.: Creative teaching and effective teaching in higher education. Int. J. Organ. Innov. (Online) **3**, 5–44 (2011)
4. Fisher, R., Williams, M.: Unlocking Creativity : A Teacher's Guide to Creativity Across the Curriculum. Routledge, London (2004)
5. Craft, A.: Creativity in Schools : Tensions and Dilemmas. Routledge, London (2005)
6. Aesthetics, D.o.E.i.A., Training, N.C.o.E.R.a.: Country Report. Art Education in India. Ministry of Human Resource Development—Govt. of India (2010)
7. Sawhney, N.: Promoting Creativity and Innovation in Education System (2011)
8. Akash Pallath, A.J.D.D.: Without Fundamental Reforms to the Education System, Indians Will Not Innovate (2019). www.fairobserver.com
9. EduTechReview.: Differences in the Perception of Creativity in the Indian Education System (2017)
10. Routhier, P.H.: The immersive experience classification system: a new, strategic decision-making tool for content creators. In: SMPTE 2017 Annual Technical Conference and Exhibition, pp. 1–13 (Year)
11. Bricken, W., Washington Univ, S.W.T.C.: Learning in Virtual Reality (1990)
12. Falconer, L.I.Z.: Situated learning in virtual simulations: researching the authentic dimension in virtual worlds. J. Interact. Learn. Res. **24**, 285–300 (2013)
13. Alahuhta, P., Nordbäck, E., Sivunen, A., Surakka, T.: Fostering team creativity in virtual worlds. J. Virtual Worlds Res. **7**, 1–22 (2014)
14. Kohler, T., Fueller, J., Matzler, K., Stieger, D.: Co-creation in virtual worlds: the design of the user experience. MIS Quart. **35**, 773–788 (2011)
15. Kohler, T., Fueller, J., Stieger, D., Matzler, K.: Avatar-based innovation: consequences of the virtual co-creation experience. Comput. Hum. Behav. **27**, 160–168 (2011)
16. Chou, P.N.: Smart technology for sustainable curriculum: using drone to support young students' learning. Sustainability **10** (2018)

17. K.G.: ICTs programmes in school education PPP models vs. integrated approach (2010)
18. Central Institute of Educational Technology National Council of Educational Research and Training, N.D.: Information and Communication Technology for the School System Curricula for ICT in Education (2013)
19. Mr. Rampravesh Gond, D.R.G.: A study on digital education in india: scope and challenges of an indian society. Anveshana's Int. J. Res. Reg. Stud. Law Soc. Sci. J. Manage. Pract. **2** (2017)
20. Anderson, L.W., Bloom, B.S.: A Taxonomy for Learning, Teaching, and Assessing: A Revision of Bloom's Taxonomy of Educational Objectives. Longman, New York (2001)
21. Kasturirangan, K., Alphons, K.J., Bhargava, M., Kureel, R.K., Kattimani, T.V., Tripathy, K.M., Asif, M., Sridhar, M.K., Gupta, R.P., Shamsu, S.T.: National Education Policy 2019 (2019)
22. Jérôme Dinet, M.o.K.: Immersive interfaces for engagement and learning: cognitive implications. In: Proceedings of the Virtual Reality International Conference - Laval Virtual, pp. 1–8. ACM, Laval, France (2018)
23. Malinverni, L., Narc, Par, S., Burgu, S.: The medium matters: the impact of full-body interaction on the socio-affective aspects of collaboration. In: Proceedings of the 14th International Conference on Interaction Design and Children, pp. 89–98. ACM, Boston, Massachusetts (2015)
24. Glenberg, A.M., Witt, J.K., Metcalfe, J.: From the revolution to embodiment: 25 years of cognitive psychology. Perspect. Psychol. Sci. **8**, 573–585 (2013)
25. Roussou, M.: Interactivity and conceptual learning in virtual environments for children. In: CHI '04 Extended Abstracts on Human Factors in Computing Systems, pp. 1049–1050. ACM, Vienna, Austria (2004)
26. Nunes, F.B., Krassmann, A.L., Tarouco, L.M.R., de Lima, J.V.: A teaching method based on virtual worlds and mastery learning. J. Virtual Worlds Res. **11**, 1–15 (2018)
27. Astuti, T.N., Sugiyarto, K.H., Ikhsan, J.: Effect of 3D visualization on students' critical thinking skills and scientific attitude in chemistry. Int. J. Instr. **13**, 151–164 (2020)
28. Slavova, Y., Mu, M.: A comparative study of the learning outcomes and experience of VR in education. In: 2018 IEEE Conference on Virtual Reality and 3D User Interfaces (VR), pp. 685–686. (Year)
29. Morgado, P.: From passive to active learners: implementing the pedagogy of "learning by doing" in a large-sized design foundation class. Transformative Dialogues Teach. Learn. J. **4**, 1–12 (2010)
30. Barsalou, L.W.: Grounded cognition: past, present, and future. Top. Cognit. Sci. **2**, 716–724 (2010)
31. Koun-Tem, S., Hsin-Te, C., Kuan-Chien, M.: Research on the application of virtual reality on arts core curricula. In: 5th International Conference on Computer Sciences and Convergence Information Technology, pp. 234–239 (Year)
32. Song, Y., Li, L.: Research on application of VR technology in art design teaching. In: 2018 International Conference on Engineering Simulation and Intelligent Control (ESAIC), pp. 343–345 (Year)
33. Dorn, C.M.a.: Assessing Expressive Learning : A Practical Guide for Teacher-directed Authentic Assessment in K-12 Visual Arts Education. Mahwah: Taylor & Francis Group (2003)
34. Beattie, D.K.: Visual arts criteria, objectives, and standards: a revisit. Stud. Art Educ. **38**, 217–231 (1997)
35. Henrik Hagtvedt, R.H., Patrick, V.M.: The perception and evaluation of visual art. Empirical Stud. Arts **26**(2) (2008)
36. Ross, A.E.: Basic and Advanced Statistical Tests: Writing Results Sections and Creating Tables and Figures. Leiden, Boston, Brill|Sense (2017)
37. NCERT.: Syllabus of Arts Education. New Delhi (2008)
38. Beattie, D.K.: Creativity in art: the feasibility of assessing current conceptions in the school context. Assess. Educ. Princip. Policy Pract. **7**, 175–192 (2000)

Chapter 61
The Influence of Industry 4.0 on Product Design and Development: Conceptual Foundations and Literature Review

Sravan Tatipala, Tobias Larsson, Christian Johansson, and Johan Wall

Abstract Since its introduction in 2011, industry 4.0 has been coined the "fourth industrial revolution" following mechanization, industrialization and IT/automation as the first three, and represents the current trend of automation technologies (cyber-physical systems, Internet of things, cloud computing, etc.) in the manufacturing industry, with their potential for disruption of the manufacturing paradigm as we know it. However, the effect and role of industry 4.0 on the design and development of the new products to be manufactured in industry 4.0 facilities is not clear. This research presents a literature review to: (1) understand the concept of industry 4.0 from an implementation (state of practice) viewpoint, (2) learn about approaches and considerations currently deployed for developing products to be produced in manufacturing plants progressively transforming into industry 4.0 environments. Results reveal that the potential of industry 4.0 is underexploited within product design and development, especially in the conceptual stages lacking methods, tools, and approaches. While later stages of the product development (production planning, ramp-up) have received some attention in regards with optimizing production operations, several publications acknowledge its potential to benefit earlier process stages.

61.1 Introduction

The fourth industrial revolution commonly referred to as industry 4.0 [1] is believed to have a profound impact on various industrial sectors, especially the manufacturing domain [2, 3]. Industry 4.0 represents focus on end-to-end digitization of all physical assets and its integration into digital ecosystems with value chain partners [4]. Industry 4.0 is characterized by emerging technologies such as the Internet of things, additive manufacturing, mobile Internet, wireless sensor networks, big data analytics, cloud computing, cyber-physical factories, nanotechnology, interconnectivity of machines, robotics and artificial intelligence [5]. Since a majority of the

S. Tatipala (✉) · T. Larsson · C. Johansson · J. Wall
Blekinge Institute of Technology, Karlskrona, Sweden
e-mail: sravan.tatipala@bth.se

A. Chakrabarti et al. (eds.), *Design for Tomorrow–Volume 2*, Smart Innovation, Systems and Technologies 222, https://doi.org/10.1007/978-981-16-0119-4_61

discussion in the context of industry 4.0 is focused around manufacturing aspects [6], the role and scope of industry 4.0 s innovation potential in the product development process is still largely unexplored, particularly in regard to product design decisions. Wettach et al. [7] assert that the digital transformation sparked by industry 4.0 is meant to be a lot more than a technology-driven change. Over the past few decades, industries have frontloaded their modeling, simulation and design exploration activities with the intention to improve knowledge early on, make informed decisions already in conceptual stages of the product development process [8] and use this knowledge in manufacturing. Manufacturing plants all over the world, in diverse sectors, are now being upgraded in accordance with industry 4.0. To take full advantage of these investments, it is essential to comprehend the interconnectedness between and the effect of industry 4.0 on the product design and development process. Thus, this paper conducts a systematic literature review (SLR) for: (1) clarifying the concept of industry 4.0 from an implementation (state of practice) viewpoint; (2) for attaining an overview of the methods, tools and considerations currently employed for product design and development in the context of industry 4.0. The governing research questions are; (1) *What are the key components of an industry 4.0-enabled manufacturing facility? and* (2) *what method, tools and considerations are design teams employing while designing products to be produced in these factories?* Additionally, the authors will reflect upon and foster the ongoing discussion of the role of product design and development in the current industrial transformation.

61.2 Research Approach

The approach employed for carrying out the SLR is based on guidelines presented in [9, 10]. The review process can be divided into two stages: *planning* and *execution*. The planning stage encompassed defining the search topic, scope, choosing and justifying the choice of database(s), formulating search string(s), defining inclusion criteria and planning analysis procedure. The execution stage involved searching and collecting literature, iteratively pruning through resultant literature, reviewing the references in the collected literature to identify further relevant studies (cross-reference) followed by classification of research into relevant categories and summarization of the review findings.

61.2.1 Data Collection and Analysis

An overview of the literature review execution is depicted in Fig. 61.1. The source for collecting literature is divided into two categories: source 1 and 2. In order to understand the concept of industry 4.0 from a state of practice viewpoint, the perspectives of global actors were gathered by reading their blog posts, opinion papers, magazines and feature articles. Company specific research literature is gathered using

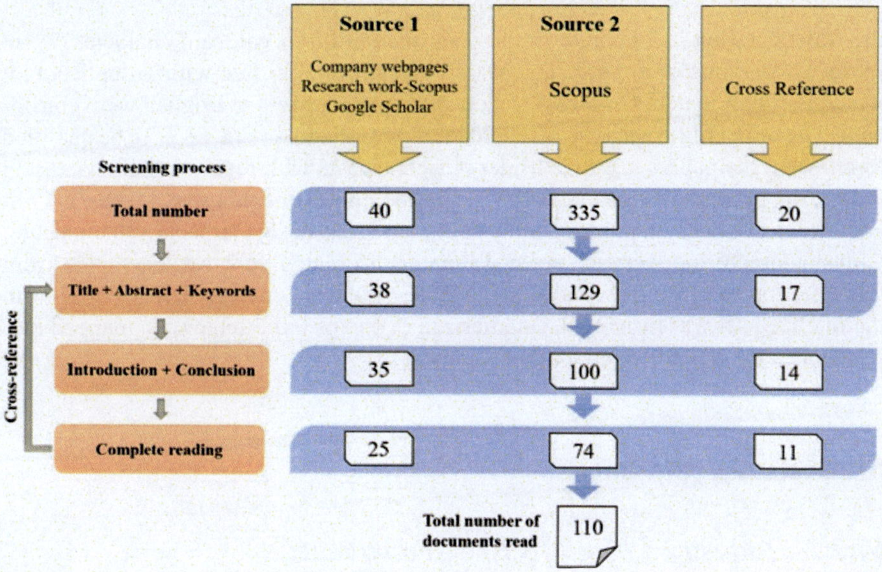

Fig. 61.1 Overview of the literature review execution (inspired from [14])

Google Scholar, Scopus and company webpages all of which is collectively termed as source 1. To learn about methods, tools and considerations currently employed while designing products for industry 4.0 environments, Scopus database is used due to the wide range of peer-reviewed scientific literature it offers and its proven relevance in the field of research which constitutes source 2. To identify further relevant studies, backward search [11] was conducted by reviewing references of the papers previously gathered (cross-reference). Other relevant databases such as Web of Science or EBSCO are excluded from the current review since the study is initial in nature. In addition, [12, 13] is referred to gather background knowledge on traditional approaches to conduct manufacturing, assembly, cost assessments during product development.

The search string used in source 2 is as follows; *(TITLE-ABS-KEY ("product development" OR "product innovation" OR "product design" OR "engineering design" OR "preliminary design" OR "concept* design" OR "concept* develop*" OR "early design" OR "initial design" OR "fuzzy front end" OR "design space exploration") AND TITLE-ABS-KEY ({Industry 4.0} OR {fourth Industrial revolution} OR {design for industry 4.0} OR {cyber physical system} OR {cyber-physical system}) AND TITLE-ABS-KEY ("manufactur*" OR "production" OR "smart manufactur*" OR "digital manufactur*" OR "flexible manufactur*")) AND (LIMIT-TO (LANGUAGE, "English")) AND (LIMIT-TO (DOCTYPE, "cp") OR LIMIT-TO (DOCTYPE, "ar")).* The search string forces the result to involve three domains namely, product design and development, manufacturing process and industry 4.0, resulting in literature with overlapping research domains. Search operators "AND" and "OR" are used due to the fragmentation of literature in the area in addition to wildcard character "*" to retrieve variations of a word. The search resulted in 335 hits with 1 duplicate hit which was eliminated. Only conference and journal papers in English were considered. The screening process involved three steps as shown in Fig. 61.1. To be qualified as relevant, the papers had to consider design aspects with regard to manufacturing during the product development process in the context of industry 4.0. Papers not addressing product or production system or related design aspects in connection with manufacturing were excluded. Literature from source 2 was exempted from this criterion to grasp the concept of industry 4.0 in its entirety. Every paper from the finalized set was fully read, summarized in one or more sentences, mapped into relevant categories followed by trend analysis.

61.3 Review Findings

61.3.1 Industry 4.0 and Its Key Components

Since the German initiative "Industrie 4.0" appeared in 2011 [1], marking the dawn of the fourth industrial revolution, it has received increased attention from several industrial sectors. In 2013, the "Industrie 4.0 working group" published first recommendations [15] for the implementation of industry 4.0, detailing its vision, enablers and some application scenarios. Despite its popularity, companies still struggle to grasp the idea of industry 4.0 [16] with diverse perceptions of the concept [17] and manifold review papers [18, 19]. Besides, very few real-life examples of industry 4.0 exist [20] with most being lab-scale or novice industrial implementations. Three key and commonly accepted components of industry 4.0 have emerged in literature as: cyber-physical systems (CPS), Internet of things (IoT) and smart factory. CPS is the integration of computation, networks and physical processes in which embedded computers monitor and control physical processes through networks while physical processes in turn affect computation through closed feedback loops [21]. CPS enables an important feature of industry 4.0 which is the fusion of the physical and virtual worlds [17]. IoT is the network used to connect smart components (e.g., sensors, actuators, logistics and tablets) in a factory allowing data transmission, interaction and cooperation for synergistic operation [17]. Smart factory is a context-sensitive manufacturing environment that assists machines and humans in execution of complex tasks efficiently [22]. Hence, industry 4.0 is an umbrella term encapsulating digital technologies and the aforementioned concepts aimed at enhancing current industrial operations [20] while meeting challenges such as growing sustainability requirements, fluctuating market demands, shorter development times as well as increased appeal for customized products and integrated lifecycle solutions like product–service systems [23]. In an industry 4.0 environment, smart factories consist of CPS that are capable of autonomously exchanging information, triggering actions and controlling each other over IoT infrastructures [17]. Smart products in these factories are uniquely identifiable, can be located at all times, are aware of their current status/history and can navigate routes to achieve target state [15]. Furthermore, industry 4.0 aims to realize end-to-end digital integration of engineering across industrial value chains and business processes, a digital model-driven process from start to end [23]. Empirical evidences for the implementation of industry 4.0 concept is presented in [20].

61.3.2 *Role of Design in the Realm of Industry 4.0*

Figure 61.2 (left) illustrates the most significant economic opportunities of industry 4.0 according to a global survey [24] interviewing experts within six nations. Amongst others, production optimization is seen as one of the main economic benefits. Furthermore, [6] acknowledges that most of the research in the context of industry 4.0 is concentrated around improving manufacturing operations while some publications [6, 7, 17, 25] admit the importance of design for the successful implementation of industry 4.0. Regarding design, both design of production systems as well as the design of products to be produced in these systems are of interest. Patil et al. [26] and Bertoni et al. [27] assert the need to alter our product design and development strategies, which, in the context of industry 4.0, are meant to support continuous improvement over the entire product lifecycle using real-time product data. For example, unlike the traditional way of developing products in discrete generations, ABB and Tesla collect data from their products in use to release new features via software updates [28]. The role of design in actualizing industry 4.0 is further emphasized by [7] which claims that "*without design, industry 4.0 will fail*" and presents six areas where design accelerates successful digital transformation in manufacturing. Likewise, [29] urges the need to combine knowledge from different disciplines and create sharing platforms since exclusive reliance on traditional engineering is no longer sufficient with design practices at the core of the digital revolution. Research also reports that investing in technology alone cannot meet the challenge of this digital transformation [30] and that industry 4.0 is meant to be a lot more than a

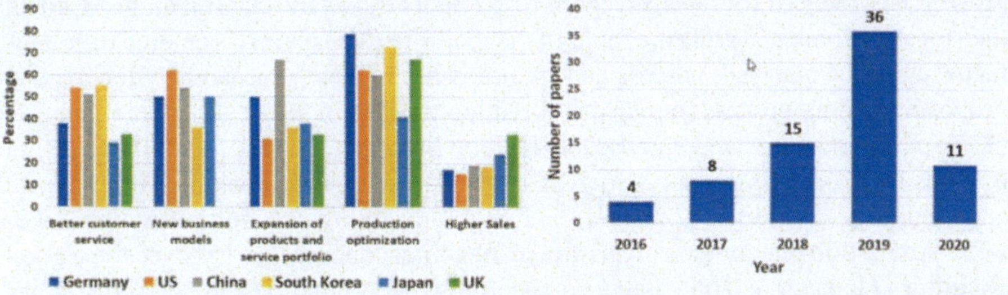

Fig. 61.2 Most significant economic opportunities of industry 4.0 based on a global survey [24] (left) and categorization of papers gathered from source 2 based on publication year (right)

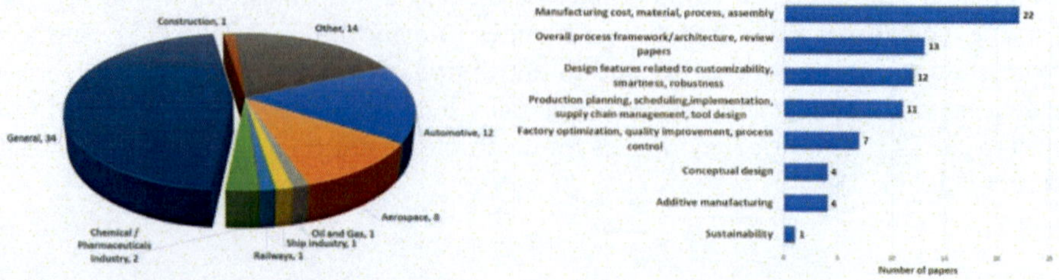

Fig. 61.3 Illustration showing the number of papers categorized based on the application domain (left) and research focus (right)

technology-driven change [7]. Since the product design and development process defines the value chain for a company to a large extent, design is believed to play a major role in fully benefiting from the fourth industrial revolution [25].

61.3.3 Manufacturing Considerations during Product Design and Development in the Context of Industry 4.0

This section presents analysis of papers from source 2. The product development process as presented by Ulrich and Eppinger in [12] is taken as a reference due to its generality. Figure 61.3 (left) shows categorization of literature based on application domain. The single most popular field is automotive (12 papers) followed by aerospace (8 papers). The majority (34 papers) fall under the general category in which research conducted does not belong to a specific domain. A deeper look reveals that these papers are published by research organizations and universities depicting the research topic's popularity. Other industrial sectors such as shipping, chemical, railways and construction industries (6 papers) are beginning to show interest in addition to electronics and food sectors (14 papers). From a time perspective, Fig. 61.2 (right) shows a constant rise in the number of publications focusing on product design and development in the context of industry 4.0, with exponential growth over the past four years. Figure 61.3 (right) shows the classification based on the "focus" of papers. While the focus of the majority (22 papers) is centered on manufacturing costs, material, process and assembly assessments, research reviewing and proposing overall process frameworks/architectures for industry 4.0 implementation (13 papers) have received significant interest. Similarly, papers focused on design features for facilitating aspects such as customizability, robustness, smartness have attracted good attention (12 papers), although presenting only initial studies. Likewise, production planning, analysis and management approaches have received good attention (11 papers). The category concerning conceptual design of products has received the least attention (4 papers) while other aspects such as factory optimization, quality improvement, additive manufacturing, sustainability were the focus of 12 papers.

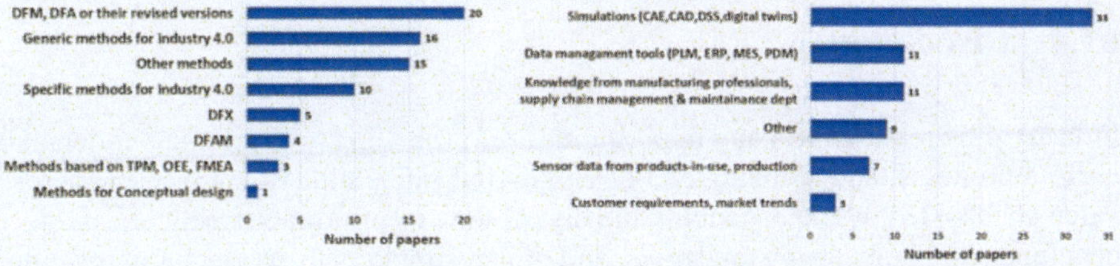

Fig. 61.4 Categorization of papers based on methods/tools/approaches employed (left) and sources of feedback (right)

Considering the methods, tools and approaches used in the analyzed papers (see Fig. 61.4, left), it is seen that a vast majority are employing traditional methods such as design for manufacturing/assembly (DFMA) (20 papers) and design for X (DFX) (5 papers), either directly or with modifications to suit industry 4.0 factory features such as flexibility [31], cyber-physical systems [32], customizability [33], recyclability [34]. Likewise, considerable papers present generic methods (16 papers) in addition to specific methods (10 papers) to support development of industry 4.0 solutions. These methods involve technologies such as augmented reality [35] virtual reality [36], digital twins [37], cloud-based platforms [38], while some of these methods support specific design aspects such as integrability between product design and manufacturing [39], design of product quality monitoring systems [40], intelligent product design with CPS feedback [41], CPS user interface design [42], IoT-based systems [43].

15 papers were assigned to "other" category, which involved methods featuring simulations, automated/smart/generative design, knowledge feedback, product lifecycle management (PLM) and manufacturing execution system (MES) tool-based approaches as well as review papers. Further categories (7 papers) involved methods based on design for additive manufacturing (DFAM), failure mode and effect analysis (FMEA), overall equipment effectiveness (OEE) and total productive maintenance (TPM). It is interesting to note that only one paper [6] presents a method contributing to conceptual design. Considering the source of feedback or information used in the presented methods (see Fig. 61.4, right), it is observed that the majority (20 papers) use insights generated through simulations (in, e.g., CAE, digital twins, CAD, DSS) or based on common data management tools (PLM, PDM, ERP, MES) (11 papers). Similarly, several (11 papers) utilize manufacturing professionals' knowledge, maintenance department logs as well as information from supply chain management to serve as feedback sources. Publications (9 papers) featuring feedback from workshops, literature review, surveys and building information modeling (BIM) were assigned to the "other" category. Interestingly, some publications (7 papers) capitalize on sensor-based data from product in use or production. Alternative sources for feedback include customer requirements and market trends (3 papers).

61.4 Discussion

Industry 4.0 creates promising opportunities for producing low-volume, on-demand custom products, and realizing end-to-end digital integration of engineering across value chains [15], while enhancing the digital flow of information between product developers, manufacturing engineers, and global suppliers to accelerate innovation, reduce risk and seek competitive advantage. Concepts such as concurrent and simultaneous engineering as well as collaborative design have existed for several years [44]. Claims like "*80% of the manufacturing costs of a product is determined by product design*" were popular amongst research publications already since the 90 s [45]. Likewise, DFX methodologies have been widely practiced by design teams for cost reduction while guiding designs to meet quality criteria related to robustness, reliability, serviceability and manufacturability, to name a few. Amongst these, DFM and DFA were the most popular methods since it directly addresses manufacturing and assembly costs [12]. The review analysis reveals the prominence of these methods even today, which are used either directly or with modifications (based on research and development) to suit industry 4.0. However, these methods do not encompass comprehensive features to foster the overall industry 4.0 "ideology." For example, they lack aspects such as, assessment of the relationship between complex industrial systems and product design features, utilization of IoT sensor data or product lifecycle data for identifying new design features [43], thorough design space exploration schemes to guide relevant design features, orchestration of industry 4.0 elements to support trade-offs, to name a few. Their role in industry 4.0 will be determined via the interoperability with new solutions based on industry 4.0. The authors recognize a need for research in this regard to deal with the complexity triggered by the fusion of physical and digital worlds within smart factories, to guide design and support decision making during the product development process. Of the publications proposing specific methods for the design of and for industry 4.0 solutions, several employ model-based engineering principles suggesting promising prospects for future research pursued in this direction. Additionally, the review highlights two promising concepts in the context of industry 4.0: product generation engineering (PGE) [46], which suggests that new products are almost always based on one or more existing product generations [6] and technical inheritance, which proposes to identify, transfer and use targeted information from product lifecycle to develop next generation products [47]. The past decades have seen a rise in the use of computer tools to virtually develop, simulate, test and validate design solutions. This is in agreement when considering sources of feedback or information in the reviewed papers where a majority utilize simulation-based sources. It is interesting to note that approaches still depend on the knowledge, experience and the intuition of manufacturing professionals. Information gathered in this manner, besides being asynchronous also consumes effort to collect and analyze. The authors argue that with increasing access to digital flow of information from different stages of the product lifecycle, there is a need for data-driven procedures to provide a quicker and synchronous stream of information for aiding product development process in real time. Singh et al. [8] exemplify the possibility of leveraging digital thread (which is defined as a data-driven architecture linking product lifecycle data [8]) to produce efficient design of the next generation of products. Most importantly, the review identifies a lack of research aimed at conceptual stages of product design in the context of industry 4.0. Although simulation-, value-, and data-driven design frameworks [23, 48–50] assist conceptual design assessments, these struggle to deal with the complexity of manufacturing process performance and industry 4.0 aspects. The authors assert a need for further research in this regard.

61.5 Conclusion

The paper presents results of a systematic literature review investigating key components of industry 4.0 and surveying its role in the product design and development process. Despite its popularity, companies struggle to grasp the concept of industry 4.0 in its entirety with many understanding the focus to be optimization of manufacturing operations. The review shows that research within product design and development in the context of industry 4.0 is scarce of which most are focused on later stages such as production system design and planning with very few in the earlier stages. The authors argue that current design approaches are not fully capitalizing on digital flow of information from different stages of the product lifecycle, which are increasingly synchronous. Furthermore, the authors acknowledge the need to consider aspects beyond manufacturing process, material and assembly procedures and focus on for example, identifying appropriate design features based on IoT sensor data, examining the relationship between complex industrial systems and product design to facilitate early trade-offs, building methods to support development of lifecycle solutions based on digital thread, and creation of designs with reduced failure rates in production to enable low-volume consumer-centric products, to name a few. The presented trends and analysis summary aim to highlight the role of design in digital transformation as well as foster the ongoing discussion on the research topic.

References

1. Kagermann, H., Lukas, W.D., Wahlster, W.: Industrie 4.0: Mit dem Internet der Dinge auf dem Weg zur 4. Industriellen Revolution. VDI Nachr. **13**(1) (2011)
2. Bartodziej, C.J.: The concept industry 4.0: an empirical analysis of technologies and applications in production logistics. In: The Concept Industry 4.0, pp. 27–50. Springer Gabler, Wiesbaden
3. Almada-Lobo, F.: The industry 4.0 revolution and the future of manufacturing execution systems (MES). J. Innov. Manage. **3**(4), 16–21 (2015)
4. Lee, M.X., Lee, Y.C., Chou, C.J.: Essential implications of the digital transformation in industry 4.0. J. Sci. Ind. Res. **76**, 465–467 (2017)

5. Wang, S., Wan, J., Zhang, D., Li, D., Zhang, C.: Towards smart factory for industry 4.0: a self-organized multi-agent system with big data based feedback and coordination. Comput. Netw. **101**, 158–168 (2016)
6. Albers, A., Stürmlinger, T., Mandel, C., Wang, J., de Frutos, M.B., Behrendt, M.: Identification of potentials in the context of design for industry 4.0 and modelling of interdependencies between product and production processes. Proc. CIRP **84**, 100–105 (2019)
7. Wettach et al.: Without design, industry 4.0 will fail: six challenges where design accelerates successful digital transformation in manufacturing. In: IXDS Human Industries Venture (2018). Retrieved from https://www.ixds.com/without-design-industry-40-willfail
8. Singh, V., Willcox, K.E.: Engineering design with digital thread. AIAA J. **56**(11), 4515–4528 (2018)
9. Biolchini, J., Mian, P.G., Natali, A.C.C., Travassos, G.H.: Systematic review in software engineering. System Engineering and Computer Science, Department COPPE/UFRJ, Technical Report ES **679**(05), 45 (2005)
10. Thomas, D.R., Hodges, I.D.: Doing a literature review. In: Designing and Managing Your Research Project: Core Skills for Social and Health Research. Sage Publications (2010)
11. Webster, J., Watson, R.T.: Analyzing the past to prepare for the future: writing a literature review. MIS Q. **26**(2) (2002)
12. Ulrich, K., Eppinger, S.: Product Design and Development. McGraw Hill. Inc., New York (2012). ISBN 9780071137423
13. Sandberg, M.: Design for manufacturing: methods and applications using knowledge engineering. Doctoral thesis. Luleå Tekniska Universitet (2007)
14. da Costa Fernandes, S., Pigosso, D.C., McAloone, T.C., Rozenfeld, H.: Towards product-service system oriented to circular economy: a systematic review of value proposition design approaches. J. Clean. Product. **257**, 120507 (2020)
15. Kagermann, H., Wahlster, W., Helbig, J.: Recommendations for implementing the strategic initiative industrie 4.0: final report of the industrie 4.0 working group. Forschungsunion: Berlin, Germany (2013)
16. Sanders, A., Elangeswaran, C., Wulfsberg, J.P.: Industry 4.0 implies lean manufacturing: research activities in industry 4.0 function as enablers for lean manufacturing. J. Ind. Eng. Manag. (JIEM) **9**(3), 811–833 (2016)
17. Mario, H., Tobias, P., Boris, O.: Design principles for industrie 4.0 scenarios: a literature review. Working paper, no. 1. Technische Universität Dortmund (2015)
18. Xu, L.D., Xu, E.L., Li, L.: Industry 4.0: state of the art and future trends. Int. J. Product. Res. **56**(8), 2941–2962 (2018)
19. Oztemel, E., Gursev, S.: Literature review of industry 4.0 and related technologies. J. Intell. Manufact. **31**(1), 127–182 (2020)
20. Da Silva, V.L., Kovaleski, J.L., Pagani, R.N., Silva, J.D.M., Corsi, A.: Implementation of industry 4.0 concept in companies: empirical evidences. Int. J. Comput. Integr. Manufact. **33**(4), 325–342 (2020)
21. Lee, E.A.: The past, present and future of cyber-physical systems: a focus on models. Sensors **15**(3), 4837–4869 (2015)
22. Lucke, D., Constantinescu, C., Westkämper, E.: Smart factory—a step towards the next generation of manufacturing. In: Manufacturing Systems and Technologies for the New Frontier, pp. 115–118. Springer, London (2008)
23. Wall, J., Bertoni, M., Larsson, T.: The model-driven decision arena: augmented decision-making for product-service systems design. Systems **8**, 22 (2020)
24. Kagermann, H., Anderl, R., Gausemeier, J., Schuh, G., Wahlster, W. (eds.).: Industrie 4.0 in a Global Context: Strategies for Cooperating with International Partners. Herbert Utz Verlag (2016)
25. Pessôa, M.V.P., Becker, J.M.J.: Smart design engineering: a literature review of the impact of the 4th industrial revolution on product design and development. Res. Eng. Des., 1–21 (2020)
26. Patil, B.A., Kulkarni, M.S., Rao, P.V.M.: New Product Development (NPD) process in the context of industry 4.0. In: 2019 IEEE International Conference on Industrial Engineering and Engineering Management (IEEM), pp. 1231–1235. IEEE (2019)

27. Bertoni, A., Larsson, T., Larsson, J., Elfsberg, J.: Mining data to design value: a demonstrator in early design. In: DS 87-7 Proceedings of the 21st International Conference on Engineering Design (ICED 17). Design Theory and Research Methodology, vol. 7. Vancouver, Canada, 21–25 Aug 2017, pp. 021–029
28. Porter, M.E., Heppelmann, J.E.: How smart, connected products are transforming companies. Harv. Bus. Rev. **93**(10), 96–114 (2015)
29. Celaschi, F.: Advanced design-driven approaches for an industry 4.0 framework: the human-centred dimension of the digital industrial revolution. Strateg. Des. Res. J. **10**(2), 97–104 (2017)
30. Synnes, E.L., Welo, T.: Enhancing integrative capabilities through lean product and process development. Proc. CIRP **54**, 221–226 (2016)
31. Karagiannis, P., Matthaiakis, S.A., Andronas, D., Filis, K., Giannoulis, C., Michalos, G., Makris, S.: Reconfigurable assembly station: a consumer goods industry paradigm. Proc. CIRP **81**, 1406–1411 (2019)
32. Ndip-Agbor, E., Cao, J., Ehmann, K.: Towards smart manufacturing process selection in cyber-physical systems. Manuf. Lett. **17**, 1–5 (2018)
33. Mourtzis, D., Zogopoulos, V., Xanthi, F.: Augmented reality application to support the assembly of highly customized products and to adapt to production re-scheduling. Int. J. Adv. Manuf. Technol. **105**(9), 3899–3910 (2019)
34. Rojek, I., Dostatni, E., Hamrol, A.: Automation and digitization of the material selection process for ecodesign. In: International Conference on Intelligent Systems in Production Engineering and Maintenance, pp. 523–532. Springer, Cham (2018)
35. Bruno, F., Barbieri, L., Marino, E., Muzzupappa, M., D'Oriano, L., Colacino, B.: An augmented reality tool to detect and annotate design variations in an industry 4.0 approach. Int. J. Adv. Manuf. Technol. **105**(1–4), 875–887 (2019)
36. Nåfors, D., Berglund, J., Gong, L., Johansson, B., Sandberg, T., Birberg, J.: Application of a hybrid digital twin concept for factory layout planning. Smart Sustain. Manuf. Syst. **4**(2), 231–244 (2020)
37. Yi, Y., Yan, Y., Liu, X., Ni, Z., Feng, J., Liu, J.: Digital twin-based smart assembly process design and application framework for complex products and its case study. J. Manuf. Syst. (2020)
38. Nellippallil, A.B., Ming, Z., Allen, J.K., Mistree, F.: Cloud-based materials and product realization—fostering ICME via industry 4.0. Integr. Mater. Manuf. Innov. **8**(2), 107–121 (2019)
39. Caputo, F., Caterino, M., De Luca, A., Fera, M., Greco, A., Lamanna, G., MacChiaroli, R., Manco, P., Manzo, M., Perfetto, D.: Product and process integrated design to enhance smart manufacturing systems. In: ECMS, pp. 213–218 (2019)
40. Shin, K.Y., Park, H.C.: Smart manufacturing systems engineering for designing smart product-quality monitoring system in the industry 4.0. In: 2019 19th International Conference on Control, Automation and Systems (ICCAS), pp. 1693–1698. IEEE, Oct 2019
41. Chen, Y.H., Ho, P.H., Chiu, M.C.: Utilizing cyber physical system to achieve intelligent product design: a case study of transformer. Transdiscip. Eng. Paradigm Shift, 1031 (2015)
42. Stern, H., Becker, T.: Concept and evaluation of a method for the integration of human factors into human-oriented work design in cyber-physical production systems. Sustainability **11**(16), 4508 (2019)
43. Albers, A.A., Bernijazov, R., Kaiser, L., Dumitrescu, R.: Internet of things canvas for ideation in model-based product generation planning. In: 2018 13th Annual Conference on System of Systems Engineering (SoSE), pp. 327–334. IEEE, June 2018
44. Polacsek, T., Roussel, S., Pralet, C., Cuiller, C.: Design for efficient production, a model-based approach. In: 2019 13th International Conference on Research Challenges in Information Science (RCIS), pp. 1–6. IEEE (2019)
45. Ulrich, K.T., Pearson, S.A.: Does product design really determine 80% of manufacturing cost? (1993)

46. Albers, A., Bursac, N., Wintergerst, E.: Product generation development–importance and challenges from a design research perspective. In: New Developments in Mechanics and Mechanical Engineering, pp. 16–21 (2015)
47. Mozgova, I., Barton, S., Demminger, C., Miebach, T., Taptimthong, P., Lachmayer, R., Peter N., Wilfried, R., Wurz, M.C.: Technical inheritance: Information basis for the identification and development of product generations. In: DS 87-6 Proceedings of the 21st International Conference on Engineering Design (ICED 17). Design Information and Knowledge, vol. 6, pp. 091–100. Vancouver, Canada, 21–25 Aug 2017
48. Bertoni, M., Eres, H., Scanlan, J.: Co-creation in complex supply chains: the benefits of a value driven design approach. In: Product Development in the Socio-Sphere, pp. 35–62. Springer, Cham (2014)
49. Wall, J.: Simulation-driven design of complex mechanical and mechatronic systems. Doctoral thesis. Blekinge Institute of Technology, Karlskrona, Sweden (2007). Retrieved from https://urn.kb.se/resolve?urn=urn:nbn:se:bth-00346
50. Bertoni, A.: Role and challenges of data-driven design in the product innovation process. IFAC-Pap Online **51**(11), 1107–1112 (2018)

Chapter 62
Digital Preservation of the Qutb Shahi Monuments: Archiving Architecture for Historical Education

Chaitanya Solanki, Deepak John Mathew, and Gino Joseph

Abstract Organized assortments of photographs have given birth to the method of photogrammetry which has earlier helped geologists in topographically mapping terrain and landscapes. Photogrammetry is a technique of measuring from photographs. It is usually associated with the making of maps from aerial photographs. When the same logic is set as an algorithm, it helps the software yield an even more accurate topographical product, but now with the additional capability of 3D projections. This can, in turn, be used to create photo accurate 3D models of any building, object, landscape desired. Along with technologies like virtual reality, photogrammetry can enable a viewer to be present in a digitized landscape that can be developed to look close to the original. This paper documents how the above set of technologies were used to virtually archive the monuments of Qutb Shahi tombs in Hyderabad. The process included photogrammetry, point cloud creation, UV mapping, texture modeling, mesh generation, etc. Thereafter, a digital setup of it was made and displayed through virtual reality. The result was a tangible, virtual landscape of the Qutb Shahi tombs, where the viewer could digitally navigate and see the monuments up close. The built artifact was then tested in a pilot study ($n = 35$) for its engagement with high school students as a tool for education in the classroom. The early results of the respondents have been positive in adopting VR exploration of historical monuments as a supportive tool for education. The future scope of evaluating this setup includes more exhaustive testing in schools of India.

62.1 Introduction

Monuments are an integral document of the country's and the world's history. Accessing and thoroughly studying it is an obligation of today's world [1, 2]. Culture

C. Solanki (✉) · D. J. Mathew · G. Joseph
Department of Design, Indian Institute of Technology, Hyderabad, Telangana 502285, India
e-mail: md19resch01001@iith.ac.in

D. J. Mathew
e-mail: djm@des.iith.ac.in

A. Chakrabarti et al. (eds.), *Design for Tomorrow–Volume 2*, Smart Innovation, Systems and Technologies 222, https://doi.org/10.1007/978-981-16-0119-4_62

and heritage have always played an important role in the collective development of society as well as the individual. Sites and monuments of cultural importance are capable of providing aesthetic and spiritual satisfaction, as well as become a center for economic growth [3, 4]. Structures of history are a testimony to the times of the past, and they help the visitor in visualizing the lives of the past. Though timeless, these monuments are prone to the adversities of time and every year, some are lost forever due to various reasons [5]. The technologies of photogrammetry and virtual reality are proposed as a preventive method that allows us to preserve the temporal evolution of the structures digitally [6, 7]. Over recent years, the technology of image analysis has grown ever more important for applications that concern the preservation and study of cultural heritage [8].

This study has emphasized the role of VR and photogrammetry for the visualization and preservation of the cultural site of the Qutb Shahi Tombs in Hyderabad. It also conducts a pilot study to test the prototype for its educational engagement with high school students.

62.2 Artifact Development

62.2.1 Case Study

At the coordinates, N17 23 43.73 E78 23 47.79, the tombs of Qutb Shahi are located close to the famous Golconda Fort in Hyderabad, India, in the Ibrahim Bagh (garden precinct) [9]. The site harbors the tombs and mosques built by the various rulers of the Qutb Shahi dynasty. Some of the tombs are single-storeyed while a few are double-storeyed. At the center of each tomb lies a sarcophagus that overlies the actual burial vault in a crypt below. A few pieces of blue and green tiles over the walls of a few domes are a testament to the fact that the domes might have been completely covered by them. Their date of building ranges between the sixteenth to seventeenth centuries [10].

Some of the major monuments include Sultan Quli Qutb Mulk's tomb, Jamsheed Quli Qutb Shah's tomb, Subhan Quli Qutb Shah tomb, Sultan Ibrahim Quli Qutb Shah's tomb, Sultan Muhammed Quli Qutb Shah's mausoleum, Muhammed Qutb Shah's tomb, Tomb of Fatima Sultana, Hakim's Twin–Tombs, Tomb of Taramati, Tomb of Premamati, Mausoleum of Kulsum Begum, and Mausoleum of Hayat Bakhshi Begum [10]. Even though the site is going through the process of restoration, all of the monuments have gone under the wear and tear of time, resulting in the loss of the original look, texture, and condition.

62.2.2 Documentation

The act of documenting a heritage site can help in preserving data regarding historical technology, lifestyle, artistic perception, aesthetics, social norms, economic levels, and building techniques. Documentation provides for the transmission of cultural heritage, visualization of the monument in contemporary manners, accurate data acquisition for future references, and knowledge about the present condition of the building. The process of documentation includes stages of data acquisition, interpretation, and production.

62.2.2.1 Data Acquisition

The acquisition of the data for this project was in the form of photographs. The task began with several visits to the heritage site for extensive reconnaissance. The geography, size, texture, elevation, ambient light, and the environment played a key role in the process.

The initial imagery was done with handheld full-frame cameras for the provision of the most quality. As the usage of handheld cameras was from ground level, it did not allow the photographers to cover the site plan of the area or the roof parts of the structure that contained the domes. To cover the complete monument, aerial photography was used through drones (see Fig. 62.1). The aerial photographs completed the documentation of all the monuments as well as provided with the site plan of the area. Following that, photographs of the environment near the perimeter were taken

Fig. 62.1 Aerial and ground level images of tombs

Fig. 62.2 Several photos of the perimeter stitched together to form the circular environment for the digital landscape

to implement later. The monuments were extensively photographed to eliminate the possibility of a lack of data for processing. This helps in the next step of interpretation and is also vital in the process of photogrammetry.

62.2.3 Development

The process to develop photogrammetric models and their subsequent deployment in virtual reality requires various steps in between; they include image interpretation and post processing, point cloud creation, mesh generation, 3D modeling, landscape creation, VR porting, and apparatus selection.

62.2.3.1 Environmental Image Stitching

Photoshop is software for photo manipulation, and in this project, it was used to stitch together the photos of the perimeter of the heritage site in order to provide for a limiting environment in the digital landscape (see Fig. 62.2).

62.2.3.2 Image Post Processing

The key usages of the Adobe Light room software were to ensure uniformity between the photographs for easy processing. Also, due to the photographs being aerial, taken with a DJI Phantom, the camera was of low quality. Light room was used to enhance the images by increasing the sharpness, contrast, and color. It was also used to remove any unnecessary flare, highlight, reflection, or shadow.

62.2.3.3 Point Cloud and Mesh Generation

After the images had been processed, they were imported into the image stitching software for their conversion into 3D models. The generation of 3D models is done one by one. When an object has been photographed from all around, reality capture can stitch the images together to export a textured and true colored 3D model. If the images are appropriate, the software first aligns them and presents a sparse point cloud with the larger icons representing the camera angles. Here, the user gets an approximate visual of the model being created. After this, the region is defined and needs to be rendered. In this case, all the tombs were rendered individually. Following that, the model can be exported in normal or high quality. Due to this project being focused for exploration in virtual reality, lighter models were preferred than heavy ones, and thus, the models were exported in normal quality (see Figs. 62.3 and 62.4).

Fig. 62.3 Sparse point clouds for four of the tombs generated in reality capture

Fig. 62.4 Generated mesh of the Hakim's tomb in reality capture

Fig. 62.5 Models of smaller monuments built in Blender

62.2.3.4 3D Modeling for Asset Creation

3D modeling software can be used to manipulate and improve the models generated from reality capture. For this project, Blender was used to eliminate unwanted foliage surrounding the monuments and to fill in for missing parts that might have been hidden in the images and thus were unable to render. Blender was also used for producing low poly environments for encompassing the high-resolution monuments. Due to the shooting permissions and duration being inadequate, the team of researchers was not able to document the smaller monuments as extensively as the bigger ones. Blender was also used to model those smaller monuments from scratch and was done so using reference images (see Fig. 62.5).

62.2.3.5 Surface Count Reduction

ZBrush was specifically used to decrease the face count of the models. When rendered from reality capture, the.fbx and.obj files can come out to be heavy and detailed. Zbrush allows smoothening of the surfaces and reduces the face count while preserving the detail. This helps in optimizing the entirety of the landscape (see Fig. 62.6).

62.2.3.6 Landscape Creation and Model Placement

After the creation of all the assets that are required for the landscape, they are imported in unity. By correlating the placement of all the monuments in respect to each other, the buildings were established onto the landscape. The plugin Steam VR was installed so that the export from unity could coherently work on the Steam VR platform and the plugin HTC Vive was installed because the final projection would have been done in the Vive. It was in unity where the interactive codes for the user to move and explore the landscape were coded (see Fig. 62.7).

Fig. 62.6 Model of the Mausoleum of Kulsum Begum being downsized and simplified in Zbrush

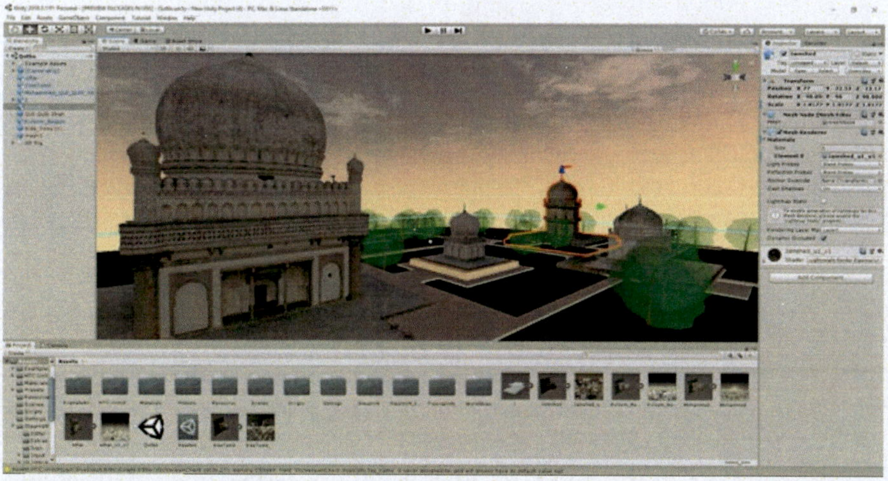

Fig. 62.7 Various tomb models being placed on the landscape in unity

62.2.3.7 Virtual Reality

After exporting the artifact as an interactive unity file, the landscape could now be viewed in Steam VR, which acts as a platform for HTC Vive. It has plugins that can handle and visualize the landscape in VR. The end result is an interactive installation in which a user could explore the landscape and the monuments of Qutb Shahi in virtual reality through HTC Vive.

62.3 Pilot Study

In order to understand whether this artifact could be an encouraging experience for the acceptance of VR landscapes as an educational tool, the researchers conducted a pilot study. The participants were guided through the experience in a controlled environment and were subsequently asked about the experience through a questionnaire. This method was used with preset guiding descriptions in-person.

62.3.1 Study Objectives

The objective of the following pilot study was to determine whether the VR exploration of photogrammetrically preserved historic monuments and landscapes could encourage students to adopt VR as a medium of learning history as well as other subjects.

62.3.2 Study Design

Convenience sampling was used for this study, where the outcome measure was a six-question written survey in which the students were asked to answer four yes/no questions, one score assigning (1–5) question, and one question that asked for the participants' suggestion regarding the improvement of the artifact.

62.3.3 Question Formation

Prior researches have confirmed that students have had positive responses to studying various school subjects through virtual reality [11]. Several studies have explored individual subjects and their fidelity in virtual reality with respect to engagement, efficiency, immersion, etc. [11–13]. While these studies have been conducted all across the world, little research is known within the Indian context. Hence, the following questions were used for this pilot study:

Q1: Was the VR experience more enjoyable than reading a textbook? (Yes/No)
Q2: Would you prefer to learn history with VR? (Yes/No)
Q3: Would you like to learn other subjects with VR? (Yes/No)
Q4: Would you like Audio Narration with VR? (Yes/No)
Q5: How enjoyable was the experience? (1–5)
With 5 being most enjoyable and 1 being not enjoyable at all.
Q6: What extra additions would you like in this whole experience? (Descriptive).

Table 62.1 Responses of pilot study sample on the experience of VR exploration of Qutb Shahi Monuments

Question	Percentage	Mean
Was the VR experience more enjoyable than reading a textbook? (y/n)	88.57	
Would you prefer to learn history with VR? (y/n)	94.29	
Would you like to learn other subjects with VR? (y/n)	94.29	
Would you like Audio Narration with VR? (y/n)	80	
How enjoyable was the experience? (1–5)		4.3

62.3.4 Demographics

The participants ($n = 35$) were students of several high schools from Kerala who were visiting the IIT Hyderabad campus on an educational trip. Response analysis showed that the group consisted of 71.43% Males, 22.86% females, and 5.71% participants who did not reveal their gender or name. The age group was between 16 and 18 years old with only one participant being 23.

62.3.5 Delimitations of the Study

This research is limited to the restrictions of convenience sampling. This study is also limited to the exploration of VR in a controlled environment; this might or might not reflect the efficacy of the technology in an Indian classroom setting.

62.3.6 Findings

Through analysis, it was shown that 88 percent of the participants found the Qutb Shahi VR exploration to be more enjoyable than that of reading a textbook, 94 percent of the participants looked forward to learning history as well as other subjects through VR exploration, 80 percent of the participants voted for the need of audio (background music as well as narration) in the setup, and average rating for the enjoyment of this experience came through to be 4.3 out of 5. In the last question where the participants were asked for their opinions on what else could be added in the experience, most of them opted for having audio and sound, whereas some replies indicated the potential for the conversion of such setups into a gaming experience in the class.

62.4 Conclusion

The landscape of the Qutb Shahi monuments was successfully archived using photogrammetry and presented in virtual reality. The project is still at the early stages of using immersive virtual reality systems for public access. The objective of the pilot study was to examine the agreeability of high school students from India in adopting VR exploration, such as the landscape of Qutb Shahi, as a means of studying school subjects in their own classrooms. The Qutb Shahi monuments that have been preserved through photogrammetry and explored through VR were shown to have a positive effect on the acceptance levels of students for the technology as an educational tool. Thus, after receiving feedback, the researchers plan to execute a more investigative approach in the future where the analysis can reveal a more generalized perception toward the artifact that was developed, potentially shedding light on whether the photogrammetric preservation and the VR exploration of other monuments throughout India can be helpful in the field of education. The study was also an indicator for the potential of a more rigorous testing that can emulate the current classroom setting and then evaluate specifics of VR exploration for education such as engagement levels, immersion, social interaction, learning efficiency, etc.

Virtual environments through the employment of photogrammetry, such as the one that was developed here, can provide rewarding aesthetic and learning experiences that would otherwise be difficult to obtain. Despite the high cost and restrictive format of these installations, it is believed by the researchers that it is well worth investigating the added value and potential that VR and photogrammetry can bring to the public domain, especially in the fields of virtual tourism, cultural heritage archiving, and digital heritage archiving, however, that is the scope of another, future study.

References

1. Georgopoulos, A., Ioannidis, C.: Photogrammetric and surveying methods for the geometric recording of archaeological monuments, archaeological surveys. FIG Working Week. Athens, Greece, 22–27 May 2004. https://www.fig.net/pub/athens/papers/wsa1/WSA1_1_Georgopoulos_Ioannidis.pdf
2. Salazar, S.D.S., Marques, J.M.: Valuing cultural heritage: the social benefits of restoring and old Arab tower. J. Cult. Herit. **6**, 69–77 (2005)
3. Bedate, A., Herrero, L.C., Sanz, J.A.: Economic valuation of the cultural heritage: application to four case studies in Spain. J. Cult. Herit. **5**, 101–111 (2004)
4. Harvey, E.R.: Política Cultural en Argentina. Unesco, Paris (1997)
5. Akinciturk, N., Kilic, M.: A study on the fire protection of historic Cumalikizik village. J. Cult. Herit. **5**, 213–219 (2004)
6. Yilmaz, H.M., Yakar, M., Gulec, S.A., Dulgerler, O.N.: Importance of digital close-range photogrammetry in documentation of cultural heritage. J. Cult. Herit. **8**(4), 428–433 (2007)

7. Gaitatzes, A., Christopoulos, D., Roussou, M.: Reviving the past: cultural heritage meets virtual reality. In: Proceedings of the 2001 Conference on Virtual Reality, Archeology, and Cultural Heritage, pp. 103–110, Nov 2001
8. Pedersini, F., Sarti, A., Tubaro, S.: Automatic monitoring and 3D reconstruction applied to cultural heritage. J. Cult. Herit. **1**(3), 301–313 (2000)
9. Callegari, F.: Sustainable development prospects for Italian coastal cultural heritage: a Ligurian case study. J. Cult. Herit. **4**, 49–56 (2003)
10. Centre, U.N.E.S.C.O. (n.d.): The Qutb Shahi Monuments of Hyderabad Golconda Fort, Qutb Shahi Tombs, Charminar. Retrieved 12 Apr 2020, from https://whc.unesco.org/en/tentativelists/5573/. Qutb Shahi tombs, 10 Mar 2020. Retrieved 10 Apr 2020, from https://en.wikipedia.org/wiki/Qutb_Shahi_tombs
11. Yang, T.H., Sun, H., Chen, C.C., Huang, C.C., Yang, Y.R.: Virtual reality applied to astronomy education in primary and middle school. In: International Conference on Frontier Computing, pp. 222–226. Springer, Singapore, July 2019
12. Munoz, A., Lewis, C., Bolling, N., Zandbergen, W., Scully-Allison, C., Le, V., Dascalu, S., Brown, B.: Virtual reality physics lab. EPiC Ser Comput **58**, 17–26 (2019)
13. Symonenko, S.V., Zaitseva, N.V., Osadchyi, V.V., Osadcha, K.P., Shmeltser, E.O.: Virtual reality in foreign language training at higher educational institutions. In: CEUR Workshop: Proceedings 2nd International Workshop on Augmented Reality in Education, 22 Mar 2019, vol. 2547, pp. 37–49. Kryvyi Rih (2020)

Chapter 63
A Novel User-Centric Assistive Device for Enhancing Luggage Security in Indian Railways

Md. Arif Ahmad, Chiranjeeb Deb, Tushar Sindhwal, and Biswajeet Mukherjee

Abstract The security system at the Indian railways mainly focuses on X-ray scanning of baggage and metal detectors for ruling out possibility of any life-threatening object. Though so, the current management does not focus on the user centric needs of safety and security as a service within the trains. The number of passengers traveling by Indian railways is ever increasing, and implanting a strong security system at all major/minor stations remains a challenge. Baggage theft in trains has remained unapprehended as a crime causing people to stay attentive by themselves which add to cognitive load for the users/passengers. To investigate this problem from a design perspective, field research was performed based on a survey of 101 participants, and 15+ interviews were conducted as a part of direct observation and analysis. Multiple user case scenarios were used to generate insights based on which a solution was proposed. This solution presents an electronic device which is paired with a smartphone. The user can place this device inside the baggage for leveraging the security of their luggage. On lifting the bag without informing its owner, the device shall detect and pass the information to the owner's smartphone through wireless technology. The smartphone receives the data and raises an alarm. The device is insensitive to the rotation and vibration of the train movement. The design of the user interface in the smartphone is based on the heuristic design principles to facilitate maximum convenience for users to operate it.

Md. A. Ahmad (✉) · C. Deb · T. Sindhwal · B. Mukherjee
IIITDM Jabalpur, Jabalpur, Madhya Pradesh, India
e-mail: 1814013@iiitdmj.ac.in

C. Deb
e-mail: 1814007@iiitdmj.ac.in

T. Sindhwal
e-mail: 1814023@iiitdmj.ac.in

B. Mukherjee
e-mail: b.mukherjee@iiitdmj.ac.in

A. Chakrabarti et al. (eds.), *Design for Tomorrow–Volume 2*, Smart Innovation, Systems and Technologies 222, https://doi.org/10.1007/978-981-16-0119-4_63

63.1 Introduction

Indian railways, one of the most trusted mode of transport in India, is quite unique and distinctive in character. Making it a safe and dependable platform is a huge challenge. The railways have the most complex interdependencies involved. Safety of 13 million passengers traveling in the railways every day is of paramount importance of the system [1]. Over the years, the network has taken a number of steps through innovative use of technology upgradation in Integrated Security System (ISS) consisting of IP-based CCTV surveillance system, access control, personal and baggage screening system and Bomb Detection and Disposal System (BDDS) [2, 3]. As per the data of 2018, a total of 473 railway stations have been provided with CCTV cameras, and the railways have planned to install CCTV cameras at all stations by the year 2021. Similarly, 1300 coaches have also been equipped by these cameras [3].

The total cases of theft in Indian railways have increased significantly over the past few years. As per data of the Indian Government, a total of 29,686 cases of theft were reported for the year 2015, 33,682 in 2016, 64,543 in 2017 and 76,474 in 2018 [4–7]. The number of cases is increased by 157% from 2015 to 2018. The responsibility to ensure security of passengers is a shared by the Railway Protection Force (RPF) and the Government Reserve Police (GRP) [1]. Under the Railway Property Act of 1966, the RPF is empowered to deal with cases of theft, dishonest misappropriation and unlawful possession of railway property. On any case of theft, the passenger can register a FIR, and their investigation and maintenance are taken care by the state governments, through the GRP. All India Security Helpline has been made functional by the RPF, which will provide passengers round the clock security assistance. Additionally, the helpline system is being upgraded to include features like auto generated SMS, computerized complaint registration, dashboard and app integration [3].

The reporting of these types of crimes is a problem of serious concern. The process of filing an FIR takes up time and, in many occasions, victim has little time to do so. In most of the cases, it has been observed that crimes are discovered when the passenger reaches their destination. It is found that only serious cases of theft are reported, and as a result, the actual data is never known. In most cases, the FIR registered has to be transferred to the Thana having jurisdiction, causing significant delay in the investigation [8].

63.1.1 Technology Mediated Support

The increase in number of cases of theft over the past few years indicate that the existing system of security, which is supported by the RPF, and the GRP leads to significant delays in investigation. There is a minimal, but emerging literature considering the use of technology. However, the works are focused mainly on theft prevention and lost luggage issues especially in airports.

U.S. Patent No. 8,742,922 B2 (2014) provides a luggage tracking device, which provides the user with real-time information of the location of his/her luggage at various stops along a journey. The device designed especially for air travel, consists of a housing configured to be attached to a piece of luggage, a location unit to determine current location, a transmitter to transmit the current location information and a controller to control operations of the location unit and transmitter [9].

U.S. Patent No. 9,439,164 B2 (2016) provides a method for locating and sharing information of checked-in luggage through a GSM device which is to be placed in the luggage. The device works in combination with a mobile device application and Bluetooth or a similar short-range RF transmitter which updates the traveler with the location of the device [10].

In another patent, an electronic luggage tag is used that uses GPS technology to track luggage. The device uses a motion detection and/or light sensing technology, which has the ability to detect whether the luggage is open or closed [11].

U.S. Patent No. 5,576,692 (1996) utilizes a nationwide paging system to track luggage items. The operation of tracking and locating is carried out by a sounding pager device or "beeper." Each beeper has its own unique number and code; in case the luggage in which the beeper is placed is lost, the beeper can be called using its number, which will cause the beeper to "beep" at whatever airport it is located [12].

In another product, a dual triggered alarm is used that includes a motion sensor alarm and a pull pin alarm. It uses a preset personal security code and designed for any bag or valuable item. If the protected item is touched, the motion sensor will trigger the alarm. The alarm than only can be offend by the personal security code [13].

In this paper, a user-centric design process was adopted, and a solution is presented in the form of a device which is paired with a smartphone. The device was designed keeping in mind the dynamic nature and challenges of Indian railways.

63.2 Methodology

63.2.1 Overview

A user-centric design process was adopted to study the problem from the user point of view and gain deeper insights. The design thinking process is an exploratory process and usually begins with defining of the problem. To walk into directed framework till the solution, the authors decided to use "double diamond" design process which involves two separate phases of divergent and convergent thinking [14]. This methodology is used because the product was not defined initially. Only after research and analysis, the product was conceptualized and developed.

The first divergent phase, known as the discovery phase begins with an initial idea or inspiration, in which user's needs are identified. Both qualitative and quantitative research is carried out using a questionnaire survey, interviews and ethnographic field study and desk research. After the first, divergent phase (Discovery), the findings were analyzed. This stage known as "define stage" included creating personas, case scenarios and framing insights.

63.2.2 System and User Study

To observe the problems and behavior of passengers, the authors traveled in two trains: one from Jabalpur to Secunderabad and the second from Jabalpur to Koderma. During observation, the authors traveled individually as normal passengers in Indian railways to observe user behavior and also understand the problems faced by passengers. Key findings were noted down, and in few cases, photographs were taken.

63.2.3 Questionnaire

The online survey method was adopted to assess the problem of theft faced by people traveling in Indian railways. For this study, a standardized questionnaire was created on the open source platform and was accepting responses for a period of 30 days. It was posted on social media and mailed to various friends and colleagues. The survey was aimed for a large variety of users in age, profession and socio-economic backgrounds. It comprised of close-ended, multiple-choice and open-ended questions. The survey did not collect any personal information like names and email addresses.

63.2.4 Interviews

Participants for the user interviews were selected randomly both while traveling in trains and during visits to railway stations. For each participant, a semi-structured interview [15] was conducted which comprised both open-ended and closed-ended questions. Expert interviews [16] of the officials like ticket collector, railway police force and CRPF were conducted. The observation and interview phase were done simultaneously, and it was done throughout a time span of 45 days. The interviews were carried out by two authors at a time. One author asked questions and the other jotted down points in a notepad.

63.2.5 User Scenarios

Scenarios are narrative stories about specific personas and their activities in a specific work situation and context [17]. Various user scenarios were developed to describe key usage situations, user goals and background, work practices, system responses, tasks, context and difficulties [18].

63.3 Results and Findings

63.3.1 System and User Study

The railways are used by almost every class of people, starting from a businessman to a ragpicker. Passengers during their travel carry many valuable belongings like wallet, Jewelry, watch and smartphones. The type of baggage that the passengers carry inside the train includes trolley, rucksack, duffle, business bag and women hand bag. Trunks, suitcase, and boxes having outside measurement of $(100 \times 60 \times 25)\ \text{cm}^3$ are allowed in the passenger compartments as personal luggage. The maximum size of the trunks/suitcase than can be carried in the AC 3 tier and AC chair car compartment is $(55 \times 45 \times 22.5)\ \text{cm}^3$. Additionally, maximum quantity permitted (free allowance) in AC 3-tier sleeper, AC chair car and sleeper class is set as 40 kg [19].

63.3.2 Questionnaire

A total of 101 participants took part in the survey. To understand the severity of the problem, participants were asked to rate the problem on a scale of 1–10 (see Fig. 63.1), where 0 is minimum severity and 10 maximum severity.

The survey was divided in two parts, one for people who had lost any personal belonging while traveling and the other for people who did not. A total of 31.7% participants reported a loss of belonging. The detailed results of these people are represented in Table 63.1.

63.3.3 Interviews

A total of 11 user interviews and 4 expert interviews were taken. The most important insights from the user interviews were:

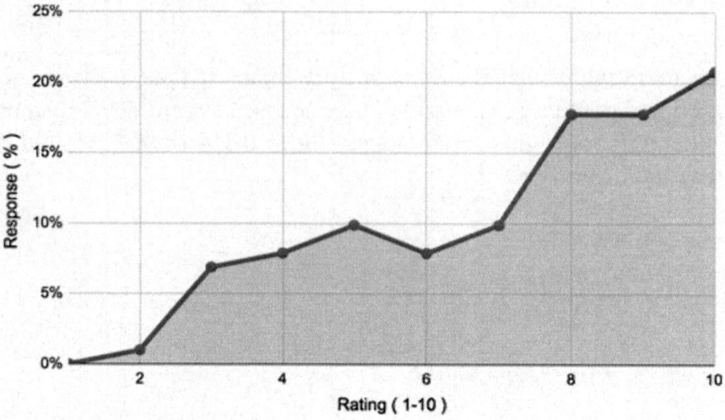

Fig. 63.1 Severity of problem

Table 63.1 Questionnaire results

Age group	Below 18	7.9%	Coach	Sleeper	67.7%
	19–30	82.2%		3 AC	6.5%
	30–45	7.9%		2 AC	9.7%
	Above 45	2%		Other	16.2%
Lost item	Cell phone	38.5%			
	Wallet	31.3%	Reported to police	Yes	38.7%
	Luggage	21.9%		No	61.3%
	Handbag	9.4%	Security measures taken	Keep bags close	84.2%
	Jewelry	9.4%		Use chain	41.6%
	Other	16%		Lock	59.4%
				Other	3%

- 72% participants mentioned that theft is one of the major tensions for passengers starting from boarding the train till the destination is reached.
- 54% participants generally use their baggage as a pillow while sleeping to ensure its safety. It was noted that the fear of theft is maximum at night.
- 45% participants fear those people who travel with no ticket or who are in waiting list. In case of any theft, reporting to RPF and filing FIR is a long process, and the chances to get the baggage back are very less.

The expert interviews however provided more in-depth insights about the system. Interview with a Traveling Ticket Examiner (TTE) revealed that thieves often travel as normal passengers, and most of the theft happens either when the passenger is asleep or when he/she goes to use the washroom. In few cases, the thieves use the small halts whenever a train reaches a small station, quickly grabs any belonging of a passenger and gets off the train, even if the train is in motion.

The interview with a Railway Police Force (RPF) thefts is mostly planned and happens mainly at night time. The railways have installed CCTV cameras, which assist them in investigating any case if reported. The investigation however takes a long time. Cases are reported in every 2–3 days. An important observation was that thefts also happen in AC coaches and also in premium trains like Rajdhani Express.

63.3.4 User Scenarios

The scenarios were build based on user observation, interviews and the qualitative data from the questionnaire survey. This is tabulated in Table 63.2.

63.4 Analysis and Insights

Following the methodology, the research concluded with a more specific problem regarding the time of the theft and the types of items being stolen. The questionnaire showed that luggage was the second highest item to get stolen, and in 96.8% cases, the stolen item was never retrieved by the traveler. The open-ended questions revealed that most of the theft happened when the passenger was sleeping. The interviews with the TTE and RPF revealed that the thefts are mostly planned and thieves often

Table 63.2 User scenarios

	Activity	Incident	Pain points
1	Sleeping (10 pm to 2 am)	Bag pack was missing when he/she woke up	To file a FIR to RPF, took 4 months to catch the thief
2	Buying food at platform	Nearby passenger stole the bag	Instant panic as they trusted strangers to keep an eye
3	Put the phone on charging and was a bit careless	A random person picked the phone	Constant vigilance on your personal belongings
4	The whole coach was sleeping	At 4 am belongings of four families got stolen	Even after reporting to RPF, the thief was never found
5	Traveling and the time was 11 pm	Train was about to start, someone snatched handbag	Unable to file an FIR, unaware of the process
6	Sleeping at night	Thief replaced the bag with another	Reporting to RPF would take a long time and is ineffective
7	Traveling with bag containing important things	Kept the bag on berth	Uncomfortable Journey lead to improper sleep
8	Sleeping on upper berth while baggage was under lower berth	In morning, the baggage was not there	Lost hope, later he found the baggage in another compartment

travel as normal passengers. This was not only observed in sleeper coaches but also in AC and more premium ones. User scenarios also revealed various incidents, where passengers have been through incidents like theft from snatching, multiple thefts in the same coach on the same night, keeping the bag as a pillow for assurance which leads to uncomfortable journeys. Most of the passengers reported to RPF but a few did not report as they were aware that it would not resolve.

The main objective was providing security for the luggage of passengers, especially in sleeper and AC class with enhanced security systems for theft protection, especially during night hours. From all the pain points and the insights gathered by the authors leads them to certain factors which the authors constantly kept in mind before ideating on the solution. The proposed concepts were designed in such a manner that it focuses mainly in the context of accessibility and reachability of users dealing with safety concerns for their heavy luggage and that too specifically during night travel so they can achieve an unstrained state of mind when traveling in Indian railways.

63.5 Prototype

63.5.1 Overview

A total of six concepts were generated, and out of them five were eliminated on the basis of cost, system failure, user behavior, complexity in use and complexity in manufacturing. The final concept consists of a device which uses a motion sensor to detect unauthorized lift/movement of the baggage. The motion sensor passes the information in two-way communication to a smartphone via Bluetooth. An android application is developed, which enables the user to control the device. The reason for opting cell phones is that cell phones are used widely by people of almost every age group. In case unauthorized lift/movement of the baggage, an alarm is raised by the app initially not the device. This is done to avoid chaos in night if the alarm is raised falsely by the user or any the nearby passenger. Figure 63.2 describes the final concept.

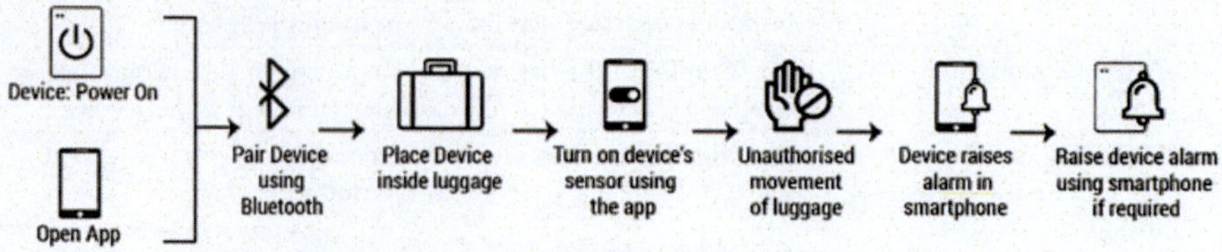

Fig. 63.2 Solution overview

63.5.2 Device Description

The device uses a motion sensor to detect unauthorized lift/movement of the baggage. The sensor neglects the movement and vibration which happens due to movement of train. The device is designed to provide instant alarm and does not involve any sort of baggage tracking. The use of motion sensors instead of GPS/GSM makes it cost effective.

The device contains a low powered RGB-LED light which provide the feedback of device's working, and its color and light intensity have chosen according to user's psychology. The LED emits blue blinking light till the Bluetooth is connected; a green light after connection is established, violet light for charging and a red blinking light to indicate low power.

63.5.3 Prototype and App Interface

The shape and form of the device is designed to meet the functional and aesthetic value of the product (see Fig. 63.3b), and a basic prototype was developed (see Fig. 63.3c). The user shall control the device with the app preinstalled on his/her smartphone. The application will enable the user to turn the device ON or OFF. The interface has been designed keeping in mind all the heuristic principles [20] (Fig. 63.4).

To use the device, the user has to first turn on the application and connects the device using Bluetooth. After connection, the device can be turned on using the toggle button on the interface (see Fig. 63.3a). In case the user wants to use the bag, he/she has to first turn the device off using the toggle button.

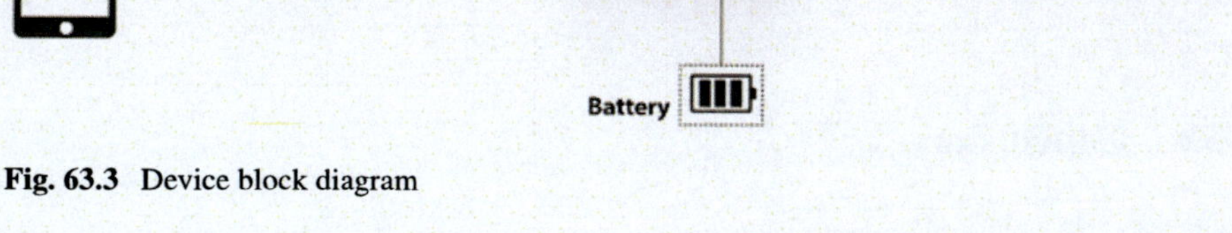

Fig. 63.3 Device block diagram

(a) (b) (c)

Fig. 63.4 **a** App interface **b** CAD and **c** Prototype

Table 63.3 Design comparison of proposed and existing product

Designs	Contextuality	Usability	Security	Cost	Aesthetic	Total
ED1	0	2	1	0	1	4
ED2	0	2	1	1	1	5
ED3	0	1	1	2	1	4
PD	2	2	1	2	2	9

63.5.4 Discussion

The final product was designed keeping the contextual constraints of Indian railways. Even though there is no product designed specifically for railway thefts, to make a comparison with the proposed design, similar products (already discussed in the literature) were taken. The existing products are labeled as ED1 [9], ED2 [10] and ED3 [13]. The proposed design is labeled as PD. The features are evaluated on a 0–2 scale (minimized Likert scale), where 0 is no presence of feature, and 2 is maximum presence of feature (Table 63.3).

63.6 Conclusion

The study helped in designing a product that targets the thefts of baggage while traveling in Indian Railways. The survey was anonymized, and no ethical clearance was needed. The proposed device is aimed to provide real-time security, and the design was made keeping in mind the dynamic nature of Indian Railways. The proposed design not only provides baggage security for any passenger, but also can be made available to the general public at a very minimal rate when compared to other products which use GPS/GSM technology.

References

1. Indian Railways. https://indianrailways.gov.in/railwayboard//view_section.jsp?. Last accessed 28 Apr 2020
2. Indian Railways Annual Report and Accounts: Security. Indian railways annual report and accounts 2010–11, pp. 70–71 (2011)
3. Indian Railways Annual Report and Accounts: Security. Indian railways annual report and accounts 2018–19, pp. 98–101 (2019)
4. Crime in India: Crime in Railways. Crime in India 2015, p. 423 (2015)
5. Crime in India: Crime in Railways. Crime in India 2016, p. 488 (2016)
6. Crime in India: Crime in Railways. Crime in India 2017, vol. 3, pp. 907–908 (2017)
7. Crime in India: Crime in Railways. Crime in India 2018, vol. 3, pp. 907–908 (2018)
8. Report No. 14 of 2011–12 (Railways): Protection of passengers and their belongings including strategy to control crime. Report no. 14 of 2011–12 (Railways), pp. 14–30 (2012)
9. Crider, E.A., Francies, V.D.: U.S. Patent No. 7,535,358. U.S. Patent and Trademark Office, Washington, DC (2009)
10. Stewart, R.J., Antoun, C.A., Naim, W., Aridi, R., Steck, H., Morgan, J.L.: U.S. Patent No. 9,439,164. U.S. Patent and Trademark Office, Washington, DC (2016)
11. Oas, D.: U.S. Patent No. 7,557,705. U.S. Patent and Trademark Office, Washington, DC (2009)
12. Tompkins, E., Hayes, W.: U.S. Patent No. 5,576,692. U.S. Patent and Trademark Office, Washington, DC (1996)
13. Doberman Security (2020) https://www.dobermanproducts.com/products/se_0304.html. Last accessed 28 Apr 2020
14. Council, U.D.: Eleven lessons: managing design in eleven global brands. A study of the design process. Design Council (2005)
15. Leech, B.L.: Asking questions: techniques for semistructured interviews. PS Polit Sci Polit **35**(4), 665–668 (2002)
16. Bogner, A., Littig, B., Menz, W.: Introduction: expert interviews—an introduction to a new methodological debate. In: Interviewing Experts, pp. 1–13. Palgrave Macmillan, London (2009)
17. Hartson, R., Pyla, P.S.: The UX Book: Process and Guidelines for Ensuring a Quality User Experience. Elsevier (2012)
18. Cooper, A.: The inmates are running the asylum: [Why high-tech products drive us crazy and how to restore the sanity], vol. 2. Sams, Indianapolis (2004)
19. Indian Railways (n.d.): Rules for Luggage. Retrieved from https://wcr.indianrailways.gov.in/uploads/files/1387190151089-luggege3.pdf
20. Molich, R., Nielsen, J.: Improving a human-computer dialogue. Commun. ACM **33**(3), 338–348 (1990)

Chapter 64
Design Strategies Enabling Industry 4.0

Vishal Ashok Wankhede and S. Vinodh

Abstract Industry 4.0 focusses on digitalization and developing intelligent products and processes. There exists need to develop robust products with enhanced process efficiency. Also, it includes integration of technologies to facilitate interoperability, modularity and so on. Design strategies would facilitate design engineer's rapid decision making for product development. In this context, this article deals with analysis of 15 design strategies. The strategies are prioritized using multi-criteria decision-making (MCDM) approach. The priority order of design strategies is derived for further deployment in terms of developing products and processes.

64.1 Introduction

Industry 4.0 (I4.0) drives the manufacturing industries through intelligentization and digitalization [1]. Various industrial revolutions made a significant transformation in manufacturing industries. The journey towards I4.0 resulted in drastic industrial transformations like steam and water powered machines, electrical-driven machines, automation and now digital automated manufacturing which makes production process more automatic, complicated and sustainable [2]. I4.0 defines a new stage of development and control on complete value chain of product life cycle which is focused towards fulfilling customer requirements [3]. The core purpose of I4.0 is to fulfil customer expectations associated in areas like research and development, order management, recycling of products, agile delivery, etc. The challenge related to I4.0 era is the complexity of new technologies which needs to adopt by the manufacturing industries for its successful implementation. The current scenario of manufacturing industry shows significant impact of this technological change with reference to

V. A. Wankhede · S. Vinodh (✉)
Department of Production Engineering, National Institute of Technology Tiruchirappalli, Tiruchirappalli, India
e-mail: vinodh_sekar82@yahoo.com

V. A. Wankhede
e-mail: vishal.wankhede06@gmail.com

A. Chakrabarti et al. (eds.), *Design for Tomorrow–Volume 2*, Smart Innovation, Systems and Technologies 222, https://doi.org/10.1007/978-981-16-0119-4_64

profitability, employment and growth in the sector [4]. I4.0 facilitates digital automated manufacturing which makes production process more automatic, complex and sustainable. I4.0 include several technologies such as cyber-physical systems (CPS), industrial Internet of things (IIoT), big data analytics, additive manufacturing and so on. The role of design methods in I4.0 enables innovation integrating various technologies and dealing with associated issues. Design strategies enable the manufacturers to visualize the adoption process of I4.0 and provide solutions needed for its transition. In this context, this article deals with 15 design strategies pertaining to I4.0 and analysed against nine criteria using Fuzzy VlseKriterijumska Optimizacija I Kompromisno Resenje or Multi-criteria optimization and compromise solution (VIKOR) approach. The quantifiable evaluation criteria for each design strategy have been obtained through literature review and finalized using expert's opinion. The priority order of design strategies is derived. Fuzzy VIKOR method would help to prioritize the design strategies pertaining to I4.0 and the prioritized design strategies would further facilitate industry practitioners to deploy smoothly for enhanced performance. The prioritized design strategies will also enable practitioners to take appropriate measures while implementing I4.0.

64.2 Literature Review

Nee et al. [5] presented augmented reality (AR) applications in manufacturing and design. They studied various AR applications in manufacturing like robotics, maintenance systems, CNC simulation, factory layout planning and operations planning. They presented the importance of AR in design phase and addressed technical challenges like accuracy, interfacing technology and latency issues. They concluded AR to be the most powerful tool in manufacturing field. Hermann et al. [6] performed literature review and provided I4.0 definition and identified design principles. From the literature analysis, they defined I4.0 as collection of various concepts and technologies of value chain organization. They also identified six design principles of I4.0, namely virtualization, real-time capability, interoperability, decentralization, modularity and service orientation. These principles help to identify the use cases and deliver guidance during I4.0 implementation. Thames and Schaefer [7] developed an architecture inspired by software defined systems using cloud manufacturing to address the challenges associated with industries practicing I4.0. The software-based cloud manufacturing architecture supports various design characteristics like programmable, manageable, interoperable, protectable and adaptable. I4.0 manufacturing systems can utilize its elements like IIoT, cloud-based manufacturing and social product development using this architecture. Future work suggested to adopt this architecture for measuring system effectiveness across various design principles and ensure its ability to security of I4.0 systems. Uden and He [8] performed a case study on application of IoT in knowledge management in automotive sector. They designed an IoT system to capture raw data from vehicles and converted into knowledge to improve decision making in transportation. The findings revealed that IoT embedded in knowledge management captures big data and so plays a vital role in quick and effective decision making. Coda et al. [9] explored the requirements of big data structures in I4.0 based on its objectives, perspectives and stake holder's expectations. Knowledge acquisition in automated specification methodology used for modelling purpose which analyses the system using objectives and requirements. Finding showed that big data systems come with different requirements and need to be modelled including desired characteristics like data procedures and security. Albers et al. [10] explored I4.0 potential on product from the perspective of its dependency on production systems. They developed a method to evaluate the risk and impact of process with respect to its production system. The model named Product Generation Engineering was developed to recognize I4.0 potential and looked for the support of approaches like design with I4.0 and design for I4.0. Further this model was integrated with model-based system engineering (MBSE) approach to enhance its capability and ensure consistency. The developed method enables product developer to address the complexity related to project and production system development. Bader et al. [11] presented a method on agile organization design for automotive manufacturing firms practicing I4.0. The authors analysed necessities which further reduced to criteria determining agility of future industries. Further, these criteria were converted into functional mechanisms for developing an approach on agile organization design. Based on the study performed, six functional mechanisms were recognized which was defined in operation design called TEAM. They found six criteria, i.e. a participative leadership, high level of transparency, small team size, flexibility, quick decision making and self-determination required for agile shop floor organization design.

From the literature review, it has been found that many authors attempted to perform study on usage of I4.0 technologies from product development and process development perspective. In connection to the above, some authors also executed mapping of design principles with I4.0 technology. But no concrete studies are reported to the best knowledge of authors on prioritizing design strategies of I4.0. Since I4.0 defines as mean of value creation, there exist a need to identify various design strategies pertaining to I4.0 and prioritize them for its successful implementation.

64.3 Methodology

Fuzzy VIKOR, a MCDM method is used as solution methodology in this study. VIKOR method was developed by Opricovic and Tzeng which computes compromise solution from the group of variables [12]. The solution which is close to the ideal solution is known as compromise solution. Though there are various MCDM methods which prioritizes variables VIKOR has some advantage over other MCDM techniques. VIKOR is an efficient MCDM method which ranks, sorts and then selects the set of conflicting variables for complex multi-attribute optimization problem.

VIKOR method can fully replicate the subjective preferences of decision makers along with individual regret minimization and group utility maximization.

64.3.1 Methodological Steps

The methodological steps are as follows [13]:

Step 1: Define the problem scope and objective for decision-making problem.
Step 2: Identification of finite set of suitable attributes, criteria and decision experts.
Step 3: Identification of suitable linguistic variables for fuzzy importance weight of criteria and rating of attributes. In this study, triangular fuzzy numbers (TFNs) have been used [14]. The linguistic terms for both importance weight and fuzzy rating are shown in Table 64.2.
Step 4: Defuzzification of fuzzy criteria importance weight and fuzzy rating of attributes. There are various methods to defuzzify fuzzy values, namely mean of maximal (MOM), alpha-cut method, centre of area (COA) method, etc. Out of these, COA method is practical and simple method as there is no requirement of including preference of any experts. The BNP score for COA method of TFN $x_{ij} = (l_{ij}, m_{ij}, u_{ij})$ is computed as follows (Eq. 64.1):

$$\text{BNP score } y_{ij} = l_{ij} + \frac{(u_{ij} - l_{ij}) + (m_{ij} - l_{ij})}{3} \tag{64.1}$$

Step 5: Computation of aggregate defuzzified importance weight and rating. The aggregate importance weight and rating is calculated as shown using Eq. (64.2).

$$z_{ij} = \frac{\sum_{j=1}^{n} y_{ij}}{n}, \quad \text{where } n \in \text{number of criteria} \tag{64.2}$$

Step 6: Evaluating the best s^* and the worst s^- values of all criteria (j = 1, 2, 3, …., n) using Eqs. (64.3) and (64.4).

$$s_j^* = \max(z_{ij}) \tag{64.3}$$

$$s_j^- = \min(z_{ij}) \tag{64.4}$$

Step 7: Evaluate values of group utility (S_i), individual regret (R_i) and VIKOR (Q_i) indices.
S_i and R_i values are evaluated using s_j^*, s_j^- and w_j using Eqs. (64.5) and (64.6).

$$S_i = \sum \frac{w_j(s_j^* - s_{ij}}{\left(s_j^* - s_j^-\right)} \quad (64.5)$$

$$R_i = \max \frac{w_j(s_j^* - s_{ij}}{\left(s_j^* - s_j^-\right)} \quad (64.6)$$

Using S_i and R_i values Q_i value is calculated using Eq. (64.7).

$$Q_i = \frac{v_j(S_i - S^*)}{(S^- - S^*)} + \frac{(1 - v_j)(R_i - R^*)}{(R^- - R^*)} \quad (64.7)$$

where $S^* = \min(S_i)$, $S^- = \max(S_i)$, $R^* = \min(R_i)$, $R^- = \max(R_i)$ and v_j is weight strategy for maximum group utility and $(1 - v_j)$ is the individual regret weight. Value of v_j is assumed to be 0.5.

Step 8: Ranking the attributes based on Q_i values. The attribute with minimum Q_i value is ranked first.

64.3.2 Proposing Compromise Solution

The attributes ranked in the increasing order of Q_i value will be considered as compromise solution only if the two conditions, i.e. $C(1)$ acceptable gain and $C(2)$ acceptable stability in decision making mentioned by Opricovic and Tzeng are fulfilled [12].

64.4 Case Study

The case study has been done in a manufacturing organization producing automotive component situated in Tamil Nadu, India. The manufacturing organization aims to adopt I4.0 technologies and in the process of developing the required infrastructure. There existed a need to adopt some design strategies for smooth adoption of I4.0 which in turn helps for value creation in the organization. Expert opinion has been taken from industry practitioner and academic researcher with expertise in I4.0. The present study aims to recognize various design strategies and further analysing it by prioritizing using a suitable MCDM method for successful deployment of I4.0 in an automotive component manufacturing industry. 15 design strategies pertaining to I4.0 are identified and analysed against nine criteria, i.e. technological readiness (C1) [15], process improvement (C2) [16], product development speed (C3) [16], design complexity (C4), workers skill (C5) [17], process quality (C6) [18], design effectiveness (C7), autonomous production (C8) [18] and responsiveness (C9). The conceptual model of 15 design strategies is shown in Table 64.1. For prioritization purpose, fuzzy VIKOR, an MCDM technique has been used. Fuzzy VIKOR aims

Table 64.1 Conceptual model of design strategies

Code	Design strategies	References
DS1	Design for agility	[11]
DS2	Design for traceability	[19]
DS3	Design for handling big data	[9]
DS4	Design for interoperability using IoT	[20]
DS5	Design for virtualization using CPS	[20]
DS6	Design for decentralization using CPS	[20]
DS7	Design for cloud-based intelligent monitoring system	[8]
DS8	Design for industrial internet of things	[7]
DS9	Design for social product development	[7]
DS10	Cloud-based design and manufacturing	[7]
DS11	Sustainable design	[21]
DS12	Augmented reality-based collaborative design	[5]
DS13	Design for service orientation using IoT	[6]
DS14	Design for modularity using internet of services	[6]
DS15	Design for real-time capability in smart factory	[6]

to rank design strategies and provides compromise solution which will help I4.0 practitioners in effective decision making. In this study, the inputs of criteria weights and design strategies ratings have been gathered from five decision experts in the form of linguistic terms and assigned respective fuzzy numbers for analysis, i.e. for ratings [Excellent 'E'-{8.5, 9.5, 10}, Very Good 'VG'-{7, 8, 9}, Good 'G'-{5, 6.5, 8}, Fair 'F'-{3, 5, 7}, Poor 'P'-{2,3.5, 5},Very Poor 'VP'-{1, 2, 3},Worst 'W'-{0, 0.5, 1.5}] and for importance weights [Very High 'VH'-{0.85, 0.95, 1.0}, High 'H'-{0.7, 0.8, 0.9}, Fairly High 'FH'-{0.5, 0.65, 0.8}, Medium 'M'-{0.3, 0.5, 0.7}, Fairly Low 'FL'-{0.2, 0.35, 0.5}, Low 'L'-{0.1, 0.2, 0.3}, Very Low 'VL'-{0, 0.05, 0.15}]. Triangular fuzzy numbers (TFNs) are used in this study [15]. An excerpt of response sheet obtained from decision experts for technological readiness (C1) is shown in Table 64.2 (Ex. # represents expert number).

The linguistic form of response obtained from decision experts is converted into respective TFNs for analysis.

The importance weights and performance ratings are further defuzzfied and aggregated using Eqs. (64.1) and (64.2). The aggregate weights have been derived from expert opinion in the form of fuzzy linguistic terms.

For Criteria 1, C1 (technological readiness), weights received from five experts are VH, H, H, VH, H. These weights are transformed into respective fuzzy numbers and then defuzzified to obtain crisp value. Further, average of crisp value resulted into final aggregated weight.

Sample Calculation

The crisp values of fuzzy number are obtained using Eqs. (64.1) and (64.2) as follows:

Table 64.2 Excerpts of inputs for design strategies against criteria

Criteria	C1				
Weights	VH	H	H	VH	H
Design strategies	Ex. #1	Ex. #2	Ex. #3	Ex. #4	Ex. #5
DS1	VG	G	G	VG	E
DS2	G	F	F	G	VG
DS3	VG	G	G	G	VG
DS4	VG	G	E	G	VG
DS5	VG	G	G	G	E
DS6	G	VG	G	VG	G
DS7	G	VG	G	VG	G
DS8	VG	VG	E	VG	G
DS9	G	F	G	G	F
DS10	VG	G	G	VG	G
DS11	G	F	F	G	F
DS12	VG	G	F	G	VG
DS13	G	F	F	G	F
DS14	G	VG	F	G	G
DS15	G	F	VG	G	F

$$\text{VH} - (0.85, 0.95, 1) = 0.933, \quad H - (0.7, 0.8, 0.9) = 0.8$$

Aggregated crisp value of weight for criteria 1 (C1)

$$= (0.933 + 0.8 + 0.8 + 0.933 + 0.8)/5 = 4.266/5 = \mathbf{0.85}.$$

Similarly, weights of other design criteria are obtained.

The aggregated crisp values of weights and ratings of decision matrix is presented in Table 64.3.

Further using Eqs. (64.3) and (64.4), the Best (s^*) and Worst (s^-) values of all criteria ratings are calculated as presented in Table 64.4.

64.5 Results and Discussions

The present study aims to rank design strategies pertaining to deployment of I4.0 using fuzzy VIKOR as solution methodology. The study focusses on ranking 15 design strategies pertaining to I4.0 against nine criteria. This is attained by computing the values of utility (S), regret measure (R) and VIKOR index (Q) using Eqs. (64.5), (64.6) and (64.7). The final ranking is done based on minimum Q_i value.

Sample calculation of S, R and Q value for design for agility (DS1):

Table 64.3 Aggregate crisp values of criteria weights and design strategies ratings

Criteria	C1	C2	C3	C4	C5	C6	C7	C8	C9
Weights	0.85	0.74	0.90	0.85	0.85	0.65	0.85	0.71	0.85
	Aggregate experts ratings								
DS1	7.6	5.9	7.6	7.1	7.1	5.6	6.8	5.6	7.1
DS2	6.2	5.6	5.9	5.6	5.9	5.6	5.3	5.6	5.6
DS3	7.1	5.9	6.2	5.9	6.2	5.3	6.5	5.9	6.5
DS4	7.6	5.6	5.9	5.9	5.6	5.6	5.9	7.1	5.6
DS5	7.4	5.6	5.6	5.6	5.9	5.3	6.5	5.9	6.2
DS6	7.1	5.3	6.2	5.6	5.9	5.3	5.9	7.4	5.9
DS7	7.1	5.6	5.9	7.1	6.8	5.9	6.5	5.9	5.6
DS8	7.9	5.9	6.2	6.5	6.5	5.6	7.1	6.5	7.4
DS9	5.9	6.5	6.8	5.9	5.9	6.2	5.9	6.2	6.2
DS10	7.1	5.6	7.1	7.1	5.6	5.6	6.2	5.6	6.8
DS11	5.6	5.9	5.9	5.9	6.5	5.3	5.6	5.3	6.2
DS12	6.8	5.9	5.6	5.6	5.6	5.3	5.9	5.6	5.9
DS13	5.6	5.6	5.3	5.6	7.1	5.3	5.6	5.6	5.9
DS14	6.5	5.6	7.4	7.1	6.5	5.9	6.8	5.9	6.8
DS15	6.2	6.5	5.6	5.9	7.1	5.9	5.9	5.6	5.9

Table 64.4 Best (s^*) and Worst (s^-) values of all criteria ratings

	C1	C2	C3	C4	C5	C6	C7	C8	C9
s^*	7.9	6.5	7.7	7.1	7.1	6.2	7.1	7.4	7.4
s^-	5.6	5.3	5.3	5.6	5.6	5.3	5.3	5.3	5.6

$$S_1 = \left[\frac{0.85 \times (7.9 - 7.7)}{(7.9 - 5.6)} + \frac{0.74 \times (6.5 - 5.9)}{(6.5 - 5.3)} + \frac{0.90 \times (7.7 - 7.7)}{(7.7 - 5.3)} + \cdots + \frac{0.85 \times (7.4 - 7.1)}{(7.4 - 5.6)}\right] = 1.77055$$

$$R_1 = \min\left[\frac{0.85 \times (7.9 - 7.7)}{(7.9 - 5.6)}, \frac{0.74 \times (6.5 - 5.9)}{(6.5 - 5.3)}, \frac{0.90 \times (7.7 - 7.7)}{(7.7 - 5.3)}, \ldots, \frac{0.85 \times (7.4 - 7.1)}{(7.4 - 5.6)}\right] = 0.60857$$

From 'S_i' and 'R_i' values, S^*, S^-, R^* and R^- are calculated as mentioned in step 6.

$$S^* = 1.77055, \quad S^- = 6.15022, \quad R^* = 0.555, \quad R^- = 0.90664$$

$$Q_1 = \frac{0.5 \times (1.7055 - 1.7055)}{(6.15022 - 1.7055)} + \frac{(1 - 0.5) \times (0.6085 - 0.555)}{(0.90644 - 0.555)} = 0.076174$$

Similarly, S, R and Q values for all design strategies are computed as shown in Table 64.5.

S, R and Q values of design strategies are arranged in ascending order for obtaining final ranking as presented in Table 64.6.

Compromise Solution

Further, a compromise solution is developed in which the study fails to satisfy condition 1 and satisfied condition 2, i.e. acceptable stability in decision making: Design strategy (1) must also be the best ranked by R or/and S. From the results, it is observed that, design for agility ($DS1$) is best ranked by $S1$ and $Q1$. Since, condition 1 is not satisfied, as indicated in methodology, set of compromise solution consists of design strategies $DS^{(1)}$, $DS^{(2)}$, $DS^{(3)}$, ..., $DS^{(m)}$ where $DS^{(m)}$ is computed by $Q(DS^{(m)}) - Q(DS^{(1)}) < (1/(m-1))$ for maximum 'm'.

Acceptable gain = $1/(m - 1) = 0.071429$. Each design strategy is checked for $Q(DS^{(m)}) - Q(DS^{(1)}) < (1/(m-1))$ and final three ranking has been obtained.

The compromise solutions consist of three design strategies i.e. DS1, DS8 and DS14 and ranked using Q value in increasing order as follows:

Table 64.5 Values of utility (S), regret measure (R) and VIKOR index (Q)

	DS1	DS2	DS3	DS4	DS5	DS6	DS7	DS8
S_i	1.77	6.15	4.29	4.80	4.96	5.06	3.56	2.35
R_i	0.60	0.85	0.68	0.85	0.85	0.85	0.85	0.56
Q_i	0.07	0.92	0.46	0.77	0.78	0.80	0.62	0.083

	DS9	DS10	DS11	DS12	DS13	DS14	DS15
S_i	3.99	3.68	5.56	5.81	5.84	2.67	4.21
R_i	0.74	0.85	0.85	0.85	0.90	0.55	0.79
Q_i	0.51	0.64	0.85	0.88	0.96	0.10	0.61

Table 64.6 Priority order of design strategies based on S, R and Q

Rank	S_i	R_i	Q_i	Rank	S_i	R_i	Q_i
I	DS1	DS14	DS1	IX	DS4	DS5	DS4
II	DS8	DS8	DS8	X	DS5	DS6	DS5
III	DS14	DS1	DS14	XI	DS6	DS7	DS6
IV	DS7	DS3	DS3	XII	DS11	DS10	DS11
V	DS10	DS9	DS9	XIII	DS12	DS11	DS12
VI	DS9	DS15	DS15	XIV	DS13	DS12	DS2
VII	DS15	DS2	DS7	XV	DS2	DS13	DS13
VIII	DS3	DS4	DS10				

$$DS1 > DS8 > DS14$$

Design for agility (DS1) > design for industrial Internet of things (DS8) > design for modularity using Internet of services (DS14). Design for agility represents quick decision making and great adaptability on shop floor which allow quick reaction time and agile design would help the industry practitioners for smooth deployment of I4.0. Design can bring a range of benefits to organization such as the variation of products from those of competitors getting competitive advantage and products complexity reduction leading to costs savings and increased profit margins.

64.6 Conclusions

I4.0 demands innovation in terms of design strategies to enhance effectiveness of product development. Design strategies ensure reducing complexity using I4.0 technologies such as additive manufacturing and IoT which can lead to improved process and production efficiency and robust products. Design strategies can be used as a tool to help decision making in new product development, but opportunities may be neglected if the project team are not attentive of all the benefits of design. In this context, this study recognized various design strategies and further analysed using a suitable MCDM method. 15 design strategies pertaining to the automotive component manufacturing industry have been analysed using Fuzzy VIKOR approach. The top priority strategies are design for agility (DS1) > design for industrial Internet of things (DS8) > design for modularity using Internet of services (DS14). The outcome of the study provides guidelines to the industry practitioners for enhancing product development effectiveness of I4.0 practices.

References

1. Vaidya, S., Ambad, P., Bhosle, S.: Industry 4.0—a glimpse. Proc. Manuf. **20**, 233–238 (2018)
2. Ghobakhloo, M.: The future of manufacturing industry: a strategic roadmap toward Industry 4.0. J. Manuf. Technol. Manag. **29**(6), 910–936 (2018)
3. Xu, L.D., Xu, E.L., Li, L.: Industry 4.0: state of the art and future trends. Int. J. Prod. Res. **56**(8), 2941–2962 (2018)
4. Laudante, E.: Industry 4.0, innovation and design. A new approach for ergonomic analysis in manufacturing system. Des. J. **20**(Suppl. 1), S2724–S2734 (2017)
5. Nee, A.Y., Ong, S.K., Chryssolouris, G., Mourtzis, D.: Augmented reality applications in design and manufacturing. CIRP Ann. **61**(2), 657–679 (2012)
6. Hermann, M., Pentek, T., Otto, B.: Design principles for industrie 4.0 scenarios. In: 49th Hawaii International Conference on System Sciences (HICSS), pp. 3928–3937. IEEE (2016)
7. Thames, L., Schaefer, D.: Software-defined cloud manufacturing for industry 4.0. Proc. CIRP **52**, 12–17 (2016)
8. Uden, L., He, W.: How the internet of things can help knowledge management: a case study from the automotive domain. J. Knowl. Manag. **21**(1), 57–70 (2017)

9. Coda, F.A., Salles, R.M.D., Junqueira, F., Filho, D.J.S., Silva, J.R., Miyagi, P.E.: Big data systems requirements for Industry 4.0. In: 13th IEEE International Conference on Industry Applications (INDUSCON), pp. 1230–1236. São Paulo, Brazil (2018)
10. Albers, A., Stürmlinger, T., Mandel, C., Wang, J., de Frutos, M.B., Behrendt, M.: Identification of potentials in the context of design for industry 4.0 and modelling of interdependencies between product and production processes. Proc. CIRP **84**, 100–105 (2019)
11. Bader, S., Barth, T., Krohn, P., Ruchser, R., Storch, L., Wagner, L., Palm, D.: Agile shopfloor organization design for industry 4.0 manufacturing. Proc. Manuf. **39**, 756–764 (2019)
12. Opricovic, S., Tzeng, G.H.: Compromise solution by MCDM methods: a comparative analysis of VIKOR and TOPSIS. Eur. J. Oper. Res. **156**(2), 445–455 (2004)
13. Vinodh, S., Varadharajan, A.R., Subramanian, A.: Application of fuzzy VIKOR for concept selection in an agile environment. Int. J. Adv. Manuf. Technol. **65**(5–8), 825–832 (2013)
14. Lin, C.T., Chiu, H., Tseng, Y.H.: Agility evaluation using fuzzy logic. Int. J. Prod. Econ. **101**(2), 353–368 (2006)
15. Samaranayake, P., Ramanathan, K., Laosirihongthong, T.: Implementing industry 4.0—a technological readiness perspective. In: IEEE International Conference on Industrial Engineering and Engineering Management (IEEM), pp. 529–533. IEEE (2017)
16. Dalenogare, L.S., Benitez, G.B., Ayala, N.F., Frank, A.G.: The expected contribution of industry 4.0 technologies for industrial performance. Int. J. Prod. Econ. **204**, 383–394 (2018)
17. Ras, E., Wild, F., Stahl, C., Baudet, A.: Bridging the skills gap of workers in industry 4.0 by human performance augmentation tools: challenges and roadmap. In: Proceedings of the 10th International Conference on PErvasive Technologies Related to Assistive Environments, pp. 428–432 (2017)
18. Makhado, M., Sukdeo, N.: Advanced Manufacturing: the key to future competitiveness in the automotive industry. In: Proceedings of the International Conference on Industrial Engineering and Operations Management Paris, pp. 687–697. France (2018)
19. Bortolini, M., Ferrari, E., Gamberi, M., Pilati, F., Faccio, M.: Assembly system design in the industry 4.0 era: a general framework. IFAC Pap Online **50**(1), 5700–5705 (2017)
20. Thuemmler, C., Bai, C.: Health 4.0: application of industry 4.0 design principles in future asthma management. In: Health 4.0: How Virtualization and Big Data are Revolutionizing Healthcare, pp. 23–37. Springer, Cham (2017)
21. Stock, T., Seliger, G.: Opportunities of sustainable manufacturing in industry 4.0. Proc. CIRP **40**, 536–541 (2016)

Chapter 65
How Real is Virtual reality—An Immersion, Interaction and Embodiedness Study

Deepak Ranjan Padhi and Sugandha Katoch

Abstract In the past few years, virtual reality ("VR") has been explored as a potential medium that provides immersive experiences. VR features are being evolved for active engagement. However, there is a lot of scope in researching how the VR experience varies across demographics. For example, how does an emergent user (education below tenth standard, hardly exposed to information and communication technologies, or "ICTs") experience VR in comparison to a tech-savvy user (graduate, highly exposed to ICTs)? We performed mixed-method research with two experiments using mobile VR headsets. Experiment-1 was conducted to study the "immersion" effect by asking the participants to watch a virtual "roller coaster" ride video. Experiment-2 was conducted to gauge the effect of "embodiedness" and the "embodied interaction" with the medium. Here, the participants were asked to play the "Moon Bird" VR game, while moving their hands like a bird's wings. Qualitative and quantitative results were compared across the two user groups. Neither had experienced VR before. We found that emergent users got engaged better than tech-savvy users when there is only a perception of the virtual environment (Experiment-1). Hand movements acted as a cognitive load for emergent users (Experiment-2) in the beginning, but eventually, most expressed themselves as the game character. On the other hand, tech-savvy users possessed a strong distinction between the virtual and real environment but were found to be more engaged when they could interact with the virtual environment, rather than just perceive it. We discuss the findings and provide recommendations for future studies.

D. R. Padhi (✉) · S. Katoch
Indian Institute of Technology Bombay, Mumbai, Maharashtra, India
e-mail: deepak.padhi@iitb.ac.in

S. Katoch
e-mail: sugandhakatoch@iitb.ac.in

A. Chakrabarti et al. (eds.), *Design for Tomorrow–Volume 2*, Smart Innovation, Systems and Technologies 222, https://doi.org/10.1007/978-981-16-0119-4_65

65.1 Introduction

The roots of the term virtual reality can be dated back to the nineteenth century when the first 360° art through panoramic murals began to appear. Nearly one hundred years later, a mechanical device, the Sensorama [1], engaged multiple senses to create an immersive VR. It was a large booth that provided a multisensory experience of riding a motorcycle, including three-dimensional, full colour film together with sounds, smells, and the feeling of motion, as well as the sensation of wind on the viewer's face.

With time, VR has evolved into a more similar representation of the real world, ushering in research in a host of domains, including computer science, medicine, education, etc. ICT and VR have become strictly interconnected as computing power increases, and human–computer interfaces have become more complete and adaptive [2]. From the host of challenges that VR faces today, one challenge remains to make it more and more immersive.

Studying the degrees of immersion is an important step in the research for VR, but studies have underrepresented the experiences of emergent users who have shown great difficulties in accessing ICTs in the past [3]. In our study, we have tried to understand the degrees of immersion and embodiedness that is felt by users across demographics. We have collated the experiences of emergent users and tech-savvy users, by conducting two experiments, based on "perception" and "perception-and-action". We have also tried to see whether embodied interactions enhance the perception of the virtual environment. The main focus of this study was to understand the reception of VR in terms of how realistic and engaged the experience it provided was, across users having different technological exposures. Qualitative data has been collected through observations, think-aloud and interviews; and quantitative data of the body movements and excitements has been recorded using Myo armband. The band can fit on the user's forearm, right below the elbow, and can record positional data using an accelerometer and gyroscope [4].

65.2 Background

The term "immersion" has widely been used in "VR", to describe the level of involvement of a person with the virtual environment. Jennett et al. [5] defined the general concept of immersion in games as the involvement in the play, which causes a lack of awareness of time and the real world, as well as a sense of "being" in the task environment [2]. Brown and Cairns [6] identified three levels of involvement where "total immersion" is referred to the highest level of involvement. This is when the users cross the barriers of empathy and atmosphere in a gameplay. When referred to VR, "immersion" is usually restricted to the meaning of spatial immersion (a perception of being physically present in a nonphysical world) [2]. To understand further possibilities, we also considered studying about embodiedness (etymologically, "em-"

meaning "in" + "body") concerning virtual reality. The term embodiment has been defined differently in different contexts due to its multidisciplinary use and application areas. One of its explanations from neuroscience is that to regulate and control the body in the world effectively, the brain creates an embodied simulation of the body in the world used to represent and predict actions, concepts, and emotions [7]. The capability of our brain of having a representation of our body results in a mental construction composed of perceptions and ideas about the dynamic organization of our own body, involving vision, touch, proprioception, interception, motor control, and vestibular sensations [8]. One could refer to this as having a "sense" of one's own body, or the underlying perception of one's bodily experiences.

When it comes to embodied cognition, Wilson [9] pointed out that the information flow between the mind and the world needs to be taken together into consideration, making the environment a part of the cognitive system. She suggested looking at this scenario from the perspective of systems. The mind itself could be one system, the environment, a system surrounding the same. How one chose to define the boundaries could depend on the purpose of the problem statement. Even without the environment, the activity of the mind would be towards the interactions with the environment, in her words, towards mechanisms of sensory processing and motor control. She also stated that cognition has its bases towards action, with the mind utilizing perception and memory to contribute towards situation appropriate behaviour. Kilteni et al. [10] used the term "sense of embodiment" or "SoE" to refer to the ensemble of sensations that arise in conjunction with being inside, having, and controlling a body, especially in relation to virtual reality applications. "SoE" consisted of three subcomponents: a sense of self-location (space where one feels present or located), sense of agency (we cause actions), and sense of body ownership (the object belongs to you).

Through VR, a person can essentially take on the role, or a perspective, of a different being through its avatar. Studies [11] have suggested that perspective-taking may lead towards reducing the mental stereotypes formed of another person, thus providing a space for changing one's negative stereotypes, even reversing racial biases [12]. In our study, when we utilize the word embodiment, we point towards the embodiment which is also enabled via perspective-taking.

In embodied interactions, we have to consider the experience of the living human body, as well as the materiality of the world we interact with and live in. This can call us to explicitly design for bodily experience, for the moving and feeling body, and to exploit our bodily intelligence [13]. In medicine especially, there lies a possibility of altering the experience of the body by designing targeted virtual environments [14].

Today we have VR systems that work at low-latency with precise tracking of movements that make usability better for the user and manage to achieve this visceral reaction [10, 13]. When technology can be transparent, users can experience an increased sense of being in virtual environments (the so-called sense of presence) [15]. Arguably, the perspective and the sense of environment enabled by VR deals mostly with the visual and the audio affordances of a device, while a person can utilize their mind and their body to interact with the virtual environment within it.

In our study, we attempt to understand the effect of immersion in VR, together with embodiedness and embodied interaction, while considering users across demographics.

65.3 Method

We performed a mix-method study with two experiments. The experiment design is shown in Table 65.1.

65.3.1 *Experiment-1*

In Experiment-1, we wanted to understand the VR experience with respect to immersion. For example, how would the participants perceive the VR experience as opposed to a realistic experience? How would this experience differ from the beginning to the end of the task? How does this immersive experience vary across two sets of users categorized as tech-savvy users and emergent users? 10 participants were recruited using convenience sampling among which 5 (5 males, 0 female) were tech-savvy users (age: 24–35 years) who were graduates and highly exposed to ICTs, whereas the other 5 (3 males, 2 females) were emergent users (age: 25–45 years) with education below tenth standard and hardly exposed to ICTs. However, none of our participants had experienced VR before. The session began with the experimenter informing the users about the objectives of the study. Participants were also briefed about the affordances of VR and consent for participation was taken. The same protocol was followed in Experiment-2. As shown in Fig. 65.1, participants were asked to stand and watch a 360 VR roller coaster ride [16] video wearing mobile VR headsets and Myo armband for 4 min. They were asked to narrate their experience while going through it. Think-aloud protocol was used to record on time user experience. Qualitative data such as excitements, expressions were collected through participant observation. Hand signals for acceleration were recorded using Myo armband for quantitative analysis. We conducted a semi-structured interview after each session.

Table 65.1 Experiment design

Experiments	Goal	User types		Type of analysis
		Tech-savvy	Emergent	
Experiment-1: VR "roller coaster" video	To understand the immersive experience	5 (5 m, 0 f)	5 (3 m, 2 f)	Qualitative
Experiment-2: VR "Moon Bird" game	To understand the embodiedness and embodied interaction	6 (5 m, 1 f)	6 (2 m, 4 f)	Qualitative + Quantitative

Fig. 65.1 Screenshot of the VR roller coaster ride and a typical user experiencing it

During the interview, participants were probed to describe the current VR experience as opposed to the real-life roller coaster experiences that they might have experienced in the past. All participants were thanked in the end.

65.3.2 Experiment-2

Experiment-2 was conducted to look at VR experience with regard to embodiedness and embodied interaction. The question we posed was whether these aspects would enhance or diminish the VR experience. Secondly, how does this experience vary between the tech-savvy and the emergent users? A total of 12 participants were recruited among which 6 (5 males, 1 female) were tech-savvy users and the other 6 (2 males, 4 females) were emergent users with the same profile mentioned in Experiment-1. None of these participants had experienced VR before. Here, the participants were asked to play the "Moon Bird" VR game [17] by wearing a mobile VR headset and Myo armband. The interesting part is that they were also asked to imitate a flying bird by flapping their hands like the wings. Among the participants, one person chose not to wear the Myo armband. So we could only collect the qualitative data from that participant. A screenshot of the game and a picture of a typical user is shown in Fig. 65.2.

We wanted to understand if the users get to interact with the VR platform through a game, how does that experience differ from just watching a roller coaster ride video. Secondly, does flapping hands bring embodiedness in the gameplay across users? We began the session by informing the users about the objectives of the study. The users were asked to relax for 5 min before participating to neutralize their mental states. Then the experiment followed a three-stage approach. We incorporated a 5 min relaxation break in between each stage to avoid the recency effect [18]. In the first stage (pre-imagination), users were asked to stand, open their arms and imagine "a bird flying in the sky" for 1 min. They were also asked to narrate whatever they were imagining (think-aloud). In stage-2 (Play the Game), users were asked to play the

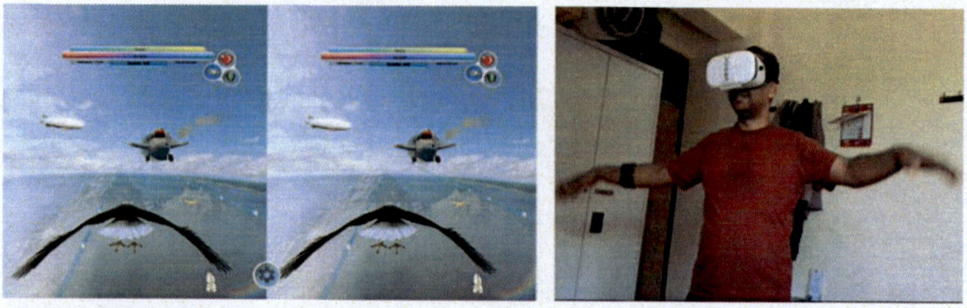

Fig. 65.2 Screenshot of the Moon Bird VR game and a typical user experiencing it

"Moon Bird VR" game wearing a VR headset for 3 min. They were initially briefed about the objectives, rules and interactions of the game. Users were asked to stretch their arms and express themselves while playing. In the final stage (post-imagination), users were asked to stretch their arms again and imagine "a bird flying in the sky" for 1 min. They were also asked to narrate what they were imagining. All narrations, body movements, excitements, and expressions were observed qualitatively. Also, the positional data and hand motion data of the users were collected through Myo armband. A semi-structured interview was conducted in the end and all participants were thanked for their participation.

65.4 Qualitative Analysis

65.4.1 Analysis of Experiment-1

We followed the grounded theory approach [19] to analyse the qualitative data obtained from think-aloud protocol, semi-structured interview, and participant observation.

VR is just like 3D; nothing more. Most of the tech-savvy users reported that while going through the VR roller coaster ride, they were very conscious about the images, picture quality, sound, and the VR platform. During the interview, one user said that "*I didn't like the animation in the ride. It would be better if it would have been a* 360° *video shot in reality*". Another user said that "*I play a lot of games. Here it felt like a 3D game. Although I felt the thrill when the roller coaster turned around but somehow I knew that it was not real*". Some users reported that in the beginning of the experiment they were not very excited, but in the middle, they got more involved and felt the fear of heights. They felt like they were taking the ride themselves. One of the users reported that "*I felt like falling when the big slope came in the ride*". But overall, tech-savvy users were less immersed, as observed by their reactions and excitements during the think-aloud protocol. They were also standing very stable during the entire session.

I was not here; I was in another world. On the other hand, emergent users were found to be more involved in taking the VR ride. One of the users said that "*I am feeling as if the world is spinning around me... Oh My God*!" Some users were found to be standing stable in the beginning but felt like falling towards the end. There were also body movements observed such as lifting hands, shaking legs, etc. while the users were taking the VR ride. Some users also took support to stand. Some users expressed curiosity about what's going to come next while taking the ride. During the interview, they reported that they felt as if they were sitting inside a roller coaster and felt the fear of falling but it's not the same feeling as reality due to the poor quality of graphics perceived through the headset.

Content is the key. We found that it is okay if the VR content is fictionalized rather than just matching to reality. The content creators could think of creative ways to develop suspense and thrill. One of the users said that "*During the ride, the roller coaster tilted several times and took a very unusual path. This usually doesn't happen in the real roller coaster ride. But that part thrilled me more. It was really exciting*". All our participants appreciated the overall content in the roller coaster ride and found it to be very interesting. However, there was a demand for creating non-animated content. Some users also complained about the creator's logo that came on the screen once. The sound effects such as the moving sound of a roller coaster, the wind flow and the change in their intensity with respect to the slopes were appreciated.

From this analysis, we found that a person's prior exposure to technology matters on how content on a new piece of technology will be perceived. Comparing both these scenarios, we could say that emergent users get more engaged than the tech-savvy users in the perception of the virtual environment although all users felt being in a virtual world. Content design plays a major role in providing a realistic experience. For example, a real 360 video is preferred over an animated one. However, there is also scope for fictionalization.

Although we collected the Myo band data for this experiment, we could not analyse it due to the difficulty in standardization. In each case, every user behaved uniquely. For example, a few users seek to hold an object in the middle of the experiment, whereas others stood still. Some users lost track during the ride and looked around a few times to get back to normal. Since experience is unique to every user, we faced difficulties in devising a controlled protocol to measure user's behaviour quantitatively.

65.4.2 Analysis of Experiment-2

Experiment-2 took place in three phases: "pre-imagination", "playing the game" and "post-imagination". We observed very interesting journeys of users from pre-imagination to the post-imagination phase.

Tech-savvy users. The journey of tech-savvy users is shown in Fig. 65.3. In the beginning, users were asked to close their eyes, stretch their arms and imagine themselves

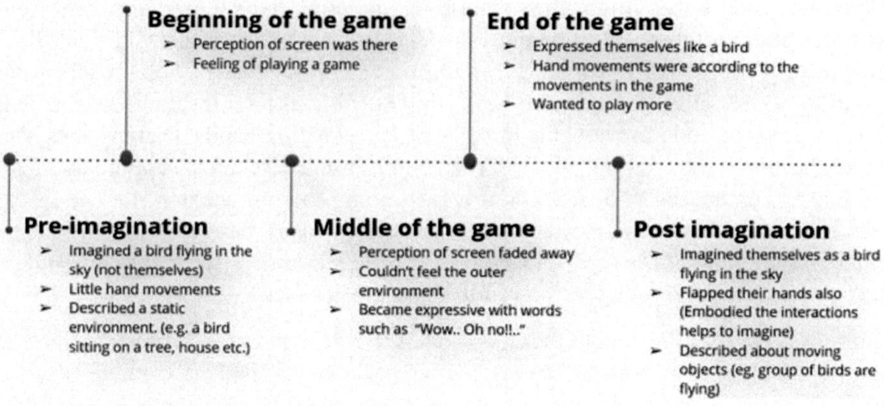

Fig. 65.3 Journey map: VR experience of tech-savvy users in Experiment-2

as a bird flying in the sky. They were asked to narrate what they were imagining. We found that all of them were able to imagine a bird flying. In most of the narrations, the users reported imagined scenarios such as "seeing a bird with trees and buildings around it", "a bird flying with other birds", etc. For example, one of the users said that when he heard the word bird he could immediately imagine "a bird sitting on a mango tree near his house and flying from tree to tree for food". In all cases, none of our tech-savvy users imagined themselves as a bird. We also observed very little hand movements.

At the beginning of the game, the users had a strong distinction between what is virtual and what is real. They felt like playing a game and there was a screen in front of them for visualization. But over time, this distinction faded away. They became a part of the virtual environment and started expressing themselves as a bird. For example, one user said, "*I am flying so high and I can feel the speed of my movement*". They got immersed in the game such that they stopped narrating and started expressing themselves with words/sentences like "*Wow*!", "*Oh no, I am flying so high now.*", "*I have to save myself from the obstacles.*", etc. All users reported that the flapping of arms helped them in realizing themselves as a bird. Although some users recognized that the flapping of arms had nothing to do with the bird's movement in the beginning, but after some time, we observed the change in intensity in their hand movements when they faced an obstacle and tried to save themselves in the game. We also observed the same when they had to take a turn. For example, if they had to take a left turn in the game, we observed more flapping on the left hand. During the interview, one of the users reported that the flapping hand interaction was looking like a hindrance in the beginning but he enjoyed the game once he was able to sync his hand movements with the bird. It gave him an immersive feeling. Another user said that this VR game is so immersive that she forgot the physical environment. All the users wanted to play more when we asked them to stop playing the game.

In the post-imagination phase, users narrated environments such as flying with other birds in the sky, clouds moving in the sky, they could see vehicles moving on the

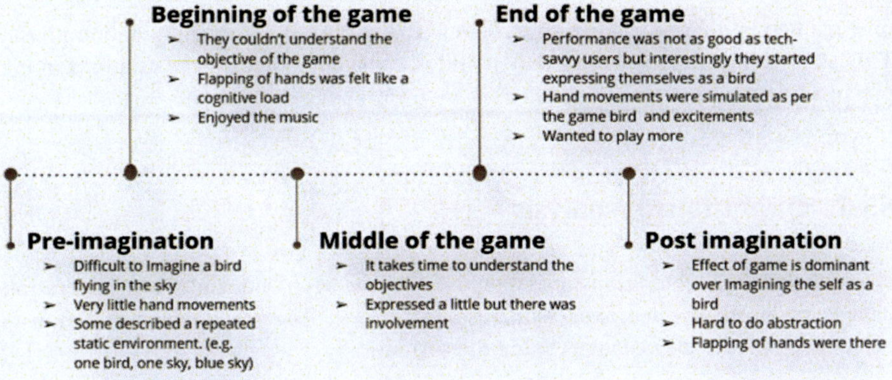

Fig. 65.4 Journey map: VR experience of emergent users in Experiment-2

road, etc. Unlike the pre-imagination phase, the users could see more moving objects in the post-imagination. This could be an effect due to the players being immersed in the game, since in the game, all the objects were moving. But the users did not mention the same set of objects shown in the game. They were also flapping their arms while narrating in the post-imagination phase. Although the flapping was not as strong as it was in the game, it was definitely more than the pre-imagination stage. Interestingly, all the users expressed themselves as a bird in the post-imagination phase.

Emergent users. The journey of emergent users in Experiment-2 is shown in Fig. 65.4. The journey took place in three phases such as "pre-imagination", "playing the game" and "post-imagination" like before. We observed that the emergent users found it difficult to imagine themselves as a bird in the pre-imagination phase. Most of the users could not even imagine a bird flying in the sky. After a few probes were given, some users started reporting repeated sets of objects such as "*I could imagine a bird, a tree, a hill… A new bird, a tree, a hill…*" Imagination was difficult for them.

In the next phase, it took a little bit of time for the emergent users to understand the objectives of the VR game and familiarize them with the interactions. However, we did not provide any training by exposing them to a VR environment since we wanted to record the first time reactions. Only verbal instructions were provided. Once they understood the objectives, they enjoyed playing the game. We observed that the flapping of arms acted as a hindrance to their perception. Some users flapped without any sync with the bird in the game. Although the flapping helped them visualize themselves as a bird, it acted as a cognitive load in performing actions in the game. However, all of them felt immersed in the virtual world.

In the post-imagination phase, the emergent users could imagine themselves as a bird. For example, one of the users said that "*I am flying with a lot of birds around. The wind speed is high*". Most of their descriptions resembled what they saw in the game. But interestingly, all the users were flapping their arms during the post-imagination stage while describing the scenario. During the interview, they said that the flapping of arms helped them visualize themselves as a bird in the post-imagination phase. This also makes it evident why they could not visualize themselves as a bird in the pre-imagination phase.

65.5 Quantitative Analysis

We analysed the acceleration data collected from the Myo armband. The acceleration data is received in three axes with respect to time. The resultant acceleration is calculated first for all instances using the formula

$$\text{Res.\_Acc} = \sqrt{x^2 + y^2 + z^2} \tag{65.1}$$

We then plotted the box–whisker plot to look over the variation in the hand acceleration across the three different phases in Experiment-2 considering the two user groups. As shown in Fig. 65.5, the acceleration in the post-imagination phase (median = 1.003, whisker range = 0.693–1.304) is more than the pre-imagination phase (median = 1.000, whisker range = 0.915–1.084) for tech-savvy users. That means the users flapped their arms faster in the post-imagination phase than the pre-imagination phase resembling our qualitative insight. The acceleration is found to be highest in playing the game phase (median = 1.004, whisker range = 0.253–1.750).

Similarly, for the emergent users, the acceleration in playing the game phase (median = 0.998, whisker range = 0.693–1.305) was found to be the highest among the other two phases. The acceleration in the post-imagination phase (median = 1.000, whisker range = 0.840–1.166) is more than the pre-imagination phase (median = 1.003, whisker range = 0.931–1.069). The variation in acceleration across the three phases for emergent users is shown in Fig. 65.6.

In all three phases, we found that tech-savvy users possessed more hand acceleration than the emergent users. However, their differences are not found to be statistically significant in the two-sample *t*-test with unequal variance ($p > 0.05$ and $\alpha = 0.05$). The comparison statistics are shown in Table 65.2.

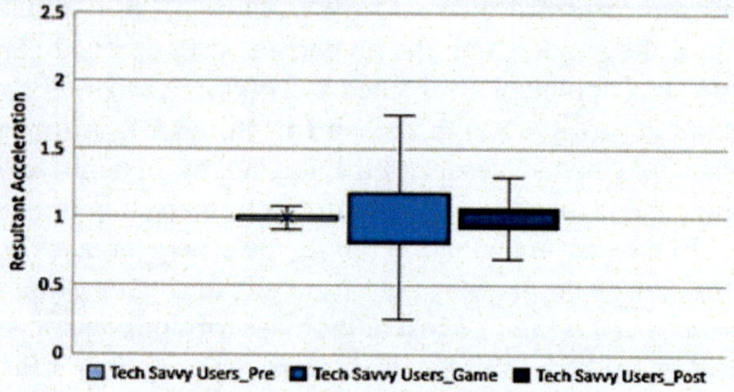

Fig. 65.5 Variation in acceleration across pre-imagination, playing the game and post-imagination phase for tech-savvy users

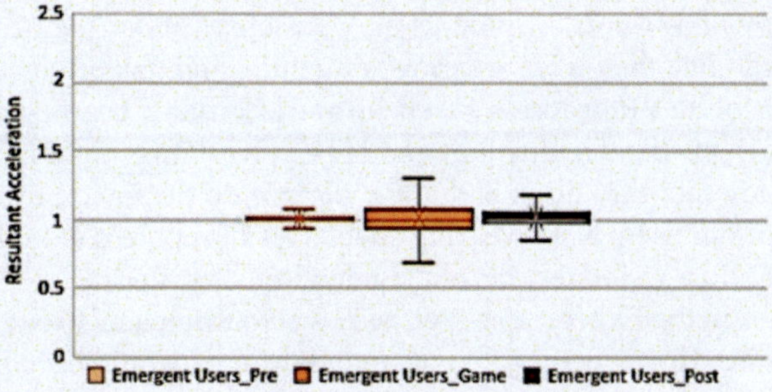

Fig. 65.6 Variation in acceleration across pre-imagination, playing the game and post-imagination phase for emergent users

Table 65.2 Comparison statistics between the two user groups across three phases

	Pre-imagination phase	Playing the game phase	Post-imagination phase
Tech-savvy users	Upper whisker: 1.084	Upper whisker: 1.750	Upper whisker: 1.304
	Upper hinge: 1.021	Upper hinge: 1.190	Upper hinge: 1.075
	Median: 1.000	Median: 1.004	Median: 1.003
	Lower hinge: 0.978	Lower hinge: 0.815	Lower hinge: 0.922
	Lower whisker: 0.915	Lower whisker: 0.253	Lower whisker: 0.693
Emerge nt users	Upper whisker: 1.069	Upper whisker: 1.305	Upper whisker: 1.166
	Upper hinge: 1.017	Upper hinge: 1.076	Upper hinge: 1.044
	Median: 1.003	Median: 0.998	Median: 1.000
	Lower hinge: 0.983	Lower hinge: 0.923	Lower hinge: 0.962
	Lower whisker: 0.931	Lower whisker: 0.693	Lower whisker: 0.840
t-test	df = 40,317, p = 0.292938, t critical = 1.96	df = 44,909, p = 0.376256, t critical = 1.96	df = 38,302, p = 0.781177, t critical = 1.96

65.6 Discussion

We conducted two different experiments to report user experience with VR (Virtual Reality). As per our analysis between emergent and tech-savvy users, we conclude that emergent users could engage themselves better when there is only a perception of the virtual environment. Perception along with action could act as a cognitive load for them. On the other hand, tech-savvy users have a strong distinction between the virtual and real worlds. Since they are better exposed to science fiction and abstraction, VR was experienced as a 3D tool for them in our perception study (Experiment-1). But they found themselves more engaged when they were able to interact with the virtual environment using their body (Experiment-2). Overall, the embodied interactions and embodiedness enhanced the immersiveness and the feeling of being physically present in the virtual world.

All our users found the VR interactions to be simple. One of the users said that "*It is so simple to play the game since no button is needed, everything is controlled by movements*". Some users were upset about the heaviness of the VR headset. It felt like they were wearing something and that affected their immersive experience. Mobile VR headsets are cheap and affordable but they don't provide high-quality VR effects like Oculus Rift and HTC Vive. This might have affected the overall result. Another limitation is that we did not do the vision test of the users. Although none of our users were wearing spectacles or reported eyesight problems, it is a good idea to do a vision test before conducting such studies.

In the Experiment-2 we asked all participants to extend their arms and resemble a bird. However, we did not study how they would have perceived the game if they did not have to flap their arms. For some emergent users, the flapping of arms acted as a cognitive load although it helped them express themselves as a bird. For quantitative analysis, we felt that the number of participants was less to generalize the results. In future, we aim to extend this study with more number of participants to better generalize our results.

References

1. Heilig, M.L.: Sensorama simulator. US Patent# 3,050,870 (1962)
2. Freina, L., Ott, M.: A literature review on immersive virtual reality in education: state of the art and perspectives. In: The International Scientific Conference Elearning and Software for Education, vol. 1, no. 133, pp. 10–1007 (2015)
3. Devanuj, J.A.: Technology adoption by "emergent" users: the user-usage model. In: Proceedings of the 11th Asia Pacific Conference on Computer Human Interaction, pp 28–38. Association for Computing Machinery, New York, NY, USA (2013)
4. Visconti, P., Gaetani, F., Zappatore, G.A., Primiceri, P.: Technical features and functionalities of Myo armband: an overview on related literature and advanced applications of myoelectric armbands mainly focused on arm prostheses. Int. J. Smart Sens. Intell. Syst. **11**, 1–25 (2018). https://doi.org/10.21307/ijssis-2018-005
5. Jennett, C., Cox, A.L., Cairns, P., Dhoparee, S., Epps, A., Tijs, T., Walton, A.: Measuring and defining the experience of immersion in games. Int. J. Hum. Comput. Stud. **66**, 641–661 (2008). https://doi.org/10.1016/j.ijhcs.2008.04.004
6. Brown, E., Cairns, P.: A grounded investigation of game immersion. In: CHI'04 Extended Abstracts on Human Factors in Computing Systems, pp. 1297–1300. Association for Computing Machinery, New York, NY, USA (2004)
7. Riva, G., Wiederhold, B.K., Mantovani, F.: Neuroscience of virtual reality: from virtual exposure to embodied medicine. Cyberpsychol. Behav. Soc. Netw. **22**, 82–96 (2018). https://doi.org/10.1089/cyber.2017.29099.gri
8. Maselli, A., Slater, M.: The building blocks of the full body ownership illusion. Front. Hum. Neurosci. **7** (2013). https://doi.org/10.3389/fnhum.2013.00083
9. Wilson, M.: Six views of embodied cognition. Psychon. Bull. Rev. **9**, 625–636 (2002). https://doi.org/10.3758/BF03196322
10. Kilteni, K., Groten, R., Slater, M.: The sense of embodiment in virtual reality. Presence Teleop. Virt. Environ. **21**, 373–387 (2012). https://doi.org/10.1162/PRES_a_00124
11. Yee, N., Bailenson, J.: Walk A Mile in Digital Shoes: The Impact of Embodied Perspective-Taking on The Reduction of Negative Stereotyping in Immersive Virtual Environments, p. 9
12. Groom, V., Bailenson, J.N., Nass, C.: The influence of racial embodiment on racial bias in immersive virtual environments. Soc. Influ. **4**, 231–248 (2009). https://doi.org/10.1080/15534510802643750

13. Hornecker, E.: The role of physicality in tangible and embodied interactions. Interactions **18**, 19–23 (2011). https://doi.org/10.1145/1925820.1925826
14. De Oliveira, E.C., Bertrand, P., Roel Lesur, M.E., Palomo, P., Demarzo, M., Cebolla, A., Baños, R., Tori, R.: Virtual body swap: a new feasible tool to be explored in health and education. In: 2016 XVIII Symposium on Virtual and Augmented Reality (SVR), pp. 81–89 (2016)
15. Cipresso, P., Giglioli, I.A.C., Raya, M.A., Riva, G.: The past, present, and future of virtual and augmented reality research: a network and cluster analysis of the literature. Front. Psychol. **9** (2018). https://doi.org/10.3389/fpsyg.2018.02086
16. VR 360 Video of Top 5 Roller Coaster Rides 4K Virtual Reality. In: YouTube. https://www.youtube.com/watch?v=hNAbQYU0wpg. Accessed 13 Oct 2020
17. Moon Bird VR Game. In: Google Play. https://play.google.com/store/apps/details?id=com.vicovr.MoonBirdDemo&hl=en_IN&gl=IN. Accessed 13 Oct 2020
18. Baddeley, A.D., Hitch, G.: The recency effect: implicit learning with explicit retrieval? Mem. Cogn. **21**, 146–155 (1993). https://doi.org/10.3758/BF03202726
19. Charmaz, K.: Constructing Grounded Theory: A Practical Guide Through Qualitative Analysis. SAGE (2006)

Chapter 66
Designing Sustainable Tourism Experiences for the Tourists of Tomorrow

Anamika Menon and Sagarika Jayawant

Abstract Tourism is one of the fastest-growing and most important economic sectors in the world with its continuously evolving trends and its potential to create millions of job opportunities. The rise of globalization, changing demographics, and the constantly evolving needs of people has widened the canvas for tourism. However, improved connectivity and an increase in the purchasing power of people have resulted in rapid and unplanned tourism which has proven to be an environmental challenge. With a desire to travel and unwind being the topmost priority among today's population, the number of tourists is predicted to increase by 3.3% per year causing tourism to become more of a hassle. Considering the current world scenario, the emerging generations (Generation Z and Millennials) are expected to contribute a significant market share in the tourism sector by the year 2040. The paper aims to identify the various pain points in the tourism experience and help advance the research and development of a sustainable tourism model. Specific attention is focused on illustrating tourism as a part of the "Experience Economy" and demonstrating how technological innovations such as Artificial Intelligence, Augmented Reality, Virtual Reality, Internet of Things have given rise to the possibility of "Smart Tourism" and can contribute to help empower the host communities and, at the same time enable a transparent, reliable, and seamless travel experience, for the tourists.

66.1 Introduction

Tourism is one of the most remarkable socio-economic phenomena that is defined by a visitors' experience away from their home environment facilitated by services provided by the host destination. It is a multifaceted industry involving a complex network of interconnected businesses and services, including various stakeholders,

A. Menon · S. Jayawant (✉)
Mumbai, India
e-mail: sagarika.jayawant@gmail.com

A. Menon
e-mail: menon.anamika97@gmail.com

A. Chakrabarti et al. (eds.), *Design for Tomorrow–Volume 2*, Smart Innovation, Systems and Technologies 222, https://doi.org/10.1007/978-981-16-0119-4_66

governmental and inter-governmental organizations, NGO networks, host communities, consumers, as well as the infrastructure required to support it.

International tourism has taken a giant leap in recent years, making it one of the most profitable sectors of the world economy. It accounts for 7% of total investment, 5% of all tax revenues, and a third of world trade in services. It employs more than 250 million people, that is, every eighth employee in the world. The travel and tourism industry contributes 3.6% of the world's GDP and is expected to rise in the coming years. The total revenue of the global tourism industry in 2019 was $2 Trillion.

Tourism being a rapidly evolving landscape, it is necessary to accord with the challenges faced at various stages of the tourism model. It causes an alteration in the traditional lifestyles of the local communities as well as tourists. Managed well tourism positively transforms the socio-cultural, economic, and political development of the host destination. On the contrary, excessive and unplanned tourism can negatively affect the social and political systems in the host destination as well as affect the natural resources and consumption patterns.

The paper focuses on considering and dealing with the areas of concern observed in the entire tourism model that can lead to envisioning a sustainable tourism experience. At the same time, determining how technological advancements like big data, AR, AI, VR, and IOTs can play an integral part in building upon the traditional hierarchies and processes of planning the tourism model, empathizing with the economic, social, and cultural concerns that lead to creating a smart tourism experience.

To better understand and achieve the above objective, the following research questions have been raised:

- What are the challenges faced by the host communities with a heavy influx of tourists?
- What are the factors that negatively influence tourist experiences?
- How can technology and innovation be leveraged in such a way that it empowers the tourism industry to deal with its economic, socio-cultural, and environmental challenges?

66.2 Methodology

The research is intended to analyze and understand the problems in the tourism industry and derive technological solutions to build a sustainable tourism model. Based on the ever-evolving nature of technological innovations and its application to sustainable tourism, an interpretative perspective was selected for this research to enable flexible and open research solutions. The study was instituted by raising a series of research questions to better understand the scope of the objectives. A comprehensive study of qualitative data was conducted to determine the chronology of the various stages of the tourism experience and the stakeholders involved at different levels. Qualitative research of the existing technologies and their applications were done to determine their relevance to solving the research questions. The data obtained was closely examined to identify patterns and facilitate value co-creation enhanced by the reconciliation and synthesis of the various domains. Work of experts in the concerned field, relevant conference papers, and journals were consulted to carry out a detailed content analysis and categorization of terms such as "Sustainability" and "multi-sensory tourism experiences" to better identify the potential of the research paper.

66.3 Analysis

The travel and tourism industry involves several pressing issues associated with social, economic, and environmental factors. The challenges faced by the tourism sector differ across geographical locations as each region has its specialties, policies, and priorities. Since it involves multiple activities in terms of host countries as well as on the tourists' end, negative consequences of tourism development are observed like, overcrowding, tourism carrying capacity, pressure on natural resources, infrastructure, preserving the local culture. Thus, these factors directly have an impact on the tourist experiences and hinder the ecosystem of the local communities and the sustainability of destinations.

66.3.1 Problems Faced by Host Nations

Economic Problems. Although tourism greatly contributes to the economy of a nation, it can sometimes prove harmful for the local communities as a high influx of money due to tourism is rarely directed to them. In regions with a heavy influx of tourists, due to high demand, there is an increase in taxes, prices of goods and services, land value, and real estate. Tourism requires heavy investments which are a concern for underdeveloped nations as foreign investors are reluctant to invest until profits are assured.

Seasonality is observed to be a significant challenge for nations as it causes wasteful use of resources as the investment in infrastructure and other activities may not be used all over the year which results in tight economic conditions. Over tourism can also backfire on the nation's economy as an increase in the purchasing power of the citizens will result in them going for vacations abroad, thus causing an outflow of money from the economy. This scenario is also known as "leakage," wherein the net income of tourism is lowered due to the local population traveling abroad.

Socio-cultural Problems. Tourism ushers a regular influx of new life and cultures. Quite often, the locals get fascinated by the "other" way of life also known as "Staged Authenticity" which is influenced by the current trend of "experiencing the authentic culture" leading to commercialization of cultural activities. Traditional festivals and ceremonies are transformed into staged performances for the tourists, which tend to lose its grace for the local communities, thus causing loss of cultural identity. Tourism development also leads to displacement of residents in certain situations. Overtourism can upset the local population due to insufficient transport facilities, communication services, safety, and security. It is commonly observed that political and economic differences often result in clashes among the local communities and tourists.

Environmental and Health Problems. Regular influx of tourists from varied geographies creates a backdrop of heightened uncertainty amongst locals concerning their health security and also causes ecological congestion. Indiscriminate efforts by host countries to manage facilities like waste disposal, water as well as unwatched development of infrastructure, which neither considers the local architecture causes loss of biodiversity. In addition, lack of initiatives by the government to develop a comprehensive healthcare model puts excessive pressure on the existing system to deal with a health crisis. COVID-19 is one very apt example of the situation. The outbreak brought the world economy to a standstill, and the tourism industry was the worst affected. COVID-19 being a contagious disease requires a high level of sanitation to improve life expectancy. The spread of disease was aggravated due to the movement of people across geographies. The inability of nations to accommodate and treat the local citizens along with the non-residents leads to an extended responsibility for the healthcare services of host countries to treat a larger population.

66.3.2 Problems Faced by Tourists

The travel experience starts right from identifying a travel destination to reflecting and sharing the trip experience with fellow travelers. Tourists today want to live the local life, engaging with locals and experiencing the authentic culture and cuisine of the place, but factors such as language, way-finding, accessibility, and infrastructure facilities prove challenging and thus could hinder the entire travel experience (Fig. 66.1).

Stage 1: Planning and Discovering

Deciding on a destination is the first milestone in a tourist's journey which is facilitated by gathering information from the Internet to connecting with tour operators. However, a disconnect between what the customer expects and what is provided and the inability to find reliable sources to gather information to create tailor-made experiences is one of the primary pain points of today's tourists.

Stage 2: Experiencing and Sharing.

Communication and mobility—Lack of sufficient infrastructure and services in the host nations creates unnecessary hassle concerning navigation and way-finding, thus resulting in unfavorable tourism experiences. Failing to understand the constraints of the local culture can keep the tourists from getting a holistic experience of the destination.

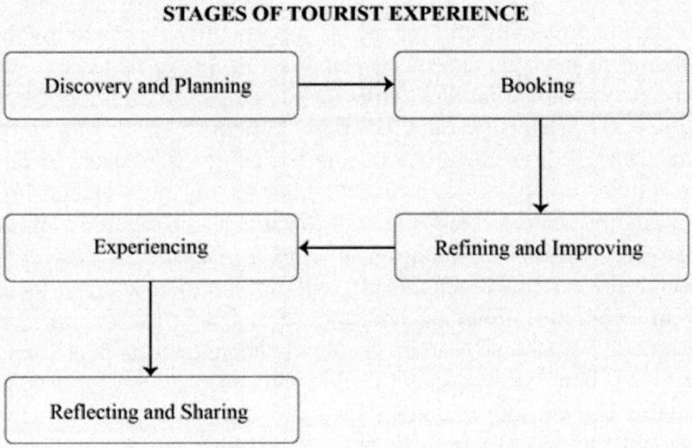

Fig. 66.1 Stages of tourism experience

Travel, health, and security uncertainties—In today's trend of solo travel, security is a pressing concern, especially for women. Unforeseen circumstances during travel, such as loss of baggage in transit, delayed or canceled flights create undue dissatisfaction in the customer journey. Tourists with pre-existing health problems are reluctant to travel to unfamiliar destinations due to uncertainty pertaining to healthcare services. On the other hand, constantly traveling across geographies makes one susceptible to contracting infections. COVID-19 has made people vary of traveling and conscious about their travel choices.

66.4 Discussion

Considering the serious repercussions of tourism, it is important to structure a tourism model that balances the needs and expectations of travelers and at the same time has a minimum negative impact on host countries. "Tourism Sustainability" determines the durability and economic profitability of tourism activity by effectively utilizing tourism resources and facilitating spatial and perception mobility. It helps attract environmentally conscious tourists. In order to achieve "Economic Sustainability" in Tourism, it is important to support the local economies. This can be done by involving the local communities in the tourism experience. This not only benefits the locals but also enhances the quality of tourism experiences. "Socio-Cultural Sustainability" reduces vulnerability, helps maintain stable health conditions, and preserves the socio-cultural authenticity of the host nations.

The world tourism organization defines sustainable tourism in the following manner: "Sustainable tourism development meets the needs of present tourists and host regions while protecting and enhancing opportunities for the future. It is envisaged as leading to management of all resources in a way that economic, social, and aesthetic needs can be fulfilled while maintaining cultural integrity, ecological processes, biological diversity, and life support systems."

Tourism, generally considered as people traveling for pleasure is nonetheless a complex activity involving various issues intersecting over several branches in the socio-economic sectors. Thus, the shift toward sustainable tourism has to be a united effort by both the host countries and the tourists in liaison with the varied stakeholders of the system. Green tourism will turn out to be a key to social change owing to imminent innovation in technology.

In this context, sustainable tourism is not only about environment-friendly travel and going green, but about various other aspects such as the local communities being profitable and causing minimal damage to the host countries. "Live like the locals" is a major factor contributing toward sustainable travel. A 2016 Expedia study derived that authenticity was a top priority for tourists of the millennial generation across the world. These services include experiences in which tourists engage in activities organized by the locals, participating in indigenous traditional festivals, cookery classes offered by locals, as well as buying local products, traveling by local transport, and staying in eco-friendly accommodation, run by the locals.

So, how do we harness the power, the opportunity, and the insights of all the raw data available and the various technologies to enhance the tourism experience?

Smart experiences aided by developing real-time monitoring, successful information systems, and a smart destination can facilitate the co-creation of experiences. The interconnection of tourism destinations and the complex structure of stakeholders through dynamic platforms, enhanced decision support systems, and knowledge-intensive flows of communication to build a sustainable model is now possible through the advent of technologies such as augmented and virtual reality, artificial intelligence, Internet of Things, and smart wearables collaborating with Web chatbots form paramount constituents of a sustainable tourism model.

66.4.1 Artificial Intelligence

Artificial intelligence has brought about a paradigm shift in what tourists want. Today, tourists wish to chart their itinerary, arrange every ancillary service throughout their journey to ensure a smooth and hassle-free travel experience. AI is a game changer in this context. AI can study consumer behavior and provide automated recommendations and services during interactions for booking tickets, creating a personalized itinerary, and even suggesting restaurants based on the users' food choices.

Recognition technology a subset of AI has immense potential in eliminating friction in a tourist's journey and enhancing seamless travel interactions. "The technology itself includes fingerprint recognition, facial recognition, retina scanning, and various other biometric identifiers. In addition, within airports and hotels, security is a key concern, and facial recognition can be used to more easily identify people, to grant specific people access to places, and to keep others out" [1].

Known Traveler Digital Identity (KDTI) "Testing and health screening at airports is difficult to achieve at scale. Under schemes like known traveler digital identity, travelers would be able to consent to share their identity and health data in advance of the journey, allowing border officials to conduct any required risk assessments in advance of the journey while avoiding queuing and bottlenecks at airports" [2].

66.4.2 Big Data

"As human beings, we create a lot of data, and it is increasing exponentially every day. The PBS program *Nova Wonders* recently shared a truly mind-boggling statistic: We now create **2.5 billion gigabytes** of data per day" [3]. Big data can be used computationally to analyze travel patterns and to understand seasonality and creating heat maps of destinations and disperse the tourists to avoid overcrowding. Before tourists arrive, a socio-demographic status is established for the knowledge of destination managers, and a plan of action can be defined to ease congestion. This system will also help effectively allocate the resources for destinations that deal with the problem of seasonality of tourism.

Environmental benefits are achieved from using integrated data to make important decisions about tourism planning in sensitive areas by identifying sites and attractions for development (McAdam 1999, [4]) and helping with tourist management techniques through zoning and identification of areas that require protection (Bahaire and Elliott-White 1999, [4]).

66.4.3 Augmented Reality and Virtual Reality

"Virtual reality technology typically involves the use of a VR headset, which helps to immerse a user in a digital environment" [5]. VR is a value addition to destinations and archeological sites that are fragile for a physical tour but can be recreated giving a virtually parallel reality of these tourist sites. The local community can be involved in this process of creating a multi-sensory extravaganza of events describing their culture and history. It can also be useful in creating an informative experience of visiting local wildlife and mesmerizing oases without disturbing the natural habitats, preserving and safeguarding the environment. VR can also prove to be extremely helpful to people with disabilities who cannot visit certain attractions due to physical limitations.

Augmented reality is a digital technology that tends to change a person's perception of the way physical surroundings which are viewed using a device. It augments the real world by an overlay of digital components, thus making their travel experience more engaging and smooth like using it for navigation or pointing at a restaurant and getting its reviews and menus.

66.4.4 IOT

An interconnected network of services and products forms the basis of IoT which facilitates seamless travel. Right from offering a smart environment for hotels that awards control to the guests to providing a seamless check-in and boarding process through beacons within airports, there are numerous offerings to this innovation. IoT is predicted to cover a $2 trillion market by 2030. Location-based services are also a major feature of IoT. Be it using sensors and sending push notifications to passenger mobiles alerting them on next available flights if one is canceled, getting directions to boarding gates, baggage arrival to providing notifications to tourists about attractions, and other services when they are most relevant based on where they are. These can also be used in-flight to monitor heart rate, anxiety level, body temperature, etc., of passengers and notifying cabin crew for assistance (Fig. 66.2).

The sustainable tourism model is designed to enhance the capabilities of the future tourism industry curating meaningful experiences for the tourists by leveraging the potential of technological innovations.

66.5 Conclusion

Technological advancements help understanding changing perceptions and expectations of tourists and have a seamless and secure journey. Collaborating and co-creating experiences act as a key to developing a sustainable tourism model and thus transform the tourism industry. Smart tourism takes into account the economic, socio-cultural, and environmental challenges faced by the host countries and the tourists and leads to a pleasant experience developing a platform to enable cooperation between the two. The tourism model proposed introduces a collaboration of systems in which infrastructure, technology, and personnel merge into a single seamless experience providing multiple personalized outputs that will not only provide a wholesome experience to tourists but also act as a catalyst in empowering the local inhabitants of the host countries helping them flourish.

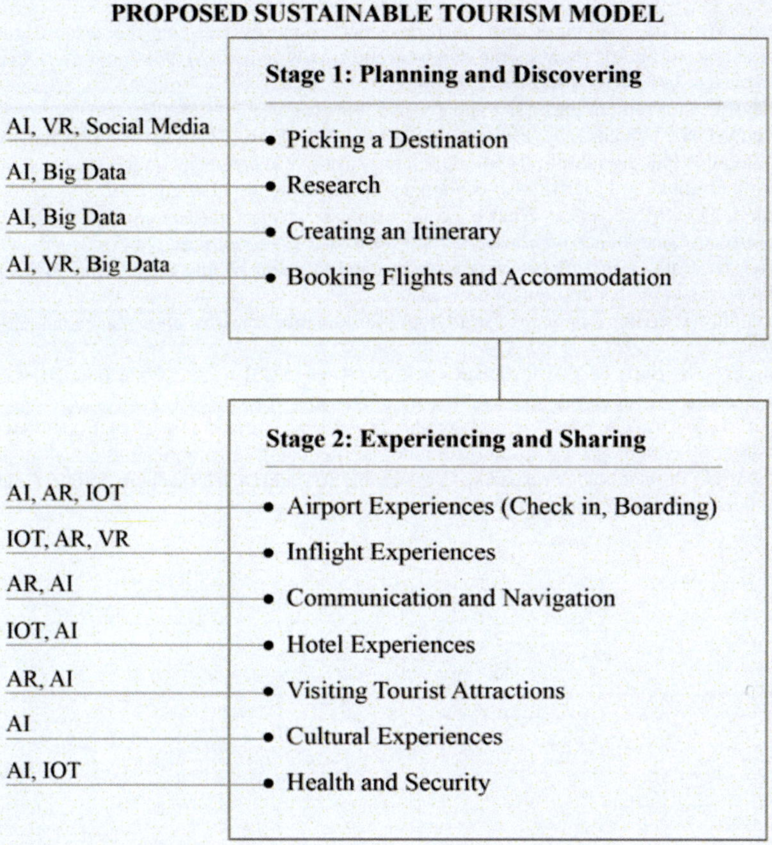

Fig. 66.2 Proposed tourism model

References

1. Revfine: 4 Ways Facial Recognition Can Be Used in the Travel Industry. https://www.revfine.com/facial-recognition-travel-industry/
2. Serra, A., Leong, C., World Economic Forum: Here's what travelling could be like after COVID-19. https://www.weforum.org/agenda/2020/05/this-is-what-travelling-will-be-like-after-covid-19/
3. Ardito, L., Cerchione, R., Del Vecchio, P., Raguseo, E.: Big data in smart tourism: challenges, issues and opportunities, 15 Jul 2019
4. Ali, A., Frew, A.J.: Technology innovation and applications in sustainable destination development (2014)
5. Revfine: How virtual reality is transforming the travel industry. https://www.revfine.com/virtual-reality-travel-industry/
6. Simpleview: How User Generated Content (UGC) Has Transformed Travel, 04 Apr 2019. https://www.simplevieweurope.com/blog/read/2019/04/how-user-generated-content-ugc-has-transformed-travel-b106

7. Ayan, M.: The challenges and obstacles that countries face on the establishment of tourism sector. https://www.simplevieweurope.com/blog/read/2019/04/how-user-generated-content-ugc-has-transformed-travel-b106
8. Essays, U.K.: Problem Faced By The Tourism Industry
9. Peceny, U.S., Urbančič, J., Mokorel, S., Kuralt, V., Ilijaš, T.: Tourism 4.0: Challenges in Marketing a Paradigm Shift, 01 Mar 2019. https://www.intechopen.com/books/consumer-behavior-and-marketing/tourism-4-0-challenges-in-marketing-a-paradigm-shift
10. Andrei, M., ZME Science: What is ecotourism and why we need more of it, 08 May 2019. https://www.zmescience.com/other/feature-post/what-is-ecotourism/
11. Ferrer, R., Caixa Bank, The sharing economy and tourism, 12 July 2018. https://www.caixabankresearch.com/en/sharing-economy-and-tourism
12. Sustaining Tourism: Resources for a Global Sustainable Tourism. https://sustainabletourism.net/
13. Kutschera, S., Soft, T.: Travel statistics to know about in 2019 and 2020, 6 June 2019. https://www.trekksoft.com/en/blog/65-travel-tourism-statistics-for-2019
14. Vidal, B., WAM: The New Technology and Travel Revolution, 5 Dec 2019. https://www.wearemarketing.com/blog/tourism-and-technology-how-tech-is-revolutionizing-travel.html
15. Dabour, N.: Problems and Prospects of Sustainable Tourism Development in the OIC Countries: Ecotourism. J. Econ. Coop. **24** (2003)

Chapter 67
Virtual Reality-Based Fire Safety Training for the Indian Context

Alan Sha, Anmol Srivastava, Madhav Haldia, Pranav Kumar, and Pankaj Badoni

Abstract A large number of fire accidents occur across India every year due to appalling status of fire safety measures and general laxity among the public. This paper explores the potential of virtual reality (VR) to educate and raise awareness amongst the masses about precautionary measures of fire safety. The VR experience is primarily targeted for non-fire fighters. A user-centered design approach was adopted to identify various problems faced at the time of fire emergencies. Semi-structured interviews were conducted, and online surveys were utilized for data collection. Based on insights gained from analyzing the collated data, VR experiences were prototyped utilizing Google cardboard and HTC vive. The prototypes educate the user about the use of fire extinguishers and behavior of smoke.

67.1 Introduction

Fire is a major disaster that threatens human safety. In India, houses and institutions are prone to fire due to lack of safety knowledge and the absence of emergency measures. According to a published report by the National Crime Records Bureau (NCRB) [1], thirty-five ($n = 35$) Indians die in a fire incident daily. Of the fire fatalities reported in 2018, 4290 were in age group of 18–30, and 3860 were in age group of 30–45. These numbers show the severity of the problem and lack of knowledge of Indian masses about fire and safety. These alarming numbers also emphasize that the educational institutions and industries must attach great importance to fire safety training. In addition, the onus also falls on the general public to learn and master the fire safety measures and necessary skills to prevent the fire and handle it properly in case of fire accidents. Fire safety education is a combination of knowledge learning and operational exercise, where teaching method plays a vital role. Until now, traditional teaching methods primarily relying on textbooks and multimedia presentations have been utilized. However, these methods are unable to capture the

A. Sha (✉) · A. Srivastava · M. Haldia · P. Kumar · P. Badoni
School of Design, School of Computer Science, University of Petroleum and Energy Studies, Dehradun, Uttarakhand, India
e-mail: alanshalimar@gmail.com

A. Chakrabarti et al. (eds.), *Design for Tomorrow–Volume 2*, Smart Innovation, Systems and Technologies 222, https://doi.org/10.1007/978-981-16-0119-4_67

essence of a hazardous environment and lack the experiential component. With the emergence of virtual reality (VR) technology, realistic three-dimensional (3D) virtual environment and immersive multi-sensorial experiences can be designed. Owing to these capabilities, VR has been utilized for many training and educational purposes. VR has also been explored extensively for fire safety training [2–4]. However, how can this technology be utilized for educating the masses especially in an Indian context? To address this question, this study adopts a user-centered design (UCD) methodology to identify the problem and design suitable solution. The aim was to understand the main problems faced in fire safety situations in Indian context through the eyes of both fire fighters and non-fire fighters. For this, qualitative data collection techniques such as semi-structured interviews and field observations were utilized. Based on the insights gained from the analysis of data and through the literature reviews, VR fire safety training experiences have been prototyped.

67.2 Literature Review

Fires are classified by the kind of fuel involved. Not all extinguishing agents are good with a wide range of fuel (e.g., water utilized on a flammable fire is probably going to build the pace of burning significantly and to scatter the fuel to cover a greater area). Thus, if a wrong fire extinguisher is chosen, the fire circumstance can be made worse, often your own personal safety. A few fire extinguishers are just more effective than others on specific classes of fire. Various classification of fire has been provided in the literature [5]. Most fire extinguishers have a pictograph label identifying the type of fuels that may be extinguished. An understanding of basic principles of combustion or fire, causes and sources of ignition, fire growth and fire spread are necessary for understanding the principles of fire control and extinguishment. Combustion usually requires an exothermic chemical reaction between a substance or fuel and oxygen. Unlike slow oxidation, a combustion reaction occurs so quickly the heat is generated faster than it is dissipated, causing a marked increase in temperature, even up to a few hundreds of degrees. Frequently, the temperature reaches so high that visible light or flame is generated. It has been shown from the triangle of fire that three factors are essential for combustion (refer Fig. 67.1). An interesting protocol to teach about fire safety is the PASS protocol. Pull the pin, aim the extinguisher nozzle at the base of flames…squeeze trigger and sweep the extinguisher or nozzle from side to side covering base of the fire [5].

A number of studies [2, 3, 6, 7] have been published on fire safety training in VR which primarily focus of fire simulation, fire safety training, smoke behavior, fire and safety in VR. Among them, Kobes and Helsloot [2] researched on the possibilities of virtual reality for studying human behavior in fires are so far scarcely adopted by researchers. The application of a behavioral assessment and research tool (BART) in virtual reality is expected to be an important addition on the existing research methods. Shamsudin [3] reviewed the history, attributes, and application of virtual reality in training and education for construction safety and health course. While

Fig. 67.1 Fire triangle

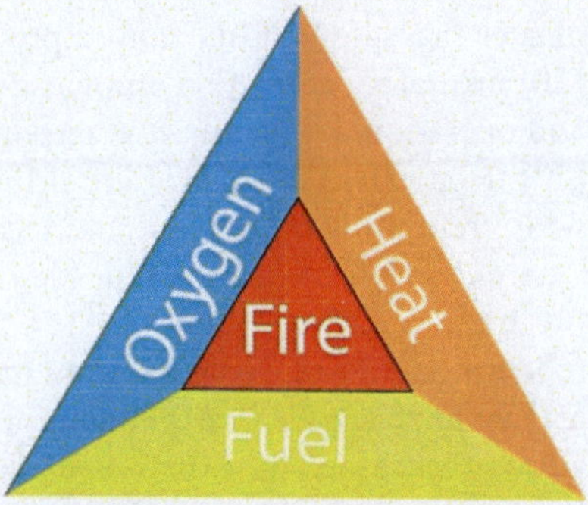

doing fire safety training in real environment, Eglin et al. [4] stated that the longer duration of exposure coupled with the inability to dissipate heat whilst wearing protective clothing can result in considerable physiological strain for fire fighter instructors [10]. Williams-Bell et al. [11] stated that the colleges and universities must attach great importance to fire safety education. Each student must learn and master the fire safety knowledge and necessary skills to prevent the fire and handle properly in the event of fire [11]. Feng and Xiao [6] suggested that a VR fire drill system has a practical significance and is instrumental in fire drills, fire safety training, and other applications [12]. Bernardes et al. [7] established virtual reality that can be a valuable tool for fire safety training, as the training can be more effective, VR reduces cost, and can increase the safety of individuals [13]. Fanfarová and Mariš [8] presented the possibility of using modern simulation tools using computer softwares. The study suggests training in VR that improve and increase the level of preparedness of fire and rescue services [3]. Yu and Guan [9] suggested that virtual training methods should be used for fighters for better performance [14]. Martínez [15]. The lack of interactivity and high equipment complexity also leads to the low practicability [15]. Moreno et al. [10] have discussed various simulations dealing with fire extinguishment actions, natural and artificial fire breaks, and various wind conditions. [16]. The authors used different fire propagation algorithms. Yang et al. [13] apply VR for gas explosion training [18]. Wang et al. [14], the paper establishes applications of fire disaster simulation and virtual fire training by using collaborative virtual geographic environment (CVGE) platform—CySim, which is developed based on open source open simulator server and second life client. García-Hernández [16], for in emergency situations, the tools described in this paper like 3D images 360° images for fighters were preferred to the already existent self-protection plans [11]. Xi and Smith [17] the success of using professional numerical fire simulation tool to support non-player characters behavior in virtual environment will provide a new way to enhance the realism of virtual environment by using high fidelity fire data sources [15]. Shamsudin et al. [3] state that VR training environment can provide a diversity of equipment operation scenarios for teaching and learning [9]. Zhang et al. [12] established that VR has enriched the teaching form of fire safety education on campus, which can effectively help students learn fire safety knowledge, master fire safety skills, and improve fire safety education effect [17]. Viant et al. [20] fire safety education emphasizes the close combination of knowledge learning and operational practice. It is important to choose the appropriate teaching method [20]. Houvouras et al. [21] fire drills can effectively help students become familiar with firefighting equipment and learn extinguishing and self-protection skills [21]. Querrec and Chevaillier et al. [22] proposed a storytelling-based approach to VR fire training [22]. Xu a et al. [18]. Rena visualization technique based on volume rendering and fire dynamics data has been especially designed to create a realistic and accurate smoke environment for the purposes of effective virtual training, which allows the trainees to experience a realistic and yet non-threatening fire scenario[7]. Itamiya et al. [19] used AR system for an experiment in evacuation drills natural hazards in elementary schools and virtual reality; it was very useful improving crisis awareness of students [20].

Based on the literature review, several directions for research in fire safety training have been identified. As the main aim of this research is to utilize VR for training the general public, which needs to have better reach in terms of equipment accessibility and cost, PASS protocol was chosen to be implemented for Google cardboard.

67.3 Methodology

This study utilizes a UCD approach and employs various design methodologies to identify the crucial needs. Semi-structured interview, field and online surveys were utilized for data collection.

67.3.1 User Interviews

The primary objective of the user interviews was to understand the problems which are faced in firefighting from the perspective of both firefighters and non-fire fighters. Semi structured interviews were conducted among ($N = 10$) participants.

Table 67.1 depicts the type of participants interviewed (Age group: 20–45, Male).

Table 67.1 List of participants and number of participants

Participants	Number of participants (10)
Fire safety manager	1
Fire fighters	2
Fire safety students	4
Fire safety instructor	1
Non-fire fighters	2

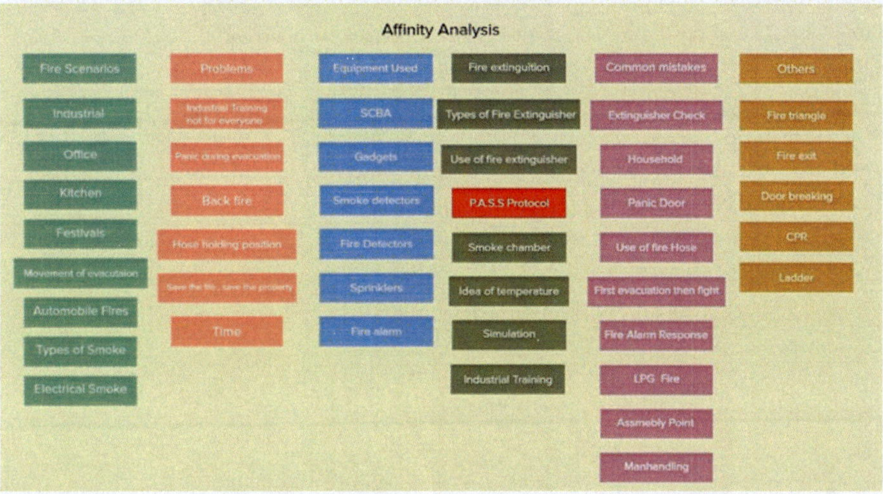

Fig. 67.2 Affinity analysis diagram from user interviews conducted

Questions pertaining to participants' work experiences and knowledge about fire training were asked. User diaries and audio recordings were utilized to document and capture interviews. The audio recordings were later transcribed for affinity analysis. Questions were asked based on the experiences of the participants regarding the real-life experiences regarding fire. Problems they faced while when fire accidents happened, what kind of training would help them in fighting fire and most common fire situations (Fig. 67.2).

From the affinity analysis, it was understood that PASS protocol is the most important protocol that should be trained for the non-fire fighters among all the other fire extinguishing techniques.

67.3.2 *Surveys*

An online survey was conducted to understand the knowledge and behavior about fire and safety. The study recruited participants in which 71.4% were male and 28.6% were female. 90.5% of the participants were in the age group 21–30. Around 57.1% were students, and 28.6% of them are employees; 14.3% are either students or employees related to fire and safety (Table 67.2).

Around 58% of the participants are not sure about the exact sound of the fire alarm. When the fire alarm is activated, 61.9% candidates feel panicked. The survey reveals that participants not associated with fire safety training lack basic understanding about fire safety and its relevant aspects—which is a cause of concern. It was also identified that these participants are also unaware of using fire extinguishers. The insights gained form user research and the literature reviews were then converted

Table 67.2 List of yes or no questions asked in the online questionnaire

Questions	Yes (%)	No (%)
Do you have any kind of previous fire safety training experience in your job or institution?	60.3	39.7
Everyone should know do's and don'ts in case of fire emergency	100	0
Everyone should be trained in fire prevention control	92.1	1.6
Do you know about fire triangle and its components?	34.9	65.1
Do you know about different types of fires?	46	54
Do you know about different types of fire extinguishers?	42.9	57.1
Do you know about "PASS" method of fire control?	22.8	22.2

Fig. 67.3 Scenario used for making fire extinguisher training

into information architecture and working prototypes. Storyboarding technique, refer Fig. 67.3, was adopted to envision how fire-extinguisher training module can be presented as a part of an immersive user experience.

67.3.3 *Basic Fire Safety Training Using Fire Extinguisher*

Methodology given in (Fig. 67.4) shows how a prototype for Google cardboard has been made for training non-firefighters in PASS protocol.

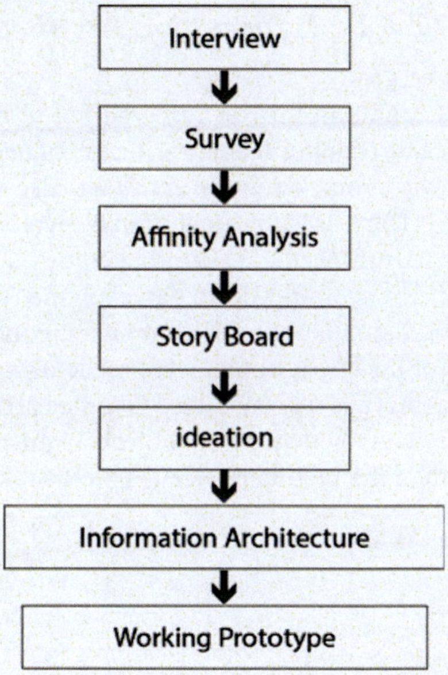

Fig. 67.4 Methodology used in making a working prototype for fire extinguisher training in Google cardboard

67.4 Ideation

In ideation part, the PASS protocol for VR fire extinguisher training is laid out stepwise (Fig. 67.5).

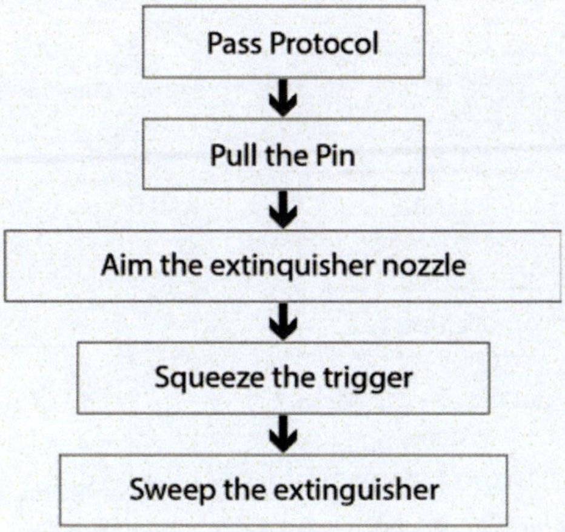

Fig. 67.5 Steps followed in PASS. protocol

67.4.1 Information Architecture

Insights from user research led to the formulation of information architecture for basic training scenarios in case of using fire extinguishers, kitchen fire accidents, fire evacuation using fire exits, and also how to react to different types of fire (Fig. 67.6).

Unity is the ultimate game development platform. Unity can be used to build high-quality 3D and 2D games, deploy them across mobile, desktop, VR/AR, and console. It is a cross-platform game engine. Which is primarily used to develop video games and simulations for computers, consoles, and mobile devices. Objects and models for the environment were made using Blender 2.80, SketchUp 2019. The interactive prototype was made for Google cardboard. Gaze-based interaction has been used to interact with different objects in the prototype since it does not need any extra gadget than the mobile device (Fig. 67.7).

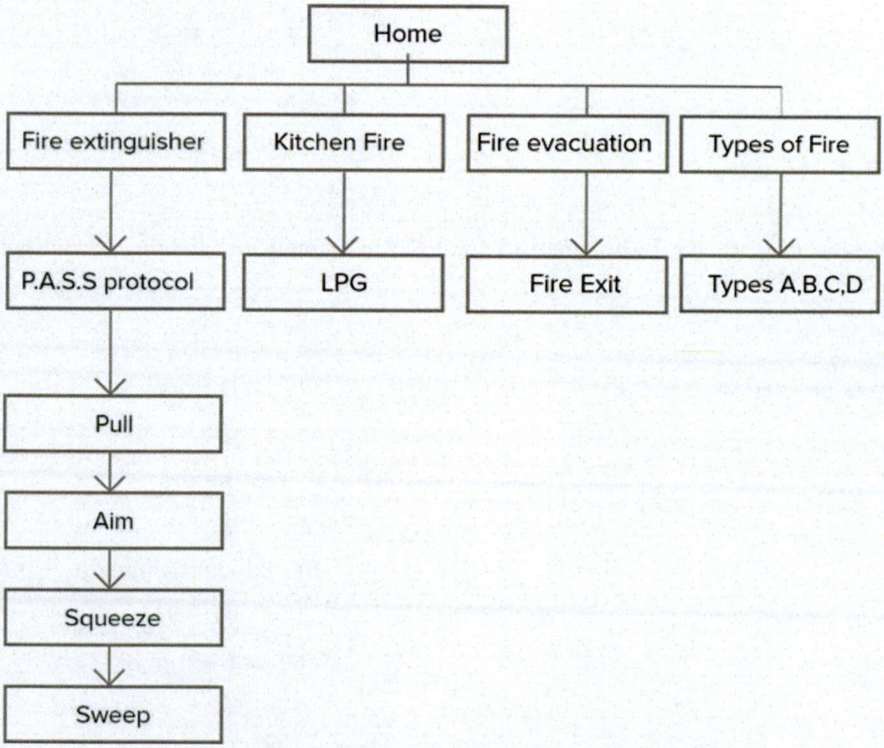

Fig. 67.6 Interaction map for VR training in fire extinguisher training

Fig. 67.7 Prototype "PASS" protocol for Google cardboard

67.5 Discussion

From the study, it is understood that there is a research gap for training of non-fire fighters in virtual reality in Indian context. Using low-cost VR headsets like Google cardboard can be solution for making masses understand the basics of fire safety training so that they can have clear understanding of different methods of fire extinguishing. Since there are limited input devices in Google cardboard, alternative input methods have to be used. Due to the COVID-19 lockdown situation of 2020, no further usability test was conducted to show how VR helps in fire training more than the existing training methods. Usability tests needs to be conducted in future so that it could be proven with help of data whether using this as a training method in Indian context is useful or not. By giving a voice interface and using commands to guide can make the prototype immersive and more functional.

67.6 Future Work

67.6.1 Fire Safety Training for HTC Vive

A virtual simulation for fire safety training in smoke chamber was developed using unreal engine for HTC vive. The smoke behavior and simulation are made similar to real-life scenarios. Smoke chamber is a common type of fire safety training given to fire fighters (Figs. 67.8 and 67.9).

Using this prototype is not practical for training non-fighters since the gadget needed for this is HTC vive. HTC vive is expensive and not easy to access in an Indian context.

Fig. 67.8 Smoke chamber simulation for HTC vive

Fig. 67.9 VR environment and smoke behavior in smoke chamber simulation for HTC vive

67.7 Conclusion

Fire safety education is important for both students and professionals. This study shows how low-cost methods like Google cardboard can be used as a method of training for non-fire fighters. Fire safety training can be further developed using VR effectively and more immersive. For Indian masses under the current situation, these types of training methods will be more time saving, and it gives better safety than exposing to real fire environment.

References

1. NCRB: National Crime Records Bureau (2018). https://ncrb.gov.in/
2. Kobes, M., Helsloot, I., de Vries, B., Post, J.: Exit choice, (pre-) movement time and (pre-) evacuation behaviour in hotel fire evacuation—behavioural analysis and validation of the use of serious gaming in experimental research. Proc. Eng. **3**, 37–51 (2010)
3. Shamsudin, N.M., Mahmood, N.H.N., Rahim, A.R.A., Mohamad, S.F., Masrom, M.: Virtual reality for construction occupational safety and health training: a review. Adv. Sci. Lett. **24**(4), 2444–2446 (2018)
4. Eglin, C.M., Coles, S., Tipton, M.J.: Physiological responses of fire-fighter instructors during training exercises. Ergonomics **47**(5), 483–494 (2004)
5. Cote, A.E.: Fire Protection Handbook, vol. 2. National Fire Protection Association (2008)
6. Feng, L.N., Xiao, Z.: A fire drill training system based on VR and kinect somatosensory technologies. Int. J. Online Biomed. Eng. (iJOE) **14**(04), 163–176 (2018)
7. Bernardes, S.M.F., Rebelo, F., Vilar, E., Noriega, P., Borges, T.: Methodological approaches for use virtual reality to develop emergency evacuation simulations for training, in emergency situations. Proc. Manufact. **3**, 6313–6320 (2015)
8. Fanfarová, A., Mariš, L.: Simulation tool for fire and rescue services. Proc. Eng. **192**, 160–165 (2017)
9. Yu, Z.F., Guan, J.L.: Fire and rescue combat technical training system construction for dangerous chemicals. Proc. Eng. **135**, 655–660 (2016)
10. Moreno, A., Posada, J., Segura, Á., Arbelaiz, A., García-Alonso, A.: Interactive fire spread simulations with extinguishment support for virtual reality training tools. Fire Saf. J. **64**, 48–60 (2014)
11. Williams-Bell, F.M., Kapralos, B., Hogue, A., Murphy, B.M., Weckman, E.J.: Using serious games and virtual simulation for training in the fire service: a review. Fire Technol. **51**(3), 553–584 (2015)
12. Zhang, K., Suo, J., Chen, J., Liu, X., Gao, L.: Design and implementation of fire safety education system on campus based on virtual reality technology. In: 2017 Federated Conference on Computer Science and Information Systems (FedCSIS), pp. 1297–1300. IEEE (2017)
13. Yang, W., Fu, G., Zhang, J., Qing, S., Shao, N., Chang, Z.: Analysis and methods of the blasters' unsafe behavior taking gas explosion accident as an example. Proc. Eng. **45**, 220–224 (2012)
14. Wang, R., Chen, B., Huang, F., Fang, Y.: Using collaborative virtual geographic environment for fire disaster simulation and virtual fire training. In: 2012 20th International Conference on Geoinformatics, pp. 1–4. IEEE (2012)
15. Martínez, D., Lawson, J.L., Molina, J.P., García, A.S., González, P., Vanderdonckt, J., Macq, B.: A framework to develop VR interaction techniques based on open interface and AFreeCA. In: IFIP Conference on Human-Computer Interaction, pp. 1–18. Springer, Berlin, Heidelberg (2011)

16. García-Hernández, C., Sánchez-Álvarez, E.J., Ubieto-Artur, P., Huertas-Talón, J.L.: Graphical tools for helping firefighters in victim rescues. Assessment during a live fire training program. Saf. Sci. **114**, 105–113 (2019)
17. Xi, M., Smith, S.P.: Simulating cooperative fire evacuation training in a virtual environment using gaming technology. In: 2014 IEEE Virtual Reality (VR), pp. 139–140. IEEE (2014)
18. Xu, Z., Lu, X.Z., Guan, H., Chen, C., Ren, A.Z.: A virtual reality based fire training simulator with smoke hazard assessment capacity. Adv. Eng. Softw. **68**, 1–8 (2014)
19. Itamiya, T., Tohara, H., Nasuda, Y.: Augmented reality floods and smoke smartphone app disaster scope utilizing real-time occlusion. In: 2019 IEEE Conference on Virtual Reality and 3D User Interfaces (VR), pp. 1397–1397. IEEE (2019)
20. Viant, W., Purdy, J., Wood, J.: Serious games for fire and rescue training. In: 2016 8th Computer Science and Electronic Engineering (CEEC), pp. 136–139. IEEE, (2016)
21. Houvouras, A.J., IV., Harvey, M.T.: Establishing fire safety skills using behavioral skills training. J. Appl. Behav. Anal. **47**(2), 420–424 (2014)
22. Querrec, R., Chevaillier, P.: Virtual storytelling for training: an application to fire fighting in industrial environment. In: International Conference on Virtual Storytelling, pp. 201–204. Springer, Berlin, Heidelberg (2001)

Chapter 68
Mobile Technologies for Ageing Population with Dementia in Singapore

William Siew, Belinda Yuen, and Arlindo Silva

Abstract Prevailing literature has shown the relevance of information to caregiving of persons with dementia. Using the case of the newly introduced Dementia Friends app in Singapore, this research examines caregivers' awareness of the availability of information resources and support services and their perception of existing app features, barriers to use and the information and support services that they most want for their caregiving. The online survey findings of 69 respondents promote the need for a systems approach towards supporting the households of persons with dementia. Active support networks can support caregivers to cope with day-to-day demands, and future app development needs to include more stakeholder discussions to address the complex needs of caregiving of persons with dementia and experiment solutions that can connect them to doctors, experienced caregivers and friends when is needed.

68.1 Introduction

One in 10 persons aged 60 and above is estimated affected by dementia in Singapore [12]. More than 100,000 individuals (spouse, children, siblings, relatives and friends) will become primary or secondary familial caregivers of persons with dementia. Despite increasing provision of community networks and information resources to support persons with dementia and caregivers, the effectiveness of support for persons with dementia and their caregivers is generally between average and below average

W. Siew (✉)
Singapore University of Technology and Design (SUTD), Singapore, Singapore
e-mail: william_siew@mymail.sutd.edu.sg

B. Yuen
Lee Kuan Yew Centre for Innovative Cities, SUTD, Singapore, Singapore

A. Silva
Engineering Product Development Pillar, SUTD-MIT International Design Centre, Singapore, Singapore

A. Chakrabarti et al. (eds.), *Design for Tomorrow–Volume 2*, Smart Innovation, Systems and Technologies 222, https://doi.org/10.1007/978-981-16-0119-4_68

[3]. This research seeks to examine caregivers' needs and barriers to accessing information and support services through mobile technologies, in particular, the Dementia Friends app. Three research questions are addressed,

- What information and support services do family members most need in their caring of persons with dementia?
- How did the Dementia Friends app help caregivers to manage their day-to-day demands of care and other family/personal tasks of work and life?
- How can mobile technologies support the needs of caregiving?

68.2 Literature

Prevailing literature has shown the complexity of caregiving and the importance of information resources to supporting families of persons with dementia [9]. Awareness of available information and support services appears lacking in Singapore even though the accessibility of information is key in determining the effectiveness of caregiving support services [5]. Information is central in self-management when caring for persons with dementia [4], reducing society's stigma towards dementia [8], handling caregivers' stress, burden of care, emotional support and psychosocial needs [11], and improving the quality of life of caregivers and persons with dementia [10].

Much has been written about the use of technologies to support caregiving, which usage depends on the 'ease of use, stability and flexibility of technology, importance of privacy, and confidentiality' [5, 13]. Since 25 October 2018, Singapore has launched the Dementia Friends app following the introduction of Dementia-Friendly Singapore (DFSG) and dementia-friendly communities in 2016. Current statistics have yet to show an active community of support and awareness of dementia for caregivers of persons with dementia [3]. There is no local research on the barriers to access informational resources and support services or the design of mobile solutions for familial caregivers. Much of the existing research is focused on physical care [7] and societal stigma towards dementia [8]. This research aims to address this knowledge gap.

68.3 Methodology

In the midst of COVID-19, an online survey was conducted from 2 March 2020 to 10 May 2020 to ascertain caregivers' adoption, perception and satisfaction with the Dementia Friends app. The survey link was sent to caregivers on the Alzheimer's Disease Association (ADA) database and posted on the Dementia-Friendly Singapore (DFSG) website and Facebook page. A convenience sample size of 300 was proposed but perhaps in view of the COVID-19 situation, only 69 respondents completed the survey; 49 respondents from ADA's caregiver database and 20 respondents through the DFSG website and Facebook page.

68.3.1 Method of Analysis

Survey data were analysed using R and Excel to study participants' awareness, usability and effectiveness of the app. The guiding framework for data analysis is summarised in Fig. 68.1. Three main analysis were conducted by profiling the caregivers to household typology and their characteristics, identifying their perceptions and opinions of the Dementia Friends app, reasons for using or not using of the app as well as understanding the ways that non-users seek informational resources and social support and the barriers and opportunities to meet caregiving needs.

68.4 Findings

The sample is predominantly female (81%). While 57 of 69 respondents have assessed themselves as digitally resourceful (83%), 20 respondents are app users (29%), and 49 are not (71%). Excluding 4 professional caregivers, 18 of 65 respondents are from single caregiver households (26%), and 47 are from multiple caregiver households (68%). The profile of the respondents is summarized in Fig. 68.2.

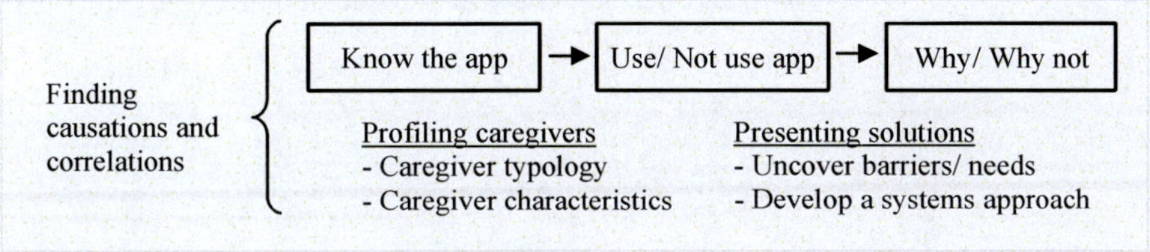

Fig. 68.1 Method of analysis for survey findings

Are you digitally resourceful? Are you registered in the Dementia Friends app?

	Not resourceful	Resourceful		Not resourceful	Resourceful
Female	12	44	**App user**	3	17
Male	0	13	**Non-user**	9	40
	12	**57**		**12**	**57**

Fig. 68.2 Profile of respondents from the survey

68.4.1 Delivery of Information and Support Services

To the question of whether they have heard about the Dementia Friends app, 40 of them have not heard, and 32 are from multiple caregivers household. The number of app user/non-users (i.e. registered/not registered in app) is shown in Fig. 68.3.

Usage of the Dementia Friends App
When asked about their usage of the app, 10 respondents indicated that they have heard about the app but do not use the app. 19 have heard and use the app, and one has not heard but uses the app. 7 of 10 who have heard but do not use are retired/unemployed/homemaker (70%). Most of them prefer searching the Internet, going to training programmes or seeking advice from organisations/institutions.

The two most frequently mentioned reasons for non-use of the app are: not aware of the app and/or no time to explore app (Fig. 68.4). The majority, 29 of 38 of those who said 'not aware' or 'no time', are digitally resourceful (76%), while 9 of 15 who have 'no time' look to other caregivers of persons with dementia for advice (60%).

One-third (5 of 15) of those who have 'no time', said that they are not likely to use the app. Half (8 of 15) of the non-users who have 'no time' are working full-time/freelance. The remaining 7 respondents are homemakers/retired/unemployed, and they use other means to seek help. 15 of 49 non-users who mentioned 'no time' also stated that app usage may cause app fatigue, app information is repetitive, and/or information can be searched on the Internet.

Have you heard of the Dementia Friends app?
Are you registered in the Dementia Friends app?

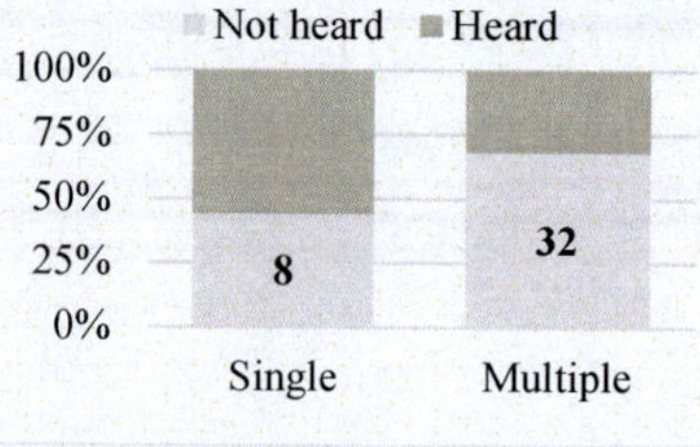

	Single caregiver household	Multiple caregivers household
App user	5	12
Non-user	13	35
	18	**47**

Fig. 68.3 Using/not using the Dementia Friends app

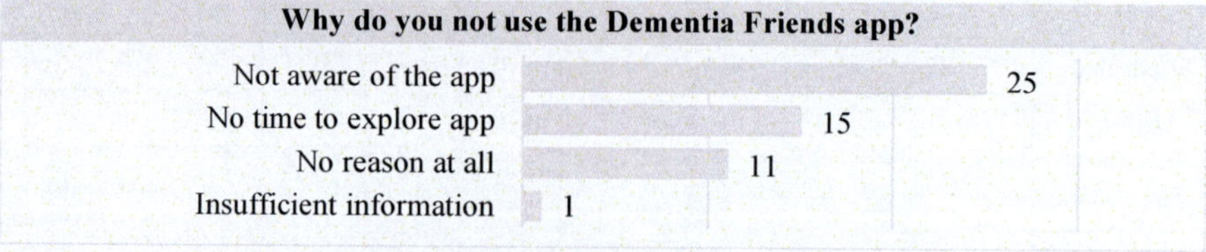

Fig. 68.4 Reasons for not using the Dementia Friends app

Figure 68.5 indicates that non-users would seek advice from doctors/nurses (76%) while app users would go to other caregivers of persons with dementia for advice (80%). This lends support to the importance of social networks in the management of care by caregivers of persons with dementia [14].

When looking into the profile of the caregivers (Fig. 68.6), singles are observed to more likely approach other caregivers of persons with dementia and colleagues for advice (41%) when compared to married respondents (25%). Married respondents are more likely to approach their family and relatives (25%) than singles (13%).

Among app users, almost half of them have not used the Dementia Friends app for more than a month (45%). When asked why, 9 of them mentioned insufficient information, and they would look to other platforms (e.g., the Internet and social media). SOS contacts, bite-sized videos (instead of informational notes) and coping strategies (e.g., exercise tips, cognitive activities and community activities) are their ideas for improvement. Generally, app users use the app to gain knowledge and resources, give support to others, find immediate help and receive updates on news/events (Fig. 68.7).

Awareness of Dementia-Friendly Communities

Even though dementia-friendly communities are being developed, 56 of 69 total respondents (81%) are not aware of the existence of dementia-friendly communities even when 16 of them (23%) are residing within these districts. This suggests an opportunity to re-examine the information and communication practice as more dementia-friendly communities are planned.

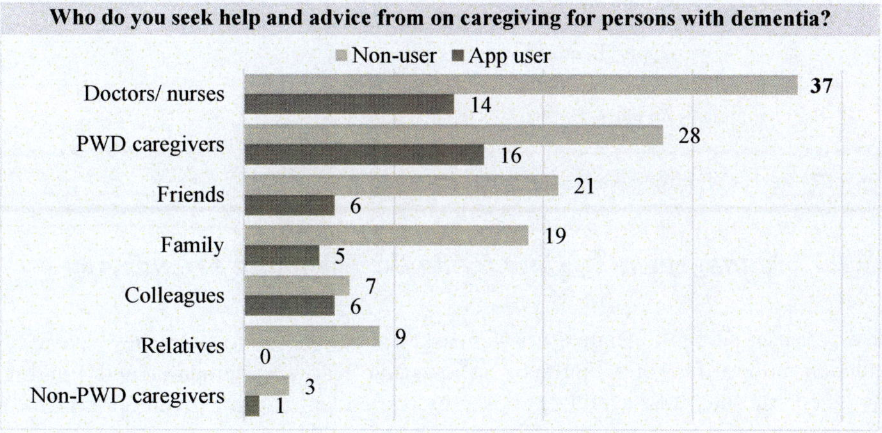

Fig. 68.5 People whom non-users/users of the app seek advice on caregiving

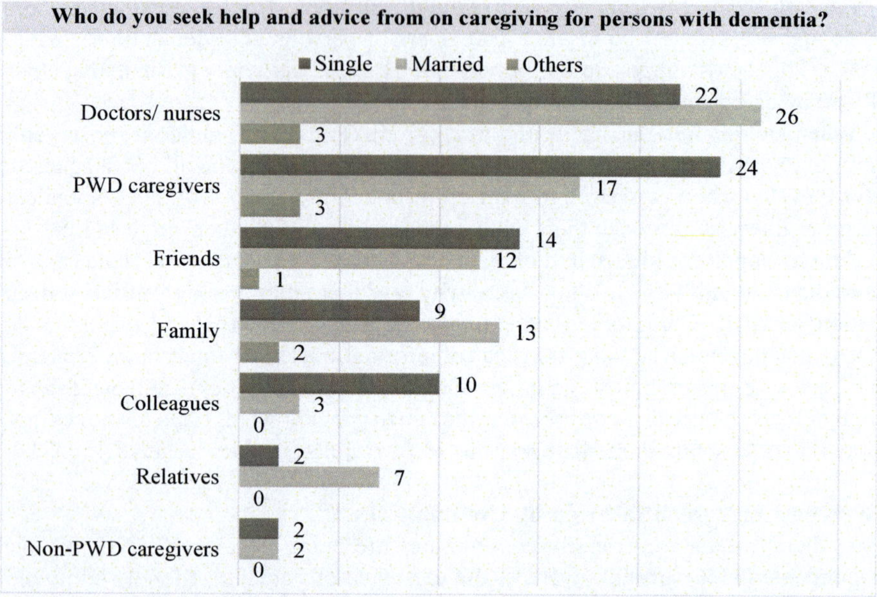

Fig. 68.6 People whom single/married respondents seek advice on caregiving

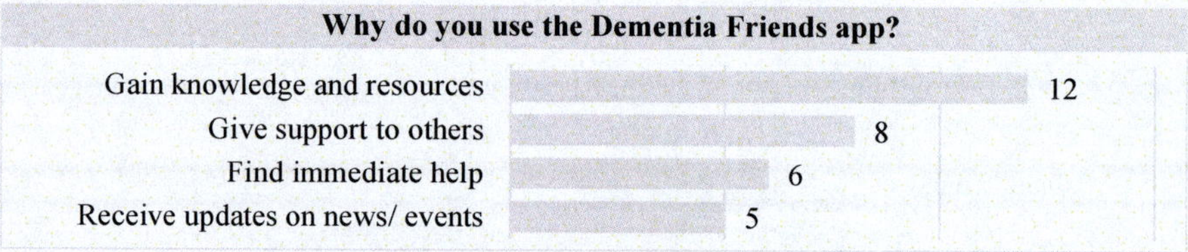

Fig. 68.7 Reasons for using the Dementia Friends app

68.4.2 *Management of Care Using the Dementia Friends App*

To assess user perception, app users were asked about the top features they have used and their ratings of the app. Figure 68.8 shows that ‘Knowing dementia’ and ‘Finding my loved one’ are most useful as chosen by respondents from a given option list of 13 features.

Respondents used the app mainly for information resources, guides and tips. Only 55% of app users agreed or strongly agreed that the app information is sufficient/the app meets their needs, indicating an opportunity for improvement (Fig. 68.9).

Figure 68.10 reinforces the findings from Figs. 68.8 and 68.9 that community support and digital resources are helpful. However, their dislikes indicate that the app is lacking features such as geo-tagging of nearest dementia go-to points, caregiver forums, bite-sized videos, and other languages. These present opportunities to

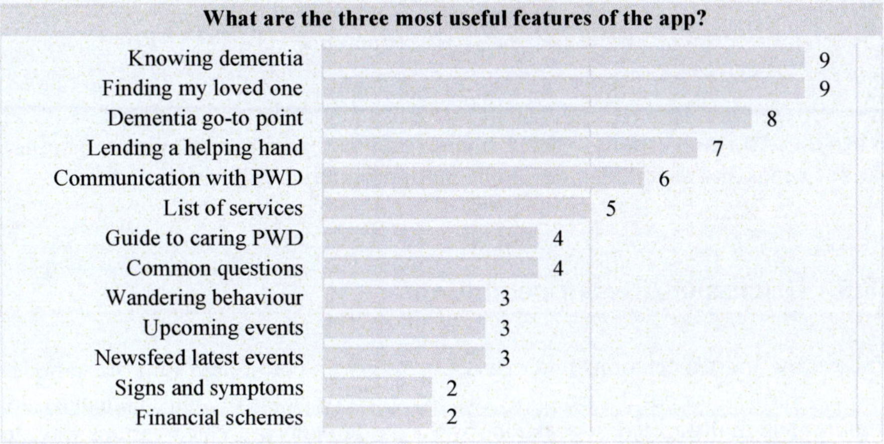

Fig. 68.8 Most useful features of the Dementia Friends app

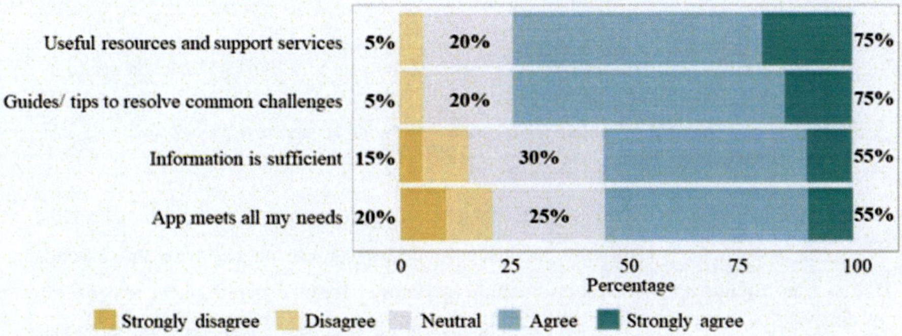

Fig. 68.9 Ratings of users on Dementia Friends app

What do you like/ dislike about the app? (Grouped by similar themes)	
Likes	**Dislikes**
Communal support to find PWD (9)	Lacking features (5)
Resources, news and events (8)	User interface not intuitive (3)
User-friendliness/ easy to use (4)	Content is shallow/ stale (2)

Fig. 68.10 Users' likes and dislikes about Dementia Friends app

provide more useful features as 40 of 49 of non-users are digitally resourceful (80%), and 28 of 40 are households with multiple caregivers (70%).

68.4.3 Mobile Technologies for Social Support and Digital Networks

When asked for suggestions to better support their caregiving, respondents identified various needs and technology development opportunities (Fig. 68.11).

68.5 Discussion/Recommendation

Uncovering the expectations and needs of caregivers could help improve mobile-based solutions for implementation [6]. As caregivers needs and challenges are complex and multifaceted, suggestions by the participants presented various ways to look at their needs and challenges as different part of the whole system to solve them. Hence, this paper proposes a systems approach through the online survey findings [2].

What type of information and support services do you look for/ wish to see in a mobile app to better support your caregiving journey for persons with dementia? Are there any other features you wish to see in the app?	
Theme	**Identified needs**
Virtual assistant/ chat-bot	Online forums/ chat groups/ support groups/ speakers to seek advice. Help from volunteer organisations for managing stress and demands.
Digital information/ resources	Medical information and knowledge on different stages of dementia. Tips to deal with behavioural issues, self-care and coping strategies.
Digital journal/ self-help tools for caregivers	Live chat with doctors, volunteers or counsellors after office hours. Get referrals or search for service providers by location. Online booking of services, e.g., respite care.
Digital activities and online learning	Online activities/ videos/ games other than informational guides. Courses for familial/ professional caregivers and domestic helpers.
E-application for financial support	Easy application process for financial assistance/ subsidy. Voice recognition for easy authentication, transaction and use. Other language selections for the app.
Assistive technology and bots for caregivers and PWD	Accompaniment and reminder sensors to help with daily routines. Pain and behavioural management, SOS contact to seek immediate support, e.g., emergency cases, mobility issues, etc. Assessment tools, modified items, e.g., appliances, etc.

Fig. 68.11 Themes for future work

	Single caregiver household	Multiple caregivers household
Digitally resourceful	1. Awareness of Dementia Friends app	2. Use of mobile app for caregiving support
Not digitally resourceful	3. Social support via digital networks	4. Awareness of dementia-friendly community

Fig. 68.12 Typology of familial caregivers to address needs

68.5.1 A Systems Approach

The study revealed that barriers to access for non-users of the Dementia Friends app were due to low awareness (51%) and no time to explore the app (31%). Yet, 80% of those who do not use the app are digitally resourceful, and most of them expressed that they are unlikely to download and use a mobile app unless it solves their direct needs. The survey responses further suggest that caregivers are experiencing information overload and app fatigue. They point to some of the gaps between provision and usage, and the dissemination of relevant information to meet caregivers' needs.

It is important to recognise that the caregivers are not a homogenous group. Resources should also be directed towards identifying and matching the functions/features that fit caregivers' needs, such as connecting them to experienced caregivers at different stages of their caregiving journey. Caregiver profiling plays a significant part in enriching the caregiving journey and can be a game changer if done right [1]. A systems approach is recommended to uncover possible enhancements to the existing app as well as the gaps in the current caregiving system architecture, policies and procedures to bring forth practical implementations that enable users to have the specific information and support services that they need at any point in time.

Learning from our findings, caregivers could be divided into single caregiver and multiple caregivers households and according to their digital resourcefulness before identifying the logical approach for each household types (Fig. 68.12).

68.5.2 Streamline-Enhance-Add-Test (SEAT) Matrix

A further tool to improve the degree of fit between technology and needs is to apply the SEAT decision matrix (Fig. 68.13). This matrix seeks to clarify the plausible recommendations to streamline, enhance, add and test using mobile technologies in meeting the needs of caregivers identified in Sect. 68.4.3.

	Does not exist*	Exist
High Need	Add	Enhance
Low Need	Test	Streamline

- Streamline: E-application for financial support
- Enhance: Digital information/ resources, Digital activities and online learning.
- Add: Digital journal/ self-help tools for caregivers, Virtual assistant/ chatbot.
- Test: Assistive technology and bots for caregivers and PWD.

Fig. 68.13 Mapping the needs analysis of caregivers using SEAT matrix

68.6 Limitations

There might be a significant proportion of caregivers who are not users of the Dementia Friends app and are not captured in an online survey. By its nature, online survey excludes those who are not technologically inclined. As the views of non-users are important to understanding the barriers to technology access and adoption, future efforts could expand the research design and methodology to reach this group of participants. In the present study, due to COVID-19 movement restriction, it is not possible to conduct methods other than the online survey. Future research can involve different profile groups of caregivers and in-depth analysis to their psychological and social needs.

68.7 Conclusion

This paper examined the usability of mobile-based technologies using the case of the Dementia Friends app, in particular, whether it can meet the demands and needs of caregiving of persons with dementia at different stages of a caregiver's journey. It is the first research that investigated the awareness, use and effectiveness of the Dementia Friends app for caregivers in Singapore. The findings revealed that more than half of the 69 online survey respondents mentioned that they have not heard of the mobile app even though most of them are generally digitally resourceful. The low public awareness of the Dementia Friends app, dementia-friendly community and dementia-friendly Singapore suggests that a review of the communication of community support nationwide is needed. More stakeholder discussions could be convened to determine the actual needs, demands and challenges of app usage. Since caregivers' needs are different, a systems approach to using mobile technologies, where solutions are mapped to caregiving needs in the SEAT matrix could offer a useful development framework. Aligning app functions with user needs would strengthen design congruence and better help families and caregivers when they encounter any difficulties during the caregiving journey. The findings reinforced the presence of help-seeking behaviour; caregivers would often seek advice from doctors, experienced caregivers and friends during times when they need information for care management, and more studies are required to assess the effectiveness of social intervention using mobile technologies for caregiving.

References

1. Brown, E., Ruggiano, N., Li, J., Clarke, P., Kay, E., Hristidis, V.: Smartphone-based health technologies for dementia care: opportunities, challenges, and current practices. J. Appl. Gerontol. **38**(1), 73–91 (2019). https://doi.org/10.1177/0733464817723088
2. Cash, B., Hodgkin, S., Warburton, J.: A transformative approach to systems theory in caregiving research. Qual. Soc. Work (2019). https://doi.org/10.1177/1473325017749988
3. Ching, R.: Alzheimer's disease and related dementias. In: Screening Statistics for Singapore (2019). Retrieved 6 Apr 2020, from https://www.screeningstatistics.com/alzheimers/
4. Huis In Het Veld, J.G., Verkaik, R., Mistiaen, P., Van Meijel, B., Francke, A.L.: The effectiveness of interventions in supporting self-management of informal caregivers of people with dementia; a systematic meta review. BMC Geriatr. (2015) https://doi.org/10.1186/s12877-015-0145-6
5. Lim, J., Goh, J., Chionh, H., Yap, P.: Why do patients and their families not use services for dementia? Perspectives from a developed Asian country. Int. Psychogeriatr. **24**(10), 1571–1580 (2012). https://doi.org/10.1017/S1041610212000919
6. Morrisby, C., Joosten, A., Ciccarelli, M.: Do services meet the needs of people with dementia and carers living in the community? A scoping review of the international literature. Int. Psychogeriatr. **30**(1), 5–14 (2018). https://doi.org/10.1017/S1041610217001491
7. Philip, Y., Seng, B.K.: Profiling the Dementia Family Carer in Singapore. Alzheimer's Disease Association of Singapore. (2017). Retrieved 29 June 2020, from https://alz.org.sg/wp-content/uploads/2017/04/Research-Profiling-Dementia-Family-Carer-SG.pdf
8. Rashith, R.: 3 in 4 people with dementia feel lonely and rejected: survey. The Straits Times, 29 Apr 2019. Retrieved 6 Apr 2020, from https://www.straitstimes.com/singapore/3-in-4-people-with-dementia-feel-lonely-and-rejected-survey
9. Setia, M., Islam, A., Thompson, J., Matchar, D.: Stakeholders' perspective on issues and challenges associated with care and treatment of ageing-related cognitive impairment disorders in Singapore. Int. Psychogeriatr. **23**(9), 1421–1432 (2011). https://doi.org/10.1017/S1041610211000846
10. Tay, K.C.P., Seow, C.C.D., Xiao, C., Lee, H.M.J., Chiu, H.F., Chan, S.W.-C.: Structured interviews examining the burden, coping, self-efficacy, and quality of life among family caregivers of persons with dementia in Singapore. Dementia **15**(2), 204–220 (2016). https://doi.org/10.1177/1471301214522047
11. Vaingankar, J.A., Subramaniam, M., Picco, L., Eng, G.K., Shafie, S., Sambasivam, R., Sambasivam, R., Zhang, Y.J., Sagayadevan, V., Chong, S.A.: Perceived unmet needs of informal caregivers of people with dementia in Singapore. Int. Psychogeriatr. (2013). https://doi.org/10.1017/s1041610213001051
12. Vaingankar, J.: IMH's well-being of the Singapore elderly study shows one in 10 elderly in Singapore has dementia. In: IMH Link, p. 6 (2015). Retrieved 18 May 2020, from https://www.imh.com.sg/uploadedFiles/Publications/IMH_Link/IMH%20Link%20Apr%20-%20Jun%202015.pdf
13. Van Boekel, L., Wouters, E., Grimberg, B., van der Meer, N., Luijkx, K.: Perspectives of stakeholders on technology use in the care of community-living older adults with dementia: a systematic literature review. Healthcare **7**(2), 73 (2019). https://doi.org/10.3390/healthcare7020073

14. Yap, W.: Nee Soon South Dementia-Friendly Community. Lee Kuan Yew Centre for Innovative Cities Webpost Notes on Age-friendly and Dementia-Friendly Cities. Note 8 (2019). Retrieved 20 Jan 2020, from https://lkycic.sutd.edu.sg/wp-content/uploads/sites/7/2019/10/Nee-Soon-South-DFC.pdf

Chapter 69
Design for Immersive Experience: Role of Spatial Audio in Extended Reality Applications

Ganesh Kailas and Nachiketa Tiwari

Abstract The incredible growth of extended reality (XR) applications will be leading us to a world beyond our imaginations in the coming decades. Extended reality is an umbrella term that encompasses different categories of immersive technologies like virtual reality (VR), augmented reality (AR), and mixed reality (MR). From the traditional applications like entertainment and training, XR has been spreading its wings into a large number of applications in health care, aerospace, product design and prototyping, e-commerce, workspace productivity, architecture, and building industries. Immersibility of the virtual reality scene into the physical world will be crucial for its acceptance by mainstream industries and future development. In addition to the virtual scene's visual perception, spatial audio is a key feature in designing truly immersive XR. Hearing is the fastest sense of humans, which makes virtual auditory display (VAD) an ineluctable part of any XR application. In this work, the importance of three-dimensional spatial audio in XR applications is explored in the user perspective approach. User experience (UX) is improved to a large extent when the applications make use of spatial audio compared to direction-less hearing experience. Spatial sound has a crucial role in giving information regarding actions in the background and beyond the field of view (FOV), and thus in making proper three-dimensional realism. However, designing user-dependent virtual audio reality is challenging because of its parametric dependence on human anthropometric features. This work also suggests the possibilities of utilizing computer-aided designing (CAD) and computer-aided engineering (CAE) tools in producing personalized virtual audio reality. While the immersive extended reality experience evolves as the next frontier in user experience designing, a sophisticated 3D audio experience will be there at the heart of it.

G. Kailas (✉) · N. Tiwari
Indian Institute of Technology Kanpur, Kanpur, India
e-mail: ganeshgk@iitk.ac.in

N. Tiwari
e-mail: ntiwari@iitk.ac.in

A. Chakrabarti et al. (eds.), *Design for Tomorrow–Volume 2*, Smart Innovation, Systems and Technologies 222, https://doi.org/10.1007/978-981-16-0119-4_69

69.1 Introduction

Interaction with sounds is an essential part of our daily life. The sense of sound elicits the fastest response in humans. But when it comes to user experience designing, typically, the visual counterpart dominates entirely, and the aural ingredient is often overlooked. However, the auditory experience can be just as relevant and informative to users as visuals are, depending on the products under consideration. Brilliant implementation of audio can provide users with a complete experience beyond visual allure. During the second half of the last decade, an immense surge in affordable extended reality (XR) products [1–3], personal assistant tools, navigation tools, and Internet of things (IoT) devices has led toward amelioration of designs and recognizing sound as an essential design component [4–6].

Encompassing virtual, augmented, and mixed reality, XR has been exploring all its possibilities in domains like health sector, entertainment, e-commerce, tourism, aerospace, product design and prototyping, workspace productivity, architecture, and building industries [7–17]. During the unprecedented pandemic situation of COVID-19, there is an increased global demand for XR technologies in many fields. Spatial audio can give an extra edge to these immersible technologies. In spatial audio, sounds can be convincingly placed in 3D space around the user in a virtual environment (VE). The effective user experience of XR is going to be the defining factor in these applications.

For satisfactory experience, products must engage the users. Designing interfaces and products with good auditory experience has started gaining attention recently. About the visual experience of XR, some functional studies have established the design parameters of virtual reality graphical UI [18, 19]. Another work [20] has developed design heuristics for dynamic VR interaction. A recent work [21] has established specific UX guidelines for head-mounted devices (HMD) XR applications. However, studies focusing on audio experiences in XR are absent. In this work, we have conducted a user perspective study on spatial audio experience in 105 users using questionnaires and feedback.

69.2 Spatial Audio in XR

Spatial audio in extended reality applications is the sound with direction and distance information in three dimensions through headphones. This category of spatial audio is called perceptually driven spatial audio. This is produced mainly using two methods: (1) Binaural recording and (2) Head-related transfer function (HRTF)-based reproduction. The binaural recording uses a dummy manikin with microphones in its two ears to produce spatial audio. HRTF is a function which gives information about how the human torso changes the sounds impinging on it before reaching the ear canal. It basically encodes different anthropometric sound localization cues such as interaural time difference, interaural level differences and spectral cues into a single entity. In

HRTF-based methods, typically, audio is recorded and synthesized using HRTF functions. HRTF is an individualized function and requires a high experimental cost to generate. HRTF measured for a particular user with unique anthropometric features is called individualized HRTF. HRTF measured for dummy models or a generic user is called non-individualized HRTF. In most applications, non-individualized HRTFs are used to reduce the cost by trading off accuracy. Localization cues in the HRTF are mixed with directionless audio to produce spatial audio.

Immersion in XR is how much we feel "real" in a virtual environment. In XR environment, simulated directional audio can draw users' attention to different directions and allows users to immerse themselves in space. A constructively designed audio can provide depth and emotions to the whole virtual experience.

69.3 A Study on User Experience of Spatial Audio

According to ISO 9241-210 (2009), user experience is "person's perception and responses resulting from the use and/or anticipated use of a system, product or a service". There is a dearth of heuristics to analyze XR's user experience, especially for the auditory experience. One of the holistic UX definitions says "User eXperience is a consequence of a user's internal state, the characteristics of the designed system and the context (or the environment) within which the interaction occurs" [22]. Structured user experience frameworks are essential for immersive technologies to develop it for better user acceptance.

Importance of *Presence* and *Flow* in immersive environments was explored in different studies [23, 24]. A combined empirical model to measure different scales affecting immersive interfaces has been developed recently [25, 26]. In these works, a unified questionnaire to measure the UX in immersive virtual environments has been proposed with reliable consistency. In our study, we have adapted this questionnaire and concentrated only on the subscales affected by auditory cues. Presence, engagement, immersion, flow, and emotion have been taken as the five subscales. These subscales as listed in Table 69.1 are not independent and are interrelated. Emotion influences flow, and engagement and immersion affect presence. Flow and presence

Table 69.1 Subscales and definitions [26]

Subscales	Definition
Presence	"Sense of being in virtual environment"
Engagement	"Energy in action, connection between the user and its behavioral, emotional and cognitive actions"
Immersion	"Illusion" that "the virtual environment technology replaces the user's sensory stimuli by the virtual sensory stimuli"
Flow	"A pleasant psychological state of sense of control, fun and joy"
Emotion	Feelings of the user in the VE

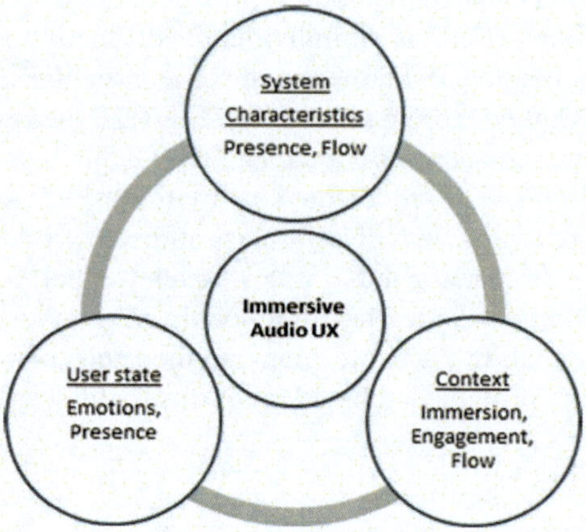

Fig. 69.1 UX scales of spatial audio

are also inter-related. The subscales can measure three components of the UX [22], as shown in Fig. 69.1.

We conducted the study to explore the user experience of spatial audio. Our research questions are (1) Do spatial audio possess a better UX compared to stereo? (2) What is the role of auditory cues in identifying direction? (3) What is the UX of spatial audio if it is considered as a holistic product? In this work, subjective methods (questionnaires and feedbacks) are followed to analyze the UX of spatial audio. The study contains mainly four parts to explore the research questions: (1) Comparison of XR experience with spatial audio and stereo, (2) Comparison of spatial awareness using visual and auditory cues, (3) UX of 360° audio interaction, and (4) Overall UX experience of spatial audio.

All four subjective experiments have been conducted with 105 participants (76 males and 29 females) in the age group of 16–73 years. The average age of the participants was 30.7, with a standard deviation of 9.8. Less than two percent of the participants used hearing aids or underwent any sort of hearing-related surgeries. Everyone used headphones with satisfactory performance characteristics. Feedbacks are also collected as comments to understand the individual perception of technology.

69.3.1 Comparison of XR Experience with Spatial Audio and Stereo

To identify the importance of spatial audio in UX of extended reality, user experience studies are conducted with two music audios without visuals; one is stereo, and the other is 360° spatial audio. Participants listened to the music audios using earphones. Both audios had an equal duration of two minutes. The updated questionnaire (Appendix) is used to assess the user experience and get feedback from

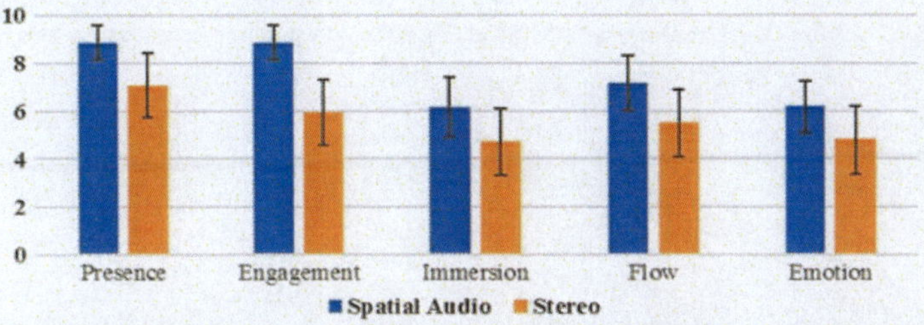

Fig. 69.2 Comparison of UX in VE with 360° spatial audio and normal stereo

the users. A 10-point Likert scale has been used for the evaluation. Participants have filled up the questionnaire after listening to each audio.

Results are obtained for all the five scales for both the stereo and spatial audios. The comparison is shown in Fig. 69.2.

It is clearly visible that all the subscales delivered an improved performance with spatial audio compared to the one with stereo. Participants were able to identify the sound sources around the space, which increased the engagement factor. Immersibility, emotion, and flow factors were also increased due to the active involvement through directional audio. All these make spatial audio as a necessary constituent of extended reality applications. Extended reality without enough deliberation on the auditory virtual environment may undergo an unavoidable reduction in its immersibility.

69.3.2 Comparison of Spatial Awareness Using Visual and Auditory Cues

A short video with a visual target moving around in different directions on a game screen with corresponding audio cues (standard stereo and spatial audio) was shown to the participants. Participants were asked about: (1) The helpful cue to find the target-audio or video, (2) Directions where spatial audio was more helpful than stereo, and (3) Directions where both stereo and spatial audio were similar. The sample game screen is shown in Fig. 69.3.

As shown in Fig. 69.4, the audio was the primary directional cue of 46% of the subjects. Visual cues were used by 36%, and 18% of the subjects used both. This indicates that the users experience exceptional spatial awareness with the directional audio even more than the visual counterpart.

When it comes to the directions in which directional audio helped identify, there is a significant variance in opinions. Technically, generalized HRTF is an important reason for this, and individualized HRTF can make spatial audio better for each user. Adaptation of faster methods to acquire individual HRTF may solve these problems to a great extent.

Fig. 69.3 Game screen sample

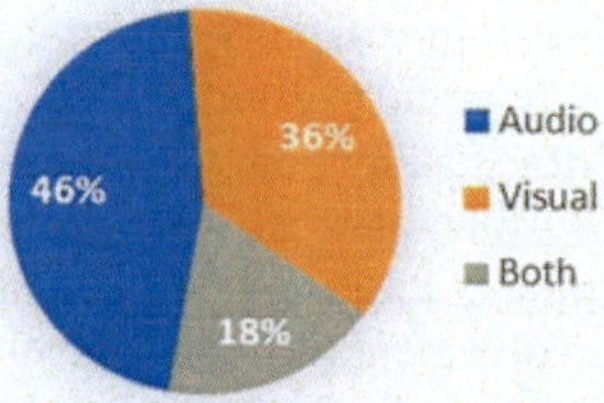

Fig. 69.4 Comparison between visual and auditory cues in identifying source

69.3.3 UX of 360° Audio Interaction

All the participants were made to listen to one interaction audio by John Tucker [27], and their user experiences in an artificial environment were examined. The same questionnaire (Appendix) was used to measure the experience. The user experience subscales are shown in Fig. 69.5. The virtual experience was rated high by participants. This study endorses the importance of directional audio cues to elevate the XR experience.

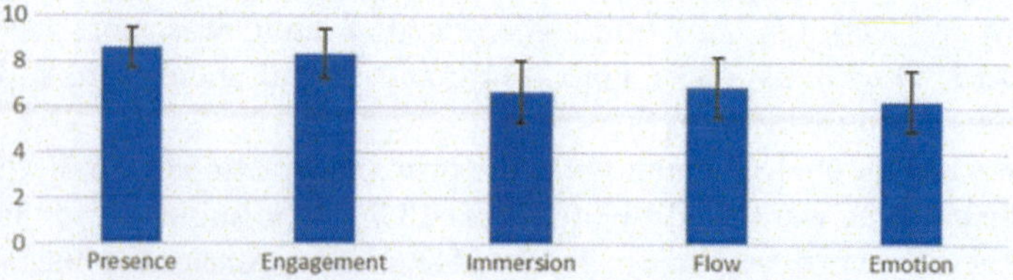

Fig. 69.5 UX subscales for interactive audio

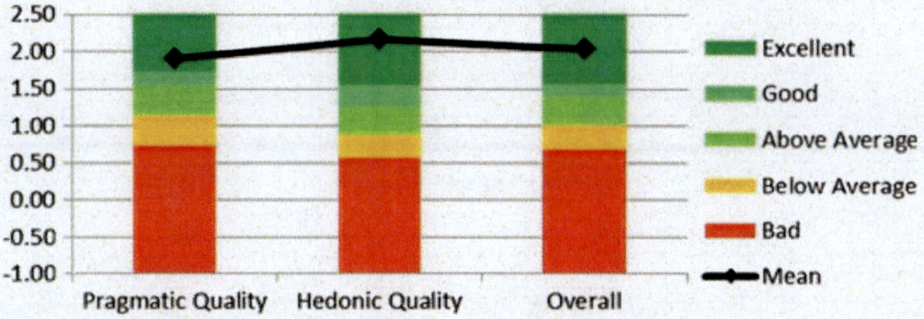

Fig. 69.6 UX performance of spatial audio using S-UEQ

69.3.4 Overall UX of Spatial Audio

Short user experience questionnaire (S-UEQ) [28] is a widely used tool to measure UX. The questionnaire has been adopted here to rate the overall experience of the technology, i.e., spatial audio. S-UEQ allows users to express their experiences of a product or technology through feelings, impressions, and attitudes. A 7-point Likert scale is used in this questionnaire. Two meta-dimensions, viz., pragmatic and hedonic qualities, are measured in S-UEQ. Figure 69.6 shows the overall spatial audio experience, and it appears to be excellent by the S-UEQ benchmarks.

69.4 Discussion

From the above subjective studies, it is very evident that spatial audio added new and more attractive dimensions to the user experience. Positive and negative comments about the experience of spatial audio have been documented. Positive comments from the participants collectively described the better user immersion and involvement they felt listening to spatial audio. Some participants felt motion sickness during some part of the audio, and others felt overuse of the directional audio compared to realistic situations. Two of the participants felt that most of the sounds were heard inside the head. These inputs can provide better insights to UX designers while designing audio in VE.

Technology advancements have also been climbing steps to improve pragmatic quality. These insights can be classified into two: (1) Improvement of technology and (2) Better audio designing.

69.4.1 Insights for Improvements in Technology

Spatial audio is still not a full-fledged technology. As discussed earlier, when it comes to accurate identification of sound sources, generalized HRTF-based spatial

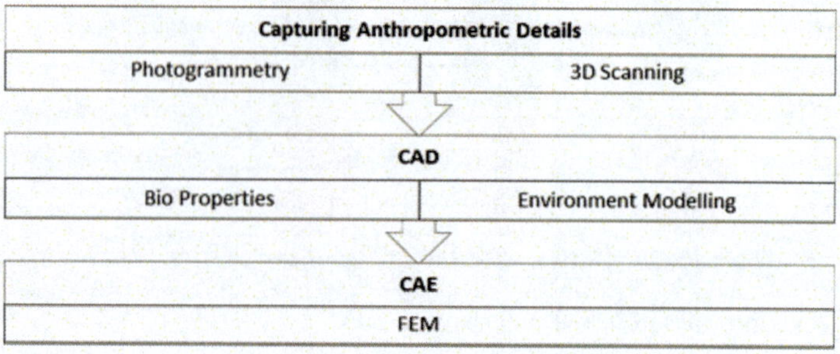

Fig. 69.7 Technical process for generating faster personalized spatial audio

audio has its limitations. HRTF is a personalized function, and it changes with anthropometric features. This difference in HRTF can be a reason for "inside-head" listening experienced by some participants. Experimental measurement of HRTF is tedious and costly. With the recent advancement of computer-aided designing (CAD) and photogrammetric tools, users' head and torso models can be generated quickly. Biomechanical properties can be applied to replicate the acoustical behavior of the human skin and body. Environment modeling is also necessary to incorporate different room acoustic properties. Using simulation methods like finite element analysis (FEA), it is easier to find an acoustic solution for the problem and obtain HRTF for the whole space around the user. The process flow is shown in Fig. 69.7. Individualized HRTFs effortlessly make spatial audio very accurate, and it will be a game-changer in the XR domain.

Consider an application like space exploration for blind people [29]. Our design process will first capture their anthropometric features using photos from different angles, and a 3D model of the user's head can be developed with acoustical properties. FEA can be performed to find the HRTF. This process flow can be employed very fast to develop an individualized HRTF, especially by using cloud services to perform FEA [30]. The HRTF can be utilized to render spatial audio and may be used in mobile applications to give spatial knowledge about different events.

69.4.2 Insights for a Better Spatial Audio Design

Sound experience design in computer interaction tools is mostly underrated. The awful audio designing at the starting of the Internet and computer animation eras also take partial blame. There was an overuse of sounds [31], and muted speakers became very common. We know that user experience is a very subjective measure, but many participants have expressed common reactions. UX designers may keep these in mind while designing audio experiences.

(a) Overuse of spatial audio: Overuse of design elements must be avoided as a general design rule. Similarly, spatial audio can also be considered as a design element. Users prefer simple and effective use of spatial audio. Style guides can be developed in audio UX similar to visual UX. Consistency and cohesiveness of the audio UX are critical.

(b) Environment design: Our natural world is not silent. Ambient noise is always there, and the absence of careful designing of the environment reduces immersibility. Some participants felt discomfort with total silence. Reverberation should also be considered, which is an essential cue for human hearing.

(c) Speed of movements: When sound moved way faster around the head, it was uncomfortable for many participants. In real life, sound rotating around the head is a rare scenario, and our virtual environment should stay close to natural audio. Motion sickness was also experienced for some subjects during the experiment. Designers must pay attention to these details while creating an environment.

(d) Quality of closer sounds: When sounds came very close to the user, most of them preferred high-quality rendering.

The above points have been derived from subjective opinion, which depended on many individual interests. Still, audio experience designers can make use of these insights while designing spatial audio and enrich the immersibility of XR environments.

69.5 Conclusion

Extended reality is going through a fascinating swell of interest from every corner. In this work, the essential role of spatial audio in immersive XR is identified, and some insights into strengthening the technology and UX are presented. As the world moves more toward remote and virtual applications, spatial audio can provide the qualities which XR applications require to surge in a promising direction.

Appendix: Questionnaire of Audio Experience in XR

The questionnaire has been adapted from ref [25, 26].

Items	Subscales
The sense of moving around inside the virtual environment was compelling	Engagement, emotion
I was involved in the virtual environment experience	Engagement
I correctly identified sounds produced by the virtual environment	Presence

(continued)

(continued)

Items	Subscales
I correctly localized (identified location) sounds produced by the virtual environment	Presence
I felt stimulated by the virtual environment	Immersion
I became so involved in the virtual environment that I was not aware of things happening around me	Immersion
I got scared/disturbed by something happening in the hearing environment	Immersion, emotion
I became so involved in the virtual environment that I lost all track of time	Immersion, flow
I felt I was experiencing an exciting moment	Flow, emotion
This experience was giving me a great sense of well-being	Flow
I enjoyed being in this virtual environment	Emotion
It was so exciting that I could stay in the virtual environment for hours	Emotion
The audio bored me to death	Emotion

References

1. Castelvecchi, D.: Low-cost headsets boost virtual reality's lab appeal. Nature **533**, 153–154 (2016). https://doi.org/10.1038/533153a
2. Daydream. https://arvr.google.com/intl/en_in/daydream/. Accessed 6 Oct 2020
3. Solutions—augmentes. https://augmentes.com/solutions/. Accessed 6 Oct 2020
4. Spatial Audio | Google VR | Google Developers. https://developers.google.com/vr/discover/spatial-audio. Accessed 6 Oct 2020
5. Audio—Facebook 360 Video. https://facebook360.fb.com/category/360-audio/. Accessed 6 Oct 2020
6. Spatial Audio—Surrounded by Sound | Delight XR. https://delight-vr.com/spatial-audio/. Accessed 6 Oct 2020
7. Andrews, C., Southworth, M.K., Silva, J.N.A., Silva, J.R.: Extended reality in medical practice. Curr. Treat. Opt. Cardiovasc. Med. **21**, 18 (2019). https://doi.org/10.1007/s11936-019-0722-7
8. Slater, M., Sanchez-Vives, M.V.: Enhancing our lives with immersive virtual reality. Front. Rob. AI **3**, 1–47 (2016). https://doi.org/10.3389/frobt.2016.00074
9. Cipresso, P., Giglioli, I.A.C., Raya, M.A., Riva, G.: The past, present, and future of virtual and augmented reality research: a network and cluster analysis of the literature. Front. Psychol. **9**, 1–20 (2018). https://doi.org/10.3389/fpsyg.2018.02086
10. Fast-Berglund, Å., Gong, L., Li, D.: Testing and validating extended reality (xR) technologies in manufacturing. Proc. Manuf. **25**, 31–38 (2018). https://doi.org/10.1016/j.promfg.2018.06.054
11. Rouse, R., Barba, E.: Design for emerging media: how MR designers think about storytelling, process, and defining the field. In: Nunes, N., Oakley, I., Nisi, V. (eds.) Interactive Storytelling. ICIDS 2017. Lecture Notes in Computer Science, vol. 10690. Springer, Cham. https://doi.org/10.1007/978-3-319-71027-3_20
12. Skult, N., Smed, J.: Interactive Storytelling in Extended Reality: Concepts for the Design, pp. 449–467. Springer, Cham (2020). https://doi.org/10.1007/978-3-030-37643-7_21
13. Guttentag, D.A.: Virtual reality: applications and implications for tourism. Tourism Management **31**, 637–651 (2010). https://doi.org/10.1016/j.tourman.2009.07.003

14. Huang, Y.C., Backman, S.J., Chang, L.L., et al.: Experiencing student learning and tourism training in a 3D virtual world: an exploratory study. J. Hosp. Leis. Sport Tour. Educ. **13**, 190–201 (2013). https://doi.org/10.1016/j.jhlste.2013.09.007
15. Mahboob, A., Husung, S., Weber, C., et al.: Smartphone as a stand-alone device for rendering, visualization and tracking for use during product development in virtual reality (VR). In: Proceedings of the ASME Design Engineering Technical Conference. American Society of Mechanical Engineers (ASME) (2019). https://doi.org/10.1115/DETC2019-97173
16. Kohler, C., Weidner, F., Broll, W.: AR training for paragliding pilots: an investigation of user experience and requirements. In: Proceedings—2019 21st Symposium on Virtual and Augmented Reality, SVR 2019, pp. 92–101. Institute of Electrical and Electronics Engineers Inc. (2019)
17. Mahboob, A., Husung, S., Weber, C., et al.: An Approach for Building Product Use-Case Scenarios in Different Virtual Reality Systems. ASME International (2018). https://doi.org/10.1115/DETC2018-85223
18. Fröjdman, S.: User experience guidelines for design of virtual reality graphical user interface controlled by head orientation input. Skripsi **75** (2016). https://www.diva-portal.org/smash/record.jsf?pid=diva2%3A939381&dswid=6783
19. Park, W., Heo, H., Park, S., Kim, J.: A study on the presence of immersive user interface in collaborative virtual environments application. Symmetry **11**, 1–19 (2019). https://doi.org/10.3390/sym11040476
20. Endsley, T.C., Sprehn, K.A., Brill, R.M., et al.: Augmented reality design heuristics: designing for dynamic interactions. Proc. Hum. Factors Ergon. Soc. **61**(1), 2100–2104 (2017). https://doi.org/10.1177/1541931213602007
21. Vi, S., da Silva, T.S., Maurer, F.: User experience guidelines for designing HMD extended reality applications. In: Lamas, D., Loizides, F., Nacke, L., Petrie, H., Winckler, M., Zaphiris, P. (eds.) Human-Computer Interaction – INTERACT 2019. INTERACT 2019. Lecture Notes in Computer Science, vol, 11749. Springer, Cham. https://doi.org/10.1007/978-3-030-29390-1_18
22. Hassenzahl, M., Tractinsky, N.: User experience—a research agenda. Behav. Inf. Technol. **25**, 91–97 (2006). https://doi.org/10.1080/01449290500330331
23. Takatalo, J.: Presence and Flow in Virtual Environments: An Explorative Study (2002). https://scholar.googleusercontent.com/scholar.bib?q=info:n449vBtFmxQJ:scholar.google.com/&output=citation&scisdr=CgVLfSDPEO_IrNB8B9Y:AAGBfm0AAAAAYC95H9bruJ03qVuLURQZNmKMa5FZ30NG&scisig=AAGBfm0AAAAAYC95H1bEFYIC3TeXu4vh31JoaujB_0Za&scisf=4&ct=citation&cd=-1&hl=en
24. Witmer, B.G., Singer, M.J.: Measuring Presence in Virtual Environments: A Presence Questionnaire (1998). https://doi.org/10.1162/105474698565686
25. Tcha-Tokey, K., Loup-Escande, E., Christmann, O., Richir, S.: A questionnaire to measure the user eXperience in immersive virtual environments. ACM Int. Conf. Proc. Ser. **10**(1145/2927929), 2927955 (2016). https://doi.org/10.1145/2927929.2927955
26. Tcha-Tokey, K., Christmann, O., Loup-Escande, E., Richir, S.: Proposition and validation of a questionnaire to measure the user experience in immersive virtual environments. Int. J. Virtual Reality 16, 33–48 (2016). https://doi.org/10.20870/ijvr.2016.16.1.2880
27. John Tucker Tucker Zone 1. In: Youtube. https://www.youtube.com/watch?v=3txhT2ncNOU
28. Hinderks, A.: Items of the short version of the User Experience Questionnaire (UEQ). https://doi.org/10.9781/ijimai.2017.09.001
29. Fg, B., Picinali, L.: Spatial audio applied to research with the blind. Adv. Sound Local. (2011). https://doi.org/10.5772/15206
30. Ari, I., Muhtaroglu, N.: Design and implementation of a cloud computing service for finite element analysis. Adv. Eng. Softw. **60–61**, 122–135 (2013). https://doi.org/10.1016/j.advengsoft.2012.10.003
31. Albers, M.C.: Auditory cues for browsing, surfing and navigating. Proc. ICAD **96**, 85–90 (1996). https://hdl.handle.net/1853/50793

Chapter 70
An Augmented Reality Application to Teach Human Anatomy to Secondary School Students

Arkoprobho Debnath, Utkarsh Pathak, and Pankaj Badoni

Abstract Students start learning about human anatomy starting from standard VI (i.e., between ages 10–15 years) and gain a basic idea about all of the important organs and their functions. However, a quick survey reveals that students find it difficult to visualize how human organs look from inside and outside. They also lack an understanding of the basic functions of these organs. Further, as most of the instructional medium is based on a two-dimensional static paper-based medium, it lacks the look and feel of actual organs. To tackle this problem augmented reality (AR) technology was utilized. AR is an interactive experience of a real-world environment that is enhanced by computer-generated perceptual information, sometimes across multiple sensory modalities. AR is an easy and feasible solution through which we can see and understand objects which cannot be seen in the real world. The proposed interactive AR application helps students learn about human anatomy in detail. The application gives the users the option to view six different organs of the human body along with their internal parts in real time. Information about each organ and its classified parts is also provided by the application. The application helps the students to learn and visualize the human organs properly. This paper presents a qualitative analysis study of the application which can help the students as a learning aid.

A. Debnath · U. Pathak (✉) · P. Badoni
Department of Virtualization, Computer Science and Engineering, University of Petroleum and Energy Studies, Dehradun, India
e-mail: pathakutkarsh2615@gmail.com

A. Debnath
e-mail: debnath.arkoprobho@gmail.com

P. Badoni
e-mail: pbadoni@ddn.upes.ac.in

A. Chakrabarti et al. (eds.), *Design for Tomorrow–Volume 2*, Smart Innovation, Systems and Technologies 222, https://doi.org/10.1007/978-981-16-0119-4_70

70.1 Introduction

The utilization of augmented reality (AR) in training is a significant subject of exploration. AR empowers the expansion of virtual items into genuine conditions to encourage constant association. The utilization of AR has gotten increasingly open as it no longer requires particular hardware and may effortlessly be utilized on cell phones. The vast majority presently own cell phones, and the utilization of these gadgets has expanded, in this manner empowering more prominent access to AR. The applications for portable AR in instruction are expanding quickly, and the possibility of versatile AR has expanded because of advances in portable innovation. AR portable applications are accessible for a few regions of instruction, and training-related AR applications are currently more regularly found on cell phones.

School is the primary stage where students learn science. Even now, with all the advances in instructive advances, the educating and learning forms in schools are as yet dependent on regular techniques. Although science course books are given in schools, beneficial learning materials are as yet required because some powerful ideas are hard to clarify in the conventional strategy for instructing. In this manner, a propelled learning material with innovation is required. Currently, teachers use the feature of videos and some applications which help the students understand the concepts. But, some concepts like human anatomy and chemistry can still not be understood properly by the students due to the lack of visualization of the organs.

Augmented reality is an innovation that can help in the viable learning process. The accompanying paper will talk about an AR Application which can assist the teachers and students with teaching and learning human anatomy all the more viably.

70.2 Literature Survey

70.2.1 Augmented Reality

AR combines real and virtual worlds, supplementing the real world with computer-generated virtual objects in real time [1, 2]. According to one of the most commonly accepted definitions, AR is said to be a technology that has three key requirements: combining real and virtual objects in a real environment, aligning real and virtual objects with each other, and real-time interaction [3]. Figure 70.1 shows Milgram's

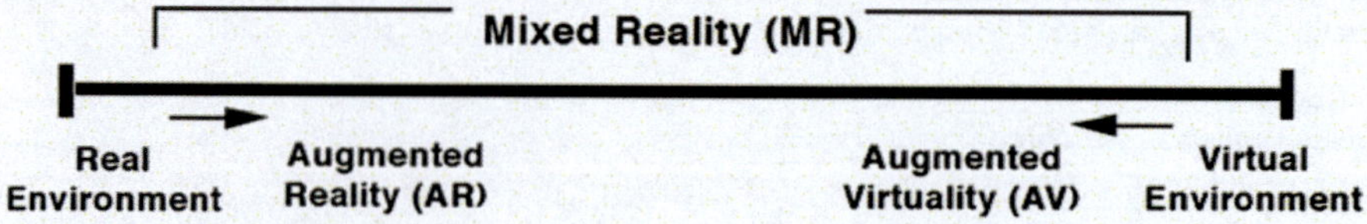

Fig. 70.1 Milgram's mixed reality continuum

mixed reality continuum which is a taxonomy of how real and virtual elements may be combined. AR lies closer to the real environment end of the continuum as can be seen in Fig. 70.1 [4].

In the case of mobile AR, technology involves the addition of digital elements to the real world through a smartphone camera. Examples of mobile AR applications include Pokemon GO, which is a location-based mobile AR game that enables users to catch various digital Pokemon creatures around their area.

70.2.2 SDK for Augmented Reality

Vuforia SDK is an AR software development kit for mobile devices launched by Qualcomm. It utilizes computer vision technology to recognize and capture planar images or 3D objects in real time and permits developers to place virtual objects through the viewfinder of the camera and adjusting the position of objects in the background of the camera. Vuforia SDK supports types of 2D and 3D objects including multiple target configurations, images with fewer symbols, and frame tags. There is an added function in the SDK. It takes advantage of virtual buttons to detect localized occlusion. Moreover, it can select and reconfigure the target image in real time and create a target set according to the scheme [10, 11] (Fig. 70.2).

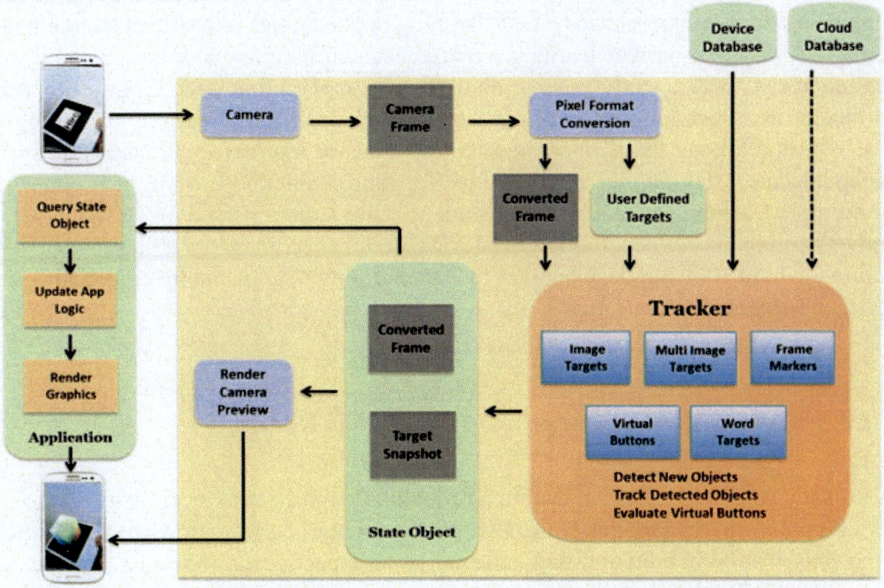

Fig. 70.2 Workflow for Vuforia

70.2.3 Augmented Reality in Education

Since its introduction, augmented reality (AR) has been shown to have good potential in making the learning process more active, effective, and meaningful. This is because its advanced technology enables users to interact with virtual and real-time applications and brings natural experiences to the user. Also, the merging of AR with education has recently attracted research attention because of its ability to allow students to be immersed in real experiences [5].

The educational value of AR is closely linked to how it is designed, implemented, and integrated into formal and informal learning environments. An important consideration is how AR technologies support and afford meaningful learning. Considering AR as a concept rather than a certain type of technology would be productive for educators [4]. The involvement of educators is important to facilitate the development of favorable AR applications for teaching, which increases the potential for AR to be incorporated in education [6]. AR applications have been developed in many areas of education.

70.2.4 AR in Healthcare Education

The effective development of healthcare competencies poses great educational challenges. A possible approach to provide learning opportunities is the use of augmented reality (AR) where virtual learning experiences can be embedded in a real physical context [7]. This study showed that AR was applied to a wide range of topics in healthcare education. Furthermore, acceptance for AR as a learning technology was reported among the learners and its potential for improving different types of competencies. AR provides opportunities for more authentic learning and appeals to multiple learning styles, providing students with a more personalized and explorative learning experience. It is believed AR will only become more ubiquitous in the future medicine [8]. But despite all of this, the number of applications for healthcare to teach students is quite limited.

70.2.5 Speech Recognition in Augmented Reality

Automatic speech recognition has been brought into normal life recently by services like Apple's Siri or Google Now. People are more accustomed to speaking into their phone or computer, and they have come to expect accurate results from such encounters [12].

Application development that combines AR technology and speech recognition is expected to improve the productive vocabulary method as it reinforces the connection between visual scripts (orthography) and audio (phonology) in the reading words.

A combination of speech recognition and AR can improve productive vocabulary methods through AR-based applications [14].

Today, implementing a speech recognition service is fairly straight forward through the use of software APIs such as Microsoft's .NET speech recognition library. Such a library provides many data structures, such as Grammars, that allow developers to define words of interest, accuracies, and strings of words that should be recognized. The use of speech recognition software for an input module for this project was a clear requirement from the beginning [13].

70.2.6 System Usability Scale (SUS)

The System Usability Scale (SUS) was used to gauge the response of the test users after using the Human Anatomy application. Based on the SUS, scale changes were made to the application.

The SUS survey included requests for demographic information from users: their name, their company, their job role, the software being evaluated, the software version, date of the user's evaluation, duration of the evaluation, and the user's experience using the software. The survey then provided the following ten standard statements with five response options (Five-point Likert scale with anchors for strongly agree and strongly disagree) [15]:

1. I think that I would like to use this system frequently.
2. I found the system unnecessarily complex.
3. I thought the system was easy to use.
4. I think that I would need the support of a technical person to be able to use this system.
5. I found the various functions in this system were well integrated.
6. I thought there was too much inconsistency in this system.
7. I would imagine that most people would learn to use this system very quickly.
8. I found the system very cumbersome to use.
9. I felt very confident using the system.
10. I needed to learn a lot of things before I could get going with this system.

There are several potential benefits of incorporating the System Usability Scale into the library Web site and system usability testing. The SUS has been in use for approximately 30 years and is a reliable, tested tool for evaluating a wide range of products and systems. The data it provides has a variety of uses. It can be used as a benchmark to measure how changes to a system or product are received by users. Or, it can be used to quantify user reaction to two alternatives, such as two different versions of the same Web page or two different user interfaces, so that they can be compared for decision-making purposes. When used in conjunction with data gathered from other sources, such as usability testing, it can be a helpful supplemental tool for researchers who need to collect a source of big picture quantitative data that can be easily communicated to library administration and stakeholders [16].

70.3 Related Augmented Reality Applications

Today's technological advancement has led to various reforms in teaching and learning systems in the classroom. The tendency for children to use smartphones has motivated researchers to study the advantages of using smartphones in children. Nowadays, AR technology in the teaching and learning environment has attracted the interest of the community as it can attract children's attention in learning environments. There are several recent AR applications in the educational field to see the effects and benefits to the students (Table 70.1). The use of AR applications is believed to be of interest to students to replace conventional and static learning environments with a more edutainment environment.

In today's life, technologies are very important and AR has become one of the most emerging technologies and started to gain attention among society. AR's ultimate goal is to provide better management and to access information wherever and whenever using a combination of interactive real-world and computer-generated world or virtual world in a coherent space. AR is a new technology that has the potential to

Table 70.1 Current augmented reality applications

AR application	Domains/uses of AR	Features of application
AR Solar System	(1) Solar system visualized in AR (2) Different planets, sun and asteroid belt can be visualized in detail	(1) Information on every planet given in great detail (2) Marker-based application
360ed's Element AR	(1) AR Technology for visualizing atomic elements (2) Periodic table is shown in 3D according to various options like ionization energy, atomic number, electronegativity, etc	(1) FlashCards used as markers for every element (2) A quiz can be taken according to the information learned in the application
Animal 4D+	(1) Animals are visualized in AR (2) Each animal produce their respective sound	(1) Information on every animal is provided (2) Flashcards of each animal are used as markers for augmenting animal
AR Tattoo	(1) Artworks and tattoos visualized using AR	(1) User-made simple markers on their body to display AR Tattoo (2) Custom made tattoo made by the user can also be displayed
AR Human anatomy	(1) Organs visualized in AR	(1) Showcases only model of the organ (2) Single marker-based application
Asthi AR - Human Anatomy in Augmented Reality	(1) The entire human body along with organs can be visualized in AR	(1) No information about the organ is given (2) Markerless application with no stability
AR Anatomy 4D+	(1) Anatomy atlas with AR elements	(1) Multiple markers for the entire application (2) All organs of the human body can be augmented (3) Interactivity present but no information provided

be used in today's world of education and it has become more widespread and practical. Researchers have found that "AR-based education media could be a valuable and attractive additional material to education in the classroom and overcome some of the limitations of text-based methods, allowing students to absorb the material according to their preferred learning style." AR can also be used to improve traditional and static learning content in the classroom. This will appeal to children with additional media content such as audio, video, graphics, 3D objects, and more.

Some of the well-known AR applications and their features are listed below.

70.4 Methodology

The causal research-based approach was used before making the prototype to understand the needs of both the students and teachers. A basic set of questionnaires was prepared for both the teachers and students differently. The target participants were students studying in secondary classes (from classes VI to X). Teachers who participated in the research were the teachers who taught biology to secondary classes.

Teachers and students were divided into different groups. Both the groups were given a defined set of questionnaires and were also asked to fill the System Usability Scale (SUS) questionnaire after using the application. A pre-usage and post-usage questionnaires were used as instruments for data collection. The questionnaires were in the form of a five-point Likert scale along with the System Usability Scale.

Questions related to the pre-usage questionnaire for the students were related to the difficulties they face during classes while studying the human body and how they generally overcome them. They were asked about the effectiveness of notes, textbooks, and the slides they use while studying. Also, they were asked about any specific organ which they felt that they were not able to understand during their lectures or were not able to visualize them properly.

Questions related to the pre-usage questionnaire for the teachers were related to the difficulties they faced while teaching about human anatomy to the students while in lectures. They were asked about which all means of communication they use while teaching the students and what all problems the students face while learning them. The teachers were also asked whether any particular organ they face difficulty to make the student visualize them better.

Questions related to the post-usage questionnaire were related to how the application is working and what all difficulties they face while working with the application and whether the application is effective or not for the students. For this, the system usability scale (SUS) was used along with an open-ended Google form to gauge the effectiveness and learn the reviews of the application.

70.5 Prototype

The prototype (Human Anatomy) AR Application was developed in Unity and using Vuforia SDK and was installed on a smartphone. Unity can be used to develop games, AR, and VR applications. After launching the application, the user has to follow a set of instructions on how to operate the application. After the instruction phase, the user has to scan the provided marker to go further (Fig. 70.3).

After scanning the marker, six organs images placed equally on the circumference of the application image are displayed (Fig. 70.4).

Each image depicts the organ that will be shown. The organ showcased is a 3D model which is shown in a real-world environment. The six organs which are shown via the application are the brain, lungs, skeleton, kidney, eyes, and heart. All the organs have a degree of realism with their animation and are constructed in such a manner that the students can easily understand the organ.

An example will be the 3D model of the lungs (Fig. 70.5). Each organ has been sub-divided into its various parts. By clicking on the organ part, only that part is shown and additional information about the part of the organ is also displayed (Fig. 70.6).

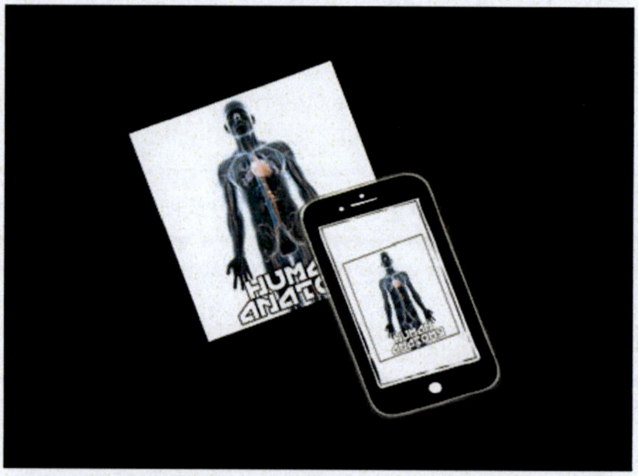

Fig. 70.3 Scanning of marker to use application

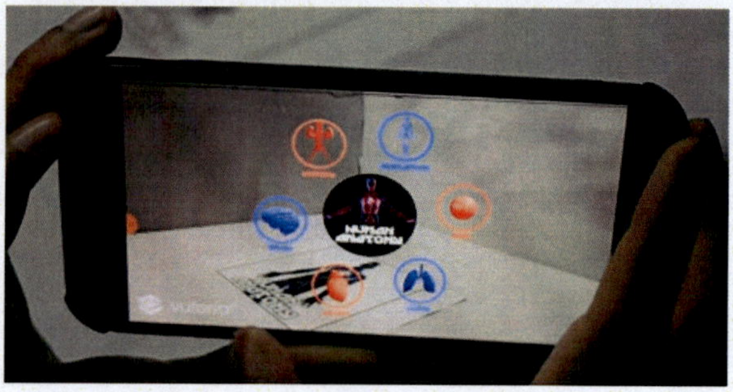

Fig. 70.4 Images of organs that can be augmented

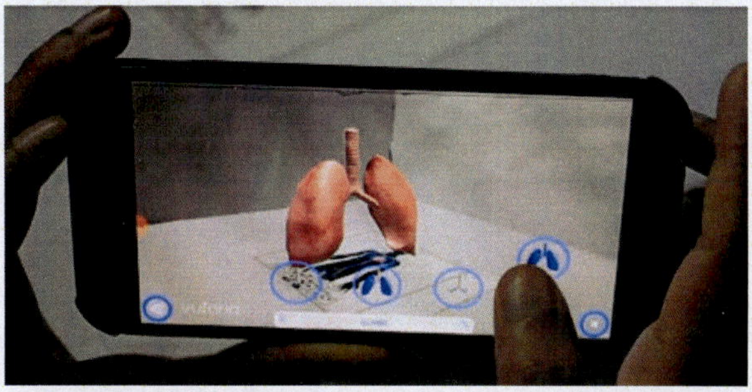

Fig. 70.5 3D model of lungs

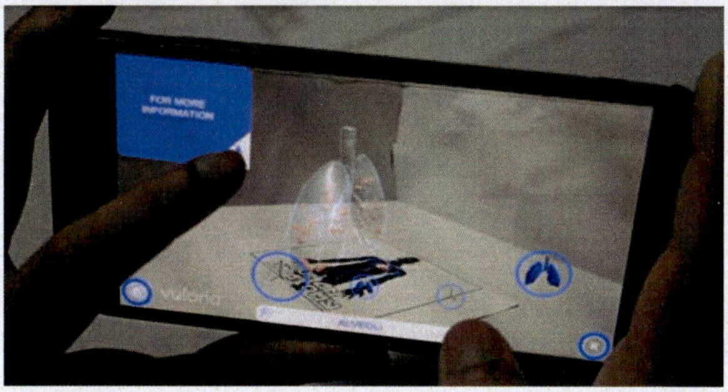

Fig. 70.6 Information of the organ

If the student wishes to acquire more information about any specific part, they can be redirected to the 'Wikipedia' page for that organ by clicking on the 'info-button' (Fig. 70.7).

Additional features about the application is the zoom-in and zoom-out feature. A student can zoom-in or zoom-out on any organ by using two fingers to get a

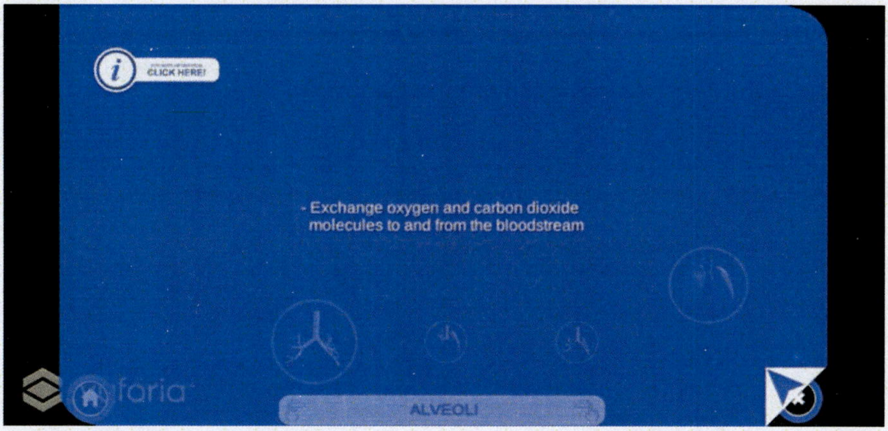

Fig. 70.7 Additional information of the organ

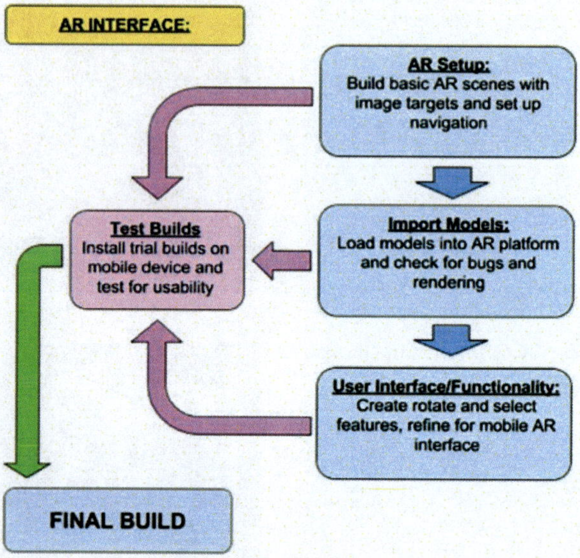

Fig. 70.8 AR Interface of the application

better look at the organ. Also, all the displayed organs rotate on an arbitrary axis by themselves which allows the user to see the organ from all angles. The user can stop the rotation momentarily when they want by holding on to the organ at any specific moment. Voice features are also available.

The workflow of the application can be seen in (Fig. 70.8).

70.6 Results and Analysis

The overall Cronbach's alpha value obtained from the system usability score (SUS) is 0.856. An alpha score greater than or equal to 0.7 indicates an acceptable value, while an alpha value of 0.8 or higher indicates a good value. Therefore, the Cronbach alpha values obtained indicate that the results obtained were reliable (Fig. 70.9).

After the post-usage questionnaire, the overall learnability score is 89.3, while the usability score is 91.5. The overall SUS score comes to a total of 91.1 which is shown in the above graph.

Using the open-ended Google form and the five-point Likert scale, we were able to test the application upon parameters like visualization, information, user interface (UI), and the overall experience.

Figure 70.10 shows that almost all the participants were able to visualize the organs better after using the application than the traditional approach of textbooks.

Figure 70.11 shows that almost all the participants were able to get adequate information needed about the various organs after using the application.

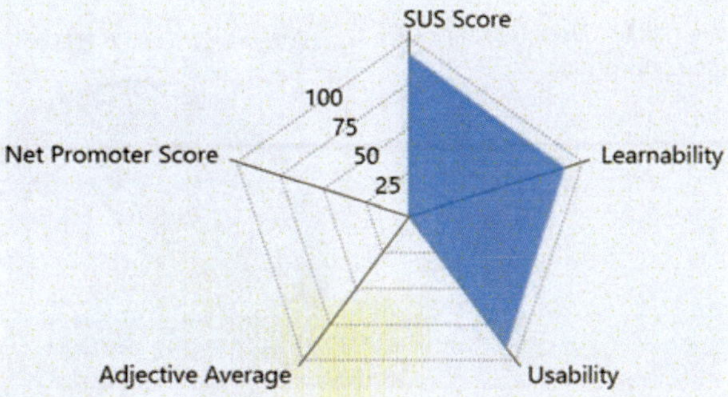

Fig. 70.9 SUS Score of the application

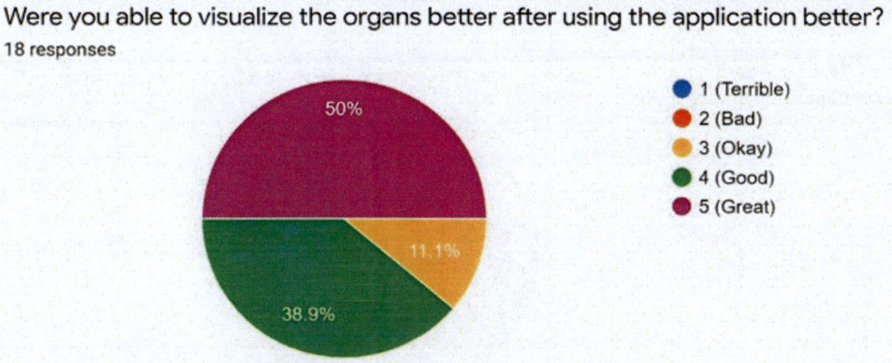

Fig. 70.10 Result for organ visualization

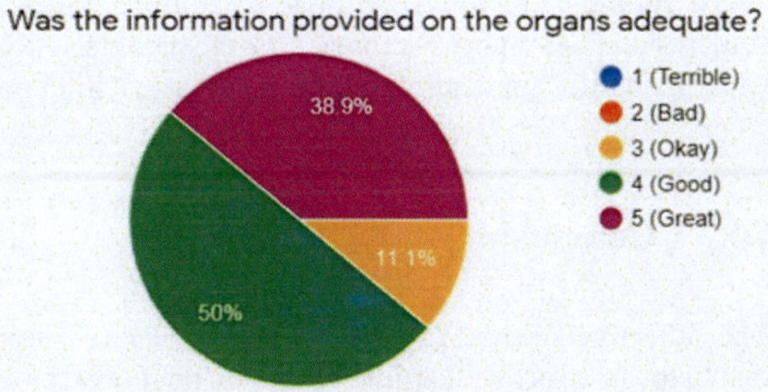

Fig. 70.11 Information provided

Figure 70.12 shows that more than 80% of the participants liked the user interface of the application and were able to use the application conveniently.

Figure 70.13 shows that the overall experience of almost all the participants was more than satisfactory. This shows that the participants were satisfied with the prototype of the application.

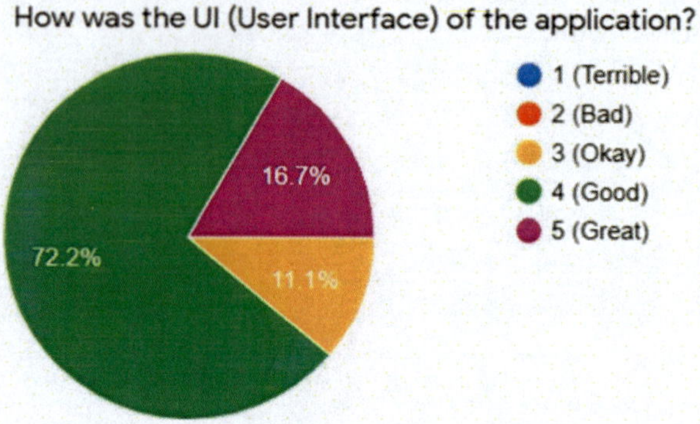

Fig. 70.12 User Interface of the application

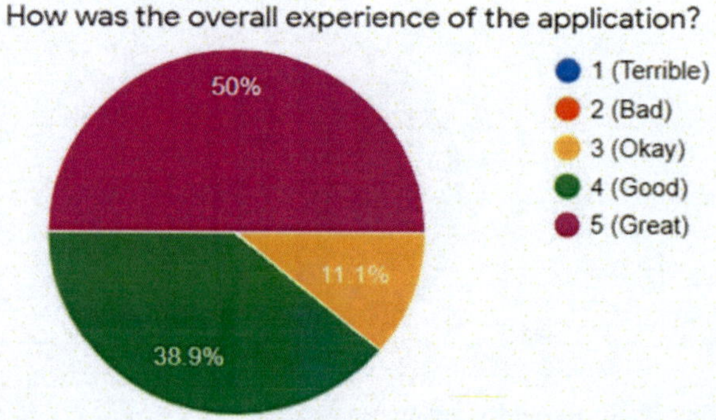

Fig. 70.13 Overall experience of the application

Therefore, using the AR Human Anatomy application has been beneficial for the students and has a positive impact. Both students as well as teachers were motivated while using the application for studying or teaching purposes.

70.7 Conclusion

The objective of this research was to showcase an augmented reality application that helps in effective learning of human anatomy for secondary school students and to understand its impact. The literature indicated that there is insufficient research on the impact of using mobile AR in education, and there is room to explore the potential of AR to improve student learning motivation and contribute to improved academic achievement [9]. Augmented reality (AR) was defined as combining real and virtual worlds, supplementing the real world with computer-generated virtual objects in real-time, and AR was explained in the context of education. Mobile AR was discussed given that AR may easily be used through mobile devices.

The literature review looked at the use of AR in education followed by an overview of some previous studies which used AR applications. AR in healthcare education was also seen where it was followed by acceptance of AR as a learning medium which helps in a personalized and explorative learning experience.

The methodology and the prototype of the AR application were discussed in detail. In-depth detail of how the application has to be handled was showcased. All the features of the application were discussed along with their pictorial representation.

The prototype of the AR application was well-received by the users. Compared to the current AR applications which are available, the AR Human Anatomy provides many more features. Currently, the applications that are available on AR usually use multiple flashcards as markers. But, this can be much more cumbersome for the user. Compared to that, our application is much more versatile and uses only one marker. Also, a lot of AR applications only show the model to the users and never give additional information to them. On the other hand, the AR Human Anatomy application gives a detailed explanation along with visualization experience to the users.

Analysis of the results shows that there has been a positive impact on the participants after using the application and is beneficial for both students and teachers. Further work has to be done to improve the application as has been said by Participant X, "Functions of the organs can also be added in short while labeling the organ name. Also, the connection between the organs should be told when the organ is displayed at the beginning."

70.8 Future Work

Future working can be done on the application in various forms. Currently, markerless technology is available for users using the Google ARCore SDK but the stability of the markerless technology is still not good for android devices. Additionally, SDK's like Google ARCore are only available on higher-end android devices which can be more expensive for the general users. Since this application is for the general public of all communities, markerless technology has not been incorporated yet but can be put into consideration for future purposes.

Additionally, for future work, we can incorporate features like body-tracking where the user has to scan a person and the scanned area will show about the organ. This way, much more models and additional features can be added to the application.

References

1. Sirakaya, M., Sirakaya, D.A.: Trends in educational AR studies: a systematic review. Malays. Online J. Educ. Technol. **6**(2), 60–74 (2018)
2. Akçayır, M., Akçayır, G.: Advantages and challenges associated with AR for education: a systematic review of the literature. Educ. Res. Rev. **20**, 1–11 (2017)
3. Di Serio, A., Ibáñez, M.B., Kloos, C.D.: Impact of an AR system on students' motivation for a visual art course. Comput. Educ. **68**, 586–596 (2013)

4. Billinghurst, M., Clark, A., Lee, G.: A survey of AR. Found. Trends Hum. Comput. Interact. **8**, 73–272 (2015)
5. Saidin, N.F., Halim, N.D.A. Yahaya, N.: A review of research on augmented reality in education: advantages and applications. Int. Educ. Stud. 8(13) (2015)
6. Wei, X., Weng, D., Liu, Y., Wang, Y.: Teaching based on AR for a technical creative design course. Comput. Educ. **81**, 221–234 (2015)
7. Zhu, E., Hadadgar, A., Masiello, I., Zary, N.: Augmented reality in healthcare education: an integrative review. PeerJ **2**, e469 (2014). https://doi.org/10.7717/peerj.469
8. Ha, H.-G., Hong, J.: Augmented reality in medicine. Hanyang Med Rev **36**, 242–247 (2016)
9. Khan, T., Johnston, K., Ophoff, J.: The impact of an augmented reality application on learning motivation of students. Hindawi. Article ID 7208494 (2019)
10. Liu, X., Sohn, Y.-H., Park, D.-W.: Application development with augmented reality technique using unity 3D and Vuforia. Int. J. Appl. Eng. Res. **13**(21). ISSN 0973-4562
11. Grahn, I.: The Vuforia SDK and Unity3D game engine. Linköping University, Department of Computer and Information Science Bachelor thesis, 16 ECTS | Computer Science 2017 | LIU-IDA/LITH-EX-G-17/059-SE
12. Maguire, B., Luz, S.: SpeechIsHard—a serious game in aid of speech recognition. University of Dublin, Trinity College, 21 May 2015
13. Lebrao, L.R., McKenna, R.E., Morrow, J.J.: Creating an Audio Game Platform for the Visually Impaired. Worcester Polytechnic Institute, Degree of Bachelor of Science, 8 May 2015
14. Hashim, N.C., Majid, N.A.A., Arshad, H., Obeidy, W.K.: User satisfaction for an augmented reality application to support productive vocabulary using speech recognition. In: Advances in Multimedia, vol. 2018. Hindawi. Article ID 9753979
15. McLellan, S., Muddimer, A., Peres, S.C.: The effect of experience on system usability scale ratings. J. Usabil. Stud. **7**(2) (2012)
16. Klug, B.: An overview of the system usability scale in library website and system usability testing, vol. 1, no. 6. Gibson D. Lewis Health Science Library. University of North Texas Health Science Center (2017)

Part X
Design Evaluation and Optimisation

Chapter 71
Part Consolidation in Design for Additive Manufacturing: A Two-Level Approach Using Complexity Metrics

Jayakrishnan Jayapal, Senthilkumaran Kumaraguru, and Sudhir Varadarajan

Abstract Design for additive manufacturing (DfAM) creates opportunities for improved product design. Part consolidation (PC) is an important design opportunity in DfAM. The existing methodologies for part consolidation in DfAM are mostly based on expert judgment or heuristic rules. In addition, they only suggest the candidates for part consolidation. They do not assess the manufacturability of the consolidated parts. This paper proposes a two-level approach for part consolidation using complexity measures at the system level and part level. At the system level, the centrality score (a complex network measure) is used to identify parts with high potential for consolidation, and at the part level, a geometry-based complexity measure—modified complexity factor (MCF)—is used to assess the manufacturability. This two-step approach is expected to improve efficiency and effectiveness in DfAM since it uses minimal information at the system level and detailed assessment of manufacturability is done only for those parts where AM potential is high. Such an approach can be extremely useful to product designers at the early stages of design. Two case studies are presented to illustrate the effectiveness of the proposed approach.

71.1 Introduction

Design for additive manufacturing (DfAM) recommends different methods and tools that help designers about the capabilities of additive manufacturing (AM) (technological, geometrical, etc.) in the design stage. In general, DfAM capabilities have been classified into two categories: opportunistic and restrictive. In opportunistic DfAM, each design potential will have a value addition and design benefit. Kumke et al. [1] have developed a semantic network of the design potentials and benefits of AM (49 nodes and 290 relations). The top three highly influential nodes in this semantic network include (a) part consolidation, (b) different material in one part

J. Jayapal · S. Kumaraguru (✉) · S. Varadarajan
Department of Mechanical Engineering, Indian Institute of Information Technology, Design and Manufacturing, Kancheepuram, Chennai 600127, India
e-mail: skumaran@iiitdm.ac.in

A. Chakrabarti et al. (eds.), *Design for Tomorrow–Volume 2*, Smart Innovation, Systems and Technologies 222, https://doi.org/10.1007/978-981-16-0119-4_71

and (c) reduced number of joints. These important design potentials represent the key drivers of complexity in any system—the parts, the content of the parts and the relationships among parts. In AM, the possibility of part consolidation is widely studied in system level, and some of the existing methods is discussed in the next session.

71.1.1 Existing Approaches to Part Consolidation in DfAM

AM has its unique process capabilities that remove the product complexity restriction because of part consolidation. Some components in the part are no longer required to be separately manufactured and then assembled; instead, they can merge into a single part to be manufactured through a single AM process. Schmelzle et al. [3] have studied the challenges that engineers face when redesigning a multicomponent assembly into a single component and fabricated using laser-based powder bed fusion for metal AM. Yang et al. [4] have proposed a new part consolidation (PC) methodology comprehensively considering function integration and structure optimization. The modular design approach is another way to perform PC. Samyeon et al. [5] have proposed a methodology to consolidate parts by considering the maintenance and product retrieval at the end-of-life stage by extending modular design approach. The modules are identified using the single component complexity index and Markov clustering algorithm. Researchers working on design structure matrix (DSM) have developed several measures and clustering approaches to study modularity in systems. For instance, Suh et al. [2] have used key drivers of complexity in any system like the number of parts, relationships and nature of relationships to develop complexity measures to guide discussions on modularity. From the literature, it is found that the existing PC methodologies are focused only on finding a suitable candidate for part consolidation in the assembly, but the manufacturability of the consolidated design is not discussed. At the same time, those focusing on manufacturability by AM have largely concentrated on geometrical complexity at the part level and not the whole.

Given that the top three design potentials in DfAM point to the key drivers of complexity in any system (number of parts, the content of the parts and the number of relations), this paper argues that there may be opportunity to improve the efficiency and effectiveness of the DfAM process in practice by using objective metrics at both the system (product) level and the part level. The paper proposes the use of complexity network measure (centrality) to identify the potential parts for consolidation from the product level design structure matrix (DSM) and an existing geometry-based approach measure for assessing the manufacturability at the part level.

The centrality score, which measures the connectedness of a node in the network, helps identify the parts with high centrality and the other parts that are directly connected with it. The manufacturability of the consolidated parts is calculated using the modified complexity factor, a measure used for assessing the manufacturability in AM [6].

71.2 Two-level Approach

71.2.1 Identification of Possible Candidates for PC

For identifying the possible candidates for PC, the overall methodology is explained in the flowchart as shown in Fig. 71.1.

The suitable candidate for part consolidation is identified by performing complex network analysis on the product architecture or design structure matrix (DSM). The product network is constructed from part interaction matrix created from the exploded view of the assembly in any CAD modeling software. Then the centrality of each component in the product network is calculated, and the component with high centrality score is selected as the candidate for PC. The centrality score is defined by the number of connections incident on each node, and it is used as an estimate of its importance in the network [7]. The centrality score obtained from the network analysis resembles the complexity metric discussed in [2], which is a function of the number of components, the number of interfaces and the adjacency matrix of the system. While consolidating the parts, the functionality of the overall design has to be preserved. So, the decision has to be taken by considering the material compatibility and manufacturability of the consolidated design in AM.

Fig. 71.1 Proposed part consolidation strategy

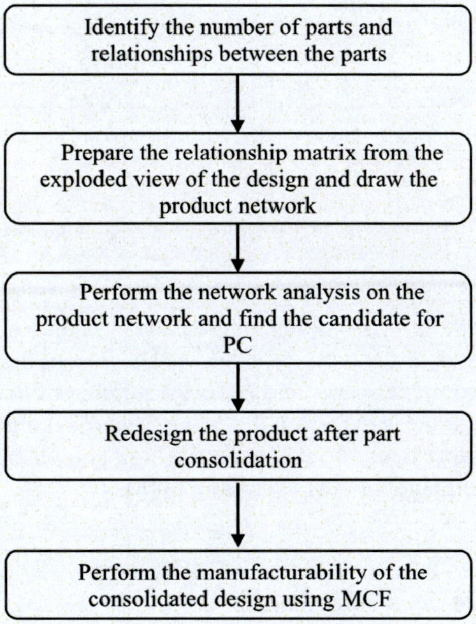

71.2.2 *Manufacturability Using the Modified Complexity Metric*

While the consolidation of the parts has been done, now we have to measure the complexity of the design to check the manufacturability. To evaluate the manufacturability of a shape for AM, Conner et al. [6] have developed a metric called modified complexity factor (MCF) which is a modified version of complexity factor mentioned in [8]. The complexity of the design, modified complexity factor (MCF), is calculated by using the geometry-related parameters like surface area, the volume of the part, volume of the bounding box, etc. [6]. The equation for calculating the MCF is shown below.

$$\text{Modified Complexity Factor, MCF} = 5.7 + 10.8C_{\text{pr}} + 18C_{\text{ar}} + 32.7C_{\text{nh}} \tag{71.1}$$

where, part volume ratio,

$$C_{\text{pr}} = 1 - \frac{\text{Volume of part}}{\text{Volume of bounding box}} = 1 - \frac{V_p}{V_b} \tag{71.2}$$

Area ratio,

$$C_{\text{ar}} = 1 - \frac{\text{Surface area of part}}{\text{Surface area of sphere}} = 1 - \frac{A_p}{A_s} \tag{71.3}$$

Hole ratio,

$$C_{\text{nh}} = 1 - \frac{1}{\sqrt{1 + N_h}}, \tag{71.4}$$

where V_p is volume of the part, V_b is the volume of bounding box, A_p is the surface area of the part, A_s is the surface area of the sphere which is considered to be a less complex shape, and N_h is the number of holes. These geometry-related parameters are measured from the mass properties of the 3D CAD model. If the MCF value is more than 44 falls under the high complexity part category, these designs are more suitable for AM according to [6].

71.3 Case Study

To understand the feasibility of the proposed approach, two case studies are carried out. Section 71.3.1 discusses an existing motorcycle steering assembly, which is already discussed in [9], and Sect. 17.3.2 discusses a throttle pedal assembly discussed in [8].

71.3.1 Motor Steering Assembly

Initially, there were seven components in the design, and Fig. 71.2 shows the initial design and components in the design. The total number of parts in the assembly is reduced to four by using the complex product network measure.

The network from the relationship matrix is drawn as shown in Fig. 71.3a, and the centrality score is calculated for each node. Based on the centrality score, the candidate for part consolidation is identified and part count reduced to four from seven. The product network after the consolidation is shown in Fig. 71.3b, and the centrality score of each part is graphically represented in Fig. 71.4.

From Fig. 71.4, upper triple clamp has the highest centrality score, so the upper triple clamp is the most important node in the product network. The next step is to consolidate the possible parts along with the upper triple clamp without eliminating the functionality of the assembly the parts need to be consolidated. So, out of seven components, the lower triple clamp, upper triple clamp and fork tubes can be considered as standard parts and without eliminating the functionality steering handle, cap, the base is combined with the upper triple clamp, and it can be made as a single part. The next step is to check the manufacturability of the consolidated part using Eq. (71.1). The consolidated design is shown in Fig. 71.5. By using the additive

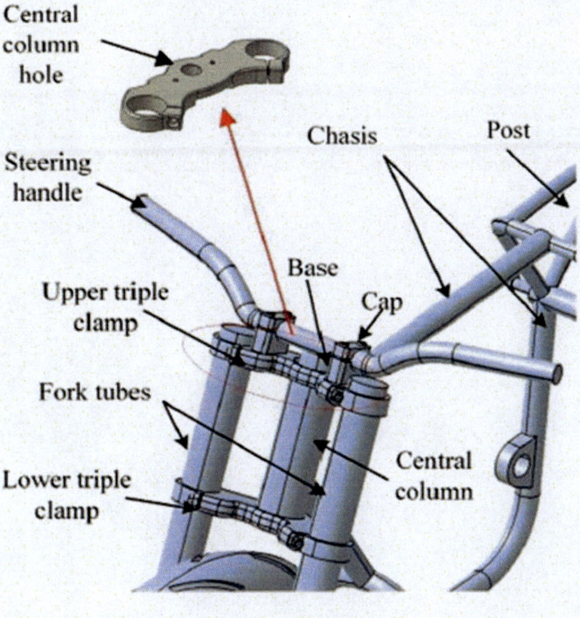

Fig. 71.2 Motorcycle steering assembly [7]

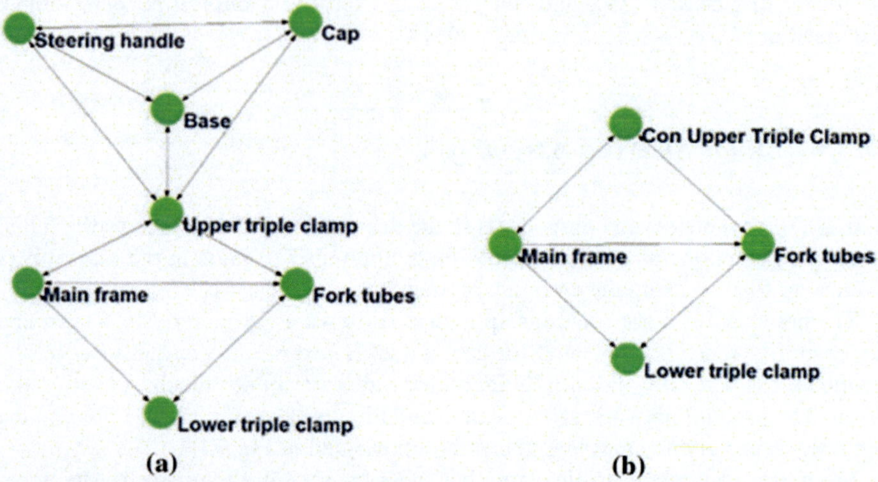

Fig. 71.3 Product network **a** before consolidation **b** after consolidation

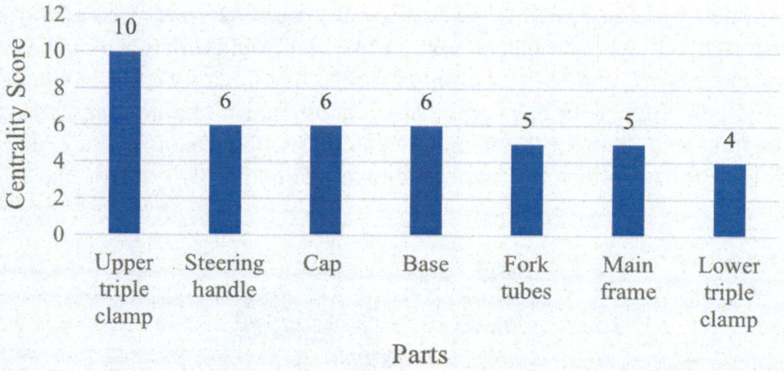

Fig. 71.4 All degree centrality of the motorcycle steering assembly

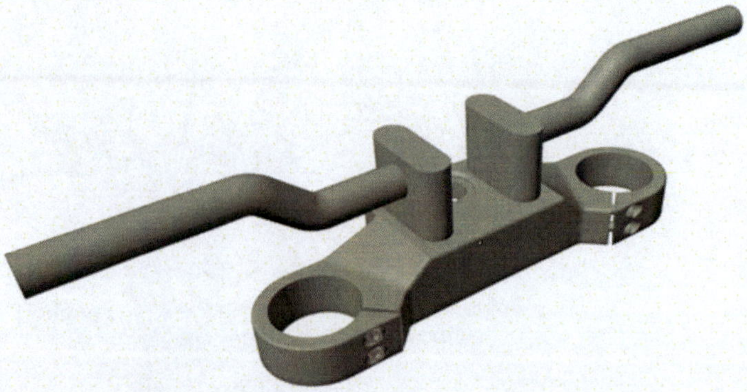

Fig. 71.5 Consolidated upper triple clamp

Table 71.1 Mass properties and parameters for calculating the MCF

Vol of the part (cm^3)	Vol of the bounding box (cm^3)	Surface area of the part (cm^2)	Surface area of the sphere (cm^2)	No of holes	C_{pr}	C_{ar}	C_{nh}	MCF
582.827	4215.809	1139.952	337.424	2	0.8617	0.7040	0.50	44.03

manufacturing-enabled part count reduction (AM-PCR) followed in [9], seven parts are reduced to five.

The details of mass properties and the modified complexity factor of the consolidated design are shown in Table 71.1.

From Table 71.1, the MCF value of the consolidated design is more than 44, so the design is suitable for manufacturing using AM [6].

71.3.2 Throttle Pedal

A throttle pedal design is selected to demonstrate the proposed framework and validate the effectiveness of the new method with the Part Consolidation Candidate Detection (PCCD) algorithm used in [8]. The original design is taken from an open-source CAD database [11]. As shown in Fig. 71.6, the throttle pedal consists of 12 parts without counting the washers and fasteners. The product network is created by identifying the physical contact between the parts. From the network, the component with the highest centrality score is identified as the candidate for part consolidation.

Based on the physical relationship between the components, the product network is drawn using the open-source network tool and is shown in Fig. 71.7. The centrality

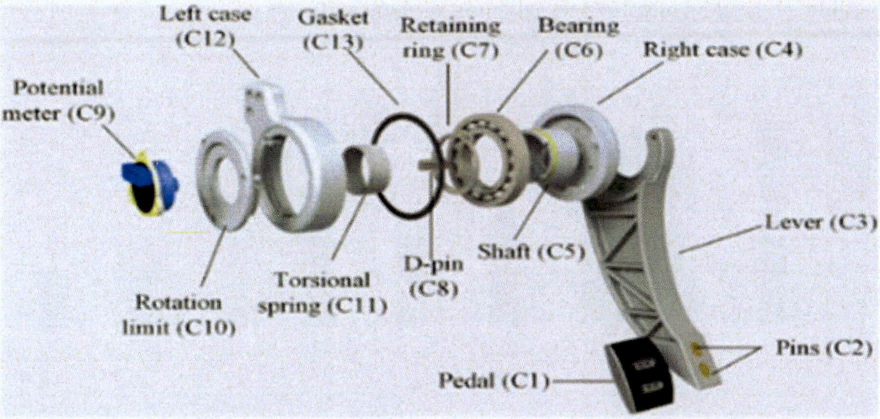

Fig. 71.6 Exploded view of throttle pedal [8]

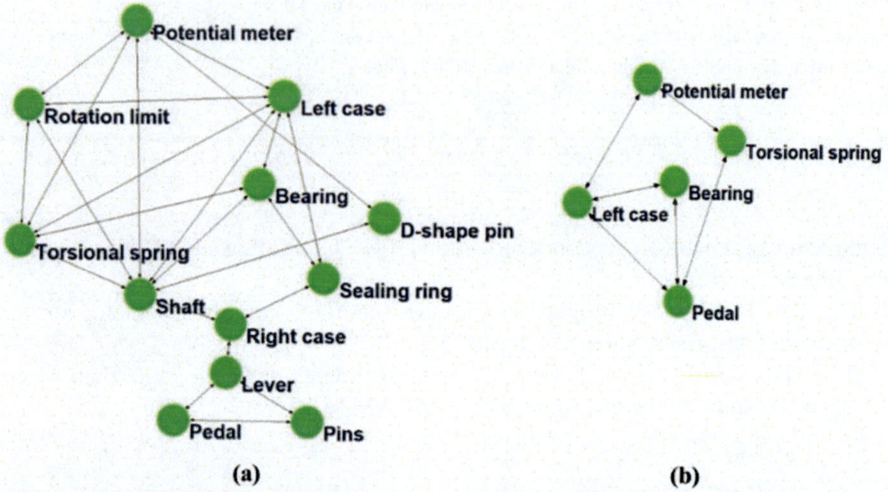

Fig. 71.7 Product network of throttle pedal **a** before consolidation **b** after consolidation

score of each node in the product network is measured and shown in Fig. 71.8.

From Fig. 71.8, the shaft has a high centrality score. Even though the shaft has the highest centrality score in the product network, all the parts around the most center node can be considered for consolidation. Out of nine parts in the assembly, some of the parts are COTS like bearing, torsional spring and potential meter. So, out of the remaining parts, we can consolidate the parts into two groups without eliminating the functionality of the product. In Group 1, the modified pedal consists of the shaft, sealing ring, the right case, lever, pin, pedal and D-shape pin. In Group 2, the rotation limit is consolidated with the left case. The manufacturability of the consolidated design is assessed with the help of MCF using Eq. (71.1), and the details are shown in Table 71.2. The consolidated design is shown in Fig. 71.9.

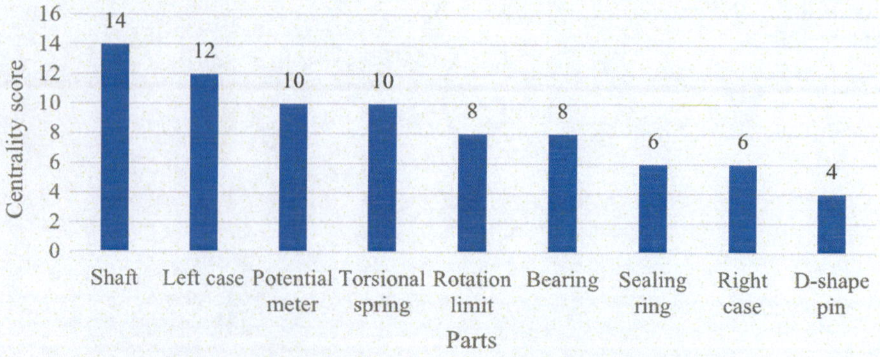

Fig. 71.8 Centrality score of the throttle pedal

Table 71.2 Mass properties and parameters for calculating the MCF

Part name	Vol of the part (cm^3)	Vol of the bounding box (cm^3)	Surface area of the part (cm^2)	Surface area of the sphere (cm^2)	No of holes	C_{pr}	C_{ar}	C_{nh}	MCF
Pedal	147.21	1307.52	631.23	134.83	3	0.89	0.79	0.5	45.79
Left case	37.63	163.55	193.46	54.31	4	0.77	0.67	0.55	44.11

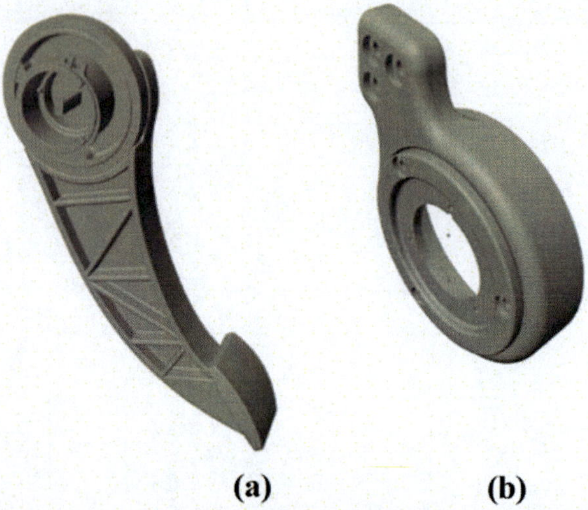

Fig. 71.9 Consolidated design **a** Group 1 and **b** Group 2

The MCF value for both the consolidated designs is more than 44, so these designs are suitable for manufacturing in AM.

71.4 Results and Discussion

The two case studies discussed in the previous section hold well with the earlier studies performed to identify the possible candidate for part consolidation using part consolidation candidate detection (PCCD) algorithm [10] and additive manufacturing-enabled part count reduction (AM-PCR) algorithm [9].

For implementing the PCCD algorithm or AM-PCR, the user has to verify the rule set for each set of component pairs in the assembly. There are seven rules in PCCD and nine rules in AM-PCR. So, the amount of information and time required to perform PC is more as the number of components in the assembly increases. But in this work, we can identify the set of components suitable for PC with fewer information and able to find the possible candidate for PC with comparatively less expertise. Also, in this proposed methodology, the manufacturability of the consolidated design is assessed using the modified complexity factor (MCF). So, if we follow the proposed two-level decision-making strategy, after part consolidation in the first level total number of parts in the assembly will be reduced and needs to check the manufacturability of lesser parts in the second stage. A detailed comparison of the existing algorithms and proposed methodology is shown in Table 71.3. The reduction in the number of parts is improved in the proposed methodology compared to the existing methodologies. So, the new two-level methodology can able to reduce the number of parts and assess the manufacturability of the consolidated parts with less expertise.

Table 71.3 Comparison of different PC methodologies

	PCCD [10]		AM-PCR [9]		Proposed Methodology	
No of parts	Before PC	After PC	Before PC	After PC	Before PC	After PC
Motorcycle steering assembly	–	–	7	5	7	4
Throttle pedal	13	7	–	–	12	5

71.5 Conclusion

An alternate approach for part consolidation using multilevel complexity measures is developed. In this work, we proposed a two-level decision-making procedure to adopt AM for manufacturing the product. In the first level from the product network, parts with high centrality scores are identified and the parts around the high centrality node are consolidated without eliminating the functionality of the product. Then the manufacturability of the consolidated design is assessed with the help of MCF in the second level. Two case studies are presented to prove the feasibility of the proposed framework for part consolidation and assessed the suitability of AM for the consolidated design. The proposed centrality score-based PC will be more effective than the modularity-based PC if the number of parts is more in the system. Further analysis can be done to check the appropriateness of the centrality-based method in future. This work can be further extended to find the suitability of the new methodology for functional integration along with part consolidation, and a view similarity-based measure for manufacturability assessment can be studied in future.

References

1. Kumke, M., Watschke, H., Hartogh, P., Vietor, T.: Methods and tools for identifying and leveraging additive manufacturing design potentials. Int. J. Interact. Des. Manuf. **12**, 481–493 (2018). https://doi.org/10.1007/s12008-017-0399-7
2. Sinha, K., Suh, E.S.: Pareto-optimization of complex system architecture for structural complexity and modularity. Res. Eng. Des. (2018). https://doi.org/10.1007/s00163-017-0260-9
3. Schmelzle, J., Kline, E.V., Dickman, C.J., et al.: (Re)Designing for Part Consolidation: Understanding the Challenges of Metal Additive Manufacturing. J. Mech. Des. Trans. ASME **137**, 1–12 (2015). https://doi.org/10.1115/1.4031156
4. Yang, S., Tang, Y., Zhao, Y.F.: A new part consolidation method to embrace the design freedom of additive manufacturing. J. Manuf. Process. **20**, 444–449 (2015). https://doi.org/10.1016/j.jmapro.2015.06.024
5. Kim, S., Moon, S.K.: A part consolidation design method for additive manufacturing based on product disassembly complexity. Appl. Sci. **10** (2020).https://doi.org/10.3390/app10031100
6. Conner, B.P., Conner, B.P., Manogharan, G.P., et al.: Making sense of 3-D printing : Creating a map of additive manufacturing products and services Making sense of 3-D printing : Creating a

map of additive manufacturing products and services. Addit Manuf **1–4**, 64–76 (2014). https://doi.org/10.1016/j.addma.2014.08.005

7. Rivette, M., Hacoët, J.Y., Mognol, P.: A graph-based methodology for hybrid rapid design. Proc. Inst. Mech. Eng. Part B J. Eng. Manuf. **221**, 685–697 (2007). https://doi.org/10.1243/09544054JEM666
8. Taylor, P., Joshi, D., Ravi, B.: Computer-aided design and applications quantifying the shape complexity of cast parts. 37–41 (2013). https://doi.org/10.3722/cadaps.2010.685-700
9. Yang, S., Zhao, Y.F.: Additive manufacturing-enabled part count reduction: A lifecycle perspective. J. Mech. Des. Trans. ASME **140**, 1–12 (2018). https://doi.org/10.1115/1.4038922
10. Yang, S., Santoro, F., Zhao, Y.F.: Towards a numerical approach of finding candidates for additive manufacturing-enabled part consolidation. J. Mech. Des. Trans. ASME **140** (2018). https://doi.org/10.1115/1.4038923
11. Miessner, H., Grab, C.A.D., Cambridge, M.A.: Throttle Pedal Design Challenge (2015). https://grabcad.com/library/pedal-one-microtechnologies-1. Accessed 7 Apr 2020

Chapter 72
A Process to Understand Products in Terms of Design Elements Using Linkographs

Ravi Lingannavar and Pradeep G. Yammiyavar

Abstract A product undergoes through a sequential process from design stage to development. The design elements and utilitarian aspects of the product need to be taken into consideration while developing /designing. As designers thought processes are converted into sketches, one has to have good knowledge on the basic elements of design. Function, aesthetics, ergonomics and usability constitute the broad elements of design. Each design will have sub-parameters, which designers need to prioritize and give more emphasis to those elements in particular. Linkographs will help in identifying the spaces, where a product can be value added to make it innovative. For all the identified design elements along with sub-parameters, linkographs have been plotted by taking a design case study. This approach shall emphasize on value parameters while designing the product and shall serve as a template for reference. Value addition can be evaluated in terms of design elements via the developed linkographs.

72.1 Introduction

A product undergoes through a sequential process from design stage to development. Initially, the problem is identified, and a typical product design cycle includes all the phases as shown in Fig. 72.1. This cycle is typically followed by almost all engineers, designers and small and medium-scale enterprises. Different stages of product development have been adopted from Chaturvedi, [1]. There are many tools and techniques followed, in achieving the product's design. There are hundreds of tools and techniques available, some example are idea generation and selection which are brainstorming, brain drawing and brain writing, fishbone diagram, How-How diagram, 5W2H and many more. These tools help in generating ideas that make the

R. Lingannavar (✉)
KLEDrMSSCET Belagavi, Belgaum, Karnataka, India
e-mail: lingannavar.ravi@gmail.com

P. G. Yammiyavar
Indian Institute of Technology Guwahati, Guwahati, Assam, India
e-mail: pradeep@iitg.ac.in

A. Chakrabarti et al. (eds.), *Design for Tomorrow–Volume 2*, Smart Innovation, Systems and Technologies 222, https://doi.org/10.1007/978-981-16-0119-4_72

Fig. 72.1 Product development stages

Fig. 72.2 Design space adopted from [5]

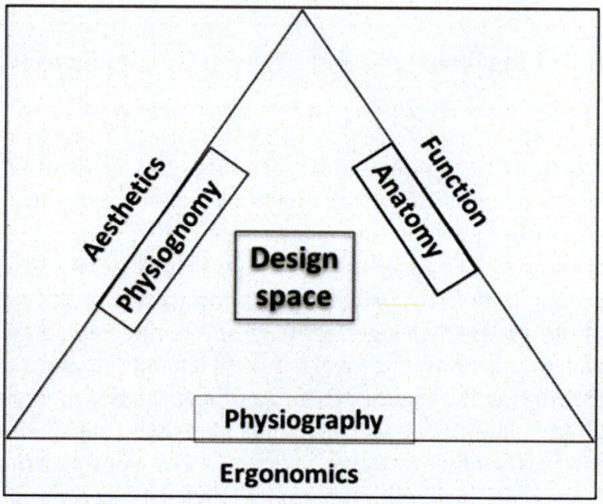

product more value added [2, 3]. It is the initial stage where designers are dealing with solution part only, not on the variables/parameters involved in design [4].

To add value to the existing product or to ideate/develop entirely new product requires knowledge of understanding design elements. Design elements are the smallest divisions of any successful product. And these are focussed more while designing the product. Hence, it becomes very important to consider them. To begin with 'design space', it is defined as 'combinatorial space generated by set of parameters, which are also the operands' (Aesthetics + Ergonomics + Function = 1, Design ratio) [5].

To achieve innovation/value addition in products, one needs to consider the design elements which starts from base, i.e., design space. Innovation in product design is defined as value addition in the product, which should be viable, feasible and desirable. And many more definitions are available in the literature [6]. It is also the foundation for the development of the proposed fractal triangle of innovation (Fig. 72.2).

72.2 Variables Involved in Design Are Complex

The criteria affecting innovation are defined in product design on the basis of the available literature. To express the variables, particularly to meet the requirements

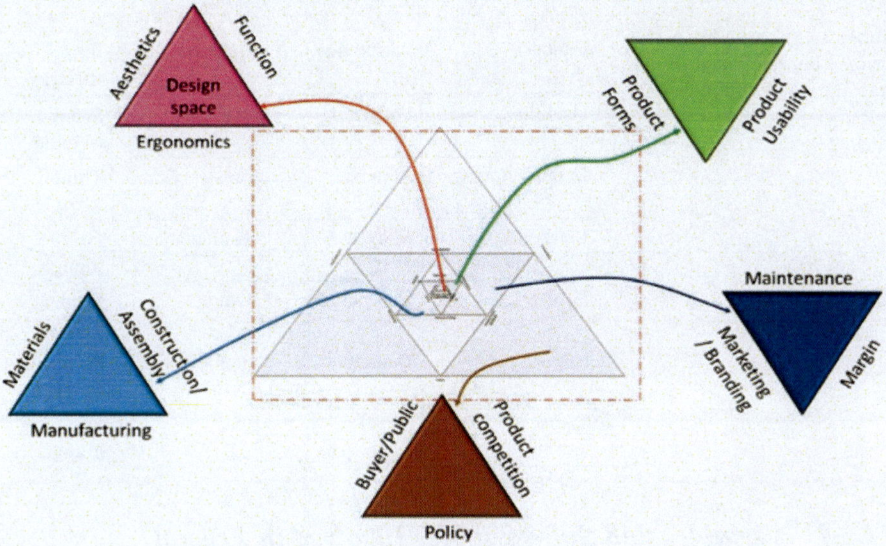

Fig. 72.3 Proposed fractal triangle of innovation

of product design, a 'novel fractal triangle of innovation' was conceptualized. It was developed as the current design morphology and did not address the questions from the perspective of the designers and engineers, as shown in Fig. 72.3. A novel fractal triangle of innovation was developed from the theory of design space. Some significant parameters which influence innovation and productivity have been established by researchers. It is posited from the literature that innovation or value addition can happen from these design elements and is discussed in the following sections (Table 72.1).

72.2.1 Variables Establishing Fractal Triangle of Innovation

There are sub-parameters for each parameter that are discussed in further sections.

The variables and sub-variables are enlisted from the available literature, keeping value addition at the center. Starting from design space as the base, first triangle constitutes function, aesthetics and ergonomics. The sub-parameters are identified based on the experience and available literature. Considering functionality as an example, the associated parameters are visual, style, color, balance, etc., which are shown in Table 72.3.

Table 72.1 Variables involved in establishing fractal triangle of innovation

Function	Aesthetics	Ergonomics	First triangle
Usability	Form	Operations	Second triangle
Manufacturing	Material	Construction	Third triangle
Marketing	Margin	Maintenance	Fourth triangle
Public	Competitors	Policy	Fifth triangle
Logistics	Demand	Supply	Sixth triangle
Tools	Novel techniques	R&D	Seventh triangle

72.2.2 *Variables and Sub-variables Involved in Fractal Triangle*

Table 72.2 depicts the main variable and sub-variable as shown. For designers and engineers, it becomes difficult in considering the higher priority or which of the following design elements have greater effect on the success of the product.

Similar way second level of triangle and its variables have been listed as shown below (Table 72.3).

Based on the literature, these design elements have been noted and few are frequently practiced in the design process by designers and engineers. The weighting factor that needs to be assigned to the design elements mentioned is the complexity that occurs. For the design elements, there is no such method of giving priorities.

Table 72.2 Variables forming the innermost triangle of design space

Function	Aesthetics	Ergonomics
Strength	Visual	Safety
Durability	Style	Efficiency
Cost/value	Color	Reachability
Purpose	Balance	Visibility
Fit	Unity	Force
Features	Symmetry	Duration
Stability	Harmony	Anthropometry
Construction	Proportion	Repetition
Ease of Use	Emphasis	Posture
	Rhythm	Direction
		Glance

Table 72.3 Second-level triangle of innovation and their design elements

Usability	Processes/Manufacturing	Operations
Functional	Human	Ease of use
Efficient	Machines	Handling
Error-free use	Tools	Installation
Enjoyable	Equipment	Number of steps
Environment	Time	Practice
Economical	Cost	Operating
	Skills	Extra fits
Construction	Power	Manual
Ease of use	Space	Maintenance
	Assembly	

Therefore, for this, a system has been created that saves time and effort as it serves as a blueprint for these kinds of products. We limit it to the first-level triangle in this paper, and it can be expanded to the third, fourth, fifth, sixth and seventh, respectively.

72.3 Current fractal triangle of value Addition proposed

The proposed innovation triangle is derived from the theory of design space and derived over the parameters identified [8–10], which contributes to innovation if implemented in the design process. Hence, the fractal triangle of innovation's building blocks is design elements. Figure 72.3 gives the overall picture of all the variable and sub-variables in triangle, i.e., fractal triangle that needs to be considered in the designing phase by the designers to add value in the product.

72.4 Connecting the Design Elements—Linkographs an Overview

Once the design elements are known to the designer, it becomes very easy to understand the importance of the same. Designer can prioritize those design elements. To better understand these elements, one can use linkography. Linkography is a technique to analyze the design process at initial levels. It can be developed on qualitative and quantitative approach. It is basically used in mapping of the designers thinking, and idea generation can be converted in a meaningful way. It is logical and meaningful to use in product design process. In fact, linkography is a matrix update, although few researchers still want to stick to matrix representation [11], specifically when we are dealing with two similar products of different company. It is the best tool to convert metacognitive thinking in design process. Each design feature is tested and marked as nodes for its relationship, and 'moves' are design moves [7]. The goal in this is to measure the innovativeness of the product in terms of design elements when two products of similar family are compared. The linkographs have been developed by using linkographer program. For the study purpose, authors have considered eight sets of product, and one of them is presented in this paper briefly.

72.4.1 Experiments Conducted

To examine the exact spaces of value addition based on the fractal triangle of innovation, linkographs have been plotted. For comparison of the products, to identify the value addition spaces, we have considered the two similar products, i.e., Vidyut laptop and Akash tablet as shown below (Fig. 72.4).

The products adopted for this study are products of IIT Bombay (Akash tablet) and belong to the same family and perform the same function. Akash tablet was designed initially to find applications in the field of education. However, the advanced version of the same was Vidyut laptop (by private firm). The user studies have proved that the latter is complicated to use—technically termed as market myopia. Countries like Japan and Taiwan are known for innovations in electronic segment over the past decades. This case study has been taken to understand the influence of design elements. The research outcomes will help the designers and manufacturers to add more value to the product and yet make it more cost effective. Vidyut laptop is one such example, as it adopts a backward integration technique, wherein the resource and technology available to design the product were used. Rest of the components like outer shell, mouse and charger, which were added to the cost of product, were outsourced. The linkographs have been plotted to identify the relative innovation index between the products. In post evaluation, we can say that the major difference between the products under study is that the laptop is just an extended version of the tablet, with components such as hinge, keyboard, mouse and casing being added

Fig. 72.4 Vidyut laptop and Akash tablet

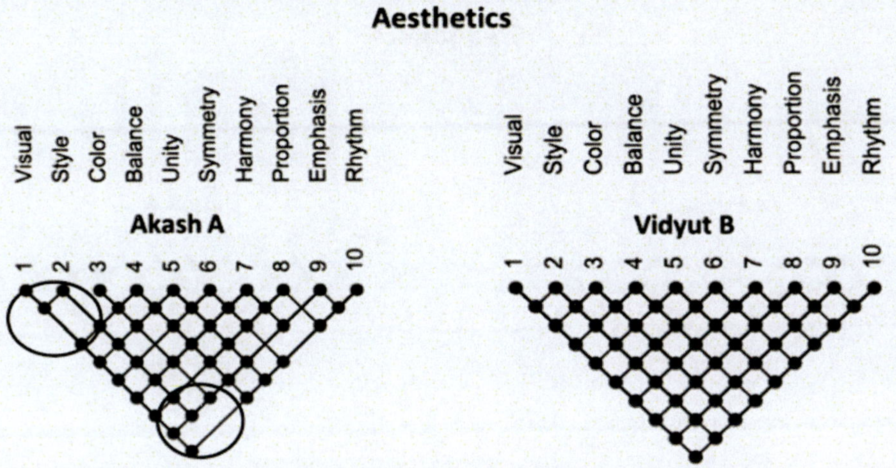

Product A

N	0	LI	%CM3	%CM 4	%CM5	%CM6	%CM7	%CM8
10	36	3.6	0.2	0.3	0.2	0.4	0.1	0.1

Product B

N	0	LI	%CM3	%CM 4	%CM5	%CM6	%CM7	%CM8	%CM9
10	44	4.4	0.2	0.3	0.1	0.2	0.3	0.1	0.09

Fig. 72.5 Linkographs for Vidyut laptop and Akash tablet aesthetics and its elements

which converts the tablet into a laptop. Therefore, we can say that innovation is doing the existing things in a new way. When the linkographs are plotted based on the design elements, it gives clear picture of where the value addition has taken place. The circles represented in Fig. 72.5 are to indicate the spaces for the value addition, i.e., Vidyut laptop is good in those places, and Akash tablet can be improved in those design elements. The plotting is done on the sheet by marking design elements and then checks for the correlation and develops the link. And this is followed over all the elements to develop a linkograph.

72.4.2 Results and Discussion

The linkographs have been plotted for both the products. In this analysis, authors have considered up to first-level triangle. Aesthetics, ergonomics and function are making first triangle. Operations, processes and usability are making second-level triangle. (Due to number of page restriction, first-level triangle to generate linkographs individual and combined has not included in this paper) (Figs. 72.6, 72.7 and 72.8).

The aesthetics and its sub-parameters are plotted for both the products. From the analysis, the relative innovation index for Akash tablet is 3.6 and 4.4 for Vidyut

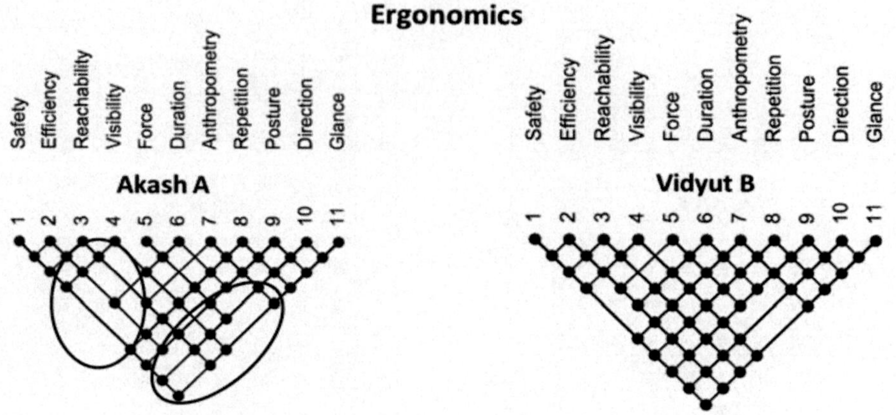

Product A	N	0	LI	%CM3	%CM 4	%CM5	%CM6	%CM7	%CM8
	11	38	3.45	0.27	0.27	.09	0.09	0.45	-

Product B	N	0	LI	%CM3	%CM 4	%CM5	%CM6	%CM7	%CM8	%CM9
	11	47	4.27	0.27	0.27	-	0.18	.27	.27	0.09

Fig. 72.6 Linkographs for Vidyut laptop and Akash tablet ergonomics and its elements

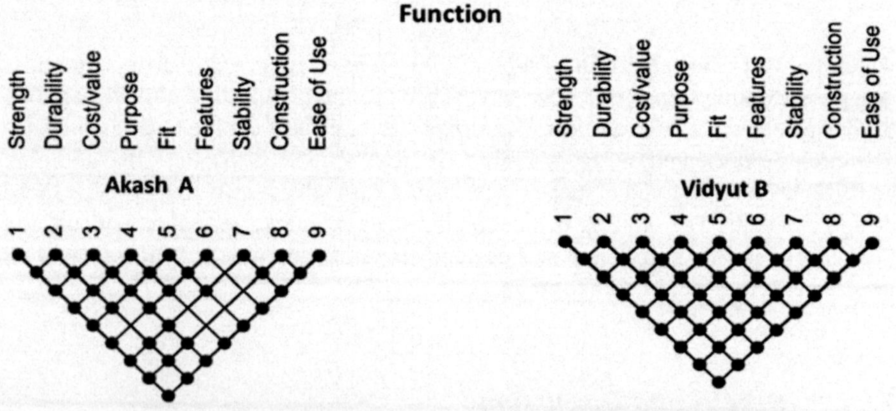

Product A	N	0	LI	%CM3	%CM 4	%CM5	%CM6	%CM7	%CM8
	9	30	3.33	.33	.44	-	.22	-	.22

Product B	N	0	LI	%CM3	%CM 4	%CM5	%CM6	%CM7	%CM8
	9	36	4	.22	.22	.22	.22	.22	.22

Fig. 72.7 Linkographs for Vidyut laptop and Akash tablet function and its elements

laptop. It leads to the relative difference of between the two products as 0.8. It can be concluded that 0.8 is the relative innovation index in terms of aesthetics, i.e., aesthetically Vidyut laptop is more value-added product than the Akash tablet. In the similar way, it has been calculated for ergonomics, function, operations, processes and usability along with their respective design elements.

When combining aesthetics, ergonomics and function, the linkograph is plotted as shown in Fig. 72.8, and it is observed that Vidyut's laptop has a higher index of innovation (AEF = 8.9). Similar way when usability, processes and operations are combined, overall link index coming for Vidyut laptop is more (UPO = 8.4) as shown in Table 72.4.

Table 72.4 is the final conclusion of plotting linkographs, which shows individual, combined link indexes which clearly indicate the design elements of improvement, in our case Vidyut laptop has higher index in almost all variables (design elements), hence making it more value added. A designer or engineer can clearly think. Hence,

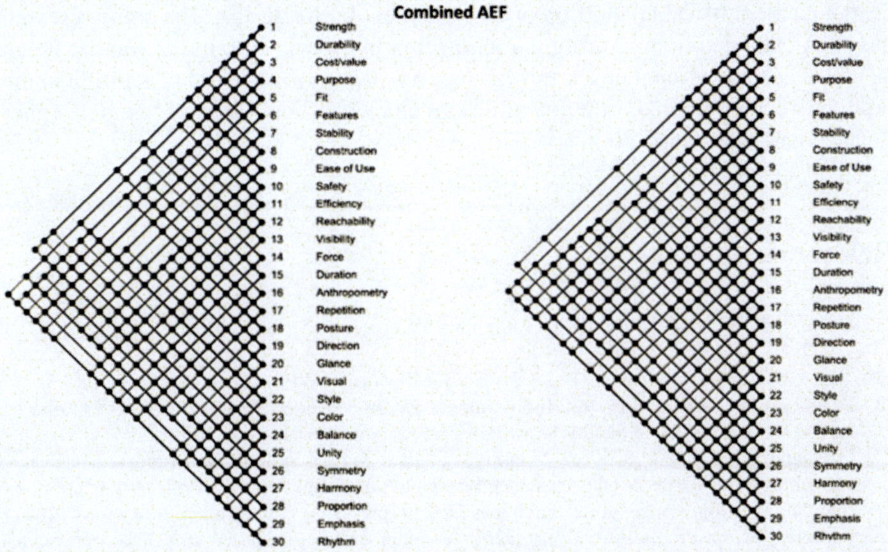

Fig. 72.8 Linkograph of aesthetics, ergonomics and function combined along with their design elements

Table 72.4 Consolidated results of link index of Akash tablet and Vidyut laptop

	A	E	F	AEF	U	P	O	UPO
Akash	3.6	3.45	3.33	8.46	2.33	2.66	2.66	6.92
Vidyut	4.4	4.27	4	8.9	2.5	4.2	3.64	8.4

A Aesthetics, *E* Ergonomics, *F* Function, *U* Usability, *P* Processes, *O* Operations, AEF, UPO Combined all

Vidyut laptop, an extended version of Akash tablet, has more value added and making it a more innovative in terms of design elements.

72.5 Conclusion

It becomes a necessary step to identify and adopt the design elements in any product's design as it leads to the value addition, focussing on design elements to be prioritized. Linkographs drawn based on design elements will help identify important variables to be considered in early stages of design stage and will help cater the value component in the product. Results of linkographs can be utilized to calculate the percentage of value addition when compared with another product of same family. Overall, the study and understanding of the design elements will lead to a more value-added product. Representation can be visualized by plotting the linkographs indicating the design element of value addition or improvement in the design. The same research has been extended in measuring the innovation in products of similar family. It can be called as innovation metrics and can measure the innovation/value addition in the product when compared, in terms of design elements. This work will help designer to catch up the value addition spaces.

Reference

1. Chaturvedi, M.: New Product Development. Wheeler Publishing. ISBN: 81–85814–92–9 (1997)
2. Tague, N. R.: The quality tool box (Vol. 600). Milwaukee, WI: ASQ Quality Press (2005)
3. Defeo, Joseph A., Joseph, M.: Juran Juran's Quality Handbook: The Complete Guide to Performance Excellence 6/e. McGraw Hill Professional. (2010)
4. Lingannavar, R., Ojha, S. P., & Yammiyavar, P. A Matrix Framework Proposal for Evaluating Innovation Criteria of a Design Process Output during Product Conceptualization. In International Conference on Research into Design (pp. 487–497). Springer, Singapore. (2017)
5. Yammiyavar, P.: Product proximity factor-A human interaction based model for optimizing parameters within a product's design space, the second, HWWE 2004 at National Institute of Training in Industrial Engineering (NITIE), Mumbai (2004)
6. Brem, A., Voigt, K.-I.: Integration market pull and technology push in the corporate front end and innovation management—Insights from the German software industry. Technovation **29**(5), 351–367 (2009)
7. Goldschmidt, G.: Linkography: unfolding the design process. MIT Press (2014)
8. Verganti, R.: Design-driven innovation. Boston, MA (2009)
9. Brown, T.: Change by design (2009)
10. Norman, D.: The design of everyday things: Revised and expanded edition. Basic Books (2013)
11. Van der Lugt, R.: Sketching in design idea generation meetings. PhD Dissertation, 2001, Faculty of Industrial Design: Delft University of Technology (2001)

Chapter 73
Daylighting Evaluation and Retrofit Strategies: A Simulation-Based Approach to Optimise the Artificial Lighting Consumption

Tarun Verma and Padmanaban Gopalakrishnan

Abstract Daylighting plays a major role in the design of lecture halls. Architects and designers take advantage of daylight in lecture halls to enhance visual comfort and energy efficiency by using daylighting simulation tools. Daylight simulations provide with the ability to compare many design parameters and optimise design alternatives to promote visual comfort and energy efficiency. Still, daylight is not commonly combined with the artificial lighting in buildings as there is a lack of information and documentation on daylighting simulation tools to assess the energy-saving potential in buildings. This study aims to bridge the gap between the use of daylighting simulation tools and evaluating the energy-saving potential of daylighting in buildings by using a single objective optimisation approach. Twelve lecture halls were selected, and five retrofitting strategies were analysed to enhance the daylighting performance and optimise the artificial lighting consumption of the lecture halls. The simulation results showed 58–95% energy savings after applying the retrofitting strategies. Similarly, the daylight autonomy increased from 40 to 89% for lecture hall G5 and 14 to 51% for lecture hall G12. The DA was found as a good performance indicator and showed a positive correlation with energy savings.

73.1 Introduction

In most of the countries, the requirement of sustainable development and energy consumption in buildings poses challenges for the engineering, construction and architecture industry [1]. For this concern, daylighting is considered as one of the essential concepts for energy-efficient, sustainable and green building design. In buildings, windows recognised as a major source of natural light which permit people to maintain visual interaction with outdoors as well as bring natural light for indoor environment with a pleasant atmosphere [2]. Daylighting is considered as a significant approach in modern architecture which has an ability to reduce the artificial

T. Verma (✉) · P. Gopalakrishnan
National Institute of Technology Tiruchirappalli, Tiruchirappalli, India
e-mail: artarunvverma@gmail.com

A. Chakrabarti et al. (eds.), *Design for Tomorrow–Volume 2*, Smart Innovation, Systems and Technologies 222, https://doi.org/10.1007/978-981-16-0119-4_73

lighting consumption and enhance visual comfort in buildings [3]. Previous studies showed that artificial lighting accounts for up to 40% of the yearly building energy consumption [4, 5]. The daylighting responsive lighting control methods are significantly impacted on the building energy use and showed 30–77% of energy saving in annual electric lighting consumption [6, 7].

Except for energy savings, the positive effects of daylighting on occupants health, well-being, comfort and productivity are also well recognised [8], that is the main reason architects and designers take advantage of daylight in their buildings. Architects and designers use different simulation tools to support their decisions during different stages of design and post-occupancy evaluation studies. Simulation-based design approach helps in optimising the energy consumption in buildings. Daylight simulations provide designers with the ability to compare a large number of design parameters and optimise design alternatives to promote visual and thermal comforts for the occupants [9]. Daylighting simulations are challenging due to the stringent requirements to represent the actual building scenario. Still, previous research works showed the reliability of simulation tools by producing the same results as obtained from the field measurements [3]. In addition, the daylighting simulation results showed a significant correlation with the occupant's subjective responses which again confirms the reliability of daylighting simulation tools [10]. Due to the stringent requirements of building performance codes, standards and green building rating systems, the daylighting simulations tools are getting more concentration in the design process [11]. However, in spite of the listed benefits, daylight is not commonly combined with the artificial lighting in building design as there is lack of information and documentation on daylighting simulation tools to evaluate the energy-saving potential in buildings [12]. The complexities involved in the daylighting simulations are also considered as one of the limitations of the application of existing computer simulation programmes [13].

This study aims to bridge the gap between the use of daylighting simulation tools and evaluating the energy-saving potential of daylighting in existing buildings. For this regard, this paper presents a single objective optimisation process to minimise the artificial lighting consumption by providing the appropriate retrofitting solutions and daylighting responsive control system. The goal of retrofitting strategies is to integrate and enhance daylight with artificial lighting to reduce energy consumption in existing building. Twelve lecture halls were selected for this study at a higher education institute of Tiruchirappalli. Five retrofitting strategies were evaluated to enhance the daylighting performance and optimise the artificial lighting consumption of the lecture halls. The proposed retrofitting strategies were selected based on building context and suitability to apply with minimal damage to the existing building envelop. Hence, minimal changes in the existing building envelop were considered a constraint to applying the retrofitting strategies.

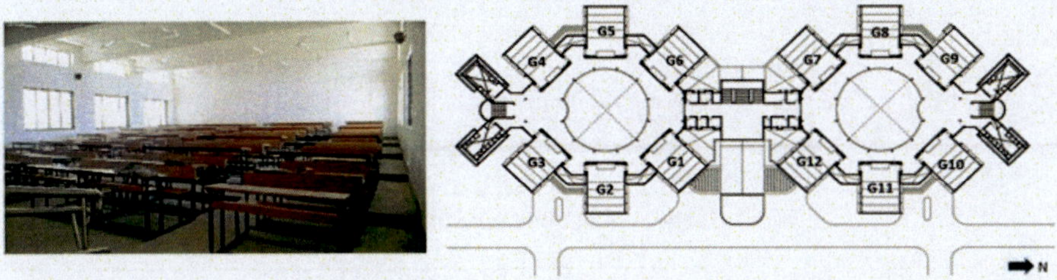

Fig. 73.1 Ground floor plan of the lecture hall building

73.2 Methodology

73.2.1 Building Details

The lecture hall building consists of 36 lecture halls distributed on three floors and oriented in various directions with the WWR (window-to-wall ratio) of 40 per cent. Twelve lecture halls (G1 to G12) were selected on the ground floor to evaluate the daylighting performance. Figure 73.1 shows the plan and lecture hall view of the selected lecture hall building.

73.2.2 3D-Modelling and Simulation Details

3D Modelling. Several tools were used to model and perform the daylighting simulations for the lecture hall building. The three-dimensional model of lecture hall building was created to represent the actual building geometry by using Autodesk Revit 2016 and then imported in Rhinoceros 5.0. DIVA 4.0 was used as a daylighting simulation tool which works as a plug-in for Rhinoceros 5.0. A grid of illuminance sensors was generated in DIVA 4.0 at a distance of 0.5 × 0.5 m for every lecture hall at a height of 1.05 m (average working plane height) from the floor level. The Tiruchirappalli weather data (.epw file) was imported in DIVA 4.0 to represent the actual weather conditions for daylighting simulations.

Simulation Details. The material properties, simulation parameters, radiance rendering settings, occupancy schedule and artificial lighting schedule (for base case) were assigned to the three-dimensional model to perform the annual daylighting simulations. The simulation parameters and radiance rendering parameters are presented in Tables 73.1 and 73.2, respectively. The opaque material (walls, roof, floor, etc.) properties were assigned based on the standards, and the glass visual light transmittance (VLT) values were assigned as per the data provided by the glass manufacturer. The material properties assigned for the simulations are presented in Table 73.3. Figure 73.2 shows the artificial lighting and occupancy schedule for the lecture hall building. The white portion of Fig. 73.2 denotes that all the artificial lights are

Table 73.1 Simulation parameters

Parameter	Value
Threshold illuminance	300 lx (NBC 2016)
Lighting set point for dimming control	300 lx (NBC 2016)
Artificial lighting operation	Dimming with occupancy on/off sensor

Table 73.2 Radiance rendering parameters

-ab	-ar	-as	-aa	-ad
5	128	256	0.15	512

Table 73.3 Material properties

Parameter		Value
Transmittance	Ventilator glass	0.89 (wbdg.org)
	Façade glass	0.37 (Provided by the manufacturer)
	Window glass	0.37 (Provided by the manufacturer)
Reflectance	Old white wall	0.55 (IES Lighting Handbook)
	Outside ground	0.20 (CIBSE Guide)
	Cement floor	0.10 (NBC 2016)
	Old ceiling	0.50 (IES Lighting Handbook)
	Door	0.50 (ECBC 2017)

CIBSE Charted Institution of Building Services Engineers
IES Illuminating Engineering Society
Wbdg.org Whole Building Design Guide
ECBC Energy Conversation Building Code
NBC National Building Code 2016, India

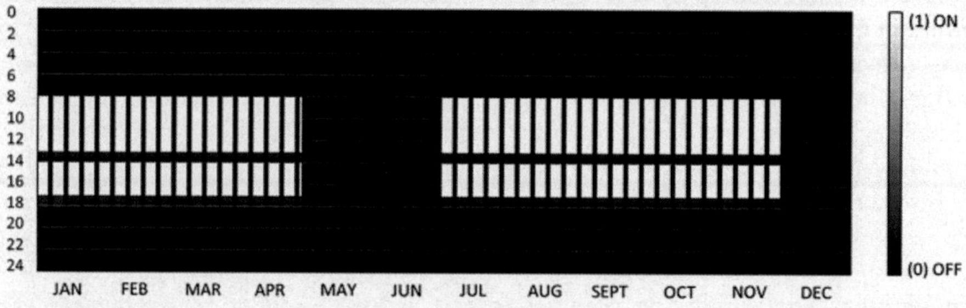

Fig. 73.2 Artificial lighting and occupancy schedule

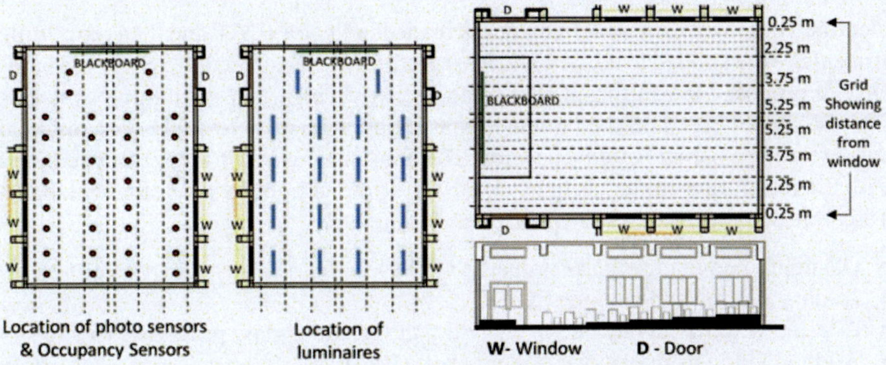

Fig. 73.3 Location of sensors, luminaires and typical lecture hall details

on (1), and space is fully occupied (1) for the respective hour of the day. The black portion of Fig. 73.2 denotes that all the artificial lights are off (0), and space if fully unoccupied (0) for the respective hour of the day. To evaluate the artificial lighting consumption for the base case, all the artificial lighting fixtures were considered switch on with the occupancy in simulations. The energy saving, due to the proposed retrofitting strategies, was compared with the base case model to identify the best retrofitting solution. Photosensors and occupancy sensors were used in simulations to reduce artificial lighting consumption. Two photosensors were placed at every luminaire in the three-dimensional model of the lecture hall building. In total, 36 photosensors were installed for every lecture hall at a height of 4.2 m (ceiling level) from the finished floor level. Figure 73.3 shows the location of photosensors, occupancy sensors and luminaires (2 × T5-28-W fluorescent tubes). The artificial lighting loads were calculated through simulations for each retrofitting strategy combing with the dimming control strategy.

73.2.3 Analysis Approach

Daylight autonomy (DA; the percentage of annual occupied hours above the illumination level of 300 lx) was used as a performance indicator to evaluate the daylighting performance of the lecture halls. The daylighting performance and energy-saving potential of retrofitting strategies were evaluated based on the annual daylighting simulation results of DA. For the analysis of simulation results, the lecture halls were divided into grids at a distance of 0.5 m, as shown in Fig. 73.3. On each grid line, the average value of DA was calculated to generate the daylighting distribution profile across the room width. The analysis of the simulation results was divided into two phases. In the first phase, the existing daylighting conditions were simulated for the twelve lecture halls and problems were identified based on the simulation results. In the second phase of the study, the best and least performing lecture halls were identified based on the daylighting performance indicator (DA), and then retrofitting strategies were applied to the best and least performing lecture halls to check the influence of proposed retrofitting strategies on energy savings. The retrofitting strategies were identified based on the building context and suitability to apply with minimal damage to the existing building envelop. The dimming control strategy was combined with the retrofitting strategies to calculate the artificial lighting load of lecture halls. The proposed retrofitting strategies are as follows:

- Changing the window glass properties (RS1)
- Adding windows in bay one (RS2)
- Adding windows in bay one + changing the window glass properties (RS3)
- Adding windows in bay one + light shelf (RS4)
- Adding windows in bay one + light shelf + changing the window glass properties (RS5)

Figure 73.4 shows the proposed architectural intervention in the existing building, and Table 73.4 shows the comparison between existing and proposed glass material specifications.

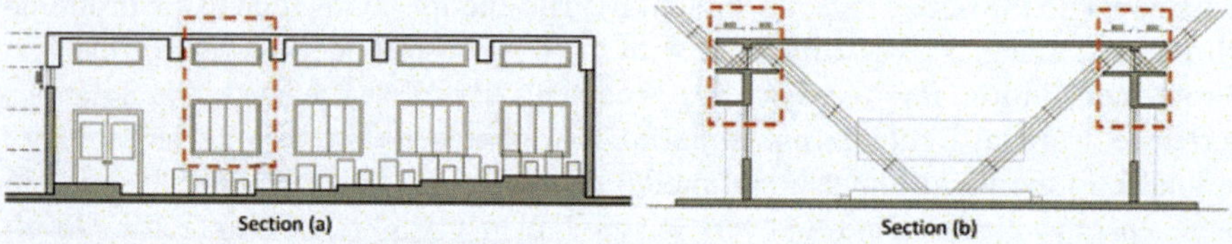

Fig. 73.4 Proposed architectural interventions for retrofitting; section **a**—adding windows in bay one; section **b**—adding light shelf

Table 73.4 Glass specifications (existing and proposed case)

Glass type	Light factors			Energy factors		
	VLT	Reflections %		Solar factor	Shading co-efficient	U value (W/sqm.k)
		External	Internal			
Existing glass	0.37	22	18	0.42	0.48	5.5
Proposed glass	0.56	15	18	0.50	0.58	5.6

73.3 Results and Discussion

73.3.1 *Phase 1—Evaluating the Daylighting Performance and Problem Identification of Existing Lecture Hall Building*

The daylighting simulation results for the existing building are presented in Table 73.5. The simulation results showed a range of average DA from 14–40 percentage. Based on the DA simulation results, lecture halls G5 and G12 were identified as best and least performing, respectively. The DA profiles across the lecture halls width are presented in Fig. 73.5. A polynomial curve fit equation ($Y = 0.29X^2 - 6.14X + 48.14$) was derived from the simulation results to represent a typical DA profile for all the lecture halls. It is evident from the simulation results that in all the lecture halls, the DA values were observed higher near (0 to 2.25 m) to the windows. The middle area (2.25 to 5.25 m distance from the windows) of all the lecture halls showed a lack of daylight. The daylight-deficient area was identified at the distance of 2.25 to 5.25 m from the windows based on the derived polynomial equation. Figure 73.6 represents the simulation result of daylighting distribution for the best (G5) and least (G12) performing lecture halls. Due to the absence of windows in bay one, most of the lecture halls show daylight deficiency in bay one and near the blackboard area. Overall, phase one simulation results showed that retrofitting strategies are needed to enhance the daylighting levels in the middle areas and bay one. The results of the proposed retrofitting strategies for lecture halls G5 and G12 are presented in phase two of this study.

73.3.2 *Phase 2—Retrofitting Strategies and Energy Savings*

The daylighting simulation results for the proposed retrofitting strategies are presented in Table 73.6. All the proposed retrofitting strategies showed a significant improvement in the daylighting levels for both G5 and G12 lecture halls. The DA results for retrofitting strategies were compared with the existing DA results. Based on the simulation results of DA, the retrofitting strategies RS3 and RS5 showed better performance compared to the other strategies. The simulation results of DA were observed in the range of 22–89%, which shows a considerable improvement in the daylighting levels of the lecture halls. The artificial lighting consumption was calculated for all the retrofitting strategies and compared with the base case. Every retrofitting strategy showed a significant reduction in the artificial lighting consumption for both the lecture halls by using the dimming control method. The energy savings are achieved in the range of 58–95%, which shows the influence of retrofitting strategies on energy savings. A correlation analysis was done to find the relationship between DA and energy savings. The DA values and percentage energy

Table 73.5 Average DA simulation results (without retrofitting)

Metrics	Lecture Hall											
	G1	G2	G3	G4	G5	G6	G7	G8	G9	G10	G11	G12
DA	14.60	38.72	17.55	22.48	40.23*	24.39	29.90	35.66	28.56	21.83	37.27	14.13**

*Maximum Value (Best Daylighting Performance)
**Minimum Value (Least Daylighting Performance)

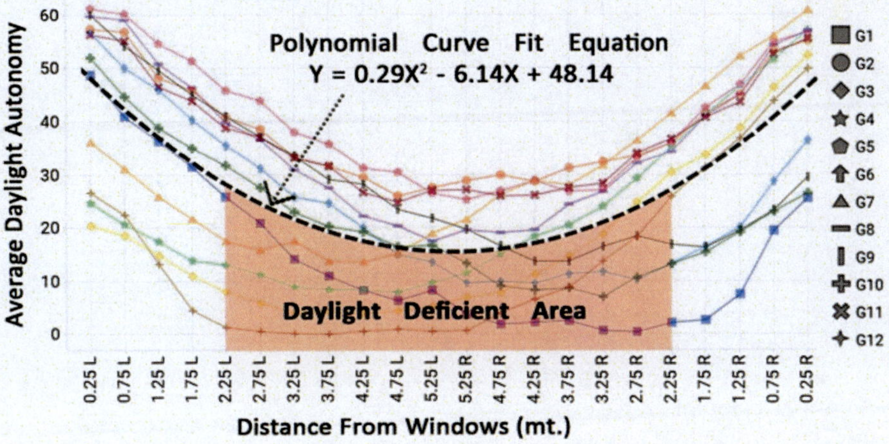

Fig. 73.5 DA profile across the lecture halls width (before retrofitting)

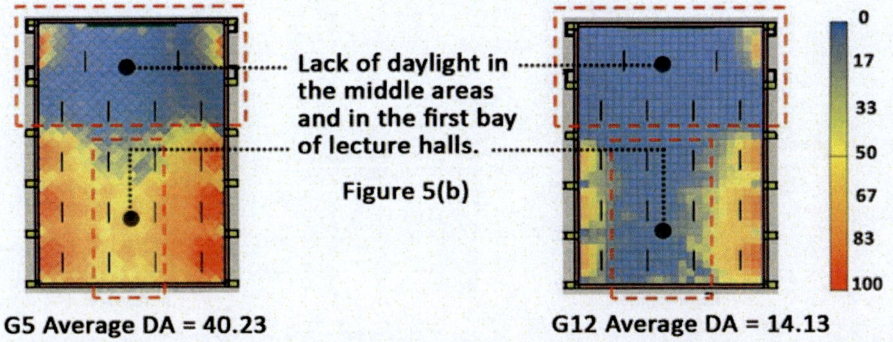

Fig. 73.6 Daylighting distribution for G5 and G12 lecture hall (before retrofitting)

savings were found positively correlated, r (8) = 0.639, p = 0.047. Figure 73.7 shows the daylighting distribution for the lecture hall G5 and G12 after applying the retrofitting strategies. The daylighting levels increased in the middle areas and bay one for both the lecture halls, which shows the positive effect of retrofitting strategies on the daylighting quantity.

The DA profile across the lecture halls width is presented in Fig. 73.8. A polynomial curve fit equation ($Y = 0.39X^2 - 7.38X + 72.77$) was derived from the DA simulation results to represent the typical DA profile for the proposed retrofitting strategies. The polynomial curve fit equation for the retrofitting strategies was compared with the existing DA curve fit equation. The highlighted area in Fig. 73.8 represents the improved daylighting band which shows a substantial improvement in the daylighting levels of the lecture halls after applying the retrofitting strategies. Looking at the polynomial curves for DA, it is evident that the minimum and maximum DA

Table 73.6 DA simulation results (after retrofitting) and energy savings

Metrics		Retrofitting Strategies									
		RS1		RS2		RS3		RS4		RS5	
		G5	G12	G5	G12	G5	G12	G5	G12	G5	G12
DA (Before Retrofitting)		40.23	14.13	40.23	14.13	40.23	14.13	40.23	14.13	40.23	14.13
DA (After Retrofitting)		58	38	51	28	75	51	54	22**	89*	40
Artificial Lighting Consumption (kwh)	Existing (Base Case)	1792	1792	1792	1792	1792	1792	1792	1792	1792	1792
	With Dimming Control	367	744	171	345	92	242	160	394	80	300
	Percentage Energy Savings	79.5	58.4**	90.4	80.7	94.8	86.4	91.0	78.0	95.5*	83.2

*Maximum Value
**Minimum Value

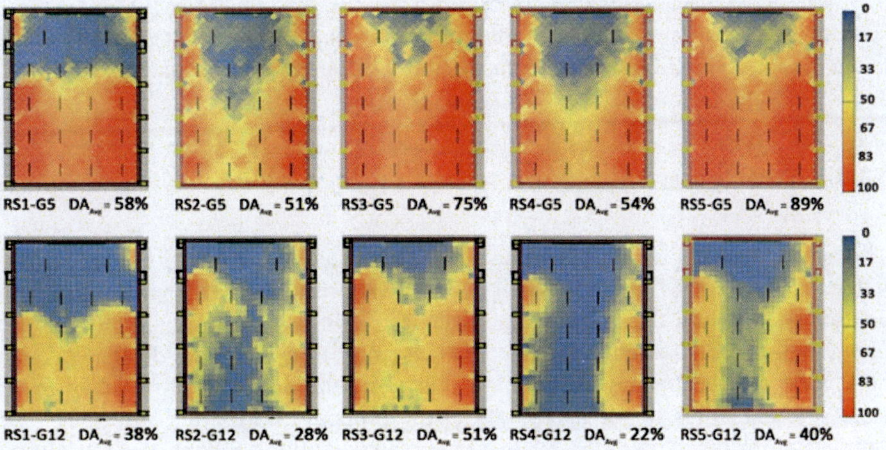

Fig. 73.7 Daylighting distribution for G5 and G12 lecture hall (after retrofitting)

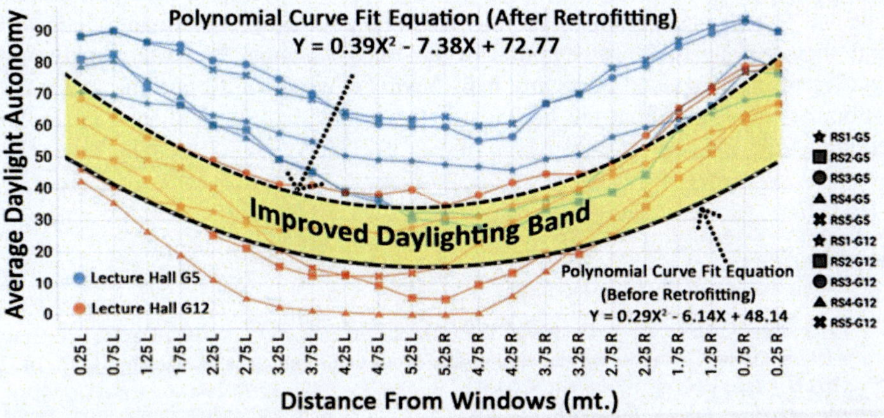

Fig. 73.8 DA profile across the lecture halls width (after retrofitting)

values increased from 15 to 35% and 50% to 75%, respectively. Looking at the individual results of retrofitting strategies, Fig. 73.8 also confirms that RS3 and RS5 performed better than the other retrofitting strategies. The RS3 and RS5 showed higher DA (>80% and >50%, respectively) values near the windows for both the lecture halls G5 and G12. Looking at the results, it is evident that both RS3 and RS5 outperformed the other retrofitting strategies. The increased VLT value for the windows was the common factor between RS3 and RS5. Before retrofitting, the VLT value for windows was taken as 0.37 (Table 73.4), which was then increased to 0.56 (Table 73.4) after the retrofitting by changing the glass specification. This finding highlights the impact of glass VLT value on the daylighting levels and energy savings. As an architect or designer, one can easily suggest a change of glass material as a suitable retrofit option to enhance the lighting levels. However, the glare should be taken into consideration while increasing the VLT value of glass material.

73.4 Conclusion

This study presents a simulation-based single objective optimisation approach to reduce the artificial lighting consumption in lecture halls by using the dimming control method and five retrofitting strategies. This study also highlights the use of simulation tools to help architects and designers to take critical design decision during the post-occupancy evaluation studies. For this regard, twelve lecture halls were selected to check the usefulness of the proposed retrofitting strategies. The simulation results showed 58–95% energy savings after applying the retrofitting strategies. Similarly, the DA values increased from 40 to 89% for lecture hall G5 and 14% to 51% for lecture hall G12. The RS3 and RS5 showed the highest energy savings; therefore, these strategies considered as best retrofitting strategies for the selected lecture hall building. The DA is found as a good performance indicator and showed a positive correlation with the energy savings. Since this study was conducted in the context of lecture halls having a WWR of 40 percent, hence the authors acknowledge that the findings of this study shall be applied to a similar building type and context only.

References

1. Li, S., Liu, L., Peng, C.: A Review of Performance-Oriented Architectural Design and Optimization in the Context of Sustainability: Dividends and Challenges. Sustainability. **12**, 1427 (2020). https://doi.org/10.3390/su12041427
2. Li, D.H.W.: A review of daylight illuminance determinations and energy implications. Appl. Energy **87**, 2109–2118 (2010). https://doi.org/10.1016/j.apenergy.2010.03.004
3. Chen, Y., Liu, J., Pei, J., Cao, X., Chen, Q., Jiang, Y.: Experimental and simulation study on the performance of daylighting in an industrial building and its energy saving potential. Energy and Buildings. **73**, 184–191 (2014). https://doi.org/10.1016/j.enbuild.2014.01.030
4. Chirarattananon, S., Chaiwiwatworakul, P., Pattanasethanon, S.: Daylight availability and models for global and diffuse horizontal illuminance and irradiance for Bangkok. Renewable Energy **26**, 69–89 (2002). https://doi.org/10.1016/S0960-1481(01)00099-4
5. Krarti, M., Erickson, P.M., Hillman, T.C.: A simplified method to estimate energy savings of artificial lighting use from daylighting. Build. Environ. **40**, 747–754 (2005). https://doi.org/10.1016/j.buildenv.2004.08.007
6. Verma, T., Valdaris, B.: Evaluating the Daylighting Performance of Lecture Halls: A Simulation-Based Approach. In: 53rd International Conference of the Architectural Science Association. pp. 303–312. Architectural Science Association, Roorkee (2019)
7. Bodart, M., Herde, A.D.: Global energy savings in offices buildings by the use of daylighting. Energy and Buildings. **34**, 421–429 (2002). https://doi.org/10.1016/S0378-7788(01)00117-7
8. Konis, K.: Evaluating daylighting effectiveness and occupant visual comfort in a side-lit open-plan office building in San Francisco. California. Building and Environment. **59**, 662–677 (2013). https://doi.org/10.1016/j.buildenv.2012.09.017

9. Nasrollahi, N., Shokri, E.: Daylight illuminance in urban environments for visual comfort and energy performance. Renew. Sustain. Energy Rev. **66**, 861–874 (2016). https://doi.org/10.1016/j.rser.2016.08.052
10. Verma, T., Gopalakrishnan, P.: Occupants' Perception versus Daylighting Simulations': A field study on lecture halls to correlate the occupant's subjective responses and climate-based daylight metrics. In: 35th Plea Conference Sustainable Architecture and Urban Design. PLEA, A Coruña, Spain (2020)
11. Wang, H., Zhai, Z.: (John): Advances in building simulation and computational techniques: A review between 1987 and 2014. Energy and Buildings. **128**, 319–335 (2016). https://doi.org/10.1016/j.enbuild.2016.06.080
12. Yu, X., Su, Y.: Daylight availability assessment and its potential energy saving estimation –A literature review. Renew. Sustain. Energy Rev. **52**, 494–503 (2015). https://doi.org/10.1016/j.rser.2015.07.142
13. Wong, I.L.: A review of daylighting design and implementation in buildings. Renew. Sustain. Energy Rev. **74**, 959–968 (2017). https://doi.org/10.1016/j.rser.2017.03.061

Chapter 74
Evaluation of Retrofitting Strategies to Optimize Thermal Performance of Naturally Ventilated Classrooms: A simulation-Based Approach

S. Niveditha and D. Kannamma

Abstract The thermally comfortable environment in classrooms is significant to improve the performance and productivity of students. A substantial number of researches emphasize energy-saving potential, especially for forthcoming projects, whereas existing buildings in the world consume 40% of the energy to meet the desired indoor comfort conditions. Passive retrofitting strategies provide opportunities for reducing energy consumptions to achieve sustainability in the existing buildings. Presently, architects and designers are taking advantage of building simulation tools that can compare and optimize design alternatives before the installation. The objective of this study is to analyse the performance of twenty-two combinations of retrofitting solutions to optimize the thermal performance of existing classrooms in the Architecture Department located in Tiruchirappalli, India, by using a simulation-based approach. The simulation results showed that retrofitting strategy 1 (RS1—expanded polystyrene + shading device option 1 + night ventilation) was performed better among all other strategies to achieve maximum comfort hours (three times more than the actual comfort hours). However, retrofitting strategy 5 (RS5—shading device option 1 + night ventilation) was found as a cost-effective strategy which does not include any insulating materials. Retrofitting strategies such as the insulating materials and the shading devices were significant only with night ventilation.

74.1 Introduction

Sustainable development in the construction field is the rising call in most of the developing countries [1]. By reducing the energy demand to operate buildings, sustainable and energy-efficient buildings can be developed for the future [2]. The emerging of

S. Niveditha (✉) · D. Kannamma
National Institute of Technology, Tiruchirappalli, Tamil Nadu, India
e-mail: niveditha14arch@gmail.com

D. Kannamma
e-mail: kanama@nitt.edu

A. Chakrabarti et al. (eds.), *Design for Tomorrow–Volume 2*, Smart Innovation, Systems and Technologies 222, https://doi.org/10.1007/978-981-16-0119-4_74

comfortable and energy-efficient buildings results as the key to a complicated optimization which involves a significant number of reliant variables [3]. On the other hand, a substantial number of researches emphasize energy-saving potential especially for forthcoming projects, whereas existing buildings in the world consume 40% of the energy to meet the desired indoor comfort conditions [1]. Retrofitting is one of the critical approaches for achieving sustainability in the existing building stock at a relatively low cost. Retrofitting solutions provide significant opportunities for reducing energy consumptions [4]. The characteristics of sustainable retrofitting should be most favourable, readily available and cost-effective. To find out the best retrofit alternatives, the role of architects, designers and technical experts is obligatory. Nowadays, architects and designers are taking advantage of building simulation tools that can compare and optimize design alternatives before the installation. Computational techniques and computer simulation tools have been efficiently used in building design stage, construction stage, operational stage, optimization and researches for eras [5]. These tools help to achieve goals such as cost reduction, energy savings and diminish environmental impacts and improve the comfortable built environment.

Public buildings became the primary source of energy consumption drastically over the last eras [6]. Educational institutions are the distinct types of buildings that have the prime objective to provide a stimulating learning/ teaching environment for the students and teachers [7]. Based on Annual Status of Higher Education of States and Union Territories in India (ASHE) 2017 report, the total number of universities in India has grown up from 602 in 2011–12 to 799 in 2015–16, growing at a compounded annual growth rate (CAGR) of 5.2%. The educational buildings use energy mainly to achieve thermal comfort. The human thermal environment, as a part of ergonomics, pacts with thermal comfort (i.e. whether an individual feel too cold, hot or sticky), particularly in their working environment. Thermal discomfort environment affects the productivity and performance of both students and teachers and hence creates unsatisfactory conditions. For this regards, the thermal comfort requirements in the classrooms of higher education institutes need serious attention.

This study is a sequence of previous research work [8] done by the authors who carried out an analysis on the thermal performance of existing classroom conditions of an architectural school at Tiruchirappalli, India, and proposed three major retrofitting solutions (design interventions) such as designing of shading devices, installation of insulating materials and night ventilation. The results from the previous study concluded that the addition of insulating materials was giving negative impacts (providing comfort hours less than existing comfort hours) in this context. Therefore, the authors further found a need to examine the performance of twenty-two combinations of retrofitting solutions to further optimize the thermal performance of existing classrooms by using a simulation-based approach.

74.2 Methodology

74.2.1 Location and Building Details

The building was chosen in Tiruchirappalli, Tamil Nadu, India (10°46′ N, 78°43′ E, 88 m AMSL), which has been classified under warm and humid climatic zone by National Building Code 2016 (NBC) of India. April, May and June represent the summer months, whereas December, January and February are relatively colder months. May is measured as the hottest of all the months with 38.1 °C as the maximum monthly mean temperature. Likewise, January is measured as the coolest with a mean monthly minimum temperature of 20.3 °C. The department of architecture of a higher educational institution, which had L configuration, was selected to perform this study. The case study building is a framed structure and consisted of four lecture halls (L1 & L2) at the first and second floor in the typical arrangements and five studio classrooms (S1 & S2) distributed on all the three floors (Fig. 74.1). Since L1 and L2 were identical in spatial arrangements, only L1 is considered for the analysis. All the classrooms and lecture halls were naturally ventilated spaces and had fans to enhance the wind movement. Table 74.1 shows the spatial characteristics of the selected classrooms.

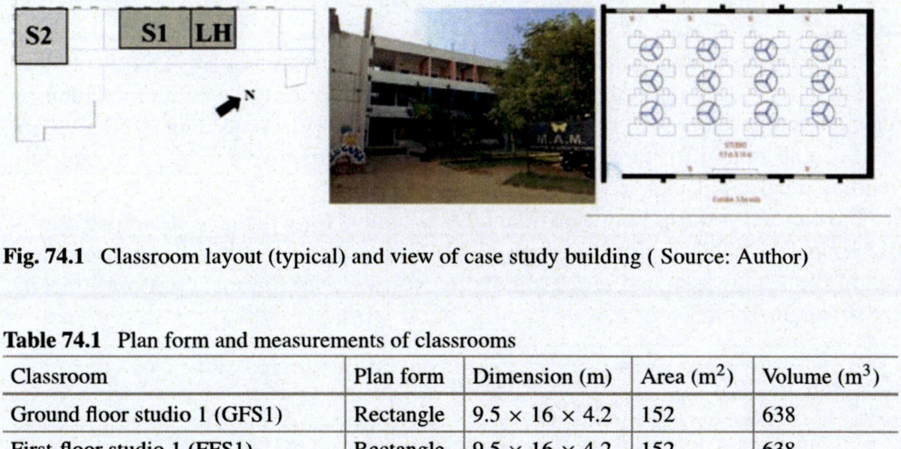

Fig. 74.1 Classroom layout (typical) and view of case study building (Source: Author)

Table 74.1 Plan form and measurements of classrooms

Classroom	Plan form	Dimension (m)	Area (m^2)	Volume (m^3)
Ground floor studio 1 (GFS1)	Rectangle	9.5 × 16 × 4.2	152	638
First-floor studio 1 (FFS1)	Rectangle	9.5 × 16 × 4.2	152	638
Second-floor studio 1 (SFS1)	Rectangle	9.5 × 16 × 4.2	152	638
First-floor studio 2 (FFS2)	Rectangle	9.5 × 16 × 4.2	152	638
Second-floor studio 2 (SFS2)	Rectangle	9.5 × 16 × 4.2	152	638
First-floor lecture hall 1 (FFL1)	Square	9.5 × 8 × 4.2	76	319
Second-floor lecture hall 1 (SFL1)	Square	9.5 × 8 × 4.2	76	319

74.2.2 Modelling and Simulations

SketchUp software (version 2017) was used to create a three-dimensional model for the case study building. Integrated Environmental Solutions software (IES-version 2016) was employed in the calculation of the indoor operative temperature and comfort hours. To simulate the results, IES used variables such as air temperature, wind speed and mean radiant temperature from the weather data file (.epw) obtained from EnergyPlus.

Building envelope template. In the IES simulation software, all the walls (both internal and external) were assigned as solid brick walls of 0.23 m thick and cement plaster of 0.012 m thick on both the sides and whitewashed properly. Along with 0.15-m-thick reinforced cement concrete (RCC) flat slab, the weatherproof clay tiles of 0.075 m thick was given as roof assemblage. The floor finishes were of lightweight vitrified tiles. The doors were made up of 0.03-m-thick plywood which has a thermal conductivity value of 0.1300 W/mK. The windows had operable steel doors and fixed grills which were single glazed having the transmittance value of 0.780.

Thermal comfort template. The study used operative temperature and comfort hours to analyse the thermal performance of all the retrofitting strategies. The formula to calculate indoor operative temperature (neutral temperature) range for the naturally ventilated buildings by using Indian Adaptive Thermal Comfort (IMAC) model is as follows:

Neutral temperature (indoor operative temperature in °C) = 0.54 * outdoor temperature + 12.83. Here outdoor temperature is the thirty days outdoor running mean air temperature (°C) ranging from 12.5 to 31 °C [9]. According to NBC 2016, the 90% acceptability range for the India specific adaptive models for naturally ventilated buildings is ± 2.38 °C [10].

The thermal comfort hours are defined as the hours for which the simulated operative temperature falls within the range of IMAC indoor operative temperature comfort band. The values assigned in the simulation software for thermal comfort template are shown in Table 74.2.

Table 74.2 Thermal comfort inputs

Factors	Values
Working hours (annually)	1841 h
People's schedule	From 09.00 AM to 05.00 PM (inclusive of break time)
People's density	3.81/m^2
People's activity—standing, sitting and writing	1.1 met
Clothing value	0.64 CLO
Ventilation type	Naturally ventilated
Comfort model	IMAC—Indian Adaptive Thermal Comfort Model

74.2.3 Retrofitting Strategies

The study analysed the indoor operative temperature and comfort hours of the existing building and identified the problems (Phase 1: Problem identification). The author proposed three retrofitting options in the pilot study to improve the thermal performance of the existing building are as follows.

Insulating materials. The primary aim of thermal insulation is to decrease the thermal conductance or U-value of walls. Table 74.3 shows the list of insulating materials of different categories from NBC 2016 used for this study.

Shading device design. The sun angles are essential to design shading devices. The author considered the solar chart of Tiruchirappalli city to design the shading devices (options 1 and 2) for western side walls (exposed mainly to solar radiation) of the building (Fig. 74.2).

- Shading device option 1 (To achieve maximum shading from April to September): HSA = -25°; VSA = 30°.
- Shading device option 2 (To achieve maximum shading from October to March): HSA = 40°; VSA = 30°.
- (Note: HSA = horizontal shadow angle; VSA = vertical shadow angle.)

Night Ventilation (NV). Night ventilation indicates the mechanism of natural ventilation at nighttime to bypass extreme heating and cooling of building envelope. A building with suitable thermal mass can reduce peak daytime temperature around 2 °C to 3 °C by adopting night ventilation. Table 74.4 highlights the macro flow inputs assigned in the software for night ventilation.

Table 74.3 Insulating materials specifications

S.No	Insulating material list	Density (Unit: Kg/m^3)	Thermal conductivity (Unit: W/mK)	Specific heat capacity (Unit: J/kg K)
1	Expanded polystyrene	25	0.035	1400
2	Glass wool	200	0.040	670
3	Polyurethane board	30	0.025	1400
4	Sheep wool	23	0.038	1300

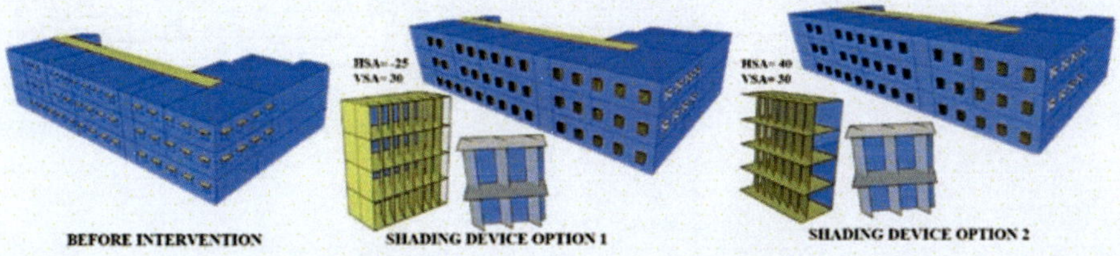

Fig. 74.2 Model of case study building with shading device options

Table 74.4 Macro flow inputs

Air exchange type		Window Open/Close Schedule
Window schedule	Without NV	Open From 09.00 AM to 05.00 PM
	With NV	Open continuously (24 h)
Window dimension		1.5 m × 1.8 m (for all the classrooms)

Combination of retrofitting strategies. Based on the results of the earlier pilot study, the authors further found a need to analyse the performance of combinations of design interventions to optimize the thermal performance of existing classrooms. Table 74.5 indicates the proposed twenty-two retrofitting strategies for this study. A colour code has been assigned to every retrofitting strategies for the analysis point of view.

Table 74.5 Colour coding details of retrofitting strategies (cluster wise)

Insulating material	Shading Device (SD)	Night ventilation	Combinations of Retrofitting Strategies (RS)	Name code	Cluster
1. Expanded polystyrene 2. Glass wool 3. Polyurethane board 4. Sheep wool	Option 1 Option 2	Changing the window open/close schedule	Expanded polystyrene + SD option 1 + Night ventilation	RS 1	1
			Glass wool + SD option 1 + Night ventilation	RS 2	
			Polyurethane board + SD option 1 + Night ventilation	RS 3	
			Sheep wool + SD option 1 + Night ventilation	RS 4	
			SD option 1 + Night ventilation	RS 5	2
			Expanded polystyrene + Night ventilation	RS 6	3
			Glass wool + Night ventilation	RS 7	
			Polyurethane board + Night ventilation	RS 8	
			Sheep wool + Night ventilation	RS 9	
			SD option 2 + Night ventilation	RS 10	
			Glass wool + SD option 2 + Night ventilation	RS 11	
			Sheep wool + SD option 2 + Night ventilation	RS 12	
			Expanded polystyrene + SD option 2 + Night ventilation	RS 13	
			Polyurethane board + SD option 2 + Night ventilation	RS 14	
			Expanded polystyrene + SD option 1	RS 15	4
			Glass wool + SD option 1	RS 16	
			Polyurethane board + SD option 1	RS 17	
			Sheep wool + SD option 1	RS 18	
			Expanded polystyrene + SD option 2	RS 19	
			Glass wool + SD option 2	RS 20	
			Polyurethane board + SD option 2	RS 21	
			Sheep wool + SD option 2	RS 22	
Before Intervention					

74.3 Results and Discussion

74.3.1 Phase 1: Problem Identification

The indoor operative temperature and comfort hours of all the classrooms were obtained from IES software as shown in Fig. 74.3. The X-axis signifies the months, whereas Y-axis (left and right) signifies the indoor operative temperature and the number of hours, respectively. The grey colour lines in the line graph indicate the indoor operative temperature of all the classrooms, whereas the yellow colour band indicates the IMAC thermal comfort range.

Based on the simulation results, the authors have identified the following problems (Fig. 74.3):

- The indoor operative temperature of all the classrooms is exceeding the IMAC thermal comfort range of Tiruchirappalli. During April, it is reaching its peak, and from March to September, the indoor operative temperature ranges of all the classrooms are increased drastically. This leads to discomfort situations to the occupants during their occupied hours.
- The comfort hours are comparatively significantly less than the total occupied hours in all the months. Especially from March to September, the comfort hours are ranging between 11 and 102 h which are approximate, twenty times less than occupied hours.

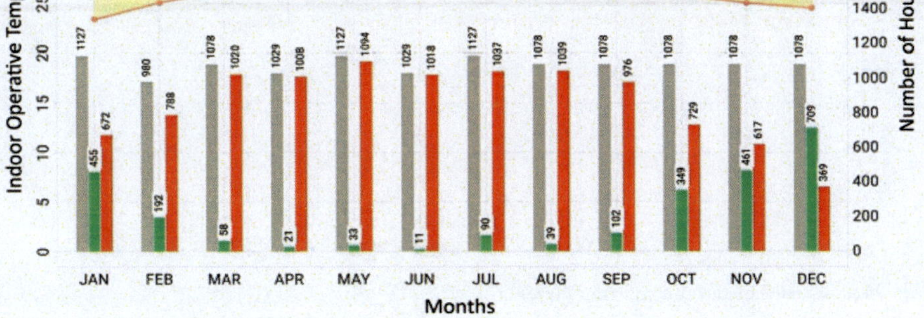

Fig. 74.3 Problem identification

74.3.2 Phase 2: Analysis of Retrofitting Solutions

To rectify the identified problems, the authors suggested the combinations of retrofitting solutions. Figure 74.4 represents the simulation results of the proposed twenty-two combinations of retrofitting strategies. The retrofitting strategies have been divided into four clusters based on the simulation results of comfort hours.

- Cluster one consists of four best-performing strategies (RS1, RS2, RS3 and RS4). All the four strategies in cluster one are achieving almost a similar number of comfort hours ranging from 1100 to 1200 h. From Fig. 74.4, it is evident that the strategies which have combined night ventilation and shading device option 1 along with insulating materials are performing the best. So, the role of night ventilation is significant to this context. In other words, the application of insulating materials is highly dependent on night ventilation. Besides, the performance of shading device option 1 is better than option 2.
- Cluster two consists of RS5 is found to be a cost-effective strategy which does not include any insulating materials. In this strategy, four classrooms are achieving comfort hours more than 1000 h, and the remaining three rooms are achieving more than 750 comfort hours.
- Clusters three and four are providing comfort hours ranging from 200 to 450 h. Explicitly, some retrofitting strategies (RS15 to RS22) have comfort hours which are less than the actual comfort hours (before intervention) because of the absence of night ventilation.

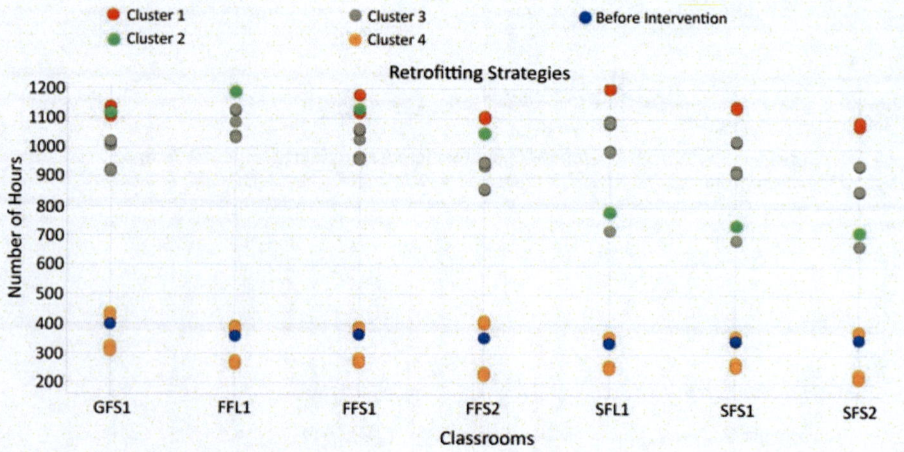

Fig. 74.4 Cluster of retrofitting strategies

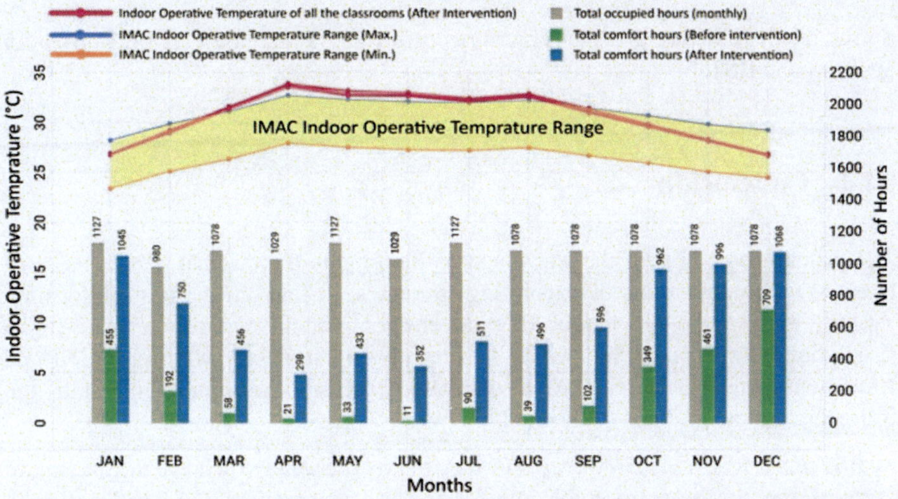

Fig. 74.5 Comparison graph of before and after retrofitting solutions for RS1

74.3.3 Selection of Optimal Solutions:

From the analysis, the authors have found that the retrofitting strategy 1 (RS1—polyurethane board + shading device option 1 + night ventilation) as the best-performing strategy. Figure 74.5 denotes the comparison graph of before and after applying RS1 in terms of both indoor operative temperature (line graph) and comfort hours (bar chart). The number of comfort hours for all the classrooms was significantly increased after optimization. The indoor operative temperature of classrooms is in between the comfort band from January to March and from August to December.

From April to July, the indoor operative temperature of all the classrooms is in between 28 °C and 33 °C, which is almost nearby in the range of IMAC comfort band. Among this temperature range, the users could achieve thermally comfortable environment by choosing their favourable adaptive options such as opening/ closing of windows and changing the seat positions.

74.3.4 Statistical Validation of RS1

A paired sample t-test was conducted to compare the monthly comfort hours of before and after applying RS1. The t-test results showed a significant difference in the monthly comfort hours of before (mean = 30, standard deviation = 32.14) and after applying RS1 (mean = 96.23, standard deviation = 41.62); $t(83) = -32.632$, $p < 0.001$. After applying RS1 (shading device option 1 + expanded polystyrene + night ventilation), the mean and standard deviation values have increased tremendously.

So the results indicate that the monthly comfort hours increased significantly after applying the RS1.

74.4 Conclusion

The primary objective of the study was to investigate the thermal performance of twenty-two combinations of retrofitting strategies, to find out the best-performing strategy to improve the indoor thermal environment in classrooms by using a simulation-based approach. Twenty-two retrofitting strategies were proposed and simulated using IES (version 2016) software. Based on the simulation results, the following are the findings of the study:

- RS1, RS2, RS3 and RS4 were performed nearly the same to achieve the maximum comfort hours and comfortable indoor operative temperature. RS1 (shading device option 1 + expanded polystyrene + night ventilations) was considered as the best in this cluster.
- RS5 (shading device option 1 + night ventilation) was considered as the cost-effective option because it achieved the maximum comfort hours in four classrooms out of seven, without the installation of insulating materials.
- Based on the simulation results, the role of night ventilation was significant to this context because the comfort hours were increased drastically (three times more than existing comfort hours).
- The addition of insulating materials without night ventilation was creating the negative impacts which were giving fewer comfort hours than the actual case. Thus, the strategies such as insulating materials and shading devices were significant only with night ventilation.

The results of this study can guide the architects, engineers and decision-makers to select the most favourable retrofitting strategies for warm and humid climates towards the enhancement of the thermal environments in classrooms. It is clear that the role of night ventilation is creating tremendous influences, but the authors are acknowledging the fact that the opening of windows during nighttime will create some safety and security issues. To overcome these issues, possible solutions such as providing iron mesh and security grills can be implemented. However, further studies on the limitations to assess other passive retrofitting strategies like the position of windows, window wall ratio (WWR), etc., can be explored. Moreover, the results of this study can be appropriate to the analogous building type and context only.

References

1. Ali, H., Hashlamun, R.: Envelope retrofitting strategies for public school buildings in Jordan. J. Build. Eng. **25**, 100819 (2019). https://doi.org/10.1016/j.jobe.2019.100819

2. Khaliq, W., Mansoor, U.B.: Energy-efficient design and sustainable buildings. In: 2014 International Conference on Energy Systems and Policies (ICESP). pp. 1–8. IEEE, Islamabad, Pakistan (2014)
3. Ferrara, M., Sirombo, E., Fabrizio, E.: Automated optimization for the integrated design process: the energy, thermal and visual comfort nexus. Energy Build. **168**, 413–427 (2018). https://doi.org/10.1016/j.enbuild.2018.03.039
4. Ma, Z., Cooper, P., Daly, D., Ledo, L.: Existing building retrofits: Methodology and state-of-the-art. Energy Build. **55**, 889–902 (2012). https://doi.org/10.1016/j.enbuild.2012.08.018
5. Nguyen, A.-T., Reiter, S., Rigo, P.: A review on simulation-based optimization methods applied to building performance analysis. Appl. Energy **113**, 1043–1058 (2014). https://doi.org/10.1016/j.apenergy.2013.08.061
6. El-Darwish, I., Gomaa, M.: Retrofitting strategy for building envelopes to achieve energy efficiency. Alexandria Eng. J. **56**, 579–589 (2017). https://doi.org/10.1016/j.aej.2017.05.011
7. Singh, M.K., Ooka, R., Rijal, H.B., Kumar, S., Kumar, A., Mahapatra, S.: Progress in thermal comfort studies in classrooms over last 50 years and way forward. Energy Build. **188–189**, 149–174 (2019). https://doi.org/10.1016/j.enbuild.2019.01.051
8. Niveditha, S., Kannamma, D.D.: An Analysis of Thermal Environment in Architectural School Classrooms of Warm and Humid Climate in Tiruchirappalli, India. In: An Analysis of Thermal Environment in Architectural School Classrooms of Warm and Humid Climate in Tiruchirappalli, India. p. 10, IIT, Roorkee (2019)
9. Manu, S., Shukla, Y., Rawal, R., Thomas, L.E., de Dear, R.: Field studies of thermal comfort across multiple climate zones for the subcontinent: India Model for Adaptive Comfort (IMAC). Build. Environ. **98**, 55–70 (2016). https://doi.org/10.1016/j.buildenv.2015.12.019
10. National Building Code. Bureau of Indian Standard, New Delhi (2016)

Part XI
Design Modelling, Prototyping, Simulation and Visualisation

Chapter 75
A Role-Based Prototyping Approach for Human-Centered Design in Industry

Nuno Miguel Martins Pacheco, Anand Vazhapilli Sureshbabu, and Markus Zimmermann

Abstract In previous work, the so-called role-based prototyping approach (RBPA) was proposed. It is for interdisciplinary student teams and puts strong emphasis on autonomy and proactivity. Progress in the project is measured by the three lenses of human-centered design, feasibility, viability, and desirability. It guides the students on important aspects of fuzzy-front product development by providing a template of the overall process and specific tools, e.g., an adapted Kanban board. This approach was limited to academia. In this paper, an application of the RBPA in an industry setting is discussed. Industry settings are different from classical fuzzy-front end scenarios because a context is given. To address this, the roles of the proposed approach were refined by assigning specific tasks to incorporate customer requirements. The result was deployed in a university course where students were given design tasks from industry. Each team is guided through this process with the help of daily coaching sessions and their progress is tracked, recorded, and evaluated. The results are then discussed in detail with an outlook for the future.

75.1 Introduction

The success of new products is crucial to the growth and prosperity of modern companies, where the ability to innovate is considered a critical lever for business success [1]. It is not easy to develop new products and to place them successfully on the market, since 80% of new products introduced do not manage to establish themselves on the market after two years [2]. New products are classified into incremental, platform, and breakthrough products by the dimensions of *extent of product change*: new core product, next-generation product, addition to product family, and derivates and enhancements; as well as *extent of process change*: new core process, next-generation process, update, and incremental change [3]. The development of new products is associated with complex processes that depend on various success factors, requiring the transformation of ideas or opportunities into tangible and viable products, known

N. M. Martins Pacheco (✉) · A. Vazhapilli Sureshbabu · M. Zimmermann
Technical University Munich, 80333 Munich, Germany
e-mail: martins.pacheco@tum.de

A. Chakrabarti et al. (eds.), *Design for Tomorrow–Volume 2*, Smart Innovation, Systems and Technologies 222, https://doi.org/10.1007/978-981-16-0119-4_75

as new product development (NPD) [2]. It is a long way in the innovation process to launch a new product and the NPD process is preceded by the fuzzy-front end (FFE) and is concluded with the commercialization [4]. The relevance of new products in a technological market [5] is forcing companies to rethink priorities due to rapid progress, global market structures, and high competition [6]. Traditional product development processes are no longer state of the art but have been expanded by important elements such as: multidisciplinary collaboration, self-organized teams, knowledge transfer, and overlapping iterative development phases [6]. FFE is not formalized and generally includes phases from Idea generation, concept development to project evaluation [5]. Product life cycles reduce and competitive market environments change rapidly, leading to FFE as the greatest weakness, but it also contains the biggest potential for improvement in product innovation [4, 7]. Zhang and Doll propose a causal model in their research, containing the necessary elements for the success of new product development as follows: *Front-end fuzziness*, *foundation elements*, and *team vision*, whereby the individual elements are further subdivided, revealing the complexity and the resulting challenges [8]. *Front-end fuzziness* stresses the following challenges: customer, technology, and competition fuzziness. *Foundation elements* emphasize the following challenges: strategic orientation, heavyweight manager, concurrent engineering, customer involvement, supplier involvement, and platform products. *Team vision* highlights the following challenges: shared team purpose, strategic fit of project targets, and clarity of project targets. This hypothesis of Zhang and Doll is further supported by the extensive literature review by Korityak and Cao, who addresses the challenges of new product development in the fuzzy-front end [9]. Therefore, a multi-dimensional complex problem in design space exists, in which boundary conditions are unknown and several factors are interdependent and influence each other.

In recent years, several academic formats based on the model of *ME310* (Stanford University) emerged, such as *Xtech Entrepreneurship* (DTU Copenhagen), *Think.Make.Start.* (TU Munich), where teams of students were introduced to entrepreneurship in the context of new product development. Recently, UnternehmerTUM GmbH started *TMS Industry (TMSI)* to offer such innovative formats in shorter 3–5-day workshops to companies that have focused on finding an idea to expand their product portfolio. In TMSI, ideas are prototyped and implemented to convince executives of the concept. Similarly, the BMW Group has also started a similar program to run its own internal sprints, where employees from different backgrounds are released for 3–5 days to work on their ideas in a set context and to implement them in a feasible prototype. Previously, a new role-based approach based on prototyping was proposed [10] that helps interdisciplinary teams while placing emphasis on autonomy and proactivity of the students as well as giving them an adequate balance of freedom and supervision. Prototyping activities form an integral part of development, and the prototyping lenses of desirability, feasibility, and viability are used for targeted development and are success factors for new product development [11]. The method guides the students through the fuzzy-front end of new product development using tools that emphasize on slow integration of artifacts via micro-logic that gets progressively complex with added methods and guidelines on prototypes and sprints. This is highlighted by the prototyping process while placing prototypes as a measure of progress, through the prototyping lens. However, the proposed method is limited to academia and is not applicable to industry because there are differences in expected output and problems between the two application fields. Development projects in academia are associated with a complete FFE, dependent on the entrepreneurship contexts of TMS. Whereas in industry, there is a given context and mostly solutions only need to be rethought for efficiency, cost or for the problem formulation due to its fuzziness with a lack of development and/or resources allotted to said problem. The objective is to connect academia and industry by adapting the previously proposed FFE method for academic contexts and using it in innovation engineering. The industry context is given through preselected problem domains, even though the problem domains are fuzzy the problems originated from presented technical problems by the OEM. Moreover, activities in the marketing and the business domain are not necessary since the result is not a startup. Through an explorative study, this paper proposes a prototyping approach linked to a role-based model, supporting student teams, and proposing a formalized prototyping methodology for industry settings. The following research questions are addressed: *(RQ1) How do novice teams embrace a formalized, goal-oriented approach to prototyping in industry settings?* and *(RQ2) How does the adapted approach affect the prototyping activities?*

In Section II of this paper, the development of the methodology for industry settings is outlined, followed by its first deployment for data collection in Section III. The paper presents the results from the deployment in Section IV and discusses the results and reveals insights and questions in the final Section V.

75.2 Methodology

75.2.1 Course Design

The basic course format is inspired from the TU Munich's Think.Make.Start. (TMS) course which aims at running student projects that have the potential to be a technology startup for standalone products [12]. Not all methodologies and processes in that format are transferable to industry settings; a new course called TMS Enterprise (TMSE) is created to address this gap in innovation process engineering for industries with student teams. The ten-week course TMSE with 20 students of TUM serves as the test-setting for this research. Current needs and problems in the industry from technological and economic systems are identified, analyzed, and validated in the interdisciplinary team and solutions are iteratively improved through prototyping activities. The students are provided an 800€ budget for project expenses, workshop access, problem owner, and coach support with methods and tools on NPD, business models and technological feasibility. The goal of the course is to create a product that is desirable (i.e., fulfilling a user need), feasible (i.e., available resources), and viable (i.e., potential customers) [11], namely a minimum viable product (MVP) [13]. The previous RBPA approach proposes three roles: business developer, tech developer, and problem expert [10]. Those roles function as enabler for collaborative work and provide a guideline but were too ambiguous for an industry context and did not provide a clear function for each role. In addition, prototypes are placed as measure of progress in the original RBPA guiding students through fuzzy-front product development. It provides a template of the overall process and tools for the case that no problem or context is defined in the beginning of the development. This approach is highly limited to academic settings and gives the students the freedom of choice to find a problem worth solving.

75.2.2 An Adapted Role-Based Prototyping Approach (Adapted RBPA) for Industry Settings

In industry, the context is given and a solution is needed. This means that process optimizations or incremental product improvements apply. It should not be forgotten that innovations are essential for industry to remain competitive, in a fast paced and changing world as defined by VUCA environments. Therefore, we propose a modified version of the RBPA approach. A holistic framework for the *adapted role-based prototyping approach (adapted RBPA)* is illustrated in Fig. 75.1.

The overall structure of the prototyping journey includes the modified team roles inspired by agile development teams. The *project manager* is responsible for meeting the project goals. He encourages the team to focus so that deadlines are met. The *function designer* is responsible for customer-relevant attributes. He interprets what people want and derives the equivalent user requirements. The *architect* is responsible to define system attributes. He defines the product architecture and therefore must interpret the user requirements from the function designer originating from users. The *component designer* is responsible for component attributes. He tries to figure out the needed technological aspects in the solution. These modified roles are

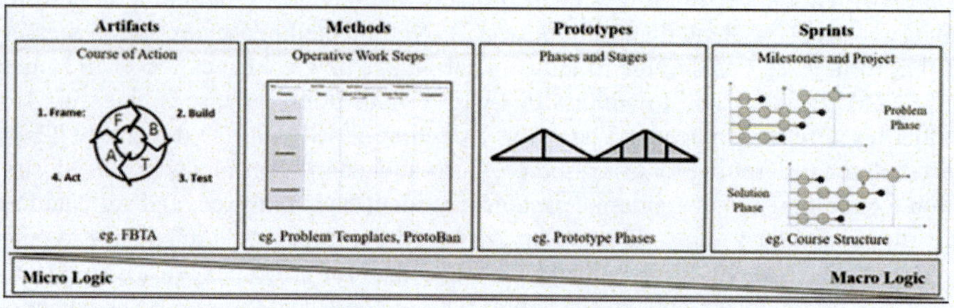

Fig. 75.1 Holistic framework for the adapted role-based prototyping approach (adapted RBPA) of TMSE consisting of a micro- and macro-logic

the enabler for collaborative work between different disciplines by clearly splitting the responsibilities in teams through four distinct roles. In the aRBPA, the prototype phase milestones are divided into three distinct metaphases: *problem*, *solution*, and *launch phase*. Those phases are subdivided into two distinct prototype phase milestones. It is only possible to continue to the next metaphase as soon as the previous is concluded. The concept of pretotyping was added to the methodology [14]. Those milestones provide tangible goals and development structure for the development teams and serve as an orientation in an iterative process. The methods and tools are not explicitly integrated in the approach as it aims at students being proactive when they face challenges. The *problem pretotype* starts the problem phase and encourages teams to understand the problem they are dealing with. It is a complete representation of the problem with all possible complications considered, for a best possible view of the problem domain. The *problem prototype* aims at defining the final set of problems that can be reasonably deduced to be solved through targeted user feedback. Those problem features help to define a singular definition for the *minimum viable product (MVP)*. The *solution pretotype* starts the transition from problem to solution metaphase. It illustrates a partially mocked-up representation of the intended product or service that can be built in minutes, hours, or days instead of weeks, months, or years. Different solution features try to solve different problem features. The *model prototype* concludes the solution phase and is mainly used to collect customer feedback and define the marketing strategy. It is a looks-like representation of the solution without features deployed on the product, but in a presentable form, which is mainly used as a demo unit, for marketing purposes and sales push. The *functional prototype* is a version of the solution with a minimal set of features deployed on a product, mainly used as demo units. The *minimum viable product (MVP)* contains sufficient features to satisfy early adopters and convince them to invest in the product. In addition to the principles of human-centered design of desirability, feasibility [11], and viability, the prototyping purposes of exploration, evaluation and communication are added [15]. This approach is not sequential, and it is possible to jump back to previous phases for pivots. If a metaphase is not concluded, it has first to be closed (Fig. 75.1).

75.2.3 Deployment of the Methodology

The methodology is deployed at a modified semester long TMS course together with an industry partner. It starts with a kick-off day that introduces production topics as the context and serves as a day to get familiar with everybody and the practical approach. In the following three weeks, an observation phase takes place in the production site of the industrial partner, where the students observe and document the most relevant topics. After the observations, the students document the encountered difficulties and create a problem wall with all problems that have been found. This problem wall contains all possible problems that were collected during the observation phase in a summarized format. The pre-event for the sprint week takes place afterward, where students and employees form teams based on the given problems. A team consists of at least three students and two employees. The employees take on the roles of project manager, architect, and function designer. In this initial phase, all the students are component designers. At the end of the pre-event, the teams create problem pretotypes that represent the problem which the team wants to solve. During the sprint week, 30 min of mandatory input on selected topics such as *design thinking, lean, agile development, user testing, pitching,* and *methodology* are given in the morning at the beginning of the day. The teams are supported in their daily sessions by coaches with methodological and technological know-how.

Progress presentations with teams presenting their current progress to their peers are held every day. The user testing for a few selected customers from the partner is also carried out. At the end of the sprint phase, the demo day marks the first milestone, where the teams showcase their products to over 50 invited guests. During the final sprint week of the sprint phase, the first day starts with a deeper understanding and specification of the problem selected in the pre-event by further subdividing the problem. The teams create problem prototype and use it to create a vision solution video. On day two, the first solution approaches are shown as solution pretotype. The third day aims to continue working on the solutions (solution pretotype) and in addition to get first customer feedback from invited guests throughout the day. On the last two days, different solutions are integrated into a functional concept called functional prototype and a model prototype is created. The purpose of the intermediate event is to get feedback on the project to decide on a continuation of the project. After the event, the teams have four weeks to integrate the received feedback into their concept and take the product to the next level. A final event at the end completes the whole process. The project is continued or gets management support, if a project is considered satisfactory.

75.3 Evaluation

The course received 35 applications from various disciplines from TUM, from which the best 20 received admission. In addition, 14 employees from the OEM participated in the sprint phase as an addition to the teams. The data was collected from the six participating teams and was considered for evaluation and analyzed. The number of prototypes in respect to the prototype phase milestones is illustrated in Fig. 75.3. In total, 5 problem pretotypes, 14 problem prototypes, 45 solution pretotypes, and 22 functional prototypes were built during the sprint phase and 7 solution pretotypes, 21 functional prototypes, and 6 minimum viable products were built during the development phase (Table 75.1).

It is evident that the teams have built more prototypes in the sprint phase than in the development phase. Additionally, most prototypes were built in respect to the solution than to the problem, which is represented by a high amount of solution pretotypes and functional prototypes. Not all milestones were covered during the prototyping activities. The teams that are highlighted in red are the winning teams of

Table 75.1 Team composition. The teams formed independently from various disciplines from TUM and employees from the OEM during the sprint phase. After the sprint week, the employees left, and the students continued the project

	Team A		Team B		Team C		Team D		Team E		Team F		Sum sprint	Sum Dev.	Sum
	Sprint	Dev.	Sprint	Dev.	Sprint	Dev.	Sprint	Dev.	Sprint	Dev.	Sprint	Dev.			
Architect (A)	0	0	0	0	1	0	0	1	1	0	1	0	3	1	4
Component designer (CD)	4	1	4	2	3	1	3	1	3	1	3	1	20	7	23
Function designer (FD)	1	1	1	1	1	1	1	0	1	1	0	1	5	5	9
Project manager (PM)	1	2	1	1	1	1	1	1	1	1	1	1	6	7	12
Sum	6	4	6	4	6	3	5	3	6	3	5	3	34	20	54

the different phases and are the one that stayed longer in the problem space. There is a clear shift from solution pretotypes to functional prototypes in respect to the sprint and development phase.

Comparing the prototype milestones in relation to the prototyping purposes, it becomes apparent that especially in the problem stage a lot of emphasis was put on communication and that the exploration took place from the moment the solution was proposed (sprint phase). Looking at the development phase, it shifts especially toward evaluation and partly stays with exploration. The relation between prototype milestones and prototyping purposes is represented in Fig. 75.4. Tracking the individual roles consisted of capturing the efforts of the individual roles, which were specified on the tasks in the Kanban board, which were directly linked to the individual prototyping milestones. Considering the engagement of the individual roles in relation to the prototype phases, it is evident that especially the component designer pushes the solution forward in the sprint phase. The project manager adds a problem focus and the other roles are more focused on early solution concepts.

During the development phase, the component designer focuses completely on the functional prototype. The project manager pushes toward the minimum viable product. The other roles are still realizing simple individual solutions within the solution pretotype. The engagement of the different roles in respect to the prototype milestone phases is illustrated in Fig. 75.5. In terms of the commitment of the individual roles in relation to the prototyping purposes, the component designer is observed to be driving the project forward with a special focus on exploration. The project manager on the other hand is largely responsible for communication within and outside the team. The other roles are more focused on early solution concepts and therefore build a bridge between communication and exploration. During the development phase, the component designer focuses completely on evaluation like all other roles. The project manager still has communication as a priority focus. The relations are shown in Fig. 75.6.

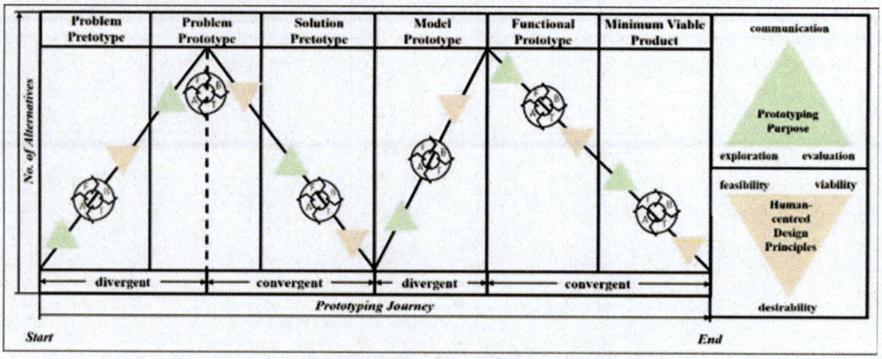

Fig. 75.2 Prototype phase milestones are split into a double diamond containing divergent phases broadening the design space and convergent phases combining previous insights into a single concept. The principles of human-centered design and prototyping purposes together with the frame, build test, and act cycles support the development activities

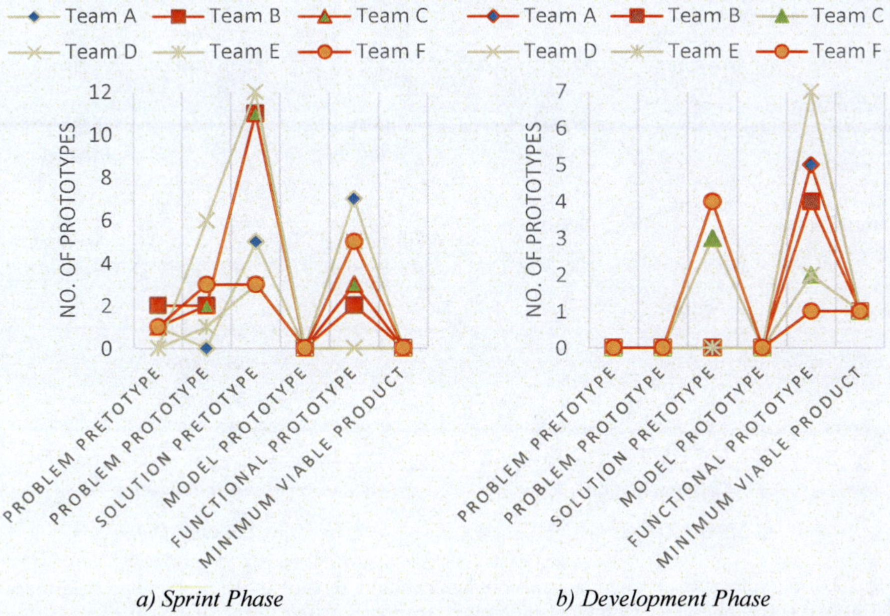

Fig. 75.3 **a** Prototyping milestones in relation to teams during the sprint phase. The prototyping toward the milestones of problem pretotype, problem prototype, solution pretotype and functional prototype were considered. Solution pretotypes received most consideration. **b** Prototyping purposes in relation to teams during the development phase. In the development phase, the prototyping activities only consisted of solution pretotypes, function prototypes the most, and the minimum viable product

75.4 Discussion

75.4.1 Methodology

The adapted RBPA methodology was successfully applied and provided the teams with the necessary support to help develop new products in an industrial context. The results show that there is a need for improvement. Student feedback was collected, and it was noted that without a standardized platform, it was difficult to collect data on retrospectives. This led us to believe that there is a lack of tools adapted to the methodology, which would have helped the collaboration of the teams to plan and reflect on prototyping activities.

In the methodology, the focus was on prototyping and the individual prototypes were developed and tested in consideration of the prototyping purposes. The roles were specified by the setting and links between the individual roles were established. This promotes cooperation and reduces uncertainty in new product development. To address this problem, several prototyping activities with different prototyping purposes and their documentation in relation to the principles of human-centered

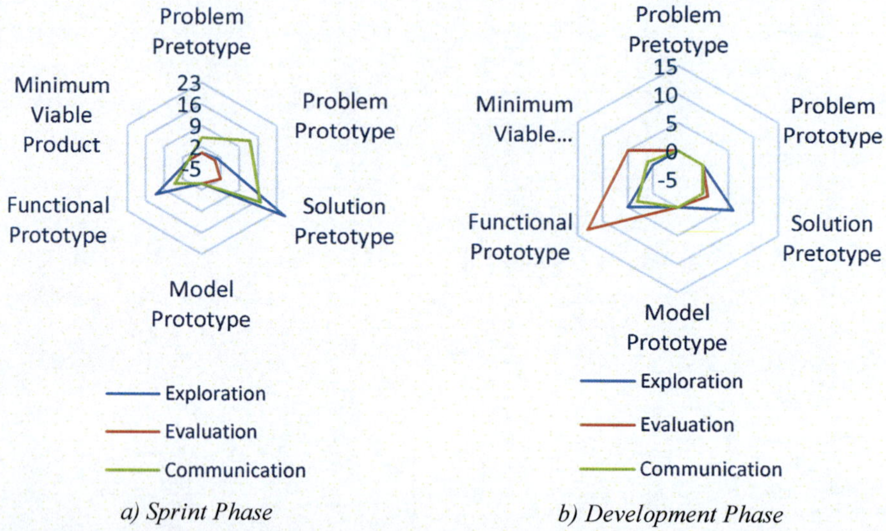

Fig. 75.4 **a** Exploration, evaluation, and communication in relation to the prototyping milestones during the sprint phase. The problem milestones are mostly used for communication while the remaining prototypes are used for exploration as well as communication. **b** Exploration, evaluation, and communication in relation to the prototyping milestones during the development phase. In the development phase, the prototyping activities shifted more to exploration and mainly evaluation, which accounts for most of it

design are necessary. This results in prototyping configurations between the human-centered design principles and the prototyping purposes. These two types serve as search vectors in the design space. This was an explorative study with the aim of adapting and testing the methodology from academia for use in industry. Regarding the application of the methodology, we can speak of a success, since most students used the approach accordingly and had a high proportion of prototyping activities. It can also be observed that teams that stayed longer in problem space and only then entered the solution space in design space were more successful in developing a minimum viable product. The impact of the methodology will have to be addressed at a later stage, but there are other challenges for improving the adapted RBPA methodology. One important observation from the aRBPA approach is the roles. This gives clear indication on the contribution of each role at any given phase or prototyping cycle. This should help in gaining insight on the kind of effort required by each contributing member of the team and how they can be best utilized. Another interesting observation is the winning teams spent a lot more time in understanding the problem completely during the sprint phases before attacking the specific attributes in the solution prototypes in the later stages. And this approach made it easier to make that observation and record it.

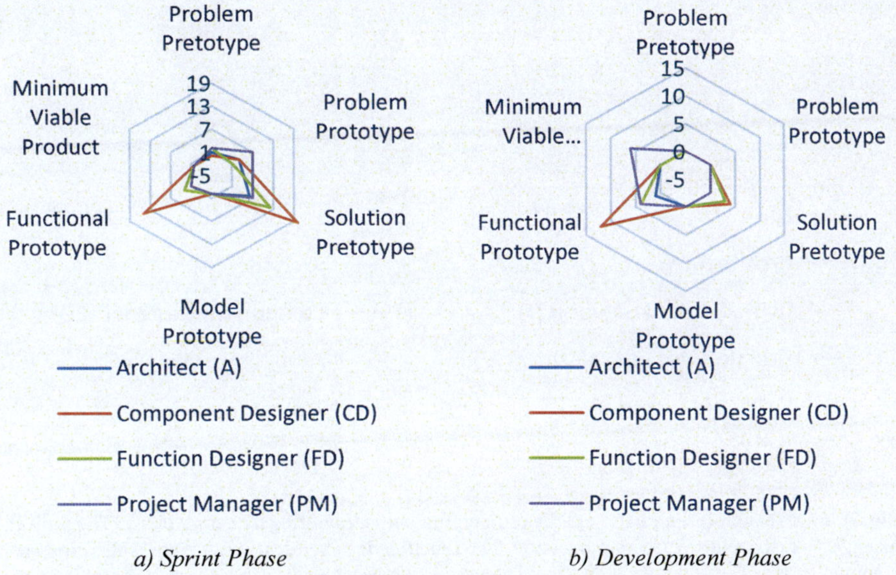

Fig. 75.5 **a** Team roles in relation to the prototyping milestones during the sprint phase. The project manager is evenly distributed and has a preference toward the problem prototype and the solution pretotype. The function designer has also a peek toward the solution pretotype. The component designer in contrast has a peek at the functional prototype and the solution pretotype. The architect, on the other hand, is balanced. **b** Team roles in relation to the prototyping milestones during the development phase. The project manager is balanced and has a tendency toward the functional prototype and the minimum viable product. The function designer has peaks at the solution pretotype. The component designer now deflects to the functional prototype. The architect, on the other hand, is balanced

75.4.2 Contribution and Limitation

This exploratory study builds on recent work in design research and represents an extension of the previously introduced RBPA methodology to support designers in the development of new products through systematic prototyping by reducing uncertainty and risk in an industrial context. The focus lies on a goal-oriented, structured prototyping approach, which facilitates the development process in conjunction with a role allocation, while adhering to the principles of human-centered design and following prototyping purposes. This approach is designed for application in student development projects at universities in an industrial context. The objective of this test was to assess the effectiveness of the methodology, but this could not be measured, as only the application and data collection were the objective. It must be emphasized that the choice of data collection using adapted Kanban boards and excel tables is not recommended. These tools were chosen to collect data, but at no time did not focus on usability for the development teams. The new more specific roles offer a direction, but the tasks of the individual roles were partly unclear and need to be

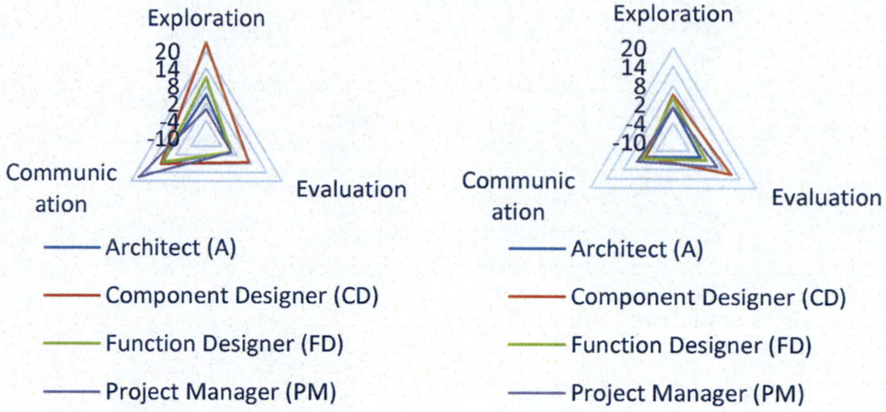

a) Sprint Phase *b) Development Phase*

Fig. 75.6 **a** Team roles in relation to the prototyping purposes during the sprint phase. The project manager is responsible for communication. The function designer deals with both exploration and communication. The component designer in contrast is focused on the exploration. The architect, on the other hand, is also responsible for the communication. **b** Team roles in relation to the prototyping milestones during the development phase. The project manager is pushing for communication and evaluation. The Function designer deals with both exploration and evaluation. The component designer is focused on the exploration. The architect, on the other hand, is balanced

communicated better the next time. A possible solution could be a tool for the development teams to track their prototyping activities and serve as a collection point for information. Prototyping phase milestones gave a good orientation by separating into problem and solution space. However, it is debatable whether these are entered sequentially or whether there are multiple loops. This was a first test in an industrial context and the next step would be to test the methodology with a company over a longer period. Overall, the aRBPA proved to a successful method when observed to the successes of the team in deploying useful solutions. The effectiveness of aRBPA was measured in terms of team successes which in turn was evaluated by efficiency of the product development and successful deployment of the developed solutions in the OEM production site. It gave very good insights on team effort by adapting a role-based approach to prototyping. The prototyping phases proved useful in self-documentation and as an unambiguous measure of progress. However, the deployment of the methodology itself needs to be done with more efficient tools and in a more intuitive manner, so that adoption is easier for novice users.

75.4.3 Conclusion and Outlook

The adapted RBPA methodology was tested and evaluated for the first time in the customized course TMSE. The adapted RBPA was extended by the prototyping purposes and delivers together with the principles of human-centered design search vectors for the design space in a non-quantifiable area. This provides a structured progress map, indicating where further activity is required. This way, the development process becomes more focused and efficient while providing sufficient freedom in early phase design work. The current evaluation is limited to the one scenario considered and only based on qualitative assessment of the development results. It is necessary to collect more and systematic data to draw a conclusion on general applicability and usefulness of the proposed approach.

References

1. Cooper, R.G.: Winning at new products: pathways to profitable innovation. In: Proceedings Project Management Research Conference, Montreal, Canada (2006)
2. Trott, P.: Innovation Management and New Product Development. Always learning, Pearson (2016)
3. Koen, P.A.: The fuzzy front end for incremental, platform and breakthrough products and services. PDMA Handbook. Pp 81–91 (2004)
4. Koen, P.A., Ajamian, G.M., Boyce, S., et al.: Fuzzy front end: effective methods, tools, and techniques. The PDMA toolbook 1 for new product development (2002)
5. Bhuiyan, N.: A Framework for successful new product development. JIEM 4. https://doi.org/10.3926/jiem.334(2011)
6. Takeuchi, H., Nonaka, I.: The new new product development game. Harvard Business Review **64**, 137–146 (1986)
7. Herstatt, C., Verworn, B.: The 'fuzzy front end'of innovation. In: Bringing technology and innovation into the boardroom. Springer, pp 347–372 (2004)
8. Zhang, Q., Doll, W.J.: The fuzzy front end and success of new product development: a causal model. European J Innov Manage (2001)
9. Korityak, A., Cao, Y.: Challenges in fuzzy front end of new product development within medium-sized enterprises: A case study on Swedish manufacturing firms (2010)
10. Martins Pacheco, N.M., Behrenbeck, J., Vazhapilli Sureshbabu, A., Zimmermann, M.: A role-based prototyping approach for human-centred design in fuzzy front-end scenarios. Nord Design 2020 14th International Nordic Design Conference, Copenhagen, Denmark (2020)
11. IDEO.: The field guide to human-centered design. Design Kit (2015)
12. Böhmer, A.: When digital meets physical-Agile innovation of mechatronic systems, Technische Universität München (2018)
13. Ries, E.: The lean startup: How today's entrepreneurs use continuous innovation to create radically successful businesses. Crown Business (2011)
14. Savoia, A.: The Right it: Why So Many Ideas Fail and how to Make Sure Yours Succeed. HarperCollins (2019)
15. Blomkvist, J., Holmlid, S.: Existing prototyping perspectives: considerations for service design. Nordes (2011)

Chapter 76
Improvement of Operator Comfort of a Vibratory Compactor

A. Gomathinayagam, S. Babu, P. Antony Stephen, R. Raja Ganesh, and D. Boopathi

Abstract Vibratory compactors are primarily used for compaction tasks in earthwork and road building. They are regularly exposed to vibrations caused by the moving elements within them like drum and engine and the unevenness of the road or soil profile. Because of these vibrations, the structures are subjected to dynamic loads and the operators are endangered to health issues. A higher prevalence of spine problem, pain in chest, abdomen, arm and shoulder are found in the drivers of the compactors. Strict norms concerning these health issues make the design of an effective operator compartment inevitable on the compactors. At present, Indian earth moving equipment manufacturers follow ISO 2631–1 and ISO 5349–1 to ensure the operator comfort in the compactors. These standards insist practical testing of the operator compartment which can be time consuming. The frequency of dynamic loading plays a pivotal role in determining the operator comfort as well as the damage of the operator compartment structures. The major vibrating frequencies prevailing in the compactor are its drum operating frequencies and engine rotating speeds. In order to protect the operator from health issues and the structures from failure, the operator compartment should be designed in a way that its natural frequencies evade the major vibrating frequencies prevailing in the compactor. In this work, an effort has been made to develop a methodology using finite element analysis to improve the operator comfort of the compactors. The FEA results have been compared with that of experimental testing, and the FEA methodology has been fine-tuned to get a good correlation. In future, this FEA approach will be used to finalize the operator compartment design prior to practical testing with an objective of reducing the development cost and time and ensuring the operator comfort.

76.1 Introduction

Vibratory compactors are primarily used for compaction tasks in earthwork and road building. They are self-propelled vehicles that use steel drums to compress the

A. Gomathinayagam (✉) · S. Babu · P. Antony Stephen · R. Raja Ganesh · D. Boopathi
Product Development Center, Larsen and Toubro Limited, Coimbatore, India
e-mail: Gomathinayagam.A@larsentoubro.com

A. Chakrabarti et al. (eds.), *Design for Tomorrow–Volume 2*, Smart Innovation, Systems and Technologies 222, https://doi.org/10.1007/978-981-16-0119-4_76

underlying soil or hot mix asphalt. They can have one or two drums. Usually soil compactors have single drum, while asphalt (Tandem) compactors have two drums. They use combined static and dynamic forces (weight and vibrations) to have a better compaction.

A typical Vibratory Tandem Compactor is shown in Fig. 76.1. It uses hydrostatic power transmission system. The drum vibration is produced by a rotating eccentric shaft and weights, and the centrifugal force created is proportional to the eccentric moment and the rotation speed. The eccentric shaft is rotated by means of a hydraulic motor which achieves desired frequencies and amplitudes by regulating speed with a variable displacement pump and by moving weights.

These compactors are often exposed to vibrations not only due to drum and engine vibrations but also due to the unevenness of the road or soil profile. Because of these vibrations, the structural components are subjected to dynamic loads and the operators are exposed to health issues. The operator comfort depends on whole body vibrations (WBV) and hand arm vibrations (HAV) experienced by them. WBV is the mechanical vibration transmitted to the whole body when operating heavy construction machinery. It is measured using tri-axial accelerometer mounted in a pad and placed on the seat surface/seat back/feet (Fig. 76.2). HAV is the mechanical vibration transmitted to the hands when operating heavy construction machinery. It is measured using tri-axial accelerometer mounted on a hand adapter (Fig. 76.3).

A higher prevalence of spine problem, pain in chest and abdomen and motion sickness are common due to WBV, while pain in arm and shoulder, reduced grip strength and vibration-induced white finger syndrome are common due to HAV [1–3]. Strict norms concerning these health issues make the design of an effective operator compartment inevitable on the compactors. At present, Indian earth moving equipment manufacturers follow ISO 2631–1 [4] for WBV and ISO 5349–1 [5] for

Fig. 76.1 Vibratory tandem compactor

Fig. 76.2 Whole body vibration

Fig. 76.3 Hand arm vibration

HAV to ensure necessary operator comfort in the compactors. These standards insist practical testing of the operator compartment which can be time consuming.

The vibrating frequency plays a key role in determining the operator comfort as well as failure of operator compartment structures. The main contributing frequencies in the compactor are its drum operating frequencies and engine rotating speeds. In order to protect the operator from health issues and the structures from failure, the operator compartment should be designed in a way that its natural frequencies evade the major vibrating frequencies prevailing in the compactor [6–10]. In this work, an attempt has been made to develop a finite element analysis (FEA) methodology using the commercial analysis package, ANSYS, to improve operator comfort of the compactors. Relative comparison of FEA results with that of experimental testing has helped to achieve a better operator comfort. In future, this FEA approach will be used to finalize the operator compartment design prior to practical testing with an objective of reducing the development cost and time and ensuring the operator comfort.

76.2 Evaluation of Operator Comfort Using ISO 2631–1 (WBV) and ISO 5349–1 (HAV)

76.2.1 Definitions

A(8): The 8-h energy equivalent vibration total value for a worker.

EAV: Exposure action value indicates the level of vibration at which action must be taken to reduce exposure.

ELV: Exposure limit value is the maximum level of vibration that should not be exceeded in any single day.

76.2.2 Procedure

As discussed in the introduction chapter, both the WBV and HAV are measured using a tri-axial accelerometer. For WBV, the accelerometer is placed inside a pad and mounted on the operator's seat surface. For HAV, the accelerometer is mounted using an adapter and held in hand or on the wrist as shown in Fig. 76.3. The coordinate system of the tri-axial accelerometer is shown in Figs. 76.2 and 76.3. For practical vibration measurements, the orientation of the coordinate system may be defined with reference to an appropriate basicentric coordinate system. In general, X-axis is aligned toward the travel direction, while Y-axis and Z-axis are aligned toward the transverse and vertical directions, respectively. Figure 76.4 shows various steps that have been followed to evaluate the A(8) value of WBV and HAV. While capturing the raw data, sampling size should be kept at least two times more than that of the maximum frequency existing in the system. The maximum frequency in the compactor is engine high idle frequency of 80 Hz, and the sampling rate considered here is 256 Hz.

76.2.3 Daily Exposure Limit Values

The daily exposure limit values of A(8) for WBV and HAV are given in Table 76.1.

76.3 Finite Element Analysis

The operator platform of a 9 T class Tandem Compactor is considered for the analysis. The operator platform is mounted on the front chassis with the help of anti-vibration

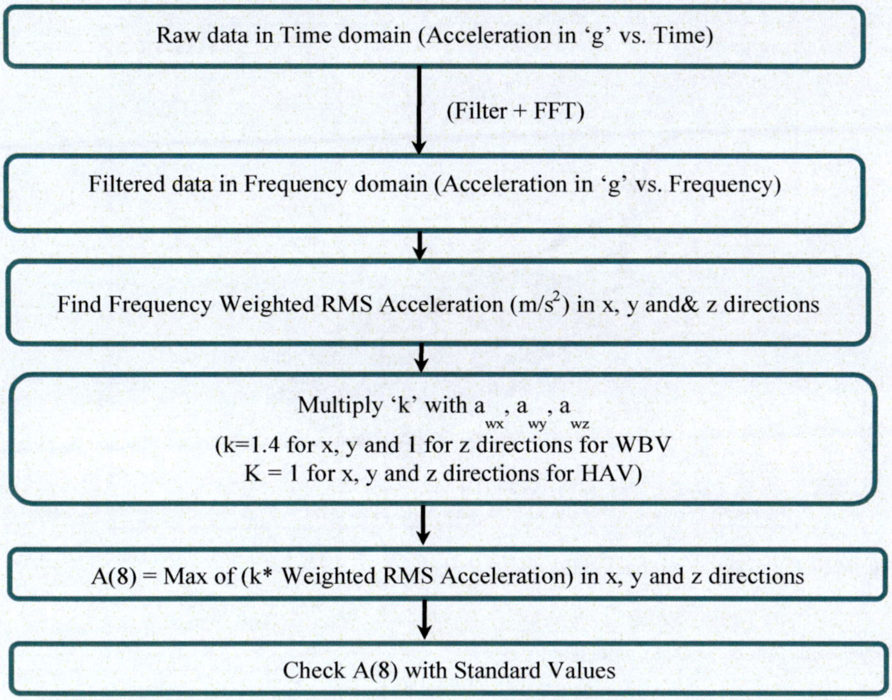

Fig. 76.4 Flow chart to evaluate A(8) value

Table 76.1 Daily exposure limit values

A(8) in m/s^2	EAV	ELV	Reference
WBV	0–0.5	0.5–0.8	ISO 2631–1
HAV	0–2.5	2.5–5	ISO 5349–1

mounts (AVM). The swivel seat is placed above the operator platform, and the steering wheel mounted on the front console is also fixed to the operator platform. The vibrations transferred from the front chassis to the operator platform and further to the seat and the steering wheel depend upon the stiffness of AVM along three directions. The stiffness of AVMs decides the natural frequency of the operator platform which needs to be away from the machine operating frequencies and engine low idle and high idle speeds.

The finite element model of the operator platform is shown in Fig. 76.5, and FEA is carried out using commercial analysis software, ANSYS. Details of the FE model like element, real constant and material properties are shown in Table 76.2. Modal analysis is carried out to find out the natural frequencies and corresponding mode shapes of the operator platform. FE model is first solved using the existing AVM stiffness values. The natural frequencies of the operator platform are also listed in Table 76.4.

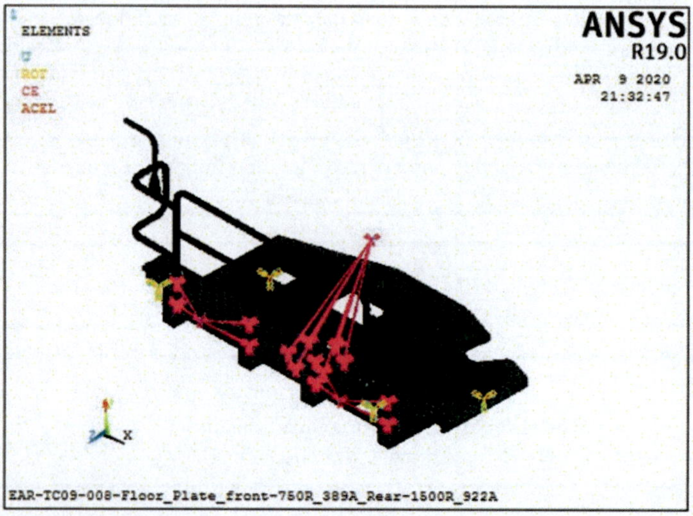

Fig. 76.5 Finite element model of the operator platform

Table 76.2 Finite element model details

Part	Element type	Real constant	Material property
Seat + operator mass Front console Tools box	MASS 21	Mass values	–
Operator platform Front chassis mounting bracket	SOLID 187	–	Elastic modulus, Poisson's ratio, density
AVM	COMBIN 14	Stiffness along 3 directions	–

It is evident from Table 76.4 that a few of the natural frequencies (highlighted) are closer to the machine operating frequencies as well as engine idle speeds (shown in Table 76.3) which shall result in the resonance causing more vibrations and discomfort to the operator. Hence, it is decided to keep the natural frequencies of the operator platform away from the operating frequencies. The simplest way is by changing the stiffness of the AVM values. FE model is solved for different AVM stiffness values, and the corresponding natural frequency values are shown in Table 76.5.

Table 76.3 Machine operating frequencies and engine idle speeds

Machine low frequency, in HZ	33~35	Engine low idle frequency, in Hz	28~30
Machine high frequency, in Hz	55~60	Engine high idle frequency, in Hz	73~76

Table 76.4 Natural frequencies for the existing AVM stiffness in Hz

6.7	19.1	42.0	61.8	83.0
10.2	21.7	52.4	65.3	89.5
10.4	25.3	**56.8**	**73.5**	90.1
16.7	**33.9**	**58.7**	**76.1**	104.8
19.1	**36.3**	**60.0**	**76.4**	108.9

Table 76.5 Natural frequencies for different AVM stiffness values

Rigid AVM		AVM set-1		AvM set-2		AVM set-3	
15.7	67.6	6.8	**56.8**	7.1	**57.0**	8.5	54.5
20.0	72.6	8.2	**58.6**	12.2	**59.0**	11.3	60.2
24.2	78.1	9.9	**59.9**	13.5	60.1	12.6	65.5
27.5	78.5	15.3	62.4	18.7	64.9	19.1	71.0
31.1	85.8	16.6	65.2	21.3	65.7	21.4	72.5
36.4	88.7	19.0	72.5	22.2	72.5	22.7	72.8
54.9	89.8	19.7	**75.9**	23.7	**76.4**	23.9	78.1
58.1	105.7	24.0	**76.4**	**28.1**	**76.6**	27.2	78.5
59.6	118.9	**32.7**	82.6	**36.3**	84.5	**36.3**	85.0
60.5	148.9	**36.3**	89.5	36.8	89.6	36.9	89.7
		42.5	90.1	43.0	90.2	49.3	90.3
		52.4	104.8	53.4	104.9	54.2	106.7

It is clear from Table 76.5 that the natural frequencies are away from the operating frequencies for the AVM set-3 except 36.3 Hz which is closer to the machine low frequency. But while checking the mode shape (deflection pattern) of that frequency, it is not contributing to the deflection of the operator platform for the machine working condition. Also, the lesser frequencies are falling closer to the disturbing frequencies for the rigid AVM. Hence, it is decided to measure the operator comfort for three cases for the operator platform with existing AVM, rigid mount and AVM set-3.

76.4 Experimental Work

In order to validate the FEA approach and to decide upon the right AVM for better operator comfort, an experimental work is carried out on the operator platform of 9 T class Tandem Compactor with existing AVM, rigid mount and AVM set-3. One tri-axial accelerometer is mounted on a pad and placed on the operator's seat to measure WBV, and another tri-axial accelerometer is mounted on the adapter fixed to operator's wrist to measure HAV.

Fig. 76.6 Experimental setup

Table 76.6 Daily exposure limit values, A(8) in m/s^2

Vibration / AVM	Existing AVM	Rigid Mount	AVM Set-3
WBV	1.29	0.68	0.4
Remarks	More than EAV	Within ELV	Within EAV
Disposition	Not Acceptable	Marginally Ok	Acceptable
HAV	4.20	2.22	1.3
Remarks	Within ELV	Within EAV	Within EAV
Disposition	Marginally Ok	Acceptable	Acceptable

The experimental setup for measuring the operator comfort is shown in Fig. 76.6. Other ends of the tri-axial accelerometer cables are connected to SVANTech DAQ system which is used to acquire the data as well as post-processing of the results. The A(8) values of WBV and HAV are calculated according to the respective standards, and those values for three cases are shown in Table 76.6. It is evident from Table 76.6 that the operator comfort values are within the EAV for the AVM set-3. Hence, it is decided to proceed with AVM set-3.

76.5 Conclusion

The WBV and HAV are used as measures of operator comfort of a Vibratory Tandem Compactor. In order to have a better operator comfort, the natural frequencies of the operator platform should be away from the machine operating frequencies. Finite element analysis is used to find the natural frequencies of the operator platform. Moreover, FEA is used to arrive at the right AVM for the operator platform in order to achieve better operator comfort. The experimental work is carried out to validate the finite element approach. In future, this FEA approach will be used to finalize the operator compartment design prior to practical testing with an objective of reducing the development cost and time and ensuring the operator comfort.

Acknowledgements Authors wish to thank the management of Larsen & Toubro Limited for the use of their Vibratory Tandem Compactor data and granting permission to publish this article. They also thank their colleagues for their assistance in carrying out the experimental work.

References

1. Bovenzi, M., Hulshof, C.: An updated review of epidemiologic studies on the relationship between exposure to whole-body vibration and low back pain. J. Sound Vib. **215**(4), 595–611 (1998)
2. Pope, M., Wilder, D., Magnusson, M.: Possible mechanisms of low back pain due to whole-body vibration. J. Sound Vib. **215**(4), 687–697 (1998)
3. Seidel, H.: On the relationship between whole-body vibration exposure and spinal health risk. Ind. Health **43**(3), 361–377 (2005)
4. ISO 2631–1:1997, Mechanical vibration and shock—Evaluation of human exposure to whole-body vibration
5. ISO 5349–1:2001, Mechanical vibration—Measurement and evaluation of human exposure to hand-transmitted vibration
6. Smets, M.P.H., Eger, T.R., Grenier, S.G.: Whole-body vibration experienced by haulage truck operators in surface mining operations: A comparison of various analysis methods utilized in the prediction of health risks. J. Appl. Ergon. **41**, 763–770 (2010)
7. Ahn, S.-J., Griffin, M.J.: Effects of frequency, magnitude, damping, and direction on the discomfort of vertical whole-body mechanical shocks. J. Sound Vib. **311**, 485–497 (2008)
8. Ren, W., Peng, B., Shen, J., Li, Y., Yu, Y.: Study on vibration characteristics and human riding comfort of a special equipment cab. J. Sens. **2018**, Article ID 7140610
9. Kuchiki, K., Hoso, H., Kawabata, M., Mori, Y.: Technology for evaluating strength, stiffness, and riding comfort of crawler crane cabins, Kobelco Technology Review No. 31, January 2013
10. Giacomin, J., Fustes, F.: Subjective equivalence of steering wheel vibration and sound. Int. J. Ind. Ergon. **35**, 517–526 (2005)

Chapter 77
Harnessing Immersive Technology with Art and Design: A Conceptual Design Procedure with the Aid of Virtual Reality

Sumana Som, Deepak John Mathew, and Kim Vincs

Abstract Art and design are increasingly crucial in Indian school curricula. Lack of specific instructional pedagogy, schools, follows a different methodology to teach art. Therefore, most students neither achieve a basic understanding of creating art nor develop design or skill. Although the article art education in India 2010 reported that it is an essential tool for equipping students to stimulate cognitive development and encourage innovative and creative thinking. Furthermore, it also emphasizes using electronic media aids to teach this subject. Nevertheless, most schools are giving less importance to teaching art and do not take the initiative to develop better pedagogy. However, education pedagogy in recent decades has struggled to develop new methods and tools for better learning. Virtual reality (VR) is one of the meaningful solutions to technology-driven pedagogy. It allows us to visualize abstract content, provide information with clarity, and actively participate in the learning environment. In this study, the researcher introduced a teaching–learning design concept using virtual reality technology so that the sixth standard student could learn art and design in the VR environment. The content is developed on elements of art, and the topics have been chosen from the National Council of Educational Research and Training (NCERT) art education program. The result of this study shows the procedure and criteria for content design with the aid of virtual reality for the study of art and design. Besides, this study examines aspects of how virtual reality helps in terms of designing the teaching strategy.

S. Som (✉) · D. J. Mathew
Indian Institute of Technology, Hyderabad, India
e-mail: md16resch11005@iith.ac.in

D. J. Mathew
e-mail: djm@des.iith.ac.in

K. Vincs
Swinburne University of Technology, Melbourne, Australia
e-mail: kvincs@swin.edu.au

A. Chakrabarti et al. (eds.), *Design for Tomorrow–Volume 2*, Smart Innovation, Systems and Technologies 222, https://doi.org/10.1007/978-981-16-0119-4_77

77.1 Introduction

It is worldwide accepted that art and design have a significant role in school education [1]. According to John Dewey, art is not only a simple utilitarian character or a mere instrument but it also transcends the legacy of culture, history, and politics and has a crucial role in the educational process [2]. Art and education relationships show diverse pathways to innovate new ideas and incubate interdisciplinary possibilities. The presence of art in education offers a dynamic teaching practice that can exhilarate knowledge, cognition, and sensitiveness of young children [3]. The art apprentice and guild system's legacy carried a vital role in the early Indian education system. It is a learning of high aesthetic sensibility, sufficient technical skill to handle the medium and blend with creative literature that invigorates the holistic learning [4]. Indeed, the subject art and design can amplify the scope of ideation and energize other disciplines by incorporating cross-curricular activities. Although there is a confutation about how to teach art, the subject itself covered a large area of creativity, innovation, and aesthetics. To create an art or design work requires intellectual concern of cultural, political, personal, and ethical aspects [5].

However, there are many significant strategies to develop good course design and curriculum for arts and design at the school level. In 1979, Chicago Board of Education published an art curriculum guide for secondary school. In which, explicitly instructed how to guide hands-on learning [6]. A well-structured course design prescribed by Efland (2002) 'line of progress' by teaching the history of art, the process of interpretation of the artwork, 'teaching formalism' or developing procedural knowledge to make artwork [7]. Clarkson [10] suggested an art curriculum that provides a model to educate the imagination and cognitive attitude, which can bridge the inner world of mythos and the outer world of logos [8]. The Art National Curriculum for England designed a two-stage curriculum strategy; stage one called attainment target 1 that deals with investigating and making to record, using resources, different materials, and settings. Stage two called attainment target 2 that encompasses knowledge and understanding in order to review, verifying, knowledge of the different types of art, different period, culture, and tradition [9]. In India, NCERT visual art program comprises handling materials, concepts, and knowledge development in the contemporary and historical context. The suggested contents are objects, people, tradition, environment, and experience-based exercises [10]. To develop a content and curriculum for art and design education, some significant factors are carried out, firstly, hands-on skill development by handling materials. Secondly, enhance knowledge and cognition by developing a sense of aesthetics, cultural concern, and historical awareness. Thirdly, engage students for the invocation of new ideas out of their experience of learning.

In the last few decades, many changes have done in curriculum. Inclusion of crafts, design, and technology are combined into an integrated art curriculum. With the advanced technology, education pedagogy has changed, and the fusion of art and technology changed the focus from the only art education to technology-aided art and design education [11]. In the new paradigm of technology-based curriculum meets the criteria of creative and collaborative pedagogy [12]. Virtual reality (VR) is one of the meaningful solutions to technology-driven pedagogy. It provides a super real, simulation-based environment for teaching and training. The teacher can explain the topic in many possible ways and make the student understand quickly. Besides, a playful interactive teaching design can enhance learning motivation, engage more deeply, and inspire the student to learn. There are a couple of studies that show the inclusion of VR in art education [13, 14] collaborative art project [15]. On the other hand, art and design study is increasingly important in school level, and the traditional teaching method needs to be reformed with efficient modern tool [16].

The subject of art itself an interdisciplinary course; it could play the role of a communicator and develop a cross-curricular activity. Indian art education system is diversified into various topics, but it is alienated from other subjects. Some of the schools follow craft practice, some theoretical practice assignments, such as a small document of Mahbubani painting and copying the style or imitate any western-style painting. Some of the schools doing mixed-method practices such as object study, crafting usable products by using eco-friendly or waste materials. Application of teaching tools are traditional such as verbal demonstration, instruction of methods, and provide an example of charts and images. For instance, students are learning the only imitation not developing their new idea. In terms of applying new technology, VR is not yet broadly implemented in the school scenario. There is not a suggested pattern on how to create content that can run into the VR environment. Therefore, this study focusses on establishing a pattern to create cross-curricular content with the aid of VR. It introduces basic workflow to develop a VR environment. The study explicitly discusses the synergy of teacher's artistic practice with VR and classroom teaching.

77.2 Literature Review

The study reviews the existing knowledge based on the application of virtual reality on art and design education, VR environment design framework, and content development with a cross-curricular approach. The study pointed on the linkages of pedagogy and practiced art and design in order to cross-curricular classroom activity and creative VR environment design.

77.2.1 Pedagogy and Practice of Art and Design with the Aid of Virtual Reality

The effects of art and design pedagogy with VR aids are two-fold. VR technology enables the effective teaching–learning with visualization of abstract content, providing information with enhanced clarity, [17]. As well as if the artist collaborates with this digital medium, it can push the limitation beyond their practice and conventional wisdom. It can open up the new horizon of inquiry. Therefore, an art educator can proceed into practice to do exploratory research with this medium as well as he/she can teach in this similar mode. For instance, an artist or an art educator creates a work of art with digital tools and uses this interactive work of art or design teacher to teach students [18]. Here, an artist's work is not only subjected as an artist's vision or thoughtful imagery expression but also it serves an open interactive space for viewer [19]. In this case, the artist works as a catalyst and connected missing link between art and design education and practice, open up the more substantial opportunity to art, design, and technology collaboration. According to Matthew J. Koehler and Punya Mishra (2006), practice of teaching requires an interweaving of different kinds of knowledge. Integrating technology with pedagogy and content knowledge is one of the critical parts of effective teaching. Teachers should know the nature of technology and how the procedure of inclusion of technology with content and pedagogy [20, 21]. In the Indian school scenario, art education pedagogy is not implemented with virtual reality technology. Besides, there are not any suggested outline to develop technological pedagogical and content knowledge for art and design.

77.2.2 Benefits of VR-Based Cross-Curricular Approach in Art and Design Pedagogy

The concept of cross-curricular teaching–learning has potential, and it has been proved in science, language, and technology education. The actual meaning of cross-curricular practice is the formation of links between seemingly separate curriculum subjects can enhance people learning and make an impact on their cognitive understanding and have a significant impact on the development of core and transferable skills [22]. This cross-curricular approach to pedagogical design is characterized by a synthesis of knowledge, skill, and understanding from the various subject areas where a number of different disciplines are applied to a single experience or idea [23]. According to Richard Harris et al. (2011–15), decision to construct cross-curriculum more or less depend and supportive of teachers' understanding of subject knowledge and identify the way to club with other subjects [24]. Application of cross-curricular approach within technology and art has outlined four principles; first, technology should be specific, and teachers need to be actively practiced with the technology to produce art. Second, the principal stated the three ways to build the bridge technology and subject by creatively, culturally, and educationally. The third principal pointed to the procedure of technology-aided cross-curricular activity that should empower and embrace the students. Forth principal says the technology embedded pedagogy must provide a different experience from existing practice that empowering the feature of contents and procedural knowledge [25].

77.2.3 Virtual Reality Environment Design Framework

Brian C. Nelson et al. (2012) described the development process of the virtual world. According to him, to create a basic design of the VR environment, commercially, people follow ADDIE (Analysis, Design, Development, Implement, and Evaluate). In some streamline, he has prescribed eight stapeses; table read, resources listing, resources gathering, recourses creation, geography creation-3D object placing, GUI creation, programming, testing [26]. To develop the VR-aided curricula, the first and foremost criteria for finding the need or problem will depend on VR-based environmental instruction to facilitate easy learning. The second criteria are the target audience, the age group, familiarity with the content, technological knowledge, and language. The third point is to focus on learning objectives through VR aided environment. In the fine stage, the focus needs to shift content organization, content description, and storyboard creation [26]. A study done by Nikiforos M. Papachristos et al. (2014) described the impact of the VR environment on educational settings. The study concluded that the environment's design needs to change with the complexity and needs of students in pursuit of the topic [27]. Pierre H Routhier [32] states that the content creator's perspective needs to be clear in which instructional problems facilitate in an immersive environment. In this context, the content creator's objective and combination of technology-based activity need to match [28].

77.3 Statement of Problem

Developing core curricula of art and design with the aid of virtual reality for middle school children is not so far practiced. To create a VR-based curricula are required to understand four significant points; first, find out a particular area of the subject requires VR implementation [29]. Second, consider the mode of interaction with the topic, whether it is based on navigation and observation or action with interaction [27].Third, figure out the time consumption of VR tools with the lesson and forth is the ease of use of these tools and applications [26]. Although some applications have developed for environmental art education [30] or game-based teaching with particular art form [31] but a VR aided syllabus for art and design for a particular class with cross-curricular activities is not so far developed. The VR-aided cross-curricular activity could incorporate the imagery influences of other subjects. For instance, choose any other subject topic and connect with a problem-based design and art study. In this case, art educators must find out what are the art lessons have possible channels to connect with the VR environment and find out the possible way to relate to other subjects.

77.4 Methodology

This study aims to identify the practical process of the creation of a VR-based art and design curriculum. A practice-based research methodology [32] has been chosen to achieve the aim, and a more natural method is described to develop the same. To support the artistic approach to developing a VR-based art and design content, the researcher used observation-based field notes, semi-structured interviews of art and craft teachers, and a detailed review of the sixth standard NCERT art, science, geography, and history syllabus.

In this research process, artistic practice is aligned with content development. This study aims to create as much as possible a more accessible mode of VR-based content creation and environment design for art and design education.

77.5 VR-Aided Content Development of Art and Design for Sixth Standard

To develop VR-aided pedagogy, first researcher has conducted semi-structured interview with ten art and craft teacher from same set of school. The interview contained an in-depth discussion on the activity usually practiced for art and design learning, the tools used for practice, and the new method they intended to do with class. Besides, researchers have scrutinized the art and design syllabus prescribed by the National Council of Educational Research and Training (NCERT). Along with history, geography, and general science lessons intensely seen to find out the possibility to intertwine with art and design syllabus. The topics are selected by evaluating its level of difficulties; those are complex to teach in traditional demonstration teaching methods. For instance, line, value, shape, and form could be taught in the 'chalk and talk' method but may not provide a clear understanding of the core concept. To make the topic more relevant, it linked with other subjects like history, geography, and science. Such as, the topic shape and form can be teached through the historical reference of stone tool, terracotta toy, reference from science 'different natural objects' and 'changes around us' and reference from geography, chapter 'globe' can be used for example or study material to teach art and design. Table 77.1 provides a glance at which particular lessons can fit the art and design study and what particular elements can come into the VR environment.

77.6 Framework for Content and Scene Design

After an in-depth discussion with ten teachers, researcher pointed out four important aspects: connectivity, image reference, mode of visualization, and translation.

Table 77.1 Application of virtual reality and topic combination of art and design with history, geography, and general science

Art and design	Technology application	History	Geography	General science
• Line • Strokes • Texture • Value • Pattern • Balance • Order	Virtual reality	**Findings from past:** Old inscription Painted pottery Coin	**Components of map:** Direction Conventional symbol, Grid Major landform of the earth (Images can be manipulated for example)	
• Shape • Form • Feasibility	Virtual reality	**Earliest city:** Stone tool Terracotta toy Stone statue Seal	**Globe:** Latitude and longitude	Different natural objects Changes around us
• Perspective • Size Balance	Virtual reality	**Earliest city:** Harappa and Mohenjo-Daro architectures and pottery	**Atmosphere and biosphere** Good example of Arial and linier perspective	

Throughout the discussion, a conceptual design has been developed that could be one way to recreate new pedagogy without changing existing syllabus.

Followed by four prescribed points and Robson's characteristics of naturalistic inquiry [32], the study finds out a strategic way that could help art educator to develop VR-based art and design content. Figure 77.1 shows the germination of a

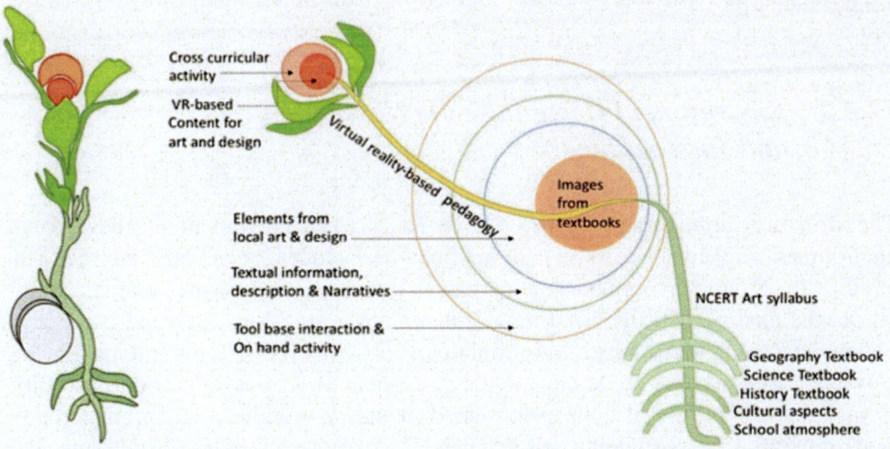

Fig. 77.1 A naturalistic approach to develop VR-based art and design content

green pea in which stem and roots represent the connectivity; the subject area of art that could be connect with geography, science, and history, and develop a VR-based cross-curricular art and design content. The middle part, germinating pea, represents image reference and mode of visualization. This part includes where from and how images will be collected and represented in 3D space. The stages are as follows: 1. the images from textbooks for 3D model, 2. local design elements (crafts and utility objects), and geographical landscape for the VR environment, 3. textual description and information to identify the problem, and 4. scripted activity with VR tool (head-mounted device (HMD) and hand controller). The cotyledon or embryonic leaf represents translation of cross-curricular VR-based art and design content.

77.6.1 Environment Design: Least of 3D Models

By following the design outline, researcher developed VR environment that contains 64 models that are majorly referred from the textbook of the sixth standard (NCERT). Reference images from the history textbook and 3D models are developed following the picture (Fig. 77.2). Likewise, the rest of the models is referenced in the book of geography and general science. Some exceptional models are developed from local art, craft, and ready-made, not directly related to the textbook. However, the students are familiar with this art forms and objects. Such as Indian brush oil lamp, block print patterns, woven fabric pattern, and folk paintings. These models are representing the images of the textbook or contemporary ready-made, but they also bridge the period-wise usability of objects from past to present, the legacy of need-based, and institutional design development. To make the 3d model, Autodesk Maya software has been chosen and Substance Painter for texturing. These two softwares are well equipped with tools and easy to plugin with each other and Unity 5 software.

77.6.2 Environment Design: Scene Navigation and Interactions

The structural arrangement of the VR scene is a crucial part of the developing environment. Space needs to be naturalistic, where students can walk, interact, and explore things with an immersive experience. To provide them with a realistic simulation, the environment required a naturalistic space division, size, and placing of objects according to life-size and formulate the pathway for their natural movement. However, art educators could change the placement or size of objects purposefully. Content-wise five virtual tables are placed in the environment. Each table represents separate topics; students and teachers can explore it with the hand controller (Fig. 77.3). Unity 5 software has used to develop the 3D environment. To look around and navigate the VR environment, HTC VIVE accessories (Hand controller

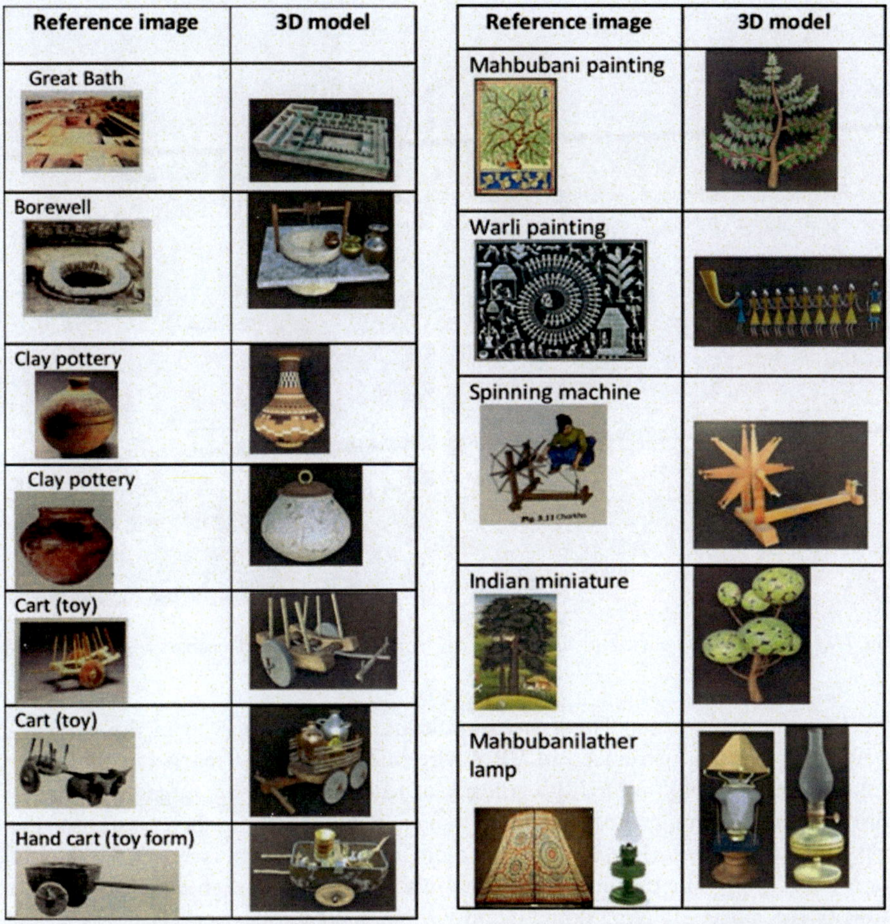

Fig. 77.2 Reference images and 3D models

and HMD) are used. Head-mounted display for users to observe surroundings and the hand controller for interaction with objects and navigation in the environment. There are a few drawbacks and limitations in this VR environment, such as teachers need good software skills to make the content. Teaching time needs to adjust to a VR-based teaching tool.

77.7 Application and Results

The VR environment tested with a small group of sixth standard students (Pilot test, sample 6). The test was based on skill, observation, and knowledge of visualization

Fig. 77.3 Virtual reality environment, content-wise interaction table and teachers are experiencing

development of students. There are significant changes observed in the student's performance after experienced in VR environments. Besides, teachers' qualitative responses mainly highlighted the efficacy of teaching–learning, engaging strategy for reluctant learners, and exploring new space without leaving the classroom. The prescribed framework (Fig. 77..1) for content and scene design is a useful guideline for art educators. The teacher's responses in this section highlighted a step-by-step strategy for conceptualizing content and visuals, image collection (Fig. 77..2), and connectivity creation.

77.8 Conclusion

Develop a synergy within the educator's artistic practice and the teaching of art and design using virtual reality; it is necessary to determine the roots of the teaching–learning objectives. To cultivate the practice of art and design education, school is the right place to introduce creative practice learning fundamentals. On the other hand, the teaching–learning of art and design curriculum is not one-sided practice. It includes artists, artisans, and designer collaboration with various mediums such as local objects, redemands, and digital technologies to develop innovative design and a new work style. The inclusion of a new style of work helps develop a new pedagogy that feeds young children's education [33].

This VR environment carries two significant characteristics; one is a close relation with four subjects (Science, geography, history, and art and craft) of sixth standard. For example, the objects in the VR environment are mostly taken from history or geography lessons. These objects, such as Great Bath and carts developed from terracotta models, are derived from history (Indus Valley Civilization), whilst small objects like lamps, bread rollers, and charkha are taken from general science. Secondly, this VR environment replicates a unique artistic utopian society that shows a balanced ecosystem, primary needs of life, a basic understanding of society's system, and development of habits. The environment shows the road with a zebra crossing, two types of dustbin for natural waste and plastic waste, and many more. The VR environment shows the form of a tree, replicating the 3D form of Mahbubani, Warli, and Indian miniature paintings.. In this endeavor, a small effort has taken to reduce alienation and give more advocacy of art and design subject (as it is called extracurricular activity) in the school curriculum. In addition, thorough out the VR-based journey with familiar objects and the environment, students will get the opportunity of minute observation. It can generate their inquisitiveness and innovative mindset.

The technology-driven pedagogy in the area of art and design is not much practiced in Indian schools. Due to the technophobic attitude of art, educators are not ready to use the digital tool and not even developed a technology-blended curriculum for art and design study. This study shows an effective way to develop art and design curriculum with VR aid. It is not only the VR-based pedagogy but also an artist or an art educator who can develop their work of art and bring into the area of school education. In this process, an artist or an art educator can make the synergy between art and design education and artist or designers' practice.

On the other hand, designing content for school or create a work of art with VR tools is not one handwork. In this case, either artist or designer needs to develop an application to create and execute the artwork or depend on the existing system [34]. This VR technology integration into art and design platform will provide a meaningful engagement of the artist, the digital system, and the viewer as a 'trialogue.' Therefore, it required an active collaboration of art, design, and technology. This type of collaborative work can make the synergy between art and design education and practice.

References

1. Maarit Mäkelä, N.N.: Tero Heikkinen drawing as a research tool: Making and understanding in art and design practice. Studies in Material Thinking vol. 10 (2014)
2. Deway, J.: Art As Experience. penguin (2005)
3. Chen, M.-H.: The formative role of art: John Dewey's art as quotidian experience. J. Literature Art Stud **7** (2017)
4. Thapr, R.: The Penguin History of Early India: From the Origins to AD 1300. penguin (2015)
5. Belluigi, D.Z.: Intentionality in a Creative Art Curriculum. J. Aesthetic Educ. **45**, 18–36 (2011)

6. Chicago Board of Education, I.L.: Design Solutions: Supplement to the Curriculum Guide for Art in the Secondary Schools. Field Test (1979)
7. Cunliffe, L.: Representing and practising meaningful differences in a well-structured but complex art curriculum. J. Curr. Stud. **42**, 727–750 (2010)
8. Clarkson, A.: Educating the Creative Imagination: A Course Design and its Consequences. Jung: the e-Journalof the Jungian Society for Scholarly Studies vol. 1, Number 2 (2005)
9. Watts, R., Cox, S., Herne, S.: Readings in Primary Art Education. Intellect Books, Bristol, UK (2009)
10. NCERT: Syllabus of Arts Education. New Delhi (2008)
11. Duppati, S.K.: Integrated Arts Curriculum At School Level: A Cultural Perspective. Scope: Contemporary Research Topics (Art & Design) 5/6, 142–150 (2010)
12. Bilyatdinova, A., Karsakov, A., Bezgodov, A., Dukhanov, A.: Virtual environment for creative and collaborative learning. In: Papadopoulos, G.A., Skulimowski, A.M.J., Kacprzyk, J., Kunifuji, S. (eds.), vol. 416, pp. 371–381. Springer Verlag (2016)
13. Koun-Tem, S., Hsin-Te, C., Kuan-Chien, M.: Research on the application of virtual reality on arts core curricula. In: 5th International Conference on Computer Sciences and Convergence Information Technology, pp. 234–239
14. Song, X.: Virtual Reality for Arts and Design education. In: 2010 IEEE 11th International Conference on Computer-Aided Industrial Design & Conceptual Design 1, pp. 679–682
15. Lu, L.: Teaching 21st-Century Art Education in a Virtual Age: Art Café @ Second Life. Art Education **63**, 19–24 (2010)
16. You-tian, Y.: Design and Implementation of Digital Art Teaching System Based on Interactive Virtual Technology. Int. J. Emerg. Technol. Learning **11**, 49–54 (2016)
17. Falconer, L.I.Z.: Situated Learning in Virtual Simulations: Researching the Authentic Dimension in Virtual Worlds. J. Interactive Learn. Res. **24**, 285–300 (2013)
18. Alexenberg, M.: Educating Artists for the Future : Learning at the Intersections of Art, Science, Technology, and Culture. Lightning Source Inc Lightning Source Inc, 1246 Heil Quaker Blvd, LA Vergne, TN, 37086 (2008)
19. Keefe, D.F., Karelitz, D.B., Vote, E.L., Laidlaw, D.H.: Artistic collaboration in designing VR visualizations. IEEE Comput. Graphics Appl. **25**, 18–23 (2005)
20. Koehler, M.J., Mishra, P., Cain, W.: What Is Technological Pedagogical Content Knowledge (TPACK)? J. Educ. (Boston, Mass.) 193, 13 (2013)
21. Huifeng, X.: An Ethnic and Folk Art Space Course based on TPACK. Int. J. Emerg. Technol. Learn. **14**, 110–121 (2019)
22. Byrne, E., Brodie, M.: Cross Curricular Teaching and Learning in the Secondary School. Science. Routledge, Milton Park, Abingdon, Oxon (2012)
23. Webb, R.: Cross-Curricular Primary Practice : Taking a Leadership Role. Routledge, London (1996)
24. Harris, R.A.: Cross-Curricular Teaching and Learning in the Secondary School... Humanities : History, Geography, Religious Studies and Citizenship. Florence : Routledge (2011)
25. Fautley, M., Savage, J.: Cross-curricular teaching and learning in the secondary school—the arts : drama, visual art, music, and design. Routledge Taylor & Francis, C/O Kentucky Dist Center, 7625 Empire Dr, Florence, KY, 41042 (2011)
26. Nelson, B.C., Erlandson, B.E.: Design for Learning in Virtual Worlds. Routledge, New York (2012)
27. Papachristos, N.M., Vrellis, I., Natsis, A., Mikropoulos, T.A.: The Role of Environment Design in an Educational Multi-User Virtual Environment. Br. J. Edu. Technol. **45**, 636–646 (2014)
28. Routhier, P.H.: The Immersive Experience Classification System: A New, Strategic Decision-Making Tool For Content Creators. In: SMPTE 2017 Annual Technical Conference and Exhibition, pp. 1–13
29. Chiou, G.-F.: Learning Rationales and Virtual Reality Technology in Education. vol. 23, pp. 327–336. Journal of Educational Technology Systems (1995)

30. Hanrahan S.S.U.C., Peter De Pietro, P.D.C., Laurie Halsey Brown, L.M.N., Alex Haw, A.Y.C.U., Julian Malins, J.M.R.A.U., Michael Milojevic, M.M.A.A.N., Martti Raevaara, M.R.T.F., Barbara Sonvilla, B.S.T.F., Jan Kenneth Weckman, J.K.W.G.C.: Interface: Virtual Environments in Art, Design and Education': A report on a conference exploring VLEs in art and design education. Arts & Humanities in Higher Education 8, 99–128 (2009)
31. Mingshao, Z., Zhou, Z., Yizhe, C., Aziz, E.-S., Esche, S., Chassapis, C.: Recent Developments in Game-Based Virtual Reality Educational Laboratories Using the Microsoft Kinect. Int. J. Emerg. Technol. Learn. **13**, 138–159 (2018)
32. Gray, C., Malins, J.: Visualizing Research : A Guide to the Research Process in Art and Design. Routledge, Aldershot, Hants, England (2004)
33. Ying, L., Jiong, Z., Wei, S., Jingchun, W., Xiaopeng, G.: VREX: Virtual reality education expansion could help to improve the class experience (VREX platform and community for VR based education). In: 2017 IEEE Frontiers in Education Conference (FIE), pp. 1–5
34. Candy, L., Edmonds, E.: Interaction in Art and Technology. Crossings: Electronic J. Art Technol. **2** (2002)

Chapter 78
Comparative Aerodynamic Performance Analysis on Modified UAV's Propeller by Using CFD

Prisha K. Asher, K. Deviparameswari, B. Feonsa Antonitta, S. Meenakshi, R. Vijayanandh, and M. Senthil Kumar

Abstract In multi-rotor Unmanned Aerial Vehicles (UAVs), the propeller and its noise generation are plays a focal position in the determination of the radar signature. Because of this detection possibility, the implementations of compact UAV in surveillance-based applications have been reduced. Therefore, nowadays, huge research is going on the noise reduction techniques in the UAV's propeller, in which propeller's profile modification technique primarily reduces the noise generation. But the research on noise reduction is the failure to include the primary consideration, which are aerodynamic parameters such as lift, drag, etc. In this work deals with the comparative aerodynamic performance analysis on various modified UAV's propeller, in which six conceptual designs of propellers are analyzed. The base propeller and five other profile modified propellers are used in this work. The finalized fundamental platform of this work is bull-nose propeller, which has a 4.5-inch pitch, 5-inch diameter and can provide 500 g average thrust. CATIA is an advanced modeling tool, which is supported a lot for all the models' construction. An advanced CFD numerical tool, i.e., Ansys fluent is used in this work for the prediction of aerodynamic performance parameters for all the models. Finally, based on exit velocity and thrust, the propeller optimization and its edge modification are obtained by using CFD's numerical tool.

78.1 Introduction About Propeller and Its Importance in the UAV

Generally, the construction of multi-rotor Unmanned Aerial Vehicles (UAVs) is comprised of two groups which are electrical cum electronics working group and the aero-mechanical group. This work deals with the aerodynamic performance analysis on one of the components under the aero-mechanical group. The aero-mechanical group includes the following components: propeller, airframe, landing gear [1]. In

P. K. Asher · K. Deviparameswari · B. Feonsa Antonitta · S. Meenakshi · R. Vijayanandh (✉) · M. Senthil Kumar
Aeronautical Engineering, Kumaraguru College of Technology, Coimbatore, Tamil Nadu, India
e-mail: vijayanandh.raja@gmail.com

A. Chakrabarti et al. (eds.), *Design for Tomorrow–Volume 2*, Smart Innovation, Systems and Technologies 222, https://doi.org/10.1007/978-981-16-0119-4_78

which propeller is more complicated than others that complexity is derived from its working and geometrical nature. Also, the propeller is the device, which has a huge platform for upcoming researches such as lift enhancement, noise reduction, design optimization, etc. [2]. The important parts of the propeller are the hub, tip, leading edge, trailing edge. Based on the positions of the leading edge toward air incident direction, the two types of propellers are available in the propulsive system of a UAV. The types are clock-wise rotational propeller and counter clock-wise rotational propeller, which are primarily supported a lot in the provision of a stable platform [3]. Aerofoil shape is the fundamental origin of this propeller, in which the aerofoil is twisted with different angles from the propeller's hub to tip. To attain variational speed throughout the section of the propeller, i.e., from hub to tip, the different pitch angles are implemented. Because of these twisting angles imposition, the useful lift force is generated under rotating conditions [4]. In rotodynamic conditions, the kinetic energy of the rotor is transferred to air fluid, and thereby, it gets energized. Due to this imparted kinetic energy, the air fluids are started to move in the pushing down manner with the help of the aerodynamic profile of the propeller. Therefore, to get upward force in a multi-rotor UAV through the propeller, the rotodynamic working nature and its advanced geometrical nature are plays a focal role. Apart from these factors, the diameter of the propeller and its pitch angle, which deals with the relationship between one RPM of the propeller with respect to its forward movement, is fundamental selection parameters used in the UAVs [5].

78.2 Literature Survey

In an experimental study done in China, a comparative aeroacoustic and aerodynamic analysis was done with normal baseline and modified propeller which had serrations on the trailing edge testing them in different conditions like hovering, and during the forward flight, the results showed that there had been a reduction of noise during testing them in forwarding flight by the serrated propellers up to 12 dBs and showed a decreased lift and drag by 5 percent but there was no change in the lift by drag ratio with bases on the advanced ratio. A computational study with serrations on trailing edge of the wing with airfoil NACA 0012 was done in [4], in which the author was analyzed the flow pattern, and vorticity was obtained also conducted a wind tunnel test and found that serrations allowed the flap to retract at higher angles of attack.

A study of an Owl Wing with Leading Edge Serrations by Shinichiro ITO [6], the jigsaw blades were inserted into the leading edge of an owl wing with different cutting teeth serrations. From this experiment, it was concluded that the airfoil had an increase in the lift with fine serrations at low Reynolds number, and the flow separation was controlled by fine serrations at the leading edge. A study of optimization of leading edge undulation of a NACA 65(12)-10 airfoil was investigated in [8], in which the optimization of LEU had been done with three different amplitudes and the wavelength. The aerodynamic characteristics were measured at the Reynolds number of 10^5. It was concluded that the maximum noise reduction is obtained with small wavelength and high amplitude, which produced more lift when the wavelength and amplitude are more. It was observed that owl species possess larger wings as compared to their body weight, which enables flight at lower velocities. As flight noise emission depends on flight velocity, the leading edge serrations naturally present at their wings change the airflow and thus reduce noise emissions. These serrations stabilize flow over a large range of velocities and angle of attack. In the Fourier domain, the noise footprint of conventional propellers consists of two main noise components, i.e., tonal and broadband [10].

The optimized geometry of the propeller with acoustic constraints resulted in noise reduction for specific operating configurations only. The saw tooth serration pattern at Trailing Edge (TE) was seen to have reduced the broadband noise component. The use of porous as well as meta-materials to reduce TE noise proved to be a highly efficient sound barrier for target frequencies also. The total noise produced by UAV has been fully relayed on airfoil self-noise, which taking a larger percentage over inflow turbulence noise. Among five types of airfoil self-noise generation mechanisms, four of them are affiliated to the flow interaction between turbulent and trailing edge. An experimental study was conducted [12], which shown that non-identical propellers with serrations at the trailing edge which are differentiated by their ratio between width of serration and half of serration height were produced the same amount of lift compared to the baseline propeller, and rotational speed of the serrated trailing edge propellers was decreased due to induced drag. With the same amount of power input, propellers with serrated trailing edge were produced the same amount of thrust compared to baseline propeller. A numerical study on leading edge tubercles [14], which conferred that propeller with leading edge tubercles with lower wavelength, came up with improved lift coefficient which leads to higher propeller efficiency and has meliorated efficiency at the larger diameter-to-pitch ratio.

78.3 Problem Definition and Solution Techniques

a. **Problem Definition [Summary of Literature Survey]**
Nowadays, lots of researchers are going on the field of the propeller and its aero-acoustic effect. Most of the scientific works are been executed on the design optimization on UAV's propeller to attain high lift with low drag, to reduce the aero-acoustic effect, etc. The major aerodynamic components such as aircraft wing, horizontal axis wind turbines, and vertical axis wind turbines have been executed the modifications on their both leading and trailing edges to increase their aerodynamic performances [7]. The curvy cut, saw tooth cut, aerofoil cut were generally used as an aerodynamic force enhancement techniques, which are been mostly implemented in the leading edge. Some of the exceptional cases such as wingtip, flaps, and hydro rotors were equipped with the techniques at their trailing edges. Apart from these modifications, a flat and thin plate-based technique has been implemented in the trailing edge of aerodynamic components to reduce the drag. All of these inputs are provided the clear path for the lift

enhancement technique and its positions on the propeller for this current work [9]. While coming to the methodology section, the two engineering approaches have been used to test the aerodynamic behavior on the relevant components. The methodologies are experimental test and Computational Fluid Dynamics (CFD)-based simulation, in which most of the aerodynamic analyses were executed with the help of CFD simulations. Because of these huge implementations, the pre-processor steps are available to everyone [11]. Especially, the boundary conditions such as type of solver, type of turbulence model, quantity and quality of turbulence model are easily available data to the new researchers. The high amounts of analyses were used velocity inlet to their problems. Thus, with these inputs, the CFD simulations are executed in these comprehensive aerodynamic analyses on various propellers [13].

b. **Solution Techniques – CFD Analysis**

In this work, analyses the comparative aerodynamic performance on a UAV's propeller by using CFD tool, i.e., Ansys fluent. The fundamental aim of this work is to select the suitable lift enhancement technique for UAV's propeller. In this regard, six different designs are modeled, in which five conceptual designs are comprised of a propeller with edge modifications, and the other one is the conceptual design of base propeller. The techniques implemented in the 5-inch diameter propeller are curvy cut saw tooth cut, aero cut, etc., and in general, the profile modifications in the UAV's propeller is executed for noise reduction. From the literature survey, it was clearly understood that the noise induced due to the abnormal environment is reduced [1, 2]. Nowadays, UAV industry needs a quite propeller so the design modifications-based propellers are suggested a lot for the construction of quite UAV. But the problem along with these types of profile modified propellers may have a chance to generate low-aerodynamic forces. Hence, the conduction of an integrated study is very important in the updated propellers to increase the implementation of the UAVs in real-time applications. The steady- and pressure-based turbulent flow is used as fundamental behavior to the working fluid for all these analyses. The aerodynamic parameters such as lift, drag, C_L, C_D are used as selection parameters for this comparative analysis [15, 16and17].

c. **Conceptual Design of various propellers**

UAV's propeller and its designs are the key role of this aerodynamic performance investigation. From the previous work, it was understood that propeller with saw tooth cut has been provided the low turbulence noise than base propellers. Thus, in this work, the aerodynamic performance of low acoustic profiled propellers is computed, in which the conceptual design of all the propellers are used from the literature survey [18, 19,and20]. Figures 78.1, 78.2, 78.3, 78.4, and 78.5 are revealed the conceptual design of propellers with saw tooth cuts, in which the locations of the saw tooth cuts are formed at the various edges of the propellers such as both leading edges, both trailing edges, one leading edge cum one trailing edge, only one trailing edge. Apart from these saw tooth cuts, the one more relevant cut is located at the leading edges of the propeller, which is v-cut.

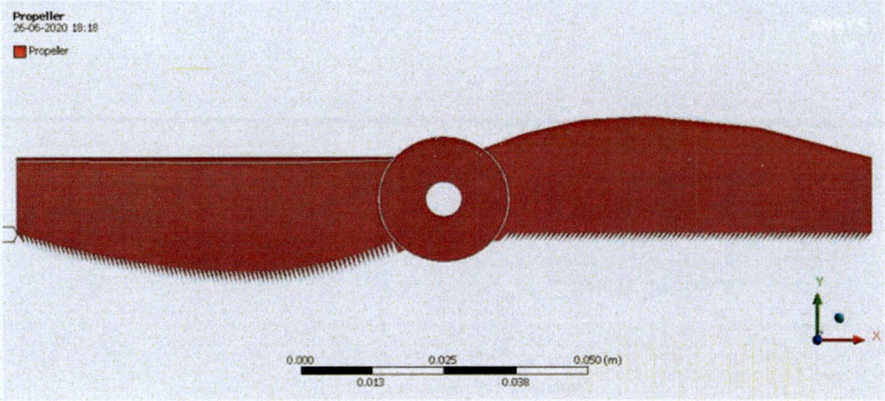

Fig. 78.1 Conceptual design of propeller with saw tooth cut at one TE and one LE

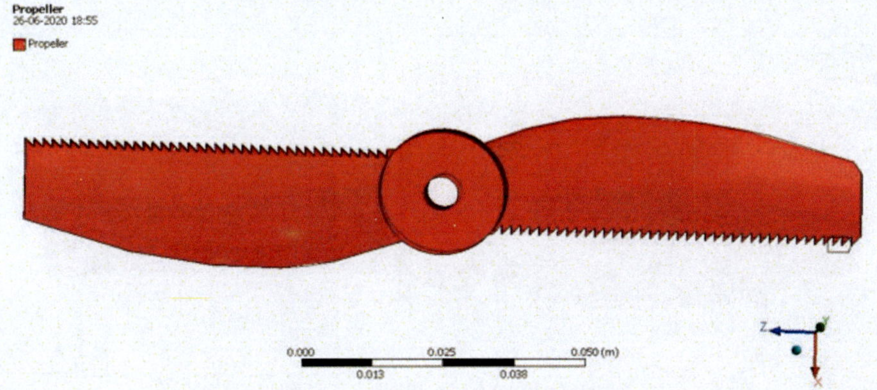

Fig. 78.2 Conceptual design of propeller with saw tooth cut at both LEs

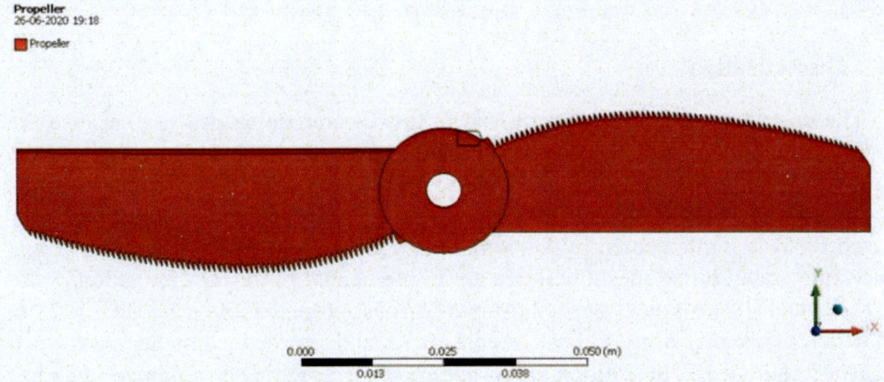

Fig. 78.3 Conceptual design of propeller with saw tooth cut at both TEs

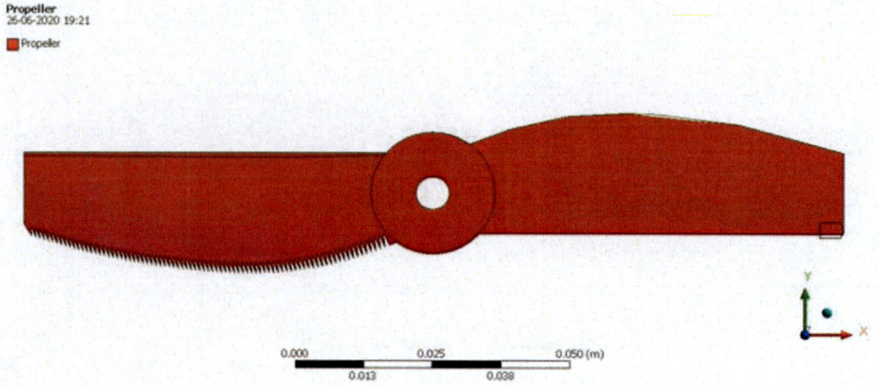

Fig. 78.4 Conceptual design of propeller with saw tooth cut at one TE and one end free

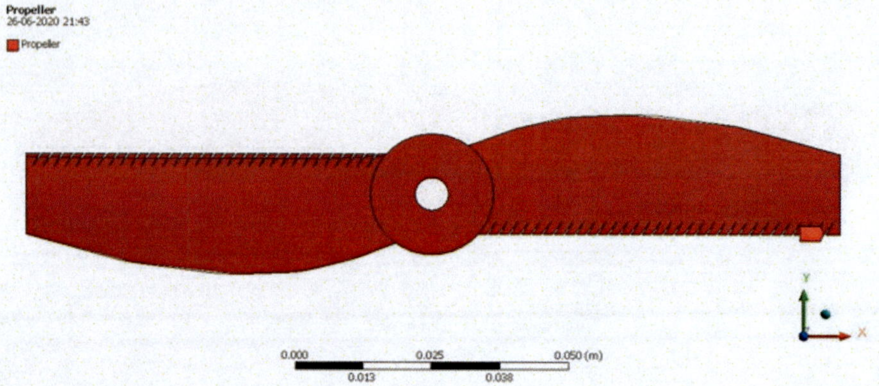

Fig. 78.5 Conceptual design of propeller with curved cut on both Les

d. **Discretization**

The unstructured grids are employed in this discretization process because the indented numerical technique for this CFD tool is Finite Volume Method (FVM). Computationally, the FVM technique is fit for both structural and unstructural grids, so in this work, purposively unstructural grids are employed, and thereby, centroid-based FVM is implemented in the computation stage. The fine proximity and fine curvature sizing based mesh facilities are implemented in this discretization phase [21, 22,and23]. The typical meshed structure on base propeller is revealed in Fig. 78.6, in which the quality of mesh is attained as 0.92, and very few elements are formed as aspect ratio of 1.2. In addition to this base process, the grid convergence study has been organized on the BOTH LE Propeller for the given velocity of 10 m/s, and the comprehensive data are revealed in Fig. 78.7.

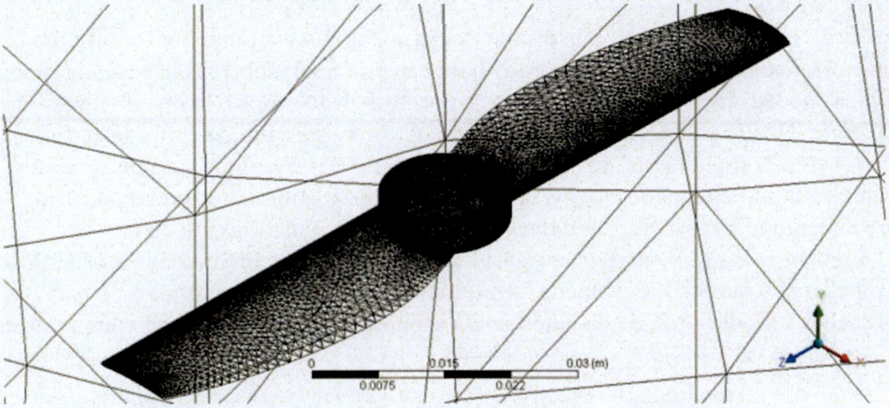

Fig. 78.6 Discretized structure of the UAV's base propeller

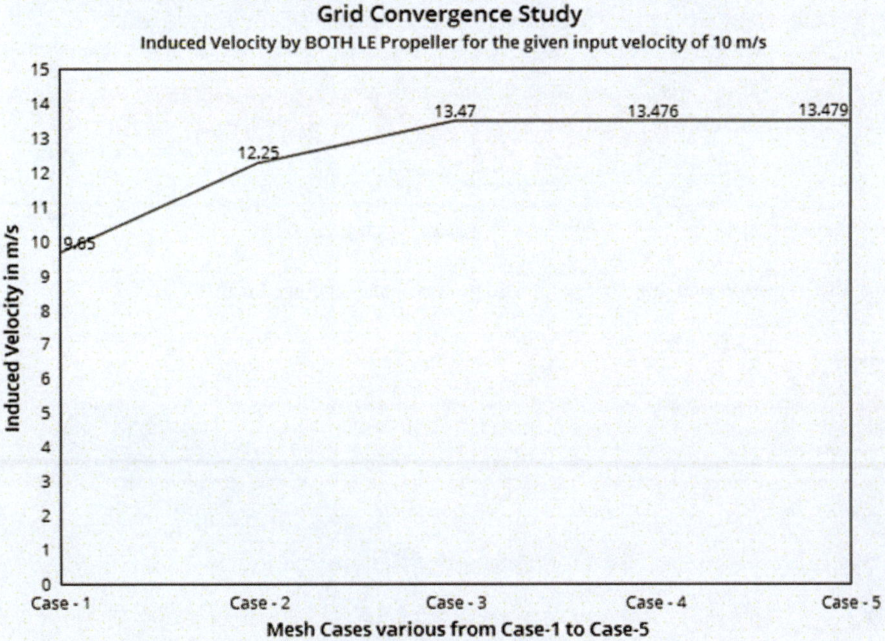

Fig. 78.7 Grid convergence test on BOTH LE Propeller for 10 m/s

78.4 Computational Fluid Dynamics (CFD) Results

The computational analyses on UAV's propeller are executed under pressure based solver circumstances, wherein the turbulent flow, velocity inlet, second order governing equations are effectively contributed and supported in computation process. The eddy formation is primary outcome of UAV's propeller for all kind of phenomenon so the prediction of eddy has been principal motto of this investigation, which guided the author to apply k-epsilon turbulence model relayed simulation on these UAV's propellers. The solution control is predominant role in turbulent flow-based computations, in which pressure–velocity coupling scheme, pressure, velocity, turbulent kinetic energy, turbulent eddy dissipation-based equations' orders are undergone utmost care, and thus, the numerical simulations are carried out.

Figures 78.8, 78.9, 78.10, and 78.11 are revealed the CFD outcomes of UAV's propellers, wherein the velocity variations over the rotor, pressure impact on propeller, and drag force generated are computed predominantly. The comparative

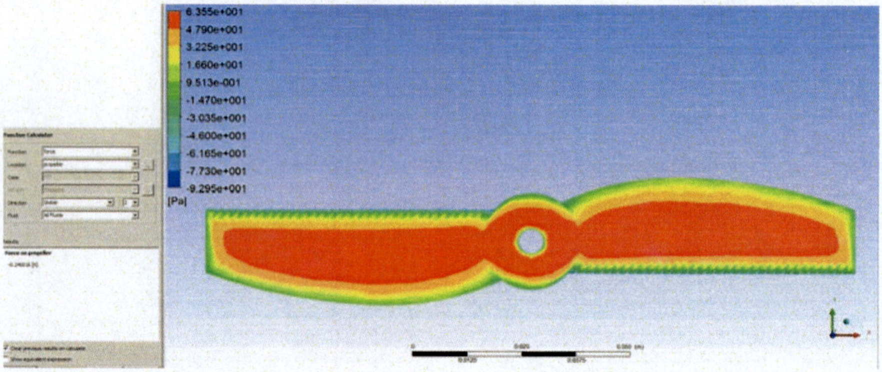

Fig. 78.8 Pressure cum drag force estimations on propeller equipped with V-Cut

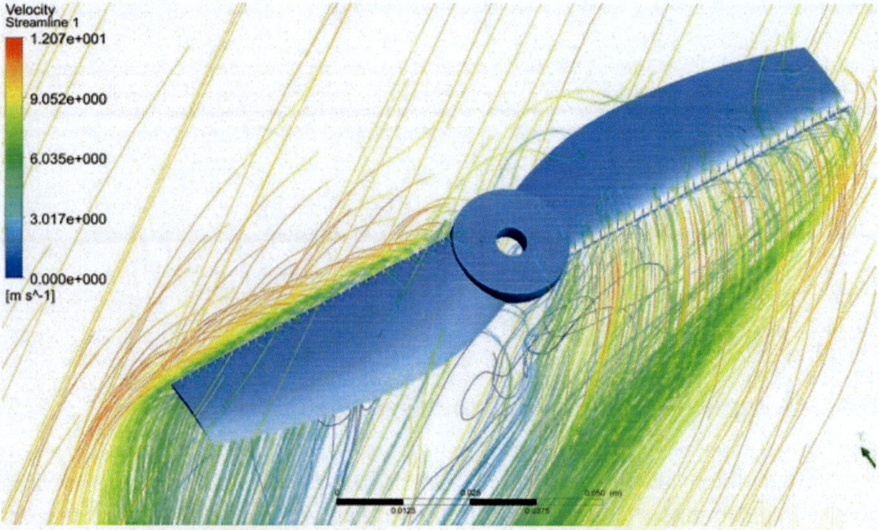

Fig. 78.9 Velocity distributions over propeller equipped with V-Cut at both LEs

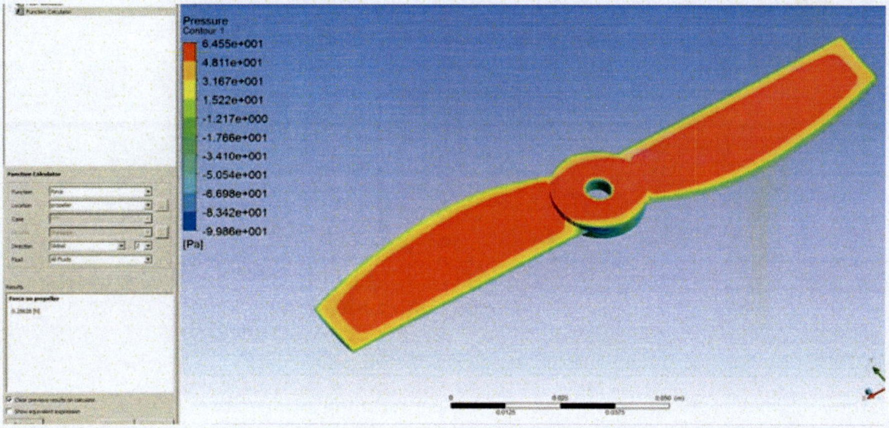

Fig. 78.10 Pressure cum drag force estimations on propeller equipped with saw tooth cut at 1-LE

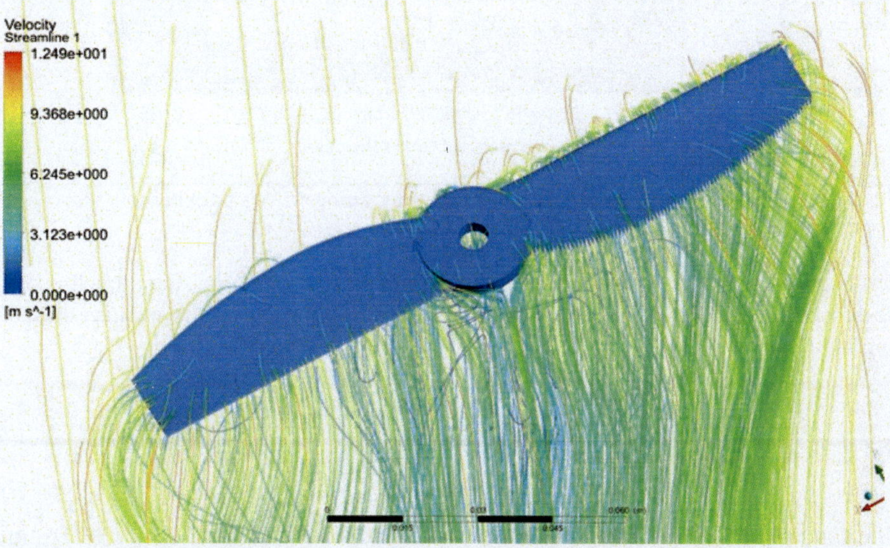

Fig. 78.11 Velocity distributions over propeller equipped with saw tooth cut at 1-LE

numerical results are shown in Figs. 78.12 and 78.13. The drag and exit velocities are considered as important selection parameters so the streamlines and planes are used for the successful capturization of aforementioned performance parameters.

From the comparative Figs. 78.12 and 78.13, it has been understood that the saw tooth cuts loaded on the edges of the UAV's propeller are performed better than base propeller. From this comparative, CFD analyses are predicted that the saw tooth cuts-based edge modification techniques have the full capability to provide good

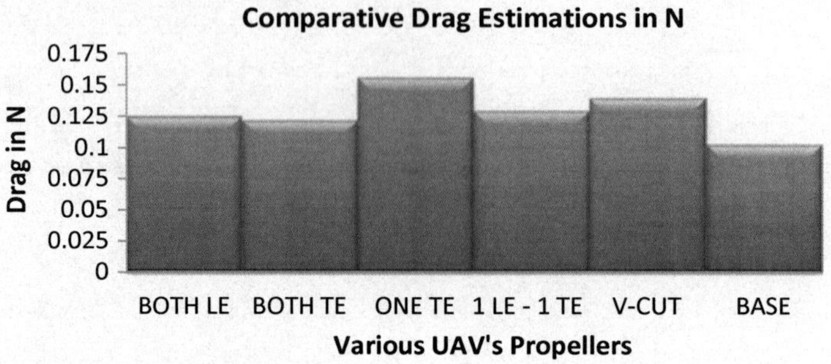

Fig. 78.12 Comprehensive drag data of various UAV's propellers

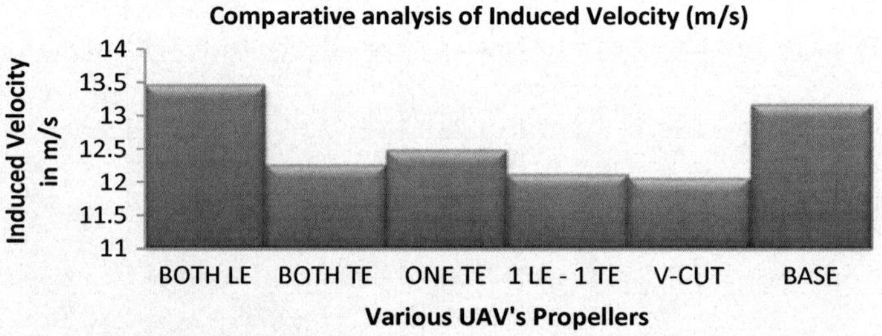

Fig. 78.13 Comprehensive induced exit velocity data by various UAV's propellers

aeroacoustic and aerodynamic performance than other propellers. Also, this kind of design modification on propeller is most suitable to survive in the military-based applications.

78.5 Conclusions

The conceptual designs of all the propellers are modeled with the help of CATIA, in which the dimensions of both propellers and the edge modifications techniques are obtained from standard literature surveys. The saw tooth-based edge modifications are primary targeted techniques of this work so based on that, four different propellers are modeled. Apart from that, one unique edge modification is also located on propeller's edge. The acceptable quality-based discretization process is constructed, which have the capability to provide robustness in the computational outcome. The standards cum reliable boundary conditions are implemented in the CFD analyses, and thereby, the numerical simulations are carried out on all the six propellers. From the results, the following conclusions are obtained: 1. the induced velocity of the fluid by propeller modified with saw tooth cuts at its both leading edges is higher than the base propellers. 2. The drag force generation is slightly higher in the edge modified propellers than base propeller. Thus, the saw tooth cut-based edge modification is capable to provide required thrust with low acoustic levels.

References

1. Vijayanandh, R., Ramesh, M., Raj Kumar, G., Thianesh, U. K, Venkatesan, K., Senthil Kumar, M.: Research of noise in the unmanned aerial vehicle's propeller using CFD. Int. J. Eng. Adv. Technol. 2249–8958 **8**(6S), 145–150 (2019). https://doi.org/10.35940/ijeat.F1031.0886S19
2. Murugesan, R., Raja, V.: Acoustic investigation on unmanned aerial vehicle's rotor using CFD MRF Approach, ASME 2019 Gas Turbine India Conference, GTINDIA 2019, pp. 1–7. https://doi.org/10.1115/GTINDIA2019-2430
3. Vijayanandh, R., Ramesh, M., Venkatesan, K., Raj Kumar, G., Senthil Kumar, M., Rajkumar, R.: Comparative acoustic analysis of modified unmanned aerial vehicle's propeller, advances in IC engines and combustion technology, Lecture Notes in Mechanical Engineering, (2020). https://doi.org/10.1007/978-981-15-5996-9_45
4. Li, Y., et al.:, Aerodynamic and Aeroacoustic Analyses of a UAV Propeller with Trailing Edge Serrations, Proceedings of Acoustics 2018, pp. 1–10, Adelaide, Australia, 7–9 November 2018.
5. Sonaimuthu, B., Panchalingam, P., Raja, V.: Comparative Analysis of Propulsive System in Multi-Rotor Unmanned Aerial Vehicle, ASME 2019 Gas Turbine India Conference, GTINDIA 2019, pp. 1–8. https://doi.org/10.1115/GTINDIA2019-2429
6. Weger, M., Wagner, H.: Morphological variations of leading-edge serrations in Owls (Strigiformes), PLoS ONE **11**(3), e0149236, pp. 1–21 (2016). https://doi.org/10.1371/journal.pone.0149236
7. Vijayanandh, R., et al.: Material optimization of high speed micro aerial vehicle using FSI simulation. Procedia Comput. Sci. **133**, 2–9 (2018). https://doi.org/10.1016/j.procs.2018.07.002
8. Ning, Z., et al.: An experimental study on small UAV propellers with serrated trailing edges, 47th AIAA Fluid Dynamics Conference, AIAA AVIATION Forum, AIAA 2017–3813, pp. 1–17
9. Vijayanandh, R., et al.: Design optimization of advanced multi-rotor unmanned aircraft system using FSI, Lecture Notes in Mechanical Engineering, Chapter 28, eBook ISBN - 978–981–13–2718–6. https://doi.org/10.1007/978-981-13-2718-6
10. Avallone, F., et al.: Noise reduction mechanisms of sawtooth and combed-sawtooth trailing-edge serrations. J. Fluid Mech. **848**, 560–591 (2018). https://doi.org/10.1017/jfm.2018.377
11. Balaji, S., Prabhagaran, P., Vijayanandh, R., Senthil Kumar, M., Raj Kumar, R.: Comparative computational analysis on high stable hexacopter for long range applications, Lecture Notes in Civil Engineering, Proceedings of UASG 2019, Unmanned Aerial System in Geomatics, eBook ISBN: 978–3–030–37393–1. https://doi.org/10.1007/978-3-030-37393-31
12. Alawadhi, H.A., et al.: CFD analysis of wing trailing edge vortex generator using serrations. EPJ Web Conf. **67**, 02002 (2014). https://doi.org/10.1051/epjconf/20146702002
13. Vijayanandh, R., Kiran, P., Indira Prasanth, S., Raj Kumar, G., Balaji, S.: Conceptual design and optimization of flexible landing gear for tilt-hexacopter using CFD, Lecture Notes in Civil Engineering, Proceedings of UASG 2019, Unmanned Aerial System in Geomatics, eBook ISBN: 978–3–030–37393–1. https://doi.org/10.1007/978-3-030-37393-15
14. Rafi Butt, F., Talha, T.: Numerical investigation of the effect of leading-edge tubercles on propeller performance. J. Aircraft **56**(3), 1014-1028 (2019). https://doi.org/10.2514/1.C034845
15. Jagadeeshwaran, P., Natarajan, V., Vijayanandh, R., Senthil Kumar, M., Raj Kumar, G.: Numerical estimation of ultimate specification of advanced multi-rotor unmanned aerial vehicle. Int. J. Sci. Technol. Res. **9**(01), 3681–3687 (2020). ISSN 2277–8616
16. Senthil Kumar, M., Vijayanandh, R., Gopi, B.: Numerical investigation on vibration reduction in helicopter main rotor using air blown blades. Int. J. Mech. Prod. Eng. Res. Dev. **8**(7), 152–164 (2018). ISSN (E): 2249–8001

17. Senthil Kumar, M., et al.: Conceptual design and comparative computational analysis of secondary inlet of rotary-wing aircraft engine. J. Adv. Res. Dyn. Control Syst. **9**(14), 1189–1209 (2017)
18. Arul Prakash, R., Sarath Kumar, R., Vijayanandh, R., Venkata Praveen, D., Raja Sekar, K., Ananda Krishnan, C.: Design optimization of convergent - divergent nozzle using computational fluid dynamics approach. Int. J. Mech. Prod. Eng. Res. Dev. **9**(1), 220–232, (2019). ISSN(E): 2249–8001
19. Vijayanandh, R., Venkatesan, K., Senthil Kumar, M., Raj Kumar, G., Jagadeeshwaran, P., Raj Kumar, R.: Comparative fatigue life estimations of Marine Propeller by using FSI. IOP-J. Phys. Conf. Series, 1473 (2020) 012018. https://doi.org/10.1088/1742-6596/1473/1/012018
20. Vijayanandh, R., et al.: Theoretical and numerical analyses on propulsive efficiency of unmanned aquatic vehicle's propeller. J. Phys: Conf. Ser. **1504**(012004), 1–10 (2020). https://doi.org/10.1088/1742-6596/1504/1/012004
21. Senthil Kumar, M., Vijayanandh, R., Srinivas, R., Arun Karthik, D., Tamil Mani, M.: Acoustic analysis and comparison of chevron nozzle using numerical simulation. Int. J. Mech. Prod. Eng. Res. Dev. **8**(7), 1089–1103 (2018). ISSN(E): 2249–8001
22. Praveen Kumar, V., Kishor Kumar, S., Sankaresh Pandian, K. R., Ashraf, E., Thanga Tamil Selvan, K., Vijayanandh, R.: Conceptual design and hydrodynamic research on unmanned aquatic vehicle. Int. J. Innov. Technol. Explor. Eng. **8**(11S), 121–127 (2019). https://doi.org/10.35940/ijitee.K1027.09811S19
23. Naveen Kumar, K., Vijayanandh, R., Bruce Ralphin Rose, J., Swathi, V., Narmatha, R., Venkatesan, K.: Research on Structural behavior of Composite Materials on different Cantilever Structures using FSI, International Journal of Engineering and Advanced Technology, ISSN: 2249 – 8958. **8**(6S3), 1075–1086 (2019). https://doi.org/10.35940/ijeat.F1178.0986S319.

Chapter 79
Prototype-Driven Innovation: Propositions Based on Challenges and Opportunities

Supradip Das and Amarendra Kumar Das

Abstract Existing new product development (NPD) methods are specification driven, where product specifications are typically made before prototyping starts. The existing method does not allow enough space for modification because of the association of sunk cost. Hence, the prototype-driven specification would enhance new product development with better prototype experience. But the prototype-driven approach does not have structured tools and methods to foster. Hence, it is very difficult to introduce and practice in design education. Systematic literature review had been done to identify the drivers, principles or guidelines and support tools in prototyping and prototype-driven new product development process. A structured and sequential process is followed to illustrate the initial reference model and impact model to identify the means that may foster the emergence of prototype-driven new product development from design education perspective. This state-of-the-art review also describes, discusses and connects the existing research to articulate propositions, which would help the future research in prototype-driven innovation.

79.1 Introduction

The capacity to do innovation and transform their own environment is one of the noteworthy accomplishment of the human species' [1]. A couple of decades back, innovation was considered as sporadic craftsmanship and selected individuals were considered as equipped for creating innovation. The innovation paradigm shifted in terms of 1. Nature: The nature of innovation changed from art to science [2]. Innovation has practical theories with associated taxonomies of the structure now, 2. Strategy: Innovation strategy changed from basic science invention driven [3] to prototype driven [4] to design driven to meaning strategy [5] to Brand driven [6]. In

S. Das (✉) · A. K. Das
Indian Institute of Technology Guwahati, Guwahati, India
e-mail: supradip.das@iitg.ac.in

A. K. Das
e-mail: dasak@iitg.ac.in

A. Chakrabarti et al. (eds.), *Design for Tomorrow–Volume 2*, Smart Innovation, Systems and Technologies 222, https://doi.org/10.1007/978-981-16-0119-4_79

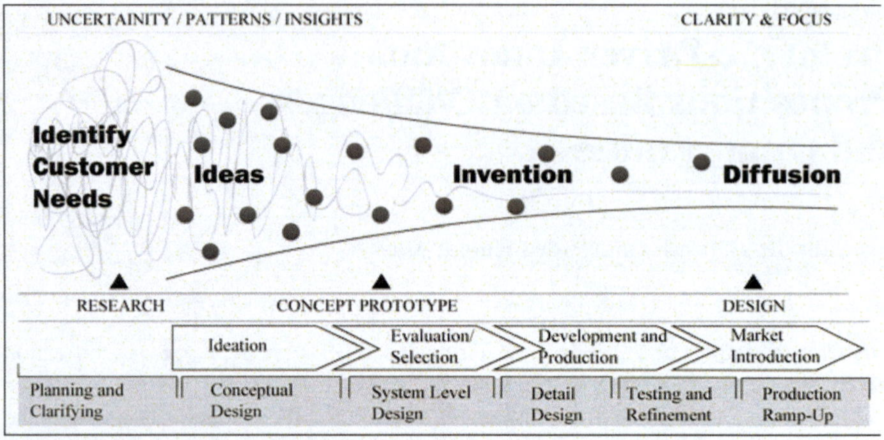

Fig. 79.1 Juxtaposition of design, innovation and new product development process

all the cases of product innovation strategy, the vantage point for the perspective is different, but the overarching goals for product innovation in all cases are the same. It follows the funnel model of the innovation process [7]. Product innovation examines the need for innovative products and processes, and so as new product development process (NPD). NPD plays in providing a strategic process to create a successful product launch [8]. Because of the similarity and natural occurrence of the steps involved in new product development [9] and innovation [7], design is implicitly part of the innovation process [10] (Fig. 79.1).

Though there are well-structured processes available, the numbers of successful innovations is less. Literature suggests that the prototype-driven approach would enhance the process of innovation and reduces the risk of failure [4]. But limited studies are available in this area to develop tools and methods to enhance the process. The following section of the paper illustrates the systematic effort to identify the challenges and opportunities for developing propositions.

79.2 Literature Review

Blessing and Chakrabarti [11] suggested a systematic and sequential process for research clarification, based on which literature review had been done. Literature was searched from the database of publishers such as Elsevier, Scopus, Inderscience, INSPEC, Thompson Reuters, Proquest, Common Ground Research Network, SAGE and also from the database of the societies related to design, such as Design Society and Design Research Society. Literature from the journals was also reviewed which are design-related or design-focused identified by Gemser et al. [12]. To make the literature review focused, areas of relevance and contribution (ARC) diagram were

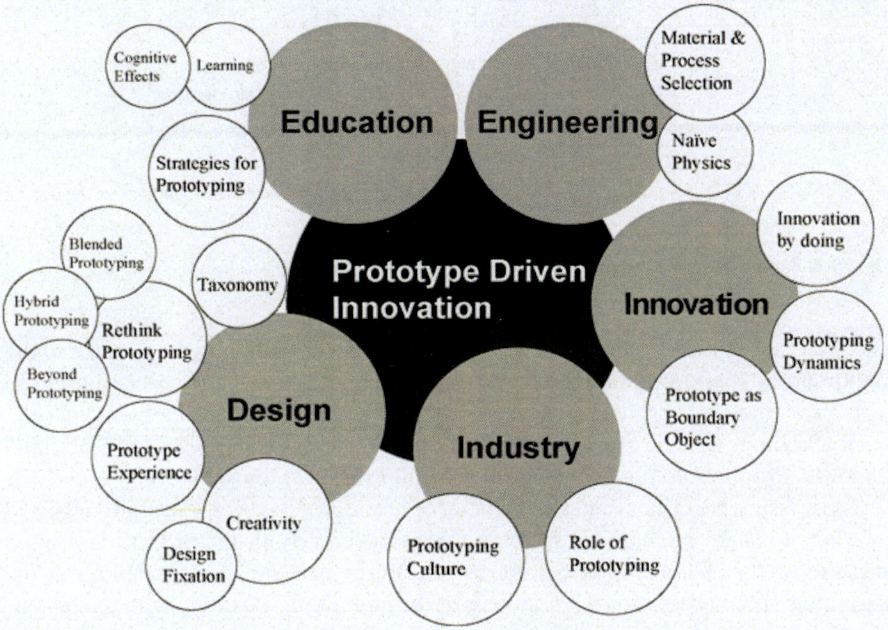

Fig. 79.2 Areas of relevance and contribution (ARC) diagram

made (Fig. 79.2), which represents and clarifies the foundation on which research is to be based and areas of contribution.

These are the following data that has been extracted from the screened articles:

(a) Definitions
(b) Drivers
(c) Principles
(d) Context
(e) Tools and techniques
(f) Prototyping in the early design phase
(g) Requirement

79.3 Prototype in Design, NPD and Innovation

According to Henry Chesbrough, "Most innovations fail. And companies that don't innovate die" [13]. This statement is an oxymoron. On the one hand, innovation is essential and on the other hand, endeavours to innovate are full of risks. Therefore, it is important to analyse the reason for failure in the innovation process. Literature review reveals these following whys and wherefores:

(A) Analysis of the innovation phases shows that undesirable project terminations are frequently occurring at two points (Fig. 79.3) in the innovation process: a) after

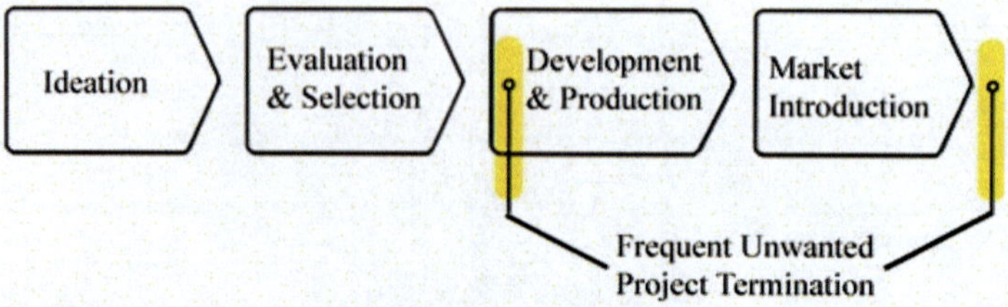

Fig. 79.3 Frequent unwanted project termination points

selections are made for product development, b) after introducing the product to the market [14]. Therefore, the innovation process should have to be re-configured in such a way that the best idea is selected to go forward for the development phase.

(B) Products fail because they do not address the user's experience and emotional response. Products do not connect to their emotional journey [15].

These issues are critical and also a matter of research. However, research suggested solutions to these issues: (A). Making a prototype early in the process is a good measure to foster innovation culture as it inspires new ideas [15]. Failing at the beginning of the process ensures success at the end. Design fixation with the help of prototypes increases the possibility of selection of best idea and reduces the risk associated with sunk cost [16]. (B) Experience prototyping [17] or collaborative prototyping [18] would lead to balanced functional and usable deliverable by engaging design team members, users and clients with prototypes.

In the literature review, it is clear that design prototypes are the boundary objects in the innovation process [19]. Besides prototypes are also an integral part of the NPD process and different types of prototypes are there to support the NPD process [20–22].

►Proposition 1: In case of prototype-driven design innovation, an exclusive process model or framework has to be proposed supported by methods and tools, where prototyping is the boundary object.

79.4 Attributes of Prototypes

Attributes of a physical model/ prototype are important because these may help to set the success criteria for prototyping. These are the elements, which can be modified for better prototype experience. Hence, these can be considered as the drivers for prototyping. To understand more about the characteristics of a prototype, analysis of the definition and objectives of the prototype are required.

According to Ulrich and Eppinger [21] "prototype is an approximation of the product along one or more dimensions". As stated by this definition, any object unveiling at least one attribute of the product that is of concern to the development team can be considered as a prototype. It means a model/ prototype (Pt) has a set of attributes (Pa…Pn) of the products (P), but all.

High fidelity prototype

↑

Low fidelity prototype

$$\text{Product (P)} = Pa_1 + Pa_2 + Pa_3 + Pa_4 \ldots\ldots\ldots Pa_{n-1} + Pa_n \quad (1)$$

$$\text{Prototype } (P_t) = Pa_1 + Pa_2 + Pa_3 + Pa_4 \ldots\ldots Pa_{n-1} \quad (2)$$

or

$$\text{Prototype } (P_t) = Pa_1 + Pa_2 + Pa_3 + Pa_4 \quad (3)$$

The more product attribute in the prototype, the more prototype quality and experience. So, the difference between product and prototype attributes (P-Pt) defines the quality of the prototype. But, how many attributes of the product to be incorporated into a prototype?

Houde and Hill [23] presented a prototype model, which illustrates a three-dimensional space which corresponds to the three attributes of the prototype such as (1) role, which give a sense of usability; (2) look and feel, which represent form and appearance; and (3) implementation, which demonstrates the working principle. Later, Broek et al. [24] provided an exhaustive list of characteristics. These are appearance, shape and design, strength and stiffness, shape accuracy, tolerance, geometric accuracy, texture, surface quality, application of features, material properties, material hardness, prototype weight, colour, conductivity, transparency.

Moreover, Lim et al. [25] describe the attributes of prototypes under two different major class, filtering dimensions, which exhibits the physical properties of the design and manifestation dimensions, which has to be considered while crafting a prototype. But, specific position in the design process would give an understanding of the purpose of the prototype. For example, in that early design phases, purpose of the prototypes is more to explore and evaluate, and later on to communicate ideas to audience [26]. While communicating to the audience, the degree of refinement (fidelity) of the prototype should be at par the ability to interpret and understand and that can be achieved by selecting the appropriate technique for crafting. Hence, an appropriate technique selection is important for representation of the prototype for manifestation.

►Proposition 2: To understand the dynamic behaviours of the filtering and manifestation dimensions for different kind of design prototype, a structured tool may be helpful to understand easily.

79.5 Prototyping for Knowledge

To understand the impact of prototyping in the knowledge domain, understanding of design cognition, i.e. human information processing in the design process is important.

To solve a particular design problem, designer travels through the external world to take information from objects (designs) and observation (context of other designs), external and the internal archive of previous experience and learning. Referring to all the sources, designers mentally encode a rich structure of knowledge, which are fractal and experiential [27].

It is important to externalize an internal mental model through multiple representations. This process helps designers' to select the best solution. Prior research expounds that prototypes help to enhance designers' incorrect mental models and lead them better ideas satisfying all the design requirements [28]. It was also observed that physical prototyping can mitigate design fixation [29].

Naïve physics' for engineering design states that engineer's mental models may not be highly precise unless recurrently verified through either extensive experience or education. That is why extensive use of physical models suggested throughout the design process to support and enhance an engineer's inaccurate mental models [28].

In contrast, teaching physical prototyping in design is not that simple because it is tacit knowledge. And the transfer of tacit knowledge is costly, difficult and less mobile [30]. It is easy to demonstrate than to articulate the tacit knowledge [31]. Learning by doing gives experience allows experimentation and helps to accumulate tacit knowledge, which lies below the surface of conscious thought [28].

►Proposition 3: Prototyping skill is tacit knowledge. To transfer the prototype-making knowledge, a structured do-it-yourself (DIY) tool is required to foster learning by doing.

79.6 Prototyping Process and Supports

A fair amount of literature is available on prototyping processes. But because of the varied meaning in different disciplines, a common process is not available. Warfel [32] introduced an iterative and evolutionary prototyping process, which is cyclical in nature. Exner et al. [33] investigated three dimensions of the prototyping process, such as objectives, dimensions and fidelity, based on which a process model was proposed. Yu et al. investigated the nature of prototyping practices between designers and engineers to suggest a prototyping process over Exner et al., which is more elaborative [34]. Research done by Exner et al. says that engineers focus on the features and functional aspects of a prototype, whereas designers explore with prototypes to understand the design space for new possibilities and explore materials and tools, particularly for lowfidelity prototypes. Later, Hallgrimsson [35] proposed a well-described model making workflow. Velásquez-Posada [36] proposed a tool for physical model making using two distinct processes adding material and subtracting material.

In the recent past, Jessica Menold and her team proposed a theoretical framework for prototyping called prototype for X or PFX to support product design during the prototyping phase of the design process only [37].

The process proposed in all the literature does not have a supporting method or tool to assist the creator of the prototype. Moreover, these processes are towards prototype creation, not ideation. In the existing process, prototyping is a one-way process (Fig. 79.4), where prototypes are made to manifest the abstract concepts generated in the concept generation phase.

Fig. 79.4 Existing process of prototyping

Also prototype is considered as merely a tool to ensure clear communication among the design team members and aids in learning to support decision making [38] (Fig. 79.5).

Towards prototype-driven innovation, where prototype is the boundary object and prototyping process is the central activity, we define prototyping as, "a process to explore and think for new concept generation".

In the case of prototype-driven design innovation, prototyping must be a two-way process to generate ideas. Thinking and making both should happen together (Fig. 79.6).

►Proposition 4: Making prototype early in the process may not work for novice designers, without any additional support of methods and tools.

►Proposition 5: During the early phases, the prototype should allow to explore, envision, and refine mental models. Hence, exploration in the early phases with 2-dimensional sketches may not work. Exploration should be done with the 3-dimensional entity.

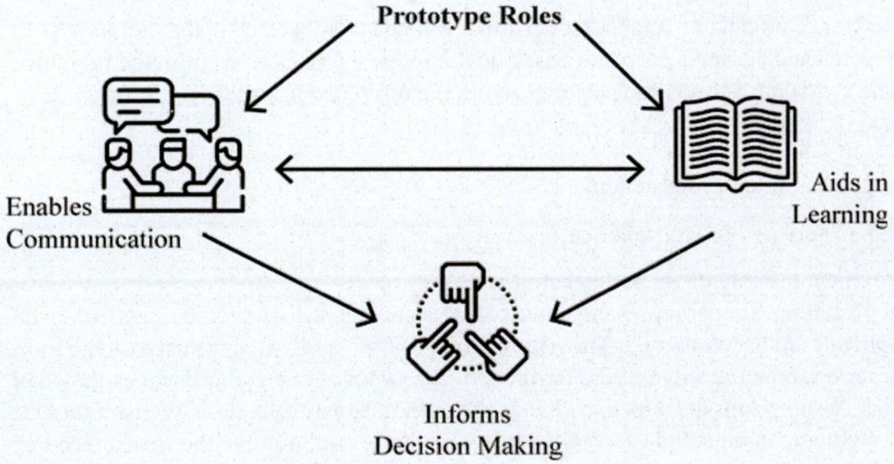

Fig. 79.5 Existing process of prototyping

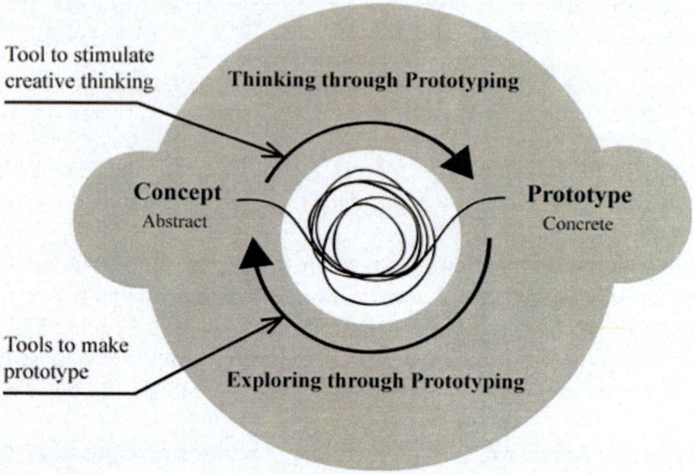

Fig. 79.6 Proposed model of prototyping

79.7 Reference Model and Impact Model for Future Direction

Reference model and impact model are developed as suggested in the literature [11]. Reference diagram is prepared based on the propositions derived from the literature. The following abbreviations were used in the reference model (Fig. 79.7):

[X]: citation number;

[A]: assumption based statement;

[E]: experience-based statement;

[O]: own investigations based statement;

Reference model represents the existing state, against which the intended developments are benchmarked. The combination of ' + ' and '−' signs used on the links describes how the value of the attribute of the factor at one end influences the other end. Assumptions are added to change the current situation. A do-it-yourself tool for prototyping is intended to foster learning by doing and improve the issues faced by novice designers in the reference model (Fig. 79.7).

It is assumed that do-it-yourself tool for prototyping would have a two-way impact. One way it would foster learning by doing. As in design discipline, students learn through learning by doing, which allows explorations, it would help to accumulate tacit knowledge implicitly. It is also easy to transfer knowledge from one person to another through learning by doing. Thus, it may influence tacit technical skill and tacit cognitive skill [30] by improving deep learning of prototyping. And literature says proper use of prototyping knowledge would lead an enhancement in innovation.

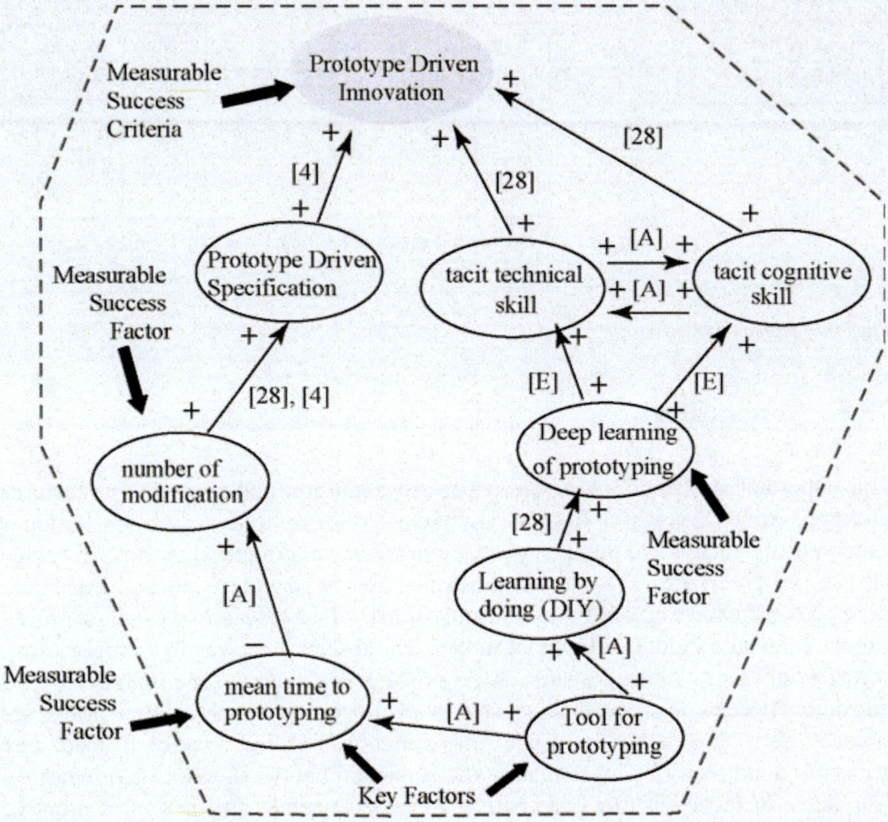

Fig. 79.7 Reference model

Another way, a tool for prototyping would help in reducing mean time to prototyping, which will help to make more number of modification within a stipulated time. And literature says prototype-driven specification demands more number of modification. Thus, the number of modification would influence prototype-driven innovation with an impact in prototype-driven specification.

Impact model (Fig. 79.8) represents the desired situation and shows the assumed impact of the support to be developed.

Reference model and impact model elucidate the opportunities and would help for future research. Impact model illustrates that the introduction of DIY support for prototyping, which would positively affect the knowledge of prototyping and increased knowledge would increase the possibility of exploration by reducing the meantime to prototyping. In the reference model, success criteria communicate the ultimate goal to which the research intends to contribute and success factors are the entity, which justify the cause-effect in the research. The term 'measurable' refers to the possibility of measuring the factors using qualitative or quantitative research methods.

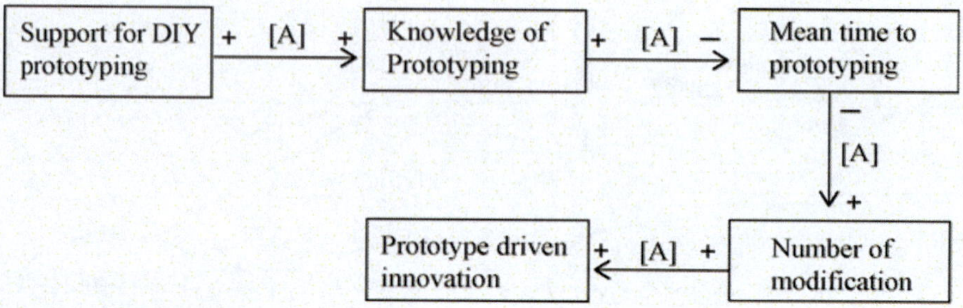

Fig. 79.8 Impact model

79.8 Conclusion

The important contributions of the paper are definition and propositions towards prototype-driven innovation from design perspective. Prototype-driven innovation is a theoretical concept and there is no concrete framework, method or tool. To bridge the gap, we have done a systematic literature review of papers from design education perspective. We have critically studied and connected the research to generate propositions, reference model and impact model. The research derives five propositions, which would guide future research. A proposition is a tentative and imaginary relationship between constructs that is stated in an allegorical form. As propositions are connects between abstract constructs, they cannot be verified directly. Instead, they are verified indirectly by investigating the connection between associated measures (variables) of those abstract constructs. The experimental construction of propositions will indicate an associations between variables to be considered as hypotheses [39]. Hypothesis can be generated from all the propositions to investigate its validity. Experiments can be designed based on the success criteria and measurable success criteria mentioned in the reference model. Furthermore, as the impact model intends to develop support in prototype-driven design innovation, a detail systemic approach required to develop the tool and test.

References

1. Youmans, R.J.: The effects of physical prototyping and group work on the reduction of design fixation. Des. Stud. **32**, 115–138 (2011). https://doi.org/10.1016/j.destud.2010.08.001
2. Hitcher, W.: The Innovation Paradigm Replaced. (2005)
3. Bush, V.: Science, the endless frontier; A report to the President on a program for postwar scientific research. (1945)
4. Schrage, M.: The Culture(s) of Prototyping. In: Bringing Design to Software. pp. 191–205. ACM Press, New York (1996) https://doi.org/10.1111/j.1948-7169.1993.tb00128.x.
5. Battistella, C., Biotto, G., Toni, A.F.D., Battistella, C., Biotto, G., Toni, A.F.D.: from Design Driven Innovation to Meaning Strategy from Design Driven Innovation to Meaning Strategy. (2012). https://doi.org/10.1108/00251741211220390

6. Abbing, E.R.: Brand-driven innovation. AVA Publishing (2010)
7. Müller-Prothmann, T., & Dörr, N.: Innovations- management. Strategien, Methoden und Werkzeuge für systematische Innovationsprozesse–Kapitel 1.1: Zum Begriff der Innovation. HANSER, Germany (2009)
8. Rainey, D.L.: Product innovation: leading change through integrated product development. Cambridge University Press (2008)
9. Phal, G., Beitz, W.: Engineering Design: A Systematic Approach. The Design Council, London (1984)
10. Utterback, J., Vedin, B.-A., Alvarez, E., Ekman, S., Sanderson, S.W., Tether, B., Verganti, R.: Design Inspired Innovation. World Scientific Publishing Co. Pvt. Ltd. (2006)
11. Blessing, L.T.M., Chakrabarti, A.: DRM, a Design Research Methodology. (2009). https://doi.org/10.1007/978-1-84882-587-1
12. Gemser, G., de Bont, C.: Design-related and design-focused research: A study of publication patterns in design journals. She Ji J. Des. Econ. Innov. **2**, 46–58 (2016). https://doi.org/10.1016/j.sheji.2016.05.002
13. Chesbrough, W.H.: Open Innovation: The New Imperative for Creating and Profiting from Technology. Harvard Business School Publishing, Massachusetts (2003)
14. Huber, D., Kaufmann, H., Steinmann, M.: Bridging the Innovation Gap. Springer International Publishing (2014)
15. Brown, T.: Change By Design. HarperCollins (2009)
16. Viswanathan, V., Atilola, O., Esposito, N., Linsey, J.: A study on the role of physical models in the mitigation of design fixation. 4828, (2017). https://doi.org/10.1080/09544828.2014.885934
17. Buchenau, M., Suri, J.F.: Experience prototyping. Proc. Conf. Des. Interact. Syst. Process. Pract. methods, Tech. - DIS '00. 424–433 (2000). https://doi.org/10.1145/347642.347802.
18. Bogers, M., Horst, W.: Collaborative prototyping : Cross-fertilization of knowledge in prototype-driven problem solving *. J. Prod. Innov. Manag. **31**, 744–764 (2014). https://doi.org/10.1111/jpim.12121
19. Rhinow, H.: Design prototypes as boundary objects in innovation processes design prototypes as boundary objects in innovation processes. pp. 1–10 (2012)
20. Otto, K.N., Wood, K.L.: Product Design: Techniques in Reverse Engineering and New Product Development. Pearson Education India (2003)
21. Ulrich, K.T., Eppinger, S.D.: Product Design and Development, 5th edn. Mc Graw Hill, Singapore (2012)
22. Ullman, D.G.: The Mechanical Design Process. Mc Graw Hill, NewYork (2003)
23. Houde, S., Hill, C.: What do prototypes prototype? Handb. Hum. Comput. Interact. 1–16 (1997). https://doi.org/10.1016/B978-044481862-1.50082-0
24. Broek, J.J., Sleijffers, W., Horváth, I., Lennings, A.F.: Using Physical Models in Design. In: CAID &CACD (2000)
25. Lim, Y.-K., Stolterman, E., Tenenberg, J.: The anatomy of prototypes: Prototypes as filters, prototypes as manifestations of design ideas. ACM Trans. Comput. Interact. **15**, 1–27 (2008). https://doi.org/10.1145/1375761.1375762
26. Voss, C., Zomerdijk, L.: Innovation in Experiential Services. Pp.390–401 (2007). https://doi.org/10.4018/978-1-4666-8216-0.ch019
27. Eastman, C.: New Directions in Design Cognition : Studies of Representation and Recall. In: Design knowing and learning: Cognition in design education. pp. 147–198. Elsevier (2001)
28. Viswanathan, V.K., Linsey, J.S.: Physical Models In Idea Generation – Hindrance Or Help? In: ASME 2010 International Design Engineering Technical Conferences & Computers and Information in Engineering Conference (2010)
29. Kershaw, T.C., Hölttä-otto, K., Lee, Y. S.: The Effect of Prototyping and Critical Feedback on Fixation in Engineering Design Experiment 1 : Prototyping Method. In: Cognitive Science Society. pp. 807–812 (2011)
30. Mascitelli, R.: From experience : Harnessing tacit knowledge to achieve breakthrough innovation. J. Prod. Innov. Manag. **17**, 179–193 (2000)

31. Dym, C.L., Little, P., Orwin, E.J.: Engineering Design: A Project Based Introduction. John Wiley & Sons Inc., Orwin, Harvey Mudd College (2014)
32. Warfel, Z.T.: Prototyping A Practitioner ' s Guide. Rosenfeld Media, New York (2009)
33. Exner, K., Lindow, K., Stark, R., Angesleva, J., Bahr, B., Nagy, E.: A transdisciplinary perspective on prototyping. In: 2015 IEEE International Conference on Engineering, Technology and Innovation/ International Technology Management Conference, ICE/ITMC 2015. pp. 1–8 (2016). https://doi.org/10.1109/ICE.2015.7438659
34. Yu, F., Pasinelli, M., Brem, A.: Prototyping in theory and in practice : A study of the similarities and differences between engineers and designers. Creat. Innov. Manag. pp.1–12 (2017). https://doi.org/10.1111/caim.12242
35. Hallgrimsson, B.: Prototyping and Model Making for Product Design. Laurance King Publication, London (2012)
36. Velásquez-Posada, A.: Model making techniques as a teaching tool in product. In: Crossing Design Boundaries. pp. 445–451. Taylor & Francis (2005)
37. Menold, J., Simpson, T.W., Jablokow, K.W.: The Prototype for X Framework: Assessing the Impact on Desirability, Feasibility, and Viability of End Designs. In: ASME 2016 International Design Engineering Technical Conferences & Computers and Infomation in Engineering Conference. , Charlotte, North Carolina (2016)
38. Lauff, C.A., Kotys-Schwartz, D., Rentschler, M.E.: What is a prototype? what are the roles of prototypes in companies? J. Mech. Des. Trans. ASME. 140, (2018) https://doi.org/10.1115/1.4039340
39. Bhattacherjee, A.: Social science research - principles, methods. and, Florida, USA (2012)

Chapter 80
Visuo-Locomotive Complexity as a Component of Parametric Design for Architecture

Vasiliki Kondyli, Mehul Bhatt, and Evgenia Spyridonos

Abstract A people-centred approach for designing large-scale built-up spaces necessitates systematic anticipation of user's embodied visuo-locomotive experience from the viewpoint of human-environment interaction factors pertaining to aspects such as navigation, wayfinding, usability. In this context, we develop a behaviour-based visuo-locomotive complexity model that functions as a key correlate of cognitive performance vis-a-vis internal navigation. We also demonstrate the model's implementation and application as a parametric tool for the identification and manipulation of the architectural morphology along a navigation path as per the parameters of the proposed visuospatial complexity model. We present examples based on an empirical study in two healthcare buildings, and showcase the manner in which a dynamic and interactive parametric (complexity) model can promote behaviour-based decision-making throughout the design process to maintain desired levels of visuospatial complexity as part of a navigation or wayfinding experience.

80.1 Introduction

The design of navigation and wayfinding systems within large-scale built-up spaces is a particularly challenging task: research in behavioural, computational, and mixed-methods interdisciplinary approaches in spatial cognition, architecture design cognition, and artificial intelligence for design have shown that diverse morphological and ecological aspects pertaining to space, perception, and human factors constitute a correlate of visuo-locomotive cognitive experience in built-up space from the viewpoint of tasks as navigation and wayfinding [3–5, 10, 17, 23, 30].

CoDesign Lab/www.codesign-lab.org > info@codesign-lab.org

V. Kondyli (✉) · M. Bhatt
Örebro University, Örebro, Sweden
e-mail: vasiliki.kondyli@oru.se

E. Spyridonos
University of Stuttgart, Stuttgart, Germany

A. Chakrabarti et al. (eds.), *Design for Tomorrow–Volume 2*, Smart Innovation, Systems and Technologies 222, https://doi.org/10.1007/978-981-16-0119-4_80

Visuo-Locomotive Experience During Navigation Visuo-locomotive experience most broadly refers to the embodied cognitive experience of environmental space through a multimodal perceptual integration, e.g., encompassing visual, locomotive and auditory exploration [4, 5]. From the viewpoint of this paper, examining the embodied experience of users through interaction with the architectural space during active locomotion includes quantitative measurements and qualitative analysis of a range of aspects including visual attention (e.g. fixation, saccades), fine-grained behaviour (e.g., stopping, looking around, interacting with other people), orientation tasks, spatial memory (e.g. sketch maps, questionnaires) etc (details in papers [6, 17]). The focus of this paper is on visuo-locomotive experience as it pertains to aspects of visuospatial and locomotive complexity as articulated in this research.

Visuospatial and Locomotive Complexity To measure the effect of the architectural space on the visuo-locomotive experience we take into consideration behavioural data of perceived visuospatial stimuli with respect to the attributes involved in the dynamic scene (e.g. form, colours, layout) (Fig. 80.1). Even though a combination of these attributes can result to complex scenes, people are able to form a coherent percept amid numerous regions and identify properties and semantics of a scene at a glance [25]. Visual complexity has been broadly defined as the level of detail and intricacy contained within the image or a scene [29]. Its effect on visual attention and spatial cognition, and corresponding measurement methodologies have been investigated in disciplines such as cognitive science, psychology, computer science, or marketing [9, 20, 22, 26]. With the term *visuospatial and locomotive complexity* we consider the combination of visual and spatial characteristics that both coexist in

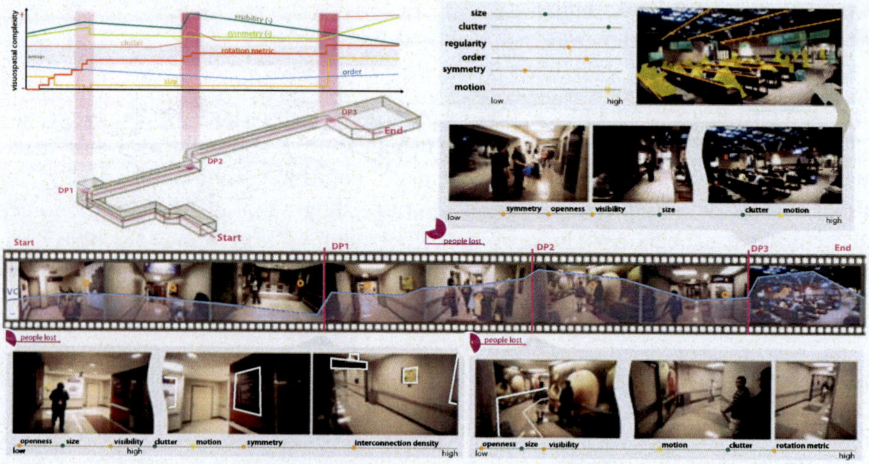

Fig. 80.1 Analysis of visuospatial complexity as it evolves along the navigation path at the Old Parkland Hospital. Analysis of decision points (DP) where participants get lost reveal the combination of attributes that compose the current visuospatial complexity level (colors correspond to the categories of attributes of Table 80.1: Quantitative (green), Structural (orange), Dynamic (yellow))

dynamic naturalistic scenes where a person navigates [17, 19].[1] Although complexity can provide an interesting and cognitively stimulating environment [8], excessive visual or spatial complexity can impact cognitive processes involved in visuo-locomotive experience of people in everyday tasks such as navigation or wayfinding [7, 11, 12].

In this paper we study the visuospatial complexity of the architectural space from the human-centred perspective of embodied visuo-locomotive experiences. We present a cognitive model of visuospatial and locomotive complexity that embeds empirical knowledge on the effect of environmental and dynamic attributes on human behaviour and we define a scale of complexity. We introduce the model to a parametric design system and demonstrate how a designer can identify *Classes* of complexity for a particular path or a segment of it, as well as modify the morphology of space along a path in order to maintain complexity to a preferable level. As currently there is no universally accepted or empirically defined scale for visuospatial complexity our method aims to organise and link the different attributes involved and their contribution to a visuospatial complexity scale, however, more empirical studies are needed to support our classification.

80.2 A Model of Visuo-Locomotive Complexity for Spatial Design

To model visuospatial complexity based on its effect on humans during active locomotion, we examine the attributes involved in the naturalistic dynamic scene and how they evolve along a navigation path. We combine empirical results from previous studies on visual and spatial complexity metrics together with the result of a behavioural study in two healthcare facilities in the Parkland Hospital [18]. We consider visuospatial and locomotive complexity as an implicit measure of cognitive load [14] and inefficient navigation, and we use the behavioural analysis along the path (through a combination of behavioural metrics such as delays, visual search performance, ask for help, loss of orientation, etc.) to identify locations with high complexity and further examine the present environmental and dynamic attributes.

Measuring Visuo-Locomotive Experience The model (Table 80.1) derives from a taxonomy of objective physical attributes from the scene analysis that affects visuospatial perception and cognitive functions (e.g. driving, walking) [19]. However, embodied cognitive experience whilst in motion along a path cannot be investigated based on a discrete analysis of a series of singular scenes. Therefore, the model is enriched with attributes pertaining to visuo-locomotive complexity such as rotation metric, visibility, interconnection density [17]. The attributes are categorised into:

[1]Henceforth, we use the terms visuospatial and visuo-locomotive complexity interchangeably. Other factors influencing embodied visuo-locomotive experience in space are familiarity, nature of the task (e.g. exploration, wayfinding), individual differences (e.g. age, spatial skills) etc. However these aspects are beyond the scope of this paper.

Table 80.1 Taxonomy of attributes for visuospatial complexity. The relationship between the individual attributes and the visuospatial complexity is defined as proportional (+) or inversely proportional (−) according to the impact they have on the visuospatial complexity scale. The (*) indicates locomotive complexity attributes

Visuospatial complexity	Proportional relationship to complexity and Description
A1. Quantitative attributes	
Size	(+/−) The dimensions of the physical space, the area covered by the visual stimulus
Clutter	(+)
Quantity	(+) No. components (objects, people, shapes etc.)
Variety of colours	(+) No. colours
Variety of shapes/Objects	(+) No. shapes / objects
Objects density	(+) No. objects in a defined area
Edges density	(+) No. edges of objects, No. of polygon vertices of a perimeter (fractal dimension)
Luminance	(−) Amount of light emitted / reflected from the scene
Saliency	(−) Prominent elements based on characteristics of colour, luminance and contrast
Target-background similarity	(+) Compare similarity in luminance, contrast, structure, orientation, etc.
A2. Structural attributes	
Repetition	(−) Recurrence of (groups) elements or characteristics on a line/a grid/ a pattern
Symmetry	(−) Resilience to transformation (reflectional/rotational/translational/helical/fractal)
Order	(−) Organised elements based on a recognised structure, fractal dimension, axial lines
Homogeneity/Heterogeneity	(−) Being all the same kind or diverse (single shape repeated - multiple distinct shapes)
Regularity	(−) Variations in a placement rule across a surface/line (polygons - abstract shapes)
Openness	(−) The ratio between empty and full space
Grouping	(−) No. elements that are part of a group
Rotation metric*	(+) Accumulated degrees of rotation angle during locomotion, No. of turns
Visibility*	(−) Visual range from a vantage point, visual connectivity between points
Interconnection density*	(+) No. of directional choices in each node (e.g. decision point, junction)
A3. Dynamic Attributes	
Motion	(+) No. people or objects moving in the scene
Flicker	(+) Abrupt changes over-time (in luminance, colours, etc.)
Speed | Direction	(+) The rate of change of position with respect to time | Move or facing towards

A1. *Quantitative Attributes.* Referring to objective environmental attributes such as low-level (edge, colors) and middle-level (corners, orientation) features of the scene. Studies on visual attention mostly on real-world static scenes, have shown how these environmental attributes can lead to a overabundance of information (known as clutter), work as distractors, and impact the visual search performance [16, 24, 26]. Clutter in architectural space may also involve the fractal dimension, referring to the number of polygon vertices of a perimeter of a space. Manifest cues, such as signage and landmarks, are part of clutter however because of their semantic significance as navigation aid objects they constitute major visual attention targets.

A2. *Structural Attributes.* Refer to the relation that the elements form due to positioning in space, or the overall distribution on the viewing scene. The presence of regularities, symmetry, repetition, or order simplifies the scene [11, 15] while randomised arrangements contribute to higher complexity classes. Homogeneity/Heterogeneity, regularity of shapes and objects, grouping of elements in space, openness of space and their relationship to complexity have been previously introduced in the fields of urban design and architecture [7, 27]. These attributes formulate the environmental structure along the navigation path, and impact the holistic legibility of the environment, including aspects of interconnection density [23, 30], visibility (e.g. visual range, line of sight) [2], rotation locomotion [17]. On the contrary, wayfinding studies show the value of architectural differentiations and breaks in the structural order as a design tool for distinguished areas that facilitate navigation [1].

A3. *Dynamic Attributes.* Studying active embodied locomotion, such as navigation, involves studying dynamic parts of the scene, as for example people or moving objects or the ratio of changes per meter. Dynamic attributes have a major impact on visual attention patterns, as clusters of attention often coincide with semantically rich objects such as eyes, hands [21]. Moreover, cortical analysis shows a selective response to moving elements on the scene, meaning that people are able to notice moving objects even if they are not looking for them [26].

80.3 An Empirical Basis for a Visuospatial Complexity Scale

Case study in Parkland Hospital In a behavioural case study conducted in two large-scale healthcare facilities (Old and New) in Parkland Hospital in Texas (USA), 25 participants performed a navigation task from the Emergency room to the Pharmacy. We employed a range of sensors for measuring embodied visuo-locomotive experience and navigation performance (e.g. eye-tracking, egocentric video, external camera based video, questionnaires, orientation task) and detect locations in the path where a number of participants were disoriented or confused (details in papers [6, 18]).

Behavioural and Environmental Analysis The correlations between the environmental and the behavioural analysis suggest that peculiar geometries (non symmetrical, not regular), narrow, long and high cluttered corridors are places where people tend to be frequently confused. Additionally, the presence of a combination of these attributes leads to low visibility and a great angle of rotation by the participant along the path that aggravate the navigation performance. For instance, in Decision Point 2 in the Old Parkland Hospital, we record 41% of participants having a disorientation or confusion event (e.g. stop, hesitation, look around, intensive visual search to both directions, etc. [17], Fig. 80.1). The place is characterised by narrow non-symmetrical and non-regular long cluttered corridors, that leads to reduced visibility. The place is also regularly occupied by dynamic attributes such as moving obstacles, pedestrian, people on wheelchairs moving in various directions. However, the questionnaires show that the landmark in the corridor (apple posters in bottom row of Fig. 80.1) was visually accessible and well recorded in spatial memory.

The effect of combined of attributes These observations suggest that the combination of attributes present in the scene can have a counterbalancing effect on the overall visuospatial complexity level, however further empirical work is needed to define the combinations of attributes that can lead to this outcome. For instance, high clutter increases the visuospatial complexity level but it can be mitigated with a well organised structural scene. Another consideration about finding balance in the visuospatial complexity scale is inspired by convex function of disorder [28] and complex adaptive systems [13], suggesting that a positive effect from the perceived visuospatial complexity is expected to reach an optimum between variety and structure (Fig. 80.2). For example, highly symmetrical spaces with repetitive features are related to low visuospatial complexity, but they have a negative effect on navigation performance as there is no architectural differentiation or distinctive object to assist cognitive mapping [1]. Consequently, in the process of developing a scale for visuospatial complexity, we consider that humans prefer to experience information at a comfortable rate (too little deprives the senses and too much overloads them [7]), and so we suggest a moderate visuospatial complexity class as a design aim (Fig. 80.2). This design aim should be adjusted based on the specific design groups (e.g. children, older adults, people with cognitive decline) in order to provide a suitable motivating environment.

Empirically model complexity Classes As an example of the modelling process we focus on selected attributes for demonstration reasons, and we define in approximation the Classes based on the behavioural analysis of the navigation study [17, 18]. The Classes are defined based on the empirical results and the proportional relation between each attribute and visuospatial complexity (Table 80.1). Further empirical work is needed for the verification of the numeric and percentage classification (Fig. 80.2). Specifically, *Rotation metric* is defined by the number of turns along the path (or along a segment of the path) as well as by the accumulating degrees of the angle the person performs while navigating. For the definition of *Size*, we combine the three dimensions of space and calculate the average along the path or the segment. *Visibility* is defined based on the percentage of the visual range a navigator has

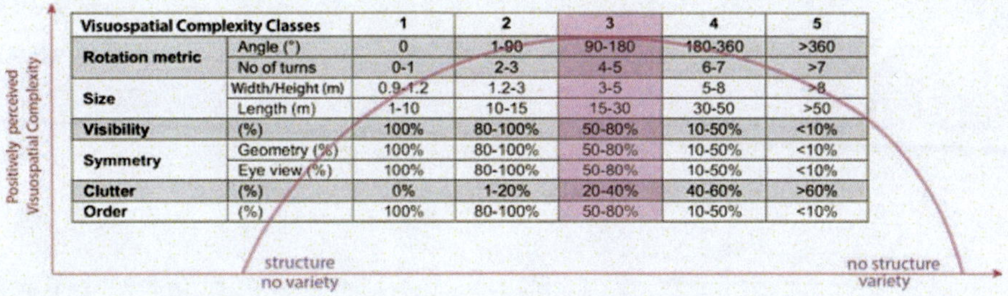

Visuospatial Complexity Classes		1	2	3	4	5
Rotation metric	Angle (°)	0	1-90	90-180	180-360	>360
	No of turns	0-1	2-3	4-5	6-7	>7
Size	Width/Height (m)	0.9-1.2	1.2-3	3-5	5-8	>8
	Length (m)	1-10	10-15	15-30	30-50	>50
Visibility	(%)	100%	80-100%	50-80%	10-50%	<10%
Symmetry	Geometry (%)	100%	80-100%	50-80%	10-50%	<10%
	Eye view (%)	100%	80-100%	50-80%	10-50%	<10%
Clutter	(%)	0%	1-20%	20-40%	40-60%	>60%
Order	(%)	100%	80-100%	50-80%	10-50%	<10%

Fig. 80.2 A simplified version of visuospatial complexity Classes developed on selected attributes for the purpose of parametric modelling (Sect. 80.4). The curve represents an interpretation of the overall positively perceived visuospatial complexity as a function of variety and structure that picks on a moderate Class 3 (Adapted from Shiner et al. [28])

while walking from the different segments towards the end of the path. *Symmetry* is calculated in respect to the layout geometry, as well as by analysing the symmetry in the 3D scene the navigator encounters along the path. *Clutter* is defined by the percentage of space covered by 3D elements in the total path (or each segment), and *Order* refers to the related organisation of these elements in space, calculated by the percentage of the area that reflects a known structure (e.g. grid, circle, rectangular, spiral, square). To calculate the visuospatial complexity class in an aggregative level for the overall path, we combine the corresponding values of each attribute and provide an average result with a numerical indication between 1 and 5 (Fig. 80.2).

80.4 Parametric Modelling of Visuospatial Complexity

By parametrising selective attributes in respect to the scale of visuospatial complexity, and monitoring the relations between them, we develop a parametric model that runs two main operations: *identification* of the visuospatial complexity class of a given morphology, and *manipulation* of the morphology to correspond to a preferred class (e.g. moderate Class 3, Fig. 80.2). We approach the parametric model from the perspective of the human-navigator, using the navigation path as the first component and the main geometrical input to the model. The parametric tool promotes interactive design decision-making based on the visuospatial complexity model (Table 80.1) that can be useful for: comparing different versions of a design in terms of the visuospatial complexity Classes, identifying the class and manipulating the morphology in relation to a selected attribute or in aggregative level for all attributes, comparing buildings in relation to their visuospatial complexity Class, manipulating morphologies to eliminate or increase the visuospatial complexity Class.

Class Identification and Morphology Manipulation The identification function receives as an input a polyline representing the navigation path, it analyses the characteristics of the path and categorises them into Classes based on the visuospatial

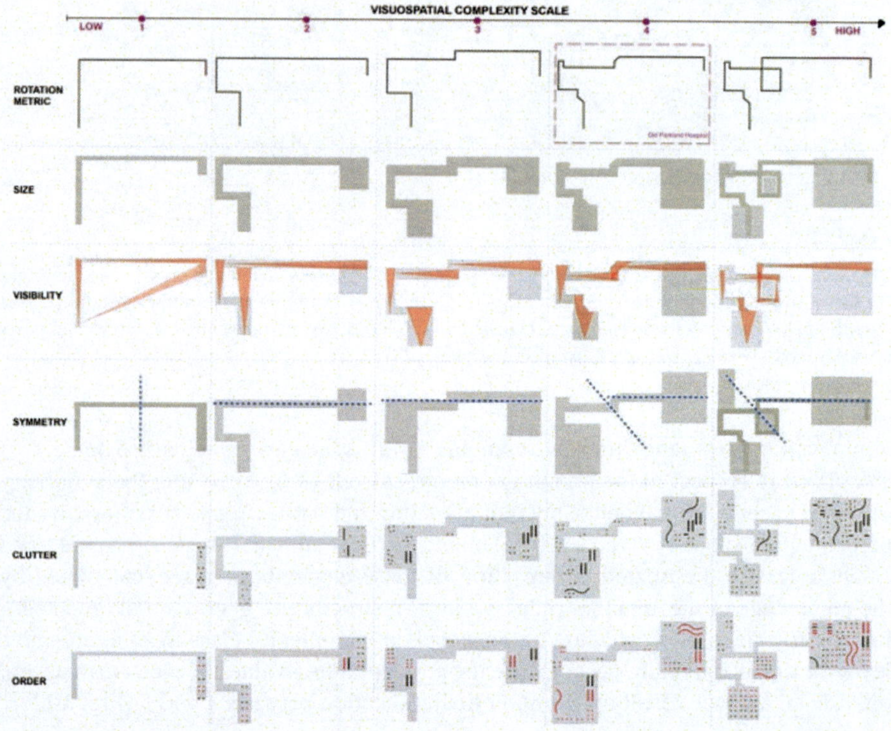

Fig. 80.3 Manipulation of the path in Old Parkland Hospital based on the scale of visuospatial complexity (Fig. 80.2) and the selected attributes that change the morphology respectively. The purple dashed rectangular is the starting point of the identification function of the given path geometry

complexity model (e.g. purple dashed rectangular in Fig. 80.3). By manipulating the morphology of space along the navigation path, we can expand or eliminate the visuospatial complexity by "controlling" the parametric "sliders" between the defined Classes of each attribute (Fig. 80.3). In the Old Parkland Hospital example (Fig. 80.4), we aim at maintaining a moderate Class 3 of visuospatial complexity along the path. The identification function gave the result of an overall visuospatial complexity Class 4. The manipulation function (based on the selected group of attributes) can be used to introduce changes in the morphology (e.g. geometry, arrangement of elements), to reduce the overall visuospatial complexity. Specifically, *Rotation metric* simplifies the geometry of the path based on the angles of turns between the segments. The *Size* function modifies the three dimensions of the segments of the path to fit the requirements of the Class. The *Visibility* reports the percentage of visible space from each segment towards the end point of the path and suggests alternative combination between visual range and Classes using an evolutionary solver (Galapagos for Grasshopper3D). The *Symmetry* function introduces new symmetry axes to manip-

21. Mital, P., Smith, T., Hill, R., Henderson, J.: Clustering of gaze during dynamic scene viewing in predicted by motion. Cogn. Comput. **3**, 5–24 (2011)
22. Moacdieh, N., Sarter, N.: Display clutter: a review of definitions and measurement techniques. Hum. Fact. **57**(1), 61–100 (2015)
23. O'Neill, M.: Evaluation of a conceptual model of architectural legibility. Environ. Behav. (1991)
24. Park, S., Konkle, T., Oliva, A.: Parametric coding of the size and clutter of natural scenes in the human brain. Cerebral Cortex (2014)
25. Potter, M.: Short-term conceptual memory for pictures. Journal of experimental psychology. Hum. Learn. Memory **2**, 509–22 (1976)
26. Rosenholtz, R., Li, Y., Nakano, L.: Measuring visual clutter. J. Vision **7** (2007)
27. Salingaros, N.: Complexity in architecture and design. Oz J. **36**, 18–25 (2014)
28. Shiner, J., Davison, M., Landsberg, P.: Simple measure for complexity. Phys. Rev. E. (1999)
29. Snodgrass, J., Vanderwart, M.: A standardized set of 260 pictures: norms for name agreement, image agreement, familiarity, and visual complexity. J. Exp. Psychol. (1980)
30. Weisman, G.: Evaluating architectural legibility: wayfinding in the built environment (1981)

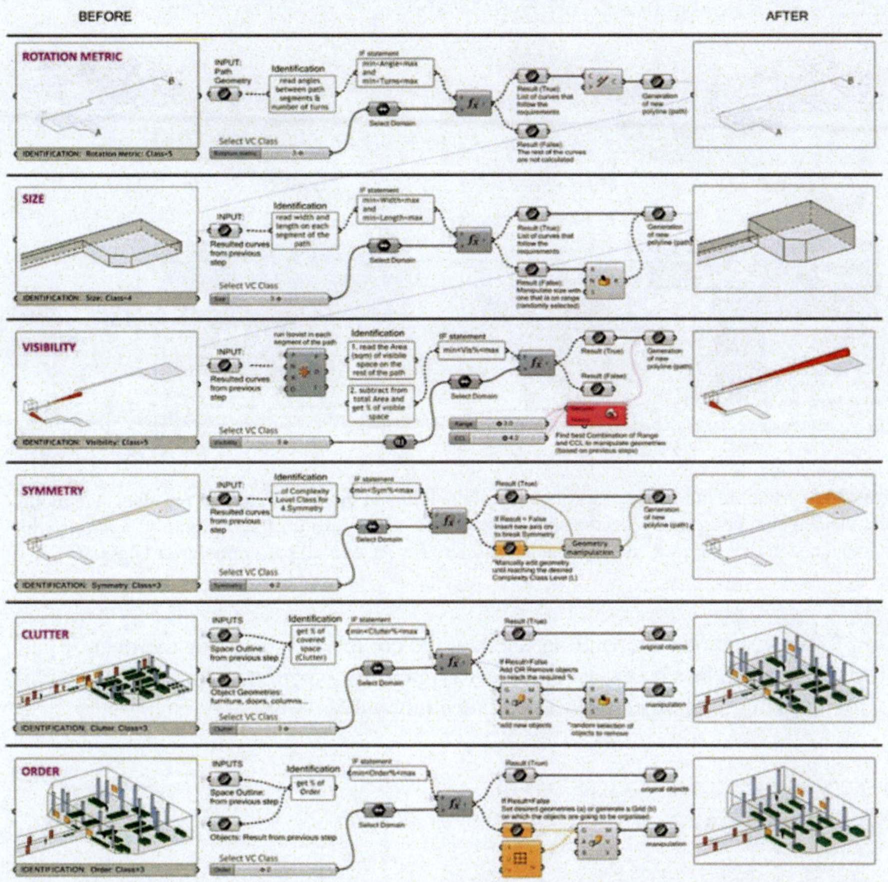

Fig. 80.4 Example of the manipulation function for the path at the Old Parkland Hospital (Fig. 80.1), implemented in Rhino CAD-Grasshopper3D. On the left the result of the identification function reporting the visuospatial complexity (VC) class per attribute, and on the right the result of the manipulation function in steps, monitoring the morphological changes and aiming at an average moderate complexity class for the overall path

ulate the geometry in order to fit the requirements of the respective Class, while the *Clutter* and *Order* functions, add or remove objects and organise them in groups.

The manipulation function can be used in the overall or in particular segments of the path to moderate, or eliminate the visuospatial complexity level. However, it can also contribute to indicating areas where visuospatial complexity can be increased to fulfil design requirements while the designer can closely monitor how these changes affect the visuospatial complexity level. For instance, the analysis of the path at the New Parkland Hospital suggests low overall visuospatial complexity (Class 2). We use the manipulation functionality for the segment of the path that involves the entrance lobby and we extent the level of visuospatial complexity from Class

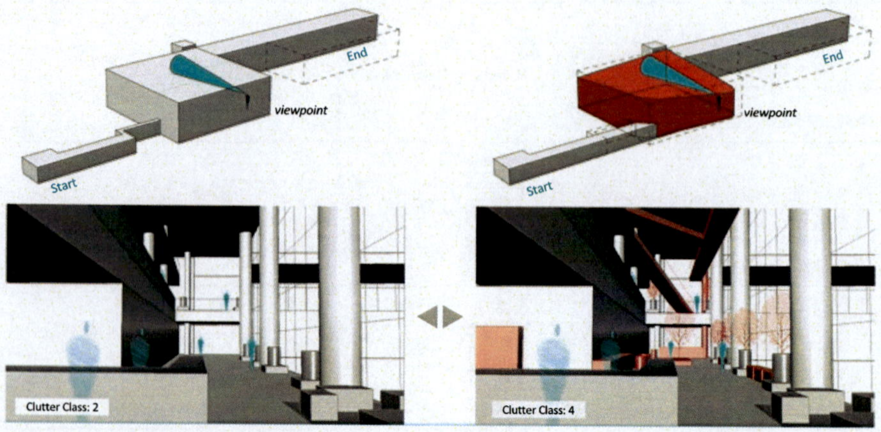

Fig. 80.5 Manipulation of the morphology of the lobby at the New Parkland Hospital. We modify the visuospatial complexity for the specific segment of the path for the attribute of *Clutter*, while we make sure that the rate visuospatial complexity for the overall path remains at Class 3

2 to Class 4 without interrupting the average complexity level for the overall path. Thus, we can reassure an acceptable navigability experience for the users while experimenting with objects at the entrance lobby (Fig. 80.5).

80.5 Summary

Architects are confronted with anticipating the effect of their design decisions on human cognitive experience—termed *visuo-locomotive experience*—in everyday embodied interaction, e.g., navigation and wayfinding [4, 5]. Towards this end, cognitive assistive technologies for design can be useful during the different steps of the design process if they incorporate empirical knowledge from behavioural studies in real-world built environments [4, 17, 18]. Parametric design systems provide the flexibility and adaptability needed, however currently their established functions are numerically and geometrically oriented. By introducing the dimension of human behaviour in parametric design systems, such as high-level cognitive design requirements, and parameters of morphological formulation emanating therefrom, we can promote people-centred architecture design process [4, 18]. The ability to analyse and manipulate the morphology of an environment (and resulting navigation paths) based on people-centred aspects, such as the scale of visuospatial and locomotive complexity and its effects on visuo-locomotive experience, is useful for the evaluation of design decisions and prediction of design performance especially for the initial stages of the design process. We have demonstrated how a parametric design system that incorporates behavioural knowledge can facilitate early design decisions

especially for large-scale built environments such as hospitals, airports, or train stations, where it is crucial to predict navigation performance of users at varying levels of visuo-locomotive complexity.

References

1. Baskaya, A., Wilson, C., Özcan, Y.Z.: Wayfinding in an unfamiliar environment: different spatial settings of two polyclinics. Environ. Behav. **36**(6), 839–867 (2004)
2. Benedikt, M.L.: To take hold of space: isovists and isovist fields. Environ. Plan. B: Plan. Des. **6**(1), 47–65 (1979)
3. Bhatt, M.: Embodied architecture design : on people-centered design of visuo-locomotive cognitive experiences. Cogn. Process. **19**(Suppl. 1), S5–S5 (2018)
4. Bhatt, M., Schultz, C.: People-centered visuospatial cognition: next-generation architectural design systems and their role in design conception, computing, and communication, pp. 207–232. Springer, Cham (2017)
5. Bhatt, M., Schultz, C.P.L., Thosar, M.: Computing narratives of cognitive user experience for building design analysis: KR for industry scale computer-aided architecture design. In: Principles of Knowledge Representation and Reasoning: Proceedings of the 14th International Conference, KR 2014, Vienna, Austria. AAAI Press (2014)
6. Bhatt, M., Suchan, J., Schultz, C., Kondyli, V., Goyal, S.: Artificial intelligence for predictive and evidence based architecture designs. In: 13th AAAI Conference on Artificial Intelligence (2016)
7. Boeing, G.: Measuring the complexity of urban form & design. Urban Des. Int. (2018)
8. Cassarino, M., Setti, A.: Complexity as key to designing cognitive-friendly environments for older people. Front. Psychol. **7** (2016)
9. Cavalcante, A., Mansouri, A., Kacha, L., Barros, A.K., et al.: Measuring streetscape complexity based on the statistics of local contrast and spatial frequency. PLOS ONE **9**(2) 1–13 (2014)
10. Devlin, A.: Wayfinding in healthcare facilities: contributions from environmental psychology. Behav. Sci. (Basel, Switzerland) **4**, 423–436 (2014)
11. Feldman, J.: How surprising is a simple pattern? quantifying "eureka!". Cognition **93** (2004)
12. Fortenbaugh, F.C., Hicks, J.C., Turano, K.A.: The effect of peripheral visual field loss on representations of space: evidence for distortion and adaptation. Invest. Ophthalmol. Vis. Sci. **49**(6), (2008)
13. Gershenson, C., Fernández, N.: Complexity and information. Complexity **18**(2), 29–44 (2012)
14. Harper, S., Michailidou, E., Stevens, R.: Toward a definition of visual complexity as an implicit measure of cognitive load. ACM Trans. Appl. Percept. **6**(2), 1–18 (2009)
15. Helm Van der, P.: Simplicity versus likelihood in visual perception: From surprisals to precisals. Psychol. Bull. **126**, 770–800 (2000)
16. Henderson, J., Chanceaux, M., Smith, T.: The influence of clutter on real-world scene search: evidence from search efficiency and eye movements. J. Vision **9(1)**(32), 1–8 (2009)
17. Kondyli, V., Bhatt, M.: Rotational locomotion in large-scale environments: a survey and implications for evidence-based design practice. Built Environ. **44**(2), 241–258 (2018)
18. Kondyli, V., Bhatt, M., Hartmann, T.: Precedent based design foundations for parametric design: the case of navigation and wayfinding. Adv. Comput. De. **3**(4), 339–366 (2018)
19. Kondyli, V., Bhatt, M., Suchan, J.: Multimodal interaction in autonomous driving. Towards human visual perception driven standardisation and benchmarking. In: STAIRS @ ECAI 2020: 9th European Starting AI Researchers Symposium (STAIRS). At ECAI 2020, the 24th European Conference on Artificial Intelligence (ECAI) (2020)
20. Madan, C., Bayer, J., Gamer, M., Lonsdorf, B., Sommer, T.: Visual complexity and affect: ratings reflect more than meets the eye. Front. Psychol. **8** (2018)

Correction to: Modular Communication as a Structure for Sustainability Within Social Enterprises

Mehak Gupta

Correction to:
Chapter 15 in: A. Chakrabarti et al. (eds.),
***Design for Tomorrow–Volume 2*, Smart Innovation, Systems and Technologies 222,**
https://doi.org/10.1007/978-981-16-0119-4_15

The original version of the book was published with missed out words in the abstract statement. The chapter and book have been updated with the changes.

The updated version of this chapter can be found at
https://doi.org/10.1007/978-981-16-0119-4_15

A. Chakrabarti et al. (eds.), *Design for Tomorrow–Volume 2*, Smart Innovation, Systems and Technologies 222, https://doi.org/10.1007/978-981-16-0119-4_81